[Enter the Past]

The E-way into the Four Dimensions of Cultural Heritage

CAA 2003 | Computer Applications and Quantitative Methods in Archaeology | Proceedings of the 31st Conference, Vienna, Austria, April 2003

Edited by

Magistrat der Stadt Wien - Referat Kulturelles Erbe - Stadtarchäologie Wien

BAR International Series 1227
2004

Published in 2016 by
BAR Publishing, Oxford

BAR International Series 1227

[Enter the Past]: The E-way into the Four Dimensions of Cultural Heritage

ISBN 9781841715926 paperback
ISBN 9781407326399 e-book

DOI https://doi.org/10.30861/9781841715926

A catalogue record for this book is available from the British Library

This book is available at www.barpublishing.com

BAR Publishing is the trading name of British Archaeological Reports (Oxford) Ltd. British Archaeological Reports was first incorporated in 1974 to publish the BAR Series, International and British. In 1992 Hadrian Books Ltd became part of the BAR group. This volume was originally published by Archaeopress in conjunction with British Archaeological Reports (Oxford) Ltd / Hadrian Books Ltd, the Series principal publisher, in 2004. This present volume is published by BAR Publishing, 2016.

BAR
PUBLISHING

BAR titles are available from:

BAR Publishing
122 Banbury Rd, Oxford, OX2 7BP, UK
EMAIL info@barpublishing.com
PHONE +44 (0)1865 310431
FAX +44 (0)1865 316916
www.barpublishing.com

Conference Organisation

Chairs_
Mr. Wolfgang Börner
Mr. Ortolf Harl

Members_
Mrs. Karin Fischer Ausserer
Mr. Joachim Ehrenhöfer
Mr. Maximilian Goriany

Magistrat der Stadt Wien-
Referat Kulturelles Erbe-
Stadtarchäologie Wien

Friedrich-Schmidt-Platz 5/1
A-1082 Vienna, Austria
kongrarchae@m07.magwien.gv.at

Conference Manager_
Mr. Andreas Felser

Editorial Staff

Mrs. Karin Fischer Ausserer
Mr. Wolfgang Börner
Mr. Maximilian Goriany
Mrs. Lisa Karlhuber-Vöckl

Magistrat der Stadt Wien-
Referat Kulturelles Erbe -
Stadtarchäologie Wien

Friedrich-Schmidt-Platz 5/1
A-1082 Vienna, Austria
kongrarchae@m07.magwien.gv.at

Graphic Design_
Mr. Maximilian Goriany

Contents

Preface

Archaeological Processes

Archaeological Prospection

Archaeometry

Computer Aided Design

Contents

Computer Applications in Osteology

Computerised methods for Rescue Archaeology

Cultural Heritage Management

Cultural Heritage Management

Cultural Heritage and Public

Database Applications

Contents

Database Applications

Geographical Information System

Geographical Information System

Integrating scientific results in archaeological processes

Internet Applications

Contents

Internet Applications

Photogrammetry

Statistic and Quantitative Methods

Statistic and Quantitative Methods

Survey and Geographical Positioning Systems

Virtual Reality

Contents

Virtual Reality

ECAI

For chapters 'ON CD ONLY' please go to this link: https://doi.org/10.30861/9781841715926_additionalchapter

Preface

[Enter the Past]

... any real body must have extensions in FOUR directions:
It must have Length, Breadth, Thickness, and Duration
There are really four dimensions, three of which we call the
planes of Space, and a Fourth, Time

H. G. Wells
The Time Machine

There have hardly been any ideas, which have fascinated human beings more than the idea of time travel, the possibility to travel into the past or into the future.

But aren't we all time travellers in a certain sense of the word? When you sit at your desk in front of your computer and move nothing but the computer mouse, time "travels" around you. The future is constantly transformed into the past; the present lasts only for a very short moment. Everything you do quickly moves into the past, which means that we travel through time (Bonsor 2000).
The idea of time travel has not only existed since H. G. Wells published his novel "The Time Machine" in 1895 but for much longer. But only Albert Einstein with his theory of relativity provided a theoretical basis for time travel. As we all know, no one has so far successfully travelled through time but at the same time no one can completely rule out the possibility either.

But what does time mean and how can we define time? You cannot see time and you cannot touch it either. You can only see its results. Time cannot exist without space and space cannot exist without time. We are talking about a space-time continuum here.

According to Albert Einstein's theory of relativity, time decelerates the closer the speed of the object is to the speed of light. Therefore many scientists thought that this would be the possibility to travel into the past or into the future. What they did not consider was that the relative mass of an object approaches infinite mass as the speed approaches the speed of light.

Most time travel theories did not involve machines. We could travel through time via natural phenomena, which should take us from one point in time to the next. These space phenomena, whose existence has not even been proven, are:

- Rotating black holes
- Wormholes (also called Einstein-Rosen Bridges) - are called star gates in science fiction literature. Carl Sagan also used them for his novel Contact in 1985
- Cosmic strings

Today all serious scientists agree that you cannot travel further back into time than to the point where the time machine has been activated ("Timeline" and the Time Machine, 2002).
So the idea to travel into the Middle Ages or even further back into the past, as described in Michael Crichton's novel "Timeline", cannot be implemented.
But now let us move from dream to reality.

The volume at hand contains all lectures held at the Congress "Enter the Past - The E-Way into the Four Dimensions of Cultural Heritage", which took place at Vienna City Hall from 8 to 12 April 2003. "Enter the Past" was jointly organised by Vienna Urban Archaeology, the Research Association Urban Archaeology and the CAA (Computer Applications and Quantitative Methods in Archaeology, UK). Further cooperation partners were AG ARCH (working group "data analysis in archaeology", Germany), AMMA (Austrian Multimedia Agency), ECAI (Electronic Cultural Atlas Initiative, USA), OCG (Austrian Computer Association), as well as UISPP Com. 4 (Union International des Sciences Préhistoriques et Protohistorique, France).

514 participants from 49 nations and all 5 continents participated in the Congress. The programme included 8 workshops, 196 lectures and 52 posters.

Dr. Thomas Klestil, Federal President of the Republic of Austria, Dr. Michael Häupl, Mayor of Vienna, Dr. Andreas Mailath-Pokorny, Executive City Councillor for Cultural Affairs and Science, and University Lecturer Dr. Ortolf Harl, former head of the Department of Urban Archaeology, opened the Congress.

Preface

Unfortunately, the Congress was overshadowed by the outbreak of the war in Iraq. In his opening address, Dr. Klestil called on the community of states to do everything possible to protect the important archaeological excavations. Every endeavour should be made to prevent the cultural treasures of Mesopotamia from becoming objects of the international antiques trade. In the course of the Congress, the participants adopted a resolution, "Vienna Declaration", calling on the warring parties to protect the World Heritage Sites in Iraq.

The Congress continued with brief speeches by Rudolf Schicker, Executive City Councillor for Urban Planning, Traffic and Transport, as well as University Professor Dr. Heinz Zemanek, the "inventor" of the first Austrian computer called "Mailüfterl".

Mr. Schicker referred to the correlation and interaction between information and communication technologies and cultural heritage. "It is fascinating to see the future from the point of view of the past".

University Professor Dr. Zemanek tried to make a connection between computer, computer language and archaeology. The building of the tower of Babel by Breughel served to illustrate the correlations.

Apart from the lectures and workshops the participants also had the possibility to visit cultural centres in the City of Vienna such as the National Library, the Museum of Fine Arts, the Prehistoric Department of the Museum of Natural History, as well as the restaurant Bieronymus . A dinner at Festival Hall of Vienna City Hall completed the Congress.

The Congress ended with a field trip to the ancient Roman town Carnuntum near Bad Deutsch-Altenburg/Petronell in Lower Austria to see virtual archaeology and theory in practice. The field trip was complete with folk music and a buffet in the newly renovated Baroque hall of Petronell Castle.

ACKNOWLEDGEMENTS

We would like to thank all participants and lecturers for their contributions to the Congress.

Special thanks go to the members of the scientific committee:

Geoff Avern, Juan Barcelo, Christian Beardah, Philip Carlisle, Francois Djindjian, Martin Doerr, Michael Doneus, Christine Finn, Lynn Dyson Bruce, Maurizio Forte, Bernhard Frischer, Irmela Herzog, Jeremy Huggett, Peter Jablonka, Elisabeth Jerem, Ian Johnson, Hans Kamermans, William Kilbride, Vu The Long, Jiri Machacek, Torsten Madsen, Branco Music, Franco Niccolucci, Clive Orton, Axel Posluschny, Maria Roussou, Nick Ryan, Apostolos Sarris, Richard Sermon, Zoran Stancic, Stephen Stead, Andre Tschan, Martijn van Leusen, Isto Vatanen, Günther Wagner, Thomas Whitley, Xiao Dong Zhu, Mikhal Zhuhovsky

The Congress was supported by:

- Vienna City Administration
- Austrian Federal Ministry for Science, Education and Culture
- Vienna Tourist Board
- City of Vienna Science Fund
- Austrian Academy of Sciences
- Cultural Department of the Province of Lower Austria
- AutoDesk GmBH, Austria
- AMMA and Filemaker
- Exhibitors (for a detailed list see www.enter-the-past.org)
- Archaeological Forum - Preconference

Preface

We would also like to thank

- Dr. Wilfried Seipl, Director of the Kunsthistorisches Museum, and
- Dr. Anton Kern, Head of the Prehistoric Department of the Museum of Natural History, for the possibility to visit the museums,

as well as

- Dr. Markus Wachter of the Archaeological Park Carnuntum for supporting our field trip to Carnuntum personally and in content.

For the technical support, we would like to thank

- the IT-Department of the Vienna City Administration,
- the employees of Vienna City Hall, and
- Telekom Austria and IVS Interactive Video Services GmbH, which broadcast the Congress on the Internet.

Finally, we would like to thank the employees of the Vienna Urban Archaeology Department, whose commitment for the Congress exceeded their usual duties by far both in terms of time and personal commitment.

Vienna, December 2003

Karin Fischer Ausserer

Wolfgang Börner

REFERENCES

BONSOR, Kevin, 2000. How time travel will work. www.howstuffsworks.com/time-travel.htm.
"Timeline" and the Time Machine, 2002.- Science in Fiction. www.3sat.de/nano/cstuecke/34308.

Opening Speech

DR. THOMAS KLESTIL

FEDERAL PRESIDENT OF THE REPUBLIC OF AUSTRIA

Owing to its location in the heart of Europe and its history, Austria and its inhabitants have been part of the manifold cultural developments and influences that have impressed their stamp on Europe.

Our country has been exposed to currents and impulses from all directions, has absorbed them and made them its own, integrating them into its very own way of life and permitting them to inspire its people to bring forth unique cultural achievements. At the same time, however, our country has passed on cultural impulses of the most varied kinds to its neighbours and partners.

As a result, Austria not only possesses a proud cultural heritage but has during much of its history also acted as something of a European cultural barometer.

To preserve this noble heritage and to make it accessible in a novel way to the largest possible number of people is the great challenge that today's society must meet as it becomes more and more aware of the immeasurable value of culture.

I am particularly pleased that Austria's capital city, Vienna, should be host to the World Congress "Enter the Past. The e-Way into the four Dimensions of Cultural Heritage". On this occasion, international scholars and scientists representing all disciplines involved in cultivating and preserving the cultural heritage of the world will deliberate on how modern information technology can contribute more effectively to networking research and deepening the general public's awareness of the great significance of cultural treasures .

In the spirit of "Sharing, Giving and Receiving", the triad that played such a pre-eminent role in promoting the birth and growth of culture, I wish our guests a stimulating and successful conference and a pleasant stay in Vienna.

Opening Speech

Dr. Michael Häupl

Mayor and Governor of the City of Vienna

The request "Enter the Past" is a challenge which I, as Mayor of the City of Vienna, can easily meet because Vienna is a city with a history that goes back more than 2,000 years. Wherever you put your feet in this city you come across the past in many ways. In the daily life of a city which has as many traditions as Vienna you enter the past on every corner. We Austrians know just too well that the past has sometimes been ignored or embellished. It is therefore essential to deal with the past and to be aware of it.

When you talk about dealing with the past and learning from history it often sounds like a stereotyped phrase. But it is legitimate and important because every generation has to deal with the past and understand it. This is particularly valid in our current climate of change. The achievements of information technologies, enthusiastically welcomed by the younger generations, force us to use new methods in dealing with the past and in imparting it. Inventions such as photography, radio, and television were major development stages and now we are also facing a new stage because with the tools of future-oriented, global technologies we can explore the past in a new way.

As Mayor of the City of Vienna I am particularly pleased that the Congress "Enter the Past", which makes a step into the right direction, is held in Vienna. We have made great efforts to provide the intellectual potential of the city with a favourable working environment. We provide focused support for creative industries such as innovative research and development projects which employ new media in the fields of art, culture, and humanities.

I hope for all participants that the Congress will be fruitful, and that Vienna and its cultural institutions will make a permanent impression on you.

Opening Speech

DR. ANDREAS MAILATH-POKORNY

EXECUTIVE CITY COUNCILLOR FOR CULTURAL AFFAIRS AND SCIENCE

Compared to just about all other fields of application, information technologies have had difficulties with culture from the very beginning and vice versa. The dilemma is huge. We could almost speak of two contrary positions. On the one hand, we have technology which has been developed to deal with homogenous and equally structured data. Its classical fields of application are the sheer endless issues of finance, administration, planning, etc.. In these areas the computer swallows, chews through and spits out one-dimensional information which can be graded with elaborate software programmes, and can be brought to a second or even third dimension.

To a large extent, culture refuses to be treated that way. What can be seen as a weakness from an information technology point of view, i.e. the enormous oscillation of the term "culture", is actually its biggest strength. Defining the term "culture" is one of the most difficult tasks. It gets out of reach just when you think you have grasped it. The term "culture" includes the highest level of creative achievements as well as the smallest purposeless designs of everyday life. Everything included under this term is further graded according to emotional, social, and/or geographic points of view, and is often scattered. So how do you want to feed a computer with such a mixture? To formulate the question even more explicitly: Why should you feed a computer with culture at all? Hasn't culture been there for thousands of years and hasn't it gotten along just fine without bites and bytes so far?

"It's the added value", is what we get to hear from the world of computers. The computer offers the chance to make cultural values generally available at any time. This reduces culture to "cultural values" but has the advantage that the achievements of information technologies cause a real democratisation process in the field of culture.

Due to the enormous width of the term "culture", the question of materials is a further issue which causes difficulties. First of all we need to analyse what kind of material qualifies for the computer, and only then can we start to process data adequately. From my point of view the concept of this Congress, hosted by the City of Vienna, is true and plausible at the same time. It is restricted to the things or phenomena which can be touched and/or fixed above ground and have been classified as cultural treasures due to their age or a convention. This is a first step to utilise information technologies at least for "cultural values". I hope that further steps will follow based on the stimuli of this Congress.

I hope that all Congress participants will be able to successfully continue on their paths, and hope for all of us that we will get to know our cultural phenomena from a new side on a day not too far in the future.

In concluding, let me make a request. Despite all the interesting issues of the Congress, please do not forget the fascinating cultural events Vienna has to offer in all fields covered by the term "culture".

Opening Speech

Dr. Ortolf Harl

Former Head of the Department of Urban Archaeology

Archaeology has its place where humanities and science fuse, where culture and technology intersect. Hence is has to face a great many challenges at the scientific, organisational, and financial levels to come to terms with discovering, documenting, conserving, mediating and destroying cultural treasures. Archaeology deals with definite objects which are usually confined to a specific place and are often several thousand years old. It is therefore a miracle that the public perceives archaeology to such a large extent.

As archeology deals with objects and places, it employs mathematical, graphic, and metrological systems which have been revolutionised by information technologies. Archeology can lead the way for an important part of culture as the objects and phenomena of all historical ages are equal before the computer. The computer distinguishes only the date and material of a Stone Age knife blade and a knife produced by Wiener Werkstätte (Vienna Workshop).

From this point of view it is consistent that the archeologists, of all people, host a congress dealing with cultural heritage in the era of information technologies. Due to its enormous acceptance our workshop "Archeology and Computers" has pathed our way on the international level since 1996, and was an important breakthrough to gain this perception. Archeology has therefore been the stimulus and motivation to hold this congress. However, it addresses everyone dealing with objects and places which form part of the human legacy in the widest sense.

Today you can only make a strong stand when you form part of a network, and we are therefore particularly pleased that we have been able to include some international organisations, first of all the CAA (University of Birmingham / UK), which celebrates its 30th anniversary during our congress. Further, the ECAI (University of Berkeley / California / USA), the Commission 4 of the UISPP, founded 130 years ago, (University of Sorbonne / Paris / France), and the working group "Data Analysis in Archeology" (University of Magdeburg / Germany). In my capacity as president of the congress I would like to thank you and welcome you most sincerely.

I would further like to thank all co-workers and colleagues of the Department of Urban Archeology and of Forschungsgesellschaft Wiener Stadtarchäologie (Research Association Vienna Urban Archeology) as well as our congress office, the exhibitors and sponsors. The harmonious co-operation with all and between all has guaranteed a very smooth organisation of the congress. However, most of all I would like to thank you, dear congress participants, for coming. I hope that your expectations will be met, and that you will be able to implement the stimuli gathered in Vienna in your own work.

The Vienna Declaration

The following declaration was initiated by the Austrian President Dr. Thomas Klestil and was passed by the Vienna Conference 'Enter the Past – the E-Way into the Four Dimensions of Cultural Heritage' of worldwide archaeological experts:

IRAQ CULTURE AT RISK
THE VIENNA DECLARATION

Vienna 11.04.03 – The conference 'Enter the Past – the E-Way into the Four Dimensions of Cultural Heritage' (Vienna, April 8th – 12th 2003) focuses on archaeology in the digital age, an increasingly significant area of cultural heritage.

This includes the virtual reconstruction of lost cities and damaged sites, such as Nimrud in ancient Mesopotamia, modern Iraq.

In his inaugural speech (Tuesday, 9th April, 2003) the Austrian President, Dr. Thomas Klestil, called for international action to prevent artefacts from Iraq becoming part of the international trade in antiquities, in the wake of developments in the Gulf.

Pledging his support, Dr.Klestil appealed to our international community 'to seek effective protection of the most important excavation sites'. He added:' Wherever this makes sense, I will support all endeavours to keep the treasures of antiquity in Iraq and to prevent any plundering of these sites. Every effort has to be undertaken that the treasures of Mesopotamia do not become the object of international trade in antiques. Furthermore, assistance must be provided at all levels to museums in Iraq to facilitate their reconstruction, wherever necessary…'.

In an Communicqué ICOMOS, the International Council on Monuments and Sites states (see attached document of ICOMOS):

… ICOMOS as an international non-governmental organisation of professionals from more than 120 countries of diverse cultures, also stresses the importance of respect for the diverse cultural heritage of the World as an essential foundation of peace and progress and expresses particular concern over the fate of cultural heritage in Iraq and the area.. .

These documents constitute the response of the international community of scholars represented at the congress, 'Enter the Past', to the President's appeal.

The extraordinary global significance of the monuments, museums, and archaeological sites of Iraq imposes an obligation on all peoples and governments to protect them.

During and after the last Gulf War, large numbers of antiquities left Iraq when museums were looted and sites plundered. Archaeologists responded by setting up databases and, more recently, specialist websites, identifying known sites and objects known to be at risk, or already on the antiquities market.

A number of cultural organisations have raised alarm about the heritage implications of the latest Gulf conflict and have made public declarations on websites. We urge all governments, institutions, and individuals in a position to act to recognize and uphold the validity of Iraq's existing, strong Antiquities Law; to secure the longterm safety of the archaeological and cultural heritage of Iraq of all historical periods, and to stop illicit digging and smuggling of antiquities that have occurred. Following a period of conflict, the staff of the Iraqi Department of Antiquities – in whatever form it might take – must be brought up to optimum strength in academic and technical fields. Most important, the number of guards for individual sites, monuments, and museums must be likewise strengthened.

As represented by the signatories of this letter, the international scholarly community is prepared, at the conclusion of the present crisis, to support the Iraqi Department of Antiquities in strengthening, training and retraining its staff, in assessing the conservation needs of artefacts and buildings, and in refitting laboratories. If asked, international archaeologists are also willing to play a role in any needed assessment of damage done by illicit digging or warfare, in salvage operations directed by the Department of Antiquities, and in repatriating stolen antiquities.

The signatories of this letter urge all governments to recognize that fragile cultural heritage is inevitably damaged by warfare, that irreparable losses both to local communities and to all humanity are caused by the destruction of cultural sites, monuments, and works of art. That is our common duty to take all possible steps to protect them.

Even in the midst of suffering history must be preserved for the future. We ask that those who would be prepared to help in this rebuilding make themselves known.

Archaeological Information
An Information Scientist looks on Archaeology

Prof Dr. Heinz Zemanek, University of Vienna

Since I saw the agenda of all your meetings and assemblies I know that the honour of being one of your opening speakers is much bigger than I thought when I accepted. But my task is proportionally more difficult. What can I tell you about archaeological information that you do not practise already? So please do not expect revolutionary news (you don't anyhow). I can give you merely a choice of philosophical considerations: what is information, how to treat it, what did the computer contribute and alter, how do I see the particular features of archaeological information and how do I see the future?

Once in a while a step back and some thoughts about what one is doing is a privilege of the human being and it can be extremely valuable for the daily work. Therefore this is a theme suited for an opening speaker.

I have witnessed the computer development during the second half of the 20th century, I was able to technically and scientifically contribute a little, and also on the international level, and I teach informatics since 55 years, two lectures a year. And you know, after five repeated lectures you begin to understand about what you are talking.

The agenda of this event calls me the "inventor" of the machine "Mailüfterl" which was the first thoroughly transistorized computer on the European continent (England was, of course, ahead). Let my slightly correct. A computer is not "invented" its structure is trivial and what is called "John von Neumann Architecture" was created by God, in the human being and already in animals. The pioneer computers were a problem of decidedness and execution strength, fighting against the conviction around the pioneer that such a thing will never work properly. And similar obstacles.

Mailüfterl: The computer designed at the Technische Hochschule Wien from 1956 to 1958, fully transistorized and with "functional bits" (forerunners of micro-programming)

My computer was unique also in the sense that I had no backing of any institution or research commission it was the personal enterprise of young university assistant, tolerated by the ministry. (During the decisive years I had no boss ...) The necessary money for somehow compensating the development team came through an exceptional channel from the Viennese Bankers Association and almost all components were gifts of the Dutch, Austrian and German industry.

The main problem was of course, seven years after its invention, the transistor. It soon turned out that only Philips in Eindhoven could and would help, but what I could get were hearing aid transistors. They would not switch very fast and so I said 1955 in Darmstadt at the First German computer conference with international participation that for our plan we can not think of a Whirlwind (the famous MIT computer of that time), but a "Mailüfterl" (a friendly Viennese May breeze) might come out. The German professors convinced me to keep Mailüfterl as the name of my computer and so I did. (We achieved anyhow a speed of 133 kc/sec.).

Electronic Archaeology

Directly I did not have to do in my life with your problems, but in the first section of my professional career I had as neighbours a kind of archaeology of information technology. The German Army had neglected Radar technology although there were some insights and indications, here in Austria (measurement of the height of rain clouds in Styria) and in Germany (Radar experiments in Silesia). When the German Air force started a programme in the fall of 1943 it was indeed too late. But I managed to get transferred into it and this enabled me to complete my diploma and to escape from the retreat of my unit through Yugoslavia. In Radar technology, the British advance could not be compensated. But there was a chance to learn from the enemy. An institution was created where all Radar technology devices found in shot down British and American planes had to be sent. They added up to a useful collection, but the devices were incomplete and most of them were damaged. The analysis of the material was not dissimilar to archaeology. Missing elements were to be added, damaged items to be repaired, at least virtually, but in many cases they first had to guess operational principles and functions. It was indeed something like archaeology of electronics.

Archaeology obviously is a lot more than information processing, but there are a number of close mutual involvements. This relation I describe today out of the view of information technology, and hope that some gain will result for you.

The title of the Congress, "Enter the Past", correctly involves the four dimensions of our cultural inheritance. There are the three dimensions of the spatial extension of which we are trained to get them reduced to the two dimensional picture spatial models would of course give better perception, but they are not so handy. And there is as the fourth dimension the time elapsed since the flourishing of the object as well as the time required for your work and your descriptions. What you do is of course information processing, in the framework of your rules, of the rules of archaeology; you do it with your experience and your derived reconstruction principles. To all of that there is a context that can not be seen at the excavation ground but which must be discovered by other considerations and which must be used the right way.

What is Information?

Information can not be defined and, therefore, must not be defined. Because definition means restriction or limitation, but on the other side of the limiting line of the "definition" is again information. Everyone has received and processed information, his body does that automatically. You know what information is. I prefer this notion in full generality: no definition.

Digitalization expresses information by 0 and 1, but even that does not restrict the generality. The big step is digital processing: the logic components of the computer restrict processing to formal model processing.

Archaeological Information Processing

Presently in the Palais Harrach here in Vienna there is an exhibition "Budapest and Vienna between Historism and Avantgarde" which shows many Hungarian treasures. One of them is a painting by Károly Ferenczy called "Archaeology" which presents your effort by its main features: two columns, i.e. surface remainders which invite for closer inspection, an excavating crew, a not yet discovered statue and the commanding scientist. One can not expect that the artist included formally the dimension of archaeology with which I will deal: the information aspect. The imagination of the spectator adds it anyhow.

In Meyer's Big German Universal Lexicon I found under the keyword archaeology the sentence: Regaining lost connections and reconstruction of only fragmentary remains are large activities in archaeology. This is a game between information and redundancy and the same game is also a means of style in architecture: the surprising elements, suddenly impressing, stand versus the repetition of already seen elements; the mighty portal for instance versus the long front of windows.

Many floors yield an even bigger number of windows, maybe ornamented systematically with different ledges. Extended to the height instead of to the width, the result is the tower. The pyramid is somewhere between, but has no windows. The information, the element of surprise, can be the entrance or the roof. (See also below: An other Distinction of Information)

When the Caliph al-Mamun (811 - 833), the first during the last 1500 years who wanted to reach the alleged hidden treasures, tried to enter the Cheops Pyramid (Zemanek 1981), he had to pierce a corridor the one by which the tourists of today get access. The Caliph found no treasures and the entrance was discovered later to be in the upper third of the same (correct) side. With entrances one may have problems even today. The architect provides for instance five of them, a typical redundancy of style. The information is: only by the most left one the visitor can get in.

Norbert Wiener, the father of cybernetics, has contributed a really new field of mathematics by a book with the title "Extrapolation, Interpolation and Smoothing of Stationary Time Series" (Wiener 1949) which excellently - but extremely theoretically - describes the information handling of archaeology:

extrapolation:
the extensions of the object must be virtually added,

interpolation:
the missing parts of the object must be virtually added, and

smoothing:
damaged parts must be (virtually) completed.

I was, by the way, impressed by the amount of statistics which, as your agenda tells, is taken into the services of archaeology. Still, I do not think that Norbert Wiener's theory of time series is of any use for archaeology. But extrapolation, interpolation and smoothing is the precondition for interpretation and virtual reconstruction of archaeological objects. You all are indeed information processors.

Archaeology like any other science applies a particular form of information processing. Information processing technology can be applied where ever routines can be made useful and this is everywhere. Consequently computers today are in every office and laboratory, in every factory and company. There are already more computers than cars.

We can be proud of what we have achieved. But I rather suggest to be modest: the possibilities and the power of modern technology invite us to underestimate the insights and the knowledge of our ancestors.

I can give an example concerning the bible, the first temple in Jerusalem. In two places 1 Kings 12, 23-26 and 2 Chronicle 4, 2-5 the bible speaks of an iron bowl called "The Ocean", with a diameter of 10 ells and a circumference of 30 ells. The wrong conclusion is that they thought that was 3. Artisans in wood and stone were able to produce very precise circles and would clearly know that n was larger. There is an interesting fact behind: the quadrature of the clock where " " is indeed 3. The bowl was placed an 12 iron bulls arranged in a dodecagon, a pattern which can be seen on each clock dial. If a square is drawn through the points 11, 1; 2, 4; 5, 7; 8 and 10, this square has precisely the area of the dodecagon and the relation between the area and the dodecagon diameter is exactly three.

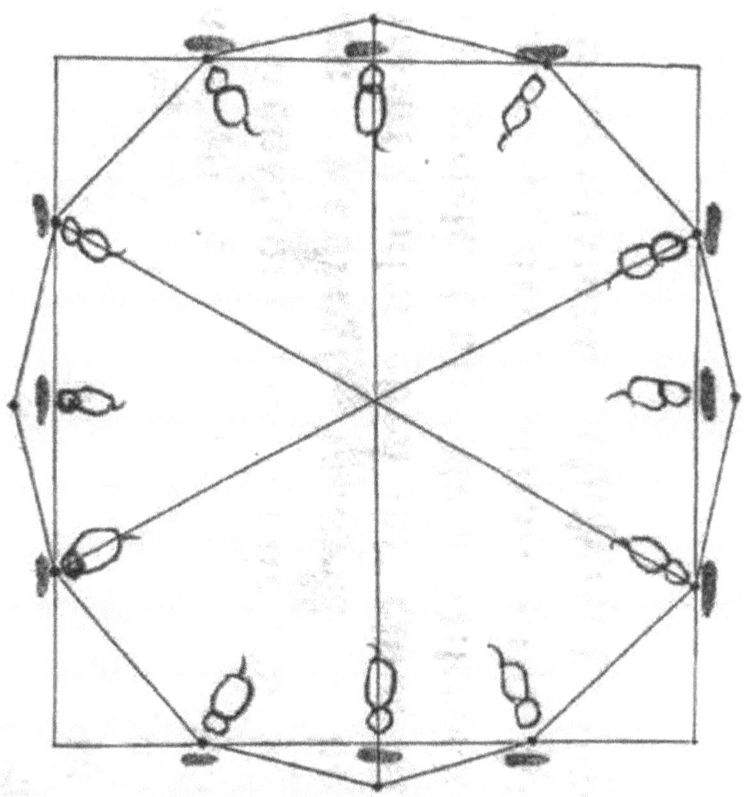

Quadratur der Uhr: This graphic, in which the square is precisely equal in area to the dodecagon, has a biblical background

My subject here is the information which archaeologists process and indeed you all are information processors. Your profession has processed information long before information technology came into existence and long before the notion of information got its modern context; originally it rather meant an aspect of education. What my profession has achieved is the development of information processing to an abstract science which deals since 50 years with many fundamental aspects of information so that now how I look on what you do might be helpful, might be inspiring for you, for your methodology. And in order to inspire your inspiring I will switch between archaeology, architecture and information technology more often than necessary.

Four interpretations of Breughel's Tower of Babylon

For my illustrations I like the Tower of Babylon by Pieter Breughel, the most famous of such paintings, which can be armired in the Vienna Museum of Fine Arts. A reproduction of this painting I had placed in the development room of my fully transistorized pioneer computer Mailüfterl, and in the conference room of the IBM Laboratory Vienna there was a special black and white big size photograph of this painting. It had three succeeding meanings to which I want to add a fourth today.

I have selected this painting because of the poor people houses which in contrast to the tower itself are "operational"; they represent in my mind the failures and weaknesses which occur in every big enterprise and were intended to remind my collaborators of the human imperfection which no enterprise can fully avoid.

The second interpretation came along internationally including pictures of the tower in computer journals and books, the Babylonian confusion in the programming languages which the bible presents as intrinsic: God needs hardly a look an the tower and does not, like in other cases, discuss with the concerned people a compromise the confusion is a logical consequence. And so it is with computers. Programming languages adapt algebra to computers, and since computers and their applications continuously change, there is an infinite need for new programming languages.

Rekonstruktion des Turmes von Breughel: The computer permits to complete the unfinished tower. Maybe the top is open - several ideas are shown

The third interpretation goes into systems design. Why does one feel when looking at the painting that the catastrophe is close? Very obviously the design is visible, the completion of the execution is merely a matter of time. Breughel has put something else into his description to carry the catastrophe message to the thoughtful receiver. The painting tells us that confusion is not a problem of design but of management. He made his work in addition a state of the art report of construction execution tools for big buildings in 1596. So he shows all the machinery in use at this time down to work by hands. Therefore all cranes are different and there is little matching between the different crews. One can feel that there is mismanagement, and with the knowledge of the bible in mind the painting predicts the soon end of the enterprise. Weak management of giant enterprises is sufficient for dispersion. You will easily find examples of our time.

The fourth interpretation of Breughel's Tower, the interpretation towards your field is a kind of virtual archaeology, the completion of Breughel's tower by computing tricks. Missing and damaged details can be taken over from completed details. That there still may be open points I want to exemplify by several roof proposals, some of them clearly only a joke, but an enlightening joke. This shows the archaeologist as a retrospective architect and an information engineer; his intention is to re-construct and he does so by applying the information he has found to imagine the incomplete as the whole. Here I need the distinction between information and redundancy.

INFORMATION AND REDUNDANCY: NOTIONS OF THE 20TH CENTURY

Like Norbert Wieners Extrapolation Theory, Claude Shannon's "Information Theory" (Shannon 1948, Shannon and Weaver 1948) being in fact a theory of the channel capacity - is of little value for archaeology and computers. But the pair of notions activated by Shannon is very helpful: the pair information and redundancy, surprise and boredom, new knowledge and what one knows already. Between the two notions is a tension, a subjective tension because not everyone knows the same, a tension important for arts and architecture in particular. As seen by an information scientist, archaeology is a fascinating game between information and redundancy, between surprise and completion or repetition, but a difficult game requiring an immense amount of context information.

First a look into the kinds of information one has to deal with. In my university lectures, I propose to distinguish five kinds of it:

compelling information:
alarms
delivered information:
  a result that had been aimed at or ordered
provided information:
  ready to be called of, prepared in bigger amounts
shielded Information:
  made invisible for the user, accessible for the service specialist
locked information:
  personal and therefore protected.

For archaeology these five kinds may be proposed slightly differently:

obliging information:
  information about new findings
delivered information:
  excavation reports
provided information:
  the pre knowledge of the archaeologist
shielded information:
  yet undiscovered, unexcavated parts
locked information:
  inexplicable parts where so far the professional knowledge
  still waits for explanation

PROGRAMMED PROCESSING - THE COMPUTER

The computer is not as trivial as its popular omnipresence suggests. And the computer is much more than a calculating device. Certainly this is how the computer its career. And since we had the watch guard mathematician, the fast development of computing went also perfectly smoothly. When you use it today, small calculations are easily inserted and large ones can be professionally executed.

Secondly, the computer certainly is the typewriter as which it appears and it is almost true that you get printed what you see an the screen. Namely: if you stay within the 26 letter the ten decadic numbers and the few other standardized symbols. Using a German "Umlaut" or a Greek letter is at your own risk, in particular if you transmit via the internet.

Apart from such minor weaknesses, the computer is a miraculous writing and storage device and has revolutionized printing. Every PC owner has become (and is sometimes forced to be) editor and press director.

Third: The computer is an organized beginning with inventory and going up beyond the system of airline systems. If your reservation and ticket scientist sits before an unwilling computer you have no chance. With your PC you have not only a dynamic address list, it also prints your envelope.

The collection of telephone books in your main post office is outdated by a much more complete and fluently updated internet access provided you know to handle it.

Four: Generally speaking the computer is a Silicium model which can be programmed to be any formal model you invent. If you design a model of intelligence or a model of satisfaction, you can have an intelligent and satisfied computer. If you really understand intelligence and satisfaction, you will know that there can be no realistic model for intelligence and satisfaction and you will refrain from a bad use of language. You will respect the limitations inherent to the notion of model.

But you have got in the computer a magnificent tool and archaeology has since a long time recognized its potential. It is not my task here to give here the certainly extremely interesting history of computer application in your field. Already the early big computers have been used in the reconstruction rebuilding of temples in Egypt. And of the present day application you know more than I do.

The Theory of Architecture

There is no reason to create a new theory of architecture: Vitruvius, the architect of the time of Julius Caesar and Augustus, has done that 2000 ago, and he did it so outstanding and generally that its application to computers and computer systems (Vitruv). He points out in detail how far the education of an architect should go. That I can not present here, but I will describe what Vitruvius considered good architecture because this has a close relationship to archaeology. Should you have interest and time: I suggest to read his first book.

What ever is designed or built has its architecture. But if the term occurs somewhere, one first thinks of the achievement of an architect: excellent design work. Clearly not all architects deliver excellent architectures; since I have studied the first book of Vitruvius, I look very critically on all modern architecture that comes before my eyes. Have they indeed read Vitruvius?

Conform to Vitruvius, the main properties of good architecture are

order: measure and proportion
arrangement: structure
eurhythmy: beauty and fitness
symmetry: agreement and matching of the components
(that is not symmetry in the geometrical sense)
propriety: conform to the elected principles
completeness: do not leave holes;
select the most general of what you include
clarity: do not complicate what can be done and said simply
orthogonality: do not interweave what is independent by nature
generality: do not restrict concepts, principles and features
more than necessary
economy: proper management of materials and sites,
(today we would add: energy and information), that includes the next keywords.
efficiency: do not waste material, energy and information
security: aim at resistance against distortions, failures and misuse
safety: no danger for environment and people in particular

I said that in the reconstruction process, the archaeologist becomes a retrospective architect like the group mentioned above who had to find out the function of the found Radar equipment. He builds up from the excavated remainders and from his historic, archaeological and general knowledge a picture, a model, a description, an imagination of the treated object. He tries to do this in the spirit of the original architect, who designed and built up the object when it was created. Finally he documents his findings and here he becomes an architect on a second level: an architect of the documentation.

One might raise the question how far the standards for elaboration and documentation should go.

Through centuries the information base of archaeology was a combination of excavation reports and library entries. The reports came then also into the libraries, and this combination, apart from history and cultural history, constituted the knowledge base.

The possibilities of information technology, memory capacity and global network formation, make information technology a new partner aside of the library, complementing it. Advantages and disadvantages of the new partner, of the medium computer, should be carefully considered before archaeology goes into the abstraction of the computer world

In the real life one typewriter after the other is replaced by a PC, and it is very practical to interconnect all of them by the internet. These are no conspicuous steps and except certain budgetary quarrels there are no objections, no obstacles. Maybe one or the other in the staff participate in a computer course, learns the required programming language.

One starts off with the authors known from the library and generalizes, using search machines. In short, in fact you all are already in our computer world. Considered precisely this is not yet really a truly scientific methodology, but it is generally accepted: other professions do not advance differently.

In fact, one could do better. But for this a particular effort is necessary and I propose one to you.

An Information Processing Competence Center

A Proposal for Archaeology and Other Fields

A study group "how to best use information technology for archaeology?" might not be a bad idea. One of its tasks would be the organization of an archaeology oriented database of history. I am sure that a lot can be found in the internet. But how reliable are the different entries? Is there any institution that checks correctness or qualifies contributors? Only the "local" specialist can evaluate the entries. As any other field you would need a generally accepted hard core distinguished from proposed addenda that are admitted but require further elaboration. And this system requires a firewall outside of which all the other, unchecked entries live.

It would be great to have a search machine which indicates for each finding to which of the three mentioned categories it belongs. The secret here is the proper selection of search key words.
It might be possible to organize this competence center as an institute, but information technology permits you a distributed solution, a center composed by sections of existing institutes. What you then need are organization and operation rules.

Actually what I have proposed here for archaeology is a structuring in the internet which would be required far more generally. But this is another story. A much more difficult one at least.

The Notion of Interface

I turn to a concept of information technology which also may be helpful here. For the investigation of an information system a suited or implied spot is selected and there a cut is applied, in reality or merely virtually, so that a look or a measurement can be made in either direction. From the measurement (or a calculation) it might be possible to conclude whether left and right are fitting or not. If not, an adjustment device may be necessary.

If the two sides of the system communicate the question arises in which language, in which code they do so or they should do it. Before the computer time that was a human problem now it becomes more and more a technical issue.

If one of the two sides is human (the user or the human "caretaker"), one faces a so called "man-machine-problem".

Man is distinguished from machine by more elementary features than most engineers think. He can adapt to machine-like behaviour but this is recommendable only in exceptional cases; normally the difference must be carefully considered in the design. The archaeologist faces a man-artefact situation and here is the analogy to the information system situation.

It might be worthwhile for the archaeologist to consider the interface between him and the object he is working on: there are models of behaviour - different ones depending of the situation, e.g. whether the object is

in existence
under design or
under reconstruction

And this invites us to consider again the steps of archaeology

Discovery, Search
something was found — coordination
how far does it reach? — extension explored
Excavation
cleaning separation of waste
Interpretation
what was it, what purpose did it have? — dynamics out of static reconstruction
how did it look when it was in order? — completion, extension

Application of Architectural Design

Architectural Design is not only required for the object as such, it is applied trifled:

to the object
to its production
to its documentation

To the documentation of archaeological work and to publication in general, I may make a few special comments. I have written more than 500 papers and I have read several thousands, many of them for evaluation. How to design a document or a publication is an art that would deserve to be the subject of a one Semester lecture for any kind of study. The main advice I can give is to spend a considerable fraction of the time for layout and list of contents (Zemanek 1980 and 1986).
One modern addendum to Vitruvius' keyword "clarity" is: be careful with abbreviations. Do not use them more than readability requires. Never use them unless they are explained in your text, in the copy or book. And in no case have an abbreviation in the title of your paper.

An opera without an overture is incomplete. The listener is not as "in" as the composer and the performing artists. He must be brought by musical means to the point where the action starts. And the same is true for any publication. Most of them neglect this. What is the prehistory of the general and of the particular subject, what is the prehistory and the view of the author?

If one documents a situation, if one writes a paper, the consideration of the architecture of the writer's product can be extremely helpful for the acceptance of the product. Has the text the right structure and the best headlines? Many such questions should be raised before giving the manuscript out of hands, before sending the document to the recipient.

An Other Distinction of Information

If one looks at the different sorts of information one sees (for instance)
1) structure information concerning the general arrangement of the parts
2) repetition information concerning typical ingredients of the design which are applied many times; important for the economy of the production
3) exception information concerning ingredients that are used one once or repeated ingredients with important deviations

That repetition information is important for the economy of architectural design is shown by the example of the Sydney Opera House. There the architect had to be dismissed because he preferred individual elements to repeated, to equal elements which correspondingly had to be fabricated individually. The budget exploded, not by the usual 25 or 30 %, but by a factor of 3 or 4. The successor was obliged by the town to use construction elements (the many bows for instance) in standard sizes so that they could be produced by few standard procedures and casting forms.

How far such considerations are valuable for archaeology and its reconstructions, is on you to judge. What I show are consideration or examples how the archaeologist can apply ideas of information technology and of architecture to his work.

3D Models, Static and Dynamic

I believe that such models are extremely helpful, not only for average visitor but also for the specialist. And here modern information technology can be of extreme service as two examples show.

The computer is a model designed for designing models.

To extend the drawing to a three dimensional model is an offer of information technology to archaeology which can be made very valuable. Because with all due respect for imagination, it is far more efficient to support the imagination of the third dimension by a model of three dimensions.

I had to evaluate such a project at the 2001 meeting of the European IST (Information Society Technologies) Prize. The project is called "Fastbuilder" and developed by the IWI Group. The subject is a three dimensional modelling tool that programs cities from maps, picture postcards, aerial views and so on. Once the model is established, one can order a certain point in the city, a height and a direction, and the program delivers a picture of the city from this point in the ordered direction.

One of the sponsors of this event is the Autodesk GmbH in Munich and in the last issue of the journal of the Austrian Electrotechnical Society this company reports its project called Munich 4D (AutoDesk 2003), simulating 400 years of the building history of the environment of the Frauenkirche and of the Karlspatz (Stachus).

The Future of Information Processing

The drive of information processing offering more and more services for a steadily sinking price makes information technology an extremely valuable tool for archaeology. And therefore it seems appropriate to conclude with a look into the future of information processing, in particular to the features of value for archaeology.

I expect for the coming decades an extension of processing to teleprocessing. What I mean is this: when you dial today from one place to another, maybe over the ocean what you trigger is an action an the other place: a switching device interconnects you conform to your instruction, namely the dialled number, to your wanted subscriber. It is easy to imagine such telecommanded processes for all kinds of purposes. The realization is in so far less easy as the execution requires a standardized code for standardized production or execution lines. And it will also need a sophisticated protection against misuse. So it will take still some time until it becomes clear in which ways such possibilities can be used successfully and economically. But I am sure that teleprocessing of material and energy will come. And archaeology will have to consider how their solutions may look like.

Already for the usual information processing one is well advised to strive for common application of advanced programs for each field of application. I therefore think that the advice I gave above to archaeology should be given to any profession: establish a competence center for the particular needs of your field, collect methods, programs and data and take care of the biggest enemy of progress: progress, progress which paralyzes prior achievements.

The main example is storage. By improvement of storage technology progress makes older carriers unreadable, their information inaccessible. It becomes necessary to translate the information from the old system to a new one. Since we collect more and more information, the costs for this translation are not going down as they do for a constant amount of information they go up. The competence center can watch the development and advise national institutions how to equip their computers software wise and can assist in the storage problems.

To this situation I tell you my closing story, not out of archaeology but out of arctic or Antarctic exploration. It was found that the administration of stored exploration results had destroyed a substantial fraction of twenty years old expedition reports. The director was heavily accused. In defence, he gave a proof that it was much less costly to repeat in certain cases an exhibition rather than to keep the complete material in storage.

This may have been before our electronic storage facilities. But somehow the story is beyond history. To store or not to store it may sound strange but it may become your question.

The speed by which the computer hardware (processor and storage) improves its parameters is as well a big chance as a heavy danger for the field of information technology, for the individual and for institutions and companies. Be enthusiastic about the chances which information technology offers you, but be also warned: the computer not only demands more discipline than can be brought up - our technology may tear you in directions you better do not take. Tools are helpful in the right hands and for the right mind.

ATTACHEMENT

As a kind of post-scriptum I want to add a few comments on Stonehenge and alignments.

Stonehenge is a perfect example for archaeological reconstruction, beginning with the question of the purpose of the site. It was with computer help that a thorough investigation of all possible alignments was carried out and the results were considered very meaningful (Hawkins 1965, Zemanek 1978). This study shows how valuable the consideration of alignments can be and the application of computer programs on found data is also remarkable.

The other two comments concern Baroque buildings, architectures still in use. One is the church of St. Charles in Vienna, a masterpiece (even a lecture in architecture) by Fischer von Erlach (who, by the way, was one of the earliest theoretical archaeologists: he reconstructed ancient buildings and the volume with his results is an admirable collection of early examples of reconstruction).

Karlskirche, Vienna, Austria

My late friend, the architect Dr. Hans Foramitti - he had, by the way, discovered Fischer von Erlach's volume in our Paris time (1948) at a bouquinist along the Seine river, bought it for a price affordable for a student, and brought it to Vienna in his rucksack - found out that the statue of St. Charles over the portal of the Vienna church gets a "holy shine" by alignment: if one takes the right position (difficult today because of the pond added a few years ago, but possible in winter time when the pond is empty) there is a front window of the lower set in line with a back window of the upper set and around the head of St. Charles is an aureole, a nimbus (moreover one of the windows necessary for the alignment has been covered for a certain purpose so that the effect did not work lately).

A different effect exists in the church of the Abbey of Melk (about 100 km West of Vienna, on the Danube). If one goes there around the day of St. Peter and St. Paul - 29 June - at 6 pm local time, i.e. 5 pm summer time - the sun rays through the back window illuminate precisely the statues of the two saint on the main altar (This I have discovered myself and published a photo in the local newspaper).

REFERENCES

AUTODESK 2003. Autodesk visualisiert München. Autodesk GmbH in Munich, Elektrotechnik und Informationstechnik 120 3, Wien:a14-a15.

HAWKINS, G.S., 1965. Stonehenge Decoded. A Delta Book, TM 755118, NY 1965:202.

SHANNON, C.E. and WEAVER, W., 1949. Mathematical Theory of Communication. Univ of Illinois Press, Urbana Ill.

SHANNON, C.E., 1948. A Mathematical Theory of Communication. Bell System Techn. Journal 27:379-423, 623-656.

VITRUV. De Architectura Libri Decem. Many translations in many languages.

WIENER, N., 1960. Extrapolation, Interpolation and Smoothing of Stationary Time Series. J. Wiley, NY 1949, 4th Edition, NY:163.

ZEMANEK, H., 1958. "Mailüfterl" ein dezimaler Volltransistor Rechenautomat. Elektrotechnik & Maschinenbau 75, Wien:453 463.

ZEMANEK, H., 1978. Stonehenge Steinzeit Computer. Elektronische Rechenanlagen 20, München:5-9.

ZEMANEK, H., 1980. Abstract Architecture General Concepts for Systems Design. In Björner, D. (ed.), Abstract Software Specifications, Copenhagen Winterschool Proceedings Lecture Notes in Computer Science 86, Springer Verlag, 1979:1 42.

ZEMANEK, H., 1981. Al Khorezmi, his background, his personality, his work and his influence. In Ershov, A.P. and Knuth, D.E. (eds.), Algorithms in Modern Mathematics, Lecture Notes in Computer Science Vol. 122, Springer, Heidelberg:1 81.

ZEMANEK, H., 1983. "Mailüfterl" eine Retrospektive. Elektronische Rechenanlagen 25, München:91 99.

ZEMANEK, H., 1986. Gedanken zum Systementwurf. Ein vom Gebäude und Computer generalisierter Architekturbegriff, der auch für Fahrzeuge und Verkehrssysteme nützlich sein könnte. In Maier-Leibnitz, H. (ed.), Zeugen des Wissens. 100 Jahre Automobil 1886-1986, Daimler Benz AG v Hase & Koehler Verlag, Mainz:XIX+1043,99-125pp.

ZEMANEK, H., 1994. The Mailüfterl Adventure. AICA Congresso Annuale Vol.1, 2 23 AICA 1994:XIV + VIII + 1874pp.

Archaeological Processes

Archaeological Typologies - an Archaeological Fuzzy Reality[1]

Sorin Hermon
PIN S.c.r.l, VAST-Lab, Prato, Italy

Franco Niccolucci
Università degli Studi di Firenze, Italy

Francesca Alhaique
Università La Sapienza, Rome, Italy

Maria-Rosa Iovino
Centro Int. di Sperimentazione, di Documentazione e di Studio per la Preistoria e l'Etnografia dei Popoli Primitivi, Siracusa, Italy

Valentina Leonini
Università degli Studi di Siena, Italy

Abstract

Since its definition in the early sixties, fuzzy logic has been successfully applied to various fields of natural and exact sciences, mostly in researches related to predictive modelling or typological classification. While fuzzy logic concepts are presently applied to predictive models using GIS applications, few typological classifications of archaeological remains have adopted concepts of fuzzy logic. Previous research has shown the problematic of archaeological typologies based on Boolean logic, whenever the definition of types is loose or a subjected interpretation from researchers is required. Moreover, uncertainties which occur in many typological classifications are obscured by the need to assign each artefact to a single, distinct type. Fuzzy logic has been successfully applied, at an experimental level to classification of lithic artefacts in the past, the research being presented in the last CAA meeting. This paper presents further experiments in applying fuzzy logic to classification of lithic and ceramic archaeological artefacts, bones deriving from archaeological sites and interpretation of use-wear analysis on lithic artefacts. Consequently, a method of classification of archaeological materials using fuzzy logic will be presented, together with its implications to archaeological reasoning.

Introduction: Fuzzy Logic and its Application to Archaeology

One of the primary goals of an archaeological research is the recognition of the cultural affiliation of the material culture under analysis (Trigger 1968), a process heavily relying upon typological research (Whittaker et al. 1998), that is classifying the artifacts into a priori types, defined by the researcher (Tixier 1967), according to a particular type-list relevant to the research area (Willey and Phillips 1958, Hours 1974), under the assumption that any culture is defined by a set of characteristic types. Boolean logic is applied in all stages of this research, even though not always all the defining characteristics of a given culture are isolated into the analyzed material, types do not always have sharp boundaries, their definition includes intuitive attributes and they are therefore exposed to different interpretations by the researchers, as shown in various occasions (for a summary, see Hermon and Niccolucci 2002). The alternative proposed in this article is to apply fuzzy logic concepts to the typological process, whenever fuzzy sets (such as archaeological types and archaeological cultures) are recognized.

Fuzzy logic (Zadeh 1965, Novák 1989, Kosko 1993, McNeill and Freiberger 1993) aims at extending ordinary deductive methods by assigning a numerical degree of truth or falsity to statements not completely true nor completely false, otherwise indiscriminately cast into the "neither" category, (varia), probably wasting valuable information (Copeland 1997). The range of truth values is the closed interval [0;1] of real numbers. The degree of fuzzy belonging of a member to its set is also given by a real number between 0 and 1 (Kosko 1993, Novák 1989).

The application of fuzzy logic to archaeological classification (for more details see Hermon and Niccolucci 2003, Niccolucci and Hermon in press) implies using a matrix A with items as rows and types as columns, as inventory catalogue of the analyzed assemblage. Its elements are the fuzzy coefficients, expressing our certainty that each item belongs to each type. It is also defined the reliability index R for each type, for each item and for the entire assemblage:

$$R(x_1,x_2,...x_m)=\frac{\left(\max\limits_k x_k\right)^2}{\sum\limits_k x_k}$$

Some case studies will hopefully clarify the methodology and its implication for the archaeological research.

Case Study I

The first example, (see Hermon and Niccolucci 2003), concerns the analysis of a sample of fifty tools originating from the lithic assemblage of the Middle Bronze Age I site of Beer Resisim (Rosen et al., in press), located in southern Israel, which is currently under analysis. The site is located along a narrow ridge facing the wadi of Nahal Resisim in the east and in close vicinity to the spring of Beer Resisim. Five areas were excavated during three seasons, which unearthed a single period settlement belonging to the end of the Early Bronze Age - beginning of the Middle Bronze Age period. It consists of a multitude of clusters of circular rooms of ca. 4 meters in diameter, some grouped into honeycomb-like buildings, and adjacent installations, such as enclosures for animals, courtyards and various activity zones. The rich material culture included a large collection of lithic artifacts.

Its classification was concluded following common standards of pre and proto-historic lithic assemblages in Southern Levant (Bar-Yosef 1970, Rosen 1997). Their analysis may contribute to a better understanding of the subsistence economy of the inhabitants in a marginal, semi-arid region. Table 1 presents the inventory list of the site, according to the traditional (crisp) methd and to the fuzzy one. The main difference concerns the relative number of scrapers, notches and denticulates, and to a lesser intensity sickle-blades, a discrepancy which may lead to different interpretations in aspects as economy of the site and cultural affiliation. Moreover, the modality of these types is suggestive and one may consider the re-definition of these types.

Type	Scraper	Tabular Scraper	Borer	Burin	Truncation	Retouched Flake	Notch	Dentic	Retouch Blade	Sickle Blade	Retouch Bladelet	Bifacial
Crisp	20%	0%	14%	2%	10%	14%	20%	4%	10%	2%	2%	2%
Fuzzy	14%	0%	12%	1%	8%	12%	18%	14%	9%	6%	1%	3%

Table 1 Inventory list of a sample (N = 50) tools from Beer Resisim

Case Study II

The second case study regards the application of fuzzy logic to the classification of use-wear traces from a sample of flint tools from the stratified cave of Corruggi, in Sicily, Italy. The cave, located two miles NW of Cape Pàssero, is situated over a rock bench that delimits the depression occupied by the marsh of Marghella, approximately 50 m from the beach of Vulpiglia, in the territory of Pachino (Syracuse, Sicily). Its origin, open along the cliff of an ancient coastal line, is by marine erosion during the Riss/Wurm Interglacial. In the successive glacial phase, represented inside the cave by a layer of red clay with Pleistocene fauna, the sea level regressed and caused the formation of a wide, coastal plain, still noticeable in the successive Mesolithic and Neolithic periods. The study of the archaeological remains from the cave is particularly important when exploring the functional differences between the Mesolithic and the Neolithic human activities and lifestyle in the area. The functional analysis of a sample from the lithic assemblage was performed by means of microscopic examination and observation of damage types (see Newcomer et al. 1986, Van Gijn 1990). Among the 100 flint artifacts (débitage and tools) selected as more suitable for the functional analysis, 60 items showed use wear along their edges, evidences of various activities performed on a relatively wide range of materials. Among these, notable is the processing of plant fibres and the working of bone/antler and of wood. Less common are the scaling of fish, hide processing and working of stone/or shell. The wear traces degree of development indicate that many of these artifacts were short term used prior their discard. The worked material was inferred with less certainty on 16% of the used edges and consequently these were simply referred to as "used". Information related only to the hardness degree of worked material was inferred on the 33% of the used edges.

The comparison of the first data interpretation (where evidence of use was interpreted only by the inference of "used" tools and/or only by the "hardness" degree and/or by a possible interpretation) with the new fuzzy methodology (interpretation of all the possible evidences under the value of a new interpretation - but with no absolute identification of use) highlighted quantitative differences in the spectrum of worked materials as well as the range of activities performed at the site. The fuzzy results presented in the Table 2 (activities performed on the site) show a higher value of non-siliceous plant processing and a high percentage of stone working versus the generic evidence of scraping activity.

Data related to the identification of processed materials (Tab.2) show high values for stone, wood and bone/antler working and a decreasing value of plant processing. The fuzzy logic interpretation enabled a classification of edge wears otherwise grouped into generic classes, such as "used" on "hard/soft" material. Thus, the additional information obtained rises up the hypothesis that stone material working and non-siliceous plant processing were underestimated in the traditional framework and should be taken into consideration when drawing the scenario of the Corruggi Cave inhabitants.

Activities	Piercing - boring	Cutting	Engraving	Hafting	Scaling fish	Stone working	Slant processing	Scraping			
Crisp	4%	10%	12%	1%	4%	1%	10%	58%			
Fuzzy	6%	13%	7%	0%	3%	17%	20%	34%			
Materials processed	Hide	Bone	Plant	Reed	Wood	Stone	Shell	Fish	Flesh	Root	Miner.
Crisp	4%	23%	34%	7%	22%	1%	3%	4%	1%	1%	0%
Fuzzy	1%	30%	18%	7%	17%	20%	0%	3%	1%	0%	3%

Table 2 Comparison between the traditional (crisp) and fuzzy methods of classification (N = 60)

Case Study III

The sample (N = 100) originates from the Bell-Beaker site of Santilario d'Enza, located in the province of Reggio Emilia, Italy. The site, excavated in the early seventies (Barfield 1974), is a small short-encampment, with refuse pits of various sizes and probably inhabited during a single phase of occupation. The ceramic material, highly homogeneous and without a doubt belonging to the Bell Beaker complex (Barfield et al. 1975, Leonini 2003), consists of 420 fragments, from which the non-decorated pottery assemblage was chosen for the experiment. The entire assemblage was initially classified according to an analytical typology based on the recognition of geometric modules, defined according to various metrical measurements, such as diameter and height and the ratios between them (Sarti 1989, Cocchi Genick 1999). Table 3 presents the inventory list of the non-decorated pottery, according to the traditional and the fuzzy method. Major differences were observed in the jars and bowls group; another, self-evident, consequence, is the disappearance of the "undetermined" (varia) group when using a fuzzy typology. Given the fact that in many cases the cultural affinity of any site is not related to statistical weights of particular tool-types, but rather on their presence/absence and on stylistic consideration, the interpretation of the observed discrepancy between the traditional and the fuzzy typology should have a different interpretation. One of the first consequences of applying fuzzy concepts is the evaluation of the reliability index R of types, which, as shown in Table 3, has low values for most types, which apparently reflects their weak definition, these tending to be often vague and easy to misinterpret. This result would invoke a re-evaluation of the relevant types' definition. Another outcome may be related to the archaeological interpretation of the results: while according to the traditional method there are some more jars than juglets, fewer bowls and a small amount of plates, the fuzzy method returns a slightly different result - the difference between the relative amount of jars and juglets is higher than before, and accordingly, less bowls and plates. Furthermore, when excluding artifacts with a low reliability index R, there is again a shift, the assemblage being dominated by juglets, and there are significantly more bowls and plates. While it is unknown which one of the interpretations corresponds better to the Bell-Beaker reality of the site, identifying "problematic" types and artifacts may add to our understanding of the past reality.

Pottery type	Jar	Juglet	Bowl	Plate	Indeterminate
Crisp	44%	32%	14%	2%	8%
Fuzzy	57%	34%	8%	1%	0%
R > 0.5 only	31%	49%	17%	3%	0%

Table 3 Inventory of the sampled pottery assemblage

Case Study IV

The site of Brega (Rosà, Vicenza) is a villa rustica excavated between 1998 and 2001. The faunal remains (Alhaique and Cerilli in press) were collected during the last season of excavation carried out under the direction of Dr. E. Pettenò (Soprintendenza Archeologica del Veneto) and coordinated by Dr. S. Tuzzato (Studio di Archeologia, Padua). All the excavated archaeological structures of the villa, three buildings and a cistern, are oriented along the axes of the roman centuriatio and located around an open space where domestic and handicraft activities probably took place. The archaeological material spans the period from the 1st to the 5th century A.D. Preliminary archaeological analyses place the faunal finds within a generic Late Antique Roman period. The osteological sample comes from the fill of the cistern and includes both food debris, represented mainly by remains of domestic animals, and residues of handicraft activities, indicated by abundant deer antler fragments with manufacturing traces. The fuzzy logic approach was applied in this sample to reevaluate the age at death of ovicaprines and the resulting mortality profiles. The method employed for the determination of the age at death of ovicaprines is based on tooth eruption and wear stages (Payne 1973). For this analysis Payne's age stages have been gathered into larger groups: very young (0-6 months), young (6-24 months), prime adult (2-4 years), adult (4-8 years), senile (>8 years). The results obtained with the "traditional" method suggest the use of these animals both as a source of meat and of live products (i.e., milk and wool) given the presence of very young as well as older and senile individuals. Still using Payne's (1973) eruption and wear stages, but with a fuzzy logic approach, it was possible to include a larger number of specimens in the "ageable" sample (N = 17 vs. N = 13 with the traditional method). The new results do not drastically change the interpretation of the assemblage, but they show a greater emphasis on young and prime adult age groups (Tab.4), which suggest that the main focus was on meat rather than on live products, therefore refining the previous interpretation. The brief example given in this paper shows that maybe one of the main advantages of using the fuzzy logic approach in faunal analysis is the possibility of increasing the study sample including also "difficult" items that would be otherwise excluded from some kind of analyses. A difficulty in the application of the fuzzy logic method to the classification of animal remains, is the need for a standardized method for determining more objectively the degree of possibility that a bone belongs, for example, to species A or to species B. Differences in the application of the fuzzy logic approach between continuous (e.g., age) vs. discrete (e.g., species) types as well as between "real" (e.g., species, age) and "artificial" (e.g., lithic artifacts or ceramic vessels) types should be also further investigated. This example shows that one of the main advantages of applying the fuzzy logic approach in faunal analysis is the possibility of increasing the study sample including also "difficult" items that would be otherwise excluded from some kind of analyses.

Age group	Very young	Young	Prime adult	Adult	Senile
Crisp	8%	30%	0%	54%	8%
Fuzzy	5%	39%	20%	30%	6%

Table 4 Age-groups of the sample

Summary and Conclusions

Since fuzzy logic was introduced in the sixties, thousands of articles and tens of books were published and international journals and societies devoted to this subject were established (for a wide list of tens of entries covering the subject from various aspects, visit (May 2003) http://www.abo.fi/~rfuller/fuzs.html). Fuzzy logic is applied in natural, social and human studies, engineering and computer sciences, covering practically most fields of modern research, a large part of our modern daily life depending on fuzzy logic based systems and machinery and the list is still long. Fuzzy logic was also applied in humanistic and natural science studies as well, such as social sciences (Ragin 2000), anthropology (Ruiz 1999), databases (Choi 2003), economy (Dolan 1994), religious studies (Raposa 1993), vegetation science (Hall 1997, Moraczewski 1993, Mucina 1997, Olano et al. 1998, Roberts 1986, 1989, Sattler, 1996), and geography (Fisher 2000, Harris et al. 1993).

But, above all, since many of our concepts in archaeology can be best defined as vague, or fuzzy, and the archaeological way of reasoning is often by means of approximation and evaluation, rather than crisp affirmation, fuzzy logic can provide the required methodological framework and the fuzzy set theory the necessary tools for research. Moreover, when data are presented in fuzzy format, the reader can have a clearer picture of the researcher's analysis process and its reliability on the analyzed material, data are more transparent and thus more accessible in raw format (lesser affected by the subjectivity of the researcher presenting the data), thus enabling a clearer distinction between "real" data (presented in fuzzy form) and the "interpreted" data, as presented in traditional form.

[1] F. Alhaique was responsible for the application of the experiment to the bone assemblage, M-R.Iovino to the use-wear analysis and V. Leonini to the pottery assemblage.

REFERENCES

ALHAIQUE F. and CERILLI, E., in press. Handicraft, diet and cult practices in the Late Antique villa rustica of Brega (Rosà, Vicenza, NE Italy). Archaeofauna.

BAR-YOSEF, O., 1970. The Epipalaeolithic Cultures of Palestine. Ph.D. dissertation, Hebrew University, Jerusalem.

BARFIELD, L.H., CREMSACHI, M. and CASTELLETTI, L., 1975. Stanziamento del vaso campaniforme a Sant'Ilario d'Enza (Reggio Emilia), Preistoria Alpina 11:155-199.

CHOI, D., 2003. Enhancing the power of Web search engines by means of fuzzy query. Decision Support Systems 35(1):31-44.

COCCHI GENICK, D. (ed.), 1999. Criteri di nomenclatura e terminologia inerente alla definizione delle forme vascolari. Atti del Convegno, Lido di Camaiore, 26-29 marzo 1998, Octavo, Firenze.

COPELAND, B.J., 1997. Vague identity and fuzzy logic. The Journal of Philosophy 94(10):514-534.

DOLAN, C., 1994. The magic of fuzzy logic. Global Investor 74:11.

FISHER, P., 2000. Sorites paradox and vague geographies. Fuzzy Sets and Systems 113(1):7-18.

HALL, A.V., 1997. A generalized taxon concept. Botanical Journal of the Linnean Society 125:169-180.

HARRIS, T.R., STODDARD, S.W. and BEZDEK, J.C., 1993. Application of fuzzy set clustering for regional typologies. Growth and Change 24(2):155-165.

HERMON, S. and NICCOLUCCI, F., 2002. Estimating Subjectivity of Typologists and Typological Classification with Fuzzy Logic. Archeologia e Calcolatori 13:217-232.

HERMON, S. and NICCOLUCCI, F., 2003. A Fuzzy Logic Approach to Typology in Archaeological Research. In Doerr, M. and Sarris, A., The Digital Heritage of Archaeology, Archive of Monuments and Publications, Athens:307-310.

HOURS, F., 1974. Remarques sur l'utilization de listes-types pour l'etude du Paleolithique superieur at de l'epipaleolithique du Levant. Paleorient 2:3-18.

KOSKO, B., 1993, Fuzzy Thinking: The New Science of Fuzzy Logic. Hyperion, New York.

LEONINI, V., 2003. La ceramica comune dell'Italia centro-settentrionale nella prospettiva di una provincia culturale europea. Università degli Studi di Pisa, Tesi di Dottorato.

MORACZEWSKI, I.R., 1993. Fuzzy logic for phytosociology II, Generalizations and predictions. Vegetatio 106(1):13-20.

MCNEILL, D. and FREIBERGER, P., 1993. Fuzzy Logic. Simone and Shuster, New York.

MUCINA, L., 1997. Classification of vegetation: past, present and future. Journal of Vegetation Science 8(6):751-760.

NEWCOMER, M., GRACE, R. and UNGER-HAMILTON, R., 1986. Investigating Microwear Polishes with Blind Tests. Journal of Archaeological Science 13:203-217.

NICCOLUCCI, F and HERMON, S., in press. La logica fuzzy e le sue applicazioni alla ricerca archeologica. Archeologia e Calcolatori 14.

NOVÁK, V., 1989. Fuzzy Sets and their Applications. Adam Hilger, Bristol.

OLANO, M., LOIDI, J.J., GONZÁLES, A. and ESCUDERO, A., 1998. Improving the interpretation of fuzzy partitions in vegetation science with constrained ordinations. Plant Ecology 134:113-118.

PAYNE, S., 1973. Kill-off patterns in sheep and goats: the mandibles from Asvan Kale. Anatolian Studies XXIII:281-303.

RAGIN, C.C., 2000. Fuzzy-Set Social Science, University of Chicago Press, Chicago.

RAPOSA, M., 1993. The Fuzzy Logic of Religious Discourse. American Journal of Semiotics 10:1-101.

ROBERTS, D.W., 1986. Ordinations on the basis of fuzzy set theory. Vegetatio 66:123-134.

ROBERTS, D.W., 1989. Fuzzy systems vegetation theory. Vegetatio 83:71-80.

ROSEN, S.A., 1997. Lithics After the Stone Age. Altamira Press, Walnut Creek.

ROSEN, S., HERMON, S., VARDI, J. and ABADI, Y., in press. The Chipped stone assemblage from Be'er Resisim: a preliminary study. In Gittin, S. and Wright, T., Festschrift in Honor of William G. Dever.

RUIZ, B., 1999. Science and philosophy. Diogenes 188:73-84.

SARTI, L., 1989. Per una tipologia della ceramica preistorica: appunti sullo studio morfologico dei manufatti. Rassegna di Archeologia 9:121-146.

SATTLER, R., 1996. Classical morphology and continuum morphology: opposition and continuum. Annals of Botany 78:577-581.

TIXIER, J., 1967. Procedes d'analyse et questions de terminologie dans l'étude des ensembles industriels du Paléolithique récent et de l'Epipaléolithique en Afrique du Nord-Ouest. In Bishop, W.W. and Clark, J.D., Background to Evolution in Africa, Chicago, The University Press of Chicago: 771-820.

TRIGGER, B.C., 1968. Beyond History: the Methods of Prehistory. Holt, Reinhart and Winston, New York.

VAN GIJN, A.L., 1990. The Wear and Tear of Flint: Principles of Functional Analysis Applied to Dutch Neolithic Assemblage. Analecta Preistorica Leidensia 22, Leiden.

WILLEY, G.R. and PHILLIPS, P., 1958. Method and Theory in American Archaeology. Chicago University Press, Chicago.

WHITTAKER, J.C., CAULKINS, D. and KAMP, K.A., 1998. Evaluating consistency in typology and classification. Journal of Archaeological Method and Theory 5(2):129-164.

ONTOLOGIES AS A REFERENCE FRAMEWORK FOR THE MANAGEMENT OF KNOWLEDGE IN THE ARCHAEOLOGICAL DOMAIN

OLEG MISSIKOFF
LUISS "GUIDO CARLI" UNIVERSITY
CENTRO DI RICERCA SUI SISTEMI INFORMATIVI
OMISSIKOFF@LUISS.IT

ABSTRACT

An ontology is a shared understanding of some domain of interest. An ontology entails some sort of world view with respect to a given domain. It contains a set of concepts (e.g., representing entities, attributes, processes), together with their definitions and their inter-relationships; this is also referred to as a conceptualisation. In other words, an ontology is an explicit, agreed specification about a shared conceptualisation.
An ontology may have different degrees of formality but, necessarily, it includes a vocabulary of terms with their meaning (definitions) and their relationships. According to [IDEF5], an ontology is a domain vocabulary containing a set of precise definitions, or axioms, that:
- provide the meaning of terms,
- enable a consistent interpretation of the terms defined in the vocabulary.
This document concerns the use of SymOntos (Symbolic Ontology System), a software prototype developed by LEKS (Laboratory for Enterprise Knowledge and Systems), at IASI-CNR in Rome, for the definition and management of archaeological domain ontologies.
The case study will be represented by the classification of Iron Age fibulae (brooches) from the cemetery of Quattro Fontanili near Veii (Rome).

INTRODUCTION

The Cultural Heritage (CH) sector is recently faced with new issues that need to be addressed quickly and competently. Among those we can list all problems related with the management, sharing and dissemination of the vast amount of digital resources now available, and the difficulties in transforming such data and information in knowledge.
For handling this ever increasing flow of information, supported by the diffusion of internet and ICTs, scholars are reconsidering studies on knowledge developed in the field of Artificial Intelligence (AI), almost 50 years ago. It is in fact in the summer of 1956 that Herbert Simon, Marvin Minsky, Claude Shannon and others meet at Dartmouth College of Hanover, New Hampshire, for discussing about the possibility of simulating human learning and reasoning processes by using rather recently invented machines: the computers. They think that time has come for this kind of studies to constitute an autonomous discipline; not without contrasts they give it the name of "Artificial Intelligence". AI provided the first, pioneer studies on knowledge and especially on its representation, producing a set of formalisms that would have paved the way for recent applications of Knowledge Management (KM) techniques in the business domain and for the state-of-art Semantic Web issues.

For more than thirty years, in fact, studies on knowledge have been confined in research labs, until the mid 1990s when the book by Nonaka and Takeushi, "The Knowledge-Creating Company" (Nonaka and Takeuchi 1995), opened to KM the doors of business and enterprise management domain. Through this contribution it became widely acknowledged that the competitive advantage of some of the world's leading companies was being carved out from those companies' knowledge assets such as competencies, customer relationships and innovation. Managing knowledge therefore, appeared a mainstream business objective as other companies sought to follow the market leaders. Technological solutions in mid 1990s, however, were still immature for permitting a successful and widespread implementation of KM techniques in enterprise management activities. After a few years of great interest, numerous theoretical publications and some failed implementation attempts, at the end of the nineties it looked as knowledge management was destined to be confined to the "management fad graveyard".

Meanwhile, the proliferation of contents and resources available on the internet has posed the problem of extracting meaningful information from an almost infinite repository: the world wide web. Again, a viable solution was spotted through the implementation of techniques and methods derived from the evolution of those AI pioneer studies on knowledge. This time the solution was called "The Semantic Web" and the proponent's name needs little presentation: Tim Berners-Lee (Berners-Lee, Hendler and Lassila 2001). The most remarkable advantages the semantic web should provide consist in the possibility to perform searches based on concepts instead of terms, therefore reducing the chances of confusion, and allowing software agents to carry out complex tasks for humans.

The Semantic Web, according to Berners-Lee, should substantially rely on well formed, interoperable and sharable contents. These conditions can be guaranteed by a recently developed knowledge organisation framework whose interest is rapidly growing in the academic community: Ontologies (Uschold and Gruninger 1996).

Archaeological Processes

The aim of this contribution is to provide CH scholars with a practical understanding of the basic principles of ontologies, and of the possible advantages deriving from their application in their domain.

Ontologies and their Use

This section will be devoted to a brief description of the nature of ontologies together with the advantages that could derive from their adoption as a domain definition reference framework.

An ontology is an explicit, agreed and shared definition of a portion of reality by means of a conceptual model. This model may exist in someone's head or be embedded in a software or information system, in an object or in a process. The task of an ontology builder is to identify the model and make it explicit. This allows the model to be accessed by, or communicated to, a wider range of potential users, be they people, organisations or software agents.

With respect to a thesaurus, an ontology aims at describing concepts, whereas a thesaurus aims at describing terms. An ontology can be seen as an enriched thesaurus where, besides the definitions of, and relationships among, terms of a given domain, more conceptual knowledge is represented. With respect to a Knowledge Base (KB), an ontology can be seen as a KB whose goal is limited to the description of the concepts necessary for modelling domains. A KB, in addition, includes the knowledge needed to model and elaborate a problem, or to answer to queries about a domain.

An ontology is composed of:

- a set of concepts (e.g., entities, attributes, processes) regarding a given domain
- the definitions ('conceptualization') of these concepts
- the relationships interconnecting entities within a given domain

Constructing an ontology implies a series of basic steps to be carried out, these are:

- examining the vocabulary that is used to describe the characteristic objects and processes of the domain
- developing rigorous definitions about the basic terms in that vocabulary
- characterizing the logical connections among those terms

For what concerns a practical use, at a higher level we can subdivide the space of uses for ontologies in the following four categories:

1. 'communication' and 'cooperation' among people
2. better institutions organization
3. 'interoperability' among systems
4. 'system engineering benefits'(reusability, reliability, specification)

Applications in the Cultural Heritage Domain

In spite of the fact that, as we all know, the CH domain is normally rather slow in responding to pressures from "the outside world", this time new factors make think that there will be a faster reaction. Apart from the already mentioned Semantic Web and the need to organise the growing amount of data available, there is an increasing effort in the conversion of traditional cultural heritage resources in digital format.[1] These activities, strongly fostered by the European Commission in 1999 with the *eEurope* initiative[2] and in 2001 with the Lund meeting,[3] are calling for new researches on the conceptual organisation of digital resources that are already available or will be soon produced.

In the CH domain, an extensive contribution has been provided by the International Committee for Documentation of the International Council of Museums (ICOM-CIDOC), and is represented by the CIDOC CRM (Conceptual Reference Model). This model provides definitions and a formal structure for describing the implicit and explicit concepts and relationships used in cultural heritage documentation:

"The CIDOC CRM is intended to provide a common language for domain experts and implementers to formulate requirements for cultural heritage information systems and to serve as a guide for good practice of conceptual modelling. In this way, it can provide the "semantic glue" needed to mediate between different sources of cultural heritage information, such as that published by museums, libraries and archives".[3]

Another valuable source of information has been provided by the "DigiCULT" FP5 IST Support Measure with its Thematic Issue No.3: "Towards a Semantic Web for Heritage Resources".[4]

In the next chapter, as a practical example, a case study will be presented: that is the classification of iron age fibulae (brooches), from the proto-historical cemetery of Quattro Fontanili near Veio (Rome).

The Methodological Foundation

This case study derives from a thesis dissertation concerning the classification of a set of archaeological objects, by means of a methodology for conceptual modelling developed in the Institute for Systems Analysis and Informatics (IASI) of the Italian National Research Council (CNR). At present a new prototype tool for symbolic ontology management, XML based, called *SymOntoX*[6] is being developed from a previous version (called Mosaico (Missikoff 1996)), and the knowledge base is being imported in this new tool.

SymOntoX is an Ontology Management System designed for supporting the construction and management of domain ontologies. It allows to manage several ontologies and supports different kinds of users with different access rights, it has multilingual capabilities and permits remote access through the Internet. SymOntoX is based on the OPAL (Object,

Process, and Actor Language) metamodel for knowledge representation (Missikoff and Taglino 2002).

A metamodel is a set of definitions and rules which allow the definition of a modelling language. In this context the metamodel allows to specify the features to be filled and the rules to be followed during the definition of an ontology. The two main mechanisms used for classifying the material are the structural specification and the hierarchical organisation. The type structural specification consists in supplying the list of properties and, for each of them, the type corresponding to the values that they can assume.

<type-name> := [<prop-name> : <prop-type>
<prop-name> : <prop-type>]

In a Knowledge Schema, it is possible to organize the type definition within an ISA hierarchy. That is basically a generalization/specialization relationship between types. For instance we can declare that: "cup ISA vase", intuitively this statement shows that all the characteristics of vase are encountered in cup as well. Besides that, the latter could have additional characteristics which are not necessarily encountered in all the vases. This principle is often referred to as principle of inheritance because the type cup inherits all the characteristics of the type vase.

In the extensional level the ISA relationship turns into an inclusion relationship between classes. The example shows that the class of cups is contained in the class of vases. These qualitative considerations are rigorously described by the system, through strict criteria that guide the building of ISA hierarchies.

The ISA relationship implies that the type being defined be a specialization of the types appearing under ISA. Beside that, the principle of inheritance is also used to obtain a more compact schema description. Inheritance can be single or multiple, if in the ISA construction appear one or more supertypes. Instead we talk of absolute inheritance when the properties of the supertype are inherited without being modified.

Having given a type, the creation of a subtype is performed through specialization. The mechanisms of specialization must be always respected in defining a type using the ISA construct. Those mechanisms are of two basic sorts, specialization by specification and specialization by restriction:

- Specialization by specification. This mechanism of specialization requires the addition of new properties to those already defined in the supertypes (which, as stated above, are inherited by the subtype). If the supertypes are two or more and have properties in common, inconsistencies can arise. This is a critic point of the multiple inheritance, the problem has been already faced in the literature and there are different ways to solve it, but their description goes beyond the purposes of this work.

- Specialization by restriction. The mechanism of restriction allows to refine in the subtype one or more properties already defined in the supertype (this mechanism is called overriding). The overriding is performed essentially operating on the two property typing tools mentioned above: the explicit listing of allowed property values in the case of categorical variables and/or the range of allowed property values in the case of continuous variables.

THE CASE STUDY

The present work aims at describing the operational steps followed in analysing a group of artefacts which have traditionally presented relevant taxonomic problems. The case study is represented by the conceptual modelling of the "Fibula" type (en. brooch), and of some of its specializations, starting from the analysis of materials from surveys and excavations in the area of the villanovan cemetery of "Quattro Fontanili" near Veii (Southern Etruria). According to the Dizionari Terminologici of the Italian Institute for Cataloguing and Documentation (ICCD), the fibula is an "object of ornament used to fix parts of clothes. It is constituted by a bow ("arco" in Italian), a needle ("ago" in Italian) generally connected to the bow by means of a spring ("molla" in Italian), and a catch ("staffa" in Italian) in which the point of the needle is inserted."

A special interest for these materials consists in the fact that they have been previously and thoroughly analysed (Close Brooks 1965:53pp, Guidi 1993, Toms 1986:41pp), thus giving a chance for useful comparisons. The present classification has been performed starting from the pioneer work of J. Sundwall in her Die alteren italienischen Fibeln, published in 1943, and developed according to a hierarchy of main attributes that generates, as a consequence, a hierarchy of types progressively more specialized. The hierarchy of main attributes is the following: 1) shape of the bow (arco), 2) shape of the catch (staffa), 3) decoration.

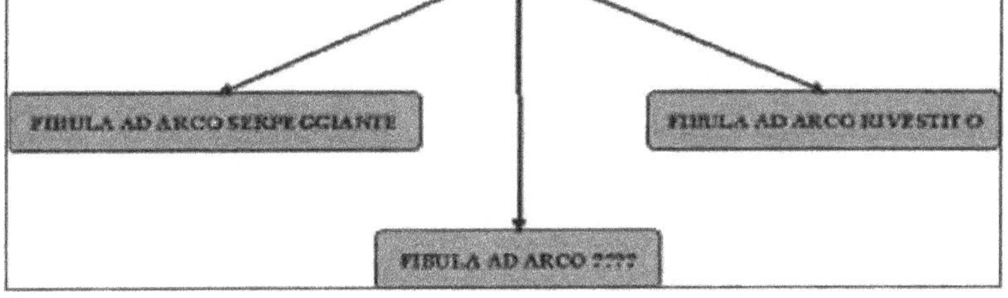

Figure 1

The use of SymOntoX and the OPAL methodology has, already at this level, stimulated a series of observations that have led to a partial redefinition of the hierarchical organisation for the Fibulae class. In fact, even if we can substantially agree in the definition of a first level based on the artefact function, that of fixing parts of clothes, in the definition of the second level a first problem arises. According to the traditional hierarchical organisation, the type "Fibula" is split, at the second level, in three subtypes: 1) Fibule ad arco serpeggiante, 2) Fibule ad arco rivestito, 3) Fibule ad arco. From a first intuitive observation it is possible to note that in cases 1 and 2 the "arco" is refined by the attributes "serpeggiante" and "rivestito", whereas in case 3 it is not refined at all. (Fig.1)

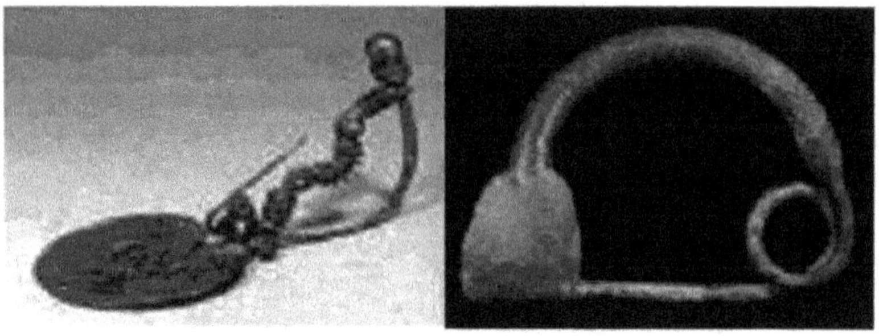

Figure 2a fibula serpeggiante Figure 2b fibula ad arco

In a mathematical/logic context and according to the classification principles previously outlined, this automatically assigns the "Fibula ad arco" entity at a more generalised level. Furthermore, from a morphological point of view, to define the body of a "Fibula serpeggiante" as a "bow" can be misleading, being rather far from a standard definition of bow or arch that can be found in dictionaries or encyclopaedias which refer always to a curved line.

This observation is reinforced by the fact that, in archaeological deposits, the "Fibule serpeggianti" show a well defined social meaning being found exclusively in male burials, whereas the "Fibule ad arco" are characteristic of female burials. Having considered the above issues, we propose an organisation of the second level, defined as "generic shape", that gathers the concepts of "Fibula serpeggiante" (Fig.2a) and "Fibula ad arco" (Fig.2b). This last concept is further refined at the third level in "Fibula ad arco rivestito" and "Fibula ad arco decorato".

Figure 3

In the third level it is therefore defined the decoration technique of the "Fibule ad arco"; here the element of distinction is represented by the type of decoration chosen for enriching the objects.

This brief example clearly demonstrates that classifications performed on an intuitive base, without relying on a formal theoretical foundation, could easily lead to inconsistencies and produce an output which shows evident weaknesses from a logical point of view. The construction of domain ontologies can certainly provide a viable solution for these problems but, so far, it is evident a lack of real case studies on which stimulate a debate. The call for both scientific communities (artificial intelligence and archaeology) is to intensify collaborations for testing methodologies and tools on real case studies and, in the long run, forming "hybrid" researchers that could rely on a more integrated background.

Conclusion

In this contribution it has been proposed an ontological analysis of a real case study as general conceptual reference framework oriented towards the support in the definition of entities in the archaeological domain. Using ontologies could also facilitate the knowledge sharing among researchers for guaranteeing a more extensive reusability of results from scientific researches. For performing the analysis it has been used "SymOntoX", an ontology management system, XML based, developed in the LEKS (Laboratory for Enterprise Knowledge Systems) of the IASI/CNR.

1 See http://www.minervaeurope.org.

2 See http://www.cordis.lu/ist/ka3/digicult/eeurope-overview.htm.

3 See http://cidoc.ics.forth.gr.

4 See http://www.digicult.info/downloads/ti3_high.pdf.

5 See http://www.symontox.org.

REFERENCES

BERNERS-LEE, T., HENDLER, J. and LASSILA, O., 2001. The Semantic Web. Scientific American May.

BERTOLAZZI, P., KRUSICH, C. and MISSIKOFF, M., 2001. An approach to the definition of a core enterprise ontology: CEO. In Luiss Edizioni (ed.), Open Enterprise Solutions: Systems, Experiences, and Organizations Workshop, Rome.

CLOSE BROOKS, J., 1965. Proposta per una suddivisione in fasi della necropoli veiente di Quattro Fontanili - Notizie degli Scavi:53pp.

GUIDI, A., 1993. La necropoli veiente dei Quattro Fontanili. Firenze.

GRUBER, T., 1993. A Translational Approach to Portable Ontologies. Knowledge Acquisition Vol. 5, no. 2.

MISSIKOFF, M. and TAGLINO, F., 2002. An ontology approach for Business and Enterprise modelling with SymOntoX. AICA XL Annual Conference, Bari.

MISSIKOFF, O., 1996. Application of an Object-Oriented approach to the formalization of qualitative and quantitative data. In Analecta Praehistorica Leidensia, Publications of the Institute of Prehistory, University of Leiden.

NONAKA, I. and TAKEUCHI, H., 1995. The Knowledge-Creating Company. Oxford University Press.

TOMS, J., 1986. The relative chronology of the Villanovan Cemetery of Quattro Fontanili at Veii. In AION Vol. 7:41pp.

USCHOLD, M. and GRUNINGER, M., 1996. Ontologies: Principles, Methods and Applications. The Knowledge Engineering Review Vol. 11, no. 2.

The Archaeological Prospection in the North-Western Caspian Region

Maria Otchir-Goriaeva
Kalmyk Inst. of Social-Economical and Legal Research, Elista, Russia

Abstract

Till today the Bronze and Iron Age Cultures of the area of interest were investigated using only the data of burials. Thousands of the burial mounds - kurgans have been excavated, the materials have been published and explored. The issues of the burial rite, dating back, social organisation, economy system and way of life of the ancient population have been mostly cleared. At the same time the numerous settlements that, due to the small portions and the absence or very thin cultural layers were traditionally considered as seasonal settlements of cattle breeders, remain virtually unexplored. The paper is devoted the first results of the archaeological settlement survey in the North-Western Caspian. The aerial photography, digital maps were used for the searching of the settlements in the steppe, and for finding the conformity to natural laws of the distribution burial mounds groups and living places in different landscape environment.

Currently, archaeological evidence for the Bronze Age of the Eurasian Steppe between the Carpathians and the Urals is represented mostly by burial mounds - kurgans. The quantity of these kurgans, which have been excavated during many years, is very large, and comes down to many thousands. The grave goods uncovered from those kurgans are very numerous, well preserved, and often undisturbed. Many of them are objects of decorative art and often are made of silver and gold. Some burial mounds have large sizes, and provide a nice possibility for reconstructions. The settlements in the steppe are not as attractive as the kurgans. Very seldom do they possess an intact cultural layer; most often the cultural layer is destroyed, and the finds are not so valuable, as those in the kurgans. Traditionally, in Russian archaeology the steppe settlements are considered to be a defective archaeological source with a little potential for valuable scientific information. They are considered as such when compared with the settlements from other geographical regions, where the sites possess thick cultural layers, remains of dwellings and other numerous finds. On the basis of very superficial investigations, and without any special analysis, many steppe settlements were interpreted as seasonal nomadic camps. In many important works and monographs archaeological material from settlements was neglected: sometimes only listed, but not analyzed and interpreted. As a result, the questions pertaining to chronology, social organisation, economic system and way of life of ancient populations of the steppe have been addressed mostly on the basis of the data derived exclusively from the burials. According to this tradition, which is still predominant in Russian archaeology, the Bronze Age inhabitants of the steppe between the Black Sea and the Urals should be interpreted as nomadic cattle breeders similar to those of the Early Iron Age.

The investigations of some leading scholars were concerned with the problem of the origin of the nomadism in the Eurasian Steppe between Black Sea and Urals, who used the evidence from the Bronze Age archaeological sites. This discussion still continues for over fifty years. It was initiated by F. Hanchar in his famous book "Das Pferd in Prähistorischer Zeit" published in 1955, in which he had come to a conclusion that "the open steppe from the Pont to the Caspian sea" was settled by the first nomads in this region. Hanchar called that part of the steppe "open" because the eastern part of the East European Steppe is not restricted by the mountains, as is its western portion skirted by the Carpathians. Hanchar suggested that the first nomads appeared in the Late Bronze Age and ascribed them to the Srubnaya Culture. Later this idea was developed theoretically by V. Shilov, N. Merpert in their fundamental works. They have extended the chronological marker for the beginning of the nomadism to the Early and Middle Bronze Age - to the corresponding Yamnaya and Catacomb Cultures (Merpert 1974, Shilov 1964, 1975, 1975a, 1985).

However, there is a serious problem with this long-accepted model of socio-economic organization of the ancient steppe populations. This problem is rooted in the tradition of interpreting ancient societies only on the bases of the burial material and living behind the materials from the so-called seasonal settlements and small habitation places.

As a matter of fact, the number of the steppe settlements is not so scarce. According to S. Pustovalov (1994), there are 245 habitation places in the Black Sea area. Out of these, 7 settlements in the Southern Bug region and 11 in the Lower Dnepr region have substantial cultural layers; and 91 settlement sites have no cultural layers. The thickness of these cultural layers varies up to 0.2 m and the area up to 100 m2 (Pustovalov 1994:86-104). V. Shilov has defined approximately 200 settlement sites in the Lower Volga, which mostly contained traces of light dwellings, ash, and numerous fragments of ceramic and animal bones (Shilov 1985:27).

The hypothesis of the emergence of nomadism during the Bronze Age was considered especially applicable to one part of the East European Steppe - the right bank of the Lower Volga. This region was noted by F. Hanchar as the driest part of the "open steppe" with the prevalence of the semi-desert landscapes. This region, namely the North-West Caspian or the Volga-Manych Steppe, by and large belongs to the modern Republic of Kalmykia in Russian Federation.

The investigations conducted so far have neglected factors that are closely related to the environmental characteristics of the steppe land. The Mesolithic and Neolithic settlements in the steppe seldom have any cultural layers, and they are also represented by small sites. For example, out of the few dozens of the Neolithic sites, which have been investigated on the territory of the Republic of Kalmykia, only three settlements have substantial cultural layer (Kolzov 1988:4, 12-14). Not a single Mesolithic site with a cultural layer has been identified there so far (Sinizyn 1933, Minaeva 1955:46-53, Praslov 1971:102-107, Malov 1989:25-37, Kolzov 1982:100-107). The Stone Age population of the Mesolithic and Neolithic Periods has never been considered as that having a nomadic way of live. I think this is a problem that has to do with the preservation of the cultural layers there. The steppe landscape is characterized by the slow soil formation properties, which include the scarcity of grass growth and strong wind erosion. In the pre-industrial times for the population, which inhabited the steppe with its scarcity of natural resources (water-grass), the most reasonable kind of settlement was a small village, containing only a few houses, the so-called "hutor" (russ). Because of such factors, the state of the preservation of such cultural layers is affected. Hence methods of field archaeology and criteria for their interpretation must be different from those used in other geographical areas (for example forest zones).

This year an archaeological Prospection project has been organized, which for the first time aimed specifically at finding settlements and occupation sites of the Bronze and Early Iron Age Periods on the territory of the Kalmyk Republic (75,9 thousands sq.km). The region of interest included three types of landscape, those from the Caspian Lowland, Ergeni Hills, and Kuma-Manych Lowland. The Ergeni Hills are the young part of the East European Plain, which are extended from North to South and are complicated by a hollow-ravines network. The hollows are extended in the lateral directions and are closely connected with the underground water system.

The Caspian Lowland is slopping from the North-West to the East-South. The elevation marks are + 12-15 to the -27 m. In the northern part of the Caspian Lowland there are few systems of valleys. The largest of them are extended along the Sarpa Lake chain. The rest of the Caspian Lowland has a regular system of limans, which are the traces of small lakes remaining from the last regression of the Caspian Sea. All these limans and limanart lowlands contain fresh water. Usually the size of the limans ranges from 0,5 to 1 km in width, and up to 3 km in length (Tashniniva 2000:12-29).

The Kuma-Manych Lowland is the valley of three rivers - the West Manych, East Manych and Kuma. The Kuma-Manych Depresion is extended from North-West to South-East. The width of the Lowland is varied from 2=3 km to 30 km. The biggest salt lake there is Manych-Gudilo (Tashninova 2000:12-29).

Our expedition conducted the archaeological prospection in the first two zones-the Ergeni Hills and the Caspian Lowland. The first question was where we should start our survey. We began to carefully investigate the maps (from 1:100,000 till 1:25,000). We have looked the space Landsat photos in searching some traces of the ancient settlements. The traces of the settlements which were leaved by population during second world war we could see very clear , but nothing more ancient. We paid attention to places where the water springs and steppe lakes or rivers were. From them we have chosen a few places, where nearby the numerous burial mounds groups of the Bronze Age were excavated. These micro regions were carefully investigated with using of the hundreds of the aerial photos, modern digital maps such as soil, water and erosion maps and two old maps, dating back last century (Using of the Land map and Water map). We tried to find the conformity to natural laws of the distribution of the burials and living places in different landscape environment. So we have learnt a lot before the beginning of the survey.

During field work we have found the rests of the four settlements, two sites with the re-deposited (due to the wind erosion) layer and numerous sites with ceramics. It is possible to distinguish three localization types. The first type of the settlement localization is characteristic of the Ergeni Hills. The sites were found on the northern bank of a small river at the bottom of the ravine, extended in the latitudinal direction. The second type is characteristic of the Sarpa Lake chain. Three settlements were found on the western shore of the Khanata Lake, and two sites on the western shore of the Tsagan-Nur Lake, which extended in longitude direction.

The last, third type, is characteristic of the Caspian Lowland, and located near by the small limans or lakes, which are spread here as a spots. So we found that the distribution of the settlement types reflects the peculiarities of the geographical distribution of the water systems meanwhile the burial mounds groups mostly follow the elevated places and hills. This result is natural and not really new. Other thing is more important, however-all our settlements are located very close to the modern water-level. They were settled within the modern ecoton zone (botanical term) - the narrow zone in which vegetation is special and differs from the surrounding flora (Otchir-Goriaeva 2002:attachment 4, Ulanova 2002: 136-140). This shows that climate and other geographic conditions, such as availability of fresh water resources, have not changed so much from those times.

There is also another peculiarity of the geographic distribution of these ancient settlements. They are located also in the proximity to the modern villages. It appears that the places suitable for settlement were not so numerous at the steppe. Maybe the best ancient settlements were destroyed by modern villages, which as a rule cover a large territory.

At the Ergeni Hills, we investigated two deep ravines-both are located near the Russian agricultural villages. The first one, Bulgun Sala, was just a few kilometres away from the capital of the Republic - Elista. Elista is located in the next deep ravine and was found by the Russian colonists in 19th century too.

The second ravine located in the central part of Ergeni contained a small river Shar Elsin. Here we found two sites with some prehistoric ceramics. Both are located not so far from the villages which were found during the second half of the 19th century.

In the southern part of Ergeni, we have investigated a large hill called Hamur with the highest point of 213m. This hill contained numerous burial mounds and numerousl springs on the hill slope. During 1937, two Stone Age settlements had been identified there, on the top of the hill, but now we could not find any traces of them-during the last 60 years their traces have disappeared. The location of these Neolithic settlements, however, suggest that during the Neolithic Period the water level was higher than now (Sinizyn 1937:15).

In the southern part of the Caspian Lowland we found numerous places with ceramics. One site was probably a destroyed grave. Two sites were settlements with the destroyed cultural layers. The settlement near the dried lake of Beloe Ozero has contained ceramics of the Catacomb Culture of the Bronze Age, and the settlement near the village of Artezian contained some prehistoric and medieval ceramics.

At the west coast of the Lake Zahanata, we found two settlements which were located very close to each other, with about 997 m between them. We made soundings at both sites (Fig.1).

The northern settlement can be dated back to the Late Bronze Age Culture - the Srubnaya one, and the second site - to the Middle Bronze Age - the Catacomb Culture. Both sites have cultural layers (15-40 cm) with ceramic and animal bones. For northern settlement it was possible to identify the animal bones as sheep, horse, cow and pig (wild?) bones (Otchir-Goriaeva 2002:attachment 2). A third settlement was found about one kilometre to the South, but it was partially destroyed by the tillage. On the surface we recorded two charcoals, a lot of ceramics, and some stone tools. These date back to the Srubnaya Culture. A fourth settlement at the lake Hanata, was located further to the south, but unfortunately, it was utterly destroyed during the construction of a dam. We have found numerous fragments of ceramics and stone tools, as well as burned animal bones. It should be outlined that these archaeological settlements are located on the place where before second world war the kalmyk sedentary village was placed.

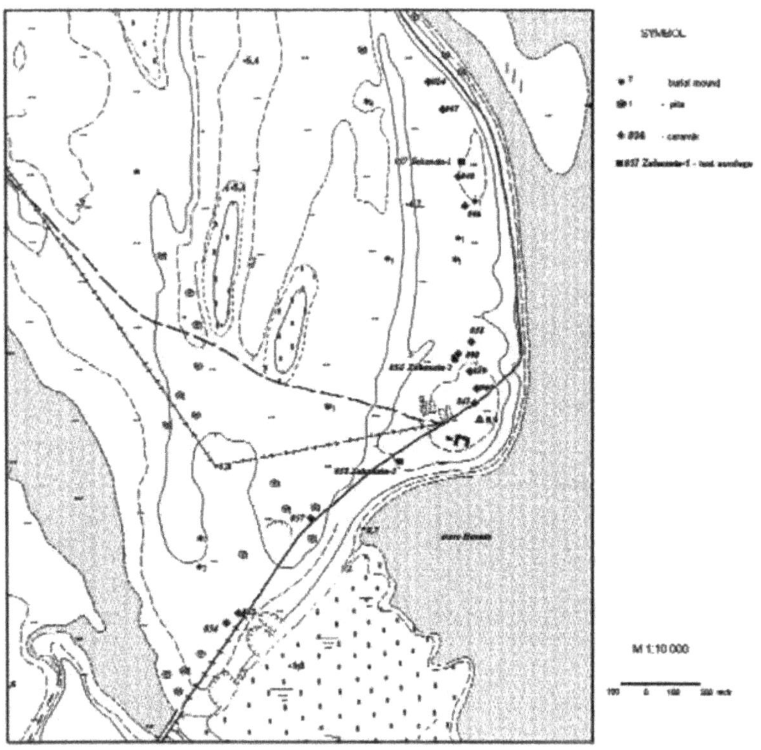

Figure 1 Zahanata

The problem of underestimation and undervalue of the settlement-sites in this region, their treatment as a defective archaeological source, and their interpretation as seasonal camps, is not only a problem of academic research. This is also a problem of the defunct Heritage Management. From 1929 until 1979 (for 24 seasons) 828 burial mounds-kurgans were excavated in the territory of the Kalmyk Republic. These produced 2630 graves (Tsutshin 1985). During the following 18 seasons (1980-1998 years) 470 kurgans more had been excavated which produced 1229 additional graves (Otchir-Goriaeva 2003). By and large, these were uncovered during various rescue excavations. During this time, however, only one settlement site had been excavated (belonging to the Neolithic Period) - settlement "Djangar (Kolzov 1988).

ACKNOWLEDGEMENT

The Project was supported by Eurasian Department of the German Archaeological Institute and Alexander von Humboldt Foundation. I would like to thank Dr. Ursula Brosseder (EA DAI, Berlin) for the creative cooperation at the start of the field work.

REFERENCES

HANCHAR, F., 1955. Das Pferd in prähistorischer und früher historischer Zeit. Wien.

KOLZOV, P.M., 1988. Neolit severo-zapadnogo Prikaspiya. Avtoreferat na soiskanie uchenoistepeni kandidata istoricheskih nauk. Moskva.

KOLZOV, P.M., 1982. Rezultaty arheologicheskih rasvedok 1976-1989 gg na territorii Kalmyzkoi ASSR. Pamyatniki Kalmykii kamennogo I bronsovogo vekov. Elista:100-107.

MALOV, N.M., 1989. Otchet ob issledovaniyah v Kalmyzkoi ASSR, v Saratovskoi I Volgogradskoi oblastyah v 1989 godu. Arhiv IA RAN, R-I, 13669.

MERPERT, N.YA., 1974. Drevneishie skotovody Volzsko-Uralskogo mezdurechya. Moskva.

MINAEVA, T.M., 1955. Stoyanka s mikroliticheskim inventarem na chernyh zemlyah. Kratkie Soobshcheniya Instituta Arheologii (KSIA), Vyp.59:46-53.

OTCHIR-GORIAEVA, M.A., 2002. Otchet ob arheologicheskih issledovaniyah na territorii Respubliki Kalmykiya v 2002 godu. Arhiv KIGI RAN.

OTCHIR-GORIAEVA, M.A., 2003. Arheologichaskie issledovaniya na territorii Respublika Kalmykia (1980-2000gg). Vestnik Instituta. Elista. In print.

PRASLOV, N.D., 1971. Pamyatniki kamennogo veka Yuznyh Ergenei. Kratkie Soobshcheniya Instituta Arheologii (KSIA) 126:102-107.

PUSTOVALOV, S., 1994. Economy and Social Organisation of Northern Pontic Steppe - Forest- Steppe pastoral populations 2750-2000 BC. Baltic-Pontic Studies, Vol.2:86-134.

SINIZYN, I.V., 1933. Drevnie pamyatniki Promorskogo raiona Kalmoblasti. Izvestiya Saratovskogo Nizne-Volzskogo Instituta Kraevedenia im.M. Gorkogo. Tom VI. Saratov.

SINIZYN, I.V., 1937. Otchet ob archeologicheskih raskopkah v Kalmyzkoi Narodnoi Respublike v 1937 godu. Arhiv IIMK. F 2, N 281.

SHILOV, V.P., 1964. Problemy osvoeniya stepei Niznego Povolzya v epohu bronzy. Arheologicheskii Sbornik Gos.Ermitaza (ASGE):86-102.

SHILOV, V.P., 1975. Ocherki po istorii drevnih plemen Niznego Povolzya. Moskva.

SHILOV, V.P., 1975a. Modeli skotovodcheskih hozyastv stepnyh oblastei Evrasii v epohu eneolita I rannego bronzoogo veka, CA. 1:5-15.

SHILOV, V.P., 1985. Problemy proishozdeniya kochevogo skotovodstva v Vostochnoi Evrope. Drevnosti Kalmykii:23-34.

TASHNINIVA, L.N., 2000. Krasnaya kniga pochv I ekosistem Kalmykii. Elista.

TSUTSKIN, E.V., 1985. Arheologicheskie issledovaniya Kalmykii. Drevnosti Kalmykii:3-18.

ULANOVA , S.S., 2002. Raspredelenie rastitelnyh soobzestv ekotonnoi zony "voda-susha" na territorii Kalmykii. Vestnik Instituta. Elista:136-140.

The Western Papaguería from the Air: Digital Imagery Using Kite and Balloon Aerial Photography

SEE THE CD FOR THE EXTENDED VERSION

K. Ahmet

Statistical Research, Inc., Redlands, California, USA

Abstract

Aerial photography has been a useful tool for archaeologists in identifying and recording sites since the early 1900s. More recently, kites and helium balloons have been used as a cheaper and more environmentally friendly alternative than airplanes. With the rapid development of compact, high-quality digital cameras, kite and balloon aerial photography has been pushing new boundaries, allowing photographs to be viewed immediately, and, more importantly, to be rectified using simple software packages. These techniques have been applied to recent surveys of the Papaguería in southwestern Arizona, USA, where geoglyphs, farmsteads, and other prehistoric rock features have been recorded digitally from the air. The photographs were then rectified and digitized as line drawings for publication. This paper explains the equipment, field methods, and results from recent fieldwork by Statistical Research, Inc. Most importantly, it shows how spectacular results can be achieved with minimal costs by university departments and cultural resource centers alike.

Introduction

Aerial photography has been a useful tool for archaeologists in identifying and recording sites since the early twentieth century. The main reason for this is that an overall view of a site, structure or feature can be gained from the air, and more importantly, can provide a more detailed view of the regional landscape. Much of this detail may not be appreciated or detected to an observer on the ground.

This paper gives a brief history of kite and balloon aerial photography and then describes the methods employed by Statistical Research, Inc. (SRI), in recent archaeological surveys in the Papaguería in southwestern Arizona, USA. Some results from that work are then presented, and, to conclude, the potential for aerial exploration in the future is discussed.

Recent Developments

The development of aerial photography in archaeology has been dominated by photos taken from airplanes. Other techniques have also been used since the last century. Reeves showed a photo of a balloon used for aerial photography in the 1930s by the Oriental Institute of the University of Chicago (Reeves 1936:Plate 4, Fig.2) and mentioned the first use of balloons to lift cameras by Major Elsdale of the British Army in 1880-1887. The first use of kite aerial photography (KAP) for an archaeological project is attributed to Henry S. Welcome, who used a kite system successfully on the Jebel Moya in the Sudan in 1911.

More recently, kite and balloon aerial photography have emerged as a cheaper and more environmentally friendly alternative. In the last 25 years, there has been resurgence in the use of KAP in archaeology. These techniques were used in the desert environment in Syria and Egypt (Anderson 1979) and during surveys of Libyan farms from the Roman period (Allen 1980). These early systems used an SLR camera rig that was attached to the kite line after the kite had already reached a stable height. As more line was let out, the camera would rise high above the feature or site. These early systems had to be fitted with a motor drive, which made them heavy and cumbersome to use in the field. The relative infancy of this technique was demonstrated in 1987, when Riley

Figure 1 Pit houses, showing four different occupation phases and construction types; Area D at Mescal Wash, Tucson, Arizona

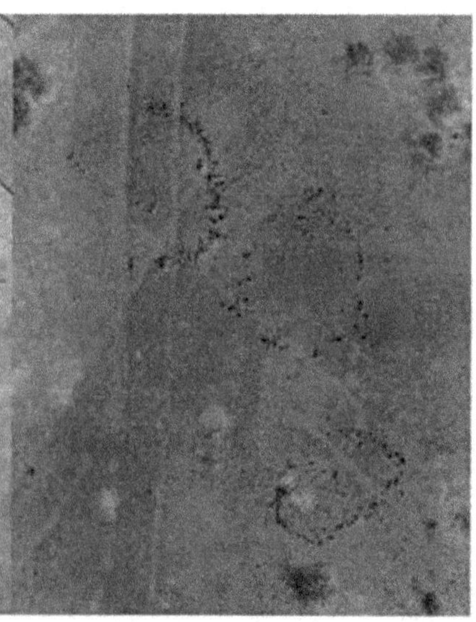

Figure 2 Aerial photograph of geoglyphs, BMGR, Arizona

made only a passing reference to the use of kites and balloons in his methods section of Air Photography and Archaeology (Riley 1987), which is dominated by photographs taken from the airplanes.

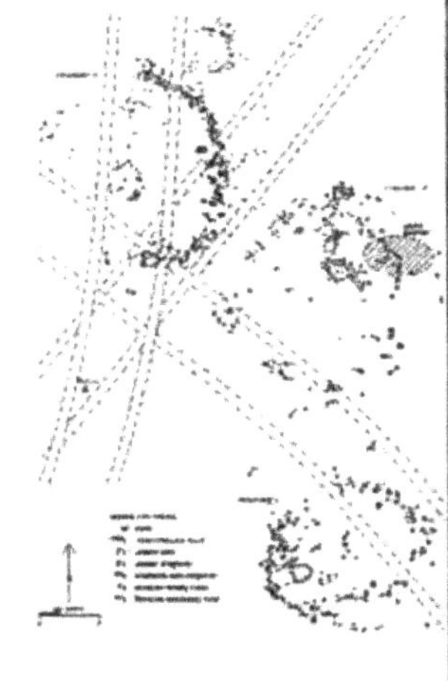

Figure 3 Line drawing of geoglyphs by SWCA, Inc., BMGR, Arizona

Figure 4 Close-up of petroglyph, BMGR, Arizona

With the rapid development of compact, high-quality digital cameras, KAP has been elevated to new heights, allowing lighter rigs to be built and producing photographs that can be viewed immediately and, more importantly, can be rectified using simple software packages. Kites and balloons have been used successfully on archaeological projects when conventional photos would be difficult, costly, or dangerous to obtain. These techniques have been successfully used by Drs. Marzolff and Ries at Johann Wolfgang Goethe University, Germany, to map and interpret erosional patterns that develop in semi-arid environments in western Africa. The team uses a kite for aerial photography when wind conditions are too strong for the hot-air balloon. KAP was used for environmental site assessment, in conjunction with geographic positioning system (GPS) equipment, to locate survey markers on the ground while mapping fluvial landforms south of the Ninnescah River in Kansas (Aber et al. 1999). Topographic maps based on photogrammetric principles have been created with KAP using stereoscopic pairs to create three-dimensional maps (Warner 1996). Kites have also been used for digital photography by Japanese research laboratories during archaeological surveys in the Philippines (Murooka 1998). A model for SRI's work has been the recent work of Bernard-Noël Chagny in the Sudan, where photos of archaeological sites, taken by kite or tethered balloon, have been rectified and incorporated into a geographic information system (GIS) or converted into stereo pairs (Chagny 1994, 2001).

Kite Versus Balloon

There are many practical problems with KAP, but the most prominent is the dependence on favorable wind conditions. The use of KAP in the field cannot be planned ahead of time because there might be light winds, which will not lift the kite, or strong and gusty winds, which make the kite impossible to control. For this reason, SRI uses a multiflare kite, which is stable and has excellent lift, and has adapted a helium balloon to lift a camera cradle in windless or low-wind conditions. Experiments using both techniques have been very successful. Our design goals were to produce a cheap, high-resolution aerial photographic system based on a cradle that could be interchanged with a kite and balloon as weather conditions changed, and that could be set up easily and used swiftly in the field to document multiple sites or features within a day. We wanted to use a helium balloon with adequate lift but with a small enough diameter to be filled with a single 1.5-m (5-foot) helium tank for easy transportation. The system also had to be easily adaptable from the wide-open desert to sites in an urban setting. Finally, we hoped to use the system not only to document sites or features from the air, but also to produce rectified photographs that could be converted to line drawings for publication or included in reports as scaled photographs. More importantly, we hoped to integrate our aerial photographs into a geographic information system (GIS) so that future monitoring of site or feature conditions could be carried out more effectively. We now routinely use these techniques in our projects, some of which are described below.

Results

The deserts of the American Southwest provide an ideal setting for aerial photography. The use of low-level aerial photography in the desert environment, and more specifically, with regard to hunter-gatherer sites, had been proposed as early as 1958 by Meighan et al., who cited contemporary experimentation with KAP and its use in archaeology by Bascom (1941) and Roy (1954). Photographs from higher elevations have been used successfully to document prehistoric irrigation systems along the Gila River but fail to pick up smaller features in the landscape. SRI's system is geared more to documenting such low-relief sites and features as trails, rock rings, pit houses, geoglyphs, and intaglios. These features are ideal for low-elevation aerial photography because they rarely rise higher than 30 cm above ground level, causing less distortion when correcting aerial views, as opposed to walled structures like mesas.

With our two systems in place, SRI can now carry out aerial reconnaissance in most weather conditions. KAP can be effectively used in conditions where winds are greater than 16 kph (10 mph), whereas balloons are suited to calmer conditions. SRI's kite system was first tested at Mescal Wash, near Tucson, Arizona, during excavations of pit houses (Fig.1). Because weather conditions were unsuitable for kiting, however, the helium balloon was used for all other aerial photographs shown.

The set-up procedure for a 2.1-m- (7-foot-) diameter helium balloon is relatively simple. First, it is anchored and filled with helium. A simply constructed cradle that houses the digital camera is then attached to the bottom of the balloon. Using a remotely controlled trigger mechanism, the camera can be operated from the ground within line-of-site range. The balloon can remain inflated for long periods and can be transported to different sites on the back of a pickup truck. If wind speeds increase unfavorably, we simply switch the cradle to the kite. On the U.S. Army's Yuma Proving Ground (YPG), SRI used both systems successfully on the same day. One of the main advantages of aerial photography is the ability to take shots of the same object at different points in time, as a monitoring tool. This repeatability allows environmental agencies to monitor the condition of sites and to measure impacts over time and also allows them to prioritize which sites or larger areas need protection. The advantage of carrying out aerial surveys from the ground, as opposed to from a plane, is that the ground crew can record those site components not visible from the air. Artifact concentrations, site boundaries, and isolated artifacts can be recorded by ground crews in combination with the aerial photographic process, rather than after it is completed.

One of SRI's first experiments with the system involved a group of three geoglyphs originally recorded by SWCA, Inc., on the Barry M. Goldwater Range (BMGR), in southern Arizona. These features were chosen for aerial survey to see if any further impacts had occurred at the site since it was last recorded. First, a photo was taken to capture all the features. It was then rotated and cropped to match the line drawing done by SWCA to monitor the condition of the site (Figs. 2 and 3).

An obvious advantage in using our system is the ability to take photographs both at close range and at higher elevations. This flexibility is best demonstrated on a trail that runs east-west that is also located on the BMGR. At ground level, the trail was barely noticeable except for a number of petroglyphs that lined the trail. Figures 4-6 show the relative positions of the petroglyph to the trail, using aerial photographs taken from three different elevations.

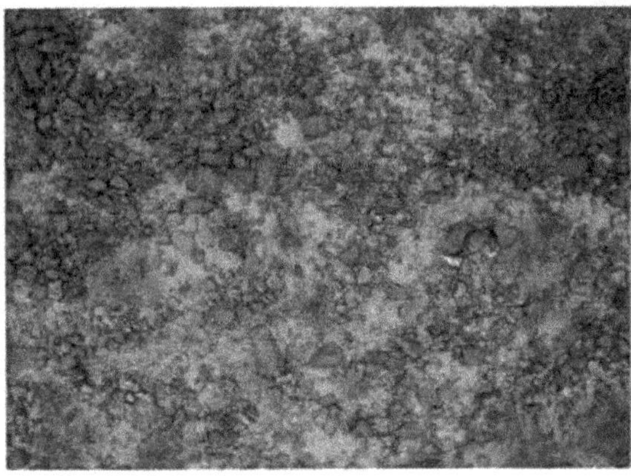
Figure 5 The same petroglyph from a higher elevation

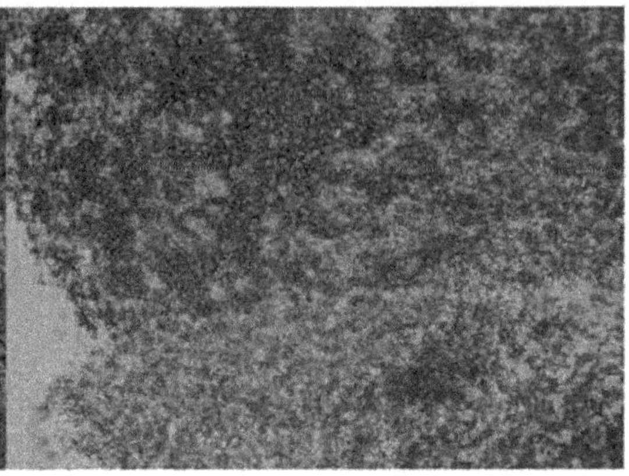
Figure 6 Petroglyph now seen as part of a trail running east-west, just above the petroglyph

A large historical-period ranching site was also recorded using a helium-filled balloon and the resulting photographs pieced together to form a mosaic that shows the entire ranch perimeter (Fig.7). These photographs have recently been used to monitor recent damage to the site by multiple vehicle tracks.

On-the-ground recording of rock features takes up valuable field time. A recent SRI project on YPG documented rock features by aerial photography that would have been impossible to document stone by stone on the ground within the allocated survey time. The accurate recording in the field of features from the air is cost-effective, and as a result, is being used more routinely on projects as a recording tool.

On YPG, a total of 22 rock features, originally classified as rock rings, were recorded by kite and balloon over a two-day period. Paper-plate markers were spaced with a measuring tape at 5-, 10-, or 20?m intervals around the features, to indicate four corners of a square or rectangle around the feature.

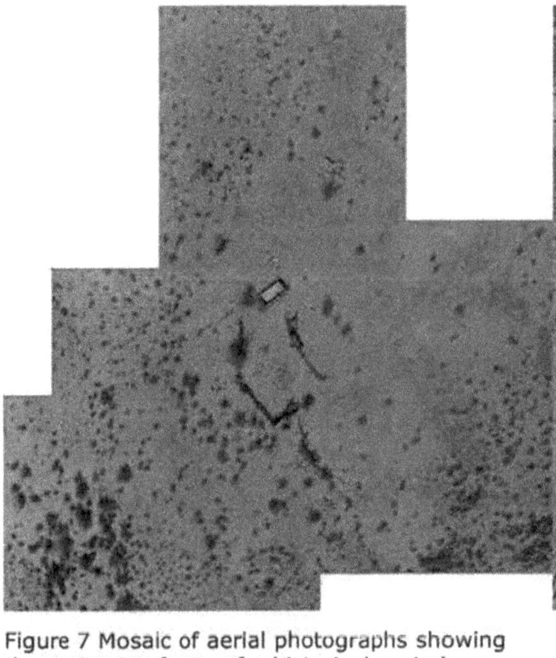
Figure 7 Mosaic of aerial photographs showing the perimeter fence of a historical-period ranching site, BMGR, Arizona

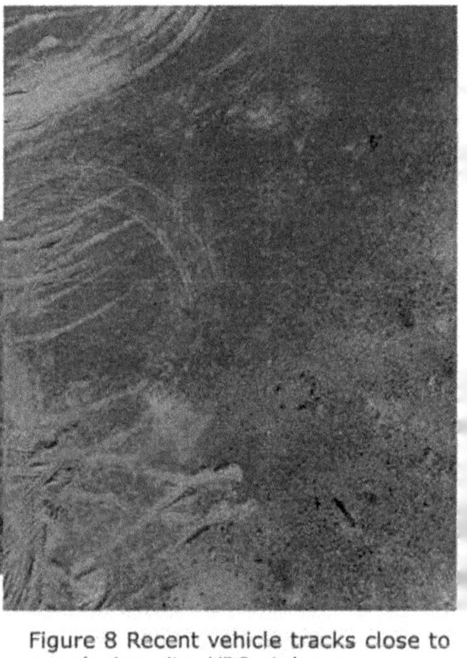
Figure 8 Recent vehicle tracks close to a rock ring site, YPG, Arizona

These corner markers were oriented to true north using a compass bearing and fixed in place using 15-cm (6-inch) nails prior to the aerial survey. The balloon or kite was then raised to an elevation sufficient to encompass all four paper plates in the picture frame. Determining the photo elevation was carried out by trial and error, and as many as 20 photos were taken at various elevations and then checked on the ground once the camera cradle was lowered. (Although a video transmitter is easy to attach to the camera so that video images from the camera can be transmitted to a handheld television, we have found our current techniques work very efficiently without adding another layer of gadgetry to the camera cradle). The images were then rectified using the paper plates as ground-reference points, and the images digitized to transform them into line drawings, which can be dropped into site maps or used as separate feature drawings. Sometimes the information provided by an aerial photograph can emphasize details difficult to portray in line drawings, especially impacts to the feature or the area surrounding it (Fig.8).

The Future

SRI hopes to continue to use low-level aerial photography to document sites in the Papaguería. We are experimenting with creating stereo-pairs to create three-dimensional topographical maps to measure quantitative changes in the landscape over time so that predictive modeling can help government agencies prioritize sites in need of protection. We also hope to adapt our camera cradle to house a multispectral scanner that can provide information about the types of materials found on landscapes and in archaeological features. The future of kite and balloon photography is assured until the resolution, cost, and accessibility of satellite imagery can match the resolution of a suspended digital camera.

References

ABER, J.S., SOBIESKI, R., DISTLER, D.A. and NOWAK, M.C., 1999. Kite aerial photography for environmental site investigations in Kansas. Kansas Academy of Science Transactions 102:57-67.

ALLEN, J., 1980. Lift off. Popular Archaeology, September:25-27.

ANDERSON, R.C., 1979. A kite-supported system for remote aerial photography. Aerial Archaeology 4:4-7.

BASCOM, W.R., 1941. Possible Application of Kite Photography to Archaeology and Ethnology. Illinois Academy of Science Transactions Vol. 34, no. 2. Urbana.

CHAGNY, B.N., 1994. Couverture photographique par cerf-volant des sites archéologiques de l'îl de Saï (Soudan). Le Lucane 68:10-17.

CHAGNY, B.N., 2001. Photographie stéréoscopique, prise de vue sous cerf-volant et visualisation des clichés. Le Lucane 96:17-23.

MEIGHAN, C.W., PENDERGAST, D.M., SWARTZ, B.K. Jr. and WISSLER, M.D., 1958. Ecological Interpretation in Archaeology: Part I. American Antiquity 24(1):1-23.

MUROOKA, K., 1998. Advent of the digital camera age. The Aerial Eye 4(1):18-19.

REEVES, D.M., 1936. Aerial Photography and Archaeology. American Antiquity 2(2):102-107.

RILEY, D.N., 1987. Air Photography and Archaeology. Duckworth, London.

ROY, E.J., 1954. Kite Takes Aerial Photos. Mechanix Illustrated October, Greenwich, Connecticut:174-176, 199.

WARNER, W.S., 1996. Kiteography. The Aerial Eye 2(2):16-17, 19.

Airborne Multispectral Remote Sensing Application in Archaeological Areas

Mercedes Farjas
E. U. I. T. Topográfica, Universidad Politécnica de Madrid (UPM), España
farjas@euitto.upm.es

J.G. Rejas

J. A. Gómez

Eduardo De Miguel

Álix Fernández-Renau

Laboratorio Central de Teledetección
Instituto Nacional de Técnica Aeroespacial (INTA), España

SEE THE CD FOR THE EXTENDED VERSION

Abstract

The Recópolis visigotic site (Guadalajara, Spain) has been largely investigated with several field surveys and laboratory analysis. As a new approach to the investigations, the area has been imaged in July and November 2002, using two airborne remote sensing instruments operated simultaneously: a multispectral scanner Daedalus 1268 and a digital camera AMDC integrated with a positioning and orientation system (GPS/IMU). The aim of this work, a joint project between INTA and UPM, is to provide high spectral and spatial resolution images to support the archaeological study. The acquired images have been used to make a preliminary site cartography, to extract current land cover/use and to test detecting and/or confirming models of possible buried archaeological structures. We describe in this paper the flight configurations, the data processing and the first results and conclusions.

Introduction

Since their early stages, remote sensing techniques have been applied to archaeological studies. A remote sensing instrument measures (usually covering a 2-D area and so rendering an image) the electromagnetic radiation coming from the study target; this radiation being either reflected sunlight, thermally emitted radiance or (in so-called active systems) energy sent by the instrument itself (microwaves in a RADAR or a laser beam in a LIDAR). Classical remote sensing is represented by aerial photography, while nowadays most applications benefit from digital data. A remote sensing system is labelled multispectral if the registered radiation is splitted in different, discrete wavelengths intervals ("bands"), and so different measurements on the same target are available.

The early works from the 60's (Donoghue 1999) showed that multispectral airborne remote sensing data were well suited to archaeological sites studies. The high spatial resolution, well suited to the scale required, coupled with the discriminant power of multi-wavelength information, make it potentially valuable for detection of buried or semi-buried features. However, technology and cost limited for many years the use of remote sensing images other than aerial photography, and few works have been published using aerial digital multispectral data as a support to archaeology.

In 2002, we have applied multispectral digital images to the exploration of the Recópolis archaeological site. Recópolis, dated 350-450 d.C., is the most important visigotic village found in Iberia. It has been subject of study of the UPM group for long time. This paper presents the characteristics of the flight campaign and the data processing applied, in order to show how to use this type of data to retrieve valuable archaeological information.

INTA Airborne Remote Sensing System

The Laboratorio de Teledeteccion at INTA owns a remote sensing aircraft and a number of remote sensing instruments, which are available to scientific or operational users. A usual set-up is to operate simultaneously two complementary cameras, a multispectral line scanner (DS-1268 "ATM") with wide spectral coverage and a multispectral digital camera (AMDC) with high geometrical quality but limited multispectrality (Fig.1).

DS-1268 is an opto-mechanical radiometer with image formation through the combination of a spinning mirror and the aircraft motion (a whiskbroom scanner in the remote sensing jergoon). It split the incoming radiation in 11 spectral bands,

Figure 1 Airborne remote sensing system of INTA, AMDC (1) , Daedalus 1268 (2) and IMU (3)

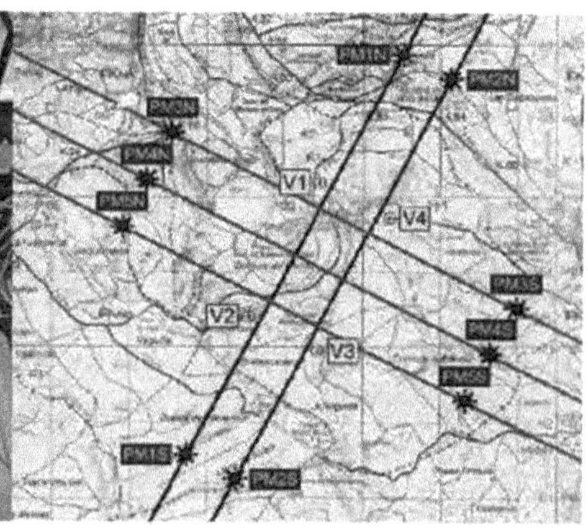

Figure 2 Study site and fligth lines

from visible through near infrared up to thermal infrared: in a typical spectral configuration the radiometer includes the same bands as the well-known earth observation satellite LANDSAT-TM. The total Field of View (FOV) is 85° and the Instantaneous FOV is 2.5 miliradians, corresponding to a spatial resolution in the range 1 to 7 meters and a swath width from 600 m to 5,000 m depending on the flight height.

The AMDC is a frame camera, i.e. an optical system imaging instantaneously a bi-dimensional area. The focal plane holds for that purpose a CCD array with 2.024 x 2.041 detectors. By the use of a filter wheel the CCD records sequentially 4 spectral images + a panchromatic one: there is a small shift between each exposure, but the imaged area is roughly the same for all 5 images. The total Field of View (FOV) is 36 x 36° and the Instantaneous FOV is 0.32 miliradians, corresponding to a spatial resolution in the range 0.5 to 1 meter and a image size from 1,000 x 1,000 m to 2,000 x 2,000 m depending on the flight height.
Both cameras are supported by a Inertial Navigation System Applanix POS/AV 410, specifically designed and installed to provide position and attitude information for the direct georeferencing of airborne images with the required resolution and accuracy (Rejas 2003).

Study Site and Acquisition Campaigns Configuration

The study site of Recópolis is located in central Spain in a small elevation facing the mid-course of River Tajo. Targets of interest are not only the visigotic city but also an Arab castle located in the same hill. The estimated area covered by the archaeological targets is around 25 Ha, from which only 1 Ha. are currently excavated while the rest extends mostly under crop fields (mainly cereals).

Flight campaigns were performed in July and November 2002, with both DS-1268 and AMDC installed in the INTA aircraft. The flight height was 2,400 m, in order to obtain a pixel size of 3.5 m for DS-1268 and 0.5 m for AMDC. Due to the resulting image width, the study area was covered by 5 different flight lines, with enough overlapping to ensure no gaps in the area coverage (Fig.2).

Supporting the image acquisition, INTA performed on site ground measurements with a field spectro-radiometer GER-1500. The output of such measurements are spectral reflectances for characterization of the surface spectral features and for eventual correction of noise or bias in the images due to atmospheric effect.

Detailed topographic information (1:500 map) is available from the UPM team.

Data Processing

Data processing techniques are used to retrieve information from remote sensing data. They usually include geometric correction, radiometric correction, and a number of analysis (image enhancement, image arithmetics and statistical analysis). The choice of techniques depends on image quality and in the required output. We present here the techniques we have used: all of them are available within the image processing package Geomatica from PCI Geomatics (also available from other packages as ERDAS-Imagine or ENVI).

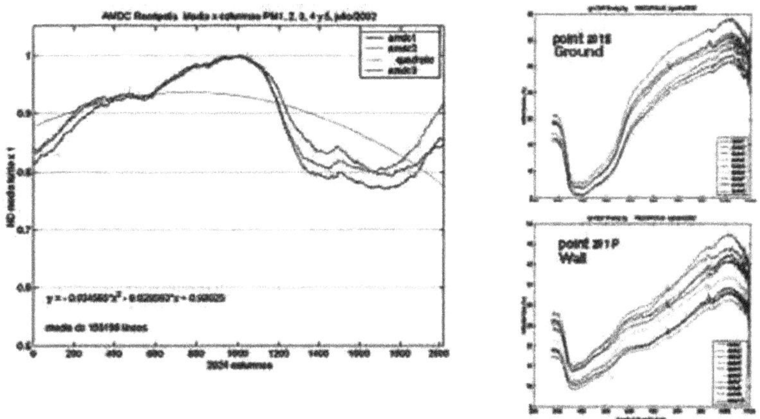

Figure 3 Samples of spectral signatures

In the AMDC images, two radiometric distortions were observed. One is a darkening of the image to the edges. The second is a local de-focussing in some bands. Both effects were caused by problems in the instrument during the campaign, and require a specific solution. For the darkening, and under the assumption that all camera detectors shall give in average the same value (200,000 image lines), a normalization function was built estimating the mean value of each image column through all the study area. The de-focussing has been minimised using a standard high pass filter. For the geometric correction, we applied the POS/AV data to the radiometrically corrected AMDC images using the commercial software Orthoengine (a module of the Geomatica pakkage). The same tool was used to build a single image per band mosaicking the individual frames. No atmospheric correction was applied. As a final step, we applied a textural analysis using well-known filters designed for enhacing image structures (for example a Sobel filter).

No radiometric correction was applied to the ATM images. For geometric correction we selected ground control points from the orthorectified AMDC images. A thin plate spline (i.e. a set of interpolating bi-dimensional polynomials) was applied to the ground control points to map the original image onto a orthophoto map.

Finally, we combined both instruments by using an image fusion algorithm implemented in Geomatica: the 11 bands

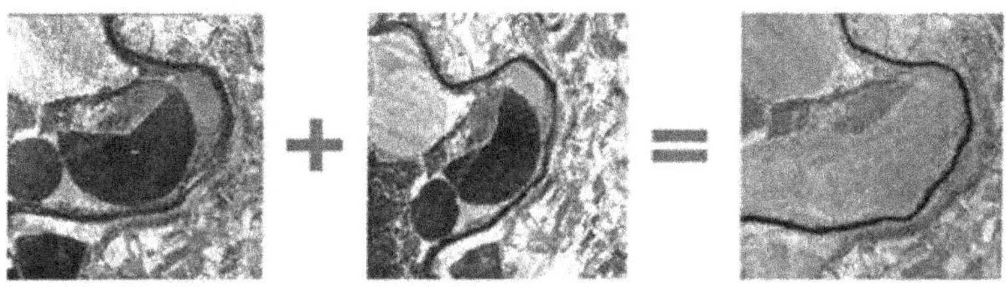

Figure 4 1:500 map and AMDC orthoimage, RGB 321

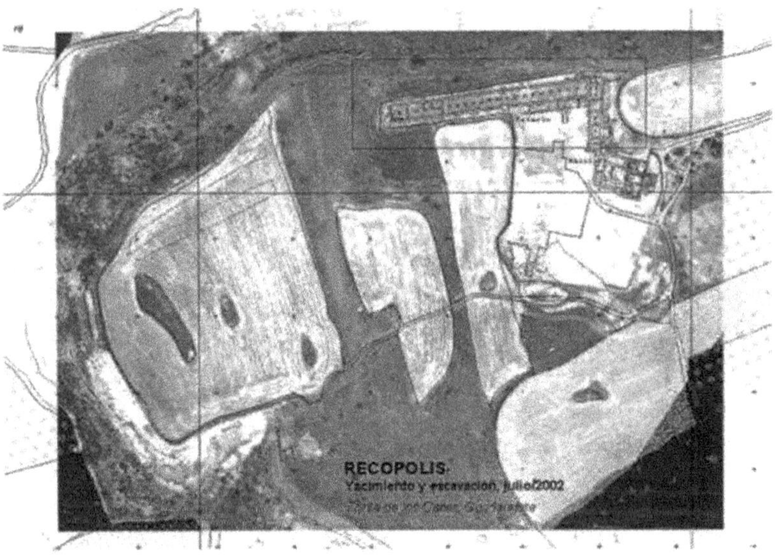

Figure 5 Sobel filter with AMDC red channel

from DS-1268, at 3.5 m resolution, were merged with the 0.5 m panchromatic band from AMDC resulting in 11 different 0.5 m images. The image fusion algorithms currently in use are intelligent tools that preserve multispectral variability while introducing spatial variability from the high resolution data (Yun Zhang 2002); however, it has to be noted that the procedure relies in data statistics and so no "new" information is created, but simply the existing one is revealed.

The combined images have been used for a statistical classification using the standard clustering algorithm ISODATA.
The GER 1500 measurements were processed using MATLAB in order to produce reflectances in bands similar to the airborne instruments bands, and then archived to be used in case the future work requires it (Fig.3).

FIRST RESULTS

The remote sensing images are currently in use as a graphic document displaying the status of the works and the actual land use through the site (Fig.4). Either direct visual inspection or integration on a dedicated Geographical Information System is possible.

The AMDC images processed with the Sobel filter have revealed some patterns which are likely to represent buried structures (Fig.5); this is the type of help that one expects from the remote sensing tools.

After different trials, the output of ISODATA classification using the bands ATM11 and ATM12 (both combined with the AMDC) was selected as the most useful. In this classified image patterns similar to the ones observed in the filtered AMDC images were apparent, patterns that are consistent with signals of man-made structures.(Fig.6).

OPEN WORK

INTA and UPM groups are working to complete the analysis of the existing images using techniques as image texture, airborne thermography, segmentation, classification and pattern recognition. In addition, it is planned to acquire new images with an instrument AHS (Airborne Hiperspectral Scanner) which can offer very interesting information.

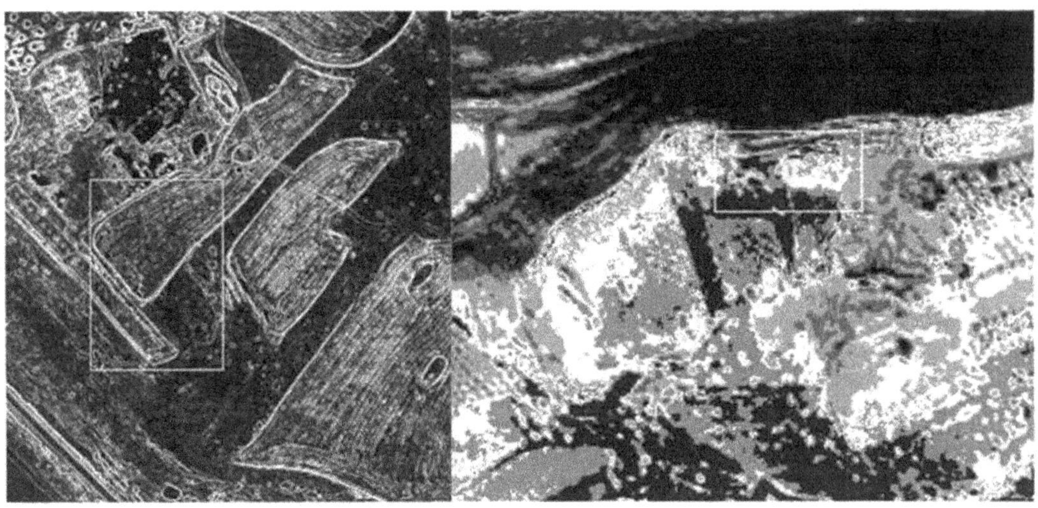

Figure 6 ISODATA classification using ATM11 and ATM12 with possible buried archaeological structures

REFERENCES

BELVEDERE, O., BURGIO, A., CIRAOLO, G., LA LOGGIA, G. and MALTESE, A., 2001. Hyperspectral MIVIS data analysis for archaeological applications. Fifth International Airborne Remote Sensing Conference and Exhibition, San Francisco, California, 17-20 September 2001.

CONSIGLIO NAZIONALE DELLE RICERCHE, 1999. Remote Sensing in archaeology. XI Ciclo di lezioni sulla ricerca applicata in archeologia, Certosa di Pontignano (Siena), 6-11 December 1999.

DONOGHUE, D., 1999. Multispectral remote sensing for archaeology. XI Ciclo di lezioni sulla ricerca applicata in archeologia, Certosa di Pontignano (Siena), 6-11 December 1999.

REJAS, J.G., GUTIÉRREZ DE LA CÁMARA, O., CALLEJO P. and GÓMEZ, J. A., 2003. Experiencias en la integración de los sensores multiespectrales aeroportados AMDC y Daedalus 1268 con un sistema GPS/INS. Sesiones Técnicas, 5ª Semana de Geomática, Barcelona, February 2003.

YUN ZHANG, 2002. Problems in the fusion of commercial high-resolution satellite and Landsat 7 images and initial solutions. ISPRS, Vol. 34, Part 4, Geospatial theory, processing and Applications, Ottawa, 2002.

Revealing the Ancient City of Lefkada through the Use of Shallow-depth Geophysical Prospection and GIS Techniques

A. Sarris
asaris@ret.forthnet.gr

S. Topouzi
stopouzi@arch.uoa.gr

F. Triantafyllidis

S. Soetens
ssoetens@yahoo.com

Laboratory of Geophysical -
Satellite Remote Sensing & Archaeo-environment
Institute of Mediterranean Studies -
Foundation of Research & Technology (F.O.R.T.H.)
Rethymnon 74100, Crete, Greece

G. Pliakou
12th Ephorate of Prehistoric and Classical Antiquities
25th March Square No 6, Ioannina 45221, Greece,
georgia.pliakou@ibepka.culture.gr

Abstract

The ancient city of Lefkas, opposite the Acarnanian coast, was inhabited from the early times and constituted one of the basic colonies of the Corinthians. The ancient city, following a Hippodamian system, expanded between the borders of the modern settlements Kalligoni and Karyotes, from the slopes of Koulmos, where some (possibly) ritual buildings and the remains of the theatre have been found, towards the coast. Parts of the city were revealed through excavations of the German Archaeological Institute at the beginning of the century and more recently by the Greek Archaeological Service.
Geophysical investigations employed vertical magnetic gradient and soil resistance mapping techniques. Processing of the geophysical data identified a number of characteristics of the ancient city plan: parallel and vertical streets, forming large building blocks, drainage pipes running to the sides of the roads, architectural remains, kilns and wall relics within the building blocks, etc. Further to the south, the density of architectural remnants is decreasing, suggesting a potential location for the southern wall of the city. Aerial and satellite images of the area were also registered to the topographic map and enhanced using image processing techniques.

Introduction

The Corinthian colony of Lefkas, founded in the end of the 7th century B.C., is located on the north-eastern coast of the Lefkada island, about 2.5 kilometres south of its modern capital city. The location of the ancient city is associated with the control of the channel, which was created between Lefkada and the Akarnanian coast by the Corinthians in the late 7th or early 6th century B.C. This passage was crucial for the navigation towards the Ionian and the Adriatic Sea (Murray 1982).

The city extends between the modern settlements of Kalligoni to the North and Karyotes to the South. The western part of the city follows the natural relief of the gentle slopes of hill Koulmos, while the eastern part extends to the offshore plain. The east and west sections of the city are divided by the modern provincial road leading from Lefkada to Nydri.

The area was not re-occupied after the depopulation of the city in the early 1st century A.D. During the Venetian period land use was limited to olive groves. The development of the area, which begun in the 1960s', was rather intensive during the last few years, due to the rapid touristic development of the island. The city walls are preserved today at the north and southwest parts of the city at a height of 2-2.5 m. Their estimated perimeter is 4.5 km. Although architectural remnants (from buildings and terraces) of the ancient city are visible at the slopes of Koulmos, those situated at the plain have been covered by shallow deposits. Thus, shallow-depth geophysical prospection survey was necessary for mapping the layout of the ancient city, in relation to the new design of the urban planning of the area.

Previous Archaeological Investigations

The first information for the remnants of Ancient Lefkas comes from the foreign travelers (Leake, Holland, Pouqueville, Goodisson, Dodwell), who visited the island during the 19th century. Systematic excavations started in 1901 by the German Archaeological Institute. Under the direction of W. Dörpfeld, excavation trenches were opened and remnants of the walls and the ancient theatre were revealed (Dörpfeld 1927). The research discontinued, since Dörpfeld's interests were focused in locating the Mycenaean Ithaka.

In the 1960s' archaeological research was re-activated in the form of rescue excavations conducted by the local Archaeological Service. The rescue excavations, which have been intensified during the last 10 years, uncovered parts of the city's cemeteries, roads and private buildings, contributing to the reconstruction of the urban planning of ancient Lefkas (Fiedler 1999, Douzougli 2001, Pliakou 2001).

According to the updated results of the archaeological excavations, the city seems to be divided in two parts: one on the hills of Koulmos to the west and one in the plain towards the coast to the east. The part that extends on the hills, where (possibly) ritual buildings and the remains of the ancient theatre have been found (Pliakou 1997, Pliakou 2001), has no specific urban planning, while the part at the plain was built according to an orthogonal system, very similar to the Hippodamian one. According to the information retrieved by different excavation trenches, the parallel streets, with average width 4.5 m (Agallopoulou 1971, Kalligas 1972, Andreou 1979), run across the city every 30 m in an E-W direction, that is from the hills to the sea. These streets intersect vertically with others of about 5.5-6 m in width (Andreou 1977, 1984), creating building plots with their long direction along the E-W axis. The study of the excavation results showed that two private houses, divided by a draining pipe, exist in every building plot (Douzougli 1993, Pliakou 1999). Other draining pipes run along the ancient streets (Agallopoulou 1971, Kalligas 1972, Andreou 1979). These pipes, being on the road surface just outside the houses, functioned as pavements.

Figure 1 Layout and codification of the geophysical grids, as seen through their overlay on the aerial image of the site

Figure 2 Results of the vertical magnetic gradient survey. The road network, structural relics, drainage pipes, kilns, and wall remains are obvious in the mosaic of the magnetic grids

The aforementioned conclusions have been compared with other known examples in the wider area, like the city of Amvrakia (also a Corinthian colony), which is in the area of modern Arta (Andreou 1993). It is worth mentioning that the orthogonal city plan system in ancient Lefkada was formed almost two centuries before the creation of the Hippodamian system.

Geophysical Prospection

The geophysical survey of the archaeological site was carried out in two phases (April 13-23 and June 16-27, 2002). In the first phase of the campaign, both magnetic and soil resistance techniques were employed in selected grids in order to test the registered signals originating by the subsurface targets. The second phase of the fieldwork activities was devoted to the extended coverage of the site through the use of magnetic techniques.

A Geoscan fluxgate gradiometer (FM36) was employed for the measurements of the vertical magnetic gradient. The gradiometer readings were able to smooth away the geological trends and diminish the external noise. Measurements had an accuracy of 0.1 nT/m. In the measurements of the soil resistance, a Geoscan resistivity meter RM15 with a Twin Probe electrode configuration was used with a spacing of 0.5 m between the mobile electrodes (Clark 1990). Emphasis was given to the wide coverage of the site with a sampling interval of 1m, while high-resolution measurements were carried out in a limited number of grids of specific importance.

The magnetic measurements in grids C, B, A and U (Fig.1), east of the provincial road from Lefkada to Nydri, indicated the existence of a network of parallel streets, at a SW-NE direction, lying at a distance of 35 m apart (Fig.2). They have a width of about 3.75-4.5 m and appear as high anomalies lying within the range of 5-15 nT/m above the background magnetic level. The dipole character of the anomalies suggests the existence of a drainage pipe running at the northern part of the roads. A road, perpendicular to the rest, appears at the northern part of grid B. Similar evidence comes from grids X and Z, suggesting that the town plan was consisting of a dense network of parallel and perpendicular roads, which divided the ancient city in building blocks.

Grids X and Z are of particular interest, since there are clear indications of architectural relics within the building blocks in both vertical magnetic gradient and soil resistance data (Fig.3). The outline of the magnetic anomalies suggests the good preservation condition of the particular structural remains. The roads appear as high magnetic anomalies within the ~3-15 nT/m range and as high resistance anomalies within the ~8-15 Ohms range. The parallel road system seems to fade out towards the east. A vertical linear anomaly, which appears at the eastern side of the particular grids, seems to be of a different nature, probably consisting of two sections, and can be considered as a candidate target for the location of the wall surrounding the town. The rest of the grids (VOR1, VOR2, TH1, TH2, TH3) towards the north-east indicate a loosen clustering of architectural remnants, which extent to the coast-line.

Streets and architectural relics are also present in grids D, E and F. In contrast, areas investigated west of the provincial road from Lefkada to Nydri (grids Q, S and T), did not reveal any significant anomalies, although it has to be mentioned that none of the above grids was laid at the projection of roads, which were identified east of the provincial road. On the other hand, north of grid T there are surface indications of a wide path leading at the top of hill to the west.

The area to the south of the SW-NE dirt-road, which divides the area of interest in north and south sections, exhibits a lower interest, as it is indicated by the decreasing density of geophysical features. The area of grids R1-R5 exhibits an elevation difference of about 3-4 m with respect to the dirt road and the north section of the surveyed region. In the projection of the dirt road to the west, large stone blocks may be correlated to the ancient city wall. Similar evidence exists along the dirt road reinforcing the hypothesis that at least a large section of the city wall is located along its length.

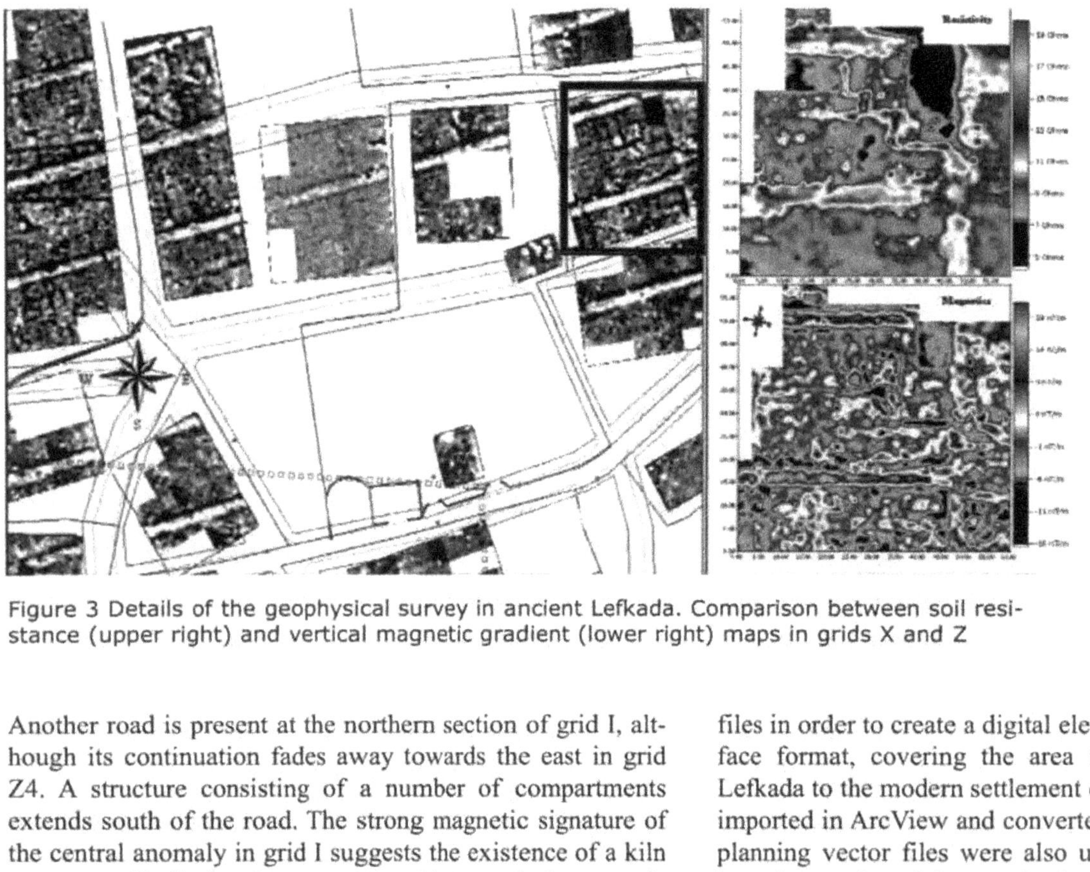

Figure 3 Details of the geophysical survey in ancient Lefkada. Comparison between soil resistance (upper right) and vertical magnetic gradient (lower right) maps in grids X and Z

Another road is present at the northern section of grid I, although its continuation fades away towards the east in grid Z4. A structure consisting of a number of compartments extends south of the road. The strong magnetic signature of the central anomaly in grid I suggests the existence of a kiln structure. Similarly, the strong negative vertical magnetic gradient anomaly to the south is probably caused by an ancient well. Other candidate kiln structures are indicated in grids P and V.

Finally, additional evidence of the extension of the city plan was provided by grids N1 and N2 to the SE. In grid N1, a cross-section is aligned in agreement to the street orientation at the north section of the surveyed region. The NW-SE road seems to be at the projection of the road registered at the NE edge of grid B, while there are also indications from grid N2 that the road continues further to the south. Due to its proximity to the coast, it is very probable that the road may lead to Lygia, where the ancient port facilities of the city were located. More specifically, in Lygia there is a mole used to connect the Eastern coast of Lefkada with the Akarnanian coast, where the fortress of Agios Georgios-Plagia is located (Murray 1982). The mole was constructed to protect the southern entrance of the canal, cut by the Corinthians and it is probably dated to the same period, namely about 600 B.C. Similarly, the SW-NE road seems to lead to the ancient cemetery, which has been located to the west of the provincial road.

A series of diagrammatic maps representing the geophysical anomalies has been overlayed on the different cartographic layers of the site (Fig.4).

Creation of a Geographical Information System

The development of a Geographical Information System was considered essential for the better representation of the results of the geophysical prospection survey in accordance with the new urban planning design. The specific Geographical Information System contains data in both vector and raster format, transformed in the same geodetic reference system, namely the Hellenic Geodetic Reference System 1987 (EGSA '87), which is used by the National Cadastre of Greece. ESRI ArcView GIS 3.2 and ERDAS Imagine 8.5 were used for the realization of the GIS.

3D topographic data in ASCII format were converted to database files in order to create a digital elevation model in a TIN surface format, covering the area from the modern city of Lefkada to the modern settlement of Lygia. Theses files were imported in ArcView and converted to shapefiles. The urban planning vector files were also used in order to determine how the results of the geophysical prospection are going to affect the proposed design plan. A number of different shapefiles was created including the road network, building plots, existing buildings, boundaries of the research area, protection zones, a.o. The outline of the boundaries of every geophysical grid was also added as a polygon. Finally, the interpretation of the geophysical maps in the form of a diagrammatic drawing (line- and polygon-shapefiles) was included and was compared to the corresponding drawings of the known archaeological remnants.

Three geographically registered orthophoto maps (date: 1998, scale 1:5,000), provided by the Ministry of Agriculture and covering the same area as the TIN model, were joined creating a photo-mosaic. This mosaic was used to georeference the rest aerial photos, provided by the Hellenic

Ministry for the Environment, Physical Planning and Public Works (Hellenic Mapping and Cadastral Organization). These 1:6,000 scale photos, which were obtained at different dates (16/05/1985 and 12/10/1994), were scanned in high resolution, so as to achieve the optimum result in the quality of every image. These photos were also processed through histogram equalization and geometric registration in order to form a time sequence of aerial mosaics.

Figure 4 Diagrammatic representation of the magnetic and soil resistance anomalies

Geophysical prospection survey grids were export in jpeg format, both as B/W and colour images, and they were geometrically corrected, using their aforementioned vector borders. Finally, scanned photos and designs of the excavations were added as hotlinks.

The available vector and raster format layers were combined in the creation of the Geographical Information System and they were superimposed on the TIN of the area. In this way, the user can choose between one of the different period's aerial photo mosaics as background information or between colour and B/W images of the geophysical grids, while at the same time he can have a direct access to the designs and pictures associated with the excavations. Most important, the user can explore which of the geophysical anomalies fall into the extents of the new urban planning design and can do the appropriate modifications. Finally, several queries related to the topographical and archaeological data can be performed.

Results - Conclusions

The geophysical prospection survey at the archaeological site of ancient Lefkas resulted in a wealth of information, providing an accurate image of the city plan and indicating the limits of the ancient city. A road network, structural relics, drainage pipes, kilns, and wall remains are included among the candidate targets that were registered by the soil resistance and magnetic techniques. The results of the geophysical survey are in perfect agreement to the surface monuments and the excavation results, completing in a large degree the fragmentary image of the ancient city.

More specifically, the city plan seems to extend to the north of the area of interest (namely, north of the SW-NE dirt-road and east of the provincial road). In contrast, the south section of the area exhibits a lower density of architectural remains and an increasing number of candidate kiln structures, suggesting workshop activities. The above observations, together with the surface monuments along the dirt-road and the elevation difference between the north and south sections in the other sides of the dirt road, suggest that part of the southern city wall is probably running along the dirt-road.

The importance of the GIS in registering the geophysical maps in a common cartographical context is obvious, due to the different areas that were surveyed. GIS contributed to the overlay of the geophysical images on the topographical layout and the aerial mosaic of the region. In this way, it was possible to have an accurate picture of the distribution of the geophysical anomalies and their correlation with the surface relics. Aerial mosaics contributed to the analysis of the evolution of the landscape. A three-dimensional representation of the research area combining the digital elevation model, draped with an aerial photo mosaic and the architectural remnants, found in the previous excavations and geophysical survey is under construction. The final system will contribute in the management of the archaeological relics of ancient Lefkas.

REFERENCES

AGALLOPOULOU, P., 1971. Archaeologikon Deltion 26, B΄:357.

ANDREOU, I., 1977. Archaeologikon Deltion 32, B1:155 (D. Lazaris plot).

ANDREOU, I., 1979. Archaeologikon Deltion 34,B1:269.

ANDREOU, I., 1984. Archaeologikon Deltion 39, B΄:187-188 (Ar. Lazaris plot).

ANDREOU, I., 1993. Ambracie, une ville se reconstitue peu à peu par les recherches. Colloque International de Clermont-Ferrand "L' Illyrie méridionale et le Épire dans l' antiquité II", Paris, 25-27 October 1990:91-101.

CLARK, A., 1990. Seeing Beneath the Soil. England.

DÖRPFLED, W., 1927. Alt -Ithaka. Ein Beitrag zur Homer-Frage:156-57.

DOUZOUGLI, A., 1993. Archaeologikon Deltion 48, B1:285-187, 293-300.

DOUZOUGLI, 2001. Nekrotafeia arhaias Lefkadas. Symposium "The capital of Lefkada", Lefkada, August 1999:45-84.

FIEDLER, M., 1999. Leukas, Wohn und Altagskultur in einer nordwestgriechischen Stadt. In Hoepfner, W. (ed.), Geschichte des Wohnens in 5000 v. Chr. - 500 n. Chr. Vorgeschichte Fruhgeschichte - Antike, Stuttgart:412-416.

KALLIGAS, P., 1972. Archaeologikon Deltion 27, B2:486.

MURRAY, W.M., 1982. The coastal sites of western Akarnania. Phd. University of Pensylvania:226-239, 243-250.

PLIAKOU, G., 1997. Nea stoiheia gia to arhaio theatro tis Lefkadas. Mia topographiki prossegissi. Hpeirotika Chronika 32:37-42.

PLIAKOU, G., 1999. to be published. Archaeologikon Deltion 54, Chronika.

PLIAKOU, G., 2001. I archaia poli Lefkas : to asti kai i evriteri periochi tou. Symposium "The capital of Lefkada", Lefkada, August 1999:30-34.

Land Use and Irrigation Works in Kazakhstan in the Present and in Historical Times. Geo-Archaeological Investigations throughout GIS and Remote Sensing

Maurizio Forte
CNR-ITABC, Istituto per le Tecnologie Applicate ai Beni Culturali, Roma

Sofia Pescarin
CNR-ITABC, Istituto per le Tecnologie Applicate ai Beni Culturali, Roma

Renato Sala
Academy of Sciences , Institute of Archaeology, Almaty, Kazakhstan

Jean-Marc Deom
Academy of Sciences , Institute of Archaeology, Almaty, Kazakhstan

Konstantine Michaelovitch
Academy of Sciences (Ministry of Sciences) Institute of Soil Sciences, Department of Genesis, Geography and Valuation, Almaty, Kazakhstan

See the CD for the extended version

Abstract

General goal of the project, coordinated by CNR-ITABC and supported by INTAS (International Association for the Promotion of Co-operation with Scientists from the New Independent States of the Former Soviet Union) is to reconstruct the present conditions and the historical evolution of the land use and irrigation works in three main areas of settlements today exposed to a sensible process of aridisation: Semirechie, South Kazakhstan, Central Kazakhstan (settlements, fortresses and ancient irrigation works from the Early Middle Ages till the rule of the Kazakh hordes VIII-XV AD). The hypothesis of research is that the three regions have always been particularly sensitive to environmental fluctuation; that proper irrigation always played a stabilizing economical role during the last two millenniums; that climatic and anthropogenic reasons have been responsible of the building, restructuring and disrupting of past irrigation works. The goal is to understand the weight of natural and anthropogenic factors behind the actual trend of aridisation and degradation in these areas; the factors that lead to the changes during history; and to compare the present and the past events. The complexity of the project embraces multidisciplinary applications: GIS, GPS, remote sensing, integrated geoarchaeological investigations. This paper constitutes the report of the scientific activity of 2002 and it concerns very preliminary results.

The Project

General goal of the project is to reconstruct present conditions and historical evolution of land use and irrigation works in three main areas of settlements today exposed to a sensible process of aridisation: Semirechie, South Kazakhstan, Central Kazakhstan.

1. Semirechie: Ili valley (Ili delta, Tianshan foothills, Chu-Ili mountains); and the delta of Talas river. There settlements and ancient irrigation works have been found from the PostSaka-Wusun period (200 BC-300 AC) till the last century.

2. South Kazakhstan, at the confluence of the Arys and Syr-Darya rivers where fortified settlements, towns and sophisticated irrigation networks were in use from the II AC till the XVIII AC.

3. Central Kazakhstan, region of Djezkazgan where have been found settlements, fortresses and ancient irrigation works from the Early Middle Ages till the rule of the Kazakh hordes (VIII-XV AC).

The hypothesis of the research is that the three regions are today and have always been particularly sensitive to environmental fluctuation; a proper irrigation always played a stabilizing economical role during the last two millenniums; that climatic and anthropogenic reasons have been responsible of the building, restructuring and disrupting of past irrigation works. The goal is to understand the weight of natural and anthropogenic factors behind the actual trend of aridisation and degradation in these areas; the factors that lead to the changes during history; and to compare the present and the past events. The project (n.699) is supported by INTAS for the period 2002-2004. INTAS is an independent International Association formed by the European Community, European Union's Member States and like minded countries acting to preserve and promote the valuable scientific potential of the NIS partner countries through East-West Scientific co-operation.

The partners of the project are: CNR ITABC; Oxford University, Dept of Environmental Archaeology; Southampton University, Institute of Irrigation and Development Studies; Institute of Archaeology of Kazakhstan; Institute of Engineering and Telecommunication of Almaty; Institute of Geology of Kazakhstan; Dept of Archaeology of Kemerovo (Russia).(Fig.1)

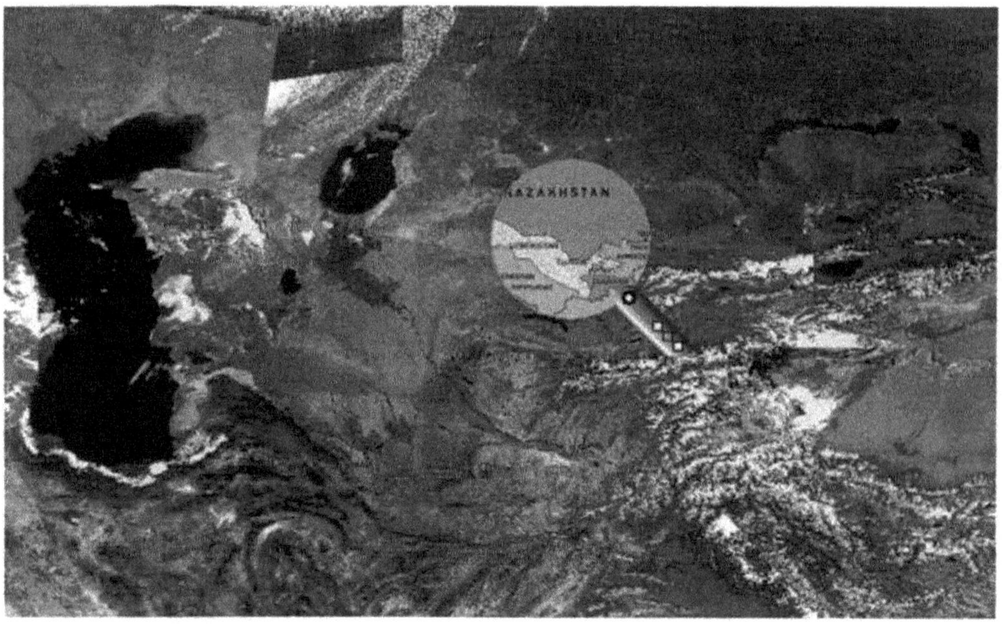

Figure 1 Satellite images and surveyed areas

[1] General goals

Main goals of the project are:

- to study the actual state of land use and the irrigation systems in the three different regions;
- to evaluate the actual process of aridisation by analyzing the stressing effects of natural factors and human negligence;
- to provide a contemporary reference (talon) for the study of the past conditions and disrupting agents in the same sites (aridisation, erosion, siltation, floods etc.);
- to create a reconstruction of environmental variations (geological, climatic, botanical, zoological) in the chosen sites, during a period going from the late Iron Age to the present time;
- to reconstruct a general historical background of the human occupation of the areas by surface archaeological surveys: detecting settlements and works and analyzing their character (temporary or permanent, etc.);
- to find out where and when ancient irrigated lands have rose and disappeared and to reconstruct the evolution of the types and techniques of water collection (wells etc.) and irrigation;
- to reconstruct the evolution in the land use: kinds of crops, type of agriculture, and their changes under climatic and historical changes.

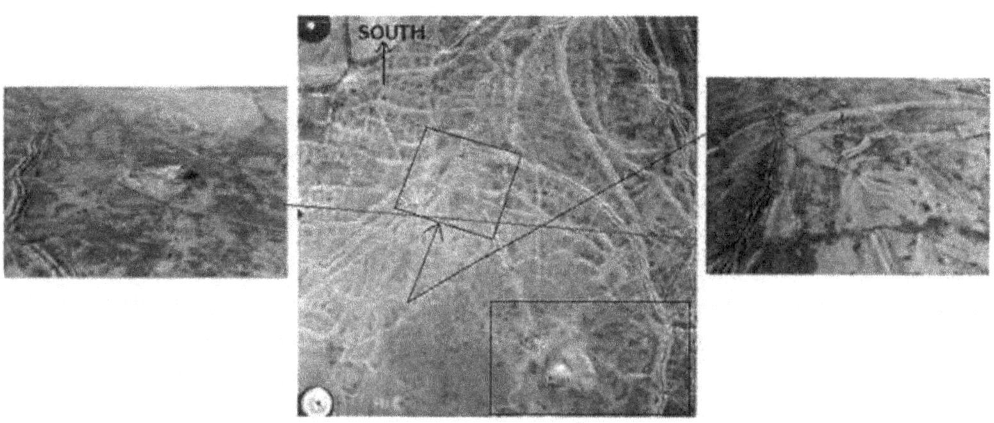

Figure 2 Altyn-Tobe's Landscape: irrigation systems

In terms of digital activities, we are working on remote sensed data (Landsat TM, 7 bands) and on GPS-DGPS survey with Arcpad installed on PDA. All the spatial collected data are publishing on line through Timemap. (Fig.2)

[2] The expeditions

In the INTAS Expeditions we pointed out three main working area:

1) Otrar oasis, in the middle part of Syrdarya river
2) Delta of Talas river
3) Western valleys of Chu-Lli mountains

As a method, every land-survey and fieldwork is followed (when possible) by aerial (microlight) survey and photo-documentation implemented by the team (Inst Arch-KZ). Georeference and rectification of the aerial photos provide an extensive digital photo-mosaic of the region, integrated with satellite data (CNR-IT).

In the season 2002, aerial-documentation has been provided for the sites of south-Kazakhstan: oasis of Otrar, oasis of Turkestan, valley of the middle course of the Arys river.

The oasis of Otrar have been documented by aerial and topographic surveys in cooperation with a UNESCO team engaged in situ for conservation works. Important information has been provided on contemporary and ancient hydrology, canals stratification, distribution of agricultural fields. In the oasis of Turkestan the aerial survey was intended to analyze general landscape features, to document the surface remains of 2 underground water galleries (karez) and to explore the possibility of undiscovered karez (eventually 2 new karez have been discovered and documented). In the valley of the middle course of the Arys river landscape, hydrology and plans of 2 early medieval towns with their surroundings have been aero-investigated.

GPS data have been collected and experiments of survey have been done. The GPS survey has been done with a Trimble Pocket PC antenna connected with ESRI - Arcpad in order to provide all the vectorial data and points in GIS format (shape file): it concerned several areas of geo-archaeological interest and first of all the identification of training regions for supervised classification of the most interesting spectral signatures. (Fig.3)

Figure 3 A phase of data collection with GPS and ESRI Arcpad installed on a Compaq PDA

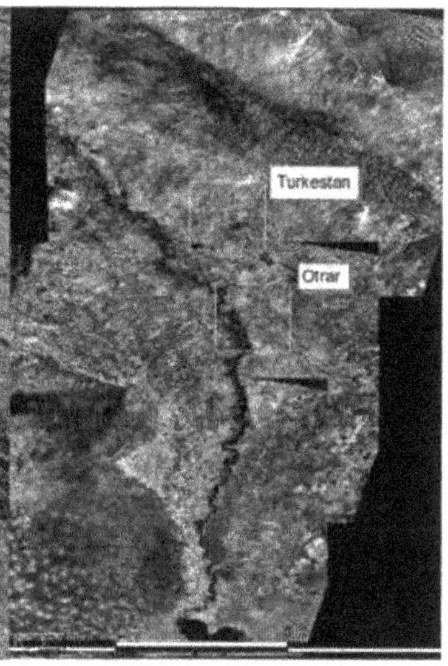

Figure 4 Remoted sensed satellite images

[3] Otrar Survey

In 2002 started Otrar expedition with the participation of the team leaders, of the specialists of five teams NIS and of the coordinator ITABC-CNR, Maurizio Forte. The work consisted in a detailed geographical, geo-morphological and archaeological exploration of the oasis (channels, tobé and fields) and of its surroundings, north and south.

We collected GPS data and we did a few survey experiments. The GPS data were collected with a PDA, a small-antenna (by Trimble) and the software Arcpad (ESRI) (Fig.2): in this way it was possible to collect the data, directly during the survey, in a vectorial format and the points, lines and polygons immediately in GIS format (such as SHP). The GPS survey was done for different areas with particular geo-archaeological interests and, moreover, it was used to identify the training regions, useful to obtain the supervised classifications of interesting spectral signature. This elaboration was done with Landsat TM satellite imagery, provided directly by NASA, whose help was possible thanks to its collaboration with CNR ITABC, started times ago for bilateral researches.

Two trenches were dug in a channel and in a field, in order to collect data useful to the paleo-climatic reconstruction and to different laboratory analysis. A general planning of the works was discussed; the planning for the second expedition in Otrar, in October 2002, was done even to keep on working on the same topics and research lines.

[4] Irrigation Systems: A Systematic Reconstruction Attempt

Up to now information on the irrigation works of the Otrar oasis has been collected by surveys limited to the surroundings of ruined towns, so that documentation was provided of settlement patterns and of the technical aspects of water management, from the size of the canals down to the kinds of conduits and containers. A systematic reconstruction of the irrigation system as a whole and of the related localization of fields and crops has never being attempted. Nevertheless this system of canals and fields is by itself the main protagonist of the history of Otrar, the one that best explains the localization of towns and villages and fortresses.

Moreover the final decay during the XVIII century of the 2000 years old oasis, synchronic with a contemporary improvement of the irrigation schemes in neighbor oases like Sairam and Turkestan, supports the hypothesis of a major management crisis.

The sites show different hydrological and climatic characters and cycles, with different irrigation facilities which predispose them to different irrigation techniques: with different chances and with different evolution patterns. Their features span from being wet by very big riverine flows likes the Syrdaria, Talas and Ili rivers, to being wet by small streams like the case of piedmonts of Tienshan, west Chu-Ili mountains and Dzhizkazgan. Moreover running in steppes or semi-desert or desert zones. Together they represent the main kinds of hydrological systems of KZ where irrigation works are possible.

THE TECHNOLOGICAL APPROACH

[1] Remote Sensing and GIS

At present we are finishing, in our digital lab at CNR, the topological overlay of the entire set of data, collected during the survey, in GIS format. A very useful result was the integration of all the spatial data collected during the fieldwork with remote sensed imagery and 1:10,000 Russian cartography. In this way we are starting to draw a new cartography of the areas of interest with a very detailed geo-archaeological mapping.

In the near future we will plan supervised and unsupervised classification in order to investigate the relations between soils, water resources, environment, canals, settlements, and oasis systems. Sharing the spatial data on line will permit to start a real cooperative work between eastern and western scientific partners. Finally, a virtual interactive reconstruction of the geo-archaeological landscape will represent the main focus of the digital processing. (Fig.4)

[2] Cooperative work: Timemap and Ecai project

Together with GIS and Remote Sensed analysis, we are also preparing the descriptive metadata for each themes (shp) overlaid in GIS (not just vectors but also raster images), in order to use a free web tool, based on Java Web-GIS interface, called TIMEMAP (in cooperation with some American researchers of Berkeley, that work in the ECAI project). This tool is quite useful indeed to share the same GIS information with all the partners of the project, through an on line data visualization, even during the phases of data elaboration, done in different and distant laboratories (UK, Russia, Italy and Kazakhstan). With Timemap, moreover, it is possible to simplify the operations of data and information exchange among the groups, and to define areas of interest for further investigations.

[2] Ecai and Timemap

ECAI (Electronic Cultural Atlas Initiative, USA, University of California, Berkely) is an international project aimed to develop and distribute geo-data based on historical and archaeological sources. It uses data, cartography and imagery of the world in connection with their content and it is based on a digital technology that combines complex data that come from different disciplines (www.ecai.org).[1] ECAI offers one of the financial support to TIMEMAP (www.TimeMap.net).

TIMEMAP is a project created by the Archaeological Computing Laboratory of Sydney University, Australia. The methodology used for TM was first developed in 1995 under the direction of Ian Johnson.

But what is Timemap? It is a free system that distributes geographical maps, updateable and interactive, together with thematic overlays connected with Time dimension and distributed through the Web.

What can we do with Timemap? Thanks to the Georeferenced data and the Time-Component, it is possible, for instance, to visualise and understand the development of a site or of a city, allowing the user, at the same time, the get detailed meta-information on the data visualised. It is possible to create and to build new hypothesis, new conceptual models just activating and overlaying different thematic layer, coming from different digital sources (vector or raster, images, web pages, databases, multimedia).

Which is at the base of Timemap? It is based on GIS to develop a clear methodological approach that allows the registration of cultural data not only in the Space (longitude and latitude values) but also in a Time-line (lower limit of date and upper limit of date). TimeMap Toolkit program provides the tools necessary to:

1. create metadata (in xml mode, based on Dublin Core);
2. register datasets with the ECAI metadata clearinghouse (just metadata will be in ECAI server [2], while the physical shape files or imagery remain in our server);
3. upload the data to a remote data server;
4. Example of registration of metadata with TimeEdit;
5. View the data in the Web with a simple user interface, based on Java programming language. (http://www.timemap.net/clearinghouse).

PRIVACY AND COPYRIGHT

We found this tool quite useful to work in a team that is quite widespread in the world. And even if the data and the digital analysis are easily available over the Web, in the project phase we decided to upload files with a password. This means that just the équipe that knows the username and the password can see data, information, have access to our database or upload new data. (Fig.5)

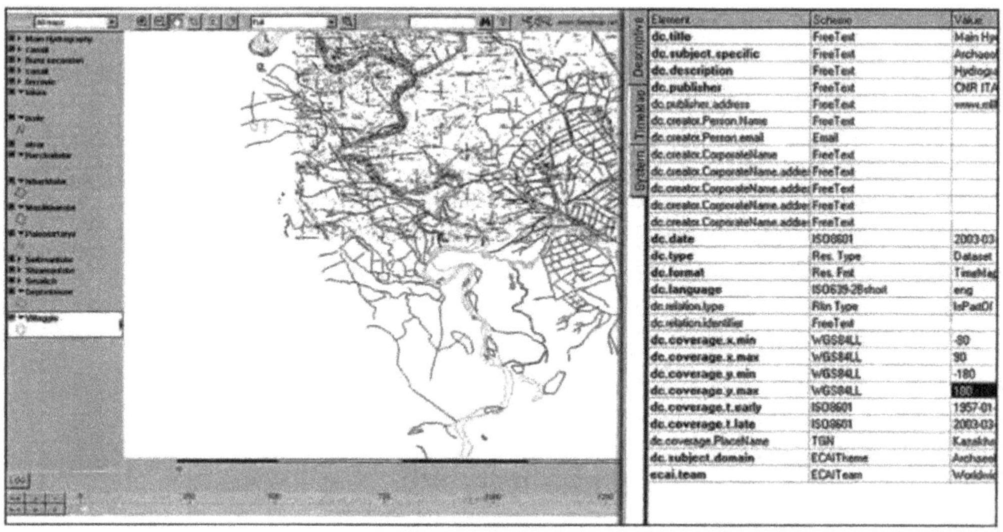

Figure 5 Example of Internet access to the data (left) and data preparation

ARCHAEOLOGICAL ANALYSES

[1] Laboratory analyses and data entry and documentation

Laboratory analyses have been done in Almaty by Inst SoilSc-KZ, Inst Geol-KZ, UN-Kemerovo and they are based on samples, collected in 4 pilot points, during the expeditions to Otrar done in June and October. They are supposed to arrange laboratory methods and procedures for the extensive researches of the second season. Digital analyses of GPS and photo-documentation are in progress by CNR-ITABC and Inst SoilSc-KZ.

The analyses have been the following:

- palinology and paleoclimatic reconstruction of the last 5,000 years in the polygon of Otrar;
- analysis of sediments (alluvial and Aeolian), for the evaluation of the geo-morphological transformation of the territory of the oasis caused by aeolian processes, by the river activity and by the implementation of irrigation activities;
- Physical-chemical analyses, for measuring the presence of minerals, salts and other chemical components(Inst SoilSc-KZ, Inst Geol-KZ);
- pedological analyses for determining the character of agri cultural soils; and micromorphological analyses for poin

ting out micro-traces of ploughing or other anthropogenic activities (Inst SoilSc-KZ);
- carpological analyses for studying traces of antropogenic organic remains (Inst Geol-KZ, UN-Kemerovo);
- analysis of pottery for the localization, stylistic and EPR absolute dating of settlements (Inst Geol-KZ);
- the topological overlay of all the data collected in GIS-Timemap is actually in progress, in the digital laboratories of (CNR-ITABC), in cooperation with ECAI of Berkeley-USA and with (Inst SoilSc-KZ), so that to share them on - line with all the participants of the project;
- Data entry of all the spatial data in GIS format in ArcView. These archives include: remote sensed imagery, Russian cartography, geo-archaeological topographic surveys, unsupervised classifications, paleo-environmental interpretations etc.

Conclusions

The project is still at the beginning but it underlines remarkable gaps in the geo-archaeological knowledge of the steppe's environment.

The study of the very articulated sequences of relations between settlements and irrigation/oasis systems will show new perspectives in the territorial analyses of these areas through the time; in this context the archaeology could give a strong reply to the modern territorial and environmental disasters of these areas. Because of these difficulties the support of remote sensing integrated with geo-archaeological surveys and with DGPS/GIS mapping is giving us important results or in the creation of predictive maps, whether in the construction of a solid geo-archaeological basis.

Finally, the next use of VR systems integrated with databases and spatial archives will contribute to visualize a very detailed virtual landscape of Kazakhstan mainly in the region of Otrar as "beta-test" for the other territories of investigation.

[1] ECAI is an international consortium of scholars developing electronic and web-based materials relating to cultural (primarily historic) information, such as texts, art, artefacts and sites, with the aim of making this information available across the WWW or through specific products such as CD-ROMs. Regional focus is primarily Asia. The Electronic Cultural Atlas Initiative is a collaborative project, which will combine global mapping, imagery, and texts. ECAI provides scholars and other users with a research resource based on digital technology which can present complex combinations of data from multiple disciplines visually and immediately. ECAI/TimeMap Project Documentation (August 2001).

[2] ECAI metadata is stored in an Internet-searchable database known as the ECAI Metadata Clearinghouse. Anyone with a web browser can search the ECAI metadata clearinghouse database. TimeMap Project Documentation (http://www.timemap.net/clearinghouse/clearinghouse_login.html, August 2001).

References

AKISHEV K.A., and BAIPAKOV, K.M, 1973. Kyarizy Saurana (Karez of Sauran). In Vestnik AN-Kaz no. 4.

BEAUMONT, P., BONINE, K. and MCLACHLEN (eds.), 1987. Qanats, Kariz and Khattara. Traditional water systems in the Middle East and North Africa. Oriental and African Studies UK.

BECKMAN, C.S., WEIGAND, P.C. and PINT, J.J., 1999. Old World irrigation technology in a new contact: Qanats in Spanish colonial western Mexico. In Antiquity 73:279.

BRIANT, P. (ed.), 2001. Irrigation and drainage dans l'antiquite. Qanats and canalizations souterraines en Iran, Egypte et en Grece. Thotm ed, Paris.

LEWIS, R.A., 1966. Early irrigation of West Turkestan. In Annals of the Association of the American Geographers Vol. 56, no. 3.

SALA, R., 2001. Reconstruction of the paleoclimate of the Late Bronze and Early Iron epochs in Semirechie. In Problemi drevniei i srednievekovoi istorii Kazakstana Vol. 3, Almaty (in English).

SPOONER, B., 1974. Irrigation and society: the Iranian Plateau. In McGibson, G. (ed.), Downing TE.

The Experiment in Applying 3D Technology of Magnetic Fields Interpretation at the Archaeological Site "Arkaim"

V. KOCHNEV
INSTITUTE OF COMPUTATIONAL MODELING SB RAN, KRASNOYARSK, RUSSIA
KOCHNEV@ICM.KRASN.RU

G. ZDANOVICH
HISTORICAL CENTER "ARKAIM", CHELYABINSK, RUSSIA
CHELYABINSK@ACES-GROUP.COM

B. PUNEGOV
HISTORICAL CENTER "ARKAIM", CHELYABINSK, RUSSIA
CHELYABINSK@ACES-GROUP.COM

SEE THE CD FOR THE EXTENDED VERSION

ABSTRACT

The technology was elaborated for solving direct and inverse problems of magnatometry. The technology offers the following possibilities:
1. Inverse problems could be solved with the help of anomaly
fields got on several layers. The number of levels is not limited.
2. The direct and inverse problems could be solved for a multi layer model. The number of layers is not limited.
The report will reflect the experience in applying this technology for defining the degree of magnet receptivity of the explored space of the settlement Arkaim, the brink of 3d - 2nd millenniums B.C. Maps and sections will be demonstrated and their interpretations will be offered to the audience. Archaeological excavations held on the explored territory gave evidence of the positive results of the offered technologies.

Prerequisites of the Experiment

Arkaim, the fortified settlement dated to 1600-1700 B.C., was discovered in 1987 (Arkaim 1995). It is situated 8.2 km north-north-west of Amurskiy, and 2.3 km south-south-east of Alexandronvskiy in the Chelyabinsk region, Russia. "The excavation revealed that the settlement had been burned and, therefore, many details were preserved. The population, however, had vacated the city before the fire and took all their possession with them. Arkaim had two protective circular walls and two circles of standard dwellings separated by a street around a central square. The external wall, 160 m in diameter and 4 m wide, was built from specially selected soil that had been packed into timber frames before being faced with adobe bricks. On the interior, houses abutted the wall and were situated radially with their doors exiting to the circular internal street" (Koryakova). The site plan is shown at Figure 1.

It is known from experience that the interpretation of magnetic fields and some of their components gives good results in the detection of archaeological objects without actual excavations. The differences in the magnetic susceptibility of the building material of walls, soil and gumus that covered the archaeological object are crucial for the successful detection. That fact was proved by the application of magnetic fields on the archaeological site "Arkaim". We made an attempt to start the construction of the model of the magnetic properties of medium under the surface. For this purpose we used the technology, developed by us for studying different geological objects (Kochnev 1993, 2003).

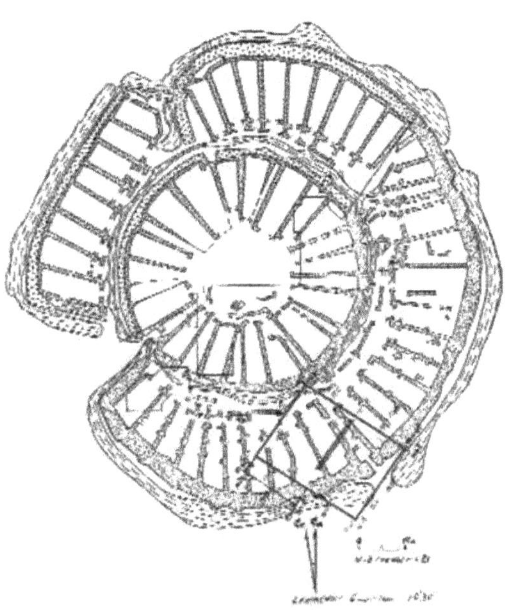

Figure 1 Archaeological site "Arkaim" (research area is shown in rectangle)

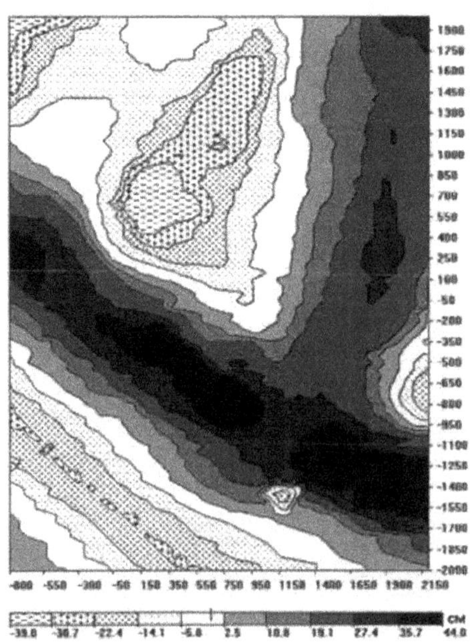

Figure 2 The relief of research area

The software package ADGM-3D is the core element of the technology that makes it possible to calculate magnetic and gravitational fields from known objects generating fields (direct problem), and the parameters of objects from the observed anomalous fields (inverse problem).

The Model of Medium and the Method of the Solution

For the solution of these problems we assumed a block-layered model with rectangular parallelepipeds as elementary objects. The lateral dimensions of parallelepipeds for the entire site are the same and are determined by the observation grid. The vertical dimensions are predetermined by the boundaries of layers. We used well-known formulas (Aleksidze 1987:-336) for the solution of the direct problem. For the inverse problem the adaptive method of solving the systems of algebraic (linear and nonlinear) equations (Kochnev 1988:-152, 1993, 1996:120-129) was applied. The method does not accumulate errors and makes it possible to solve the systems with a large number of equations and unknowns quantities.

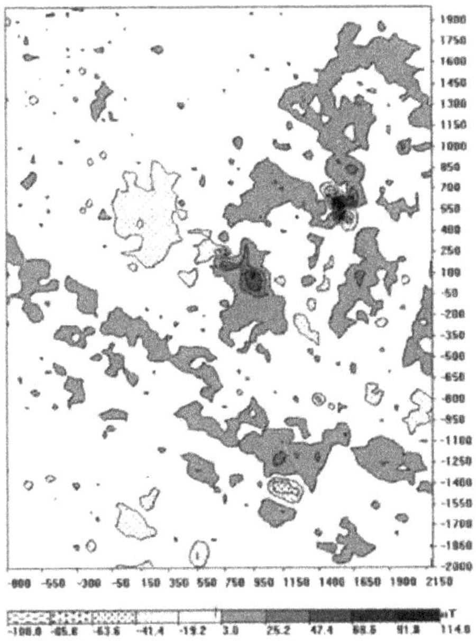

Figure 3 Initial anomalous magnetic field observed at the surface

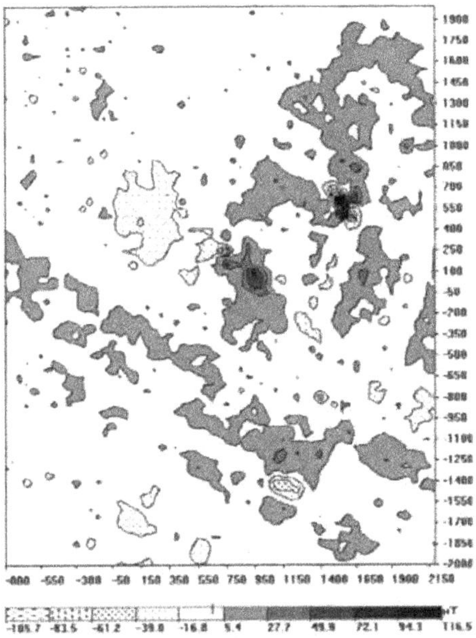

Figure 4 Modeled magnetic field

Initial Data

Let's move on to the object under study. In Figure 1 we can see the model of "Arkaim" site and the fragment which we used for the experimental interpretation (red lines). The Northwestern segment was excavated earlier, but the eastern part was not, including the area of our research. Figure 2 shows the area relief of the studied fragment. The survey of magnetic field was conducted on the grid of 0.5 x 0.5 m. The dimensions of the studied fragment are 30x40 m. The number of points is 61 x 81=4941.

The initial magnetic field (in 10-colors palette) is shown in Figure 3. Negative values are in red colors, positive - in dark-blue. We can see two strips of positive anomalies which coincide with the direction of convex relief forms. The limits of the field vary from -116 to 123 nT. Rectangular forms of slight positive anomalies can be discerned in the background (in dark green) We can also single out two strong anomalies of the complex form. The first is located in the central part (profile 0), the second one is situated to the north. (It should be taken into consideration that the anomalies go from South to North almost bias from the bottom left corner to the top right corner) On the profile -1500 we can see a negative anomaly, caused by a natural deepening. Two areas with the calm field (yellow spots in the lower and central parts of the figure) should also be focused at.

Solution of the Inverse Problem

In order to solve the inverse problem it is necessary to assign the initial approximation of model. The solution can be found while applying the problem to the single-layer model, but it proves to be unstable. Minimum discrepancy occurs at the first iteration, but it increases at the iterations following the first one. (Discrepancy is the difference between the original and modeled values of field). After a number of experiments, we accepted the three-layered model of medium with the boundaries of the layers imitating the surface relief. The layers were 30, 60 and 120 cm thick. Thus, the total thickness of model was 210 cm. It was assumed that magnetic susceptibility for all layers is 50*10-5 SI, and the initial approximation error is 10*10-5 SI.

The following discrepancies were obtained in the process of solving the inverse problem:

9.7   5.2   3.9   3.1   2.6 nT.

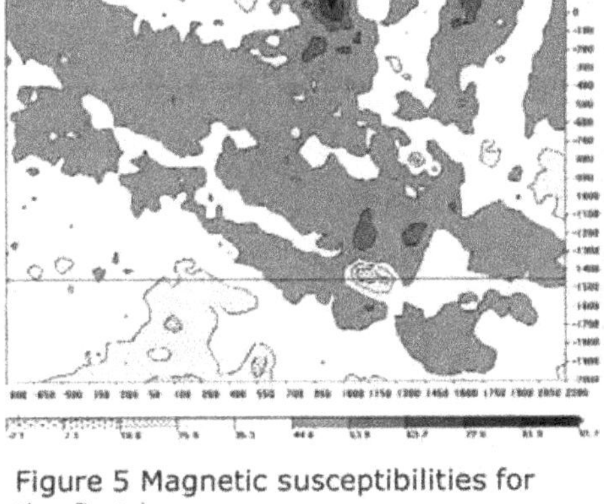

Figure 5 Magnetic susceptibilities for the first layer

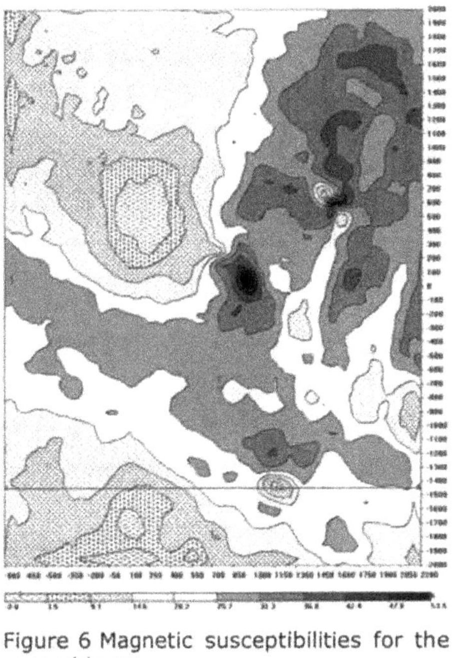

Figure 6 Magnetic susceptibilities for the second layer

In 1 hour we obtained the 3-layer model of magnetic susceptibility for the entire area of study. Figure 4 shows the magnetic field, calculated from the model. It is almost undistinguished from the initial field. The proximity of the initial and the model magnetic fields suggests that one of the possible solutions is obtained.

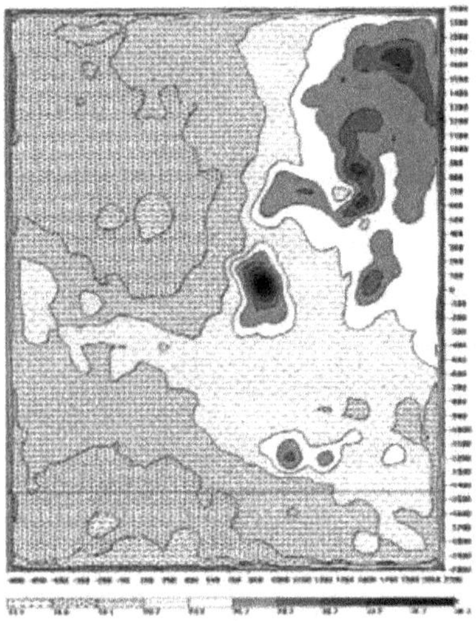

Figure 7 Magnetic susceptibilities for the third layer

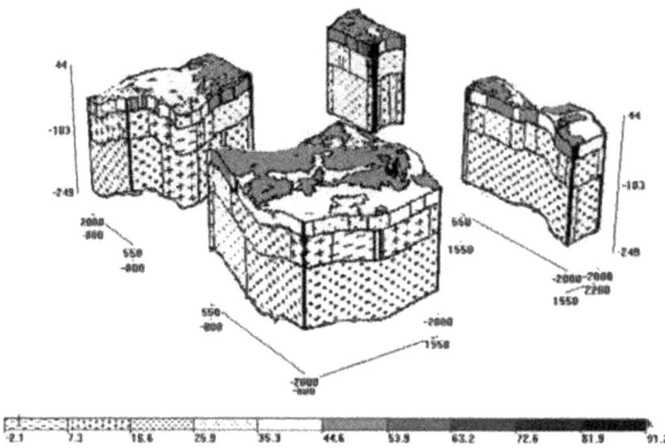

Figure 8 3D model (block variant). Vertical scale is 8 times enlarged. Vertical dimensions in cm, horizontal in meters

Analysis of the Results

In Figure 5 we see the magnetic susceptibility of the first layer. It changes from 2 to. 91*10-5 SI. At the picture we can distinctly see those special features of the forms which were visible in the magnetic field. Emphasized parts are regions of calm (not technogenic) field.

The distribution of the magnetic susceptibility in the second layer is shown at Figure 6. One can trace here the same features noted before but in a smoother form. The magnetic susceptibility of the third layer (shown here in the individual local palette) we can see in Figure 7. The range is from 13*10-5 to 38*10-5 SI. Figure 8 show the three-dimensional model of magnetic susceptibility. Vertical scale is 8 times enlarged.

Let us further look at the most interesting sections first in the left-to-right, and then in the bottom-to-top direction. On the upper sequence of Figure 9 we see a section along the profile -1500, passing through a negative anomaly, which is located to the south of the bank.

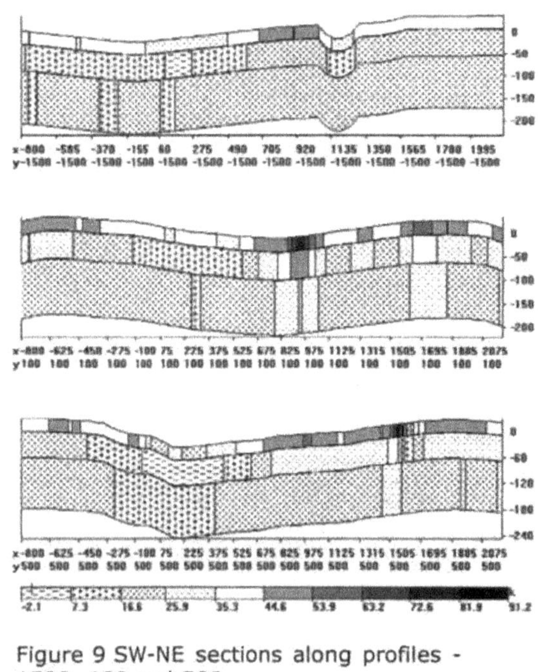

Figure 9 SW-NE sections along profiles -1500, 100 and 500

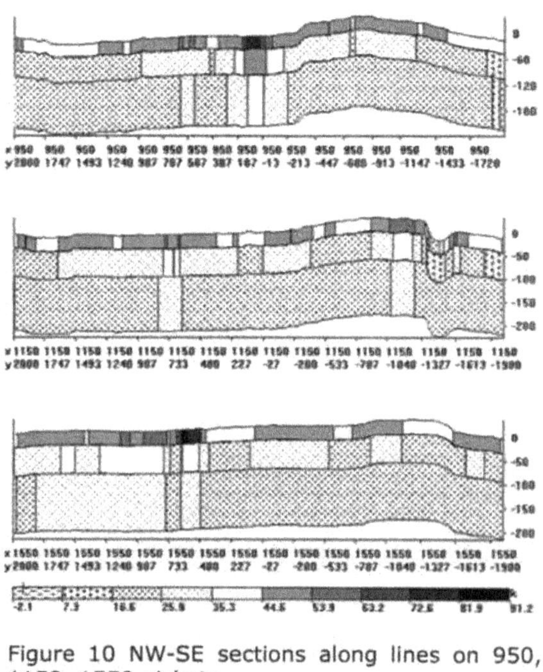

Figure 10 NW-SE sections along lines on 950, 1150, 1550 pickets

Note two specific features. The upper layer has high magnetic susceptibility while the lower level is characterized by low magnetic susceptibility. This is especially visible in the area of the decline in the relief.

The second sequence presents a section along one of the positive anomalies. Two parts with the increased magnetic susceptibility are noted here: in the center and on the right. Moreover the first one corresponds to the negative form of relief, and the second one - to the positive. The third sequence shows a section along profile +500. In the cavity on the left a reduction in the value of magnetic susceptibility can be seen. Note that the lowered regions of relief both beyond the boundary of the fortress (sequence 1) and within the limits of the fortress (sequence 3) have lowered values of magnetic susceptibility with smooth changes. Does that indicate that in this part of the fortress there are no walls of dwellings or some other constructions? Could it be a small reservoir inside the fortress, surrounded by trees?

Figure 10 shows profiles that pass from the lower part of the section to the upper. All special features, indicated on the left-to-right profiles, are visible here as well.

The Experimental Processing of 4-Level Observations

In the small section of 20x1.5 m geophysicists carried out the survey of magnetic field on 4 levels: on the surface, 5, 15 and 25 cm above the surface with a sampling interval 50 cm. The initial and modeled magnetic fields are shown in Figure 11. The upper frame is a section. Let's look at the section and compare them with the results obtained with level 1of the survey (Fig.9). They somewhat differ.

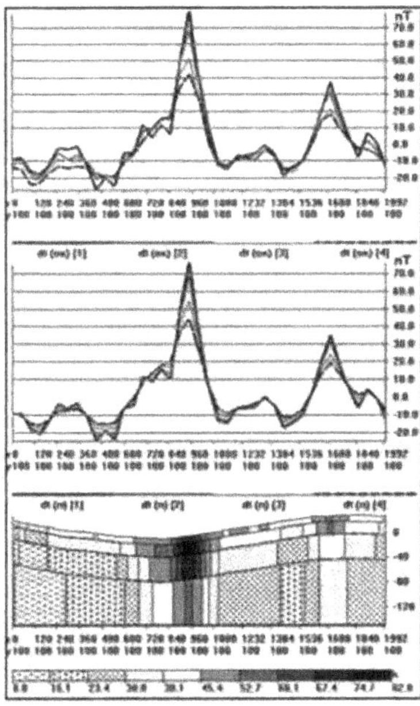

Figure 11 Four curves observations. Plots of initial dt field (up), modeled dt (middle) and section (bottom), visualized block-style

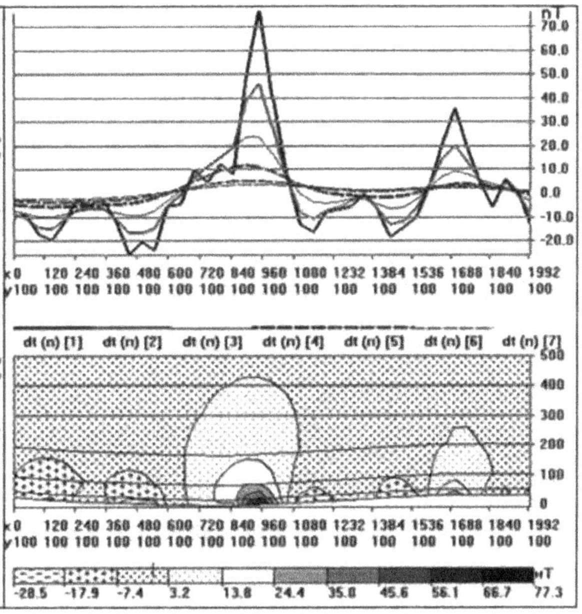

Figure 12 Magnetic fields modeled at profile 100. Plots (up) and section of field modeled on 7 levels

Let's do the calculation of the magnetic field at 7 levels, imitating the surface. The first is on the surface, and the last is at height of 3.5 m. We can see that the magnetic field at latter two levels is close to 0. Analyzing the results of simulation we come to the conclusion that the optimum level of observation could be 0, 15, 30 or 60 cm.

Brief Description of the Technology

The software package ADGM-3D serves as the basis of the technology. The number of profiles, points, layers and the number of observation levels are unlimited. The rapid and stable method of solving inverse problems in combination with the graphic possibilities gives the possibility to solve complex problems. All illustrations in this paper were obtained with the graphic means of the package.

Conclusion

We obtained a 3-layer, 3-dimensional model of magnetic susceptibility of a near-surface part up to 210 cm in depth (30x40 m) of the archaeological site "Arkaim". This approach enabled us to compose maps of each layer and sections along different paths. We show 6 sections, crossing characteristic anomalous zones. Zones with strong and weak fluctuations of susceptibility were identified on maps and sections. 3 strong anomalies were found. First - negative, beyond the fortress limits, correlated with local lowered form of relief and probably connected with the burial site. Two other ones are positive, inside the fortress, probably connected with ancient metallurgy items.

Data from 4-level survey of 20x1.5 m part were processed. These data turned out to be higher. We would recommend to use this technology advantages for choosing an optimal survey scheme.

The first experiment of the application of this technology showed that there are many interesting tasks in the field of archaeology to be worked at by geophysics and archaeologists in cooperation. We should not forget that magnetometry is a rather precise and highly productive method and at the same time comparatively simple in application. Topographic survey and mapping are the most laborious part of such works. One of the tasks for archaeologists is to analyze the results of experimental interpretation with the aid of our technology, which was created for solving forward and inverse problems in geology. We worked on this problem with great interest and enthusiasm.

Special thanks

The authors thank the co-authors of the package D.Vasil'ev and V.Sidorov for the collaboration, and also I.Goz., M.Starikova for the help in preparing and translating the work.

REFERENCES

ALEKSIDZE, M.A., 1987. Approximation methods for solving forward and inverse gravimetric problem. Nauka, Moscow.

ARKAIM, 1995. Studies and explorations. Chelyabinsk.

KOCHNEV, V.A., 1988. Adaptive methods of interpreting seismic data. Nauka, Novosibirsk.

KOCHNEV, V.A., 1993. Adaptive methods of solving inverse geophysical problems. Comp. Center of SB RAN, Krasnoyarsk.

KOCHNEV, V.A., 1996. Khvostenko V.I - Adaptive methods of solvinginverse gravimetric problem. Geologiya i geofizika 7:120-129.

KOCHNEV, V.A., VASILIEV, D.V. and SIDOROV, V.Y., 2003. The technique of solving 3-D gravity problems. SEG International Exhibition, Salt Lake City.

KORYAKOVA, L. Sintashta-Arkaim Culture (http://www.csen.org/koryakova2/Korya.Sin.Ark.html).

Archaeometry

A Dacian Sacred Enclosure in the Carpathian Mountains - Pietroasele - Gruiu Darii (Romania)

Valeriu Sîrbu
Brailei, Piata Traian, nr. 3, 6100 Braila
valeriu_sirbu@yahoo.co.uk; vsirbu@MuzeulBrailei.ro

Dan Stefan
Garibaldi, nr. 5, Ap. 23, Sect. 2, 7000 Bucuresti
dstefan@net4u.ro

Cristina Garganciuc
Serban Bogdan Stan, nr. 3, bl. 311, ap. 17, sect. 6, 7000 Bucuresti
cristal@net4u.ro

Sebastian Matei
4 Muzeul Judetean Buzau, Bd. Nicolae Balcescu, nr. 50, 5100 Buzau
sebastianmatei@yahoo.com

Abstract

The archaeological inventory discovered is from Eneolithic, Bronze Age and La Tène period.
The essential question is - what type of monument is Gruiu Darii: a settlement, a fortress, a necropolis or a sacred enclosure?
The methods used in the research of this enclosure were the following: a) Digital archives from excavation and fieldwork; b) Mapping the past - binding cartography and archaeological databases using GIS; c) Digital Photography and Video recording; d) Interpretation of imagery using Photogrammetry corrections; e) Virtual Reconstruction of the Past using 3D Virtual Reality.

Given its truncated form, with three abrupt sides, the plateau (534 m high, situated in the Buzaului Subcarpathians) was easy to defend and fortify, it provided great visibility over the hills and the plain and was surrounded by high mountains from West, North and East.

Enclosure

The fortification includes an enclosure approximately 2,500 sq. meters wide today and at least 3,500 sq. meters in the Dacian era. A total of 1,000 sq. meters of the site have been excavated, mostly the Dacian layer. The 140 meters in length of vestiges of the fortifications are still impressive, as the ruin at the base was 8 -10 meters wide and 3-4 meters high in the central area. In the last stage in the use of the enclosure (1st century A.D.), the wall, made of two faces and emplecton was 2 - 2.20 m large; today, this wall is 1.10 -1.20 m high. On the Southern and Eastern sides, the wall and a part of the enclosure have been destroyed over the centuries by the limestone quarries (Dupoi and Sîrbu 2001:13-15).

The Votive Deposits

In practice only two types of complexes were discovered in situ in the Dacian layer:
a) the votive deposits, mostly with rings and fireplaces at the bottoms
b) isolated fireplaces.

The votive deposits consist of agglomerations of archaeological materials and soil with a semi-oval profile, marked at the bottom by round or oval rings. Sometimes, the rings are extremely well outlined, with one or two lines of stones on top of one another. In other cases though, the stones are much more scattered, but one can make out their round or oval trajectory. Sometimes, stones cover the entire surface of the rings. There is a great variety of situations insofar as the shape, size and position of the rings are concerned: round or oval, ranging in diameter from 1 to 2 meters. There are also concentric rings, "mother" rings with smaller ones beside them, compact surfaces of stones (Dupoi and Sîrbu 2001:18-20). In the most part of the rings the deposits seem to be the remains of dwellings and their inventories, but we could not say why they did it that way. Some of the objects found here may have come from remote places, which leads us to the conclusion that communities living on a very wide geographical area had brought offerings. There are significant differences between the inventories of the complexes - some are "richer", others are "poorer".

Archaeological Material

Since numerous items - mostly ceramic vessels, in one piece or broken in situ - were found in the complexes, it is obvious that they did not get there by accident, but they were put on purpose (Dupoi and Sîrbu 2001:20-46). As we are dealing with items in one piece, sometimes of a high inner or utilitarian value, it is clear that the Dacians would not have given them up but for strong reason, of a cult nature.

How can we find an explanation for the great variety of the inventory found here? It could be the consequence of varied offerings requested by gods, as well as of the depositors' different professions, given that they asked for protection and wealth for their trades.

It is obvious that there is a certain chronological sequence of the votive deposits - emphasized by the vertical and horizontal stratigraphy - but their dynamics will be clear only after detailed analysis of the inventories in all the area. Dating the complexes with votive deposits in the 1st century B.C. - 1st century A.D. relies on various categories of objects: coins, fibulae, Hellenistic imports, tools, weapons and adornments, Dacian vases etc. When the Romans conquered the Dacia, in 101/102 or 105/106 A.D., the enclosure has been destroyed; after this date there is no other proof of its sacred function.

Modern Techniques Used at Pietroasele - Gruiu Darii

The research of the Dacian sacred enclosure of Pietroasele - Gruiu Darii has been taking advantage of the modern technologies only since last two years. The main goal of using new technology in support of traditional archaeological research was Mapping and Recreating the Past.

Our first objective was the development of a good digital model, not only for the plateau where the site is located, but also for the surrounding landscape. The spatial (geographical) data and the information obtained during the research and digging campaigns were integrated using the GIS technology connected with a data base engine.The objective we aimed for was to develop a complex set of tools. This approach provided the necessary support for the integration of all relevant data in archaeological research of the Dacian sacred enclosure of Pietroasele - Gruiu Darii.

For the topographical measurements made in the close proximity of the plateau where the sacred enclosure is located, traditional topographical equipment was used, e.g. theodolit and total station. The error of the determinations in this area was under 3 cm. For special points (landmarks, measurement axes, and so on) the measurements were compensated for the elimination of errors, and the accuracy obtained was of 1-2 mm.

Although the classical topographical equipment gives us high precision measurements, in some situations the use of other methods was required. Photogrammetry was the method that allowed us extremely efficient data collection, using digital photography, obtaining at the same time a good accuracy for complex situation recordings. The precision of these determinations was in all cases better then 1/20,000.This becomes an efficient work rule in case the team deals with complex spatial situations like object contours and structure traces, when recording only the points location, even if very accurate, is not sufficient. This is such an example: (Fig.3)

The need to collect spatial information from vast geographical areas (25 sq. km. or more) and the requirement that the data collected should be as accurate as possible, at least for those locations where the research takes place, forced us to use a differentiated approach, with the use of different equipment for different situations.

Figure 3

Figure 4

For the general recording of the landscape we chose the GPS technology. After the elimination of the Selective Availability by Department of Defense of USA (DoD) inserted in the system for the civilian application, a precision of 4-6 m for surfaces and 10-15 m for altitudes become possible even using standard GPS equipment, on sale for just a few hundred dollars. Using an altimetry barometer connected with the GPS equipment and by making use of the differential corrections provided by the WAAS-EGNOS technology, the precision can be further improved up to 2 m both in plan and elevation axes. We have used Garmin eTrex Vista GPSs with altimetry barometers and a WAAS-EGNOS correction technology: (Fig.1, Fig.2)

Here we needed a top drawing for the complex 25, but a usual photograph distorts the reality due to perspective and spherical deformation. Applying the necessary corrections, we obtained a good image for the drawing.(Fig.4, Fig.5)

The use of digital photography corrected by photogrammetry is well suited for complex situation recording but only for extremely small areas (typically a few square meters). Creating execution plans and drawings becomes very easy. The digital land model was created using AutoDesk applications (AutoDesk Map and AutoDesk Land Desktop) and animations and renderings were made using AutoDesk Viz. We combined traditional cartographic instruments with modern tools like total station and GPS to collect data while in the lab all of these were computed using GIS technology and a lot of companion tools. The geographical information was completed using also old drawings and maps which were computed and digitized with Auto Desk Raster Design.The space information was

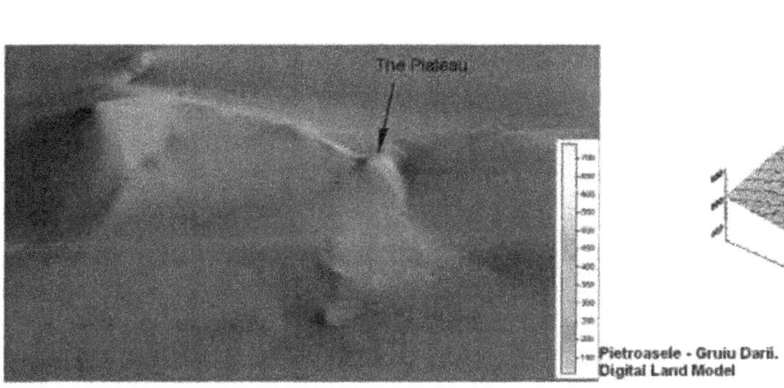

Figure 1

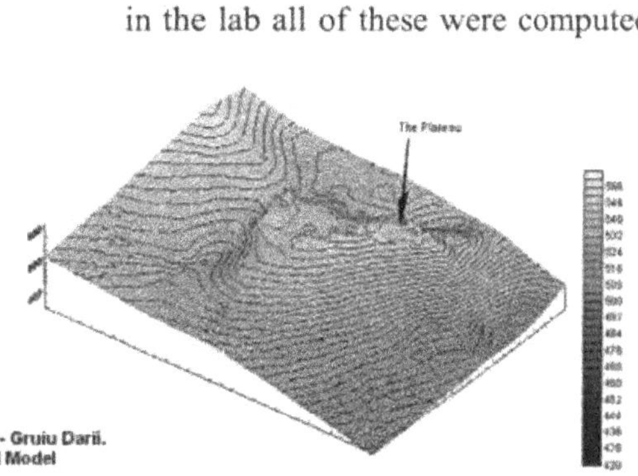
Figure 2

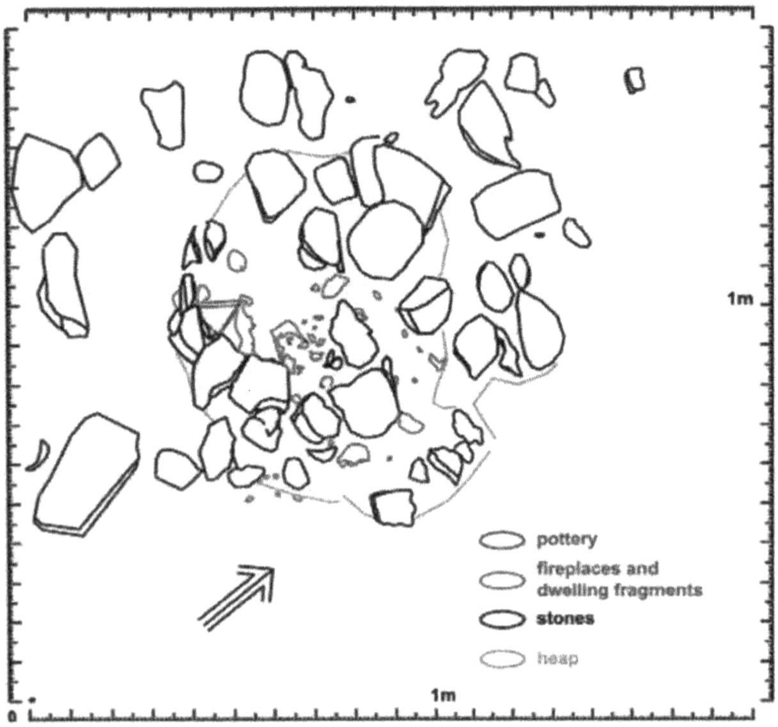

Figure 5

connected with the inventory databases and many other records and can be queried using a powerful search engine, Inter base SQL Server. The front-end was developed in Borland Delphi.

With the help of Virtual Reality technology we tried to recreate some aspects from the past of the Dacian sacred enclosure. Renderings and animations were made in AutoDesk Viz. while 3D objects were exported to Macromedia Flash to minimize the size of the files and making them accessible in Multimedia Presentations and on the Internet.

The discoveries made so far could let us imagine the following sequence of the rituals for the complexes with ring and fireplace. First, they made stone rings and lighted fire in the fireplaces inside them, thus marking and consecrating that space by certain rituals. Then, they put in the rings fragments of dwelling walls, complete vases, vases broken in situ and a lot of vases fragments that cannot be completed, tools and utensils, jewellery and clothing accessories, weaponry and coins, anthropomorphic and zoomorphic figurines, toiletries and harness gear etc. In the end, the mound type deposit was covered, more or less compactly, with stones and soil. It is also obvious that depositing entire items - vases, tools, weapons and adornments etc. - namely giving up useful goods, sometimes with an intrinsic value, cannot be explained without a cult motivation.(Fig.6)

Final Considerations

The essential question is - what type of a monument is Gruiu Darii: is it a settlement, a fortress, a necropolis or even a sacred enclosure? We will start with what the site of Gruiu Darii could not be.

It is not a fortress with a strictly military function because these ones had totally different features. It is not a fortified settlement, as no dwelling complexes, no household annexes, no workshops were identified. Why it is not a necropolis? Because: a) non-cremated human bones have been found only in 4 of the 170 complexes (some of these findings may suggest human sacrifices); b) neither cremated human bones, nor cremation pyres have been identified; c) in some vases with a lid there were only fragments of fired clay walls or some remains that could suggest some food offerings food, drinks, boneless meat). Consequently, we couldn't consider them to form a necropolis (Sîrbu 1993:39-40).

On the contrary, the votive deposits and the inventory discovered here let us assume that we are dealing here with a sacred enclosure.

What reasons made them act like this? Which deities were invoked? Who were the deposits' "recipients"? Was the entire enclosure meant for votive deposits or only a part of it? What are the dynamics of the deposits? What was the ceremonies scenario? There are a lot of questions here that cannot be answered until the entire enclosure is excavated and more written information is found. Doing otherwise is overspeculating.

What is also beyond any doubt is that, given its monumentality and the richness of the votive deposits, Gruiu Darii was an important sacred centre (Sîrbu 1995:313-330) and that analyzing and interpreting the vestiges will substantially enrich our knowledge about the spiritual life of the Getae-Dacians (Crisan 1986, Sîrbu and Florea 2000).

Figure 6

REFERENCES

CRISAN, I.H., 1986. Spiritualitatea geto-dacilor. Bucuresti.

DUPOI, V. and SÎRBU, V., 2001. Incinta dacica fortificata de la Pietroasele-"Gruiu Darii". jud. Buzau I, Editura Alpha, Buzau.

SÎRBU, V., 1993. Credinte si practici funerare, religioase si magice în lumea geto-dacilor. Editura Porto-Franco, Braila/Galati.

SÎRBU, V., 1995. Un nouveau type de monument sacré chez les Géto - Daces. Acta Musei Napocensis 32, Cluj/Napoca:314-329.

SÎRBU, V. and FLOREA, G., 2000. Les Géto-Daces. Iconographie et imaginaire. Fondation Culturelle Roumaine, Cluj/Napoca.

A New System for Interactive Vessel Reconstruction and Drawing

Francisco Javier Melero
Dpt. Lenguajes y Sistemas Informáticos,
Universidad de Granada
fjmelero@ugr.es

Alejandro J. León
Dpt. Lenguajes y Sistemas Informáticos,
Universidad de Granada
aleon@ugr.es

Francisco Contreras
Dpt. Prehistoria y Arqueología,
Universidad de Granada
fccortes@ugr.es

Juan Carlos Torres
Dpt. Lenguajes y Sistemas Informáticos,
Universidad de Granada
jctorres@ugr.es

Abstract

We present an easy, automatic and efficient full vessel reconstruction and drawing system, which works using a mesh model of the fragment taken from a 3D scanner.
To get the appropriate orientation, the archaeologist selects some areas of the border. A genetic algorithm is then applied to the points on these areas obtaining the plane that best fits the border of the fragment.
The profile of the vessel is also interactively chosen by the expert, who draws it on the fragment as in a graphical editor. This profile is also editable, it being possible to eliminate, move or insert points. From this information, the system generates the unknown part of the profile using splines.
Finally, the system makes it possible to create a 3D virtual view of the complete vessel and the traditional 2D drawing, such that it can be edited (i.e. by positioning the original fragment or placing measurement lines).
The software is very efficient, enabling the creation of the drawings in a few minutes.

Introduction

One of the more tedious tasks after an archaeological excavation is to manage the thousands of small fragments of pottery that may appear on the site, trying to select those which offer more information so as to be able to reconstruct the original shape of the vessels, to draw them and take the measurements necessary to classify the pots. Once the sherds from a vessel have been selected, the process of reconstructing it can take about fifteen minutes, and involves five steps:

1. Orientation. The sherd is placed on a table, with its rim completely in contact with the table surface. In this position, the vertical projection of the sherd is measured; the sherd is oriented by placing it in such way that the vertical projection is the same as the previously measured (see Fig.1a).

2. Diameter estimation. Carried out using a bordimeter (Fig.1b), i.e. is a set of concentric circles drawn on a sheet of paper. The diameter of the vessel at the rim is the diameter of the circle that best fits the external arc of the rim.

3. Profile estimation. The profile is the cross-section of the fragment in the direction of the rotational axis of symmetry (Kampel and Sablatnig 2000). To extract the profile, the expert uses a slide caliper, and measures the width of the sherd at different positions (Fig.1c)

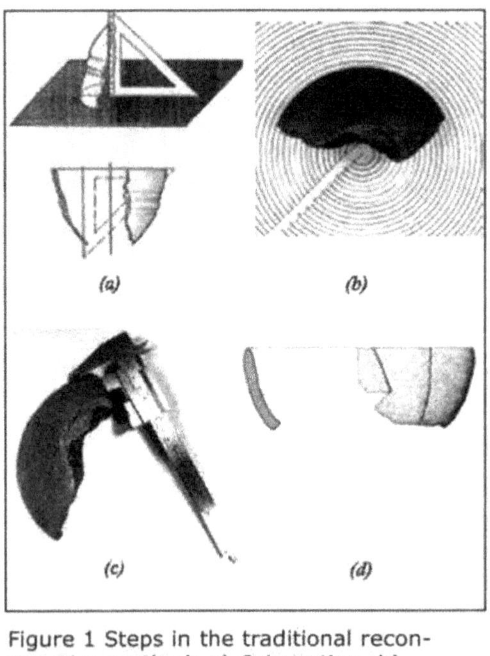

Figure 1 Steps in the traditional reconstructing method: a) Orientation, b) diameter estimation; c) Profile extraction; d) Drawing

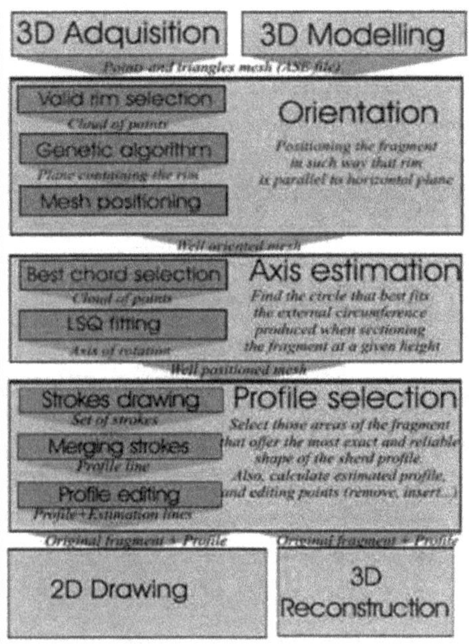

Figure 2 Scheme proposed in this paper

4. Drawing. Based on the measures taken previously (diameter, vertical projection, profile), an artist makes a pen-and-ink drawing, following a standard that establishes the appearance of the drawing: lighting and shading, stippling lines, textures, etc... (Fig.1d)

5. Additional measurements. For example, the diameter at different heights, rim angle, etc. that help the archaeologist to classify the vessel.

The error made via this procedure has been estimated at ± 15% (Leonardi and Penello 1991).

OUR APPROACH

We have designed a software tool to carry out this task in a semi-automatic way. The software has been developed under the Windows platform, using C++ as programming language and OpenGL as graphic library (Davis et al. 1993).

One of the aims of this software is that the archaeologist should feel comfortable working with it. The goal was not to build "perfect" software which would leave no room for the expert to interact in the reconstruction process, giving the solution without taking into account the archaeologist's experience and knowledge.

Therefore, our system interactively follows the same steps as in the traditional procedure (Fig.2).

[1] Data acquisition

Our goal is to be able to use a 3D scanner to capture the polygonal surface of the sherd, and then manipulate it with the software. However, although scanner technology has improved in both accuracy and speed, it is still very expensive and the error margins are unacceptable when small sherds are scanned.

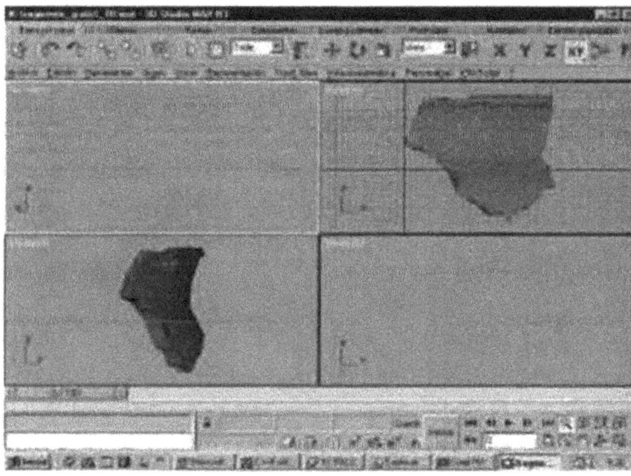

Figure 3 Constructing the synthetic models

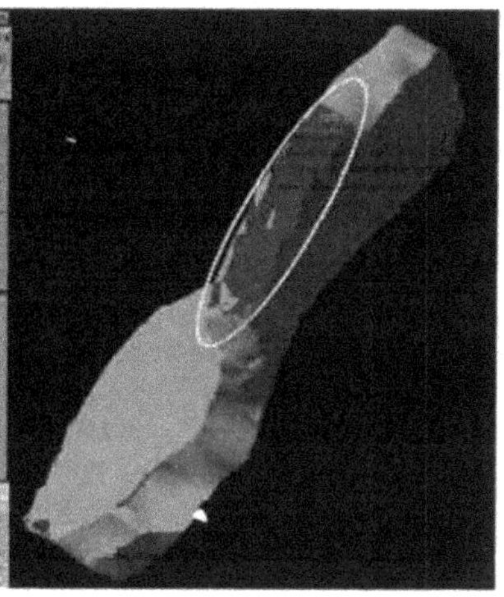

Figure 4 Area of the rim selected as valid

In fact, these errors are a consequence of the high accuracy demanded, but from our point of view, no more than one point per millimetre is necessary to get a good mesh.

Due to these technical problems, during development phase we used synthetic models generated with 3D Studio Max, based on real profiles and with irregularities like real ones (Fig.3). Thus, we were able to test the method without the problem of acquiring tri-dimensional data.

[2] Orientation

Orienting a sherd merely involves finding the plane that contains the rim and rotating the sherd in such way that the plane is parallel to the XZ plane.

This problem may be solved via Least Squares Fitting or other statistical or mathematical methods, but we have chosen a more flexible approach: Genetic Algorithms (GA) (Michalewicz 1996, Reynoso and Jezierski 2001). These algorithms are easy to implement and allow one to change the goodness criteria, it being possible to make the method more robust by adding constraints to this criteria (e.g. the incident axis of the surface normals).

A GA is based on a set of individuals, each one defined by a chromosome, which is a solution for the problem at hand.

In our case, we have a chromosome composed of three genes, each one corresponding to a point of the selected rim. More precisely, the rim is divided into three zones (rightmost, central and leftmost) by a k-means algorithm, and the chromosome contains one gene for each zone. These three points define a plane, and our aim is to evaluate the goodness of this plane.

The goodness criteria for our chromosome (plane) is to consider the number of points of the rim contained by the plane and the number left to one side of the plane. That is to say, it's not sufficient to have a plane containing a lot of points, but it's also necessary for it to lean on the rim, just touching it, not crossing through the middle of the fragment.

The goodness of a chromosome c can be expressed as:

Where Rc and Lc means the number of points at right and at left respectively, of the plane defined by c, and Nb is the number of points selected as a part of the rim. Pc is the number of points whose distance to the plane is below a threshold. The rest of the parameters of the GA (e.g. cross and mutation probability) can be modified by the user.

In our software, the user must select the area of the rim that is useful for this step (Fig.4), seeking to avoid noisy parts of the rim like bubbles, breaks and other irregularities. This area is used for generating the population of the GA and finding the plane.

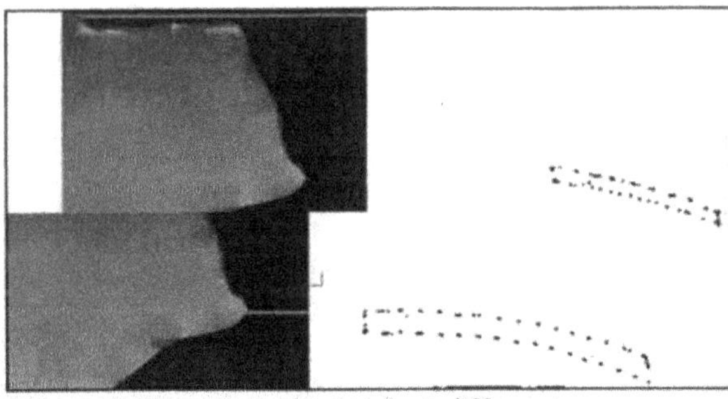

Figure 5 Two arcs obtained at different heights of the sherd by cutting it with a horizontal plane

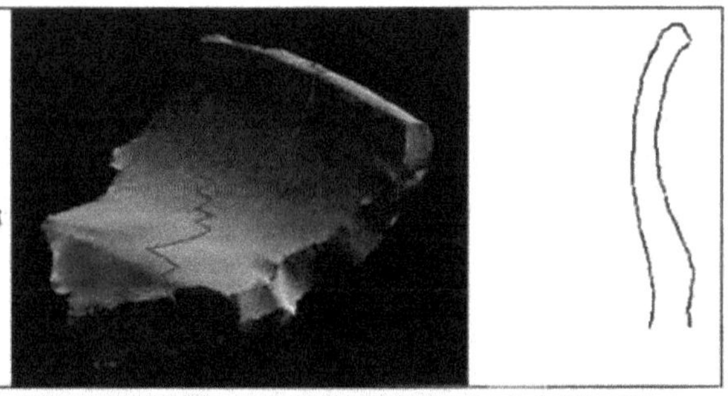

Figure 6 Profile extraction. Note at left the strokes on the surface of the sherd, and at right the profile obtained

[3] Diameter estimation

Estimating the diameter of the original vessel is analogous to finding the rotational axis of the pot. As we work with prehistoric ceramics, made without a potter's wheel, its surfaces are very irregular, and it is not easy to find the axis by calculating the incident axis of the surface's normals (Halír 1997).

We use a procedure similar to the traditional one:

- The expert selects a horizontal cut through the sherd by moving a horizontal plane with the mouse (Fig.5). The expert may have selected the longest arc or the one with fewest irregularities.
- From this arc (in fact, its a double arc: the external and the internal), the problem is now that of finding the centre of the circle that best fits the external arc.

To fit the circle to the arcs, we follow a LSQ fitting algorithm. First, we fit a circle using both arcs, so it will pass between them. Then, we remove the internal points and repeat the procedure with the external ones, so the new circle fits by minimizing geometric distance with the sequence of points that define the external arc.

The rotational axis is the vertical line that passes through the centre of the calculated circle.

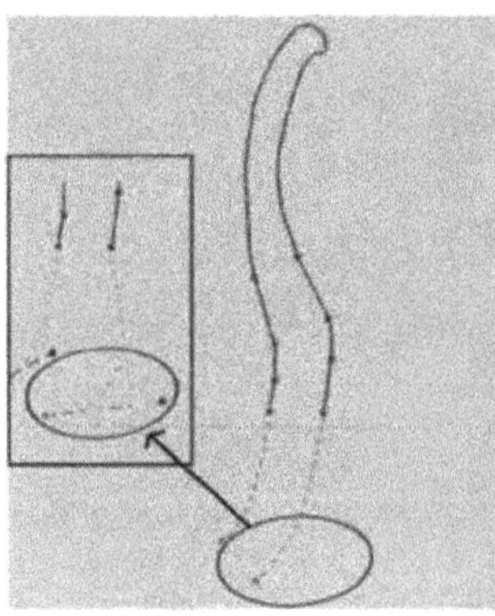

Figure 7 (left) Editing the estimated profile. Note the tangent has been moved and, due to this, the estimated profile has changed.

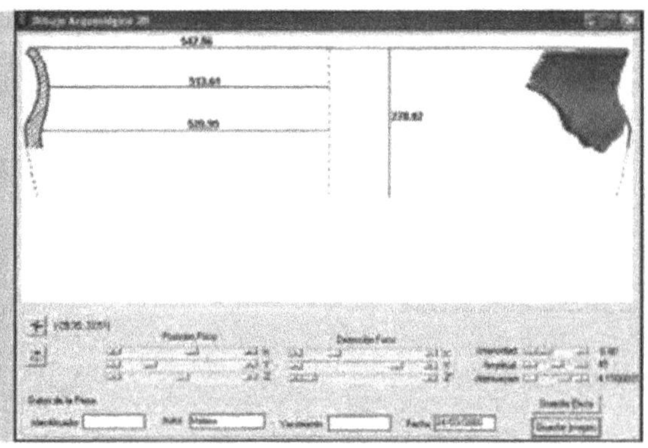

Figure 8 (right) 2D Drawing following the standards and with the measurement lines

[4] Profile extraction

From the tri-dimensional model of the sherd, the expert selects those areas considered to best define the shape of the profile. This step is done by drawing lines over the sherd, using the mouse, as if it were a pen over the real sherd (Fig.6), thus obtaining a set of 3D lines that are projected over the planes containing the previously calculated rotational axis. This can be done because we have previously translated the point coordinates into polar coordinates; when selecting a point, we take the module and the azimuth of its polar coordinate, and so it is easily translated into 2D coordinates.
Given the sequence of 2D lines, the system fixes them in an unique stroke, detecting overlapping and irregularities. This profile is fully editable, being possible to add, eliminate or move points, so, if the software does not remove all irregularities, the expert can do it in a few seconds.

As described above, the drawing of the vessel includes an estimated profile, that indicates how the profile should have continued if the whole vessel had been available. The length of this estimation differs from one sherd to another, because it depends on the typology and the experience of the expert, but standard practice is to extend it by about two centimetres.

The software generates a first approach to this prolongation by a cubic spline that follows the shape indicated by the lowest 15% part of the profile. This spline is calculated by taking 4 points separated by 5% of the height of the profile obtained.

As typologies in prehistoric ceramics are very hard to establish, and very restricted to small areas, we consider that the experience of the expert is very important at this point, so we also allow him/her to edit the estimated prolongation, by moving the points or the tangents at the end of the prolongation (Fig.7).

[5] Drawing and 3D reconstruction

Once the profile and the rotational axis have been obtained, it is an easy matter to obtain the traditional drawing of the ceramic (Fig.8).

This step of the reconstruction process is also interactive, because the system allows the expert to:

- Decide the position of the sherd in the drawing, depending on the details he wants to show. As in manual drawing, a false perspective is used, to avoid hiding part of the surface when placing the sherd at left or rightmost position.
- Modify the lighting, to improve the visualization of several details of the sherd.
- Take measurements over the drawing, such as the diameter at several points, height of the sherd, etc... These measurements can be translated, modified, etc.

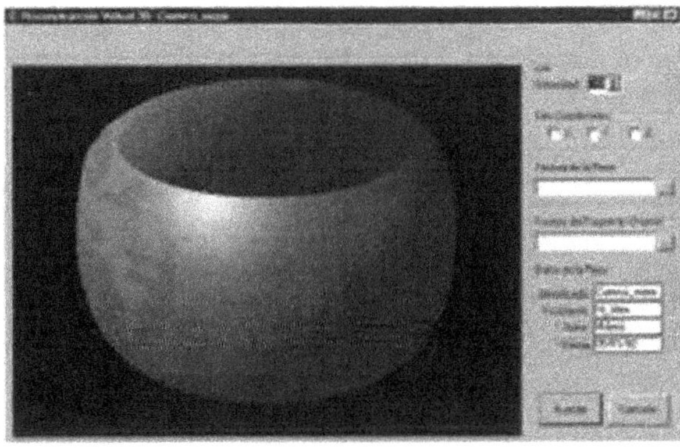

Figure 9 Virtual reconstruction of a pot. Note the original sherd and the lighting

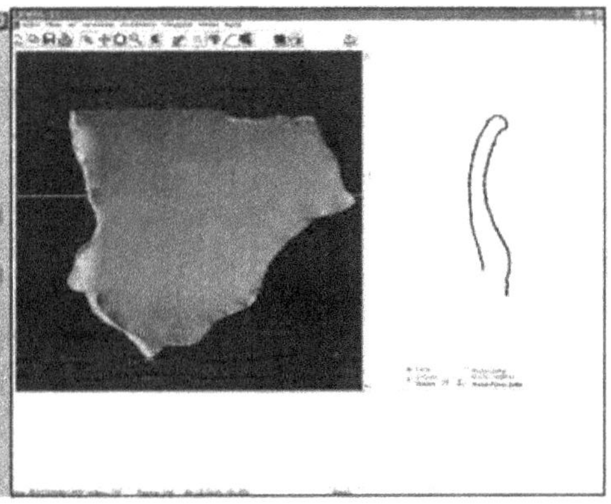

Figure 10 The software working at the profile extraction step

The image can be saved as a GIF, JPEG or BMP image, and so it can be edited by any image editing software.

The tri-dimensional reconstruction of the pot is generated by rotating the profile (both the real one and the estimated prolongation) around the rotational axis. The surface corresponding to the estimated part of the pot is shown with a semi-transparent texture, in order to identify clearly both parts of the reconstruction. The original sherd is shown in the vessel, as in real reconstructions, so the goodness of the reconstruction and the exact place where the sherd was located can be appreciated (Fig.9).

The reconstruction can be rotated and zoomed to obtain different views of the vessel, each of which can be saved also as an image.

Conclusions and Future Works

We have proposed a software that reduces the time required to draw and measure a sherd by 80 %.

We have tested it with several synthetic models, and the error margin is reduced by 60% with respect to the traditional method. It has been estimated by reconstructing some pieces several times, and computing the average and standard deviation of measurements on each reconstructed piece. We have got an error margin of +/- 5%. So, we believe this system can be very useful for archaeologists.

The most serious problems were encountered in scanning the sherd, and so the software won't be fully operative until the scanning procedure is capable of providing sufficient accuracy and speed.

Our intention is to add other capabilities to the software such as:
- morphological measurements, such as the angle of the rim, volume, etc...
- realistic textures for the tri-dimensional reconstruction
- pen-and-ink textures for traditional drawing
- the addition of a database to register all measurements for later statistical studies.

More detailed information about the software can be found at http://lsi.ugr.es/ ~fjmelero

Acknowledgments

We thank the Dpto. de Informática of the University of Jaén for their help with the scanner.
This work was partially supported by the MCYT (Science and Technology Ministry) by project TIC2001-2099-C03-02.

References

DAVIS, T., NEIDER J. and WOO M., 1993. OpenGL Programming Guide. Addison-Wesley, New York.

HALÍR, R., 1997. Estimation of rotation of Fragments or Archaeological Pottery. In Proceedings of the 21th Workshop of the ÖAGM, Hallstatt, Austria.

KAMPEL, M. and SABLATNIG, R., 2000. Computer Aided Classification of ceramics. VAST'00.

LEONARDI, G. and PENELLO, G., 1991. Il disegno archeologico della ceramica. Saltuarie dal Laboratorio del Piorego 2, Padova.

REYNOSO, C. and JEZIERSKI, E., 2001. A Genetic Algorithm solver for Archaeology. 29th Conference of CAA: Pushing the Envelope (BAR International Series 1016).

MICHALEWICZ, Z., 1996. Genetic Algorithms + Data Structures = Evolution Programs. Springer-Verlag.

Application of 3D Modeling in Ship Reconstruction and Analysis: Tools and Techniques

George Indruszewski
GWZO Leipzig/NMF Roskilde

Gerald Farin
PRISM, ASU, Tempe

Anshuman Razdan
PRISM, ASU, Tempe

Arleyn Simon
ARI, ASU, Tempe

David Van Alfen
ARI, ASU, Tempe

Jeremy Rowe
PRISM, ASU, Tempe

SEE THE CD FOR THE EXTENDED VERSION

Abstract

Using different data capturing techniques (manual drawing, laser pen-aided drawing, MicroScribe and FARO arm drawing, adjustable 3D pin and cardboard models) the hull of a vessel is reconstructed for further manual and/or computerized analysis (I-ship program). The application of 3D computerized modelling techniques has hitherto been used for visualization purposes only (Rhinoceros software). The application of 3D scanning coupled with the specific use of customized software (by PRISM) reveals the potential for analysis of symmetry in reconstructed hulls. This pilot study used a scaled hypothetical model of a partially preserved Viking Age vessel found in the Roskilde Fjord, Denmark. Through the symmetry analysis of the hull of this vessel we can 'rediscover' hidden design aspects of eye-built hulls, identify technological assembly processes, and facilitate accurate reconstructions of scaled and full-sized models. The effectiveness of 3D scanning for data capture and analysis suggest its applicability in ship reconstruction and analysis.

Ancient ships pose complex problems in all aspects of their research. Archaeological investigations of shipwrecks are conducted in stages, starting with data capture and ending with object display and/or reconstruction. Although much data capture is obtained through elaboration of 2D drawings, photographs, and manual measurements of 3D objects, idealized models are drafted with AUTOCAD or other drawing programs. Recent advances in computerized visualization have resulted in an ever-increasing use of 3D models for presentation of reconstructed or preserved objects. We summarize the advantages and limitations of data recording techniques used in marine and nautical archaeology, focusing on the potential of 3D scanning technology to capture, archive, and analyze accurate, detailed 3D geometric models of ship hulls.

The hull of a vessel may be reconstructed using different data capturing techniques (manual or laser pen-aided drawing, MicroScribe and FaroArm digitizing, and adjustable 3D pin and cardboard models) for further manual and/or computerized analysis (I-Ship program) or 3D visualization (Rhinoceros software). In this study, we used a scaled hypothetical model of a partially preserved Viking Age vessel (from Roskilde Fjord, Denmark, http://vikingeskibsmuseet.dk) and applied 3D scanning and customized software (developed for the 3DK project (Shurmans et al. 2002) by PRISM, http://3DK.asu.edu) to examine the potential of symmetry analysis for reconstructed hulls. Through such analyses, we can rediscover hidden design aspects of 'eye-built' hulls, identify technological assembly processes, and facilitate accurate reconstructions of scaled and full-sized models. The effectiveness of 3D scanning for data capture and analysis supports its further use in ship reconstruction and analysis.

Recording Technology and Techniques. Viking Age ship finds recovered from archaeological contexts are a major information source on shipbuilding and seafaring (Crumlin-Pedersen 1997), but the challenges of data collection increase the complexity of reconstructing the maritime past. Hulls are often only partially preserved and without major structural timbers such as stems and sternposts as the result of a deliberate action (dismantling) or of formation processes. Northern European ships from the first millennium AD are often found without accompanying artifacts, which make reconstruction of the ship's hull imperative. These ships were built by eye and 'rules-of-thumb' and have no surviving written documentation. In order to rediscover the shipbuilder's knowledge of that period, the researcher has to piece together the recovered ship timbers to best resemble its original shape. The technological process is assessed to rediscover the mental template(s) used as a blue print in shipbuilding (Indruszewski 2000). For this purpose, expert knowledge and accurate measurements of preserved timbers are vital for the reconstruction of complex 3D ship shapes. Thus, nautical archaeologists are involved in 3D modelling and reconstruction from the very beginning and by the nature of the objects of interest.

But to do this, each recovered ship timber has to be measured and drawn by edge tracing; each timber is laid horizontally under a clear Plexiglas plate positioned parallel with the surface; edges and all other features pertaining to use-wear, functionality, and material are drawn on clear film affixed to the plate. Although this method is reliable, it also contains some difficulties; superimposed edges in the same plane are difficult draw in 2D. Tracing irregular and deformed edges and shape-ends has proved to be tedious, requiring many hours of effort. The amount of information rendered on a surface, encompassed between four inter-related edges or surrounded by a contour line, is a representation of the visual perception of the draftsperson at the time of drawing.

In 1998, an improvement was made by attaching a permanent marker (used in free-hand tracing) at an angle to a laser-pen set perpendicular to the drawing plane. By moving both pens as a unit, the operator is able to simultaneously trace the edge or any other feature on the object and on clear film. The result was improved drawing accuracy obtained with less time and effort. The drawings are scanned, scaled down, and printed for building a 3D cardboard model, which facilitates the reconstruction of the hull itself, the relocation of each timber, and importantly, defines the sequence of the assembly process. Once finished, the model is used for recording the ship lines from which the final ship drawings are elaborated. These drawings constitute the template used in the construction of full-size replicas and scaled models utilized in the study of practical navigation, technological processes, and analysis of hull hydrodynamics. The analysis is carried out at present by utilizing a software package called I-ship and NMF-ship, designed to construct a 3D hull and calculate several parameters such as main and additional dimensions, hydrostatic data, and form coefficients.

The MicroScribe and FaroArm devices recently replaced the manual edge tracing method and are used to digitize various ship timbers and/or entirely preserved hulls. By dragging the FaroArm pointer on the digitizing surface instead of marking points at regular or random intervals, the operator can produce computerized feature lines with less time and effort (3 hours for the Kolding cog knee and 1.5 days for the Hjortspring vessel hull). These lines of individual ship timbers are assembled into a hull 3D view with Rhinoceros software. However, the data capture is the electronic version of the manual edge-tracing method and is dependent on the operator's perception of the artefact; subtle surface deformations or deviations in planar shape may be missed.

Differences between a 3D computer model and a 3D cardboard model become obvious while assembling ship timbers into a hull. Electronic edge tracing methods can't be applied in special cases where hulls, such as the 10th-century AD Ladby graved vessel, have been totally decomposed and where the only remaining evidence is rivet imprints and stains. In this case, a 3D model made of movable pins was required to obtain the lines of an otherwise invisible ship shape.

The Rhinoceros program is problematic for the recording and documentation stage and as an end-process for 3D visualization. Attempts to tie Rhinoceros to NMF-ship that would result in visualizing the hull and completing calculations on it have been unsuccessful. The latest version of Rhinoceros has enhanced interface capabilities and can carry out some hydrostatic calculations; but these standard calculations are not appropriate for Viking Age vessels because of the way these ships were built. Rhino cannot calculate displacement for hulls with 'naked edges' (that is, clinker-built hulls with strakes that overlap each other) below the designed waterline (DWL). The issue is further complicated by the fact that these hulls were not always symmetrically built, thus posing more calculation stress for Rhinoceros and also NMF-ship software (Jensen 1999).

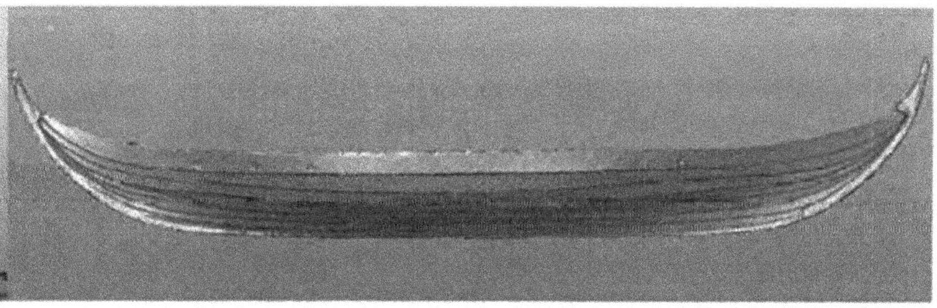

Figure 1 (left) Scanning of the scale model Viking ship using a Cyberware 3030 scanner at PRISM
Figure 2 (right) 3D model side view of scale model Viking ship

Scanning the Scale Model Ship. In 2002, the 3D laser scanning method was introduced in an effort to document sequential working stages in the recording, analysis, and reconstruction of ship hulls. A scaled model of the Skuldelev 3 replica, the Roar Ege, was built as a finite hull. Because the stern of the original sunken ship was not preserved, the dimensions and shape are approximated and mirrored. The scale model serves as a proxy and is used to assess hypothetical variations in symmetry of various sections of the ship and to examine how the shipbuilders made adjustments for the placement of the rudder on one side of the stern.

The ship model was scanned at the PRISM lab with a Cyberware 3030 Laser Scanner (with RGB colour) to capture three-dimensional geometry (Fig.1).

The 3D model was built from a total of 48 scans (4 sets of 12 rotational scans) to capture the detail of the interior and exterior surfaces. The individual scans are merged to create one file. The resulting coordinates (a 3D point cloud made up of individual x, y, and z coordinates) are fitted with a triangulated mesh using the Cyberware software. This produces a full-rotational 3D model of the object (Fig.2), not just the appearance of 3D obtained by animation of a series of 2D images (QuickTime VR).

As in Rhino or I-ship, the 3D model may be rotated in all directions for close examination of features on the interior and exterior surfaces. An oblique view of the inside illustrates the internal structure of the hull (Fig.3) and the end-view

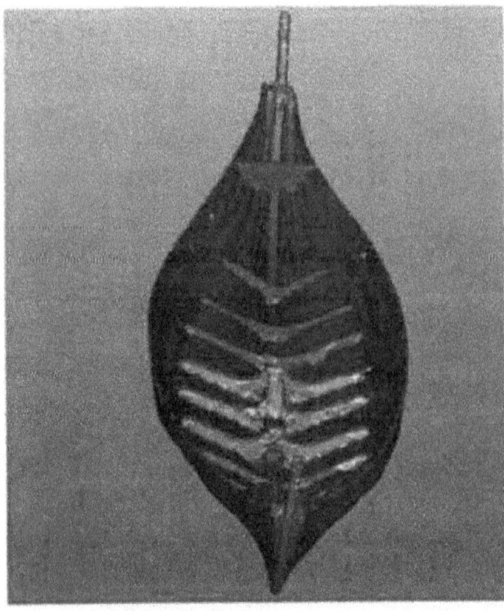

Figure 3 3D model oblique view of scale model Viking ship

Figure 4 3D model end view of scale model Viking ship

offers a profile of the ship and the hooding ends of the strakes (Fig.4). The 3D model file is reduced in size through a process called decimation (reducing the number of triangles to those essential to defining the object) to make it more manageable in the analysis software, but retaining the necessary data to define the 3D form in detail.

Curvature and Symmetry. The methods used in this pilot study are drawn from those developed by PRISM (http://prism.asu.edu) for prehistoric ceramic vessels (Simon 1998, http://archaeology.asu.edu). As most ceramic vessels have a vertical axis of rotational symmetry, a profile curve is obtained by cutting a vertical cross-section through the vessel. The archaeologist controls the placement of three orthogonal planes to create a cutting plane best representing the profile of the vessel (Simon et al. 2002). The resulting string of points is fitted with a least squares cubic (degree 3) B Spline curve within a certain tolerance. The number of control points and the parameterization of the B Spline curve are automatically computed (Bae 1999) to smooth out the noise in the data and present the profile curve as a mathematical representation.

The curvature plot represents how the curve bends and flattens (Farin 2001, Razdan and Farin 1998); given the curvature plot one can recreate the curve (barring position and rotation in space). Since the profile curve is a planar curve, its curvature plot is also two-dimensional; hence we can assign a sign to the curvature; i.e., positive or negative. Where the curvature plot crosses the zero line is called an inflection point, and these tell precisely where the curve changes from being convex to concave or vice versa. Inflection points and corner points (points of high curvature) are two important diagnostic features that are more easily detected from the curvature plot than the original curve.

Symmetry can be examined by mathematically comparing the two halves of the curvature plot. If the vessel is highly symmetrical, the two halves will not differ, but if the vessel is uneven, there will be considerable differences between the two. Symmetry is scored on a standardized scale (1 = high and 0 = low):

Let $f(x)$ be a (nonzero) function defined over $x \in [0,1]$. If f is symmetric with respect to $t = 1/2$, then $f(x) = f(1-x)$. A measurement σ for the symmetry of f is given by

$$\sigma f = \frac{\int_0^1 [f(x) - f(1-x)]^2 dx}{\int_0^1 [f(x) + f(1-x)]^2 dx} \quad (1)$$

Further, a unit measurement σ' for the symmetry of f is defined by $\sigma' f = 1.0$ (2)

Thus $\sigma' f = 1$ for a symmetric function.

Using this approach, the software developed by PRISM allows the quantification of the symmetry of 3D forms based on user-defined cross-sections. The symmetry measure is based on a difference in area between the two halves of a two-dimensional graph of the magnitude and direction of curvature (i.e. the curvature plot). This symmetry measure has been tested extensively on prehistoric pottery collections from Arizona and Mexico and yields interpretive results that are used in archaeological studies of ancient pottery technology.

Although the ceramic analysis software has not yet been modified for objects with longitudinal (i.e. mirrored) symmetry, the user-defined sectioning with profile curve and profile plot can be successfully applied to the hull model example (Fig. 5). In this hypothetical case, we sampled cross-sections from the mid-ship, near the bow, and near the stern and these were evaluated against the two possible extremes (1 = symmetrical and, 0 = asymmetrical). While we did not expect perfect symmetry due to cumulative variation from the individual strakes, it is informative to examine the variation in the scaled model as a proxy for full-sized ships. The section near the bow is the most symmetrical (.7095), the mid-ship section is slightly less symmetrical (.6204), whereas the section near the stern has extremely low symmetry (.0467).

The resulting symmetry measures indicate that in a hypothetical scaled model of Roar Ege, the bow of the ship is the most symmetrical, whereas the stern is the least. If such asymmetry was documented in the actual archaeological specimens and was an intentional part of ship building technology, such asymmetry in the stern could potentially be related to the attachment and hydrodynamic role of the side-rudder against the hull. The challenge has been to obtain direct and accurate 3D data with which to quantify asymmetry to study such questions. Further research with 3D scanning of highly accurate reconstructions where the bow and stern are more completely preserved will be needed to fully test this hypothesis. However, this pilot study demonstrates the proof-of-concept that highly accurate scanning provided by 3D technology and customized analytical software has the potential to address these questions.

Conclusion. Access to 3D scanners, equipped to capture data

on different size ranges of objects, is becoming more economically and technologically feasible for researchers in many disciplines. Recent advances in 3D scanning applications and CAGD modelling include the capturing of detailed and accurate point cloud data (x, y, and z coordinates), a surface texture map, and curvature modelling. With analysis using customized software, quantitative measures such as volume and symmetry can be obtained. A further benefit is the ability to make rapid prototypes of the scanned models at different sizes, for study and visualization. Although the current analysis software, designed for rotational symmetry, does not allow the generation of longitudinal profiles (buttock line sections) of the ship, PRISM researchers are developing software for the analysis of other types of symmetry.

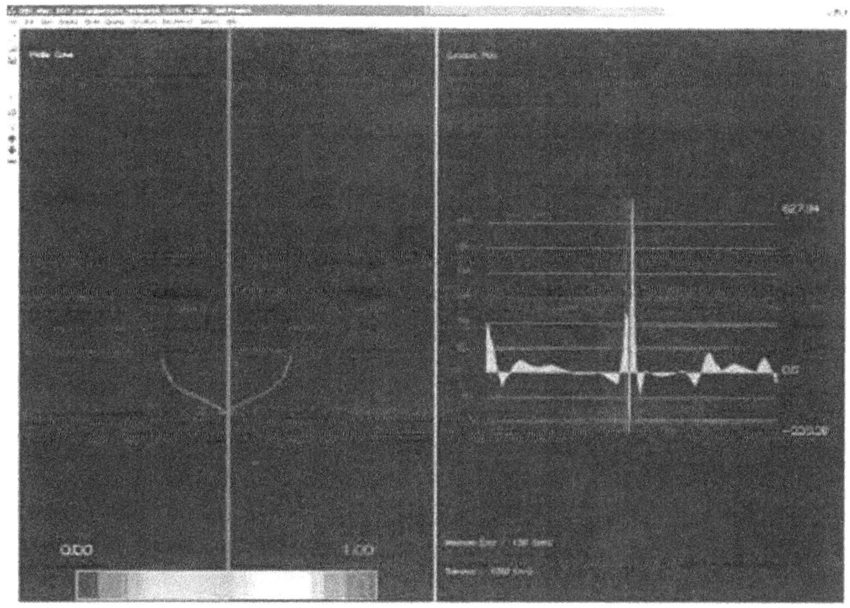

Figure 5 Analysis software view of mid-ship section illustrating profile curve (left) and curvature plot (right) of the scale model Viking ship

Our experience with this pilot project indicates that the transformation and grouping of sequential working stages (from recording and documentation to display and analysis) from a manual to an essentially electronic process is far from complete. There are advantages to various methods, but there are still major issues to be addressed. For example, using a FaroArm to edge trace brings the documented ship timber closer to assembly, 3D display, and analysis, but a cardboard 3D model seems irreplaceable as a tool for understanding the assembly processes used in shipbuilding. However, the ultimate goal for 3D visualization is not display alone, but should include additional calculation capabilities that are sufficiently flexible to address specific problems arising from the nature of the artefact itself. Such interdisciplinary research expands the concepts and the means by which we learn more about 3D objects and the cultural and craft traditions that developed them.

References

CRUMLIN-PEDERSEN, O., 1997. Viking Age Ships and Shipbuilding in Hedeby/Haithabu and Schleswig, SBN 2. Schleswig & Roskilde.

INDRUSZEWSKI, G., 2000. Man, Ship, Landscape: Ships and Seafaring in the Oder Mouth Area 400-1400 A.D., A case study of an ideological context. Ph.D. dissertation, Københavns Universitet Det Humanistiske Fakultet.

JENSEN, K., 1999. Documentation and Analysis of Ancient Ships. DTU, Lyngby.

KDI,1999 - 2003. 3D Knowledge: Acquisition Representation & Analysis in a Distributed Environment, Arizona State University PRISM (NSF IIS-9980166), 1999-2003 (http://3dk.asu.edu, http://prism.asu.edu).

SCHURMANS, U., RAZDAN, A., SIMON, A., MARZKE, M., MCCARTNEY, P., VAN ALFEN, D., JONES, G., ZHU, M., LIU, D., BAE, M., ROWE, J., FARIN, G. and COLLINS, D., 2002. Advances in Geometric Modeling and Feature Extraction on Pots, Rocks and Bones for Representation and Query via the Internet. In Burenhult, G. and Arvidsson, J. (eds.), Archaeological Informatics: Pushing the Envelope CAA 2001. Computer Applications and Quantitative Methods in Archaeology. Proceedings of the 29th Conference, Gotland, April 2001, Archaeopress, Oxford, (BAR International Series 1016):191-202.

SIMON, A. (ed.), 1998. Salado Ceramics and Social Organization: Prehistoric Interactions in Tonto Basin: The Roosevelt Archaeology Studies 1989-1998. Roosevelt Monograph Series 11, Anthropological Field Studies 40, Office of Cultural Resource Management, Department of Anthropology, Arizona State University, Tempe.

SIMON, A., RAZDAN, A., FARIN, G., ROWE, J. and VAN ALFEN, D., 2002. Ceramic Vessel Morphology Analysis and Archiving using 3D Modeling: A Case Study from the American Southwest. International Symposium on Archaeometry, Amsterdam, 22-26 April 2002.

Definition and Construction of Parametric Archetypes of Archaeological Artefacts by means of CAD Techniques

SEE THE CD FOR THE EXTENDED VERSION

Giuseppe Di Gironimo
Dipartimento di Progettazione e Gestione Industriale, Università degli Studi di Napoli Federico II, Italia

Bianca Ferrara
Dipartimento di Discipline Storiche, Università degli Studi di Napoli Federico II, Italia

Stanislao Patalano
Dipartimento di Progettazione e Gestione Industriale, Università degli Studi di Napoli Federico II, Italia

Abstract

The present work represents a contribute to improve the classification and the recording techniques of archaeological artefacts. With a particular reference to the class of clay artefacts of Magna Grecia, a new method for the CAD representation of artefacts, based on the definition of 3D archetypes, has been developed. At first the method allows to use the same CAD archetype to build the virtual model of any type of artefacts, in particular of any vase with different shapes and dimensions; then, the method allows to upgrade the geometry of the model and to converge toward a representation of the real artefact more and more close to that experimentally obtained.
The proposed procedure is a valid substitute to the one, actually in use, to put the archetypes in the inventories that is essentially based on the manual development of the drawings through the use of shapes reproducing the object geometries. The great reduction of times in reproducing the artefact images and the possibility to operate classifications based on the similar geometric features used in the solid modelling activities, represent a reason of interest and of innovation for the Archaeology. These data can contribute to the solution of the problem of archaeological artefacts classification and recording.

The Sanctuary of Hera at River Sele Mouth: Typological Layout and Classification of the Black Varnish Pottery

"... beyond river Sele mouth, Lucania and the Sanctuary of Argive Hera, settled by Jason and nearby, 50 stadia far, Poseidonia..." (STRABO, Geography, VI, 1, 1). This way the Greek geographer Strabo, describing the Lucania district, mentions the Sanctuary of Argive Hera at river Sele mouth about 9 km far from Paestum, for many centuries one of the most important religious sites in Great Greece. In ancient times the river flow marked a natural boundary between the land occupied by Etruscan people settled in the Picentini Mountains area and the land where Greeks settled, at the beginning of VI century b.C., Poseidonia town and the Sanctuary of Hera at river Sele mouth. The river Sele area was initially sacralized through a mere altar made of earth and ashes. At the end of the V century b.C., Lucanians arrival brought drastic transformations: two new reception spaces and a squared building. After the latest roman era interventions and the successive neglecting, the sanctuary turned into a fenland.

In 1987, the Archaeological Bureau of Salerno, Avellino and Benevento Provinces starts up a new excavations season and a massive rearrangement work of these materials, coordinated by professor Giovanna Greco. A team of young archaeologists, first from University of Salerno and then from "Federico II" University of Naples, is given the exciting task to put in order what is left and reopen the famous cases; findings are still wrapped up in '30s and '50s newspapers or kept in tins and cigarette boxes, together with notes and lists drawn on bits of paper while excavating. The heap of objects is huge and a multitude of finding classes is identified: Corinthian pottery, black-figure pottery, red-figure pottery, black varnish pottery, coroplastic. These findings, when their origin context can not be identified, have been rearranged in conformity with classes and organized according to a well structured typology. In this respect, particularly interesting is the study carried out on the black varnish pottery, which is one of the best supported classes within the Hera Argiva Sanctuary at river Sele mouth and attested by many thousands of fragments and hundreds of complete or restored items.

The significance of a typology arises from the necessity of creating logic series which, once chronologically organized, will allow the reconstruction of certain pottery productions, in order to define their specific peculiarities and identify their real characters. A typology is a fixed reference system and must be explanatory, descriptive and absolutely not interpretative. According to ceramist Ann Shepard (Shepard 1965), the study of vases shape can be faced from the aesthetic, functional or taxonomic point of view. In the first instance used criteria are too much subjective, in the second too much uncertain, therefore the most obvious approach is the geometric one. But even the geometric method offers a range of several possibilities resolving between two extreme solutions. The first one - not more than a mere procedure of first approximation - consists, for instance, of classifying truncoconical, bitrunconical, tritrunconical etc. shaped vases into different groups. The second one aims to decompose each vase, mainly its shape, into elements which can be codified and in case numbered through an analytic method. This last method has been followed by J.P. Morel in his "Céramique campanienne: les formes" (Morel 1981), a work that constitutes a turning point for the statement of black varnish studying and is not only an irreplaceable tool for this class settlement, but above all a fundamental methodological base. The typology built by Morel, sole instance within classic pottery, lies on the hierarchisation at morphological level of evidences into categories, genres, species, series, types. Such system is an open one, that is any moment increasable thanks to several degrees of investigation allowing the insertion of new items into hierarchic groups. Therefore, this open method of classification allows not only a whole evidence arrangement without any forcing, thanks to the various hierarchic levels, but even a clarification and pointing out of significant similarities and differences.

Groups description constitutes the typology base, for its aim is reporting pertinent criteria - primary or secondary - opposing different groups within the given set (that is a hierarchically superior group). The preliminary definition of this set, itself a part of a superior one (and so on until the category, first subdivision within the set of black varnish pottery), let us forget repeating, for each constituent group, the elements that such group shares with the other groups in the set. For instance, the factor distinguishing each category from the others is the presence/absence of handles. But this is the only identifiable and absolutely definable characteristic; in fact the other levels - genre, species, series - are characterized and differentiated by formally dissimilar factors for each category. "Type" is instead a model shape which one or more "samples", associated by some morphological characteristics, have been related to: two vases will never be identical, but what matters is the ideal model followed by their creator. According to Morel, a typology's subject is then the items shape, while a classification consists of organizing items within classes, and a class is the set of technical characteristics associating vases created by the same workshop or by a group of similar workshops; consequently, class stands for Morel as production stands for other scholars. Following a precise theoretical choice in distinguishing types, he does not take into account either decorations or any other factor but morphological. In consequence, the whole typological system lies on the "similarity" criterion and consists of gathering morphologically similar vases or distinguishing them according to significant dissimilarities. Particularly relevant is calculating proportions, which can be extremely important and interesting in order to define a series within a site production and, above all, in the view of its formal and chronological evolution.

Therefore such a project, aiming to recreate the whole vase from the single fragment, can only arise the strongest interest in an archaeologist tempting to restore through traditional methods the shape of vases from fragments, in order to put them into a wider typology pursuing the reconstruction of production centres and historical contexts (Caputo 1999, Di Gironimo 2001, De Napoli 2001, VV.AA 2003).

Definition of Drawing Archetypes Belonging to Classes of Selected Artefacts

The classification of archaeological artefacts, that today is essentially based on the manual development of the drawings, can be improved thanks to the availability of parametric CAD systems. In fact, the use of CAD techniques and the knowledge of significant geometric characteristics of the artefacts have induced to formulate a new method for the CAD definition of virtual archetypes representing classes of selected artefacts.

According to the present method the virtual archetypes have to satisfy two Functional Requirements. At first the possibility to use the same archetype to build the CAD models belonging to any type of artefacts, in particular of any vase with different shapes and dimensions. The variability taken into account is always related to that one of the black varnish pottery examined in the previous section of this paper. Besides, the choice of using a unique typology of archetype is related to the possibility to take full advantages of a parametric-variational CAD modeller (Patalano 2001).

The second Functional Requirement of the virtual archetypes consists in the possibility of upgrading the geometry of the model and converging toward a representation more and more close to that experimentally obtained. This requirement is related to the starting data set (clouds of points) necessary to build the CAD models, but also to the specific aim foreseen for the models. If the data acquisition from the archaeological finds is manual then the number of datum points is not so high. On the opposite, if it is possible to perform an automatic acquisition, a more detailed profile could be favourable. Finally, if the CAD model is due to perform a comparative evaluation between different reconstruction techniques is better to use the maximum level of detail; if the main goal is the visualisation and the model exchange by internet is better a lower level of detail with lighter files.

The first Functional Requirement implies the need of using reference planes, whose positions are modifiable, to locate the encumbrance of the artefact and the more significant geometries (the edge, the bottom, the foot). Therefore the dimensions used to locate the reference planes have to be taken into account during the CAD modelling of the archetypes.

The second Functional Requirement implies the need of using a set of datum points, whose number is modifiable, to "well represent" all the significant features of the artefact. This need imposes the use of a CAD tool that allows to change, in real time, the number of datum elements in use.

Construction of the Archetypes by Means of a 3D Parametric Modeller

The suggested construction of the CAD models foresees the creation of the principal datum planes corresponding to the longitudinal section plane and to the most significant features of the artefact: the top plane, the lower plane, the bottom plane, the foot plane and, finally, the edge plane. These planes are located from a reference depth conventionally established to the top plane. Then it is possible to create two series of initially equidistant datum points. The first series deals with the definition of the external profile of the meridian section; the second series with the internal one. The number of points depends on the availability of points coming from the acquisition of real artefacts. Two tables control the position of datum points through a couple of dimensions: the depth and the diameter of a datum circle. These tables can be filled in using text external files. The parametric characteristic of the modeller in use allow to regenerate different CAD model through the changes of numeric values belonging to the tables.

Then the construction is increased through the profile sketching of the meridian section. The sketch uses spline curves passing through the existing datum points. A spline curve is a smooth, freeform curve that connects a series of control points. Changing any single control point will result in a change in the curve, so that the curve can pass through the new point (Bertoline 2002). This property makes this kind of curve able to satisfy the functional requirements above mentioned. The archetype arranged by the authors is able to automatically upgrade the profile once the datum points have been positioned using the tables. Once the profile is drawn, the system automatically upgrade a solid revolved feature that uses as axis of revolution the connecting line among the centres of the datum circles.

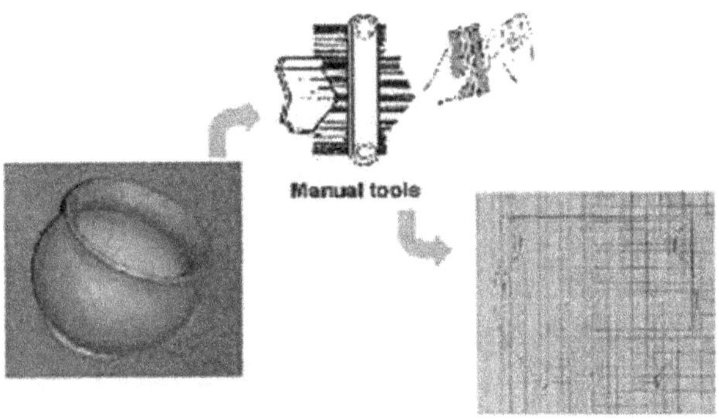

Figure 1 Profile acquisition with manual tools

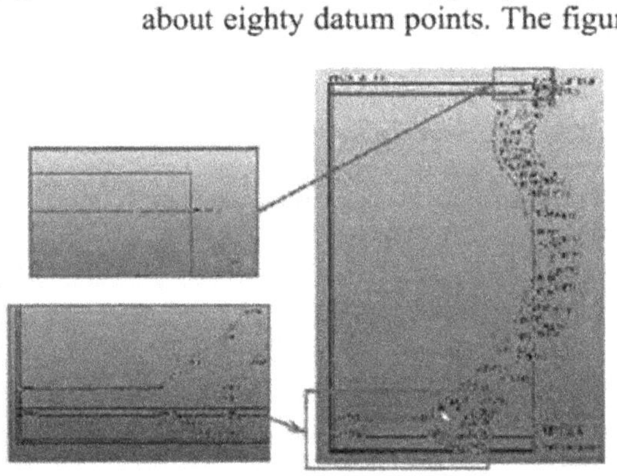

Figure 2 Profile description

CASE STUDIES

A first case study deals with the reconstruction of a vase whose typology is similar to one of the artefact described by Morel (Fig.1). The profile acquisition has been performed in a manual way with a profilometer. The bidimensional and usual representation of the vase has been drawn on a grid squared paper with only an half section to see the internal profile. At first, the archetype presents five datum planes in a default position. From the paper drawing it is possible to extract a certain number of relevant points for the external and the internal profile and to insert their coordinates in their respective tables. The relative distance between each point is function of the curvature of the profile. For example, when the curvature is small it is possible to give a relative big distance between adjacent points because the spline algorithm, in this case, works well. On the opposite, when the curvature is more accentuated it is necessary to reduce the distance between the points. Then it is possible to build the profile by means of an unique spline curve that passes through both the external and the internal points. In particular there is a little straight line between the first external point and the first internal point, connected with the rest of the curve by imposing a continuity condition not only on the tangency but also on the curvature. In the lower part there are similar continuity conditions (Fig.2). The resulting solid model is shown in figure 3.

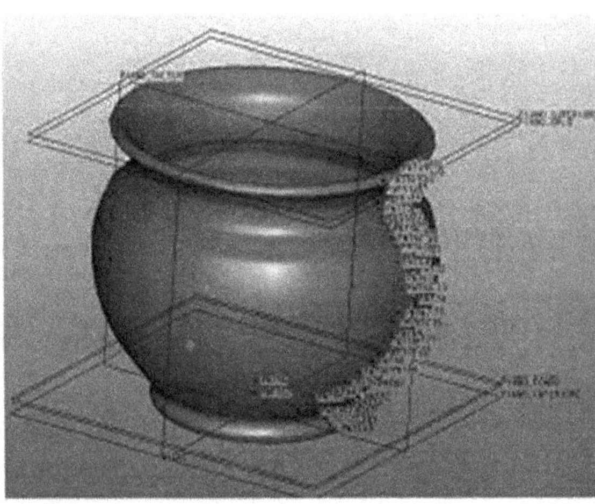

Figure 3 Solid model

Figure 4 Parametric modification

In order to show the potentiality of the modelling technique used the figure 4 shows the effect of a parametric modification on some datum points of the profile using the values contained in the guiding tables. A second case study deals with the reconstruction of the artefact in figure 5. In particular it is the one classified by Morel (Morel 1981) with the code number "1123a1". The CAD model comes from the characterisation of the same archetype by means of five datum planes and about eighty datum points. The figure 5 shows also the axonometric views of the CAD model.

CONCLUSIONS

The present work deals with the representation of archaeological artefacts. In order to improve the recognition and filing of geometries a new CAD modelling procedure has been proposed as a substitute of the one that is essentially based on the development of the manual drawings. The great amount of types have suggested the authors to take advantages of the modern parametric CAD system characteristics in order to foresee a procedure able to give a three-dimensional description of the artefact geometries. In particular, it has been proposed the construction of only one parametric archetype to give the availability of a solid model of all the artefacts belonging to a class or to a similar classes. At first the procedure allows to set the position of some datum planes in order to define the bounding box of the object and other characteristics of interest, used by the archaeologists for the classification. The initial reference planes, that are the planes of interest for the archaeologists, give the possibility to immediately visualize the height of the model and other significant positions. The construction of the datum points that control the profile's shape occurs in two steps: individualization of the number and characterization of the corresponding coordinates. This procedure allows the construction of a database using 3D virtual models of the vases. The database will be used to perform the automatic recognition of partial fragment of artefacts. In fact, the mathematical knowledge of geometries can allow the use of searching algorithm based on gauss curvature comparison.

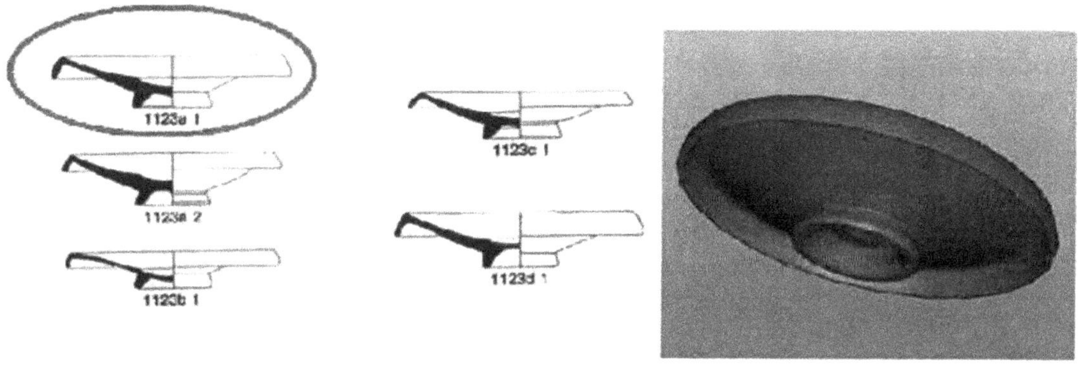

Figure 5 CAD reconstruction of the 1123 a1 from the Morel Classification

ACKNOWLEDGEMENT

The authors, who have carried out the present work with a same contribution, thank to Prof. Ing. Francesco Caputo and Prof. Giovanna Greco for their encouragement and suggestions. The present work has been developed with the contribute of MIUR-Italy performing the activities of National Project: "Classification and Restoration of Archaeological Finds by means of CAD-RP Technologies".

REFERENCES

BERTOLINE, G. and WIEBE, E., 2002. Fundamentals of Graphics Communication 3rd Edition. McGraw Hill.

CAPUTO, F., GRECO, G. and DI GIRONIMO, G., 1999. Sull'impiego di tecniche CAD nella modellazione e nella ricostruzione di reperti Archeologici. Proceedings of 11th ADM International Conference, Palermo, 9-11 December 1999.

DE NAPOLI, L., LUCHI, M.L., MUZZUPAPPA, M. and RIZZUTI, S., 2001. A Semi-Automatic Procedure for the Recognition and Classification of Pieces of Archaeological Artefacts. Proceedings of XII ADM International Conference, Rimini, 5-7 September 2001.

DI GIRONIMO, G., FERRARA, B., GERMANI, M. and MARTORELLI, M., 2001. Reverse Engineering Techniques in the Reconstruction of the Virtual Shape of an Archaeological Find. Proceedings of XII ADM International Conference, Rimini, 5-7 September 2001.

MOREL, J.P., 1981. Céramique campanienne: les formes. Roma.

PATALANO, S., GERBINO, S. and DI MARTINO, C., 2001. Complex Product Development by Parametric-Associative CAD Systems: Modeling Car Body Components by using Surface Features. Proceedings of International IFIP Conference on Feature Modeling and Advanced, FEATS 2001, Valenciennes, France, 12-14 June 2001.

SHEPARD, A., 1965. Ceramics for the Archeologist. Washington.

VV.AA., 2003. Classification and restoration of archaeological finds by means of CAD-RP techniques. Naples (www.dpgi.unina.it\archeocad).

COMPUTER AIDED RECONSTRUCTION OF HUMAN CRANIA

P. GUNZ
VIENNA INST. FOR ANTHROPOLOGY, VIENNA, AUSTRIA

P. MITTEROECKER
VIENNA INST. FOR ANTHROPOLOGY, VIENNA, AUSTRIA

FL. BOOKSTEIN
VIENNA INST. FOR ANTHROPOLOGY, VIENNA, AUSTRIA
MICHIGAN CENTER OF BIOLOGICAL INFORMATION,
UNIV. OF MICHIGAN, USA

GW. WEBER
VIENNA INST. FOR ANTHROPOLOGY, VIENNA, AUSTRIA

ABSTRACT

We present statistical and geometrical techniques to reconstruct incomplete human crania using techniques that formalize the biologist's prior understanding of the considerations that govern form: continuity, symmetry and integration. The modern morphometrics of landmarks and curves makes it possible to blend statistical and biological reasoning in this domain. Factors such as size allometry or sex and also directional asymmetry whether zero or nonzero can be explicitly incorporated into the data estimation by way of the corresponding covariance structures. For tasks of estimation based on very small samples we show a variant based on the continuity assumption of the thin-plate spline. When complete specimens are adequate in number our estimation can be regression-driven instead. All the missing points can be estimated at once by maximizing the likelihood of the resulting configuration in a reduced-rank model of a multivariate Gaussian distribution. Whatever integration the form posses-ses is automatically exploited in the course of these regressions. We demonstrate the accuracy of these approaches using a dataset of 388 anatomical landmarks and semi-landmarks on 52 complete H. sapiens crania. After deliberately deleting regions of landmarks we estimate the missing data and compare the estimated specimens to the originals. Our results indicate that the accuracy of estimation is sufficiently close to the precision of measurement.

Life after death is not kind to paleo-anthropological specimens. Forms are distorted, parts break off, and the textural cues we use to locate landmark points may be effaced or covered by matrix. Distortion shifts landmarks away from their original context; we will not discuss that form of data damage here. The effect of the other two types of processes, breakage and effacement or encrustation, is to render the landmark locations actually unobservable in the single specimen; such points are coded as missing. The topic of the present paper is the subsequent treatment of points coded as unobservable in this way.

Across the general run of modern computational statistics there is one core algorithm for handling missing data, the EM ALGORITHM (Dempster et al. 1977). E stands for EXPECTATION, and M for MAXIMISATION. In any EM analysis, an incomplete data set is treated as having arisen from a complete data set by random knockout of observed values, and the data set is completed as a whole (i.e., not specimen by specimen but all at once) in order to maximize some overall probabilistic likelihood, such as a multivariate Gaussian one. The algorithm is iterative: a covariance structure is approximated, then missing data is estimated by regressions datum by datum, then the covariance structure recomputed, the regressions redone, and so on until convergence. The procedure is robust and reliable in many settings within the natural and the social sciences (Allison 2001, Little and Rubin 2002).

However, anthropological questions involving missing data do not necessarily suit any of these typical settings. Data might not be missing at random: Geologically older specimens might be missing more of their landmarks, infant or small individuals have fragile bones that break easily - and if not the discrete landmark points then certainly the semi-landmarks (Bookstein 1997, Mitteroecker et al., in press) that are missing tend to cluster on forms. We may not be intending to optimize any sort of a likelihood for the sample as a whole, but only some descriptive functional for the reconstruction of one single form; and the purpose of the reconstruction might be to understand gross aspects of size and shape, or instead details of local modeling.

The difference between a concern for spatial position per se, versus a concern for aspects of shape, is closely related to the distinction between two quantities that are familiar separately in the literature of geometric morphometrics: Procrustes distance (sum of squared separations of shape coordinates regardless of spatial position) and bending energy (weighted sum of partial warp scores that take spatial contiguity into account in a very fundamental way) (Bookstein 1991, 1997, Rohlf and Slice 1990).

We will show that either of these concerns can be set at the core of an algorithm for estimating missing data that yields reasonable results in realistic simulations. One approach is using the thin plate spline for GEOMETRIC RECONSTRUCTION while the other uses multiple multivariate regression for STATISTICAL RECONSTRUCTION. The choice between the methods is not, at root, a matter of algebra, but of science: the methods are pertinent to two different paleo-anthropological contexts. We will return to this distinction in the final discussion, after notating and demonstrating the two methods.

At this point we want to completely dismiss a method that is often found in the literature: the method of MEAN SUBSTI-

TUTION borrowed from the social sciences. In a mean substitution, missing entries are substituted by the mean of the non-missing entries over the full sample data set. When data are Cartesian coordinates or shape coordinates this procedure makes no sense either as statistics or as science - because of the method's prevalence, the test results of mean substitution are shown nevertheless for comparison.

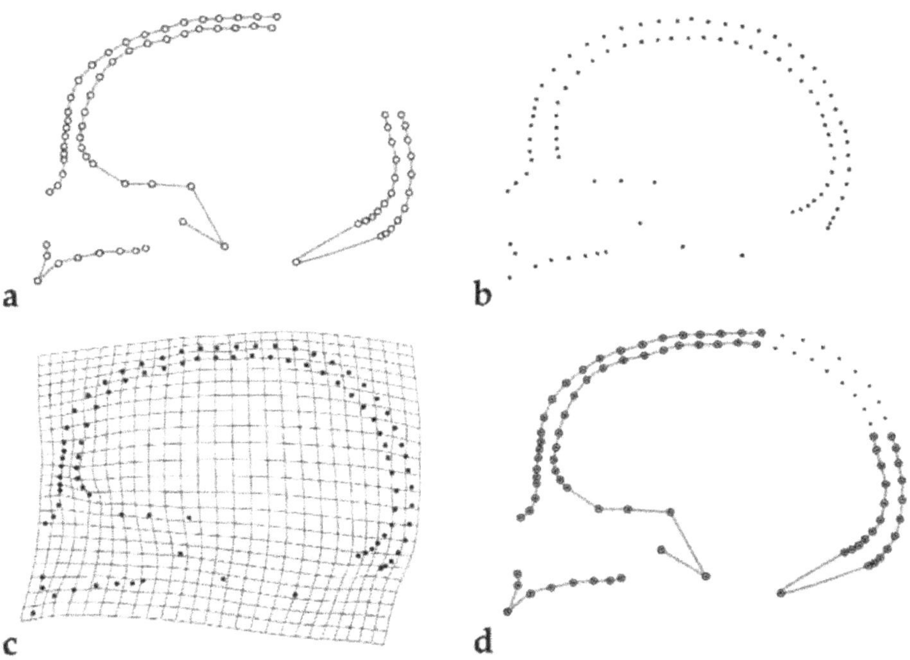

Figure 1 GEOMETRIC RECONSTRUCTION. (a) The target specimen is missing landmarks in the neurocranium. (b) Complete reference specimen. (c) The thin plate spline deformation grid between the specimens in (a) and (b) is computed from the subset of all landmarks available in (a). (d) Missing data of the target specimen is substituted according to the deformation grid shown in (c), which maps the landmark locations of the reference specimen to the target specimen

THE METHODS

Imagine a data set of many landmarks or semi-landmarks over many specimens. Consider, first, the very simplest case: exactly one landmark is missing for only one specimen. There is a good deal of information available to help us to reasonably estimate its location. In one approach, the STATISTICAL RECONSTRUCTION, we think of this point as correlated with all the other landmark locations of the landmark set. Using the other - complete - specimens, we work out the prediction function that predicts it with the minimum sum-of-squares given the other data.[1]

But we might just as well predict the missing parts based on geometric properties of the single specimen, such as continuity information of curvature - GEOMETRIC RECONSTRUCTION. For this purpose we use the thin plate spline interpolation function: We predict the missing data mapping the average of the complete cases to the specimen with missing landmarks - using the thin plate spline interpolation based on the subset of observable landmarks.

These two alternatives yield different locations, in general, but for reasonable numbers of reasonably distributed landmarks the discrepancy is virtually undetectable.

If more than one form is missing landmarks, then whichever figure of merit we choose, we need an iterative approach, because in practical applications not all specimens are missing the same landmarks. First we estimate the missing points, but then using the regression method we have to recompute the covariance matrix that gives us the prediction formulas (because now it is using all the forms, as they have all been tentatively completed), and in the thin-plate method we have to re-compute the average form that gives us our spline, because now it averages over the entire sample combining both the originally complete forms and the tentatively completed. In either version, the iteration converges rather quickly for reasonable data schemes.

At the conclusion of either of these algorithms, we have estimated all of the missing data anywhere in the data set in order to optimize the quantity that is encoded in the estimation step. For the first option, that quantity is the net regression residual sum-of-squares for prediction of landmarks by the others; for the second option, it is the net bending energy of the completed sample around its grand mean.

SIMULATIONS

We demonstrate the accuracy of these approaches using a dataset of 388 anatomical landmarks and semi-landmarks on 52 complete H. sapiens crania. After deliberately deleting regions of landmarks we estimate the missing data and compare the estimated specimens to the originals. Figure 1 shows the 4 different KNOCKOUT-SETS that were used to compare the accuracy of the two estimation methods.

First the landmarks in the shaded regions were deleted on one of the 52 specimens, then this specimen was reconstructed using the two methods introduced above. This was done for each individual. As in each calculation-cycle only one form is missing landmarks, there is no need for iteration here. Figure 2 summarizes the results as the mean of the summed squared residuals per landmark for each knockout-set and estimation method. A semi-landmark carries only shape-information perpendicular to the curvature, so when a semi-landmark was deleted, only the residual normal to the ridge or surface was used.

Each bar's length represents the total error of estimation. Mean substitution always performs worst. With the exception of the first knock-out set, the regression method is always better than the thin plate spline warping.

These differences result from intrinsic properties of the compared methods: The spline performs best only in the case of

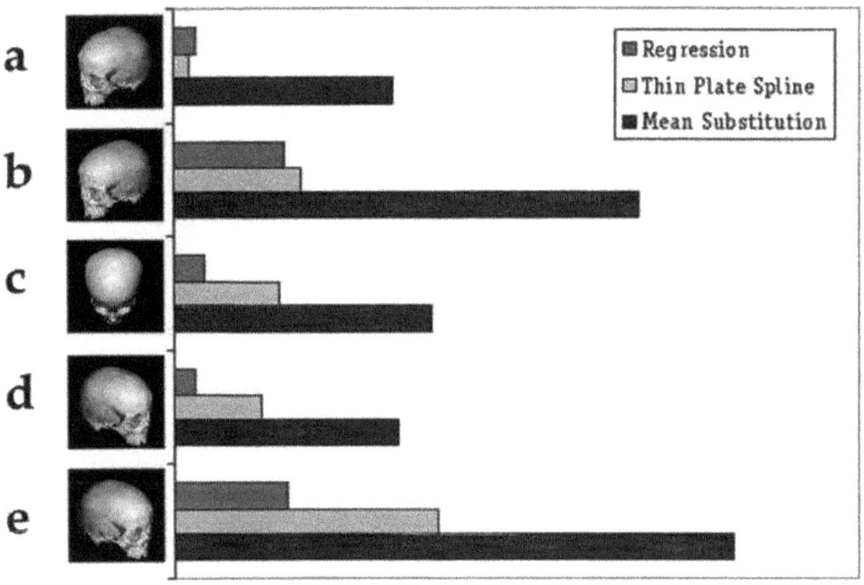

Figure 2 plate spline warping and regression and mean substitution. Each bar's length represents the total error of estimation. Mean substitution always performs worst. With the exception of the first knock-out set (a), the regression method is always better than the thin plate spline warping

Figure 2a, where only a small part of the neuro-cranial surface is missing. The thin plate spline computes the deformation that is least bent, which turns out to be the best method for the smooth curvature of the neuro-cranial surface. Particularly elucidating is the contrast between Figure 2b, where regression and thin plate spline perform almost equally well and 2d and 2e, where the regression is more than twice as precise as the spline. In the former case, the whole face has to be estimated, in the latter, information is missing on only one half of the cranium. While it is equally hard for both methods to reconstruct the face when only neuro-cranial information is available, the different results for the knockout-set of 2d and 2e demonstrate that only the regression has access to symmetry information. But not only symmetry - the regression exploits all information about biological factors present in the reference population like allometry or morphological integration (Bookstein et al. 2003).

Principal Components

We performed principal components analysis (PCA) of the Procrustes coordinates (Rohlf 1993) of the original and the estimated specimens. Figure 3 shows the first two PC's of the knockout-set of Figure 2e for each method. Each complete and reconstructed specimen is connected with a small vector, where the arrowhead points towards the reconstruction. The arrows of the thin plate spline reconstructions exhibit no obvious patterning and the estimation errors of the regression are almost too small to be seen in the plot of the first two PC's.

Discussion

As we have already argued, MEAN SUBSTITUTIONS should always be avoided. We have reviewed two methods for imputing missing data that are similar in many ways (requiring iteration, ending up with a sum of squares) and different in one very important matter. For one, the sum of squares is of a Procrustes distance, and for the other, a bending energy. The choice between the methods is thus, in fact, a choice between these figures of merit. Under what circumstances would one wish to minimize one of these, or the other?

Thin plate spline interpolation can be considered a reasonable method in the absence of a reference population, as it requires only a single reference form. This single specimen however conveys no information about population variance and covariance; in some cases it may be plausible to match reference and target specimen on some single quantity (for example sex or age).

When specimens are adequate in number, the missing data estimation can be regression-driven instead: The likelihood of the population that includes the resulting landmark configuration is maximized, subject to the proviso that the range of dimensions one intends to impute is actually present in the reference population (e.g. growth allometry).

In the presence of a reference population of which the reconstructed form is likely to be a part, the decision rule we suggest depends on the use to which the completed forms will be put. If you're going to be talking about cranial capacity, distances between landmarks, or other large-scale properties, you want the most precise landmark locations, and so you use the regressions as best you can. If you want to talk about features of shape of the single form, like bumps and bends that might be characters, you use the bending energy.

Take care that the assumptions of the reconstruction must not overlap with the hypothesis you are testing. Statistical reconstruction exploits all information

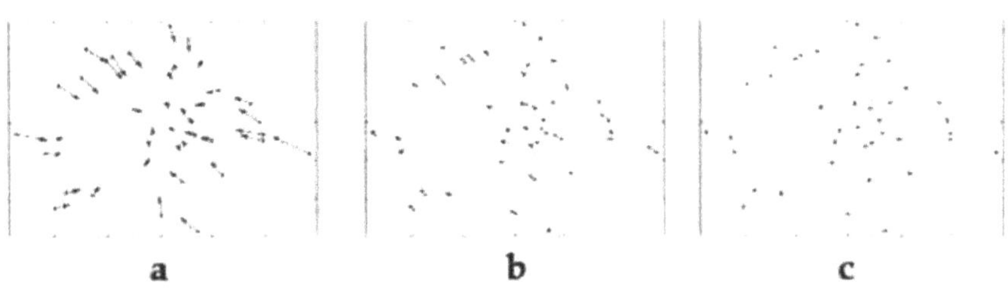

Figure 3 First two principal components of shape of the 52 complete original and the reconstructed specimens for the knockout-set of Figure 2e. The vectors point from the original towards the reconstruction. (a) Mean subsitution. Note that the reconstructions are biased towards the mean. (b) Thin plate spline. No obvious bias. (c) Multiple multivariate regression. The error is almost too small to be drawn.

present in the reference population; hence applying any procedure that involves a covariance matrix (eg. principal components or singular warps [Bookstein et al. 2003]), the statistically reconstructed specimen is 'overestimated'. In such cases it makes sense to use geometrical reconstruction instead.

ACKNOWLEDGEMENTS

We want to thank Horst Seidler, Katrin Schäfer, Dennis Slice, Bence Viola, Markus Bernhard and Simon Neubauer for their advice and support. Our research is supported by the AUSTRIAN SCIENCE FOUNDATION Project P14738.

[1] In practice, given the predictions by each other shape coordinate separately - typically we cannot invert covariance matrices among shape coordinates, so we just add up the separate predictions or the predictions by the first few principal components of those other coordinates.

REFERENCES

ALLISON, D.P., 2001. Missing Data. Sage University Papers, Thousand Oaks.

BOOKSTEIN, F., 1991. Morphometric tools for landmark data: geometry and biology. Cambridge University Press, Cambridge (UK)/New York.

BOOKSTEIN, F., 1997. Landmark methods for forms without landmarks: morphometrics of group differences in outline shape. Med Image Anal 1:225-43.

BOOKSTEIN, F.L., GUNZ, P., MITTEROECKER, P., PROSSINGER, H., SCHAEFER, K. and SEIDLER, H., 2003. Cranial integration in Homo: singular warps analysis of the midsagittal plane in ontogeny and evolution. J Hum Evol 44(2):167-87.

DEMPSTER, A.P., LAIRD, N.M. and RUBIN, D.B., 1977. Maximum Likelihood from Incomplete Data via the EM Algorithm. Journal of the Royal Statistical Society 39:1-38.

LITTLE, R. and RUBIN, D.B., 2002. Statistical Analysis With Missing Data. John Wiley & Sons, Hoboken, New Jersey.

MITTEROECKER, P. and GUNZ, P., BOOKSTEIN, F., in press. Semilandmarks in three dimensions. In Slice, D.E. (ed.), Modern Methods in Anthropolgy, Kluwer.

ROHLF, F.J., 1993. Relative warp analysis and an example of its application to mosquito wings. In Marcus, L.F., Bello, E. and García-Valdecasas, A. (eds.), Contributions to Morphometrics, Monografias, Museo Nacional de Ciencies Naturales, Madrid:131-159.

ROHLF, F.J. and SLICE, D.E., 1990. Extensions of the Procrustes method for the optimal superimposition of landmarks. Syst. Zool. 39:40-59.

NEW MORPHOMETRIC METHODS IN PALEOPATHOLOGY: SHAPE ANALYSIS OF A NEOLITHIC HYDROCEPHALUS

P. MITTEROECKER
VIENNA INST. FOR ANTHROPOLOGY, VIENNA, AUSTRIA

P. GUNZ
VIENNA INST. FOR ANTHROPOLOGY, VIENNA, AUSTRIA

M. TESCHLER-NICOLA
MUSEUM OF NATURAL HISTORY VIENNA, AUSTRIA

GW. WEBER
VIENNA INST. FOR ANTHROPOLOGY, VIENNA, AUSTRIA

ABSTRACT

We provide a shape analysis of a Middle Neolithic (4900-4300 BC) human cranium introducing a new tool to geometric morphometrics: semilandmarks on surfaces.
Two hydrocephalics (34 and 12 years old) and a control group of morphologically regular formed crania (N=20) ranging from 2 years of age to adults are compared against an adolescent middle Neolithic specimen which is presumably hydrocephalic.
42 anatomical landmarks and several thousand points on the neurocranial surface were either digitised using a Polhemus device or extracted from CT-data. Our algorithm automatically places 336 semilandmarks on the neurocranium and relaxes them against the Procrustes average preserving just information perpendicular to the surface. These three dimensional semilandmarks offer the opportunity of incorporating information about curved forms that lack traditional landmarks into statistical shape analysis.
While all 3 hydrocephalics possess very prominent parietal bosses, the Neolithic hydrocephalus is different in frontal bone shape, lying close to the variability obtained in the control group. This suggests that the Neolithic specimen could represent a case of acquired hydrocephalus where the sutures at the parietal and occipital region of the skull still had enough degrees of freedom to compensate the intracranial pressure, whereas the frontal suture had already fused.

INTRODUCTION

Hydrocephalics from prehistoric time are rare findings (Teschler-Nicola et al. 1986, Schultz and Teschler-Nicola 1989) and - in addition - questions concerning differential diagnosis and causality are often hampered by their poor preservation status and fragmentary character (Teschler-Nicola and Gerold, in press). A completely preserved skull of a subadult individual excavated in Kamegg, Lower Austria and dated to the middle Neolithic is assumed to be hydrocephalic (Teschler-Nicola et al. 1986): Not only the enlarged neurocranium but also the results of a detailed analysis of pathomorphological features predominantly manifested at the internal cranial base gave arguments to support this hypothesis. Nevertheless, the Kamegg skull is distinct to most other known hydrocephalics, since it exhibits a moderately large and triangular shaped cranial vault (Fig.1a-c) whereas the "classical" hydrocephalics possesses a typical large rectangular shaped neurocranium (Fig.1d). According to histological and detailed investigations of pathological features an inflammation in the region of the right temporal bone (Otitis media), or a trauma of the cranial vault during early childhood could be responsible for the characteristic features in this case of a probably acquired hydrocephalus.

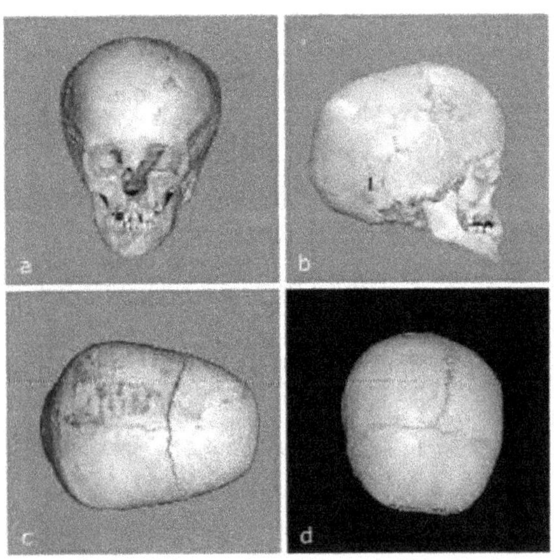

Figure 1 Photographs of the Neolithic specimen (a, b, c) and the 32 year old hydrocephalic (d)

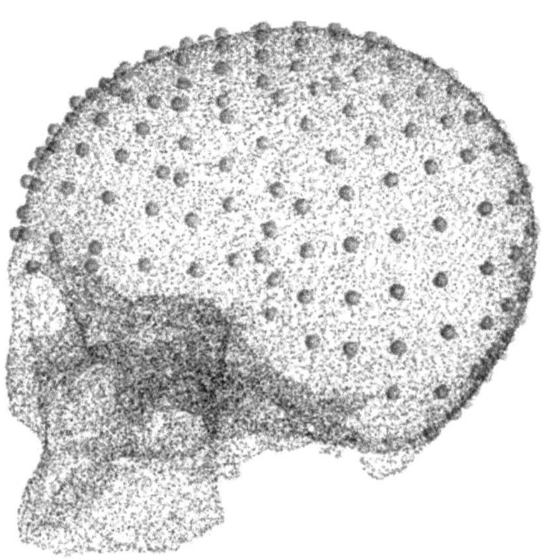
Figure 2 Dense surface points extracted from a CT-scan and semilandmarks on the neurocranial surface (larger spheres)

The use of traditional metric analysis for differential diagnostic purpose lacks reliability since they do not include landmarks of the most diagnostic regions on the skull, e.g. the cranial bosses. We therefore applied a rather new morphometric method, semilandmarks on surfaces (Mitteroecker et al., in press), to compare the Neolithic skull - which is representing the most ancient hydrocephalic finding in Europe - to a set of anamnestically well documented pathologic specimens as well as a set of inconspicuous modern reference skulls for the improvement of the hypothesis of representing an acquired pathology.

Material

During the systematic archaeological investigation in the middle-Neolithic concentric ditch system in Kamegg, Lower Austria, carried through by the Institut für Ur- und Frühgeschichte, University of Vienna in 1984 (Trnka 1986), a well preserved skeleton of an immature individual was found at the base of a settlement pit. Based on fragmentary pottery the finding can be attributed to the Lengyel-culture, 4900-4300 BC. By using dental eruption and mineralisation the age of death of the individual was restricted to 10 - 12 years. Regarding this subadult age, non-conventional methods such as dental dimensions and relative proportions of tubular skeletal elements have been used to determine the individual's sex: The majority of the metric dimensions of that features are in favour of a female.

The most conspicuous morphological feature in this individual is the enlargement of the neurocranium in length and width and an 40% extension of the cranial capacity compared to the reference skulls. Impressive and characteristic is the triangular shape of the Neolithic specimen in the vertical (and the trapezoidal shape in the occipital) view, which is resulting from the considerable development of the parietal bosses.

For a comparative morphometric analysis we used two recent hydrocephalic skulls (age at death: 34 and 12 years) and a control group of morphologically regular formed crania (N=20, age of death: 2 years to adult).

The two hydrocephalics belong to the collection of the Federal Pathologic-Anatomical Museum Vienna (diagnosis were made by former pathologists and recorded in ancient hand written inventories), all other specimens are part of the osteological collections housed at the Department of Anatomy, University of Vienna and the Department of Anthropology, Natural History Museum Vienna.

Methods

We digitized 42 clearly identifiable anatomical landmarks on the whole skull and several thousand random points on the neurocranial surface either using a Polhemus device or a CT-reconstruction of the specimens. Based on this surface scatter our algorithm automatically places 336 semilandmarks on the neurocranium and relaxes them against the Procrustes average to decrease bending energy between them (Fig.2). This algorithm omits information about shape deformation along the surface (that is an artefact of landmark spacing) while preserving information perpendicular to the surface (that is due to morphological differences). The resulting semilandmark positions can thus be treated as homologous in the subsequent statistical analyses (Bookstein 1997, Mitteroecker et al., in press). The 23 sets of 378 landmarks were superimposed using a GLS Procrustes registration (Rohlf and Slice 1990) which partials out position, size, and orientation. The resulting Procrustes coordinates contain shape information only and overall size is measured explicitly as centroid size (the square root of the sum of squared distances from all landmarks to their centroid). For the explorative study of shape we use relative warps analysis (i.e. principal components of Procrustes coordinates, see Bookstein 1991, Rohlf 1993).

Results

Centroid size as a measurement for overall scale for each individual is plotted in Figure 3. While the Neolithic specimen is larger than all skulls of the control group, it is also clearly smaller than the two hydrocephalic skulls. In a relative warps plot of all neurocranial landmarks the hydrocephalics and the Neolithic specimen are completely separated from the control sample (Fig.4a). When only the semilandmarks on the frontal bone are analyzed (b), the Neolithic specimen does not cluster together with the other hydrocephalics though it is distinct from the reference population. With respect to the first Relative Warp its shape is completely different from the other two pathologic cases. If only the posterior part of the cranium was analyzed, the two hydrocephalics and the Neolithic specimen are clearly separated from the control group but cluster together (not shown). Even when the specimen's size is considered in the analysis by re-multiplying the shape coordinates by their respective centroid size before performing the relative warp analysis (c, d), the frontal bone of the adolescent Neolithic specimen lies very close to the normal variation.

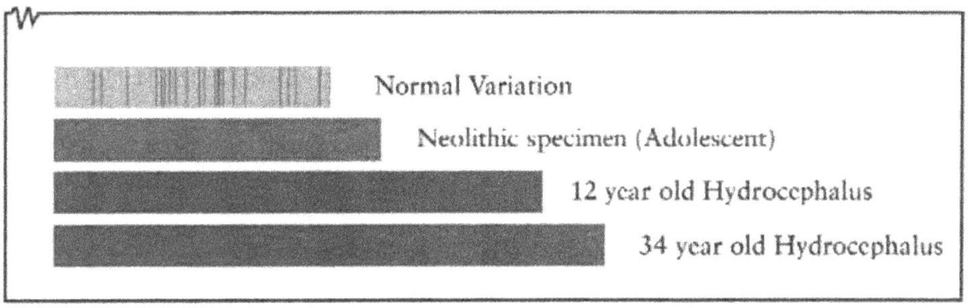

Figure 3 Centroid size as a measure of overall scale. The upper bar shows the size range of the control sample, the black lines marking single specimens. The adolescent Neolithic cranium is larger than all non-pathological cases

Figure 4 shows the actual differences of the neurocranial shape of the hydrocephalics to the mean form of the control sample. Both known hydrocephalics exhibit enlarged frontal and parietal bosses (a, b) while the Neolithic specimen's neurocranium, though generally enlarged, mainly deviates in its posterior regions.

Discussion

The shape analysis confirms the impression that in the Neolithic specimen almost only the parietal bosses seem to be enlarged while the form of the frontal bone lies close to the normal adult variation (Fig. 4). Richards and Anton (1991)

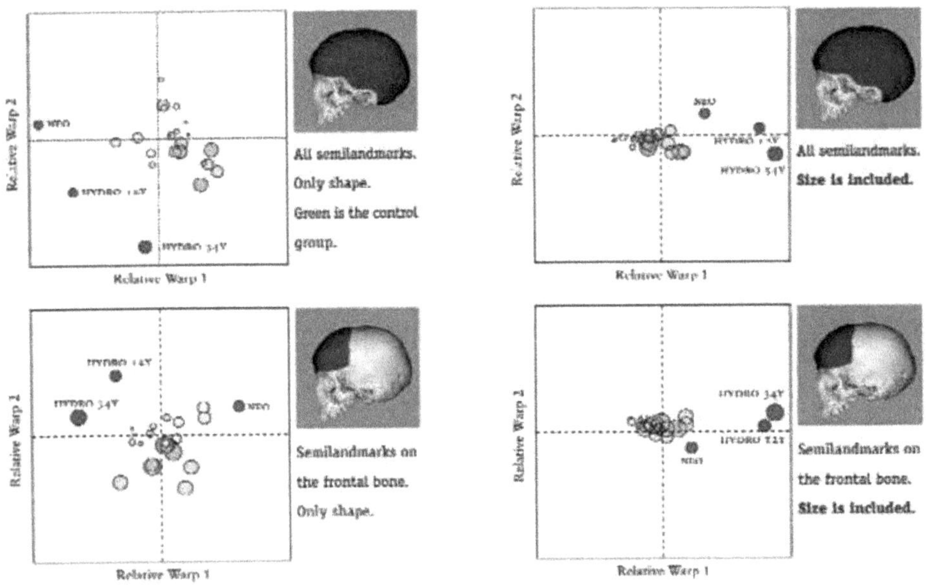

Figure 4 Relative warp analysis (principal components of Procrustes coordinates). Age is coded by disk size; the control group is unlabeled. The hydrocephalics and the Neolithic specimen are separated from the control group. When only the semilandmarks on the frontal bone are analyzed, the Neolithic specimen does not cluster together with the other hydrocephalics but is still distinguishable from the control group. This keeps true also when the specimen's size is considered in the analysis by re-multiplying the shape coordinates by their respective centroid size

distinguish between rectangular and more triangular shaped hydrocephalics and suspect the reason therefore in a communicating versus non-communicating form of hydrocephaly. We add another explanation: The different onset of the cause for the enlargement. If the intracranial pressure rises because of an increase of cerebrospinal fluid (CSF), the hydrostatic pressure against the skull vault is likely to be more or less uniformly distributed over the whole interior cranium. The head does not expand isometrically, though; the bones can compensate the pressure only in certain areas, depending mainly on the cranial sutures. In congenital (or very early acquired) hydrocephalus, none of the sutures have fused-the intra-cranial pressure causes the head to expand rather uniformly resulting in a rectangular shape. When hydrocephaly is acquired later in life, the frontal suture is already fused and the cranium can only expand in the parietal and occipital region, leading to a triangular shape.

The frontal suture normally fuses until the age of two, so the Neolithic skull could represent a case of acquired hydrocephalus with an onset of intracranial pressure later than two years of age. This hypothesis is supported by the fact that the skull's endocranial lamina exhibits vestiges of an inflammatory process presumably caused by meningitis or meningoencephalitis. The latter can lead to obstructions of CSF flow and/or resorption defects-causing a dynamic block and increasing CSF pressure (Teschler-Nicola et al. 1986).

Traditional craniometrics offers only very few reliable landmarks on the neurocranium while the face posses-ses a lot of unequivocally defined point locations (Type I and II, as defined by Bookstein 1991). Landmark definitions based on maximal extensions like Eurion (Type III landmarks) are not suited for geometric morphometric analysis because they are well defined in just one dimension and lack the other two coordinates. In this study we have approached this problem by applying semilandmarks on surfaces (Mitteroecker et al., in press). With this method we could show differences in neurocranial form and shape between the Neolithic specimen and the control group as well as the two hydrocephalic skulls.

In the current example it would be hard to diagnose hydrocephaly on the frontal bone only, but we clearly could distinguish the Neolithic individual from the others in the relative warp plots (Fig.3b, d). The methods shown in this study can provide a quantitative and transparent way in the identification of anomalous morphologies.

Acknowledgements

We thank F. Bookstein, J. Kim, B. Patzak, H. Prossinger, K. Schaefer, H. Seidler, D. Slice and T.B. Viola for access to material and helpful discussions. This study was supported by the Austrian Science Foundation P14738.

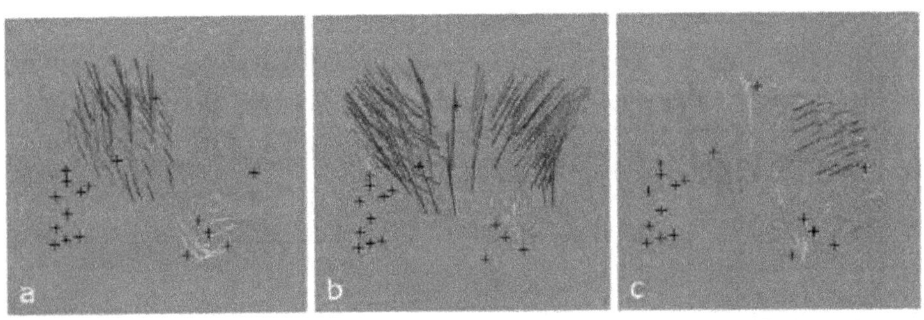

Figure 5 Shape residuals of the semilandmarks after the real landmarks (black crosses) were warped to the consensus. The little lines visualize direction and magnitude (length and color) of deviation from the mean form of the control sample. The two hydrocephalics (a, b) are easy to recognize because of their enlarged frontal and parietal bosses. The Neolithic cranium (c) shows some differences compared to the other two hydrocephalics - the parietal bosses are very enlarged while the frontal bosses exhibit only small expansion

REFERENCES

BOOKSTEIN, F., 1991. Morphometric tools for landmark data: geometry and biology. Cambridge University Press, Cambridge/New York.

BOOKSTEIN, F., 1997. Landmark methods for forms without landmarks: morphometrics of group differences in outline shape. Med Image Anal 1(3):225-43.

MITTEROECKER, P., GUNZ, P. et al., in press. Semilandmarks in three dimensions - Modern Methods in Anthropology. D.E. Slice, Kluwer.

RICHARDS, G.D. and ANTON, S.C., 1991. Craniofacial Configuration and Postcranial Development of a Hydrocephalic Child (ca. 2500 B.C. - 500 A.D.): With a Review of Cases and Comment on Diagnostic Criteria. Am J Phys Anthropol 85:185-200.

ROHLF, F.J. and SLICE, D.E., 1990. Extensions of the Procrustes method for the optimal superimposition of landmarks. Syst. Zool. 39:40-59.

ROHLF, F.J., 1993. Relative warp analysis and an example of its application to mosquito wings. Contributions to Morphometrics. In Marcus, L.F., Bello, E., and García-Valdecasas, A. (eds.), Monografias, Museo Nacional de Ciencies Naturales, Madrid:131-159.

SCHULTZ, M. and TESCHLER-NICOLA, M., 1989. Osteologische Untersuchungen an bronzezeitlichen Kinderskeletten aus Franzhausen, Niederösterreich. Verh. Anat. Ges., Leipzig:82, 407-409.

TESCHLER-NICOLA, M., SCHULTZ, M., KILLIAN, W. and LUBEC, G., 1986. Osteologische Untersuchungen an dem Kinderskelett aus der mittelneolithischen Kreisgrabenanlage von Kamegg, Niederösterreich. Archeologia Austriaca 70:113-136.

TESCHLER-NICOLA, M. and GEROLD, F., in press. Zur Palaeoepidemiologie des mittelalterlichen Gräberfeldes von Gars/Thunau.

TRNKA, G., 1986. Vorläufige Ausgrabungsergebnisse der Kreisgrabenanlage von Kamegg (NÖ). Archeologia Austriaca 70:87-111.

THREE-DIMENSIONAL IMAGERY : A NEW LOOK AT THE TAUTAVEL MAN

GASPARD GUIPERT

MARIE-ANTOINETTE DE LUMLEY

HENRY DE LUMLEY

BERTRAND MAFART

LABORATOIRE D'ANTHROPOLOGIE, FACULTÉ DE MÉDECINE,
SECTEUR NORD, MARSEILLE, FRANCE
GASPARD.GUIPERT@TIMONE.UNIV-MRS.FR

ABSTRACT

In 1981 was shown an empirical reconstitution of the incomplete skull of a fossil man called the Tautavel Man. The recent advent of digital picture gave us a second opportunity to study the fragmentary remains (Arago 21 & 47) of the 'Oldest Frenchman' (450,000 years old). After researches on the original fossil records as well as on X-ray digital acquisitions, we made a 3D model by virtually replacing the fossils. A preliminary work was realized on solid cast then, in a second step, the considered placements were analyzed virtually. The correct middle face was mirrored, different deformations were balanced, the right parietal was placed differently. The lack of some bones was counterbalanced by a substitution with others more or less contemporary European skulls (Sima de los Huesos,...) digitalized as well. The different hypotheses can be tested directly, preventing a long and expensive work by means of cast reconstitution, which often proves difficult to handle. This work will allow us to better know the hominids of the Middle Pleistocene by resituating the Tautavel Man more accurately, while waiting for the discovery of the missing fragments during future excavations.

INTRODUCTION

In 1971, a human face was discovered in the "Caune de l'Arago", a cave in Tautavel, France, in layer "G" dated to 450,000 years BP (Lumley 1973, Yokoyama 1991). It was the 21st human remains found in this place, therefore called "Arago 21" (Fig.1). In 1980 was discovered an incomplete right parietal bone, "Arago 47", which could be joined with the face. Those two bones were baptized "Tautavel Man", a Homo heidelbergensis. Other discoveries were realized in the laboratory after studying isolated small bone fragments. We have two other parietal fragments corresponding to the right and left sides of the bregma area. They concern the same individual. At the moment other fragments are under consideration : one left parietal bone and two foramen magnum parts. A first reconstruction was realized in 1981; nowadays, with new technologies, we can improve this modelling.

MATERIALS AND METHODS

[1] Specimens

We used 4 human remains : Arago 21, Arago 47 & 47A, Arago 3A. Arago 21 is a deformed face, due to a lateral pressure after depositing. This face is complete, including malar, maxillary, nose and frontal bones. There are many deformations, structural and superficial. The most distorted part is the left side of the face. The external surface of the right malar bone is deep, without internal deformations. The right parietal bone (Arago 47) is incomplete. The posterior and temporal parts are totally preserved with a large angular torus and a portion of the sagittal suture. We have 2 mm of coronal suture. It is the principal clue that allows us to say that it is the same individual. Arago 47A and 3A are two parietal bregma fragments fitting together with the frontal and permitting us to reconstruct the anterior part of the sagittal suture.

The first reconstruction, quite acceptable in those days, was achieved by M.-A. & H. de Lumley and R. David with an empirical method using those fragments and parts of mouldings of other skulls (Swanscombe for occipital, Sangiran 17 for temporal). Improving of this reconstruction was however necessary. In-

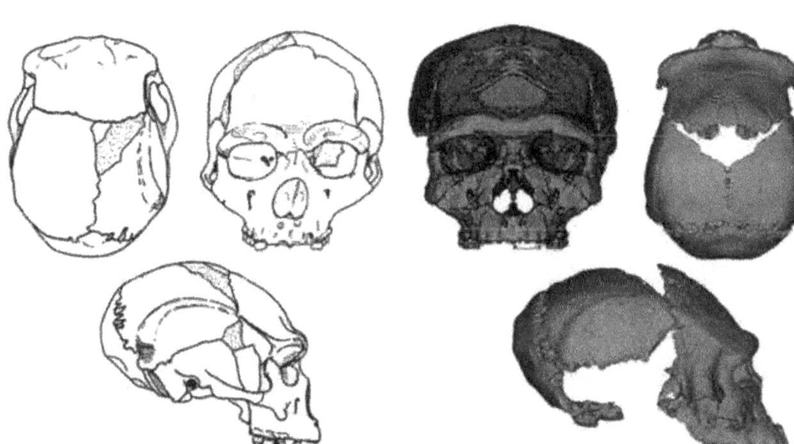

Figure 1 (left) Frontal view of Arago 21

Figure 2 (right) Different views of former (left) and newer reconstitutions (right) of the Tautavel Man. We can see the junction point between the face and the parietal

deed, the posterior curvature between the two parietals (original and reconstructed) did not correspond to an anatomical structure, creating an unusual protuberance. The face was distorted, not balanced, always asymmetric (different orbital cavities, slantwise nose,...). The sagittal suture was not in line with the lateral curvature. Also, the landmarks were not aligned. Finally, the asterion-lambda-asterion angle was too short for a Middle Pleistocene Hominid. It was more like that of the present Homo sapiens, with anatomical characteristics incompatible with Homo heidelbergensis (i.e: torus angularis, parietal thickness, supra-orbital torus, prognathism).

[2] Computer analysis

In collaboration with Initial Society and P. Corbex, we worked on computers with different software's (Mimics 2001, Magics 2001) and 0.5 mm thick CT scans of the original bones. We wanted to use virtual reconstruction because of the brittleness of the original fossil material and to prevent physical disassembly.

[3] Methodology

After researches on the original fossil records as well as on X-ray digital acquisitions and by taking into consideration the different points of view (Grimaud 1982, Spitery 1982, Vlcek 1986), we performed a computerized reconstruction of the skull. We used external anatomical features (parietal line, coronal suture) to position the bone fragments on the computer screen. The inferred effects of general deformation due to the compression of the face that occurred during fossilization were corrected by mirroring.

In the first place, we sought to obtain a more accurate version of face: we corrected the frontal bone (width, positioning of fragments, curvature), we modified the "nose" so as to have a less distorted nasal cavity and a smaller rise of the right malar bone (the external aspect of the left malar bone is correct, but not its location) and finally, we finished the face reconstruction by mirroring to preserve the right side, which is the less deformed. In the second place, we had to position the parietal bone (Arago 47): we used the portion of the coronal suture to connect the parietal and the frontal bones, like in the former reconstruction. Nevertheless we rose the parietal right posterior part, using measurements obtained by preliminary work on a moulding. It allowed us to test different arrangements. After that, we did a reconstruction of the midvault by mirroring the right parietal bone. Only for comparison (Fig.2), we adjusted on our rebuilding the occipital of Swanscombe, but not the two bregma fragments which were not scanned at that time.

Results

The consequences of the new bone arrangements are the following. We observe a growth of the missing pterion part and a reduction of the missing sagittal suture. The asterion-lambda-asterion angle is growing too. In the lateral and the sagittal view, we can see the connection of parietal and the frontal bone, with a continuity of the parietal line. There is no more break anymore between the face and the parietal in superior view (Fig.2, on the left). The sagittal suture is now straight, without curvature. The midvault is more rounded in frontal view.

The occipital bone and the midvault fit together without any modification. That was not the case with the former reconstruction which required an occipital adjustment by cutting off the occipital suture. It gave a value of 73° for asterion-lambda-asterion. The new value is 90,7°. For comparison, the average of Homo sapiens is 85° and, for Homo neanderthalensis, it varies between 70 and 99° and between 87 and 108° for Homo erectus (Spitery 1984). This new measurement corresponds more to an old hominid.

Table 1 indicates intermediate results, estimated and direct measurements taken from the newer and former reconstructions, 4 Homo neanderthalensis and 3 other European Hominids. These measurements were realized on mouldings.

	Max Midvault Width	As-As Width	La - G Line (M3)	La - Na Line	La - B line	B - G	B - Na	Min Frontal Width (M9)	Max Torus Width	Max Frontal Width (M10)
Ceprano	156	128	174	175	101	102	106	106	130	118
Arago 2003	**159**	**122**	**183**	**181**	**98**	**105**	**110**	**114**	**132**	**124**
Arago Swanscombe	142	112	181	175	100	103	105	103	125	107
Petralona	143	119	186	180	104	108	109	110	133	119
Sima de los Huesos 5	139	112	168	169	98	99	102	104	124	114
Neandertal	147	119	186	184	105	112	117	108	121	121
La Chapelle-aux-Saints	152	113	187	182	100	105	105	108	122	123
La Ferrassie	154	112	189	188	104	118	121	107	120	120
Monte Circeo	153	116	178	176	93	109	112	110	122	123

Table 1 Comparative table (Italicized numbers represent estimations. All the values are in mm.)

Discussion and Conclusion

The work on the Tautavel remains will be finished within a few months with the addition of the two bregma fragments. We will be able to correct definitively and precisely the curvature of the superior part of the frontal bone and the coronal suture. Our work will consist in levelling out the small excrescences created by frontal deformation. After that, we will have the sagittal outline from the rhinion and the nasion to the bregma, 2 cm of sagittal suture, a small gap (2 cm) and the rest of the sagittal suture to the lambda.

Anatomical characters look more archaic than later Middle Pleistocene Homo erectus and Homo heidelbergensis. Comparison with the skulls of other fossil men allows us to place this reconstitution closer to an older skull, the one from Ceprano, Homo cepranensis (Mallegni et al. 2003). Our reconstruction shows many similarities with this skull: a broad nasal bone, a torsion of the supraorbital torus, a bilate-

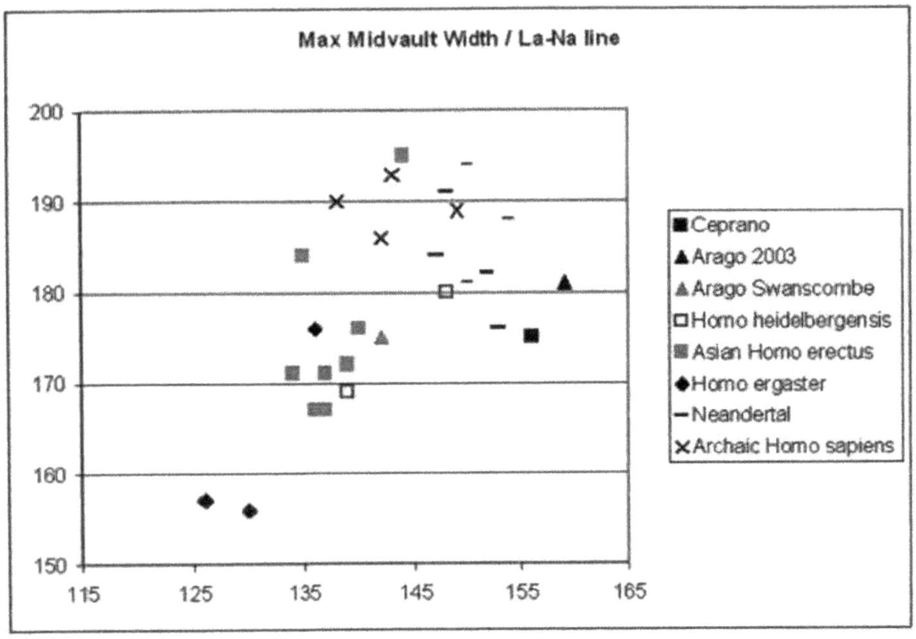

Figure 3 Comparison of different Hominid populations with the newer reconstruction of Arago

ral discontinuity of the supratoral sulcus, a pronounced post-orbital constriction, a frontal keel, a flattened parietal, a maximum breadth across a prominent angular torus, and a low cranial vault. Anatomically, the Ceprano Man shares many features with our reconstruction and the dimensions of his temporal bone correspond to the gap in the Tautavel Man. Figure 3 illustrates the closeness between them. A comparison with this man who died 900,000 or 800,000 years ago (Mallegni et al. 2003) points out the archaic characteristics of Arago 21-47. The dating back to 450,000 years BP is an estimation, but we could imagine that this man is even older. We will develop this question when this work is totally finished.

We will continue our reconstruction by assembling this one with the two bregma parietal fragments and by filling the missing parietal anterior parts. Another part of the work will consist in finding similar skulls in order to reconstruct the missing parts. After scanning those skulls, we will "carve" them to take the interesting fragments (temporal, sphenoid and occipital bones). Presently, the most interesting skull is Ceprano. Once this stage has been achieved, we will be able to make resin models by stereolithography (Zollikofer 1998, 2002).

With digital technology, we will suggest a new reconstitution after we have directly tested different hypotheses, preventing a long and expensive work, by means difficult to handle cast reconstitutions. This work will allow us to have a new vision on the Hominids of the Middle Pleistocene by better situating the Tautavel Man, pending further discoveries of the missing fragments during future excavations.

Acknowledgement

Initial Society and Patrice Corbex for their help. We are deeply grateful to Francis for translation and to Valérie for her assistance.

References

GRIMAUD, D., 1982. Evolution du pariétal de l'Homme fossile - Position de l'Homme de Tautavel parmi les Hominidés. Doctorat de 3ème cycle: "Géologie des formations sédimentaires", option: Géologie du Quaternaire et Préhistoire.

LUMLEY, H. and LUMLEY, M.-A. de, 1973. Pre-Neanderthal Human Remains from Arago Cave in South-eastern France. In Buettner-Janusch, J. (ed.), Yearbook of Physical Anthropology, Washington.

MAGICS 7.5., 2001. Materialise. Leuven, Belgium.

MIMICS 7.1., 2001. Materialise. Leuven, Belgium.

MALLEGNI, F., CARNIERI, E., BISCONTI, M., TARTARELLI, G., RICCI, S., BIDDITTU, I. and SEGRE, A., 2003. Homo cepranensis sp. nov. and the evolution of African-European Middle Pleistocene hominids. Comptes Rendus Palevol 2:153-159.

SPITERY, E., 1982. L'occipital de l'Homme de Tautavel. Essai de reconstitution. In CNRS (ed.), 1er Congrés international de Paléontologie Humaine - L'Homo erectus et la place de l'Homme de Tautavel parmi les Hominidés Fossiles, Nice:89-109.

SPITERY, E., 1984, unpubl. Evolution de l'os occipital chez les Hominidés. Doctorat de 3ème cyle: "Géologie des formations sédimentaires", option : Géologie du Quaternaire et Préhistoire, Université de Provence.

VLCEK, E., 1986. Les Anténéandertaliens en Europe Centrale et leur comparaison avec l'Homme de Tautavel. L'Anthropologie 90, Paris:503-513.

YOKOYAMA, Y., NGUYEN, H.V., FALGUERES, C., BIBRON, R. and LEGER, C., 1991. Datation directe par la spectrométrie gamma non destructive des restes humains: comparaison avec d'autres méthodes. In CNRS, Raynal, J.-P. and Miallier, D., (ed.), Datation et caractérisation des milieux Pléistocènes, Bordeaux:11-17.

ZOLLIKOFER, C., PONCE de LEON, M.S. and MARTIN, R.D., 1998. Computer-assisted paleoanthropology. Evolutionary Anthropology 6:41-54.

ZOLLIKOFER, C.P.E., 2002. A Computational Approach to Paleoanthropology. Evolutionary Anthropology Suppl1:64-67.

3D Modelling and Database of Indian Archaeological Pottery Assemblages-HNBGU-NDSU Collaborative Initiative

Vinod Nautiyal

Sudhir Nautiyal

Mohan Naithani

Juyal Sanjiv

Archaeological Computing Laboratory,
Department of History and Archaeology,
HNB Garhwal University (HNBGU),
Srinagar-246174, Uttaranchal

Jeffrey T. Clark

James E. Landrum III

Aaron Bergstrom

Richard Frovarp

Justin Hawley

Derrick Eichele

Archaeology Technologies Laboratory (ATL),
North Dakota State University (NDSU),
Fargo, ND, USA

See the CD for the extended version

Abstract

Indian archaeological ceramics have been considered the most useful cultural resource for reconstructing various aspects of ancient society. Scanning technology has not been used effectively in India for rapid documentation of collections and there is no centralized or easily accessible archive of pottery available for comparative analysis. With this in mind, the authors have initiated collaborative work for the development of computing technology for 3D modeling and an Internet-accessible database of Indian archaeological pottery.

Introduction: Indian Pottery Collections and Indian Heritage

Pottery constitutes the major proportion of all material recovered from archaeological excavations in India. The mosaic of Indian history and archaeology, spanning 5,000 years, is reflected through the different and diverse pottery traditions, commencing in the Neolithic and continuing throughout the Indian archaeological sequence. The different ceramic traditions and cultural phases are reflected through specific, diagnostic pottery types. In addition, pottery defines the cultural, social, economic, and technological status of the people who developed the different cultures and phases over time and space. In short, pottery collections in India serve as one of the primary sources of historical reconstruction. Therefore, a large number of studies focusing on the importance of pottery have been done in India (e.g. see Pant and Jayaswal 1997). Indeed, the cross-cultural associations of the different Indian protohistoric sites have been reconstructed based upon the relationships of pottery types (Srivastva 1979). However, the pottery assemblages from a wide range of sites are stored at distantly located institutions throughout India. Access to these preserved materials and the associated documentation is often difficult and expensive. No centralized or otherwise easily accessible archive of the corpus is available for comparative analysis. Also, in spite of the well-established classification of Indian pottery collections in terms of shape, color, fabric, paintings and designs, etc., there has not been a concerted effort to develop or introduce an automated system for fast documentation of pottery shapes. Nor has there been an effort to exploit the advances in computing, software, 3D modeling, and Internet technology to develop a Web-based database to access the detailed information on Indian archaeological pottery collections.

In order to fill this gap, the first major challenge has been to introduce some form of automated device and to make use of computers and other graphics tools to develop the technique further. To some extent, Dr. Nautiyal, along with students and staff, at HNB Garhwal University (HNBGU) has been able to address partially that goal (Nautiyal et al. 2000, 2001), but there is still great need to advance and develop their work on 3D ceramic modeling. Due to limitations of equipment and technological expertise, however, that advancement has been difficult. Therefore, a collaborative project was initiated by Drs. Nautiyal and Clark and their teams that will link the efforts at HNBGU with the work being done at the Archaeology Technologies Lab at North Dakota State University (NDSU). With joint funding from the DST (India) and NSF (USA) programs, the digital database of Indian pottery being developed at HNBGU will be made accessible through the Digital Archive Network for Anthropology and World Heritage (DANA-WH). Furthermore, NDSU will assist HNBGU with the application of the latest 3D scanning technology and a host of 3D graphics and database software. The present paper summarizes the work at HNBGU and describes the preliminary results of the collaboration between HNBGU and the NDSU.

Related Research

Pottery is one of the most conventional sources of data for understanding and reconstructing the past. Typically, a ceramic assemblage from an archaeological site is collected, catalogued, and documented through traditional means for research purposes as well as for museum display. However, in recent years, with the advancement of computing technology, 3D graphics software, and innovative imaging and analytical techniques, there has been a shift towards the investigation of the various complex attributes of pottery through the application of geometrical 3D shape modeling, 3D reconstruction of axially symmetric pots from fragments, texture mapping, and more. While a comprehensive review of recent literature about computer based visualization of ceramics and virtual reconstructions of vessels cannot be presented here, there is clearly an increase in the number and sophistication of such efforts, as illustrated by several papers at the CAA 2003 and other publications (e.g. Schurmans et al. 2001, PRIP 2000, Cooper et al. 2001).

HNBGU System: Optical Plotting - Extraction of 2D and 3D Profile and Diameter

In view of the latest advances in the field of computers and various software applications, HNBGU has exploited the tools of AutoCAD for manipulating 2D outlines generated through an optical plotter (called Reflex Metrograph). These 2D outlines are then used to produce 3D rendered images of the archaeological ceramics (Nautiyal et al. 2000). The development of this AutoCAD-based technique opens many new possibilities in the areas of visualization, shape modeling, and use in digital archives for dissemination of pottery collections and other cultural heritage resources via the Internet. However, a lack of more advanced technology at HNBGU has hindered further developments in these areas.

The optical plotter that is being used by Dr. Nautiyal was designed for orthodontics. However, its simple application was also found to be useful in archaeology by Scott (1981). At HNBGU, an optical plotter was used to make drawings of pottery fragments and full vessels to demonstrate the inaccuracies present in the manual method of pottery drawing (Nautiyal et al. 1995). The details of the system have been described elsewhere (Scott 1982), therefore only a brief description is given here.

The optical plotter (Fig.1) is used to digitize the x-, y-, and z-coordinates of an object through the semi-reflecting mirror in ASCII format. The light source is moved over the reflection of the object to generate an outline of the object. Rplot2 software is used to convert the RPT files into DXF format, which allows the HNBGU team to manipulate and modify the files using more recent releases of AutoCAD, such as R14 and 2000. A 3D virtual reconstruction of the vessel to the lowest point of the vertical profile line is then generated by a series of AutoCAD commands: Rimdiameter, DView, Mirror, Resurf, and Render. This technique has been successful for generating 2D images and rendered 3D models of pottery.

Texture mapping, or the projection of a surface image onto an object, has been added to the HNBGU system as a result of the collaboration with NDSU. In addition to shape reconstruction, the surface texture (color) of pottery is as equally important for generating a realistic impression. In earlier work at HNBGU (Nautiyal et al. 2000) attempts had not been made to texture map the pottery assemblage. Following discussions between our teams, the materials library tools in AutoCAD have now been effectively used for texture mapping at HNBGU. These tools provide a workable solution to generate the closest possible photo-realistic model of the vessel. In this study, the digital image (JPEG and BMP) of the pottery fragment is imported into the materials library and the color captured during this stage is wrapped to the object. It is then rendered to create a textured full model.

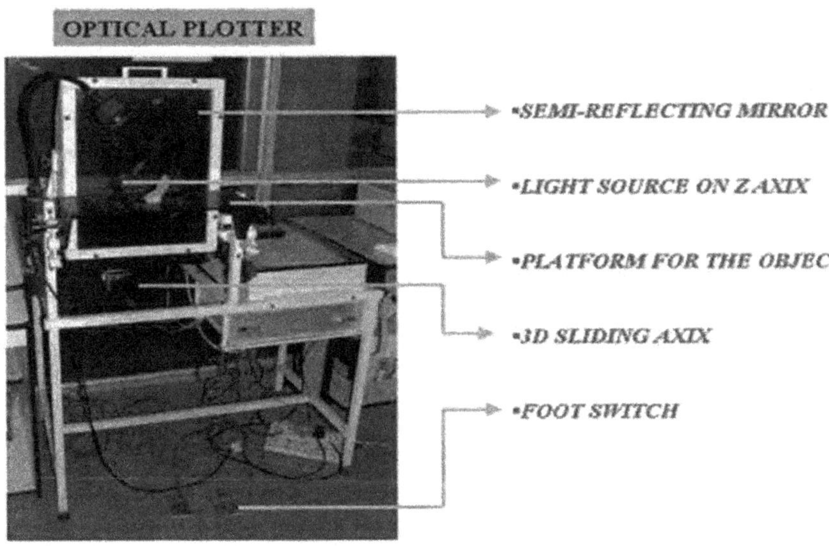

Figure 1 Optical plotter used by HNBGU

Thus, optical plotting hardware coupled with AutoCAD software has provided an inexpensive technique for drawing archaeological sherd profiles and for creating partial virtual reconstructions of vessels. To these reconstructions, textures can be added based on digital photographs of the original sherd. These reconstructions significantly improve efforts to classify pottery from archeological collections. This technique opened up new possibilities for Dr. Nautiyal to begin working with Dr. Clark on laser-based techniques of modeling, texture mapping, and Web-enabled digital archives.

ATL- NDSU

In November 2001, the Archaeology Technologies Laboratory (ATL) (http://atl.ndsu.edu) at NDSU received funding from the National Science Foundation's Digital Libraries (NSDL) program, for a two-year project designed to create and implement a Digital Archive Network for Anthropology (DANA). This network is being designed to function as a federation of distributed, interoperable databases, each with specific content of value to anthropology (Clark et al. 2002). With the realization that a network such as DANA could grow larger than first conceived, the scope of its content grew to cover all domains that have as their focus the study of cultural and biological human heritage. This new vision also included expanding the content beyond US-based institutions to include international participants. As a result, DANA was enlarged in name as well as content, and is now the Digital Archive Network for Anthropology and World Heritage (DANA-WH, http://www.dana-wh.net). DANA-WH is also accessible through the NSDL portal (http://nsdl.org) and the ATL Web page (http://atl.ndsu.edu/archive/).

DANA-WH differs from other digital heritage archives; in addition to textual and two-dimensional (2D) graphical (digital photos, sketches, etc.) content, it includes accurate, 3D models of artifacts and fossils. The 3D models are surrogates of the real objects that can be variously manipulated to be viewed from all angles, and they are sufficiently precise to allow for a range of detailed measurements and analyses. The ATL has developed a customized 3D viewer for DANA-WH that includes virtual tools that enable users to take and record basic measurements of the 3D objects for individual research and analysis or group studies. As an online distributed network, DANA-WH will allow "anytime, anywhere" access to content and services. Users can take advantage of the digital library for formal research, as an educational resource, or simply to satisfy curiosity.

The ATL is equipped with three non-contact laser digitizers/scanners and an array of software for 3D modeling and graphics. Each of the scanners captures a point cloud that represents the surface of the object. Point clouds are merged using third party software, either PolyWorks (http://www.innovmetric.com/) or Geomagic (http://www.geomagic.com/). One of these digitizers, a Minolta Vivid 900 (http://www.minoltausa.com/vivid/default.asp), captures color-per-vertex information, thus eliminating the need for a separate step for texture mapping. The 3D surrogates are available for viewing and a range of measures through DANA-WH, which also presents a range of metric and non-metric associated data.

As a further result of the HNBGU-NDSU collaboration, we are now working to integrate knowledge and methods to develop a more efficient system for vessel reconstruction from single sherds. Rather than optical plotting, the ATL employs its laser scanners to create highly accurate 3D models of the entire sherd. From this sherd surrogate, we can define a profile by a plane through the sherd. We are working to further define the vessel profile, to the extent possible from size of the sherd, based on a series of slices through the digital surrogate. The advantages of this technique over that used at HNBGU is that the basis for the vessel reconstruction then becomes the entire sherd rather than a single linear profile of one slice of the sherd, and therefore the profile is more precise. Furthermore, the Minolta Vivid 900 scanner allows for the automatic texture mapping of the sherd, which greatly facilitates the simulation of full-vessel surface color.

Certainly there are limits to how much one can reconstruct, or model, based upon one sherd; for example, if the lower portion of the vessel is missing, particularly the base portion, then reconstruction will typically be most problematic. This limitation can be ameliorated somewhat, however, in situations where the ceramic assemblage is reasonably well known for a given temporal and cultural period. In such a situation, a series of templates of characteristic vessel forms can be developed to serve as the basis for shape comparisons. That is, the sherd form is used to reconstruct the vessel shape as far down as the sherd size allows; extensions of the walls are then projected beyond that area. New software will be developed that will be used to compare the simulated form to the set of vessel templates for a best-fit match to some specified level of confidence. The template form parameters can then be used to create an idealized model of the full vessel shape. We are still far from developing the shape comparison capability, but software for similar types of shape analysis is being developed at Arizona State University (e.g. Schurmans et al. 2002) and elsewhere. Whatever the extent of the vessel reconstruction, though, the sherd can be viewed in an appropriate place on the whole pot, and the real sherd can be distinguished from the proposed reconstruction.

Digital Database of Archaeological Pottery from India

The primary objective of our collaboration is the development of a pilot program and proof-of-concept for the Indian ceramic comparative collection delivery through DANA-WH. The inclusion of the HNBGU database of Indian pottery in DANA-WH will allow it to be networked with other databases and to make it available to a global audience of researchers. In return, DANA-WH will be enriched in terms of its content. Furthermore, HNBGU will be the first international participant in DANA-WH. In addition, our collaboration will provide mutual benefits for distributed database network development, 3D modeling, and pottery vessel reconstruction.

To start this work we have selected a well documented and published pottery collection from Atranjikhera, an Indian archaeological site in District Etah, Uttar Pradesh, excavated by R. C. Gaur in 1964. The most noteworthy feature of this published material is that pottery from the earliest Ochre Color Pottery (OCP), to Painted Grey Ware (PGW), Northern Black Polished Ware (NBP), Black and Red Ware (B&R), through late Red Ware comes from a well-stratified deposit and comprises almost all the typological shapes of Indian pottery (Gaur 1983). In the development of the database,

Gaur's work has been used to demonstrate the real and future potential of digital archive technologies in archaeological research. In this case, the long-range goal is to establish an online Indian comparative ceramic collection for analysis and refinement of the Indian pottery typological classification system.

In our initial work, the HNBGU content was converted to more generic database formats that are compatible with a broader range of JDBC-compliant relational database management systems. Oracle 9i (http://www.oracle.com/) and PostgreSQL (http://www.postgresql .org/) were used for the initial implementation. A prototype of the Indian Ceramics Comparative Typology, comprised of 200 ceramic specimen records and associated 2D images of pottery profiles (adapted from RC Gaur 1983), is now available through DANA-WH (Fig.2).

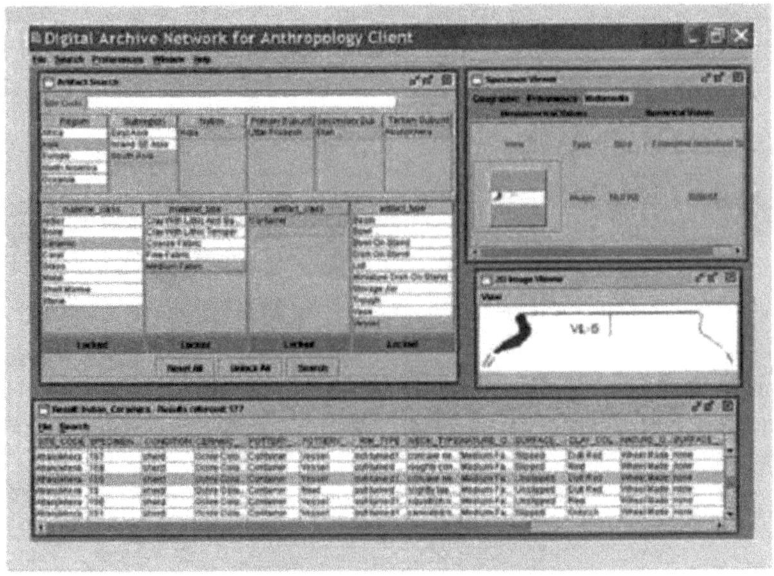

Figure 2 Compilation of screen captures from DANA-WH showing search interface with geographic and artifact type query (top left), list from query results with specific object selected (bottom), multimedia options for object (top right), and profile view f sherd and vessel (right center). Indian pottery vessel from HNBGU database

A major challenge to our goal has been the insufficiency of the technology infrastructure that serves HNBGU, which is in a mountain valley in northern India. Due to a current lack of IT infrastructure at HNBGU, the ATL is hosting the Indian database and image archive during the pilot phase. Through the assistance of Dr. Franco Niccolucci, however, plans are currently underway to secure financial assistance from the Italian government to upgrade the IT infrastructure to HNBGU. When that upgrade is completed, the content will be relocated to HNBGU where it will continue to be accessible through DANA-WH.

Conclusion

The application of computer technology in Indian Archaeology has been little exploited compared to other scientific studies. Therefore, computer applications in archaeology, especially with regard to visualization of artifacts, has remained an undeveloped domain in India. The Indian team at HNBGU has tried to fill the gap by developing an AutoCAD based technique for pottery drawing in archaeology. In this paper we have given a detailed discussion on this technique, and preliminary results of work on 3D modeling, texture mapping, and the development of a web-enabled database of Indian pottery from the archaeological site of Atranjikhera under the DST-NSF collaborative project with the ATL.

The collaborative initiative will be extremely useful in facilitating Web-based access to a distributed network of database and multimedia archives with contents on Indian ceramics, through DANA-WH, for use in the comparative classification and typology of specimens. This will not only benefit Indian archaeologists but also various other scholars interested in South Asian Archaeology.

Acknowledgements

The present work has been done under a collaborative research project sponsored by the Department of Science and Technology (DST), India, and the National Science Foundation (NSF), USA. The senior authors are thankful to the DST and NSF for providing the financial support to undertake the work in the USA and India.

REFERENCES

CLARK, J.T., SLATOR, B.M., PERRIZO, W., LANDRUM, J.E., III, FROVARP, R., BERGSTROM, A., RAMASWAMY, S. and JOCKHECK, W., 2002. Digital Archive Network for Anthropology. Journal of Digital Information 2:1-16.

COOPER, D.B., WILLIS, A., ANDREWS, S., BAKER, J., CAO, Y., HAN, D., KANG, K., KONG, W., LEYMARIE, F.F., ORRIOLS, X., VELIPASALAR, S., VOTE, E.L., JOUKOWSKY, M.S., KIMIA, B.B., LAIDLAW, D.H. and MUMFORD, D., 2001. Assembling Virtual Pots from 3D Measurement of their Fragments. In Proceedings VAST 2001: Virtual Reality, Archaeology, and Cultural Heritage, Glyfada, Nr Athens, Greece, 28-30 November 2001, Association for Computing Machinery, New York.

GAUR, R.C., 1983. Excavations at Atranjikhera-Early Civilization of the Upper Ganga Basin Delhi. Motilal Banarsidass and Center for Advanced Study, Department of History, Aligarh Muslum University.

NAUTIYAL, V., NAUTIYAL, K.P., KHANDURI, B.M., NAUTIYAL, S.C. and NEGI, J.S., 1995. Optical Plotting Through Reflex Metrograph: A New Technical Approach In Pottery Drawing. Man and Environment 20:91-6.

NAUTIYAL, V., NAUTIYAL, S. and NAITHANI, M., 2000. Optical Plotting and AutoCAD® for Drawing Pottery. CSA Newsletter Vol XII no.

PANT, P.C. and JAYASWAL, V. (eds.), 1997. Ancient Ceramics: Historical Enquiries and Scientific Approaches. Agam Kala Prakashan.

PRIP, 2000. Computer Aided Classification of Ceramics (http://www.prip.tuwien.ac.at/Research/ArcheologicalSherds/lit.html, 12 September 2000).

SCHURMANS, U., RAZDAN, A., SIMON, A., MARZKE, M., MCCARTNEY, P., VAN ALFEN, D., JONES, G., ZHU, M., LIU, D., BAE, M., ROWE, J., FARIN, G. and COLLINS, D., 2002. Advances in Geometric Modelling and Feature Extraction on Pots, Rocks and Bones for Representation and Query via the Internet. In Burenhult, G. and Arvidsson, J. (eds.), Archaeological Informatics: Pushing the Envelope, CAA 2001, Proceedings of the 29th Conference, Gotland, April 2001.

SCOTT, P.J., 1981. The Reflex Plotter: Measurements without Photographs. Photogrammetric Record 10:435-46.

SCOTT, P.J., 1982. Drawing and Measurements of Finds: A Reflex Action. World Archaeology 14:191-96.

SRIVASTAVA, K.M., 1979. Community movement in Protohistoric India. Agam Kala Prakashan, New Delhi.

Computerised methods for Rescue Archaeology

Between Soil and Computer - Digital Registration on a Danish Excavation Project

HENRIK SKOUSEN
MOESGAARD MUSEUM, DK-8270 HOJBJERG, DENMARK[1]

SEE THE CD FOR THE EXTENDED VERSION

ABSTRACT

The starting point for my paper is a major excavation project in progress under the administration of Moesgaard Museum, previous to the construction of a highway on the east coast of Jutland in Denmark. On this project we have tried to introduce systems for electronic registration of field data throughout all levels of archaeological investigation.

First a few words about the project and the project area illustrated by the use of digital maps.

In connexion to the project there has been established a formalised co-operation with the local authorities in charge of the construction plans - the county of Aarhus. In this co-operation we exchange digital maps and other data, which facilitates our communication and in a wider perspective the communication with landowners and other persons who might be affected by our excavations and who may wish the possibility to gain an insight into the project.

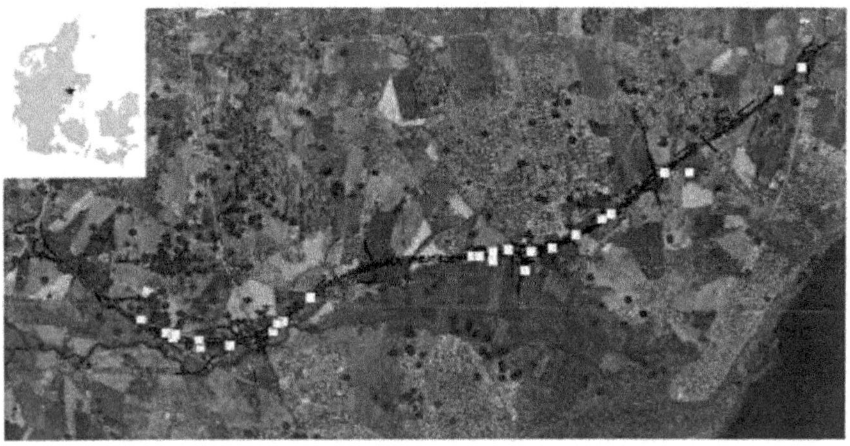

Figure 1 Administration of archaeological data. Map showing the project area with the construction plans of the highway inserted and linked with information from The Danish National Record of Sites and Monuments. White rectangles represents known registrations on the 13.6 km highway tracé combined with newfound sites from field reconnaissance in the area

In this way the GIS maps have become an integrated part of the project from the very beginning and through all subsequent levels of archaeological investigation.

At the map - Figure 1 - we see the highway inserted on the basis of the GIS material we have received from the local authorities constructing the highway.

Our first task was to give an account of the archaeological interests and to this purpose we had a splendid basis in The Danish National Record of Sites and Monuments which is an electronic database containing information on all registered finds and monuments nationwide. At present, The National Record contains data on almost 150,000 finds. Our digital maps could easily be combined with information from this database. At the map we see a plot from the Danish national record containing registrations in vicinity of the projected highway. Each dot contains information about the archaeological object, state of investigation and status. On this basis we were able to give an impression of the archaeological interests in the area. After this operation status was that 20 sites must be excavated within two years by 5 excavation teams employing all in all 40-45 archaeologists and students. To carry out this task and to avoid a situation by which we would end up with an overwhelming mass of finds and paper recorded data and no energy left to take hand on the final reporting, we formulated a set of wishes and preconditions for a system that could ease and rationalize the task of data recording and reporting.

At the end of an excavation all field data should be electronic. What we needed was...

- A system for electronic data registration
- A system for digitising ground plans
- A registration scheme following a uniform procedure

This leads to the second part of my paper, which will be dedicated to what I call the digital excavation, the use of electronic field registration and the systems we have introduced in our daily work. Illustrated by the...

- Recording of field data on handheld computers
- Data collection in the Archaeolnfo database
- From field pantographs to digital area plans in MapInfo
- And the linking of collected field data

In our procedure for recording field data we have tried to combine existing equipment with new technology.

As basic elements in the fieldwork we have introduced: Handheld computers, field pantographs and GPS survey equipment combined with software from Microsoft Access and MapInfo. As handheld computers we have tried working with Psion 5mx PDA's, which are functional, require low battery supply, have a handy keyboard as well as a touch screen at a price of about 200 Euro a piece.

The field pantograph is a drawing machine which works by simple mechanics and quick and easy to use. The GPS survey

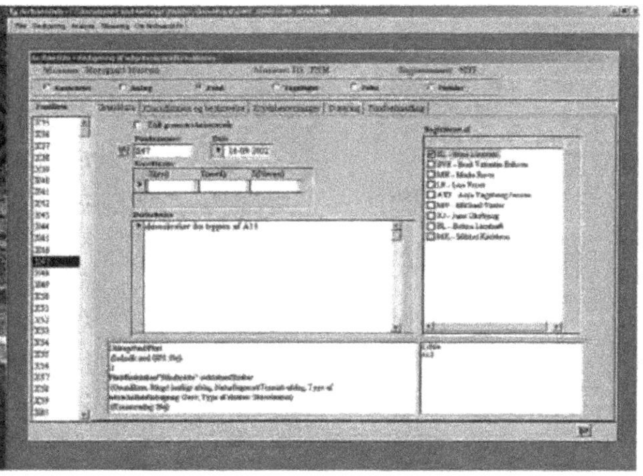

Figure 2 (left) "Hardware" for fieldwork. Working with a field pantograph, handheld computer and GPS survey equipment
Figure 3 (right) ArchaeoInfo - an object oriented information system for archaeological excavations

equipment is an advantage when working in GIS-systems. It is important to notice that all elements are safe and stabile for outdoors working.

Field data is registered during excavation on handheld computers. All observations regarding contexts, constructs, finds, drawings, photos etc. are registered and classified in a predefined, uniform data structure. In the mens wagon these data are regularly transferred to the main excavation database on portable PC's. In this way data are secured and checked simultaneous to fieldwork.

The handheld computers have shown to be surprisingly stabile, functional and useable under any weather condition. In case you have a malfunction on the hand-held PDA data are secured on a memory card.

Never the less experiences during our first excavation season shows that they do have clear limitations. Many people have great difficulties give up the paper. The greatest difficulty is probably the mental adaptation as the mental process is different from paper recording. The record does no longer have any immediate physic dimension. The immediate clarity is different. And some persons simply feel insecure with computers.

For the recording of excavation data, we use ArchaeoInfo. ArchaeoInfo is an object oriented and flexible Access based database created by Torsten Madsen lecturer at the Institute of Prehistoric Archaeology in Aarhus[2]. Data are recorded on a structured form, which you can define on your own and all data are internally cross connected.

Collected field data are transferred from the handheld computers to ArchaeoInfo by an automatic converting to Access. In the process of transference, data are checked for errors such as double entries etc.

By Access we get the opportunity to link and exchange data between the various recording systems used at the excavations. In this way we obtain the requested and improved analytic and reporting possibilities.

The linking of several independent databases hasn't been without problems. Working on more databases, which are subsequently combined, demands that you are very conscious about the potential and limitations of the systems. Although functional, one of the problems is the integrity to the overall system of the very simple database on the handheld computers, which isn't optimal. After a phase of introduction these problems have been eliminated.

Recording of area plans and other digital map material is done in MapInfo. From ArchaeoInfo you can link to MapInfo.

Ground plans are drawn by the use of field pantographs and fixed by GPS measured UTM reference points. By these the drawings can be geographically orientated and digitised in MapInfo. In this way it is possible to carry out excavations without any local coordinate system. Digitising is done as part of the fieldwork, which means that area plans are quikkly secured for further fieldwork, printed on waterproof, transparent paper.

MapInfo is a GIS like program with geographic objects linked to information in tabular data. It has been introduced to Danish archaeology by the national archaeological authorities as a standard to be used in the digital recording of archaeological area plans.

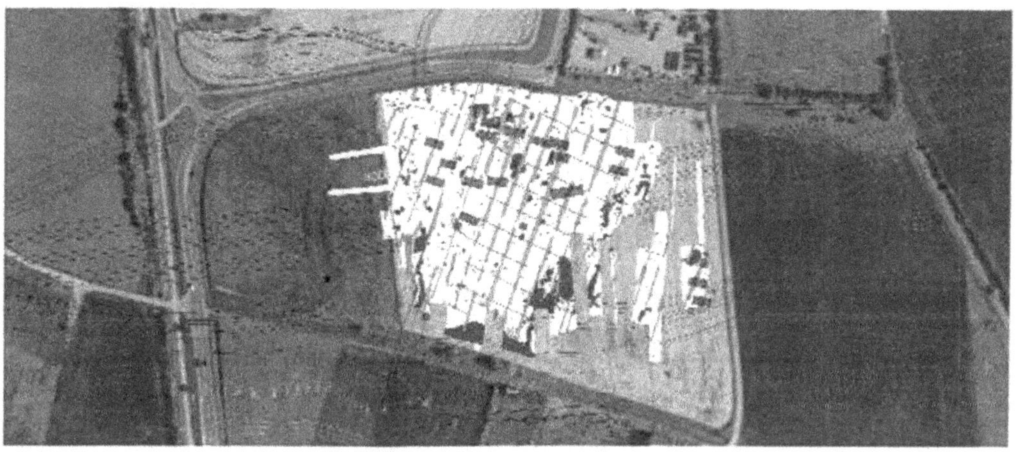

Figure 4 Digital area plan. Example of an excavation plan digitised in MapInfo superimposing an ortophoto

In MapInfo objects are fixed and registered in different layers within the global UTM coordinate system. You define the structure of layers as appropriate. This gives us the opportunity to collect all relevant map material in a structure of layers containing administrative as well as archaeological maps. Behind the digital maps you have the tabular data linked to the objects in the layers. By these data you get the opportunity to carry out advanced search and analysis.

All finds are recorded during fieldwork with a pre-printed label containing barcode information.These pre-printed labels prevent double entries and the barcode facilitates further work and registration of the objects.

Further registration is carried out at the headquarter where all finds are washed, numbered and finally classified. Besides all characteristic finds are photo registered in an arrangement with a camera connected to a computer.
Photo registration is carried out in an independent photo database, but the photos can as well be linked to ArchaeoInfo containing all the recorded data relevant to a given photo.

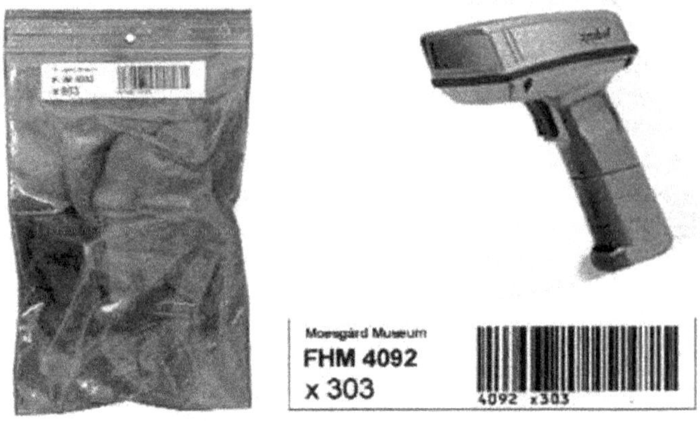

Figure 5 Registration of finds using pre-printed labels

[1]Visit our site on www.moesmus.dk and see more pictures from the excavation project.
[2]ArchaeoInfo was presented at the CAA-2003 by T. Madsen. See: ArchaeoInfo. An Object Oriented Information System for Archaeological Excavations.

Digital Recording of Stratigraphic Excavations

Michael Doneus

Institute for Prehistory, University of Vienna,
Franz-Kleingasse 1, A-1190 Wien, Austria
Michael.Doneus@univie.ac.at

Wolfgang Neubauer

VIAS -Vienna Institute for Archaeological Science,
Franz-Kleingasse 1, A-1190 Wien, Austria
Wolfgang.Neubauer@univie.ac.at

Abstract

Beside the creation of a stratigraphic sequence (Harris matrix) the stratigraphic excavation method is based on the recording of every single unit of stratification. Conventionally it is done by analogue drawings. GIS is a perfect tool for the digital recording of stratigraphic excavations.

During the last years we developed a GIS based procedure for the digital documentation of stratigraphic excavations. To be able to fully reconstruct the part of the site destroyed by excavating, the surfaces of the excavated deposits have to be fully documented in 3D ("single surface mapping"). Every single surface is photographed using a digital camera and rectified on site. The boundary polygon, the topography and the location of finds are recorded using total stations. All data are imported into a ArcView GIS, where the georeferenced data are immediately available for further on-site analysis. Import, terrain modelling, and contouring, of the surfaces is automated by the developed extension ARCDig 1.1. Further modules of the extension allow to calculate cross sections along arbitrary defined lines and to calculate the volume of any deposit. All additional attributable data as descriptions of surfaces and deposits, the find database etc. are integrated in the GIS for further analysis of the data.

Introduction

For a long time in its history, archaeology was considered a science dominated by pragmatism and adapted its methods or techniques from other disciplines, mainly natural sciences. Methodological criticism was and still is mainly based on the methods of interpretation but not on the processes to gain the primary facts or the archaeological record.

One of the most important realizations of the last centuries, that changed archaeological research, was the fact that any archaeological site is stratified and any archaeological stratification is unique (Harris 1989). Stratigraphy, the description and interpretation of stratification is the main key for any further analysis of any archaeological finds. It defines the relative sequence of events marking any site's unique history. This most important attribute of stratigraphy made it necessary to base the recording of archaeological excavations on the principles and needs of stratigraphy. The recording process which takes place during excavation aims at splitting the site into its component parts, the units of stratification. This is done by removing them in the reverse order to which they were formed. During this process, their physical, spatial, and stratigraphic properties are recorded, and finds and samples are collected.

As GIS provides the ability to store, visualise and analyse graphical information in combination with descriptive information, it is a perfect general tool for the visualization and analysis of excavation results. With GIS, archaeologists are able to reproduce the topographical development of sites in an efficient manner that was almost impossible to carry out before the invention of GIS and computers (Harris 2001).

During the last years we tried to develop a GIS based procedure for the digital documentation of stratigraphic excavations. This paper will outline the recording of stratigraphic excavations of the type that directly relates to stratigraphy and GIS, namely plans or maps and related databases.

The Stratigraphic Record

The stratigraphic excavation method, as defined by E.C. Harris (Harris 1989) makes it possible to record the single units of stratification (i.e. deposits and surfaces) along with all its attributes and relations, and to create, a stratigraphic sequence or a Harris matrix from this data, named after its inventor. As Harris points out, every stratification is formed by material (deposits) and immaterial aspects (surfaces or interfaces) that have to be found and recorded by the excavating archaeologist. In the first instance these two aspects are the main objects to be recorded on a stratigraphic excavation. Any finds, samples, or other information and observations have to be related to the deposits and surfaces, i.e. the units of stratification.

Single deposits are 3D volumes of material that are destroyed by excavating them in the reverse order of their deposition. Deposits can only be recorded partially as most of their mass is simply excavated. The material aspects of stratification, the deposit's data, can only be captured by sampling. For the stratigraphic record the deposit is reduced to a unique number in the stratigraphic sequence. It imparts this number to all of the portable finds and samples found within its volume during the process of destruction. The finds or samples are isolated from the stratigraphic unit while excavating it. Their 3D position can be easily defined upon discovery or extraction. The finds and samples are recorded in 3D space as three dimensional points or small volumes, thus reduced to a set of coordinates. All further data derived from the finds can be stored in a database, photographs and find drawings.

In contradiction to deposits, surfaces, the immaterial aspects of stratification, can be recorded entirely. Single-surface recording provides the ability to virtually reconstruct the excavated volumes in three dimensions. Therefore, 3D recording of the top and the bottom surface of any single deposit is necessary to fully reconstruct the part of the site that was destroyed during the process of excavation. The recording of a single surface can be done by giving it a unique number and documenting its boundary polygon as well as its topography. Therefore it is reduced to a set of coordinates, too. The collected sets of coordinates from surfaces and deposits are the primary raw data for the GIS.

Contrary to the deposit, a surface can be recorded in its entirety and that is the reason for its outstanding importance to stratigraphic analysis. Any deposit is enclosed by its top surface, once exposed to air, and its bottom surface which is in superposition to the top surfaces of older deposits or interfaces. Interfaces are surfaces in their own right as the surface of a pit, ditch, wall etc. Thus it becomes clear that surfaces always comprise more than 50 % of any stratigraphic record! Surfaces represent the use periods of a site and thus account for far more time in the site's history than deposits do. It is the single surfaces that can be recorded and have to be recorded entirely. They are the primary element for the reconstruction of the topography of the site through time.

Stratigraphic recording ends in the creation of a stratigraphic sequence based on the spatial relationship of the units of stratification. The stratigraphic sequence provides the ability to visualise time. It is the important diagrammatic representation of time and provides a relative calendar, the testing pattern for any further analysis.

The Stratigraphic Record and GIS

The most important concept to come out of the fore said is the need for single surface planning (Harris 2001). The outstanding value of a GIS is its ability to reproduce the complete record of a stratigraphic surface as well as any related descriptive information. The GIS functionality provides the ability to visualize surfaces as contour plots or triangulated irregular networks. It makes it easy to combine the boundary polygon of surfaces or deposits as once exposed to air with rectified digital images. The finds can be mapped as registered within the volumes defined by top and bottom surfaces of the corresponding single deposits, classified by stratigraphic position or material aspects.

The way GIS functions, it permits the dynamical mapping of single surfaces or the creation of composite maps (phase or period maps, sections at any position etc.), based on the recorded data. The decisions on how to compose the necessary maps is derived from the analysis of the stratigraphic sequence. The secondary data dealing with the various aspects (location, material, date etc.) of the finds uncovered is stored in the spatial database of the GIS. There, it can be combined with the graphical visualisations, analysed and counterchecked.

The Recording Process

At stratigraphic excavations it is of major interest, to check the recording during the process of excavating. Therefore, we developed techniques of recording, where boundary polygons and topography of surfaces, finds and the photographic record of each unit of stratification are recorded digitally and immediately interfaced to the GIS. The resulting GIS-based system consists of a total station, a digital camera and a PC/Laptop with ArcView GIS (Version 3.2) and Monobild (or a similar software for rectification). For the creation of the stratigraphic sequence we use ArchEd http://www.ads.tuwien.ac.at/arched.

The standardised process of recording is divided in several subsequent steps, that are repeated for any excavated deposit:

- Defining the limits of the top surface of the uppermost deposit
- Unique numbering of the top surface and the deposit
- Photographic record
- Measuring ground control points
- Measuring boundary polygon and topography of the top surface
- Excavating the deposit, taking samples, measuring find and sample coordinates, describing deposit, describing finds
- Unique numbering of the bottom surface
- Photographic record of the bottom surface
- Measuring ground control points
- Measuring the topography of the bottom surface

Figure 1 Graphic raw data of stratification unit s059 in Schwarzenbach after import.

The process of recording of interfaces is as follows:

- Unique numbering of the interface
- Photographic record of the bottom surface
- Measuring ground control points
- Measuring the topography of the bottom surface
- Describing the interface

The stratigraphic relations of the top surface of the deposit the deposit and its bottom surface are trivial. The stratigraphic relations of any top surface, bottom surface or interface have to be observed and recorded. Find and sample records, descriptions and stratigraphic relations are recorded on preprinted sheets or directly in a database on site referenced by the unique number of the unit of stratification.

INTERFACING TO THE GIS

To be able to use the digital photographic record in the GIS, it has to be rectified and georeferenced. We had been running extensive tests to find a fast and easy way to rectify the digital image within an accuracy that would not exceed +/- 5 cm. During these tests, we realized, that to stay within this range of accuracy, a combination of vertical photographs and a projective transformation of the image would be sufficient in most cases. The use of a projective transformation assumes, that the topography of the surface is relatively flat. The plane of the surface is defined by at least 4 ground control points, which must be well distributed around the surface to be recorded. They are measured in 2D space and define the parameters of the transformation.

This simple method of rectification could yield reasonably good results as long as the surface has minor differences in height or is evenly sloping up (Fig.1). The issue gets more problematic, when the topography of the unit of stratification is rough, having large height differences on several locations. In this case an accuracy within +/- 5 cm cannot be guaranteed. To solve this problem, more control points have to be added defining smaller plane units within the surface, which then are rectified individually. The transformed parts are mosaicked together again in GIS. Another possibility would be to use software capable of differential rectification using a DTM, as we did in our tests. So far, the procedure of rectification is done using the software "Monobild" (http://www.geosi.at/produkte/monobild/monobild.htm). It is an easy to use software designed for our needs.

The spatial dimensions of the single surfaces can be recorded in its entirety in 3D using a total station. To allow a continuous recording of the excavation, the total station is set up during the entire duration of the excavation. All measurements are stored on the module using the internal data format of the total station. In front of each new feature class a codeblock is stored containing the number of the unit of stratification, a code, and a possible comment. The code defines the type of feature and is used for the handling of the following coordinates.

To be able to automatically import and analyse the spatial data in ArcView GIS, the extension module ArcDig was programmed using Avenue, the object oriented programming language of ArcView 3.2. The import routine reads the native ASCII data file format of the geodetic instrument and converts the data into the native shape file format of ArcView. The measurements of the boundary polygon are stored as a 2D polygon shape and as a 3D polyline shape. The 3D mass points of the surface and optional break lines are imported as points respectively lines. Finds and samples are imported as points, where the numbers of the finds are automatically added to the database connected with the shapefiles. Ground control points are written to ASCII text files - one file per unit of stratification. The format is fitted to the needs of the import function of the rectification software.

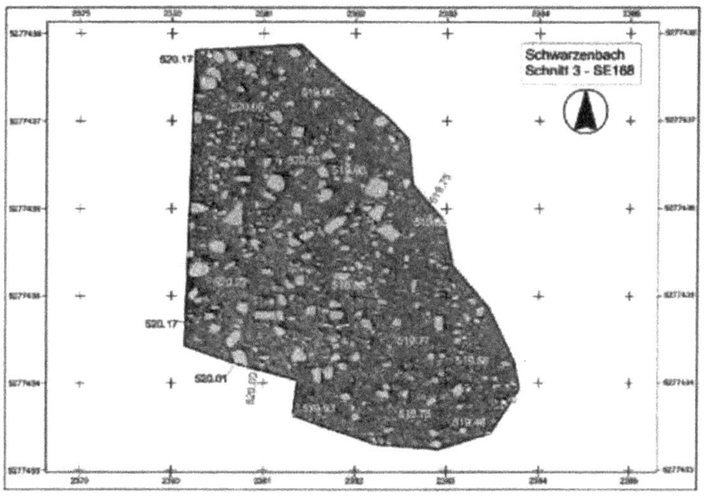

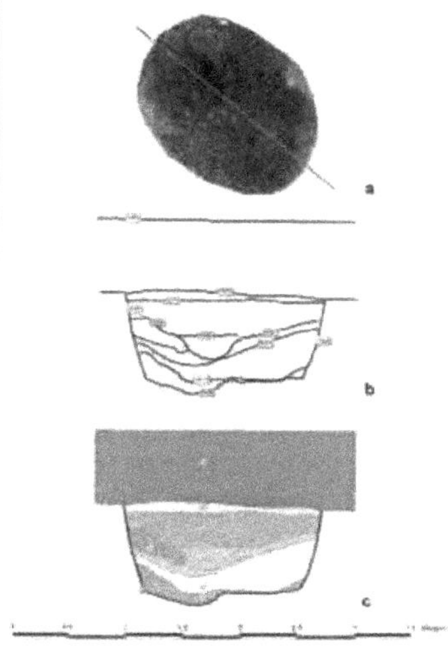

Figure 2 (left) Stratification Unit 168 of Schwarzenbach.
On top of the clipped, rectified image, details were drawn using ArcView GIS

Figure 3 (right) Procedure of creating a cross-section. (a) Defining the line of the cross-section. (b) Calculation of intersection lines with each underlying surface. (c) Cross-section after editing

Points, break lines and 3D boundary are then automatically used to create a digital elevation model of the surfaces (Fig.2). During this process of surface creation, contour lines are automatically interpolated from the triangulated elevation models of each unit. The interval can be chosen interactively.

POSTPROCESSING AND FIRST ANALYSIS

After the importation of the entire graphic record, it is available in the GIS and can be checked for errors. The descriptive information stored in databases can now be linked to the graphic record.

The rectified images are added. Experience showed, that in many cases, the outline of a deposit can be better recognized in the digitally enhanced image than on site. Therefore, each boundary polygon is displayed on the rectified image, chekked and - if necessary - corrected. Details like skeletons, layers of stones, ceramics, pavements etc. can be drawn directly on screen using the rectified images, while the next deposit can already be excavated. During this process, only those parts of the image, which lie within the control points can be drawn with the desired accuracy. The accuracy decreases largely outside of the area enclosed by the control points. To prevent unintentional drawing in the outside areas, the images can be clipped to the boundary polygon of the corresponding stratification unit (Fig.3).

For the production of arbitrary sections out of the three dimensional spatial record we programmed another script, which enables the user to interactively calculate sections at any part of the excavated area. To do so, a line has to be drawn, which is then intersected with the triangulated irregular network of each underlying surface. The resulting intersection lines are then compiled into a new drawing representing the section. The advantage is clearly to be seen: sections can be reconstructed automatically at any line of interest (Fig.4).

Future Aspects of Digital Recording

Terrestrial 3D imaging laser scanning forms a method to acquire a large number of precise data points in 3D space representing the surface of objects under investigation. These scanners are an effective tool for the collection of data to create a digital elevation model of the topography of a site as well as of the surface of a single archaeological deposit. Together with RIEGL LMS GmbH (www.riegl.com) the excavation campaign 2002 at Schwarzenbach could be monitored for two weeks using the RIEGL LMS-Z360 and LMS-Z210 devices. The scanners were used for a high resolution 3D recording of the surface of the open 900 m2 trench, so far done using a total station.

Figure 4: 3D laser scanning for the documentation of single surfaces at the excavation Schwarzenbach. Upper left: Laser scanner in action. Lower left: Intensity image of a RIEGL LMS-Z210 laser scanner showing the excavation. Upper right: comparison between data and TIN using a total station (left) and a laser scanner (right). Lower right: Visualization of the surface

The primary data of a 'scan' is mainly a so-called 'point-cloud' in the sensor's local coordinate system. In general it will be necessary to obtain multiple 'scans' from different positions to overcome the problem of occlusions that will be present in any scan from a single point of view. In order to obtain a quite complete representation of the surface by a point cloud, in general numerous scans have to be merged into a common coordinate system ("registration").

For the registration of the scans 11 reflectors were mounted on fixed positions. Right after the digging out of a single deposit, the surface of the trench or parts of it were scanned, using the highest resolution of the specific instrument. The collected and registered point clouds were resampled on a 5 or 10 cm regular grid and converted into a native ArcView shape-format.

For further data processing we used the function of the GIS ArcView. The single surface of a deposit was clipped from the triangulated topographic scan data using the outbound polygon of the stratification unit, measured with the total station. Rectified digital photographs were mapped on top of the triangulated surface (Fig.6). The 3D laser scan devices showed a high reliability and efficiency for topographic single surface recording in every day archaeological work. The scanner, as used here, could do the same recording job, done so far by two people, in only 20 % of time collecting up to 50 times more data. This would save at a typical 1 month excavation up to 100 man hours. The 3D laser scanner can be seen as a future standard tool for the high resolution 3D recording of single surfaces on a stratigraphic excavation.

References

HARRIS, E.C., 1989. Principles of archaeological stratigraphy. 2nd edition, London.

HARRIS, E.C., 2001. The only way to see. In Harl, O. (ed.), Workshop - Computer and Archaeology (CD), Vienna.

New Achievements on Pottery Reconstruction[1]

Martin Kampel

Robert Sablatnig

Pattern Recognition and Image Processing Group
Institute for Automation, Vienna University of Technology,
Favoritenstr. 9/183/2, A-1040 Vienna, Austria
Fax: +43 (1) 58801 183 92
email: {kampel,sab}@prip.tuwien.ac.at

Abstract

We are developing a documentation system for archaeological fragments based on their profile, which is the cross-section of the fragment in the direction of the rotational axis of symmetry. Hence the position of a fragment (orientation) on a vessel is important. To achieve the profile, a 3D representation of the object is necessary. The main technical goal of our project is to perform an automated classification and reconstruction of archaeological fragments by using the profile section of the oriented object and additional attributes (type of clay, dimensions, type of vessel and the site) associated with the fragment. The final aim is to provide a tool that helps archaeologists in their formation of their archive of records. This paper gives an overview about new achievements on our potter reconstruction system.

Introduction

Archaeology is at a point where it can benefit greatly from the application of computer vision methods, and in turn provides a large number of new, challenging and interesting conceptual problems and data for computer vision (Martinez 2001:9-15). In particular, a major obstacle to the wider use of 3D object reconstruction and modelling is the extent of manual intervention needed. Such interventions are currently extensive and exist throughout every phase of a 3D reconstruction project: collection of images, image management, establishment of sensor position and image orientation, extracting the geometric detail describing an object, and merging geometric, texture and semantic data. Improvements in rangefinder technology, together with algorithms for combining and processing 3D data allow us to accurately digitize the shape and surface characteristics of physical objects.

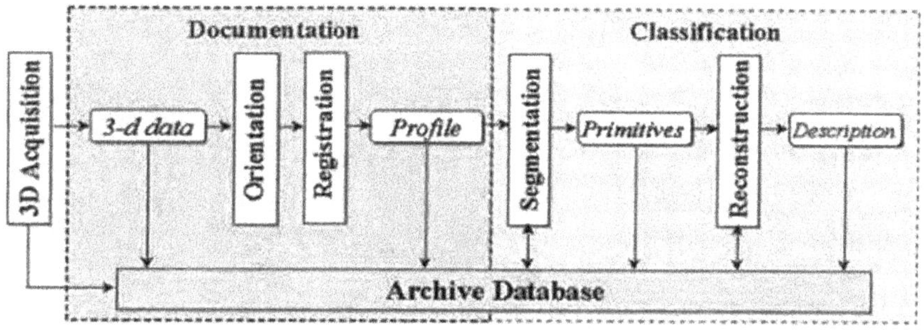

Figure 1 Automated Archivation Process

The range- and pictorial information of a pottery fragment recorded by the acquisition system serves as the basis for the further classification and reconstruction process. The profile of a sherd has to be determined by orientation. The term orientation describes the exact positioning of the fragment on the original vessel with the help of the axis of rotation. To automate this process, the profile has to be determined in the same way as in the manual documentation. The generated profile is used to perform the reconstruction and retrieval of fragments of the same type. The reconstruction procedure works if the size of the fragment covers a large part of the original vessel in the vertical direction. The profile is rotated by the original axis of rotation, thus measurements like volume can be estimated. Figure 1 shows the automated archivation process schematically, giving an overview of the intermediate steps.

Furthermore, the process of documenting a fragment is improved, by the automation of the procedures of measuring, drawing, and description. With the help of 3D data, the profile (a cross section in the three dimensional model) of the fragment is constructed. The frontal view is represented with the help of the pictorial information of the surface of the fragment and the surface model. This representation can be used for publication or for retrieval from the database, put on the Internet by other users. This will enable the publication of both the profile of the sherd and a virtual reconstruction of the whole vessel.

Data Acquisition

The ShapeCam Technology developed by Eyetronics[2] consists of a Sony TVR-900E digital camera and a Leica slide projector. Figure 2 illustrates the ShapeCam: a digital camera and a specially designed flash device are mounted on a lightweight frame. The flash device projects a predefined grid or pattern onto an object or a scene, which is viewed by a camera from a (slightly) different point of view. The camera also grabs the texture information, which can be mapped on the resulting 3D geometry of the object.

The ShapeCam technology is a commercially available technique, which is a based on the structured light approach (Klette et al. 1996). It allows the generation of 3D models based on the use of a single image taken by an ordinary camera. As this system is a handheld device, the shapes can be recorded in situ. Within the 3D MURALE (Cosmas et al. 2001) project we carried out on site tests to capture 3D pots-

Figure 2 Eyetronic's ShapeCam

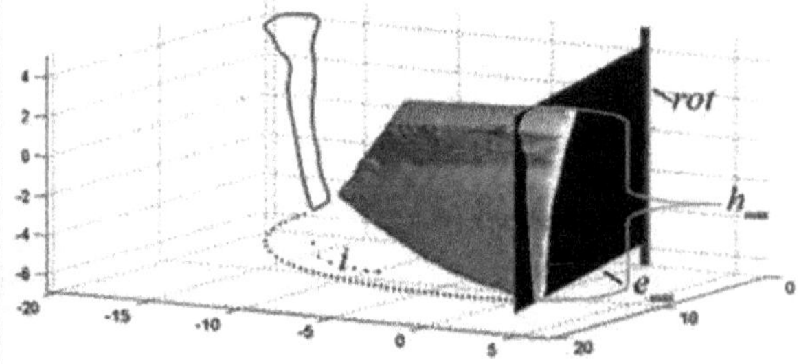

Figure 3 Orientated sherd, rotational axis, intersecting

herds and other finds from the excavation site in Sagalassos (Cosmas et al. 2001). The ShapeCam hardware has been adapted to facilitate such work.

Data Processing

Archaeological pottery is assumed to be rotationally symmetric since it was made on a rotation plate. With respect to this property the axis of rotation is calculated using a Hough inspired method (Sablatnig and Kampel 2002:90-103).

To perform the registration of the two surfaces of one fragment, we use a-priori information about fragments belonging to a complete vessel: both surfaces have the same axis of rotation since they belong to the same object. Consequently we register the two views by calculating the axis of rotation and by bringing the resulting axes into alignment. A detailed description of the registration algorithm can be found in (Sablatnig and Kampel 2002:90-103). The registration of front- and back-view together with the axis of rotation provide the profile used to reconstruct the vessel.

Figure 3 shows the 3D-model of a sherd and its rotational axis rot as a vertical line along the z-axis. The black plane represents the intersecting plane emax at the maximum height hmax of the sherd. The longest profile line is the longest line along the surface of the sherd parallel to the rotational axis rot. The radius r is the estimated mean radius of the profile line. The extracted profile line is shown in the xz-plane. Our algorithm for the estimation of the longest profile line is fully described in (Kampel and Sablatnig 2003).

Results

The resulting 3D reconstruction of fragments depends on the correct orientation of the profile section. The evaluation of the 3D representation is rather complicated since verifaction is not available due to the fact that there is no 3rd dimension in archaeological archive drawings and the object is no longer complete. The description of shape is subjective and is not standardized, by archaeologists.

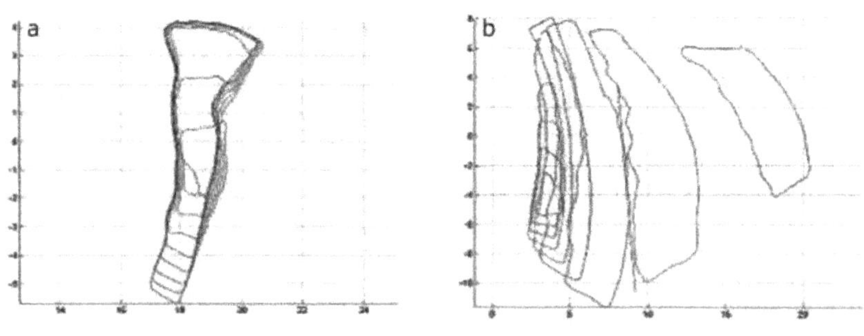

Figure 4 Multiple profile lines: (a) using a correct estimated rotational axis (b) using a incorrect rotational axis

Experiments were done on all fragments of our pottery database. The success rate for correct extraction of the profile line and consequently the percentage of sherds used for further reconstruction is around 50 % of the data selected at the excavation site. This should be compared to manual archivation done by archaeologists (Orton et al. 1993): for coarse ware around 35 % (Degeest 2000) and for fine ware around 50 % (Poblome 1999) of the findings are used for further classification. It heavily depends on the shape of the fragment (e.g. handle, flat fragments like bottom pieces, small size, etc.). 18 fragments have been excluded from reconstruction due to incorrect estimation of the axis of rotation.

Experiments with synthetic data have shown that the correctness of the reconstruction depends on the correct estimation of the axis of rotation and on the resolution of the 3D scanner used. The number of vertices in the data used ranges between 4,000 and 15,000, leading to a profile line with 200 to 300 points. The execution time using a prototype written in MATLAB running on a Pentium III, 1 GHz is less than a minute per sherd. It depends heavily on the computation of the axis of rotation (70 % to 80 % of the execution time).

In order to demonstrate the correctness of the computed profile lines, Figure 5 shows a recorded sherd (dark object) and its computed profile section (vertical line). The computation of the virtual fragment (grey object) is based on the profile section. One can see that the recorded fragment fits into the virtual fragment, which indicates that the computation is correct. Following multiple cross-sections along the perimeter of the virtual fragment - Figure 4a - one can see hardly any deviation from the original fragment. Some deviations are caused by the bumpiness of the surface, as the surface is not exactly rotationally symmetric since it is hand-made pottery. If the fragment was orientated incorrectly, the recorded fragment does not fit into the virtual object and multiple cross-sections along the perimeter of the virtual fragment show large deviations from the original object (see Fig.4b).

Figure 6 displays a reconstructed pot (grey object) out of a rim- fragment (dark object) based on the profile line (light line) and its axis of rotation (dashed line).

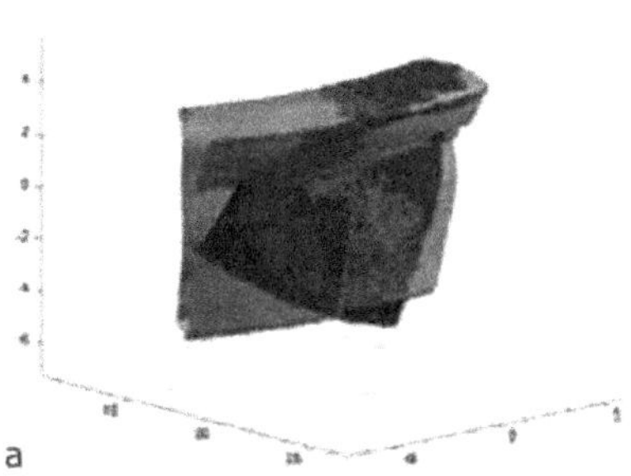

a

Figure 5 Reconstructed fragment, profile section and recorded fragment

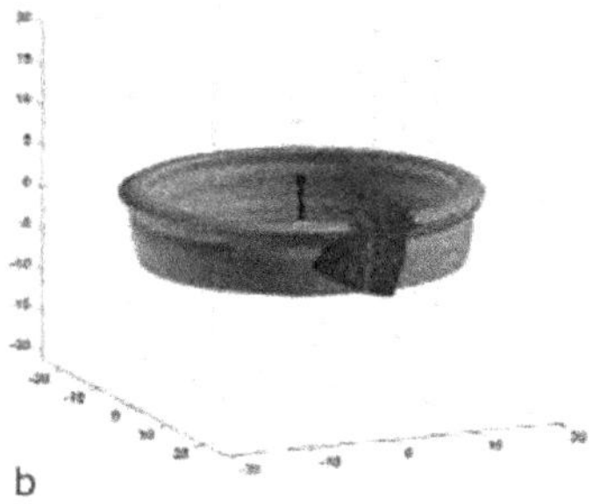

b

Figure 6 Reconstructed pot

The ceramic documentation and reconstruction system described is currently under further development to be integrated into the virtual excavation reconstruction project 3D MURALE. Future work will be directed towards setting up a pottery database with more then 1,000 fragments. Currently we are working on the classification system based on the profile in order to classify all profiles and to find matching fragments from similar vessels and eventually from singular vessels.

Conclusion and Outlook

We have proposed a reconstruction-method for pottery based on its profile section. The profile, which is the cross-section of the fragment in the direction of the rotational axis of symmetry, represented by a closed curve in 2D, is computed by finding the correct orientation of the fragment. Comparing the performance of our automated technique to the manual, archaeological approach, the archival process of pottery is accelerated and the quality is improved.

Acknowledgments

The authors would like to thank Prof. Marc Waelkens and Roland Degeest from the Katholieke Universiteit Leuven, Eastern Mediterranean Archaeology, and Kristina Adler and Martin Penz from Vienna University, Institute of Classical Archaeology for their archaeological support and contributions in evaluating the profile representation approach and for helpful and inspiring discussions.

[1]This work was partly supported by the Austrian Science Foundation (FWF) under grant P13385-INF, the European Union under grant IST-1999-20273, the Austrian Federal Ministry of Education, Science and Culture and by the innovative project '3D technology' of the Vienna University of Technology.

[2]Eyetronics HQ, Kapeldreef 60, 3001 Heverlee, Belgium, www.eytronics.com.

REFERENCES

COSMAS, J. et al., 2001. 3d Murale: A multimedia system for archaeology. In Proceedings of Intl. EuroConference on Virtual Reality, Archaeology and Cultural Heritage, Athens, Greece.

DEGEEST, R., 2000. The Common Wares of Sagalassos. In Walkens, M. (ed.), Studies in Eastern Mediterranean Archaeology III.

KLETTE, R., KOSCHAN, A. and SCHLÜNS, K., 1996. Computer Vision. Räumliche Information aus digitalen Bildern. Vieweg Verlag, Braunschweig/Wiesbaden.

KAMPEL, M. and SABLATNIG, R., 2003. Profile based Pottery Reconstruction. In Proceedings of IEEE/CVPR Workshop on Applications of Computer Vision in Archaeology, Madison, Wisconsin, USA, 17 June 2003.

ORTON, C., TYERS, P. and VINCE, A., 1993. Pottery in Archaeology.

MARTINEZ, P., 2001. Digital Realities and Archaeology: A difficult relationship or a fruitful marriage. In Arnold, D. (ed.), Proceedings of the International Symposium on Virtual Reality, Archaeology and Cultural Heritage, Athens:9-15.

POBLOME, J., 1999. Sagalassos Red Slip Ware. In Walkens, M. (ed.), Studies in Eastern Mediterranean Archaeology II, Brepols.

SABLATNIG, R. and KAMPEL, M., 2002. Model-based registration of front- and backviews. Computer Vision and Image Understanding 87(1):90-103.

Information System and Computerised Methods for Rescue Archaeology

François Djindjian
University of Paris
1-Pantheon-Sorbonne & CNRS UMR 7041

Abstract

In Rescue Archaeology, where the control of both efficient diagnostics, scientific control of operation, budget and planning are major constraints, computerised methods are the key of improvement of quality and productivity of the projects. Various computerised methods are described here and integrated into a global methodological framework for Rescue Archaeology. Four phases in Rescue Archaeology are described : Diagnostic, Evaluation, Operation and Publishing. For the operation phase, nine steps have been defined and detailed : specifications of the operation, complete documentation concerning the area under survey, logistic support, G.I.S. implementation, zoning, intrazone field survey by sampling, kriging & predictive modelling, intrasite operations by surface survey, geophysical survey, core surveys & excavations, sampling artefact collections, post-excavation studies and publishing preparation.

Introduction

The historical links between Rescue Archaeology and Sampling

If the concept of archaeological sample was recognised since the beginning of XX° century by many archaeologists, it is certainly Vescelius (1960) who the first recognizes the link between an archaeological sample and a statistical sample. But the success of the concept of statistical sampling in the context of the North-American Archaeology is certainly due to both G.L. Cowgill (1964) and L. Binford (1964), the first proposing to sample in large shard collections, the second using a probabilistic sampling to optimise surveys of archaeological sites. The development of the concept of sampling were then strongly associated with the first rescue archaeological projects, both in North-America between 1965 and 1975, remembered by the book "Sampling in Archaeology" (Mueller 1975) and in UK during the seventies : "The role of sampling in contemporary British Archaeology" (Cherry et al. 1978). The difficulties met by the application of probabilistic sampling, needing high rate of sampling for delivering results with low reliability and bad precision, obliged to test new other techniques, supposed to be more efficient, like Kriging (Zubrow and Harbaugh 1978) and Predictive Modelling (Scholtz-Parker 1982), but without the expected success (Djindjian 1991, 2001). The use of sophisticated methods for archaeological surveys of large rescue archaeological projects restarts during the nineties with the success of the Geographical Information System (G.I.S). The data management and cartography functions of G.I.S. packages offer then both the framework and the environmental information needed to implement easily simultaneously various techniques like probabilistic sampling, kriging and predictive modelling, alimented with the large cartographic data provided by the G.I.S. implementation (Wescost and Brandon 2000).

Principles in Rescue Archaeology

The archaeological workflow of Rescue Archaeology is very characteristic of a project-oriented approach. Generally, the four following phases are considered :

Phase 1 : Diagnosis

The diagnosis phase, made by the contracting authority, must estimate, with the help of experts, the archaeological impact of projected works upon the concerned area.

Phase 2 : Evaluation

The evaluation phase, which is a further step of diagnosis for large operations, made by the contracting authority or subcontracting consulting specialists, allows obtaining the information needed to make the specifications of the operation, a budget and an agenda.

The evaluation phase is including in-depth diagnosis, audit by experts, preliminary documentation by use of various retrieval information systems, global surveys including aerial photography and satellite image, preliminary zoning, test surveys by random sampling, global specifications, use of configuration tools (templates using time and expenses ratios by categories of sites) for budget estimation, budget negotiations with third parties, etc.

Phase 3 : Operations

The operation phase realizes a rescue archaeological project. The operation is directed by the contractor, after a negotiation between the contracting authority (a public authority in Culture Resource management), the third party (generally a civil work company) and the contractor (a public or private company specialised in rescue archaeological projects).

Phase 4 : Publishing

The publishing phase is based on the definition of a standard framework for editing archaeological publishing, multimedia databases, CD-ROM and even Virtual reality.

The 9 Steps of a Rescue Archaeological Project

A Rescue Archaeological Project may be split in several steps, which will be detailed below:

1. Detailed Specifications of the project,
2. Complete documentation concerning the area under survey,
3. Logistic support to install,

4. G.I.S. implementation,
5. Zoning the area,
6. Field survey : finding archaeological sites by sampling, kriging & predictive modelling,
7. Intrasite operations: surface survey, geophysical survey, core surveys & excavations,
8. Sampling artefact collections,
9. Post-excavation studies and publishing preparation.

STEP 1 : DETAILED SPECIFICATIONS

Starting from the specifications made during the evaluation step by the contracting authority, the contractor must control and detail the received specifications, and before starting any operation, has to obtain a validation of the new specifications he has produced by the contracting authority :

1. To define precisely the area of the rescue project,
2. To obtain digitalized various maps of the area,
3. To get a general view of the area, by various means,
4. To obtain information about the archaeological sites known in the area (archaeological map),
5. To define the various constraints about the rescue project area,
6. To design an initial zoning of the area,
7. To validate the budget,
8. To elaborate a detailed agenda of the project,
9. To list the resources and skills needed.

STEP 2 : COMPLETE DOCUMENTATION

The complete documentation is the existing documentation concerning archaeological information of the discovered artefacts or archaeological sites and the archaeologist activities in all the places where the information may be stored.

- Museum documentation,
- Archives,
- Toponymy,
- Various maps, including altitude (DEM), hydrography (at a level n), vegetation, soil, modern infrastructure, etc.
- Aerial photography and satellite images,
- Geomorphologic studies,
- Archaeological map,
- etc.

STEP 3: LOGISTICS

Logistic support has a major role to improve the productivity of field works. Such a logistic support is listed below, as an example to emphasise the importance of an in-house prepared operation, before any field survey :

manpower (human resources), tool kits, machines, facilities (water, electricity, telecommunications, mobile home, etc.), food supply, storage, laboratory services, transport services, computers (hardware, software and customisation), recording procedures, methods (manual), security rules, training, insurances, code of practice, organisation & tasks of the project team, etc.

STEP 4: G.I.S. IMPLEMENTATION

Computer services need a complete organisation including resources, hardware, software and customisation, prepared before the rescue operation in a dedicated computer department, offering services for all operations.

- Dedicated G.I.S. resource,
- Computer dedicated for project : installation & customisation,
- Software products for the project management
    - Project Management (Pert or Gant diagram) : MS-project,
    - Budget control management : Excel or dedicated package,
    - Time and expenses management : Excel or dedicated package,
    - logistic support management : Access or any other DBMS,
    - G.I.S.,
    - AutoCAD,
    - Statistical package,
    - Dedicated sampling algorithms
- Import existing digitalized maps,
- Digitalization of needed published maps,
- Design of the data model,
- Import existing archaeological data,
- DEM model,
- Zoning at different scales,
- Sampling procedures,
- Recording survey data,
- Special software for sampling techniques,
- etc.

STEP 5: LOCATION & ZONING

The step of location and zoning is one of the most critical step of the operation.

Location may be realised by two techniques :

- Field localisation by GPS,
- Field marker.

Zoning is realised at different scales, first from the landscape map, and then at a macro and micro scales where intrazone sampling surveys are applied.

- Landscape Zoning (more than 1 km2)
- Macro-Survey Zoning (from 0.01 to 1 km2)
- Micro-Survey Zoning (from 100 to 10,000 m2)

STEP 6 : INTRAZONE SAMPLING SURVEY

Intrazone sampling survey are realised with three main statistical techniques used with an increasing efficiency : probabilistic sampling, kriging and predictive modelling.

The main characteristics and reliability of these different techniques are listed above :

Probabilistic Sampling
- Method
    - non straight random sampling,
    - clustered sampling,
    - Shape & size of the sampling unit : transect (l = 20 m, l = 1 m),
    - Sampling rate : 5 to 30 %
- Performances of surveys and excavations:
    - high rate, bad accuracy,
    - rare structures not discovered,
    - limited to initial sampling.

Kriging (Spatial Interpolation)
- Method
    - Initial Random Sampling,
    - Computation by Spatial Interpolation of a three--dimensional Graph estimating the probability of density of archaeological sites,
    - Priority of Survey of Higher Density Units
- Performances
    - Random Sampling Survey up to 25 % rate ?
    - 50 % of Sites discovered by Surveying 50 % of the Zone,
    - 100 % of Sites discovered by Surveying 75 % of the Zone
    - Potential of Gain 25 %

Predictive Modelling
- Method
    - Initial random sampling & kriging
    - Site & non site differentiation,
    - G.I.S. recording of topographical, ecological and
    - settlement variables,
    - Discriminant analysis between sites & non sites (logistic regression, correspondence analysis),
    - Establish a list of the most explaining variables for the location of sites,
    - Produce the highest probability zones for location of sites, for surveys,
- Performances
    - predictability 75 %,
    - gain 50 %.

Step 7: Intrasite Survey

Intrasite surveys are methods oriented to find and locate the settlements.

- Surface collection techniques : structure or artefact location, walking procedure, collecting procedure, artefact typology, cartography, multidimensional spatial analysis,
- Sampling strategy : unit size, depth, sampling procedure : sampling & kriging, accuracy, simulations,
- Excavation strategy : adapted methodology to various fields, surface exhaustivity principle, recording exhaustivity principle

Step 8: Sampling Artefact Collections

When numerous artefacts have been collected and stored, it is often useful to design a sampling procedure to avoid to study all the artefacts when a representative sample is only needed.

- Cluster sampling	unit/artefact
- limited to artefacts more than	1,000 units
- rate	down to 5/10 %
- good accuracy	depends on the sampling procedure
- gain	up to 50%

Conclusions

Rescue archaeology is an engineering application of archaeological science in a special context where diagnostics, scientific control of operation, budget and planning are major constraints. In such a case, a rigorous methodology helped by computerised tools is the only way to obtain the level of efficiency and of reliability needed by the project. We have tried to propose here a methodological framework to improve it.

References

BINFORD, L.R., 1964. A consideration of archaeological research design. American Antiquity, no. 29:425-441.

CHERRY, F., GAMBLE, C. and SHENNAN, S. (eds.), 1978. Sampling in contemporary British Archaeology. Oxford (BAR Int. Ser. 50).

COWGILL, G., 1964: The selection of samples from large shert collections. American Antiquity, no. 29:467-473.

DJINDJIAN, F., 1991: Méthodes pour l'Archéologie. Armand Colin, Paris.

DJINDJIAN, F., 2001: Prospection archéologique au sol : de la région au site. In Miskovski, J.P. (ed.), Géologie de la Préhistoire, Nouvelle Edition, Geopre, Paris.

MUELLER, J.W. (ed.), 1975. Sampling in Archaeology. The University of Arizona Press, Tucson.

SCHOLTZ-PARKER, S. and LIMP, W.F., 1981. The application of multivariate logistic model to the prediction of site locations. In X° UISPP Congress, Conference n°5, Mexico 1981, Preprints:236-255.

VESCELIUS, G.S., 1960. Archaeological sampling : a problem of statistical inference. In Essays in the science of culture in honour of L.A. White, G.A. Dole & R.L. Carneiro (eds.), T.Y. Crowell co., New York: 457-470.

WESCOST, K.L. and BRANDON, R.J. (eds.), 2000: Practical applications of G.I.S. for Archaeologists: A predictive modelling kit. Taylor & Francis, London.

ZUBROW, E.B.W. and HARBAUGH, J.W., 1978. Archaeological prospecting: kriging and simulation. In Hodder, I. (ed.), Simulation studies in archaeology, Cambridge University Press, Cambridge:109-122.

ArcheoPackPro!
A Software System for Digital Data Management for Archaeologists

Nenad N. Tasic[1]

Department for Archaeology, Univ. of Belgrade

Vitomir Jevremovic[2]

Abstract

The authors are presenting the software package that will unify all the elements of computer usage during archaeological campaigns and provide an interactive research tool for data analyses after campaign ends. It uses a unique graphics user interface in 3D environment, allowing simultaneous and interactive work with the 3D terrain model, Databases, Statistical and mathematical analysis, 3D models of archaeological finds, etc. It will open a new chapter in archaeological documentation and interactive fieldwork, offering an ultimate context preservation tool for versatile field documentation, its safekeeping, flexible presentation and publication possibilities for a new age archaeology. 3D Scanning technologies will also be introduced in function of better site preservation techniques and material 3D scanning options.

Until recently there was a big gap between the potential offered by computer technologies and the traditional methods of archaeological field work, recording and data processing. Data collected and recorded in archaeological excavations are a combination of images, videos, drawings, and various kinds of numerical data and textual descriptions. In the pre-computer era, browsing through field documentation was a complicated and time-consuming task.

When computers appeared in archaeology during the late 70s and 80s it seemed we had finally found the way to manage the data. However, they were mostly used for statistical analysis of finds and other numerical operations. While databases flourished it was almost impossible to link them to the archaeological context and make them interactive. GIS systems offered a significant breakthrough in this field and managed to integrate maps and plans with databases. Using computers in the "traditional" manner as text, image and data base processors has revealed few serious limitations. However, there has been no real interactivity between data scattered in different archives. Digital data from sites were rarely linked to the plan or drawings of the sites. Images, videos and drawings of finds were accessed through different file directories and not directly linked to the data bases.

Computer experts and archaeologists working together at the Vinca site in over the last 6 years have produced a software system able to equip archaeological teams with a complete data management system for archaeological fieldwork. Our primary aim was to establish an interactive system allowing researchers to move easily from one database to another focusing on the data rather than the underlying system. ArchaeoPackPro! has been successfully tested in both a grid and unit system at Vinca, a prehistoric, Late Neolithic site with 10 meters of cultural deposit. The system has proved efficient, easy to use and reliable in data input and management and for work with complex and demanding methodology at this stratified site.

The System

ArchaeoPackPro! is designed as a single software package that communicates with the SQL database on a server system. The focus of the design was to build a link between different data structures used in archaeological field documentation. The key concern is a successful correlation of scattered archives. The archaeological data (texts, images, videos, databases) are ideal for this purpose since all the data in archaeological documentation systems represent logically defined and organized structures. Archaeological material and objects which are excavated and analyzed have their physical coordinates. Textual and numerical descriptions of the material and objects are stored in different, interrelated structures. Excavated material is spatially static so 3D coordinates can be taken as a basic organizational element. These 3D coordinates (or EDM points) are the starting point for linking databases of specific archaeological material such as pottery, polished stone, flint, etc; they contain both 3D coordinates from the field and the exact time and description that can be used to link them to other data structures. EDM points are the basic structural elements of the system and everything is

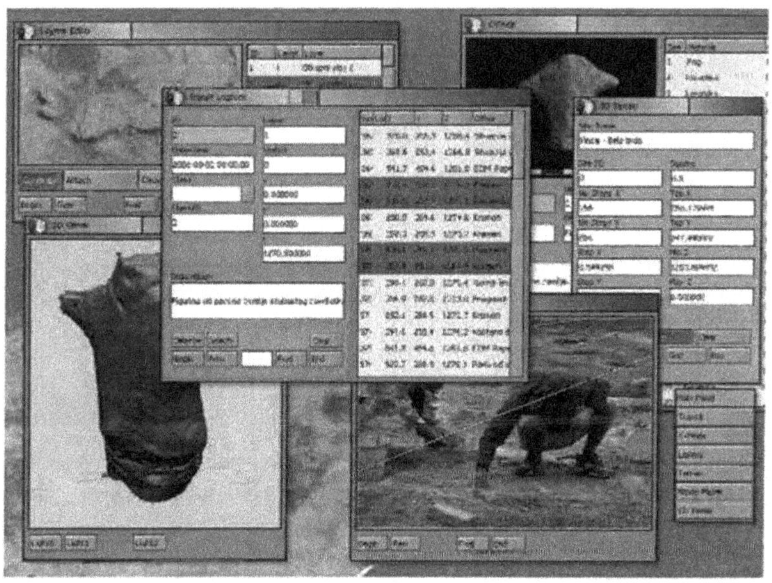

Figure 1

appended to them. Each database involved in archaeological field work is linked to a specific EDM point.

THE PROCEDURE

The basic procedures, which are fundamental to every methodology of archaeological excavation, run as follows:

3D coordinates input is necessary for the 3D configuration of the terrain, small finds, architecture, photo grid and other classes of coordinates. Careful thought is required to determine the amount of points for topography. These will later be intersected with the texture or orthogonal photograph. The larger the number of points, the better and more accurate the 3D mapping of the terrain.[3] The data can either be downloaded directly from the EDM in the form of a table or entered data manually. The latter is suitable for digitalization of archive field documentation or when working with "traditional" levelling equipment such as theodolite. In both cases all three values must be entered. If an artefact or other specimen has been discovered out of its primary context, i.e. in a flotation process or in the sieved earth, it will have a "virtual EDM". In this way all finds from an excavation can be displayed on a 3D terrain with a clear mark which indicate which of them have a "virtual EDM".
Attributing the class of EDM entry allows a researcher to map and distribute each class of artefact and other finds. The operator - usually the same person that operates the EDM - selects a point and attributes it to one of the classes. Classes contain basic categories of archaeological material and objects and can be easily upgraded to match the exact needs of the excavator.

Unit managing. The unit system is a method of excavation which is directly reflected in the structuring of EDM points.[4] Units correspond to specific features of the excavated surface and can be represented by shapes and/or 3D objects. They can be inter-connected to form different horizontal or vertical relations (stratigraphy, parent-child relationship or horizon). Organized into a full graphical and logical representation of an excavated site, they are a tool for organizing EDM points, materials, artefacts, excavated objects and features.

Determining horizontal and vertical relationship of units. One implementation option of the unit system would be a Harris matrix which is fully supported and additionally expanded. By offering many different types of relations between unit elements the dendrogram becomes more elaborate, offering new options for unit management and organization. Structures, objects, features and units can be organized in a dendrogram based not only on stratigraphy but also on parent-child hierarchy (e.g. house - trench - post) or horizons cross-linked with parent-child hierarchy. If possible, this procedure should be performed during the day in the field or at the end of each day in order to obtain updated and accurate relations between units.

Images and video input. The system contains the database for images and videos from a campaign. The process of input is simple - downloading the photographs or video material to a specific folder. The next step is classification and attributing the image to the unit/square or other EDM entry such as specific find, sample for analysis etc. The database for images contains classes related to the type of image (field, orthogonal, journal entry, pottery, flint, studio photograph etc). Once images are downloaded, attributed and cross-linked with EDM points and the unit system a search can then be performed according to the classes, description in the photo diary or the position on a 3D terrain. This makes search and browsing the photo diary fast and seamless. The same applies to video material. The description for the image is standard and includes reference to the unit/square, find, orientation etc.

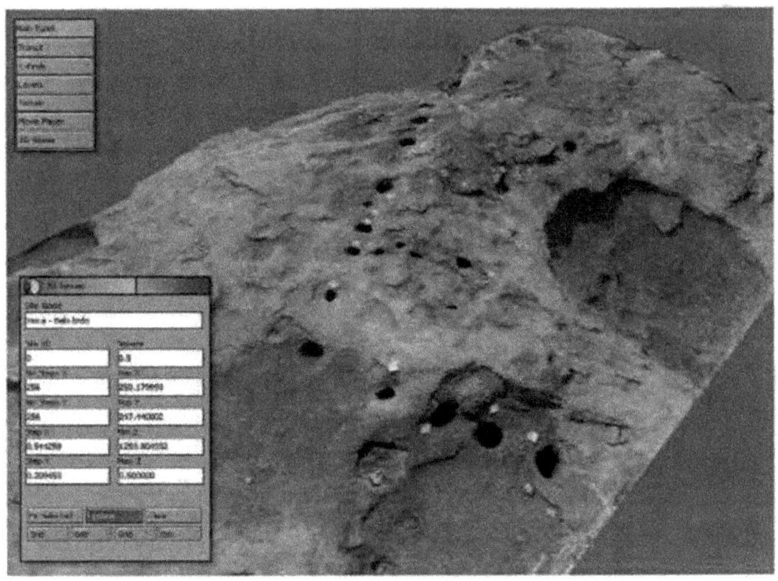

Figure 2

When these basic procedures have been completed it is possible to browse through the field documentation and carry out different search actions including mapping, statistics, unit managing, drawing and printing. All the elements of field documentation can be displayed on the monitor and analysis becomes just a click away. Since the 3D plan of the site and every database and log book is instantly available for the researcher, spatial analysis of finds becomes routine. For example, if one wishes to map flint discovered during the excavations it is enough to select entries with this description (or a code) in the EDM log book and the next click will show its exact position on the 3D terrain model. These points obtained can be highlighted or hidden. It is also possible to analyse more than one category: e.g. the relation of flint to a house floor, an oven or the site. One can then overlap graphical displays of different analysis performed and compare or print them. The diversity of such spatial analysis is, of course, related to the amount of information recorded and EDM points taken during the excavation.

The module "research organizer" that will save the process of analysis and prepare elements for a report is in preparation and will be available soon.

The System

ArchaeoPackPro has been developed in C++ with usage of OpenGL. OpenGL is a standard graphics library for managing and drawing 3D objects using modern graphic card hardware. New generation of graphic cards allow extremely detailed 3D models to be processed in real-time, thus allowing our system to work simultaneously with different 3D objects such as terrain model, 3D scanned artefacts, User Interface elements, etc.
As both C++ code and OpenGL code were developed to be easily portable we hope that in the near future ArchaeoPackPro will be available for all platforms and operating systems that support 3D hardware. These would include Windows, Macintosh, UNIX, Linux, BeOS or any other system that has hardware support for OpenGL.

One important issue not to be forgotten is that not all institutions and archaeologists have computer systems capable of running ArchaeoPackPro. 3D graphics cards are extremely cheap and affordable now but we are aware that many older computers are still in use. As ArchaeoPackPro! is connected through server and SQL database located on the server, we are developing a special package feature that will allow data to be entered the using older computers - the option of accessing databases through standard Internet web pages. Designed in pure HTML and DHTML, web forms will be an interface with databases used by ArchaeoPackPro. Through local network or internet connection it is possible to access and manage the data even using old 486 or Pentium I computers and Internet Explorer, Netscape Navigator or any other compatible browser.

Introduction of digital technologies in archaeological fieldwork does not only simplify and speed up the process. It also influences the basic methodological techniques and practices. Unforeseeable possibilities of a 3D representation and context preservation of archaeological material and objects resulted in a radical shift in the field documentation. In previous systems the backbone of the system was the field journal with all the details concerning the dynamics of the fieldwork, context of finds and links to the technical documentation and databases. The abundance and diversity of textual input in field journals has been constantly narrowing the search so within this software package the starting point for linking became the EDM log book i.e. the 3D image of the data. The EDM enables archaeologists to acquire both the absolute height for the find or object and the X and Y values instantly. It makes 3D representation possible and extremely suitable for linking information about the exact position of small finds and objects, photo and video documentation, contexts and other information from various databases for every activity undertaken during an archaeological fieldwork. This is probably the most important change in the field documentation and indicates the most likely direction of its development. There are no methodological obstacles in applying this system to any type of excavation or site. It is equally efficient at excavations using either the grid or the unit system and for emphasizing either the vertical or horizontal dimension. The accuracy of the 3D representation depends exclusively on the number of x-y-z points taken during the excavation - which is in the hands of the site director. With the grid system the minimum requirements would be to take the points of each square, position of the finds and the contours of objects and features. With the unit system more EDM points should be taken if greater accuracy and precise position and contours are required. Hence, regardless whether one is digging mounds, graves, ditches, pits, prehistoric subterranean, medieval donjon or any other structure, whether one is doing trenching, gridding, open stripping, Wheeler box-grid or open-area excavation the procedure of recording will be identical: entering the data in the EDM log book, attributing the points and linking with the unit/locus/square log book, field journal, photographs, plans etc. ArcheoPackPro! is designed to replace the old-fashioned procedure of data input with the numerous possibilities offered by modern computer systems. It will speed up fieldwork, improve the quality of the documentation and introduce new methods of data input, processing and analysis.

[1]Author is assistant professor at the Department for Archaeology, Univ. of Belgrade.

[2]Author is a computer programmer, author of numerous computer games and graduate of Archaeology at the Univ. Belgrade. Web address is www.online-archaeology.com.

[3]This, however does not apply when working with a 3D scanner.

[4]It should be noted that units can be also treated as squares for those that apply that methodology of excavations.

Europe's Electronic Inheritance: The ARENA Project and Digital Preservation in European Archaeology

Jonathan Kenny
Archaeology Data Service,
Department of Archaeology, University of York

William G. Kilbride
Archaeology Data Service,
Department of Archaeology, University of York

Abstract

As long ago as 1994 Torsten Madsen warned that as we step 'into the steep road of electronic information, we should worry about one thing only, how do we ensure that what is electronic information today remains a stable and permanent source of information tomorrow as well'. This paper will consider the progress, or otherwise, that we have made in ensuring the stability and permanence of digital archaeological archives. To consider this topic the paper concentrates on the work of the Archaeology Data Service in the United Kingdom and of the ARENA project on a European scale.
The ARENA project is hosting a day workshop on the practical process of digital preservation and considering some of the issues surrounding this work. The author of this paper will be taking part in the ARENA workshop and will then present a 'school report' on the current state of preservation work for digital archaeological archives.

Introduction

The vulnerability of data stored in digital formats has become apparent to an increasing number of archaeologists and archivists over the last ten years (Madsen 1994, Richards, Miller and Wise 1999, Richards 2002). Richards et al. (1999) illustrated awareness of the fragility of archaeological data: Although computers have been widely used in archaeology for the last 15 years or more, as recently as 1999, an extensive survey of professionals in the UK demonstrated that "little attention" was being paid to the long-term preservation of digital data.

The Archaeology Data Service (ADS) was founded in 1996 specifically with the task of providing a digital preservation service for archaeologists in UK universities (Richards 2002). That mission has included a range of research and development activities, as well as guidance and awareness raising activities (Gillings and Wise 1999, Bewley et al. 1999, Richards and Robinson 2000, Schmidt 2002, Eiteljorg et al. 2003, Fernie and Richards 2003). The work of the ADS is intended to support researchers based in the UK, wherever they chose to work. Consequently, extending expertise on preservation to partners outside the UK is important for the ADS's own mission, and it is consistent with the work of heritage agencies the world over. In 2002 an opportunity arose through the European Union through the Culture 2000 programme, to bring together a small network to research and extend expertise in digital preservation. The resulting ARENA (Archaeological Records of Europe: Networked Access) project is a path-finding initiative working on issues of data preservation and access (Kenny, Kilbride and Richards 2003). This paper summarises the work of the ARENA partners to bring forward unexpected themes and issues in digital preservation. The experience of the partners has raised issues that will become key areas of debate in the preservation of digital archives in archaeology. These issues include data quality, migration, data from "live" projects, blurring interface and preservation, possibilities for e-publication, digitisation and indexing, and advocacy. It also raises issues about the boundaries between Europe's diverse heritage agencies.

Value: Why do We Preserve Archives?

The need to create an archive containing the recorded activity of an archaeological excavation is treated as a 'given' by most archaeologists. After all we work in a field that, unlike many sciences, destroys the evidence and doesn't allow for exact experiment replication. But in themselves there is no value to an archive, the value comes from its use. The first use to which an archive is put is to substantiate the interpretive claims made by the archaeologists who carried out an excavation. This has led to a close relationship between publication and archive (Richards 2002:354-357). As Richards points out this has contributed to a publication 'back log' problem, as authors attempt to integrate huge archives into publications (Richards 2002).

The problems of archaeology are unusual, but preservation is a characteristic of all viable scholarship. Research values in all disciplines dictate that scholars hand on their results to future generations that may use and refine their own work. If scholarship moves to digital forms - and the growth of e-journals, e-science and e-publishing shows that some disciplines are already very advanced in this respect (RSLG 2003) - then so there is a pressing need for that information to be preserved and transmitted in digital form too. It is the archives that keep archaeology a dynamic research pursuit as well as a development driven activity.

Richards (2002) has recently reemphasised the opportunity that is presented by the computer age to expand the value of archaeological archives, by making them accessible and searchable to a wide audience. Partners in the ARENA project have each recently released a set of archives (available online at http://ads.ahds.ac.uk/arena/), which will be useful for researchers in each of the partner countries - but which are also research in their own right. The most striking feature is that there is not a single model for archaeological archives. These archives presented through ARENA fall into a number of categories, representing in part the nature of the organisations that presented them. Six broad categories can be identified:

- Archaeological excavation archives. The ADS, the Danish National Agency for Cultural Heritage and the Institute for Archaeology (the FSI) in Iceland all made archives from excavations available. The archives were Danebury and Cottam in the UK (ADS), Tarraconensis landscape survey in Spain (ADS), Vorbasse, Dankirke and Helme in Denmark (The Danish National Agency for Cultural Heritage) and Hofstaðir in Iceland (FSI).
- Antiquarian archives. The Institute for Cultural Memory (cIMeC) in Romania is making available an extensive digitised archive the Archaeological Repertory of Romania (RAR).
- Catalogues. cIMec has also made available an online catalogue, the Chronicle of Archaeological Research in Romania
- On Line Publication. The partners from Poznan Archaeological Museum in Poland have made available a summary version of the paper publication of excavations at Kowalewko.
- Historical Archives. The Poznan Archaeological Museum is also hoping to make available a set of images recording the excavations at Biskupin in the 1930s
- Landscape archives. The Norwegian partners at the Museums Project have generated a resource holding all of the documentary records of discovery and archaeological intervention for Norway broken down by individual farmsteads. This resource is being presented in relation to the archaeological landscape, allowing researchers access to documents for each farmstead in the landscape arranged in date order. For the purposes of ARENA the resource is being pioneered using the Egge and Hegge landscape.

It is clear from the archives above that the opportunities highlighted by Richards (2002) are broad and exciting. The simple, but vital, value of archaeological archives is vested in them by their use, not by their simple existence. Despite this however, archives must be preserved before they can be used. The experiences of the ARENA partners, gained in making the above archives available illustrate some of the issues involved in preservation.

Data Migration

The Danish National Agency for Cultural Heritage has worked on three archaeological sites that have run between 1970 and 2000. At all three sites (Vorbasse, Helme and Dankirke), there was no digital recording on site. The digitising of the records has been office based and has taken place over a long period of time. It soon became clear that there were data structure and quality issues as well as a variety of data formats. The process of making data available in useful formats was very time consuming,ensuring that open formats were used and that these were in the a format that was likely to be readable by users. The approach to making data available was two fold; one that the data did not need to be supplemented to make it useable and two that the format issues were best dealt with by migration. The latter is a key debate in preservation activity, migration or emulation. For all of the ARENA partners faced with this issue, migration has been the preferred option, it is easier to achieve and far more convenient for the user.

Ongoing Projects and "Live" Data

For FSI in Iceland, working with a project that was still "live" posed a problem but also an opportunity. The development of the Hofstaðir archive has been interesting because excavation is continuing. New approaches to digital archives, including policy for preservation, have to feed into working practices. The archive covers many seasons of excavation raising a number of problems; there is a mix of materials in varied formats and documentation is fragmented and inconsistent. Because it is "live" the archive is constantly changing and lastly the archive needs to be connected to the post excavation process. The answers to these problems lie in the management of the archive, once more a time and resource consuming activity, ensuring the application of data standards and quality assurance. The importance of the management role on a "live" project is emphasised by the Hofstaðir experience. The FSI experience highlighted the need to remember the user when preserving data; but it is difficult to know what the user will want.

Blurring the Line between Interface and Archive: E-Publication Possibilities

The experiences of FSI and the Danish National Agency for Cultural Heritage connect preservation issues with user needs. As the Danish partners found, there was a need to decide on how to present archives. The decision to make available archives that are preserved and migrated onto useable formats, backed up by brief descriptions of the site, leave the use of the archive to the user. But attention must still be given to the user; they need some presentation of the site to decide what use they may make of the data. This can be provided online, either as a brief summary or as a more extensive online publication.

In the case of the ADS archive preparation the interface became a key issue for the preservation of the archive. The Cottam archive gives a brief overview of the project and makes preserved data archives downloadable for reuse, the basic service, properly preserved. The digital archive from the Ager Tarraconensis field survey project was created in the late 80s. It was used to create the paper publication and was then left on disc for storage. As it transpired the map data could not be migrated into a useable format. Without the maps the rest of the data made little sense, its reuse value was lost. Simply making available the data files that were migrateable would not make sense to the user, even though the data might be properly preserved. A user interface was required to demonstrate the potential of the archive. This was achieved by first redigitising the map data from the paper publication. Second the scanned maps were enhanced using simple html to show the relevant distributions and spatial relationships. It is then possible to consult the data online through the maps, giving a browsable archive that demonstrates its content. From here the user can then decide to download the preserved data, including digitised maps, for their own use. This approach, inspired by necessity, blurs the line between simple preservation and developing the user interface. By serendipity the ADS experience suggested an answer to the issues raised by Aldred at the FSI, making the potential of the archive clearer to the user. Enhancement is particularly pertinent for archives that relate to a paper or electronic publication, allowing for researchers to download and work with data whilst also allowing publications to be made shorter or interactive and more accessible. The third archive presented by the ADS, the Danebury project, archive highlights this association between preservation and publication enhancement still further. The archive is linked, through simple html, to online copies of four of the five Danebury publications.

Digitisation, preservation and archive enhancement.

The same blurring between preservation and user needs was illustrated by the experience of the Romanian partner. CIMeC found that the preservation and presentation of the catalogue of the Chronicle of Archaeological Research was relatively straight forward as there was relatively little digital data involved. The digitisation of the Archaeological Repertory of Romania (RAR) raised more issues however. To preserve the scanned documents there was also a need for documentation and indexing. It is not enough to ensure that the images are in a preservation format; they must have relevant documentation that makes sense of the content and indexing. Because this is a historic archive many of the objects or sites recorded must be identified in contemporary terms to give them a value to the user. The documentation and indexing of the digital archive in a database brings very real value to the archive, ensuring its public use and thus its preservation. It also raises an issue of investment, how deep do we go in enhancing the archive? Costs and time are the delimiters, as well as the need to define at which point to leave the use of the archive up to the user.

The archives used by the Norwegian partners at the Museums Project are documents that have been scanned, enhanced through OCR and extensively indexed using XML. Preservation work has been carried out at the same time as enhancement is added through detailed indexing. The Museum's project is taking the enhancement of the user interface still further by relating particular documents to specific landscapes. The intention here is to provide a kind of "glue" for more users to make sense of the archives in a spatial setting.

Advocacy

One of the fundamental points that applies to those working with preserving digital archives is the degree of awareness amongst archaeologists themselves. In Poland, partners have found that the adoption of information technology has been a slow process. The use of IT and thus the value of IT has been the preserve of a few experts. They are aware of the preservation needs of data, but their advocacy role remains important. This is illustrated by the presentation of the Kowalewko data. This important archive has very few digital components, leading to the publication of a digital version of the paper publication. This raises a question for all who attend CAA, how far has the need for digital preservation facilities been recognised? Is the value of digital data and its concomitant need for preservation still only recognised by a small expert group?

Breaking Down Boundaries

When looking at the experiences of the ARENA partners in the round, one may observe the breaking down of boundaries. An understanding of the place of standards in preserving digital data breaks down national boundaries, if standards are going to work for data preservation in a global sense they cannot afford to be localised. There are other boundaries to be broken down here to. The perception of the importance of preservation issues has to be spread throughout the archaeology community. This must be done hand in hand with the emphasis on the value of such archives, both as a record of an otherwise destroyed data set (the archaeology itself) and as a research tool.

Conclusion

Europe's digital inheritance is fragile and its longevity is not assured. All organisations, governments and individuals that decide to engage with digital technologies will have to develop strategies to maximise the lifespan of key digital assets. By providing an overview of the challenges faced and processes adopted by key cultural heritage agencies in Europe, the work of the ARENA partnership represents an important contribution to the preservation of digital data, and also provides a further opportunity to raise awareness of the problems which all will face.

REFERENCES

BEWLEY, R., DONOGHUE, D., GAFFNEY, V. VAN LEUSEN, M. and WISE, A., 1999. Archiving Aerial Photography and Remote Sensing Data: A Guide to Good Practice. AHDS Guides to Good Practice Series, Oxbow (http://ads.ahds.ac.uk/project/goodguides/apandrs/).

EITELJORG, H., FERNIE, K., HUGGETT, J. and ROBINSON, D., 2003. CAD: A Guide to Good Practice. AHDS Guide to Good Practice Series, Oxbow (http://ads.ahds.ac.uk/project/goodguides/cad/).

FERNIE, K. and RICHARDS, J., 2003. Creating and Using Virtual Reality: a Guide for the Arts and Humanities. AHDS Guides to Good Practice Series, Oxbow (http://vads.ahds.ac.uk/guides/vr_guide/index.html).

GILLINGS, M. and WISE, A. (eds.), 1999. GIS: A Guide to Good Practice. AHDS Guides to Good Practice Series, Oxbow (http://ads.ahds.ac.uk/project/goodguides/gis/).

RSLG, 2003. Research Support Libraries Group Report to Higher Education Funding Council for England (http://www.rslg.ac.uk/).

KENNY, J., KILBRIDE, W.G. and RICHARDS, J.D., 2003. Enter the ARENA: preservation and access for Europe's archaeological archives. In Doerr, M. and Sarris, A. (eds.), Computer Applications and Quantitative Methods in Archaeology 2002, Archive of Monuments and Publications, Hellenic Ministry of Culture:349-353.

MADSEN, T., 1994. IT-based information processing in archaeology - a key to progress? Paper delivered at unpublished seminar on The Problems and Potentials of Electronic Information for Archaeology, British Academy, 1994, cited by Richards, J.D. et al., 1999 op cit.

RICHARDS, J. D., MILLER, P. and WISE, A., 1999. Digital Archives in Archaeology. In Hansen, H. and Quine, G. (eds.), Our Fragile Heritage:123-130, Copenhagen: National Museum of Denmark.

RICHARDS, J. and ROBINSON, D. (eds.), 2000. Digital Archives from Excavation and Fieldwork: Guide to Good Practice Second Edition. AHDS Guides to Good Practice Series, Oxbow (http://ads.ahds.ac.uk/project/goodguides/excavation/).

RICHARDS, J.D., 2002. Digital preservation and access. In European Journal of Archaeology Vol. 5(3):343-366.

SCHMIDT, A., 2002. Geophysical Data in Archaeology: A Guide to Good Practice. AHDS Guides to Good Practice Series, Oxbow (http://ads.ahds.ac.uk/project/goodguides/geophys/).

Cultural Heritage Management Based on Information Science in China

Xiaodong Zhu
State Administration of Cultural Heritage
Beijing 100020 CHINA
drxdzhu@263.net

Abstract

The State Administration of Cultural Heritage(SACH) is one of the department of the State Council, aimed to protect, utilize and manage the cultural heritage of China. Digital and network technologies had been used in cultural heritage management since 1991. We are carrying out cultural heritage information construction, gradually realize the digitization of cultural heritage resources and the networking of industrial management and make use of internet to collect and propagate accurate, authoritative and abundance information about cultural heritage is an important guarantee for us to enhance the overall strength of cultural heritage.
The SACH has established a governmental website (http://www.sach.gov.cn) for releasing information about cultural heritage in 1998, and has started the construction of some databases about cultural heritage protection and management projects. Our major task is to formulate, improve, promulgate and implement the standard of cultural heritage information, thereby laying a foundation for realizing interconnection and information share in China.

As the world famous country with an ancient civilization, China has a long history and very rich cultural heritage. There are 28 sites of the World Heritage List by UNESCO in China, 1271 most important old buildings and monuments and archaeological sites, more than 12 million movable relics. The policy about Chinese cultural heritage protection is "protection utmost, rescue first, reasonable use and efficient management". Chinese government lay particular stress on the protection of scientific, historical, cultural heritage. The State Administration of Cultural Heritage (SACH) is responsible for protecting, managing and utilizing these cultural heritage, has an irreplaceable role in the information construction of cultural heritage. To carry out cultural heritage management based on information science, gradually realize the digitization of cultural heritage resources and the networking of industrial management and make use of internet to collect and propagate accurate, authoritative and rich information about cultural heritage is an important guarantee for China to enhance the overall strength of cultural heritage undertakings as well as the only way for the future development of the said undertakings.

Cultural heritage management based on information science takes a special position in Chinese information construction and is an important measure for publicizing the ancient Chinese civilization and the characteristics of the traditional Chinese culture in the space of digitized development.

What we did

According to the progress of the development of network technology throughout the world, to put the information about domestic traditional cultures, museums, natural landscapes and colleges and universities on Internet has become a strategic choice of all developed countries. In order to achieve survival and development in the network world, some developing countries have also established cultural heritage-focused websites of traditional cultures.

Since 1996, a series of achievements has been made in cultural heritage information construction in China:

a. Achievements of the current stage have been made in the front-end work of national cultural heritage information construction. The building of the Database of Treasured Chinese Cultural Relics, a basic research job approved by the Ministry of Science and Technology, is under smooth progress. The formulation of Provisional Regulations of China on the Management of Information about Museum-held Objects and the Provisional Regulations of China on the Management of Information about Cultural Heritage Preservation Units has been basically completed.

b. Various websites concerning cultural heritage and museums have been established. The SACH has established a governmental website "http//www.sach.gov.cn" for releasing information about cultural heritage and museums, briefing on the relevant knowledge and publicizing the relevant rules and regulations. Nearly one hundred units such as Shanghai Museum, the Chinese National Museum, Henan Museum have successively established public professional websites or web pages for providing information about cultural heritage, treasured museum-held objects, academic researches, news about cultural heritage and the relevant exhibitions to be held.

c. Certain achievements have been made in the development and application of museum-held cultural relic management system. A group of museums represented by the Palace Museum, Shanghai Museum and Tianjin Historical Museum have completed the building of LANs and put them into operation with self-raised funds. The management system and digital image data of cultural relics developed by the Palace Museum take a leading position in the cultural relic management systems across China.

d. China has cooperated with some foreign countries in the development of virtual reality technology for the manage-

ment of museum-held relics and historical heritage preservation units. Research on digitized mural preservation technology jointly carried out by Dunhuang Institute and the US Northwest University is under smooth progress.
e. The SACH has started the construction of a database of cultural relic protection projects and the development and application of information technology for general investigation of cultural relics in the whole cultural relic and museum system of China.

WHAT WE ARE DOING AND GOING TO DO

We will strengthen the research on the standardization of cultural relic and museum information and, while enhancing the basic work of cultural relic and museum management, preliminarily realize the digitization of cultural relic resources and the networking of cultural relic management and information propagation, expand foreign exchange in respect of cultural relic information and try hard to increase our country's cultural relic and museum information construction to an advanced level.

a. To formulate, improve, promulgate and implement the standard of cultural heritage information, with particular stress laid on the standard of the information about museum-held objects and the standard of information about cultural heritage preservation units, thereby laying a foundation for realizing interconnection and information share in China.

b. To construct the Database of Cultural Heritage Information Construction in China, with the contents mainly covering the policies, laws, rules and regulations on cultural relics and museums, the places in World Heritage List, national key cultural heritage preservation units, famous historical and cultural cities?towns?districts, large archaeological sites, sites of archaeological excavation, museums and warehouses of cultural relics, treasured museum-held cultural relics and important objects unearthed, propaganda on cultural relic protection, exhibitions, public collection and cultural relic circulation, cultural relic-related cases, scientific cultural relic protection, safety control projects, foreign exchanges, the relevant institutions, experts and talents, academic studies, education and training, etc.

c. To build a Cultural Relic Investigation and Cultural Relic Protection Projects Management System, including various special-topic databases concerning the construction and maintenance of national key cultural relic preservation units, the management of archaeological excavations, the maintenance of key museums and cultural relic warehouses, the safety control projects of cultural relic preservation units under grade A risks, the collection of public-held cultural relics, the scientific projection of treasured museum-held cultural relics and important objects unearthed.

d. To build a network system that is composed of three websites including Network of National Cultural Relic and Museum System, Network of Institutions under the SACH and Chinese Cultural Heritage and Museum Network in horizontal direction and four levels of nodes covering "state-provinces-prefectures (cities)-cultural heritage management units" in vertical direction. To build a Database of Cultural Heritage Infrastructure Construction Projects and Environmental Management Projects, including the national key cultural heritage preservation units and national key museums, large archaeological site protection zones and etc. Last August, Mr. Zujin Wang and I advanced a new idea: According to the database, the SACH can utilize the 3S technology(RS,GPS,GIS) and management information system to supervise the old buildings & monuments and archaeological sites (the national key cultural heritage preservation units) in China. We already finished the project proposal.

e. To establish a national cultural heritage and museum administration and service management system with the institutions under the SACH as the centre for realizing automated administration in such aspects as the circulation of official documents between the SACH and the units under it and provincial cultural heritage administration departments as well as the approval and management of projects.

f. In accordance with the requirements of the relevant international conventions that China has joined, to construct such special-topic databases as National Cultural Relic-related Cases and Lost Cultural Relics Registration System for comprehensively, timely and accurately master the relevant situations and provide information service for severely cracking down and effectively preventing cultural relic-related crimes and recovering lost cultural relics.

g. To carry out the relevant front-end work for starting the construction of China Digital Museum and make it a representative of the image of Chinese cultural heritage in the digitized virtual world.

MAJOR POLICIES

The SACH established a cultural heritage information construction leading group with it's head acting as the group leader. The executive office of cultural heritage information construction is responsible for the cultural heritage information construction and carry out centralized coordination. Every province's cultural heritage administration designates the relevant departments and special persons to be responsible for this work. The Chinese Cultural Heritage Information Consultation Centre will undertake the relevant work of cultural heritage information construction under the authorization of the SACH.

a. To speed up the training and introduction of talents and promote the popularization and application of achievements in IT-related scientific researches. To complete the program of training on computer and network for cultural heritage staffs, to take effective measures to introduce talents of information and network and make use of the existing achievements in IT-related scientific research in the cultural relic system.

b. To strengthen the building of the legal system and perfect the relevant laws, rules and regulations. To strengthen the research on intellectual property and information security for realizing the effective protection, reasonable use and efficient management of information about cultural heritage. To encourage persons from all circles to contribute their efforts for cultural heritage information construction.

c. To strengthen foreign exchange and cooperation in respect of cultural relic and museum information construction. To make full use of the foreign successful experience, adopt world advanced technology, participate in the formulation of international standards and regulations on cultural relic information construction, send professional and managerial personnel abroad to study advanced cultural heritage information network construction technologies.

International Reference Collections

A. G. Lange
Netherlands National Service for Archaeological Heritage

Abstract

Whatever paradigm is guiding our research and analyses, in whatever specialist niche we are interested, archaeological theory and practice will always be founded upon the interpretation of physical remains from the past. Such remains can be tangible remains (finds, movables, artefacts) or recognisable traces in the soil of past human activity (features, crop marks, chemical composition). The interpretation of remains takes place on three levels: a) by analysing the intrinsic characteristics of the remains themselves, b) by analysing the (extrinsic) context of the remains in the local setting and c) by comparing the intrinsic and extrinsic characteristics of the remains with similar finds or features from elsewhere. Comparable remains are dispersed all over Europe (and beyond), as are the specialists and the literature on the specific remains. Contemporary specialists together with the available literature covering finds and features form the basic archaeological knowledge base.
In the Netherlands, many developments in society at large and in the discipline itself are responsible that the archaeological knowledge base has become widely scattered, fragmentary and, partly, out-of-date. The erosion of knowledge continues at a high pace and threatens the quality of archaeological research, heritage management and education. In the Netherlands the National Services for Archaeological Heritage (ROB) studies the possibilities of developing an instrument for acquisition, storage and access to expert knowledge on archaeological artefact and feature types (Nieuwhof and Lange 2003). This so called National Reference Collection is considered an essential part of the quality control in the Dutch system of archaeology, as this has developed after signing the Treaty of Valetta (Malta, 1992) (Willems 1997).
It is proposed here to develop this knowledge management system further on a European scale as a) similar changes in the archaeological "setting" are recognised in many other countries, and because b) the dispersion of culture, material or immaterial, has never been constrained by national borders.

Knowledge Base

Any interpretation of material remains from the past needs the ability to identify archaeological finds and features, which is the very basis of the discipline of archaeology. In order to do further archaeological research, to protect sites, to teach archaeology and to tell the story of our past, we first must identify what was found. To identify the remains, references are used as pointers to knowledge of similar remains found elsewhere. Interpretation of the identified finds will lead to models of explanation. For over 200 years professional archaeologists have continuously accumulated knowledge on material culture (Brongers 2002). At the universities, this knowledge passed on from generation to generation by lectures and through the literature. The published corpus together with the insights of contemporary specialists, form the knowledge base on which archaeological theory and practise are constantly being (re-) built.

Changes, Threats and Opportunities

Recent developments in European society as a whole, and the archaeological discipline in particular, threaten the accumulation and accessibility of this knowledge base, in particular on material culture. With this knowledge base under pressure throughout Europe, the identification and interpretation processes, that are considered the core-business of archaeology, are being threatened as well.

This is illustrated by the following phenomena:

First, when evaluating the educational system, only part of the archaeological knowledge has been made explicit in publications and therefore a proportion is not accessible, except by direct contact with the specialist. In recent years, this so-called tacit knowledge has had little chance of finding its way to a younger generation of students through the educational system, due to a combination of reasons:

- The ready availability of absolute dating techniques, made the classification and use of type series much less important as a dating tool. Material culture studies lost their indispensable place in archaeology.
- Repeated budget cuts at the universities in recent decades are a major reason why older staff has not been replaced and there is little capacity for material culture studies pro grammes.

- The recent reduction and the limitation of the period of study have decreased the capacity for the time-consuming mate rial culture studies as well.

Secondly, in the practice of excavation, the need for access to information on material culture is growing fast, due to an enormous growth of the number of projects and participants. Large infrastructural projects have involved in many investigations a great number of young archaeologists. At the same time, with the arrival of commercial archaeology in a few big, but many small companies, the participants are often removed from, and have no ready access to, information-rich resources.

Additionally, the equally fast growing numbers of archaeological reports sometimes have a limited distribution. With low levels of availability, these reports reside in the realm of "grey literature", and contribute little to the archaeological knowledge base. The chances are that the expected level of scientific sophistication is no longer attainable.

Although we see a tremendous growth in the number participants, for the commercial archaeologists the possibility of participating in the conventional archaeological discourse and the scientific network is often restricted due to the physical distance, difficulties of planning or for budgetary reasons.

Finally, in the broader context of environmental planning, since the signing of the Treaty of Valetta (Malta, 1992) the number of people involved from outside the discipline has grown tremendously also. Today, civil servants, architects, landscape designers, environmentalists, contractors, solicitors, planners, etc. are professionally and structurally occupied with archaeology. Archaeology is taken seriously and plays a role in landscape and town planning in its early stages. However, the knowledge base, on which decisions are made, is often far from explicit and remains a black box to the uninitiated. Consequently, archaeological motivations are rarely objectively verifiable, while the consequences for society might be considerable.

To summarise, the "old" expert knowledge base is incomplete and widely scattered, and the accessibility of the new literature is low. This impedes the proper accumulation of new knowledge, while the demand for knowledge is growing at a high speed. This implies the need for measures to safeguard the very basis of our discipline.

A rather obvious solution is sought in the wide-ranging use of the various possibilities of the Internet. Web-applications can be developed as a means for storing, distributing and accessing the knowledge base. Web-based communication allows professionals to take part in the scientific discourse, independent from place and time.

Dutch Model

The Netherlands National Service for Archaeological Heritage (ROB) has submitted applications to national and international organisations for the co-funding of the development of a prototype for the web site of the National Reference Collection (NRc). The NRc will be a central site where the visitor finds state-of-the-art knowledge on material culture, in the form of images and descriptions or is referred to the latest literature and is presented with information on where the physical remains can be studied. In a fully developed NRc, there will be a discussion and publication platform, while a users are invited to pose questions to the system. Where the technical and functional infrastructure of the NRc is centrally organised, the content is provided by specialists and organisations scattered over the various regions. The actual data may reside at different web sites, but is harvested into the central site ("information hub"). The visitor need not necessarily has to be aware of this process.

The prototype of the NRc will present two data sets. The first is a description of a type series of late medieval glass. It will present profile drawings and photographs of the material as well as a strict, but widely accepted, nomenclature. The ROB maintains the described collection. The second collection to be presented is the Corpus Middeleeuws Aardewerk (Vlierman 1983). This is a series of descriptions of medieval pottery types from particular well-dated closed finds. As such, this "collection" exists only on paper, while the objects themselves are scattered among different archaeological archives. In the Netherlands, the provincial governments maintain these archives where, in principle, all excavated material is deposited. The archives are now developing from warehouses, closed to members of the public, towards public information centres (Ploeg 2001) where it is possible to study the actual remains and original documentation of excavations. The archives all use the same software package to register the remains. The web site of the NRc will play a role in providing information about location, content and accessibility of the different elements of the Corpus Medieval Ceramics. The ability to provide a visitor of the web site with a location of a particular collection and give him/her information on the content of this collection, is an asset in its own right, next to the virtual collection that is present on this site integrated with the virtual collections of other sites in the network.

European Added Value

From early prehistory onwards, material culture (artefacts, techniques, raw material) has been transported over large distances across Europe (and beyond). It is of vital importance for researchers to know what specialists in other countries contribute to the knowledge base, in terms of, for instance, the provenance of raw materials, production methods, cost calculations, trading routes and distribution patterns in geographical, socio-economical and cultural sense.

The development of a European network of national or superregional and interoperable web sites, each presenting reference collections relevant to the own region seems only a natural extension of the Dutch model, and certainly would serve as a significant step forward in conserving and updating the European knowledge base. Such a development could be described as "Moving from idea to practise in E-Science", building on earlier EU-funded initiatives like Aquarelle, ArchTerra and ARENA, etc. It will consist of a network of national and/or super-regional "information hubs". In turn, these central hubs will be the access points for the regional and local sites. Together this network forms the European Reference Collection (eRC). The wider archaeological discipline is currently operating at quite different levels of sophistication. The eRC builds bridges for all to the same knowledge base. This process is sometimes referred to as the "democratising" of knowledge.

At the same time, the eRC can be an instrument to unify European archaeological descriptive concepts, while preserving scientifically idiosyncratic "dialects", to indicate gaps in the knowledge base, to formulate general hypotheses, etc.

Open Expert Network Infrastructure

The eRC requires the co-operation of many, and the agreement and acceptance on many issues, such as technical protocols, procedures of communication and coding, to name but a few. Building an eRC is more than just presenting cultural heritage on the Internet's World Wide Web. As scientific insights and theories change, the content of the eRC should change accordingly. The eRC should be a state-of-the-art body of knowledge, kept up-to-date by groups of specialists. Throughout Europe national and international specialists groups are operating, some of these formally organised. Actually, the eRC should evolve towards a virtual meeting point for these specialist groups: a place to exchange ideas, to develop theories, to publish articles and to make suggestions for reformulating typologies and thesauri. The eRC also makes specialist knowledge available to all who are interested: discussions will be open to specialists and students, to professional non-archaeologists and to non-professional archaeologists, from inside and outside the European Community.

The involvement of many content providers involves a reorientation of the communication practices of the specialists groups. Content providers will be invited to participate in discussions on the Internet. Such discussion could take place in the framework of, for instance, Internet research communities. Internet communities are the virtual counterpart of assemblies of people who are interested in a special topic. Access to a particular community might be restricted, but generally everyone can join. The advantage over conventional means of communication is that in virtual communities discussions can be followed independently from time or place. Changes in behaviour are often difficult to realise. The investment in behavioural change would be easily paid back by the much wider audience and the speed of the new media.

The eRC will further enable the identification of gaps in our knowledge and will, consequently, stimulate multidisciplinary material culture studies and the international exchange of students and scientists. The archaeological heritage will become also easily accessible to a wide public with an interest in human history and technological developments. The eRC will thus contribute to the insight and public support for cross-cultural heritage, which adds to the coherence of the multicultural community by showing the influences of, similarities and flux between technological and cultural developments in various regions in Europe (and beyond).

Next to providing access to the European knowledge base on archaeological heritage, the eRC serves also as knowledge base on the more technical issues involved in developing and maintaining the network. Examples of current issues are; international standards for the exchange of metadata, machine to machine protocols for data transmission, standardisation, personalisation, data enrichment, generating presentations and the protection of Intellectual Property Rights (IPR).

The Semantic Web is another example of an innovative technology to which cultural heritage is presently contributing very little, in terms of solutions, but the technology is also yet to present its challenges. So it may happen that this technology grows without taking into account the particular needs of cultural heritage, notably archaeology, e.g. in terms of validation, scientific credibility and IPR. The eRC can bridge the gap between technological innovation like the Semantic Web and as yet implicit requirements of archaeological research concerning new technologies. An example of the latter could be the development of Content Based Image Retrieval methods. These will be crucial for an optimal service to the user of the image-oriented site of the eRC. The eRC will seek alliance with national and international technical research programmes to develop together with ICT-scientists new tools for archaeology. Another positive contribution of the eRC would be to balance the differences in awareness and starting conditions about such topics in the different European countries.

Conclusion

The development of the eRC will be a long-term commission involving many participants from a number of disciplines to build a network along the contours as sketched above. The eRC will deliver a knowledge base, which is preserved, maintained and updated, and accessible to anyone. The content will consist of the foundation of archaeological theory and practice: the knowledge on archaeological material remains and features. The eRC will promote high standards of the evaluation of archaeological remains throughout Europe. By unifying identifications and nomenclature the long-term accessibility and usability of the archaeological record is guaranteed.

The costs of the development will be high, both in time and money, but the reward will be a solid, sustainable foundation for the wider archaeological discipline and historical science at large. Therefore, the commitment from European partners is of vital importance.

After a presentation at the EAA conference in Thessaloniki in 2002 (Lange 2002) a number of participants had already taken sincere interest into the matter, notably from England, Denmark and Belgium. Since then, this group expanded with representatives from institutes from Spain, France, Italy, Romania, Finland,

Norway and Iceland. From this group a task force has been formed to develop plans for the realisation of the eRC and to formulate proposals for funding by national and European bodies. A first task will be to arrange for funds to enable the participants to meet and discuss the co-ordination of activities and prepare a time schedule.

REFERENCES

BRONGERS, J. A., 2002. Een vroeg begin van de moderne archeologie; Leven en werken van Cas Reuvens (1793-1835). ROB, Amersfoort, The Netherlands.

LANGE, A.G., 2002. Co-operation in the development and exploitation of National Reference Collections. The European Archaeologist 18.

NIEUWHOF, A. and LANGE, A.G., 2003. Op weg naar een Nationale Referentiecollectie Archeologie; Eindrapport Haalbaarheidsonderzoek NRc. ROB, Amersfoort, The Netherlands, in press.

PLOEG, F. van der, 2002. Brief aan de Tweede Kamer 27 432 nr. 54. State secretary of the Ministry of Research, Culture and Science, 27 mei.

VLIERMAN. K., 1983. Aardewerk uit de inventaris van twee laat-middeleeuwse vrachtschepen OK73-74 en ON5. Corpus van middeleeuws aardewerk uit gesloten vondstcomplexen in Nederland en Vlaanderen, 1-2. Rijksdienst voor de IJsselmeerpolders / Museum voor Scheepsarcheologie, Lelystad/Dronten.

WILLEMS, W.J.H., 1997. Archaeological Heritage Management in the Netherlands: Past, Present and Future. Archaeological Heritage Management in the Netherlands: Fifty Years State Service for Archaeological Investigations. Willems, W.J.H., Kars, H. and Hallewas, D.P.(eds.), Van Gorcum/ROB, Amersfoort, The Netherlands, 1997.

Optimizing Digital Library for National Heritage As a National Culture Managing Device

RULY DARMAWAN
PADI ARTGROUND
40135 WEST JAVA - INDONESIA
RDARMAWAN@YAHOO.COM

ABSTRACT

Accessibility and connectivity is a significant issue in a recent networked world. This accessibility and connectivity are taken as benefit to gain and deploy information and knowledge as much as people need. Both information and knowledge have a broad spectrum in which cultural issues are also considered. Cultural issues are significant for countries and/or communities to promote their cultural potentials and, of course, cultural identity. When technology and culture issue are met, a proper national culture management could eventually be achieved.

INTRODUCTION

Culture is ensemble of social processes by which meanings are produced, circulated and exchanged (Thwaites et al. 1994). People live with their culture and they will always be. Since culture played an important role in civilization development, people and its culture are inseparable. Managing the culture means also managing people, a community, and even a nation for better future sustainability. Therefore, each attempt towards better national culture management and promotion should be at the top level of consideration.

POTENCIES OF INDONESIAN CULTURE

Indonesia has approximately 13,000 islands, which occupied by various tribes. As aforementioned above that, most people whether they lived individually or in a group have and produce their own culture. Tribes in Indonesia are widespread on islands. Some places are hard to access because of their geographical condition.

Within the tribe's community, culture is manifested into some mediums, visualized in various forms. For example, the faith on transcendence between life and death is expressed on either temple (candi) architectural design or dancing. As should be understood here that culture provides more than artistry and/or aesthetical values. Wisdom of life, faith is some inserted sense by which culture become a precious legacy for the next generations, as well as the root of their national identity.

Indonesia has both tangible and intangible culture. Tangible culture is manifested on several traditions in Indonesia, which would generally be, not limited to, as:

1. Oral, Performance Tradition,
2. Writing, Drawing Tradition

Other tangible culture could also be found as Temples or Candi, Manuscripts (on bamboos, stone carvings, barks, palm-leafs), Ornamental Designs, Stone Carvings, and/or Musical Instruments.

Meanwhile, manifestation of intangible culture would generally be, not limited to, as:

1. Folklore,
2. Mores, and/or
3. Community System

Nowadays, some institutions attempt to develop some culture manifestations. As an example, PADI ARTGROUND develops a number of manuscripts in which the ideas are taken from the origin Indonesian culture manifestation.

Examples shown above prove that existing culture manifestation could be developed. Because art has played such an important role in the development of culture and human civilization, artistic development can, therefore, be equated with cultural development (Darmawan and McGlynn 2003). The result of such research and development may contribute a new insight for local people and may enrich the trove of national culture.

ACCESSIBILITY TO INDONESIAN CULTURE RESOURCES

The lack of accessibility to any cultural material resources is the common obstacle of the most Indonesian scholars who interested on their own national culture development. Furthermore, most Indonesian scholars are also facing a problem of limited fund while they wish to travel in order to research and gather cultural material directly from the origin place. This might also experienced by the foreign scholars when they intended to do the same thing about Indonesian culture.

Culture is like an unfinished story, which should be continued (Peursen 1988). In order to keep national cultural potencies sustainable, accessibility to the some, if not all, Indonesian cultural potencies should be provided and developed.

DIGITAL LIBRARY FOR NATIONAL HERITAGE: DEVELOPMENT AND OPTIMIZATION

As a product of Information Technology and Communication (ICT) development, digital library provides opportunity to unlock the access to any knowledge resources anywhere in the world. Though most regions in the world, in particular developing countries, are experiencing digital divide, there are still plenty of alternatives, which might be helpful for them to gather and disseminate knowledge.

Indonesian Digital Library for National Heritage tends to be developed on interactive multimedia version. Because of the inappropriate bandwidth capacity, this interactive multimedia remains a better solution to bridge the digital divides, which are the most obstacles for Indonesian. Nevertheless, the on-line version remains a primary consideration regarding to convenience accessibility to all resources of Indonesian cultural potencies.

This project was planned to be publish on series. These series based either on institutions or on type of traditions. Whatever its form, this national culture management demands a serious effort that in turn could affect on national cultural identity retrieval. As a standard, information is categorized in collections and research-and-development, and it might be deepen and modifiable if necessary. It is quiet simple, but might be able to cover all essential of Indonesian cultural potencies and its developmental efforts.

One of outstanding cultural potencies of Indonesia is about its drawing and writing tradition. As a part of Indonesian institutions in which cultural potencies are collected and developed, PADI ARTGROUND begins to promote Indonesian cultural potencies on drawing and writing tradition.

The following picture below is an example, which shows visualization and content management of PADI ARTGROUND's collections within digital library interactive multimedia format. Activity in PADI ARTGROUND, an art institution, is now focused on the research and development of Indonesian drawing and writing tradition in which external influences such as Hinduism, Buddhism, Islam, and Chinese are penetrated. All these activities and collections may contribute something significance as a best practice to any national culture development effort in Indonesia. Moreover, another institution who would like to do the similar thing might like to see and review this project as a developmental reference.

Collections and situations in PADI ARTGROUND, which presented on this digital library interactive multimedia format reflects some effort of the national culture management in Indonesia. Moreover, this media could be an evidence for a developing country who wish to achieve better national development and sustainability based on its cultural contents.

Conclusion

Developing a medium for national culture management, for instance, is an excellent effort towards civilization sustainability. Even some obstacles occurred on the way of national development such as social gaps and low human resource capacity, there would always a way out. New technology development, in particular Information and Communication Technology (ICT) development, might provide a solution, such as information management system, software engineering for information and communication device, and multimedia technology. Those technologies offered a plenty of opportunities to be taken as benefit for purposes such as national culture management. However, both local content and local potential should be consider as main priority rather than the technology itself. This is an idea which will enable people to promote and preserve their own culture for their own next generation, and, eventually, for their own next civilization.

Figure 1 Visualization and Content Management of PADI ARTGROUND's Collections within Digital Library for National Heritage Interactive Multimedia Format

References

DARMAWAN, R. and MCGLYNN, J.H., 2003. Establishing a National Network for National Heritage. Proceeding: International Symposium and Workshop on Managing Heritage Environment in Asia. Yogyakarta: Gadjah Mada University.

PEURSEN, C.A. van, 1988. Strategie Van De Cultuur. Translator: Dick Hartoko. Yogyakarta: Kanisius:12.

THWAITES, T., DAVIS, L. and MULES, W., 1994. Tools for Cultural Studies. Macmillan Education Australia Pty Ltd., South Melbourne:1.

Integrating Web and GIS Services into Archive and Collection Management Systems

S. Rußegger

H. Zeiner

H. Mayer

Institute of Information Systems and Information Management
Joanneum Research Forschungsgesellschaft mbH
Austria

Abstract

Archive and collection management systems based on information technologies are nowadays widely in use and have proven to provide valuable support for the management of objects in the cultural heritage domain. At the same time the Internet and the world-wide-web have become one of the most important channels to provide access to these objects next to their physical presentation in museums and archives. However, very often there is no direct connection between the databases holding the scientific description of all objects and their public, visitor centred presentation on web sites. In this paper we discuss the technical possibilities and their implications to modern collection management systems provided by modern Web technologies and also how to integrate geographical information to such solutions. Based on a concrete example, our collection management solution IMDAS-Pro, we will demonstrate how easily it is possible for an ordinary user to manage collection objects together with geographical information and to publish information about objects on the Web.

Introduction

The Institute of Information Systems and Information Management develops new and integrated methods designed to organise and manage the constantly growing flood of data and to selectively filter out the desired information. For this purpose, the institute's experts combine Internet technologies with collection management systems to create extensive Web information systems, developing tools for analysing and processing content. The great strength of the Institute is the tight intermeshing type of media - for today the digital world no longer consists of numbers, text and pictures alone, but also increasingly offers audio, video and map data. The result leads us to new technologies and methods that help us search for and, above all, find content both on the Web and in multimedia museum and archive data. The systems are tailor-made to the client's needs and stand out due to their clear structure and a high level of user friendliness. The possible applications range from libraries, museums and television archives to computer-assisted training and e-technologies.

During the last years archive and collection management systems based on information technology found their way into the field of cultural heritage. More than 10 years ago the Institute of Information Systems and Information Management started analysis projects in the field of cultural heritage. Aim was to find out and document the work flow within cultural institutions regarding the cultural objects they are dealing with. Results of these projects were very detailed documentation documents including not only the main work processes but also concrete descriptions of necessary functions and user interfaces.

Before a museum object has been registered, inventoried and scientifically documented, it must go through many departments. Within a "perfect" museum the first station of an object is the incoming department where a rough description of the object is done (registration). Further on the object comes to the scientists where it is going to be researched, analysed and inventoried. Especially the scientific treatment and documentation must be the "living" part of a software program that should assist the museum staff in their daily work. This main part of a "museum program" will increase by the usage and possibilities of state of the art technologies. Starting with "only" textual descriptions some years ago modern museum programs are including multi media descriptions (pictures, films, graphics) nowadays as well.

Service Oriented Collection Management System

The IMDAS-Pro software package (Rußegger, Preininger and Müller-Straten 2000) - that was developed at the Institute of Information Systems and Information Management and is based on very detailed research work in the cultural heritage sector - can greatly speed up these time-consuming organisational procedures. The great strength of the documentation system for museums is its flexibility - it can be used to record dinosaur bones just as well as to catalogue Picasso's paintings.

The development concept of IMDAS-Pro is to provide a very strong documentation system in combination with a service oriented architecture (SOA) by using add-on services such as web content management system or a GIS service. The combination and not the complete integration of services support the user with a set of very powerful and well integrated set of tools. Therefore the user may choose tools and software fitting well into his work flow. The offered add-on services can be easily fed with the data of the IMDAS-Pro database, so that it is not necessary to know much about data structure and storing mechanism. This concept is a very time and cost saving one because you do not have to switch the software tools.

Living in a world and time that seems to grow together by the usage of Internet on the one hand and where the interests of people increase on the other hand the necessity arises that even cultural institutions must open their archives and have to appear before the public. Therefore the Internet and world-wide-web have become one of the most important channels to provide access to these objects next to their physical presentation in museums and archives. Based on the knowledge and experience of the members of the Institute of Information Systems and Information Management in the field of cultural heritage and in the field of information technology the following concept was developed to offer a flexible and easy to use system to present data - stored in a database - on different output devices. One of these devices might be for instance a web page with the museums data.

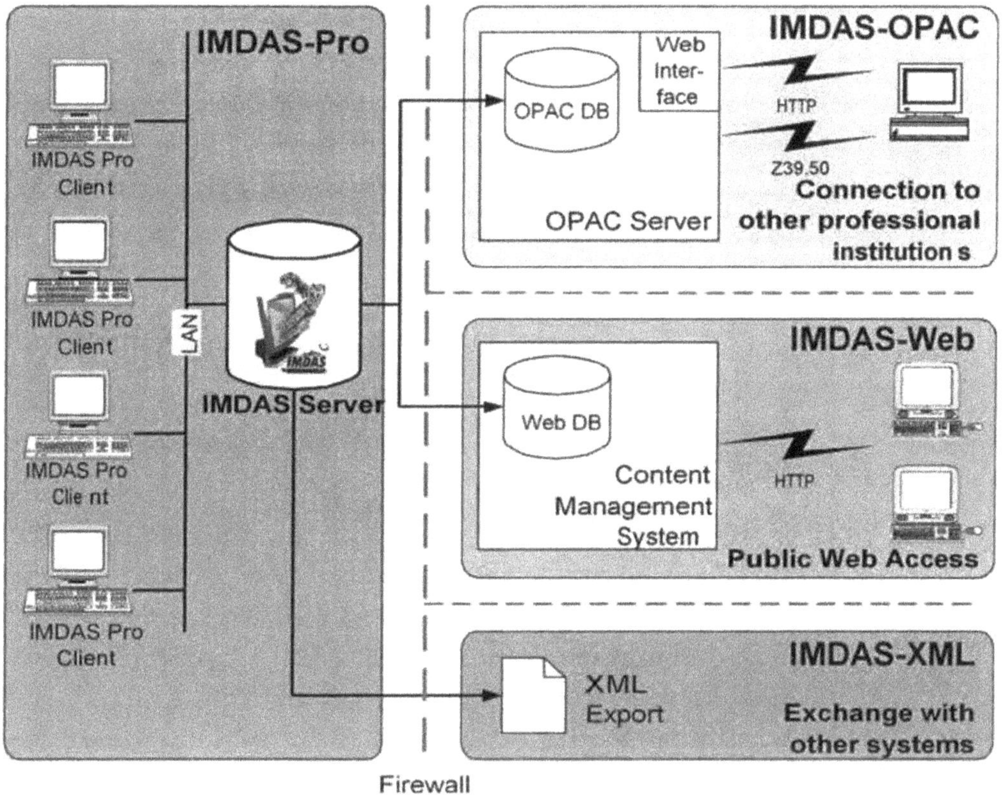

Figure 1 IMDAS-Pro components

The concept shows an add-on service to any data entry software that is able to export data in a structured, defined way. These exported local data are stored in a server database (in the graphic: IMDAS server) from which the data can be disposed to the output devices. One of these devices is an OPAC server that can be used for very special, scientific and detailed search and retrieval functionalities. Another device could be the presentation on web in the form of web pages with administrative and cultural information or even virtual exhibitions to different themes. An example for the usage of this web concept you can have a look on the project web site http://iis.joanneum.at/ dominico/.

Incorporating Geographical Information

Another important concept is used in the field of cultural information systems in combination with Geographical Information Systems (GIS). Geospatial data is part of the data flow. It is a known fact that geographical information is identified in about 74% of the data flow of the workflow process. Organisation such as OGC (Summary Report [...] 2000) increased the integration of geospatial data information in the last few years. OGC is a not-for-profit association dedicated to new geospatial technical and commercial approaches to interoperable geo-processing (Open GIS Consortium, Homepage).

The Geographical Mark-up Language (GML) specification (OpenGIS® Geography Markup Language (GML) 2002) defines the features and syntax that GML uses to encode geographic information in XML. GML supports the definition and data exchange in services. GML is a well established format, so sites do not need to support proprietary data formats and GML is extensible and XML-based, which makes it easy to manipulate, change, add to its contents and integrate it in a (web) service infrastructure. GML will have a significant impact on the ability of organizations to share geographic information with one another, and to enable linked geographic datasets. The format can embed links associated with features from other domains such as cultural heritages. Geospatial technologies applied and used in collection management tools such as IMDAS-Pro provides an additional spatial view of the collection itself. The power of GIS lies in the ability to link cultural objects on a map to a content management system, combining the graphical display of data with content query and analysis functionality. This helps the scientist to analyse, maintain and deduce information from the objects of the collections. One of main advantages is the integrated of the GIS component within the collection management tools such as IMDAS-Pro.

The IMDAS GIS module includes basis GIS features such as zoom in, zoom out, panning, measure the distances between two points, layer management, import and export of geographical data. Advanced features are e.g. the generation of objects on the map by using the geographical coordinates of the underlying thesauri and time related search are supported too. The result of this type of query returns different coloured objects according to their age. So it is possible to show the development progress of e.g. cities. Museum relevant data can be displayed along with the objects (e.g. excavation area boundaries). A selection of one or more objects in the map is equivalent to the selection of one or more objects in IMDAS-Pro. Afterwards these objects can be combined to a special IMDAS-Pro group.

Conclusion

The application and service structure described in this paper is an example the new service oriented approach. The sound combination of visual representations (images, symbols, maps) and intelligent collection management based on IMDAS-Pro improves the usage and value of the cultural heritage objects. A sometimes underestimated issue in the area is that this additional local information, or more precisely the semantic information which is implicitly contained in the local information, can simplify the usage of this content by humans, e.g. for navigational purposes.

IMDAS-Pro offers different components (in a service like manner) which are independent from the local documentation systems of the institutions. This concept enables a customized solution for each user and user group. However, we believe that this solution of combining already existing types of information systems into easy to use applications will lead to new type of content and collection management systems.

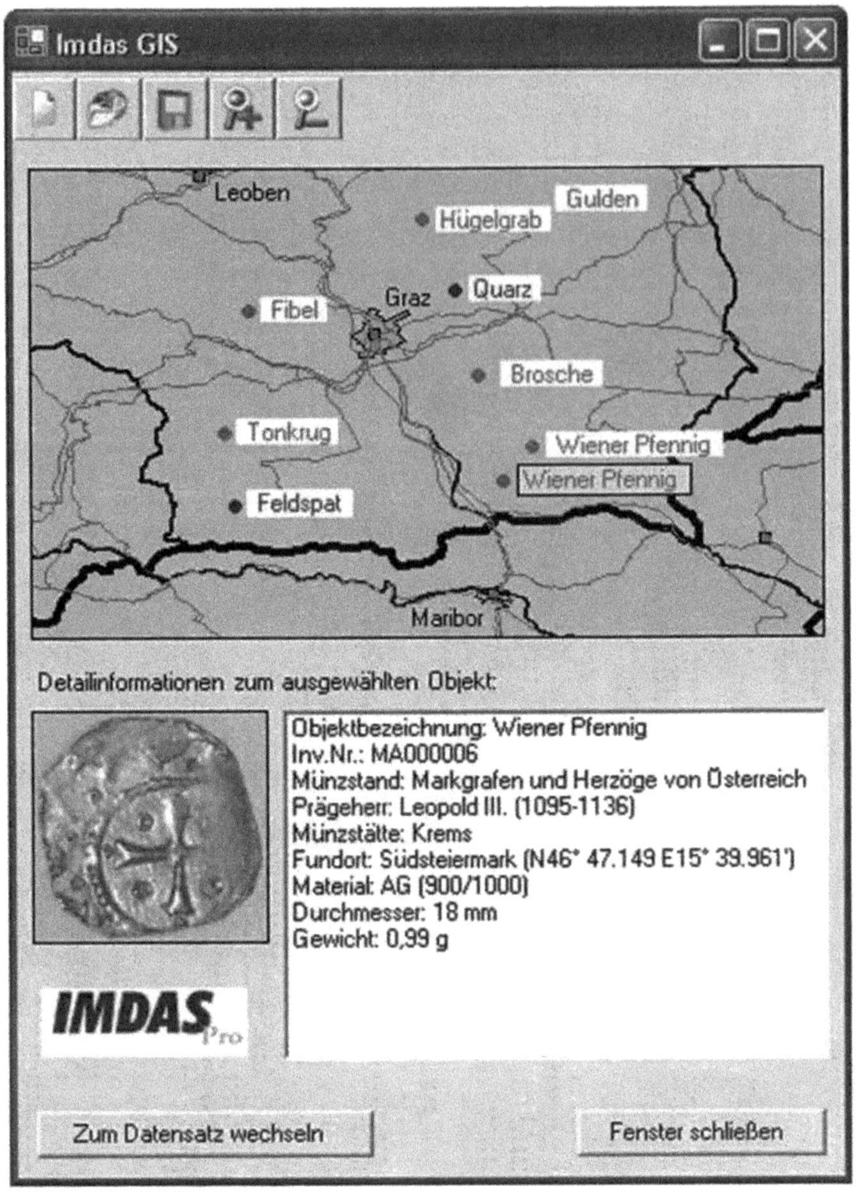

Figure 2 Screenshot of GIS component within IMDAS-Pro

References

OPENGIS® GEOGRAPHY MARKUP LANGUAGE (GML), 2002. Implementation Specification, 2.1.2 (http://www.opengis.org/techno/implementation.htm, last access: June 2003).

RUßEGGER, S., PREININGER, W. and MÜLLER-STRATEN, C. 2000. Die Version 1.3.0 von IMDAS-Pro. In MUSEUM AKTUELL.

SUMMARY REPORT of Version 1.0 of the Bureau Architecture, 2000. Bureau of Land Management, 31 March 2000 (http://www.blm.gov/ba/documents/BA_Summary_v1.pdf).

OPEN GIS CONSORTIUM (OGC) (http://www.opengis.org, last access: June 2003).

ODENSE BYGIS - ODENSE URBAN ARCHAEOLOGICAL GIS

KARL BRIX ZINGLERSEN[1]
MA STUDENT, DEPT. OF MEDIEVAL ARCHAEOLOGY,
UNIVERSITY OF AARHUS, DENMARK
KARLBRIX@SOL.DK

SEE THE CD FOR THE EXTENDED VERSION

ABSTRACT

The Danish town Odense is not only the birthplace of the famous writer of the fairy-tales Hans Christian Andersen (The Little Mermaid, The Ugly Duckling, etc.), but also has a heritage of Viking and Medieval origin.
Odense City Museums have the latest years developed an online database and GIS system handling the archaeological excavations in the county (www.arkdata.dk), primarily those most recently completed. However a full geographical overview of the excavations, test pits etc. from the previous decades in the medieval city centre do not exist at present time. Consequently in 2001 the author finished his MSc dissertation from University of Newcastle, UK, dealing with the subject of mapping these investigative activities. The ambition was to reach the most low cost and basic way of recording needed.
At home at the University of Aarhus a new dissertation for the MA is now prepared, coping with the modelling of the deposits and mapping of the medieval cultural heritage still left beneath the modern city surface. The different methods will be discussed and the problems and drawbacks accounted for, as well as the possible achievements. This is the initial stage of the project, so pieces of good advice are very much appreciated.

INTRODUCTION

The archaeological resource is a valuable element of the urban heritage, and excavations in urban environments are exciting and complex, but expensive projects are damaging to the sponsors purse and the archaeological resource.

The author's studies deal with the urban deposits in the medieval city of Odense, Denmark, by applying GIS deposit modelling, terrain analysis and spatial statistics to the material. The prime aim is to encourage the Danish medievalists to better understand the 3D development of the medieval town and be aware of the situation of the archaeological resource for the improvement of future strategies for its preservation and study. The result will be a deposit terrain model to support better-informed decisions and priorities prior to intensive investigations and planning schemes.

MANAGING AND PRESERVING THE CULTURAL RESOURCE

[1] Handling the pressure of development on urban archaeology

Until recently archaeological excavations have been carried out prior to construction works with substantial sacrifices for both the developer's finances (or government funds) and the archaeological remains - particularly in urban locations. Though knowledge of the town's history feels natural to the archaeologist, often only little is known about archaeological survival on-site until the digging begins. Well-informed decisions concerning the archaeological remains prior to development planning and constructions are crucial. The local curators need to have strategies for the urban archaeological resource and the strategies should be shaped by archaeological knowledge and research priorities based on assessments of the archaeology and maps or models of the deposits' conditions. The strategy, assessment and deposit map form the framework of the local site evaluations, producing a more detailed picture ahead of excavation works for the benefit of the archaeological judgments and efforts.

[2] Estimating the archaeological resource

Prediction of archaeological settlements and interpolation of archaeological deposits in the open landscape and the crowded city is not novel. Particularly the landscape-wide applications are widespread in North America and to some extent in Europe and other continents, e.g. Highway projects in Slovenia (CAA 2000). Volume computation of urban archaeological deposits is slightly more rare, although introduced early in United Kingdom and France on an analogue basis. The first projects by British researchers include The Future of London's Past (Biddle et al. 1973) and research by Martin Carver (Carver 1983), both emphasising the importance of the three-dimensional body of deposits representing the archaeology rather than isolated monuments and the usual two-dimensional plane recording approach. The development in the recent decades of the GIS software has emphasised the three-dimensional methodology. As Paul Miller points out: "The data we model do not exist on a two-dimensional plane, but in a three dimensional world where the extremes of topography have a significant effect upon the use of space both now and in the past" (Miller 1997:100).

Estimation of urban deposits usually has a different ambition to prediction modelling on landscape level; the aim is often to assess the deposit distribution and accumulation and the topography of a known urban settlement - e.g. Roman London or York. Landscape predictive modelling is often based on the calculation of probability of the location of certain phenomena and deduction from known situations to more or less unknown wider contexts (Andresen, pers.comm.). Modelling in urban environments is not less environmentally deterministic than traditional landscape predictive modelling: the predominant data sources are known archaeological excavations and boreholes, and the aim is the calculation via interpolation of deposit volume, accumulation and survival. The detailed knowledge of depth and quality of deposits and structures is important to measure the expected impact on archaeology from developments, prior to urban site evaluation (Richards 1990:32, Carver 1983).

Archaeological predictive modelling is often viewed with scepticism offering no alternative for archaeological investigation and research whatsoever (Carver 1983:341). Prediction and interpolation is naturally a best guess depending on data and use of algorithms and one can never produce an exact result. Particularly if the model avoids areas of archaeological importance, scepticism is provoked - often rightly - and as Martin Carver highlights "emphasis of these exercises is always on the positive rather than the negative results" and "no archaeological map, can or should be taken as complete" (Carver 1983:369). However, deposit interpolation and computation in urban environs does have its justification when used consciously for the benefit rather than to the detriment and negligence of the archaeological resource, particularly as a tool for assessing the resource and formation of strategies for its preservation and investigation. Examples from United Kingdom will illustrate this approach in the following chapters.

Mapping Urban Archaeology in England: UAD and YAA

[1] The Urban Archaeological Database programme

The necessity of strategies for urban archaeology has been recognised at national level in England through the English Heritage (EH) programme Urban Archaeological Databases (UAD), which is still in progress. A number of towns of chronological depth, good preservation and clear development pressures have been mapped and recorded intensively or extensively (English Heritage 2001, Thomas, pers. comm.). The GIS and databases form the starting point for new and repetitive assessments of the urban archaeology preceding the development of strategies for preservation of the archaeological resource and its role in the modern urban environment.
In the UAD structure the urban archaeology is considered as a collection of monuments following an Event-Monument partition (Thomas, pers.comm., Hopkinson 2002:18). The UADs incorporates from time to time relatively simple maps of the distribution of deposits, but seldom by a full-scale deposit model.

[2] The York studies

The York Archaeological Assessment (YAA) project was a pilot project of EH and York City Council prior to the UAD programme, but the approach of YAA was different. The town was perceived in a more holistic way as an undivided and integrated entity with focus on the distribution and quality of deposits rather than a number of separated monuments - a concept based on research by Martin O.H. Carver (Oxley, pers.comm., Carver 1983, Richards 1990). The YAA embodied a collection of data and a series of map layers displaying the archaeological development and characteristics, including a 'four dimensional' model of the topography of deposits. With his PhD thesis in 1997 from University of York Paul Miller enhanced the value of YAA adding supplementary technical and conceptual discussions on the capabilities of applying GIS methodologies to study the multi-dimensional and multi-temporal sequence of deposits in the city (Miller 1997).

Recent studies in York by Guy Hopkinson have addressed the deposit analysis and modelling from a task-based perspective in the hands of the contracting archaeologists in archaeological site evaluations (Hopkinson 2002). Data from the urban database were enhanced with supplementary borehole data and a context-based data structure for a more detailed analysis. Hopkinson points out that any new site evaluation might ask new questions of the material, which the existing data set might not be suitable for. Conversely much of the detailed site evaluation data will be superfluous to the scope of the urban database and GIS (Hopkinson 2002:19).

The Odense Studies: Odense ByGIS - Urban GIS and Deposit Model

The Danish city of Odense has a rich Viking and Medieval origin and the archaeology is managed, conducted and archived by the Odense Bys Museer (Odense City Museums) archaeological unit.

The author's MSc dissertation 'Odense ByGIS' was the first project to deal digitally with the full body of archaeological excavations in Odense producing a relatively straightforward database and GIS handling of the excavations and their basic information for curatorial use (Zinglersen 2001). The topic of my current MA dissertation includes the issues of deposit modelling and mapping of the urban archaeological resource.

[1] The first project: Mapping the antiquarian events basically - the Newcastle Dissertation

The initial stage of the Odense ByGIS (or urban GIS) dealt with the antiquarian events: boreholes, watching briefs, excavations, trial trenches etc. and their data of an administrative and interpretative nature, approximating to what in England would be termed a UAD, though just slightly more basic. The system was intended to assist the curator in managing the archaeological investigations by providing him with the basic information and a geographical overview for the initial site evaluation prior to examination of the extensive analogue archive.

The system consists of a MapInfo and an MS Access application - the software packages in use at the archaeological unit. The Access database is holding, managing, modifying and sorting the records with their information and values, and then distributes the results as flat tables to MapInfo automatically. The data included describe the archaeological investigations and the listed buildings of the city. MapInfo displays the data connected to the digitalised spatial entities, together with a series of map overlays.

The system is designed for basic input and retrieval of information and relatively fast and cheap implementation in the organisation using existing data sources. The project is currently only submitted as a dissertation from the University of Newcastle, being a pilot study for further implementation, hopefully to be realised within a short time. It is planned to expand the system with a database on written sources including estate tax registers.

[2] The new project: Deposit Modelling - the Aarhus Dissertation

Inspired by a Danish national scheme on mapping areas of cultural heritage, including the medieval urban centres, and by the English Heritage and York projects my MA thesis from University of Aarhus deals with the matters of archaeological resources in the field of urban archaeological deposit modelling and preservation using GIS.

The prime ambition of the project is to boost the discussion in Denmark on the priorities of urban archaeology and the understanding of the three- and four-dimensional development of the medieval town. Initially the result of the project provides the Odense Bys Museer with an enhanced GIS tool for mapping, recording and managing the archaeological resource. The project will identify the advantages and limitations of Deposit Modelling in urban environments and the technical and theoretical methodologies to be applied.

A detailed study answers the questions on how to digitally record, map and model the resource using different sources; on how to employ the results in the research, the management and preservation of urban archaeology in practice, particularly in the estimation of deposit volume and character in site evaluations.

Data acquisition and estimation of data quality is the initial concern. The types of sources are excavation plans and data, boreholes, watching briefs and other surveys, modern digital maps, geological maps and if possible cellar surveys. Each type of data is handled individually including its own variety of inaccuracies, biases and oddities recorded as metadata for mapping their effect on the result.

The next issue is modelling of the deposits. A range of interpolation methods (Triangulations, Krigings, and Inverse Distance Weighting etc.) will be considered and tested for their usefulness on the data set and the results analysed with spatial statistics.

The resulting assessment, models and maps should preferably display the calculated capacity and state of preservation of the urban medieval archaeological deposits, and give a general presumption of the correlation between the distribution of deposits and the topography and development of the medieval town as outset for more detailed future studies.

For a fuller explanation of the project and for reviews of Danish projects on mapping the cultural heritage areas as well as on the UAD and YAA projects, please read the CD-ROM.

ACKNOWLEDGEMENTS

My gratitude is expressed on this occasion for the support provided by Jens Andresen and Mette Svart Kristiansen, University of Aarhus both supervising my work on the MA thesis. Many words of thanks also go to Penny Spikins, University of Newcastle, supervisor on my MSc thesis 2001 and Jakob T. Christensen, medieval archaeologist of Odense Bys Museer.
I wish also to acknowledge Roger Thomas (English Heritage), David Jamieson (MoLAS), Julian Richards (University of York/ADS) and John Oxley (York City Council) for their kind help and sound advice. I also owe a dept of gratitude to Dominic Powlesland and Guy Hopkinson (Landscape Research Centre) for their excellent hospitality and pleasant and interesting company during my visit to England in spring 2003.

[1]The author is currently an MA student at the Department of Medieval Archaeology at the University of Aarhus and holds an MSc in 'GIS and Archaeology' from the University of Newcastle upon Tyne. The MSc dissertation dealt with the construction of a straightforward GIS tool for managing the excavations of Odense; the MA dissertation in progress deals with deposit modelling and preservation in Odense.

References

BIDDLE, M. and HUDSON, D.M., 1973. The Future of London's Past: Survey of the Archaeological Implications of Planning and Development in the Nation's Capital. Rescue Publications no. 4, Worcester.

CARVER, M.O.H., 1983. Forty French Towns: An essay on archaeological site evaluation and historical aims. Oxford Journal of Archaeology 2(3), Oxford:339-378.

ENGLISH HERITAGE, 2001. Archaeological Review 1998-99, London.

HOPKINSON, G., 2002, unpubl. Archaeological Deposit Modelling: An investigation into the techniques and potential of archaeological deposit modelling, focusing on the Hungate area of York. MSc Dissertation University of York, York.

MILLER, A.P., 1995b. The York Archaeological Assessment: computer modelling of urban deposits in the city of York. In Wilcock, J. and Lockyear, K. (eds.), Computer Applications & Quantitative Methods in Archaeology 1993, Tempvs Reparatvm, Oxford:149-154.

MILLER, A.P., 1996. Digging Deep: GIS in the city. In Kammermans, H. and Fennema, K., (eds.), Interfacing the Past: Computer Applications and Quantitative Methods in Archaeology, Analecta Praehistorica Leidensia 28, Leiden: University of Leiden, pp. 369-376.

MILLER, A.P., 1997, unpubl. The York Archaeological Assessment: an investigation of techniques for urban deposit modelling utilising Geographic Information Systems. Doctoral thesis, Department of Archaeology, University of York.

OVE ARUP & Partners and York University, 1991. York Development & Archaeological Study, York.

RICHARDS, J.D., 1990. Terrain Modelling, Deposit Survival and Urban Archaeology. Science and Archaeology no. 32:32-38.

ZINGLERSEN, K.B., 2001, unpubl. Urban Archaeological GIS in Denmark and Odense ByGIS. MSc Dissertation, University of Newcastle, Newcastle upon Tyne

Some Potentials of Digital Literacy in Archaeology of Sri Lanka

Sameera Liyanage
Department of History, University of Ruhuna, Sri Lanka.
Email. sameera_liyanage@hotmail.com

Abstract

The use of computer in the field of archaeology in Sri Lanka has been introduced in the last decade. Archaeologists use computers for data acquisition, analysis and presentation at basic level. The Department of Archaeological Survey has taken steps towards establishing a National Data Base of Archaeological Sites (SMR) for Sri Lanka since 1998 and has planned to link SMR to a Geographical Information System (GIS). Two universities have introduced archaeological computing courses. The Postgraduate Institute of Archaeology is carrying out research work with use of GIS.
Some general constrains of archaeological computing in Sri Lanka are non-use of expert advice and Incompetence of researchers and lecturers who are interested. The undergraduate courses are limited to giving practice in word processing, data analysis and drawing programs at basic level. The improvement in this field is affected due to financial difficulties too.
A scheme for collaboration of available expertise in Sri Lanka with foreign colleagues is required to improve of research work and teaching affects. It should ultimately bring together archaeologists, computer specialists and geologists, which would certainly enhance efficiency and productivity of cultural heritage management of Sri Lanka which has a growing global scale demand.

Introduction

Use of computer technology is a new trend in the field of archaeology in Sri Lanka. The early attempts of such can be traced to mid 1990's. In a broader perspective, the emergence of the use of computer technology in the field of archaeology in the country can be considered as a technological implantation, rather than a historically derived phenomenon.

This paper seeks to review the existing nature of the computer usage of the subject. It also attempts to identify the potentials of digital literacy in Sri Lankan archaeology and so outline the future prospects in the field.

Computers in Archaeology: A Retrospect

The pioneer of computer application in the field in Sri Lanka is the Postgraduate Institute of Archaeology (PGIAR) of the University of Kelaniya, The early attempts were confined only to a kind of architectural reconstruction of a ruined building in the royal complex belonging to 5th century A. D. at Sigiriya in the Central province. AutoCAD software has been used and the results of the simulation were incorporated into the interpretation of the excavation results of the particular building.

Several academic departments in the universities have also contributed their skills and energy to the development of the field. Most of them introduced computer-oriented syllabuses into their degree programs. Such syllabuses help to trace the basic elements of the relationship between computer technology and archaeology. The analytical procedures of the subject are not well addressed by those elementary teachings. The Departments such as the Department of Archaeology at the University of Kelaniya, the Department of History at the University of Ruhuna, and the Department of Archaeology at the University of Peradeniya are conducting such teaching programs.

Several government institutions such as the Department of Archaeology and the Cultural Triangle that come under the Ministry of Cultural affairs are also using computer technology in the field of cultural heritage management. A major project initiated by the Department of Archaeology is to integrate the data available for the archaeological sites and monuments of the country. Spatial and thematic data of each monument have been encoded using GIS applications. This project is moving slowly due to both lack of proper personnel and lack of adequate technological backing.

Recent archeological research on ancient settlements conducted by the Postgraduate Institute of Archaeology under the technological guidance of the Uppsala University in Sweden proved the importance and value of the computer application in the field of archaeology. A limited area about 230 sqkm was surveyed to uncover the distribution pattern of the ancient settlements in the lower Kirindi Oya basin in the Southern province of Sri Lanka. The spatial data of that survey has been processed in GIS. The spatial diversity of the ancient settlement distribution in several time episodes has been simulated. The methodology followed in undertaking this computer simulation is the amalgamation of the spatial information and the archaeological data. The reconstruction of the ancient city including its suburbs, very clearly shows, the potentials of artificial intelligence that archaeology bears in such area surveys.

The Potentials

The growing research in the Sri Lankan archaeology is now demanding the use of computer technology. The rapid development of the information technology, the efficiency and sophistication made by the use of artificial intelligence, creates a new demand among the new generation of archaeologists in the country. However, the viable sub-fields of archaeology of Sri Lanka that has a vast potential of computer application can be classified into the following major groups.

1.Field Surveys: The landscape approach of modern archaeology emphasizes a high priority for using compute programs in the landscape analysis. As a developing country, this is urgent as mass-scale accelerated development programs are taking place. For example, during the construction work of the Mahaweli river diversion project, considerable extent of lands, with great archaeological value, were destroyed. No records were left behind. The unawareness of the methods of artificial intelligence to be utilized to record such widespread spatial entities, and lack of proper personnel for such a task is explicit in the circumstances alike.

2. Excavations: each year at least one excavation is conducted somewhere in Sri Lanka. So those excavations produce enormous amount of data pertaining to human existence in different periods of the past. Computer usage provides a wide opportunity to organize such data in an accessible way. For example, the excavations of the Jethavana stupa site during the last decade have uncovered nearly 600,000 ancient beads

The excavation section of the Department of Archaeological Survey has already developed an archaeological data management system: a suitable catalogue and a storage system for archaeological objects recovered from scientific excavation. It is 'Archaeological Data Visualizer': a special computer software program written and encoded with MS Access and MS Visual Basic program interfaces. However it is limited to store data collected from previous excavations. Developing a system to store data at any excavation process has become a necessity.

3. Record of monuments: in digital format by using AutoCAD, GIS and VRML. For instance, in publications and presentations, virtual reality and simulation could be used to acquire maximum results. To reconstruct past for public presentation projects under the cultural triangle could be presented as virtual tours through internet and at the sites.

4. Database management: Creation of databases for the use of researchers and pubic has become an essential need. A national database for inscriptions is an example.

The library of the Postgraduate Institute of Archaeology has already catalogued all books using software ISIS and catalogue searching is available to anyone.

The National Museum of Colombo is also establishing a database named 'Project Object ID' to record and manipulate the information regarding the artifacts at the museum guided by the Royal Tropical Institute of Amsterdam, Netherlands.

Constraints

On one hand one should appreciate the researchers and lecturers who work under minimum facilities and trying circumstances to develop archaeological computing in Sri Lanka. They hardly have previous experience or training. On the other hand in some institutions facilities are not being used properly because of the lack of enthusiasm towards archaeology.

In this situation, the main constraints to archaeological computing in Sri Lanka can be summarized as follows:

1. Inadequate academic response due to inexperience and under-training of the researchers and lecturers in archaeological computing.

2. Lack of proper communication between administrators and those in the archaeology field in respect of policy making and implementation.

3. Lack of teaching materials including books and essential software and hardware.

4. The lethargic and disinterested attitudes of the professionals in archaeology and the computer specialists towards a common goal.

5. Non-availability of vast international knowledge and experience in computer applications in archaeology in Sri Lanka.

Probable Future Prospects

It is vital as responsible institutes that the above-mentioned contributors in archaeological work, must rethink, plan and organize their activities to get maximum use of computer to the field of archaeology.

An impressive national or international conference and a series of workshops in archaeological computing should organize to make aware the applicable personalities and institutions about the importance of the previous and present research work that make use of digital literacy in Sri Lanka.

At least selected suitable personalities from each institute should be highly trained in archaeological computing.

For undergraduate courses at universities proper syllabuses and teaching materials should be prepared. Adequate practical knowledge should also be provided to upgrade the archaeological computing courses.

It should ultimately bring together archaeologists, computer specialists, geologists and mathematicians to enhance the efficiency and productivity of the cultural heritage management of Sri Lanka.

ACKNOWLEDGEMENTS

I express my kindest thanks to Raj Somadeva, Postgraduate Institute of Archaeology attached to the University of Kelaniya for his input to this paper in the form of good advice. I would also like to thank to Prof. Sudarshan Seneviratne,the University of Peradeniya, Anura Manathunga and Mangala Katugampala, the University of Kelaniya, Gamini Kulathunga, the National Museum of Colombo and all the officers at the Department of Archaeology, Sri Lanka who gave their kind assistance to collect data. Any mistakes and omissions remain my own.

REFERENCES

DAWSON, T., 1998. A Simple Introduction to use of Computers for Archaeologists. Welipila Vol. II., Central Cultural Fund, Colombo.

KATUGAMPOLA, M. and RESTA, F., 2000. Computer applications in Archaeology: a review. A paper presented at the Conference held in the Department of Archaeology, September 2000, Sinhala.

PEREA, NIMAL, DELDENIYA and RUCHIRA, 2000. An Introduction to the Archaeological Data (Movable) Recording and Storage. Gaveshana I, Department of Archaeological Survey of Sri Lanka, Colombo.

SAMARATHUNGA, S., 2000. Geographical Information Systems and Other Computer Applications in Archaeological Exploration: Case Study of Pinwewa-Galsohonkanatta Archaeology Program 1997. Gaveshana I. Department of Archaeological Survey of Sri Lanka, Colombo.

SOMADEVA, Raj., 2002. Computer Applications in the archaeological research in Sri Lanka. A paper presented at a conference held in the Faculty of Economics, University of Uppsala, 21 November 2002.

A Geo-Archaeological Model of Holocene Landscape Development and its Implications for the Preservation of Archaeological Sites

Benjamin Ducke
Projekt Archäoprognose
Archaeological Heritage Management Brandenburg,
Wünsdorf 15838, Germany
benducke@compuserve.de

See the CD for the extended version

Abstract

This paper presents a geo-archaeological approach to estimating the potential state of preservation of archaeological sites in the Northern European Lowlands where the Holocene shaping of the landscape has been dominated by processes of soil erosion and deposition. Over time, as determined by topography and soil properties, protective soil layers are either removed (erosion), thereby exposing sites to an increased risk of being destroyed or built up (deposition), preserving the sites but making them hard to find. With the help of a geo-physical model and the integration of archaeological data, it is possible to locate and assess the potential locations of eroded and covered sites.

Assessing the Invisible

It has been understood for some time, that archaeological research on a landscape scale must pay close attention to the intimate relationships between cultural sites and landscape development (e.g. Davidson 1985, Holliday 1992). Besides conserving visible monuments of our past, we must also strive to preserve the true, unseen extent of our cultural heritage as it lies buried in the ground, hidden from the senses of the archaeological prospector[1].

Primarily, it is the strength of the geomorphological processes of erosion and deposition that determine the state of conservation of the embedded sites. If the protective soil cover has been reduced by erosion, the archaeological site's remains will be exposed to human activities (ploughing, construction work), and gradually destroyed. On the other hand, sites that have spent the last millennia under a thick, protective cover of deposited soil, can be expected to produce well-preserved features. Knowing where to expect the latter provides important hints for the allocation of resources. The planning and conduction of rescue excavations in particular could benefit immensely from the additional information offered by a robust model of erosion and deposition. These considerations led to the design of an erosion model for Holocene landscape development in the Northern European Lowlands. The geographical and archaeological data to support the research presented in this paper is taken from a study area west of the German capital city Berlin in North-eastern Germany ("Havelland" in Fig.1). With its extent of about 3 x 17.7 km, the region can be considered large enough to justify a landscape scale approach on a detail level of 10 x 10 m ground resolution.

Soil Erosion and Landscape Development

As far as the Northern European Lowlands are concerned, the most recent large scale shaping of the landscape by natural geological and geomorphological processes can be dated back to the final glacial phase of the last Ice Age ca. 12,000 years ago. The geological epoch that followed, the Holocene, has so far been one of stable geological conditions. The fact that archaeological stratigraphies in the area under discussion usually display a complex, heterogeneous depositional pattern can mostly be attributed to human impact. The European Plains as we perceive them today are much different in appearance from what they were like when the first farmers settled there in the 6th millennium BC. In fact, what we see is only a flattened, smoothed reminiscence of the natural Holocene landscape (Fig.2). Claiming one patch of land after another for millennia in succession, the farmers of the Northern European Lowlands have been stripping the landscape of its natural cover, hacking, burning down and up-rooting the vegetation and replacing dense forests with open, unprotected agricultu-

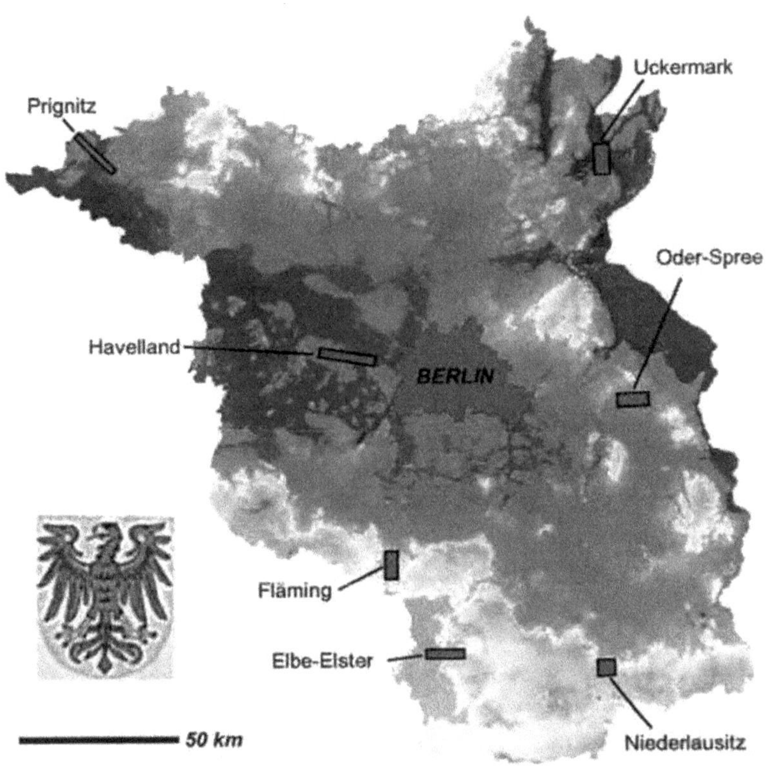

Figure 1 The state of Brandenburg in North-eastern Germany. The study area discussed in this paper is designated "Havelland"

ral fields. With the natural vegetation gone, there is nothing that could prevent the fertile, volatile cover of soil from being washed off and transported down the slopes by intense rain and from being blown away by the wind, whenever fields lie bare of crop. This process, called soil erosion, results in severe soil degradation and a flattening of the landscape. The archaeological literature abounds with stratigraphical and geomorphological observations that indicate events of disastrous erosion by water and wind. Some recent publications include examples from Central Europe as well as virtually all other parts of the world (e.g. Sandor 1992, Verhagen 1996, Zeidler and Isaacson 2001). However, precise quantifications of the soil volumes involved are hard to find. Fortunately, some recent publications with geo-archaeological study cases from North-eastern Germany (Bork et al. 1998) have considerably alleviated the stand-still situation. These recent studies have shown:

1. In the area under study, soil erosion has been and still is the dominant geomorphological process.

2. Soil erosion started as soon as agriculture was introduced in the area under study.

3. The intensity of soil erosion processes has seen several dramatic peaks; provably in the iron age (ca. 8th century AD to 4th century BC) and in the 14th century BC.

4. In the long term, soil erosion leads to a flattening of the landscape that makes geomorphological features harder to assess.

Figure 2 A typical landscape in North-eastern Germany

Figure 3 Stratigraphic cross-section of Dyrotz 37 (courtesy of L.A.N.D. ltd.)"

Modelling Soil Erosion

Although having been smoothed by soil erosion, the landscape's relief still retains its ancient topographical properties. It is possible, therefore, to model the movement of soil masses through the landscape in the past based on presently available data and from such a model learn about the locations of soil depositions that might still cover largely undisturbed archaeological sites in the landscape today. The last decade has seen the emergence of refined erosion models, many of which go far beyond the mere practical needs of calculating erosion risk. Process-based models such as the Water Erosion Prediction Project (WEPP, see WEPP 2001) or the European Soil Erosion Model (Eurosem, Morgan et al. 1998) allow for detailed studies of the processes involved in erosion to gain a better understanding of landscape development. While in theory being excellent research tools, the practical drawback of these models is that they require lots of difficult-to-obtain, precise quantifications for their numerous parameters.

An optimal model for archaeological erosion risk assessment would therefore combine easily accessible and robust parameters with the predictive capabilities of a more advanced model. Fortunately, just such a model exists. USPED (Universal Stream Power Erosion Deposition) calculates rates of erosion and deposition according to a simple geophysical model (based on original concepts by Moore and Burch 1986). It assumes a steady state overland water flow with uniform rainfall excess conditions. Steady state water flow can be expressed as a function of upslope contributing area per unit contour width which in turn can easily be calculated from a DEM. The net erosion/deposition rate is then estimated as divergence of sediment flow (Mitasova and Mitas 1999). Instructions for several popular GIS platforms and an in-depth explanation of the mathematical methods have been compiled by Mitasova and Mitas (1999). Armed with a high quality DEM and the USPED model, one only needs to supply a few parameters. All of them are stable enough in time to allow incorporation into a model for the entire Holocene and can be represented adequately on a landscape scale level.

1. Rainfall intensity can be approximated from histograms of local rainfall capacity.

2. Vegetation cover is a very complex parameter but can be roughly estimated using standard relations for vegetation classes (see Mitasova et al. 2001).

3. The erodibility of soil depends on many physical properties, such as particle size and cohesion, proportion of organic material, aggregate sizes and permeability. Fortunately, these micro-properties of soil can often be mapped to the most common classes of soil types (such as sandy, loamy etc.) via standard estimates.

4. A topographical parameter is derived from the DEM of the study area.

An Archaeological Study Case

The Havelland is a typical landscape in North-eastern Germany which is located west to the urban area of Berlin (Fig.1). Until modern times, the landscape's appearance was characterized by a dense network of streams that made the region a much-frequented passage way and established its status as an important centre of prehistoric settlement activity from the Neolithic to the Iron Age. The reorganisation of agriculture on an industrialised level after WWII brought the ruin of the historical landscape and subsequently gave way to the modern, minimised topography which is composed of broad and gentle slopes that are predominantly used for growing crops (Fig.2). As a case study for the local impact of the global erosion/deposition pattern, we will take a look at the prehistoric settlement site of Dyrotz 37. Artefacts indicate that several prehistoric settlements have been erected on the site over a period of approximately 4500 years: 1. Middle Neolithic (ca. 4600-4300 BC), 2. Late Neolithic (ca. 3500-3100 BC), 3. Late Bronze Age (ca. 1300-1100 BC), 4. Iron Age (ca. 550 BC-0 AD), 5. Germanic Age (ca. 0 AD-500 AD). Figure 3 shows the stratigraphic situation in those parts of the site that were covered under soil deposits: several colluvial layers can be distinguished by their darker shades. Undisturbed, natural soil layers of lighter shade in-between them are evidence of prolonged periods without settlement activity on the site and its immediate neighbourhood. The features in the western part of the excavated area lay buried under more than 2 metres of colluvial depositions and were accordingly well preserved. Roughly two thirds of the excavated area is situated up-hill where erosion has been pre-dominant and has essentially erased all of the prehistoric layers. The bottom parts of settlement pits were essentially the only surviving archaeological features in this area. In ignorance of the geomorphological situation, the site's importance was well underestimated and the resources allocated to its excavation were insufficient leading to a great loss of valuable archaeological information. It should be noted that the erosion/deposition model depicted in Figure 4 has been colour-coded for easy recognition of erosion/deposition patterns. In the natural landscape, the "steepness" of the slope on which the site is located is around 2 degrees and hardly recognizable at all. Even this small gradient, however, does make a decisive difference for the site's stratigraphic structure and the level of preservation of its archaeological features. The USPED-based model accurately predicts the depositional pattern of the site. In the case of Dyrotz 37, the individual depositional layers are up to 1.5 meters in height. The chronological determination of each erosion/deposition event and the accurate quantification of its strength requires geo-archaeological data of a quality that the entire study area presently lacks. For future work on the subject, many more site stratigraphies will have to be examined before a statistically valid relationship between USPED-calculated values and actual soil volumes can be established.

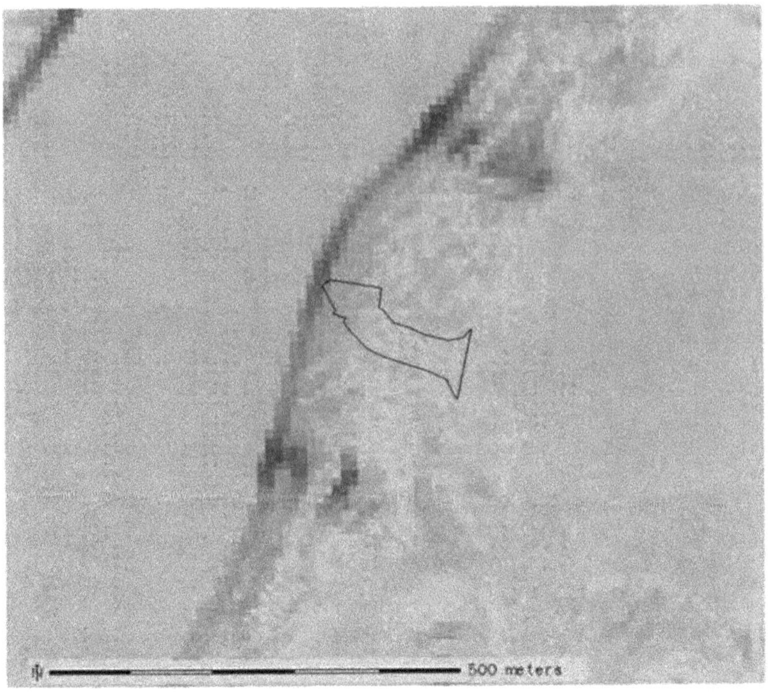

Figure 4 USPED-based erosion/deposition model of the archaeological excavation site "Dyrotz 37" (lower right, corner, red outline) and its surroundings. Orange to yellow colours (light) indicate erosion; greenish to blue colours (dark) indicate deposition

Results

Erosion and deposition are factors of primary importance when trying to predict the potential locations of well-preserved archaeological sites in the landscape. The method described in this document yields valid results only for a restricted class of landscape types where erosion and deposition caused by run-off are the dominant processes. The model can thus be applied to vast areas of the Central and Northern European Lowlands from the Netherlands to Poland. In other types of landscapes, Aeolian depositions, stream meandering, littoral processes etc. may be of importance.

[1] The work presented in this paper is partly derived from the author's M.A. thesis (Ducke 2002, in German language, yet unpublished) and is also part of "Projekt Archäoprognose Brandenburg" , a research project aiming to provide a complete predictive map for archaeological sites in the state of Brandenburg, North-eastern Germany (Müller and Kunow 2002, http://www.uni-bamberg.de/~ba5vf99/index.html.

REFERENCES

BORK, H.-R., BORK, H., DALCHOW, C., FAUST, B., PIORR, H.-P. and SCHATZ, T., 1998. Landschaftsentwicklung in Mitteleuropa - Wirkungen des Menschen auf Landschaften. Gotha, Stuttgart.

DAVIDSON, D.A., 1985. Geomorphology and Archaeology. In Rapp, G., and Gifford, J.A. (eds.), Archaeological Geology, New Haven/London:25-55.

HOLLIDAY, V.T., 1992. Soil Formation, Time and Archaeology. In Holliday, V.T. (ed.), Soils in Archaeology - Landscape Evolution and Human Occupation, Washington/London:101-188.

MITASOVA, H., MITAS, L., 1999. Erosion/deposition modeling with USPED using GIS. (http://skagit.meas.ncsu.edu/helena/gmslab).

MITASOVA, H., BROWN, W.M., HOHMANN, M. and WARREN, S., 2001. Using Soil Erosion Modeling for Improved Conservation Planning: A GIS-based Tutorial (GMSLab and USA CERL). (http://skagit.meas.ncsu.edu/~helena/gmslab/reports/CerlErosionTutorial/denix/denixstart.html).

MORGAN, P.C., QUINTON, J.N., SMITH, R.E., GOVERS, G., POESEN, J.W.A., AUERSWALD, K., CHISCI, G., TORRI, D., STYCZEN, M.E. and FOLLY, A.J.V., 1998. The European Soil Erosion Model (EUROSEM): documentation and user guide Version 3.6. Silsoe College, Cranfield University, Silsoe, Bedford.

MOORE, I.D. and BURCH, G.J., 1986. Physical basis of the length-slope factor in the universal soil loss equation, Soil. Sci. Soc. Am. J. 50:1294-1298.

SANDOR, J.A., 1992. Long-term Effects of Prehistoric Agriculture on Soils: Examples from New Mexico and Peru. In Holliday, V.T. (ed.), Soils in Archaeology - Landscape Evolution and Human Occupation, Washington/London:217-246.

VERHAGEN, P., 1996. The use of GIS for modelling ecological change and human occupation in the Middle Aguas Valley (S.E. Spain). In Kamermans, H. and Fennema, K. (eds.), Interfacing the Past. Computer Applications and Quantitative Methods in Archaeology CAA95 Vol. II (Analecta Praehistorica Leidensia 28), Leiden:317-324.

WEPP, 2001. Water Erosion Prediction Project (http://topsoil.nserl.purdue.edu/nserlweb/weppmain/wepp.html).

Information Management for the Conservation of Archaeological Sites - Suggestions for a Site in Western Anatolia[1]

B. Nilgun Oz
KA.BA Conservation of Historic Buildings and Architecture Ltd

Nimet Ozgonul
East Technical University, Department of Restoration

Abstract

In the world of today, regarding the increasing number of threats to archaeological sites and the complexity of problems that are faced, management of archaeological sites has become one way to ensure their conservation. An important aspect of managing an archaeological site is to evaluate the information on the site and to suggest methods for the proper management of information. Especially in archaeological sites where excavations are continuing, new information is reached continuously and is added to the already existing data, which can be in forms such as plan drawings, inventories, photographs, etc. The decisions made during the management planning of a site and those that will be made in the following years depend on the available information on the site; thus, employing wise, systematic means and a holistic approach in collecting, classifying and disseminating all the information on the site is critical. The importance of information management increases in providing the optimum circumstances for understanding the site. In this respect, utilization of computers and other digital media are all the more favoured. Within this framework, current circumstances in an archaeological site in Turkey can be discussed to evaluate conditions for the sustainability of the site.

Introduction

The complex circumstances in which the works for the conservation of archaeological sites have to be accomplished, necessitate methods that allow the responsible people / institutions to act on as much information as possible about the site. The need to make decisions based on detailed information, so as to avoid making mistakes, requires employing management systems both for the overall conservation of the site and also the storage, access, dissemination and review of information about the archaeological site in question. In this context, management of an archaeological site means "...to approach a defined objective, or objectives, for that site in a disciplined way which, with given constraints and resources, will most easily allow that objective to be realised" (Evans 1986:10). In short, management is about planning the future of an archaeological site with conservation aims, defining guidelines for matters such as visitor management, consolidation and restoration of ruins, interpretation and presentation and maintenance

To understand a site, a range of information is necessary, from archaeological and architectural data to social aspects, legal administration, land ownership of the site, etc. since the site is considered as an integrated whole with its surroundings. At present, the use of digital technology in the studies on archaeological sites lessens the amount of time obtaining information but also increases the amount of data to be stored, re-accessed and updated. On the other hand, multi-disciplinary studies have increased in the scientific works carried out at archaeological sites, and as a consequence, many kinds of information exist for understanding an archaeological site, other than those of only archaeological interest. Consequently, systems where the data is recorded, classified, accessed, disseminated and updated are important. Applying appropriate digital technology can aid decision-making for conservation in the management of archaeological sites, through cross-relating the stored and updated data.

In this respect, the following study contemplates on the necessity of an information management system for an archaeological site in the Aegean Region of Turkey, in order to develop a holistic perspective in the decision making process for the sustainable conservation of the site.[2]

Archaeological Site Conservation in Turkey

Turkey, having a total of 4920 listed archaeological sites and archaeological research projects (incl. scientific excavations and surface expeditions), is showing an increased interest in the conservation of its archaeological sites. The number of archaeological excavations, which are conducted by either Turkish or international teams or the local museums, increased from 71 to 282 in the last 20 years. Similarly, developments in the field of conservation are observed: the sites themselves, as a whole, gained importance since the 1990s, upon which conservation and development plans for

Management	Management of Archaeological Sites
- Understanding resource through analyses - Evaluating data to find means of achieving the aims - Foreseeing changes in the resource and its sorroundings - Reconciling conflicts	- Holistic and Integrative Approach - Determination of values - Planned decisions opposed to crisis decisions - Gathering of related interest groups - Peridov reviews of the management plan and its decisions based on updated information

Figure 1 - Management for the conservation of archaeological sites can be useful in providing the above

Figure 2 - Location of Magnesia ad Maeandrum (National Geographic Society, Washington DC, September 2002)

archaeological sites, among them Hierapolis/Pamukkale and Perge, were prepared. Previously, the activities of conservation in archaeological sites in Turkey were mainly oriented towards consolidation, repair and restoration of single monuments. Lately, management as a system for enabling comprehensive site conservation is being debated and a site management plan for Hierapolis/Pamukkale has been prepared.

However, there are many factors common for all archaeological sites threatening the future of archaeological sites in Turkey. Especially worth mentioning are the lack of integration of conservation works with development policies and the disregard of the importance of cooperation of key interest groups of the archaeological sites. A step forward could be to collect all information about an archaeological site, and to form a medium to share it with the determined key interest groups and most importantly the responsible institutions, so as to make decisions for the sustainable conservation of the site. At a smaller scale, the information collected through documentation and research and other means, should be gathered in a database system, enabling evaluation by considering a variety of information, resulting in the definition of priority actions and long and short term plans for the conservation of the site.

Suggestions for the Use of Information Management in Magnesia

The ancient city of Magnesia is located in the Germencik district of Aydin Province, on the western part of the Great Menderes (Meander) Plain near Gumusdag (Mount Thorax) (Fig.2). It is situated in a rural area surrounded by villages but is close to urban centres such as Kusadasi, Aydin and Izmir. The city was an important religious centre from the archaic period until it was abandoned in the 1300s. The most important historical information on Magnesia is the epiphany -a divine manifestation- of the Goddess Artemis that took place in the city and the Isitheria festivals that started to be held in honour of this event. Magnesia attracted many travellers before the 20th century, especially because its Temple of Artemis was mentioned in Vitruvius' 'Ten Books on Architecture', but the first excavations took place in the 1890s, carried out by C. Humann.

Magnesia is a registered archaeological site since 1990, with visible archaeological remains belonging to the Hellenistic, Roman, Byzantine and Principalities Periods. Major aboveground remains are the Theatron, Gymnasium and the Roman Baths/Barracks as well as the Altar and Temple of Artemis, the Market Basilica and the Propylon of the Agora, Mosque and Cemetery. However, some of the ruins, among them the Temple of Zeus, were transported to Europe prior to the 20th century and today are in the museums of Louvre in Paris and Pergamon in Berlin.

Currently, archaeological excavations are continuing under the supervision of Prof. O. Bingol from Ankara University, who has been working at Magnesia since 1984. The excavations are carried out on the expropriated areas; the larger section of the site is privately owned and used for agricultural facilities. The works accomplished during the scientific excavation consist of digs in many locations, environmental investigations, restoration of building remains, material conserva-

Strenghts	Weaknesses
- Registered archaeological site status provoding compulsory protection - Ongoing archaeological excavations - Representative of history of archaeological research in Turkey - Local support and interaction - Cultural and socio-economic values of the site	- Registered archaeological site boundaries not covering the whole ancient city - Inability to work in larger areas within the site - Less known compared with sorrounding sites such as Ephesus and Miletus - Use of predominantly manual methods in documentation and archive keeping - Lack of integrated information for evaluation - Inefficient holistic decisions for conservation - Lack of finacial sources
Opportunities	**Threats**
- New information reached continuously - Sponsors aided the scientific works - Situated in a higly attractive tourism area close to many famous archaeological sites - Well-informed visitors - Location in well-known region - Located near a frequently-used highway - Transition to digitised methods in architectural documentation and archive keeping	- Lack in legal provisions for the security of the excavated ruins - Lack og communication between legally responible organizations - Uncontrolled tourism development in the vincinity - Expansion of new construction areas towards the archaeological site boundaries - Natural hazards - Illicit excavations within the boundaries - Site separated by a highway

Figure 3 Analysis of current conditions in Magnesia (SWOT Analysis)

tion, architectural documentation and environmental organization. This leads to a variety of information about the site, which has to be evaluated to understand the site's past and to plan for its future. A detailed analysis of the current conditions (Fig.3) reveals that there are multi-faceted conditions calling for necessary experience of a multi-disciplinary effort. There is a need for an information management system that enables comparison and interaction of information and a mechanism to monitor the condition of the site. Considering the fact that Magnesia has several problems to be solved, including natural, legal and financial ones, there is a need to make comparative analysis of the information relating to these problems, to define strategies for conservation.
In this context, the main goal for the conservation of the site should be to find means for establishing a site management mechanism, working not only for Magnesia but also for other sites in the region. Similarly, the short-term objectives for managing Magnesia for its conservation should include aims about establishing a digital information management system for the scientific archaeological excavations that are being carried out.

- Archaeological Site Boundaries
- Cadastral plan
- Roads, Railway and Paths
- Topography
- Geological Characteristics
- Landownership
- Land use
- Flood Areas
- Area Protected with Fence
- Excavated Areas
- Architecturally Documented Areas
- Buildings(ruins, new blds. etc)
- Periods of Ruins
- Condition of Ruins
- Locations of Ruins Abroad
- Locations of Small Finds
- Favourite Tourist Routes
- Location of Information Panels

Figure 4 Possible layers of information for a GIS application in Magnesia

In this way, users of information about the archaeological site, including the excavation directorate, local and central government and the public, would be connected. The objectives can be:

- Establish a digitised database of all information on the site (incl. socio-economic conditions, legal framework) providing evaluation of data in relation to one another
- Establish a system enabling integration and interaction of information that gives way to making appropriate decisions for the conservation of the site
- Transform previously finalized architectural documents into a utilizable database
- Prepare action plans for archaeological excavation and architectural documentation
- Ensure periodic revision of action plans and update information for decision-making

Considering the current conditions at Magnesia, there are some priority decisions that need to be made, which at best should be decided on through utilizing methods such as GIS technology. These decisions are:

- Designation of zones showing physical expansion directions/areas and reserve areas of archaeological work
- Designation of areas that need to be expropriated for the continuity of the excavation
- Designation of ruins in need of immediate intervention

In the case of GIS applications in the archaeological excavations at Magnesia, a number of layers of information need to be studied (Fig.4). The correlation between these layers is essential in coming up with the optimum solutions to problems. As an example to decision-making, the designation of expansion zones depends on the cross-relation of most of the layers printed above, among them, a relation between land ownership status showing lots that can be expropriated in the short term, and the location of ruins in need of urgent intervention, can be mentioned.

CONCLUSION

The conditions at Magnesia and the methods to provide its sustainable conservation can be common for many archaeological sites in Turkey. This means that, there is a need for a holistic system in Turkey. At this point, it is essential to understand that, one of the key aspects (what should also be the priority aim) in archaeological site conservation in Turkey is the develop information systems on archaeological sites and to provide their integration into the national development systems. Within this framework, a selected part of the data in the information system, that is to be based on the site's scale and characteristics, which are at best created by a multi-disciplinary team working at the site, should be made public and accessible to a wide range of users, from academic researchers, to the local community. This could provide related persons and institutions, such as scientists working on the site and the local and central governments, with the data required and form the basis of an integrative approach in conservation. In the future, a transition from information systems of archaeological sites based on geographical regions to information systems based on various themes could be put forward. A sample further step could be to devise an information system that could present information about the Roman archaeological sites in the Aegean Region.

[1]The study presented here is a further examination of a part of an MSc Thesis prepared by B.N. Oz, to the Faculty of Architecture, Department of Restoration, Middle East Technical University, Ankara, Turkey, under the supervision of Inst. Nimet Ozgönül of METU and the excavation director of Magnesia, Prof. O. Bingol of Ankara University. The author (B.N. Oz) would like to thank Prof. O. Bingol for his permission to study at Magnesia and to use related documents for the preparation of the thesis.

[2]This study reflects primarily the views of restoration architects. It is acknowledged that most of the information and suggestions supplied here can be more comprehensive if studied with a multi-disciplinary group.

[3] Data taken from related reports of the Ministry of Culture for the year 2001.

References

BINGOL, O., 1998. Magnesia on the Meander. Ankara.

EVANS, D.M., 1986. The Management of Archaeological Sites. In Huges, W. and Rowley, O. (eds.), The Management and Presentation of Field Monuments, Oxford University Dep. for Exterior Studies, Oxford.

FEILDEN, B.M. and JOKILEHTO, J., 1993. Management Guidelines for World Cultural Heritage Sites. ICCROM UNESCO ICOMOS, Rome.

OZ, B.N., 2002, unpubl. Management of Archaeological Sites; Case Study Magnesia ad Maeandrum. MSc Thesis, Department of Restoration, Faculty of Architecture, Middle East Technical University, Ankara.

SULLIVAN, S., 1997. A Planning Model for the Management of Archaeological Sites. In Torre, M. (ed.), The Conservation of Archaeological Sites in the Mediterrenean Region, Getty Conservation Institute, Los Angeles.

RETHINKING THE CULTURAL HERITAGE DOMAIN: TOWARDS AN ECONOMY OF THE VIRTUAL

MAURIZIO FORTE
CNR-ITABC, INSTITUTE OF TECHNOLOGIES
APPLIED TO THE CULTURAL HERITAGE
00016 MONTEROTONDO ST., ROMA, ITALY
MAURIZIO.FORTE@MLIB.CNR.IT

OLEG MISSIKOFF
LUISS "GUIDO CARLI" UNIVERSITY,
CENTRO DI RICERCA SUI SISTEMI INFORMATIVI
00198 - ROMA, ITALY
OMISSIKOFF@LUISS.IT

ABSTRACT

This contribution represents a proposal to extend the CH research area towards disciplines able to provide a strong support in the creation of Value Added Services and so stimulate an interest in this domain from potential industrial and financial partners. The corollary is that CH (and, even more, VH) management should be considered as any other productive activity, and therefore deploy adequate organizational models and technological solutions for building (and managing) sound and well structured value chains. The "Libera Universita' degli Studi Sociali" (LUISS) in Rome and the CNR/ITABC (Institute of Technology Applied to Cultural Heritage) intend to propose, starting from a state-of-art analysis, the implementation of Business Process Re-engineering models and techniques in the CH/VH management. Projects clearly aimed at pointing out the ROI possibilities that could derive from this approach will surely provide better chances of raising funds for improving the quality of services, (and preservation) in sites, archives, museums, and exhibitions.

INTRODUCTION

The Cultural Heritage (CH) domain, in Italy as well as in Europe, has recently been interested by a deep rethinking of its intrinsic essence. The socio-economic context in which CH institutions were created has dramatically changed, and the pressure for undergoing a substantial restructuring is constantly increasing. Such restructuring, in order to produce results, should indeed affect most aspects of CH institutions: from the mission itself to inner organisational configuration, from human resources to customer relationship management, from technological solutions to marketing and promotion. This phenomenon is generated by two key factors: 1) the need to raise financial resources independently from public funds, and 2) the increasing request for high quality, value added, services from a more and more demanding cultural tourism.

As an answer to these issues, the present contribution aims at providing a theoretical reference framework for supporting the formulation of general standardised requirements on Cultural Heritage Information Systems (CHIS) development. A CHIS is intended as an integration of organisational models and technological solutions, customised for providing the various categories of CH institutions with a support for designing a sustainable evolutionary roadmap and a management model for facing the new challenges.

The final goal is represented by the transformation of cultural institutions in "state of art", networked organisations, here defined as Cultural Service Providers (CSP). The roadmap towards innovation will therefore proceed along three main streamlines: 1) Organisational models (designed for cultural institutions); 2) technological solutions (for supporting innovation and management activities); and 3) Knowledge management and dissemination (by means of a Domain Ontology).

VIRTUAL REALITY BETWEEN SCIENCE AND ART: EPISTEMOLOGY AND NEW "CONSUME" OF THE DIGITAL CULTURE

A very complex subject such as the case of the virtual reality, suggests several considerations from both a technological and epistemological perspective. The first depends strongly on the second which represents a fundamental base of discussion for interdisciplinary projects and for a correct interpretation of the aspects related to cybernetics and of all the applications in this field. According to Howard Rheingold (1991) we can imagine virtual reality like a magic window opened to other worlds, from the molecules to the mind. According to Wooley (1992) virtual reality is the technology used for creating a specific interface between human beings and digital images. Pierre Levy (1994) considers virtuality and actuality as two different ways of being: the Virtual is not the opposite of the Real but of the Actual. The Real is diluted in the Virtual and represents its conceptual thought, visual and abstract, but anyway communicative. The artificial world cannot simulate the Real, but it can increase it, decrease it, code it, interpret it. It is a common opinion that the search of photorealism is a primary goal of the virtual reality, but the Virtual does not aim at substituting the Real, but at changing its representation. The digital photoreality does not necessarily increase the perception, but simply fits the description and illusion of the reality to the artificial worlds. In the digital photoreality while an illusion of modelling the Real is created, at the same time a new context is created: the new Real.

Therefore, virtual reality can suggest new grammars of communication and inter-connection, directed to the exchange of information, events and behaviours, apart from principles of likelihood to the "true" or to the "presumable". The rules of meta-representation of the Virtual also concern the artistic thought: with regard to this, Arnheim writes that: "If we consider famous pictures of the past - Raphael's, Rembrandt's [...] - without a deep attention - they would have seen expressly precise copies of natures, landscapes, insides, still life, human beings. Was it probable that they knew presence of totally abstract patterns and, as images, they meant something completely different"? (Arnheim 1972).

The same principle ruling construction and perception of abstract patterns, of sensible informative geometry, much different from the Real, can be found in the virtual reality applications and they represent a visual thought. Virtual Reality is itself symbolic, because it communicates through abstractions: "An image acts like a symbol when it figures things at a higher level of abstraction than the symbol itself is. According to the cybernetics of G. Bateson (1972), the learning is through difference, the perception acts only on the difference. Receiving, grabbing information means always and necessarily to receive news of difference, and perception of the difference is always limited from a threshold. Differences too slight or presented too slowly are not perceptible: they do not feed the perception. Science doesn't prove, it tries..." (Bateson 1972). According to the thought of Korzybski "the map is not the land", in the virtual reality the "map is the land". In the Virtual the "difference" represents the cognitive value of the artificial environment, of the logic of the learning; in the dynamic of events and interactive environments, a difference is created, it is the preamble of any cognitive acquisition. This is the "cybernetic transformation", the principal scientific aim of the virtual reality.

On the basis of these considerations and on the complexity of the concept of "Virtual" it is important to understand the epistemological content of the VR systems and applications in a revolutionary and key scenario of the digital world. Where is the information? In which way VR and different epistemological contents/representations can modify and suggest new consumes and learning of the cultural heritage (scientific, didactic, tourist, etc.)?

The conversion of a real world resource in digital information, augmented reality, interactive and perceptive act or process, increases the value of the resource as a purely physical object, disseminate it, spreads out its content and message. In this increase of value it is possible to identify the true economics of the virtual, of which we are maybe not yet fully aware but, nevertheless, we should evaluate its impact both in the society and the community: the capital resides in the information and communication, be it in the physical, direct fruition or in the virtual. The Virtual, because it puts together, multiplies and re-assembles new contexts, can add new cognitive and communication geometries thus opening up different and broader learning processes. The problem consists in the fact that very few are yet the spaces or environments designed for an advanced fruition-interaction with respect to so called linear didactic tools (books, audio guides, catalogues, panels, etc.) and, generally, these few seem to be destined only to restricted scientific communities and not to the public in a broader sense. For this reason it is justified the infrastructural need of Virtual Heritage Centres.

The current Scenario

The first and, probably, hardest issue to be addressed is represented by a profound revision of the mission of CH institutions. In fact, traditionally, their task has always consisted in protecting the heritage from all possible dangers of being damaged, destroyed, stolen and so forth. At the time when interests towards antiquity arose, culture (just like sports, tourism and most of leisure activities) was enjoyed only by a limited elite. The socio-economic changes that took place from the 1960s, however, created the base for a mass access to CH resources. Not unlike what happened, for example, in tourism services, previous management models soon appeared unsuitable to deal with the requests posed by the new scenario. The CH domain however, mainly because it could always rely on public funds, has proven almost completely untouched from these new pressures until recently, when the international economic situation has made it necessary for governments to reduce the resources destined to cultural heritage.

Moreover, the recent diffusion of ICTs has created expectations in users, who find the average cultural institutions' communication generally old fashioned and sometimes even unintelligible. Meanwhile, the "digital revolution" is having a strong impact on CH domain and the number of activities concerning digitisation is rapidly increasing. In fact, the European Commission has fostered researches in this area by launching, on the 8th of December 1999, the eEurope initiative. A first offspring of the eEurope initiative has been the Lund meeting, organised on the 4th of April 2001 under the auspices of the Swedish Presidency for gathering representatives and experts from all European countries. The major outcome of the meeting was represented by the definition of the so called "Lund principles":

"Europe's cultural and scientific knowledge resources are a unique public asset forming the collective and evolving memory of our diverse societies and providing a knowledge basis for the development of our content industries in a sustainable knowledge society."[1]

The Lund meeting triggered a series of activities among which is important to quote the "Minerva" FP5 Network of Excellence[2] and the "DigiCULT" IST Support Measure.[3] Both of these resources provide a huge quantity of information concerning application of technologies, and more specifically digitisation activities[4] , to the cultural heritage domain.

Theoretical Background to the Research

The epistemological foundation of this contribution is derived from studies on cultural economics (Baumol and Bowen 1966), ontologies (Uschold and Gruninger 1996), and cybernetics.[5] In this paper an integration of these three fields of research, opportunely adapted to the characteristics of the CH domain, is proposed as a viable solution for the complex scenario outlined in the previous section.

[1] Cultural economics

The relationships between economics and culture have traditionally been of reciprocal suspect. In fact, as it always happens when two disciplines start interacting for addressing new problems, in a first phase scholars just try to export their paradigms onto the other discipline forcing old models in totally unusual scenarios. Most of the times this approach does not work and it is necessary to wait until the second phase, when some scholars complete the "hybridation" process and more organic scientific outputs begin to appear. The first such output can be recognised in the seminal work by William Baumol and William Bowen (1966) that, after its publication in 1966, stimulated an increasing flow of papers and books on the subject. This work, however, was oriented more towards the performing arts, which present different characteristics compared to the cultural heritage. This sort of subtle distinctions represent the typical pitfall for economists trying to perform a straightforward application of standard economic models to a complex field such as the culture in general and the heritage in particular.

[2] Ontologies

The first step to be taken, in order to produce a usable and interoperable output, shall be represented by the construction of a Domain Ontology. Ontologies are defined as "a shared understanding of some domain of interest which may be used as a unifying framework" for "facilitating knowledge sharing and interoperability between independently developed subsystems" (Uschold and Gruninger 1996). In the CH domain, a valuable and extensive contribution has been provided by the International Committee for Documentation of the International Council of Museums (ICOM-CIDOC), and is represented by the CIDOC Conceptual Reference Model (CRM). This model provides definitions and a formal structure for describing the implicit and explicit concepts and relationships used in cultural heritage documentation:

"The CIDOC CRM is intended to provide a common language for domain experts and implementers to formulate requirements for cultural heritage information systems and to serve as a guide for good practice of conceptual modelling. In this way, it can provide the "semantic glue" needed to mediate between different sources of cultural heritage information, such as that published by museums, libraries and archives".[6]

[3] Cybernetics

The fruition of culture has a strong "addictive" nature. In fact, according to Throsby[7] , the approach to culture is incremental and current fruition, if integrated with deeper knowledge, increases the wish to replicate it in the future. However, for activating this feedback process, it is necessary to accurately design clear and well segmented access paths to heritage and the integrative knowledge. Such strategy implies a sustainable entry level and a series of incremental loops aimed at stimulating the user's desire to go deeper and, at the same time, make him or her feel comfortable with the level of knowledge required by each phase of the cultural experience. The model underlying this strategy is derived from the cybernetic approach described in Gregory Bateson's "Steps to an Ecology of Mind" (Bateson 1972).

A viable Evolutionary Model

The idea underlying the model proposed in this contribution has emerged during the course of two projects: 1) the FP5 "E-Culture Net" project[8] and 2) "TransICT", a research project financed by the Regione Lazio for guiding small and medium enterprises in the transition towards the Net Economy.[9] The proposed roadmap foresees the constitution of enabling structures, the Virtual Heritage Centres[10], for providing cultural institutions with the all the support needed in order to safely pursue the mentioned innovation process. This support should consider all "sensitive" areas, from training the staff to helping managers in achieving the "vision" required for reaching the target in terms of business models, organisational configuration and technological solutions. From a practical point of view, results are expected in the value production in terms of both knowledge dissemination and financial resources raising. The target scenario will activate a virtuous circle: increasing the quality of services would attract more visitors and, consequently, an interest from investors; on the other hand, more resources at disposal would provide chances for further improving knowledge dissemination capabilities, quality of services (on and off line), and the range of products available for customers.

Operational Steps

The re-engineering activity starts from an ontological analysis of the Cultural Heritage domain, providing a better accessibility to information, on a semantic base, for visitors/scholars and the basis for interoperability among the stakeholders composing the scenario. The following step consists in designing a set of basic business models for each of the various categories of memory institutions, to be further customised for meeting the needs of specific situations.

For the implementation of business models it is important to be aware of the substantial lack of accountability in Italian cultural institutions. In fact, according to Normanton, "to be accountable means (...) to give reasons and explanations of what one does" (Forte 2000). This condition is due to two key points which are far from being addressed: besides the already mentioned weakness in knowledge production, management and dissemination, it is also necessary to develop a "cultural" governance. Supervisory boards of cultural institutions should in fact empower their managers for allowing them, within the framework of the institution's mission and budgetary constraints, to freely make all the necessary strategic decisions (Zan 1999).

Besides operating at the micro level, it is also important to stimulate inter-organisational cooperation. The "Virtual Enterprise" model is proposed as a basis for building Digital Cultural Districts. Case studies are provided by the FP5 "Minerva" project[11] and a series of interviews with cultural institutions' managers.

[1] See http://www.cordis.lu/ist/ka3/digicult/lund_principles.htm
[2] See http://www.minervaeurope.org
[3] See http://www.digicult.info
[4] See also http://www.minervaeurope.org/publications/globalreport.htm
[5] See http://cidoc.ics.forth.gr
[6] See http://cidoc.ics.forth.gr
[7] See http://www.transict.org
[8] See http://www.mmi.unimaas.nl/eculturenet
[9] See http://www.transict.org
[10] See http://www.virtualheritage.net
[11] See http://www.minervaeurope.org

References

ARNHEIM, R., 1969. Visual Thinking. University of California Press.

BATESON, G., 1972. Steps to an Ecology of Mind. Chandler Press, San Francisco.

BAUMOL, W. and BOWEN, W., 1966. Performing Arts: The Economic Dilemma. The Twentieth Century Fund (EVL), New York.

FORTE, M., 2000. About virtual archaeology: disorders, cognitive interactions and virtuality. In Barcelo, J., Forte, M., and Sanders, D., (eds.), Virtual reality in archaeology, ArcheoPress, Oxford (BAR Int. Ser. S 843).

LEVY, P., 1994. L'intelligence collective. Pour une anthropologie du cyberspace. Coll. Science et société, La Découverte, Paris.

NORMANTON, E.L., 1966. The accountability and audit of governments - a comparative study. Manchester University Press, New York.

RHEINGOLD, H., 1991. Virtual Reality. Summit, New York.

THROSBY, D., 1994. The Production and Consumption of the Arts: A View of Cultural Economics. Journal of Economic Literature Vol. XXXII, no.1.

USCHOLD, M. and GRUNINGER, M., 1996. Ontologies: Principles, Methods and Applications. The Knowledge Engineering Review Vol.11, no. 2.

WOOLEY, B., 1992. Virtual Worlds. Blackwell, Oxford.

ZAN, L. (ed.), 1999. Conservazione e innovazione nei musei italiani. Etas, Milano.

REVELATION: PRACTICE, TECHNOLOGY, DISSEMINATION AND THE DESIGN OF A FIELD RECORDING SYSTEM

K. MAY
ENGLISH HERITAGE, ARCHAEOLOGY DEPARTMENT,
LONDON, UK

S. CROSS
ENGLISH HERITAGE, CENTRE FOR ARCHAEOLOGY,
PORTSMOUTH, UK

ABSTRACT

On site computer recording of archaeological information has been a reality for 30 years and yet it is still seen as an experimental or at least innovative approach. Why? While most archaeologists use computers in some aspects of life, field recording is still done largely on paper. What is the challenge in field recording that makes it resistant to change? The issues may be as much cultural - to do with the way in which we see 'the field' in relation to the rest of our work, our colleagues and our audiences - as they are technological. The Revelation project has undertaken a comprehensive review of information systems at the English Heritage Centre for Archaeology in the context of the broader profession. Our aim has been to understand how we use data throughout the life of an archaeological project so that we can design a field recording system that is used by the majority of our field teams. In the process of this assessment we have had the chance to reconsider how our working practice feeds our understanding and how it can be supported and improved by better designed systems.This paper presents the results of the assessment and plans for implementation.

INTRODUCTION

Revelation is a Centre for Archaeology (CfA) project to provide a coherent digital information system that will make the capture, analysis and dissemination of CfA research faster and more effective. The project team expects that this will include on-site digital recording based around spatial data as well as complete integration between site data and off-site data. We are not trying to build a single system that meets all the needs of archaeologists since the needs of archaeologists can be remarkably variable. The CfA need a system which:

- allows for flexible and powerful access to the broadest range of data from any investigation.
- facilitates integrated working between dispersed teams of specialists, all of whom have responsibilities to their sub - disciplines as well as to the team.
- builds a manageable data library for the work of the CfA, particularly enhancing the understanding of those sites we have statutory or commercial responsibility for.
- allows for broad scale comparative work examining detailed research questions on the condition of the archaeological resource for issues such as agricultural damage, erosion and climate change.
- speeds up extended work programmes where individual member's input varies and key people's time is limited.

BACKGROUND- THE NEED FOR REVELATION

The Revelation project has recently completed a review of CfA systems. It shows that the CfA and its predecessors over the last 30 years have developed at least 83 separate computer information systems. Development began in the mid seventies with the creation of one computerised system for fieldwork data (Delilah) and slightly later a computerised system for collections management (Labfile). These two systems provided the main computer records up until the early nineties. With the advent of more user-friendly software, there has been a proliferation of small systems developed in increasing numbers over the last ten years. The reasons for this vary according to the complexity of the systems. Most common is the growing expectation of users for more functionality from a system, which results in frustration with inevitably ageing software and hardware. Another factor is the growing level of user skills and familiarity. A further significant reason is the increasing availability of better technology that allows more people to consider developing their own mini-systems. Time and again when users were asked about the main strengths of their systems, the recurring reply came back that "it does what I need", but with far less regard to the possible needs of others.

Although archaeological research is a collaborative exercise it is conducted by a disparate group of individuals. Some of these people work quite closely together during or immediately after fieldwork, but over the many years typically pass between fieldwork and publication, people on the project team have many different needs and requirements. It appears that as software becomes more user-friendly, many of the users decide to develop systems to meet those needs as they go along. The challenge therefore that faces the CfA is not a simple case that we haven't got enough IS, but rather more complexly that the current systems are 'organic'. Each time a person runs across a new problem, they develop (or in rare cases ask someone to develop) a solution for that problem.

There are some advantages to this approach, which is why it has gone on for so long. The most immediate is that it provides a quick and usually workable solution for the individual - or in some cases project - concerned. If the primary concern is to complete work to a fixed timetable this can often become an over-riding factor. But there are longer-term disadvantages. Most of these systems don't speak to each other. As more of them develop the fragmentation of information increases. The individual systems will often replicate the data that is held in others and this may lead to a further duplication in analysis work. Worse still it serves to preserve, if not encourage, the sense of isolation that is the nemesis of collaborative interdisciplinary work.

The aim of the Revelation project is to identify and address the longer-term requirements for archaeological projects. But, as already suggested, getting such requirements from archaeologists can prove remarkably difficult. The approach that Revelation has adopted is to undertake a major assessment prior to the design phase.

As the first element in this assessment the Revelation team have looked at the existing systems described above, to establish what functions they serve and to learn the lessons from previous developments. Since drawing (as opposed to spatial recording) is at the heart of the current process, we needed to scope the difficulties attending it. On-site digital drawing trials were our second step in research. Thirdly, we held workshops with fieldwork project teams to establish what projects need from systems and will back this up with a comprehensive questionnaire. Finally, we are conducting a review of sectoral practice - both through a literature review and through talking with other people implementing systems. Initially the interviews and hence the results presented below, have all been within Britain. We now hope, through the presentation of papers at conferences such as CAA, to take in experience from other archaeological organisations that have developed digital systems for archaeological recording, analysis and dissemination & archiving in recent years.

Existing Practice

Looking at the British experience, including CfA, we see many systems that aim to have a digital record before finishing fieldwork, but generally they still rely on some, or all, of the recording being done on paper and transcribed during the excavation (Roskams 2001:275). While such systems are an advance on the delay involved in digitising during post-excavation work, they have drawbacks. Not only are there time costs to the continued double handling, but it also leaves the door open to transcription errors. More fundamentally, however, such an approach separates the process of analysis from the process of recording and overlooks the advantages of digital systems over paper for recording.

The principle advantages of paper recording are seen to be flexibility and transparency. The flexibility is largely required because, during excavation, we may not fully understand what we are doing and so we like to keep our options open. However, this flexibility of a paper record can become meaningless if the data is entered for analysis into a digital system that cannot cope with it. If the data is going into a digital system at some point, that system should be designed to take the whole data set, in the manner it is recorded. Also on closer examination the transparency of the paper record is often somewhat murkier. In reality the paper record, while having room for authorship, has no room for multi-vocal and dynamic representations of the recording process because everything gets put on one sheet and annotation, if it happens at all, may not be attributed. Often even changes made many months later are simply inserted into a sheet without comment.

In the face of these concerns, digital systems can make recording easier and better - which is why commercial companies in many domains use such systems for all sorts of complex work practice. Digital systems provide immediate validation and quality checking. They can also provide feedback to the recorder about the quality of the record they are creating and about how a record contributes to broader research questions and the overall interpretation of the site (Andrews et al. 2000). A digital system can act as crib sheet, checker, and reference manual. Waiting until after the record is created before using a digital system means you are always dealing with legacy data which may have quality problems and which may then need to be shoehorned into a system that is not designed to deal with it.

Identifying the Requirements - Revealing the Archaeological Processes

The word Revelation means "knowledge disclosed" or more straightforwardly, "revealing some fact". There is also a sense in the word of the intervention of some divine or supernatural power that reveals this knowledge or fact. This is the 'black box', which describes the view held by many archaeologists of how a computer delivers information to their desktop (Hugget 2000). The 'black box' might also be a metaphor for the processes required to transform the action of an excavators trowel on a piece of soil into the publication of a monograph about the archaeology of a site and the landscape within which it exists.
People working with archaeological Information Systems (Stewart 1997) often describe the archaeological process in a linear fashion. The following stages are commonly identified from a data driven perspective:

- data collection,
- data management
- data manipulation and analysis and finally
- dissemination (of both data and interpretation)

But this sequence rather quickly glosses over the core of the process and the most important element in the design of the system - data manipulation and analysis. Contemporary archaeological theory is less keen on linear processes than earlier incarnations and most archaeologists accept that interpretation and analysis are also part of data collection (Hodder 1999).

An alternative model might propose five recurring analytical tasks, present at all stages of research from trowel to monograph and beyond, which need to be supported by an IS.

- Atomisation - the construction of the units of analysis, is a choice, a process that often takes place before data collection begins, although repeated through analysis. Key elements in atomisation are the data structure itself, controlled vocabulary, and the spatial definition of boundaries.
- Quantification - can be as simple as measuring the diameter of a posthole or more complex as in estimating the mean level of a particular element in an assemblage of slag.
- Ordering - this may include classification, categorisation and sequencing.

- Integration - regrouping, also bringing together different classes of data (and interpretation).
- Explanation - where the other processes are incorporated with broader understandings and belief.

The key difference in the second model is that the processes are iterative rather than linear. Data isn't atomised, or explained, or subjected to any other process only once and therefore the IS needs to be able to track or audit these processes and their more complex relationship(s) to each other.

Supporting existing Practice

Existing practice in archaeology has developed over the last 150 years and has many elements that are work-arounds for the technology we've been using. Some people have been expecting that the Revelation project will recommend a revolution in existing practice - opting for 3D recording, for the replacement of plans with photos, for the replacement of the 'Cranbourne Chase model'. While some or all of these developments may well take place in the next few years - and we need to take these directions into account - the requirement to produce a useful system (one that is used) must be the prime focus rather than any "technopolistic imperative" for innovation (Huggett 2000:19).

Our approach has been to identify current good practice that needs to be supported with better systems, rather than simply an attempt to create new practice. We have produced workflow diagrams of our existing archaeological practice and will use these to build upon the elements that are strong and need supporting while also identifying the gaps and problem areas that need improving.

On most English excavations a single person is responsible for the excavation, drawn record, text record and to a degree the photographic record of each individually identified context. This approach is seen as the best way to get the most integrated record and maximises the excavator's engagement in the recording and analysis process, thereby increasing both the quality of the record and the excavator's skills and experience.
One possible implication of maintaining this practice is that it will require tools for digital drawing on site, rather than off-site digitisation of the pencil plans or remote survey drawings created using a total station. For this reason we have already carried out some preliminary testing of digital field drawing systems to examine the viability of digital drawing in the field more closely. Our testing of digital drawing with pen computers demonstrated that it is clearly feasible to draw scaled drawings directly on a pen computer. Further, this achieves the same levels of accuracy of depiction as achieved by drawing on paper and digitising later, which is the method that the CfA and many other units currently use.

Improving existing Practice

One area that can be improved by a well-designed system is the feedback given to people recording on site. With existing paper-based systems, people who complete context sheets in the field are very rarely involved in the later analysis or interpretive work using the information they have recorded. If people do not understand why they are being asked to record a piece of information they either do not record it correctly, or simply do not record it at all. A typical example is that excavators are asked to give sedimentological descriptions of each deposit but they never find out what the geoarchaeologist does with the information. Hence they have little idea of why they are recording it and many people are very lax about entering this information. A similar lack of feedback mechanisms occurs with finds and sampling information, and indeed with most information on the context sheet. Information about the future use of data can be presented in context sensitive help files on a digital system. Further, the front-loading of analysis into the excavation season allows site staff to understand the process better (Andrews et al. 2000). These improvements rely on the integration of project team and on having in-house specialists involved in the design of the system.

Conclusion

The Revelation project assessment is a product in itself. Most archaeological units have neither the money nor the scope to conduct this kind of assessment, which can mean that systems are developed in a piecemeal manner. When this stage is complete, further consultation will be needed on technical aspects. We are already planning work with a European consortium called SEMKOS to look at issues of cross-domain mapping and 'future proofing' our work through the use of XML and the Semantic Web.

The final outcome of the assessment will be a plan for implementation. This will probably be modular and involve adoption of modules from other systems where appropriate. The key aim is to provide something where all aspects of the project, from start to finish, are managed together and that all project team members have access to each others work.

Acknowledgements

The views expressed here are principally our own, but they are based on the work of the entire Revelation team. We would therefore like to thank Brian Attewell, Paul Cripps, Vicky Crosby, Tom Cromwell, Karla Graham, Jen Heathcote, Manny Lopez, Eddie Lyons, Andy Payne, Sarah Reilly, David Robinson, Kirsty Stonell, and Mary Walkden for their hard work and advice.

REFERENCES

ANDREWS, G., BARRETT, J. and LEWIS, J., 2000. Interpretation not record. Antiquity 74:525-530.

HODDER, I., 1999. The archaeological process: an introduction. Blackwell, Oxford.

HUGGET, J., 2000. Computers and Archaeological Culture Change. In Lock, G., and Brown, K. (eds.), On the Theory and Practice of Archaeological Computing, Oxbow books, Oxford.

ROSKAMS, S., 2001. Excavation. Cambridge University Press, Cambridge.

STEWART, J., 1997. Has 25 Years of Computing Provided Greater Physical and Intellectual Access to Archaeology? In Computer applications and quantitative methods in archaeology.

THE OFFICIAL LIST OF ARCHAEOLOGICAL SITES IN THE CZECH REPUBLIC - AN INFORMATION SYSTEM OF ARCHAEOLOGICAL SITES IN THE CR

LENKA KRUŠINOVÁ
CZECH REPUBLIC

SEE THE CD FOR THE EXTENDED VERSION

ABSTRACT

The project handles the specification of the territories with archaeological findings (i.e. archaeological monuments or sites) according to the state of knowledge for purposes of archaeological heritage management which is executed by governmental authorities. The collection, automatical elaboration, keeping, updating and amending of the data about the archaeological findings for the open information system is secured by implementation of the application SAS CR 2.0 end the application ARCHAEOLOGIE for the inter/active utilization for the graphical data and information from the d-base in the enviroment GIS ArcView in the version 2.0 being newly accesable also via Internet. So far 53 district offices have been participating in the project on contractual basis. In connection with the reform of public administration the district authorities ceased to exist. Prior to this contracts have been signed with 6 of the new larger administration units in 2002 and with the remaining 8 contracts are under preparation. Tens of archaeological institutions and archaeologists have been contracted for processing of archaeological data and nearly 50 regional archaeological institutions are maintaining, updating and using the data today.

THE DEVELOPMENT

The archaeological heritage and its management have been defined in the Czech Republic only during the nineties of the last century. The dynamic changes in the legal system caused by the major societal changes after 1989 gave rise to emphasis on the focusing of archaeological interest in the territory. In the area of state protection of historical monuments archaeologists are now trying to establish a system of protection of archaeological monuments as a consequence of both the enormously increased requirements of building contractors on execution of archaeological salvation researches and requirements of the public administration authorities on the recording of areas with archaeological findings in the preparation stage of construction work. The position of specialized archaeological institutions is determined by the changes in the system of public support of research as well as by changes in the establishing function of public administration authorities as a result of their reform.

The management and preservation of archaeological heritage as a whole can only be possible through establishing an open information system of the presently identified archaeological heritage, i.e. both of the movable findings already excavated from identified archaeological sites and of the immovable part, covering still unexcavated movable archaeological findings and undiscovered immovable archaeological findings. The information on relations between the above parts of archaeological heritage represents a limited and unique source base for the research of the ancient history of humanity and its historical memory. Therefore the archaeological sites so far undisclosed either for reasons of insufficient archaeological research capacities or due to unsystematic approach to research deserve the maximum efforts regarding their recognition and preservation. Especially this part of the archaeological heritage is most endangered and destroyed due to ignorance and disinterest and it is desirable to focus the attention on its recognition in the territory.

THE MALTA CONVENTION.

An important step of the Czech Republic was the ratification of the "European Convention on the Protection of the Archaeological Heritage" (revised) of the Council of Europe Nr. 143 (signed in La Valetta on January 1, 1992) in 2000. In order to fulfil the principles of Malta Convention it is necessary to reinforce the administrative and practical aspects of a more effective management of the archaeological heritage both in the specialized archaeological and in the public administration institutions evenly covering the entire territory of the CR. Regarding the domain of archaeological heritage, the Ministry of Culture of the CR, being responsible for the state preservation of monuments, refers in its "Conception of more effective conservation of monuments of the CR" (whose proposal was approved by the government on April 22, 1998) to the intention "to finalize the identification of territories with archaeological findings in the Czech Republic and develop a specialized central archaeological list". The Ministry of the Environment of the CR considered the archaeological heritage both in the wording and fulfilment of the Act No. 244/1992 on environmental impact assessment and by focusing one of the research and development programs on "Applications for a complex digital description of archaeological sites and mining areas in connection with the execution of public administration" in 1998. The Ministry for Regional Development of the CR is cooperating with the Department of Archaeological Heritage Management (DAHM) of the National Heritage Institute (NHI, until December 31, 2002 the State Institute for the Preservation of Monuments) in the area of sharing and utilizing the data on digital archaeological zones (derived from the data of the SAS CR project) for the needs of town and country planning in the geographical information system (GIS) environment.

THE SAS PROJECT

The identification of areas with archaeological findings for the purpose of their protection and salvation is the subject of a long-term research project "The Official List of Archaeological Sites in the Czech Republic" (in Czech: "Státní archeologický seznam Ceské republiky", SAS CR). The conception of SAS CR stems from a thorough analysis of the situation, needs, and possibilities of the state administration authorities (former district authorities) and of the archaeological institutions processed during the years 1991-93 at NHI. The precondition for a viable cooperation of public administration authorities with specialized archaeological institutions is the identification of areas with archaeological findings in connection with property rights (on a resolution basis allowing the link to cadastral maps). The information funds hitherto collected by the Institutes of Archaeology of the Academy of Sciences of the CR turned out to be too incomplete especially with regard to the areas of potential archaeological findings. Furthermore, the levels of data processing in the territories of Bohemia, Moravia and Silesia were too different to comply with the needs of public administration authorities and archaeological institutions both on the regional and national level for producing an information system for protection and salvation of the archaeological heritage. For these reasons it was not possible to unify the development of a technical instrument covering both the registration of areas already archaeologically researched (i.e. the already destroyed areas with archaeological findings) and the purpose of identification of areas with archaeological findings.

The standard for the scope of data for collecting and processing of information for the purpose of archaeological heritage management was set by a research team of DAHM led by principal investigator Lenka Krusinova. The team has developed and implemented the methodology enabling the solution of the problem of identification of areas with archaeological findings based on the present state of archaeological knowledge by developing the instruments for archaeological data processing in GIS environment. Decision making of public administration based on the use of such information systems is in accordance with the EU recommendations in support of application of GIS and data harmonization and standardization. The classification of areas with archaeological findings is split into four categories: I - areas with a positive proof of findings, II - areas with a 51-100 per cent probability of findings, III - areas presently without identified archaeological sites, cultural landscapes with the probability of findings less then 50 per cent, IV - areas with no real probability of archaeological findings - all excavated areas.

The start of the research project SAS CR was made possible by a financial donation from the Embassy of the UK in the years 1993-1997. Only starting from 1995, the Ministry of Culture also granted financial contributions to the project. Moreover, the planned duration of the project was changed to 2003 from the original 2005. The implementation of the project will continue from 2004 on as an integral part of the internal research activity within NHI.

The aim of the SAS CR project is to generate a digital map of areas with archaeological sites in the Czech Republic in a GIS environment interlinked with the database of data about archaeological sites and provide the mechanism for updating and amending the archaeological information. This will establish an open information system operated on user stations allocated throughout the territory of the Czech Republic. The basic relations of the project are created on contractual basis between NHI, namely DAHM, the public administration authorities, and the archaeological institutions licensed to execute archaeological research. NHI secures the collection of archaeological data and their central management; it also secures the digitalization of graphical data into the basic archaeological coverage. Numerous regional archaeological institutions as well as single archaeologists have been participating in the collecting of standardized archaeological data, transferring positive archaeological findings into the database, and preparing expert maps in 1:10,000 scale, showing the locations with archaeological findings. Figure 1 shows the state of processing of archaeological data gained in cooperation with archaeological institutions until December 31, 2002.

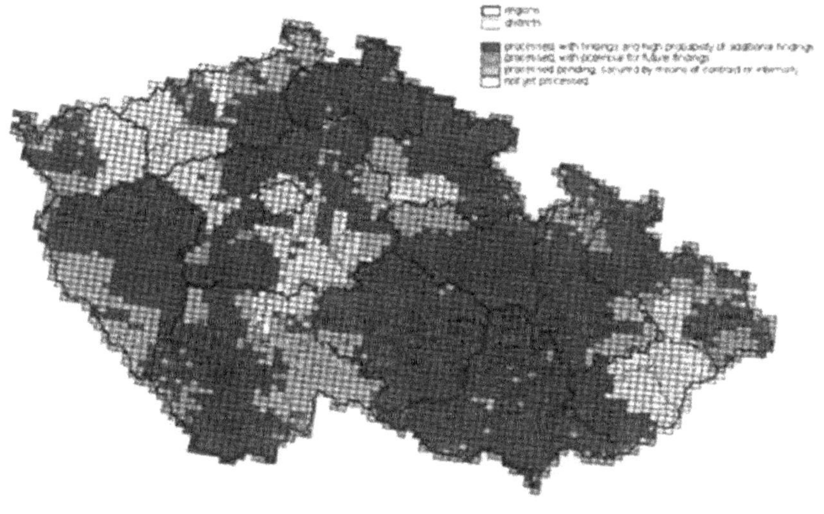

Figure 1 SAS CR - state of processing of maps on 31.12.2002

The regional management, completion, and updating of archaeological data is being executed (also on contractual basis) by 40 archaeological institutions, cooperating with public administration bodies in the protection and salvation of archaeological heritage. Archaeological institutions are utilizing the data from the regional SAS CR projects in GIS for solution of their professional tasks (see Fig.2). The database is operated within the SAS CR version 1.5 application (in the environment of Paradox for Windows) developed to allow automated archaeological data collecting and management and currently contains 22.000 records in total. New SAS CR version 2.0 application has recently been developed to improve the professional quality of data (supplemented by some more detailed items such as the information on the reason for entering of each specific site, obligatory links between selected items, etc.) and for better comfort of the users. Presently the extended database is being completed, the data are being transferred into the new structure and prepared for distribution, and user manual for the new application is under completion.

Figure 2 SAS CR - state of contractual cooperation with the old districts and the new regional authorities and the licensed regional archaeological institutions on 10.04.2003

The graphical coverage contains 20,000 graphical representations of digitalized sites. Out of the total of 4,754 maps in 1:10,000 scale, 3,959 maps (corresponding to 86 per cent of the territory of the Czech Republic) have been processed until end of 2002. Presumably more than 90 per cent of the territory will be covered before the expiration of the project at the end of 2003. For the remaining part analytical instruments in GIS environment can be used for modelling the possibilities of occurrence archaeological findings.

The function of public administration authorities consists in utilizing the archaeological information for protection and salvation of archaeological findings in the area of state preservation of monuments. Before the former district authorities terminated their activities on December 31, 2002 (as a partial step of the public administration reform), contracts have been signed with 53 of them (out of then existing 77 in total). They supported the utilization of archaeological information and project data in their offices in GIS environment. They also supported the regional management of data together with specialized archaeological institutions and provided their own reference (positional) data needed for the project. The cooperation with the newly established regional administration authorities in fulfilling the aims of the project started already during 2002. By April 10, 2003, contracts have been signed with 8 of the newly established 14 regions, with the remaining contracts being under negotiation and/or preparation.

Application software Archeologie 1.1 and 2.0

Archeologie version 1.1 application in the environment GIS Arc View 3.x or T-Map Viewer 2.x, enabling effective work with data in interaction between database and graphical data, was installed at 46 out of then existing 77 district authority offices and at 36 regional archaeological institutions. NHI was sharing the positional data with 49 district authorities in total. However, due to the termination of their existence the data have been returned. DAHM has secured raster data in 1:50,000 scale for the entire territory of the CR and, based on a contract with the Ministry of Defense of the CR, also vector data in 1:25,000 scale with the necessary number of user licenses for the cooperating regional archaeological institutions. New Archeologie version 2.0 application in GIS environment (presently under completion) can be installed locally and data can be accessed and worked with also via Internet. Figure 3 shows a sample of the use of Archeologie version 1.1 application, showing in red the location of a selected archaeological site at Svucice (part of Misovice) and in the window the corresponding data on findings at this location. Software versions 1.0, 1.1, and 2.0 of Archeologie application have been developed for NHI by T-Mapy, s. r.o., see http://www.tmapy.cz. The application can be used on two levels. On the first one, the user can use all the basic functions for the work with data regarding archaeological sites such as viewing, sorting, selections, graphical presentation, preparation of output for printing etc. The second level represents extended access for system managers. In addition to the above it also enables the access to detailed information regarding each site as well as to the instruments for a more detailed sorting of information and setting of passwords. Detailed description is available in the manual prepared by the software supplier.

It is obvious that public administration authorities are inter-

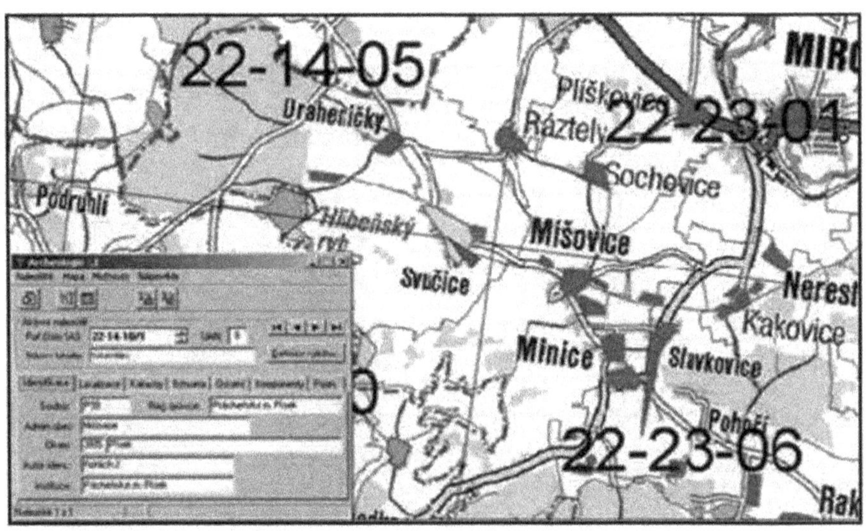

Figure 3 SAS CR - Aplication Archaeologie 1.1 - sample of interactive work with project data in GIS

ested to participate in taking care of the archaeological heritage in the area of protection and salvation of archaeological findings in cooperation with the archaeological institutions. The regional administration authorities, which are also responsible for the founding of new specialized archaeological institutions since 2003, are cooperating with NHI on contractual basis. This is a fundamental precondition both for the stabilization of the professional archaeological activities in the regions and the search for an adequate model of the system for the protection of archaeological monuments fulfilling the aims of the European Convention on the Protection of Archaeological Heritage, a system enabling the utilization of updated specific archaeological information in the decision making processes of the public administration authorities fully considering the requirements for preservation, protection and conservation of the archaeological heritage.

ACKNOLEDGEMENT

The author would like to express his thanks to Mr. Becvar and Mr. Volfik from NHI for the technical preparation of the figures.

References

BAŠTOVÁ, D., KRUŠINOVÁ, L., SKLENÁROVÁ, Z. and VOLFÍK, P., 1997. Státní archeologický seznam Ceské republiky - informacní systém archeologických nalezišt. In Machácek, J. (ed.), Pocítacová podpora archeologie, Masarykova univerzita v Brne, Brno:115-123.

BAŠTOVÁ, D., KNECHTOVÁ, A., KRUŠINOVÁ, L., SKLENÁROVÁ, Z., VOLFÍK, P., KRECMER, T., KAMENICKÝ, J., TRHON, P. and VACULÍK, J., 1998. Aplikace Archeologie 1.1 - prírucka uživatele.

BECVÁR, L., CEJPOVÁ, M., ERNÉE, M., KNECHTOVÁ, A., KRUŠINOVÁ, L., SKLENÁROVÁ, Z., VACHUT, P. and VOLFÍK, P., 2003. Státní archeologický seznam CR, verze 2.0 - uživatelská prírucka. Praha.

BECVÁR, L., DOSTÁL, L., HULEK, T., KAMENICKÝ, J., KNECHTOVÁ, A., KRECMER, T., KRUŠINOVÁ, L., VACHUT, P., VOLFÍK, P. and VACULÍK, J., 2003. Aplikace Archeologie 2.0. - prírucka uživatele.

BLAŽEK, J., 1998. Ein willkommenes Provisorium. Bodendenkmalpflege in der Tschechischen Republik. In Archaelogisches Nachrichtenblatt Band 3, 2/1998:123-128.

CEJPOVÁ, M., 1997. Archeologická památková péce v podmínkách okresního úradu. In Archeologická památková péce - legislativa a skutecnost, MENHIR - spolecnost na podporu archeologických aktivit, Olomouc:45-50.

http://www.EUROGI.org.

Koncepce úcinnejší péce o památkový fond v Ceské republice do roku 2005. Praha 1998.

KRATOCHVÍL, L. and KRUŠINOVÁ, L., 1997. Právní hlediska porizování a prenosu archeologických informací. In Archeologická památková péce - legislativa a skutecnost, MENHIR - spolecnost na podporu archeologických aktivit, Olomouc:23-36.

KRUŠINOVÁ, L., 1992. Postavení oboru archeologie v systému památkové péce. In Zprávy památkové péce 2:20-23.

KRUŠINOVÁ, L., 2002. The Czech Republic. In Fairclough, G. and Rippon, St. (eds.), Managing Europes Archaeological Heritage, Europes cultural Landscape, Archaeologists and management of Change, EAC Occasional Paper 2:182-185.

KUNA, M., KRIVÁNKOVÁ, D. and KRUŠINOVÁ, L., 1996. Archiv 2.0 - Systém archeologické databáze Cech. Uživatelská prírucka, Praha.

SKLENÁROVÁ, Z., KRUŠINOVÁ, L., BAŠTOVÁ, D. and VOLFÍK, P., 1999. Státní archeologický seznam CR, verze 1.5. Uživatelská prírucka, Praha.

KRUŠINOVÁ, L., 2000. Archeologické dedictví v CR z pohledu státních úradu a archeologických institucí. In Cáslavský sborník, Cáslav.

The author would like to express her thanks to Mr. Becvar and Mr. Volfik from NHI for the technical preparation of the figures.

Deconstructing the Crystal Ball: the State of the Art in Predictive Modelling for Archaeological Heritage Management in the Netherlands

Hans Kamermans
Leiden University, The Netherlands

Jos Deeben
State Service for Archaeological Heritage Management, The Netherlands

Daan Hallewas
State Service for Archaeological Heritage Management, The Netherlands

Martijn van Leusen
University of Groningen, The Netherlands

Philip Verhagen
RAAP Archeologisch Adviesbureau BV, The Netherlands

Paul Zoetbrood
State Service for Archaeological Heritage Management, The Netherlands

See the CD for the Paper

Abstract

This paper presents the first results of a three-year study into the application of predictive modelling techniques in archaeological heritage management in the Netherlands. Predictive maps play an increasingly important role in the decision making process for planning schemes on a municipal, provincial and national level but at the same time the validity and reliability of the models that form the basis of predictive modelling have been questioned internationally. In the Netherlands a national research team recently started a project called "Strategic research into, and development of best practice for, predictive modelling on behalf of Dutch Cultural Resource Management".

The goal of the project, which runs until the beginning of 2005, is a thorough analysis of the various models and methods used in current predictive modelling practice, the exploration of possibilities for methodological improvement and greater efficiency, and the formulation of recommendations for the Dutch Handbook of Archaeological Quality Norms (http://www.cvak.org/).

The first phase of the project is now completed. This paper presents a review of the current practice of both commercial and governmental predictive modelling in the Netherlands. In some ways the conclusions are remarkable.

Building Community-based Information Management Systems within Aboriginal Cultures of the Kimberley Region, Australia

Greg Mules
APR Geo Information Services, Brisbane, Australia

See the CD for the Paper

Abstract

With the passing of each generation of Australian Aboriginal people, cultural knowledge is endangered and in threat of being lost. For the traditional owners of the Kimberley Region of northwest Australia, cultural is bound up in the inter-relationships between people, language, land (and sea), spirituality and history. Community-based information management systems can provide the tools for Aboriginal people to access knowledge about their living culture and cultural heritage. Access to knowledge can empower communities to meet the challenges of self-determination, land rights and healthy country.

Integrated (database, GIS, GPS and www) techniques are being developed to manage knowledge-based systems. These systems include cultural archives, art and artifact catalogues, plants and animals surveys, traditional land use and occupancy maps and language web sites. The processes and challenges of implementing Aboriginal community-based information management systems in the Kimberley are discussed. A review of the nature and scope of such systems across the region is presented along with examples of projects currently being implemented.

JERG RATGEB AND THE HERRENBERG ALTARPIECE - A PROJECT OF INTERDISCIPLINARY DOCUMENTATION OF HISTORY AS A CULTURAL HERITAGE

WOLFGANG DIETZ
HISTORICAL INSTITUTE, DEPARTMENT OF LOCAL HISTORY, UNIVERSITY OF STUTTGART

LEIF SCHEUERMANN
HISTORICAL INSTITUTE, DEPARTMENT OF LOCAL HISTORY, UNIVERSITY OF STUTTGART

ABSTRACT

Based on the Multimedia CD "In den Wirren des Bauernkieges - Jerg Ratgeb und der Herrenberger Altar", this article shell outline ideas, how a subject can be approached, using thoughts and viewpoints of some diverse scientific disciplines.and introduce some thoughts on the aesthetic and didactic virtue of historical reconstruction and representation as well as their value for historical research.

The project as a whole was initiated by the Artentity foundation.[1] It started with some conceptional ideas in the mid 90's and is not expected to be finished for some years to come. Besides the Historical Institute - Department of Local History - at the University of Stuttgart, the Archive of Herrenberg, the Bauernkriegsmuseum Böblingen, the Fraunhofer Institut für Arbeitswirtschaft und Organisation and the Wilhelm Schickard Institute for computer science and graphical-interactive systems at the University of Tübingen were involved in the project.

The main objective was using the possibilities of the New Media for an integral vision of art and history. The multiple connections of Jerg Ratgebs altarpiece to the contemporary art- and cultural history form the main focus on it.

During a 3 year project a group of students under the scientific lead of Dr. Gerhard Faix were able to produce a CD Rom based encyclopaedia over 2 CD Roms (Burgbacher and Faix 2000). These CDs content as topics the Herrenberg Altar by Jerg Ratgeb, the "Gesamtkunstwerk Chorraum", the choir stalls of the collegiate church at Herrenberg, the building history of the collegiate church at Herrenberg, the Brethren of the Common Life, the Twelve Articles of the peasants and the Peasants War in Württemberg.

Beneath this historical and arthistorical analysis the possibilities of the multimedia applications for the arts became a focus of attention: the structural possibilities of hypertext, the reconstruction of complex spatial structures the visualization of dynamic processes an the use of audio and video sequences. The CDs were created using the ToolBook software and contain about 1000 pages of code.

The CDs were awarded the Prix Möbius Germany 2001 in the category Science and Communication and had a special mention at the world final in Peking. It was also awarded the seal of quality by the Europrix 2002 in the category knowledge, discovery and culture.

In the following some contents of the CD Roms will be presented.

The first CD Rom is dedicated to the altarpiece and is original location, the Choir of the collegiate church at Herrenberg. The second CD Rom allows an insight into the events of the Peasant War in the duchy Württemberg - located in the southwestern part of Germany - and presents Jerg Ratgebs disastrous part in it. The history of the "collegiate church at Herrenberg" are illustrated by the representations of the different phases of the building process during the 800 year history of the churchbuilding. Apart from such static 2dimensional pictures the CD also contains some three dimensional reconstructions.

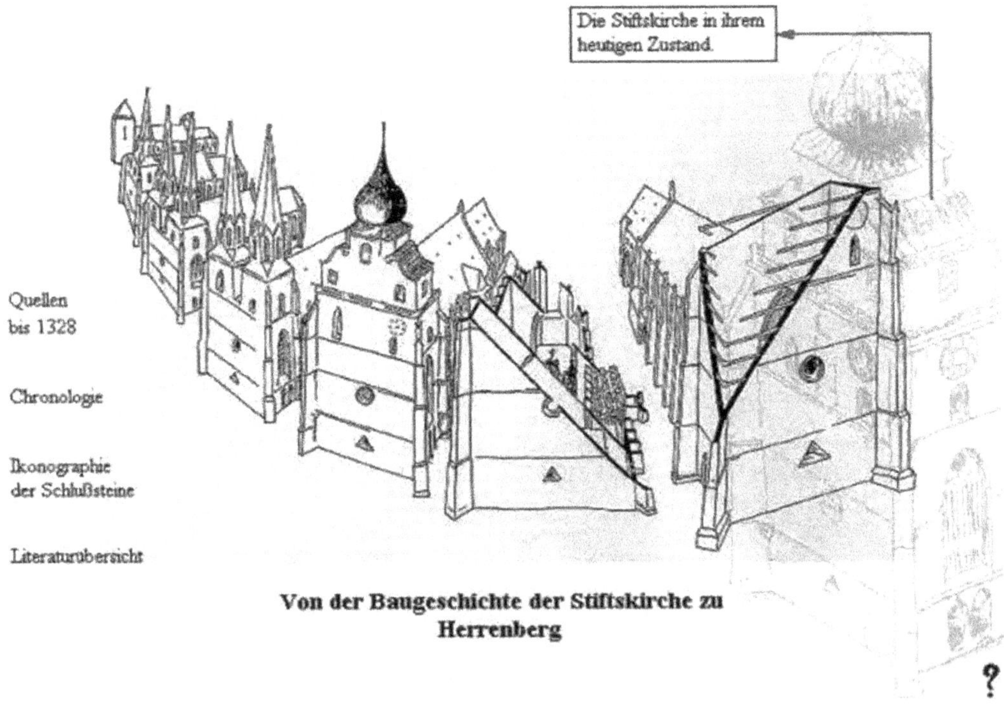

Figure 1 The representations of the different phases of the collegia church building

In an animation the user can see a reconstruction of the inside of the "collegiate church at Herrenberg" as it was at the beginning of the 16th century. This animation was generated using a so called CAVE based at the Fraunhofer Institut für Arbeitswirtschaft und Organisation. CAVE stands for Computer Aided Virtual Environment and is a three dimensional projection space. The user can enter this space and can navigate the projection and interact with it using a Joystick. This way it is possible to move freely through the inside of the church and explore it. As well it is possible to explore the architecture in Jerg Ratgebs altarpiece, during a virtual Walk trough them. The the paintings were created in the late 15th century by Jerg Ratgeb and can only be seen today at the Staatsgalerie Stuttgart, wich is out of its original context.

The CD provides multiple options to position the altar tables in relation to each other and compare them. This way an onlooker is better able to understand Jerg Ratgebs "pictorial agenda". Furthermore the CD gives one the ability to precisely observe the various pictures via a zoom function.

One of the most important expressions of the time line between Middle Ages and the early Modern Times in Europe is the Peasants' War in the German speaking regions of the year 1525. This, in the modern historical research called "Revolution of the Common Man" (Blickle 1993), was intended by the growing dissatisfaction about the social, economical and religious changes through the whole society.

To show the reasons and the development of the peasants campaign in the duchy of Württemberg a interactive map was designed. The user is given the possibility to explore this in different levels, more or less detailed. (Fig.3)

Beginning with summary the user gets short information about every important place or station during the march of the peasants through the duchy. In further levels the user can go in to greater detail about the locations, the separate occasions or the exact way, the peasants marched in April and may 1525 shown in an animation in a three dimensional topographical model of Württem-berg.

Figure 2 Homepage of the pictorial agenda of the altarpiece by Jerg Ratgeb

To secure these information in a scientific way the CDs are not only designed to simply impart such knowledge but also to enable the user to reflect on the historical events and draw his own conclusions. The contemporary sources used in the production of the CDs were scanned and embedded in order to empower the user to question our interpretations. In order to facilitate navigating the sources a system of footnotes was designed.

The relationship between the altar and the choir benches as it was planed by the "Brethren of the common life" is presented in an other animation.

These Animations and reconstructions are didactic as much as artistic elements, designed to lead to onlooker from the visual reality towards the historical questions.

On the CD Roms another focus was given on the altars wide relationship to the culture and art history of the period. The contemporary social movements in the 15th and 16th century and the rapid expansion of the art of printing marks a time of upheaval which has been methodically consolidated and presented.

Finally some words to the experiences made during this project. The most exciting examination for historians with the possibilities of New Media is to overcome the deep gap between the "Old School", which widely refuses using it, and the needs of modern education. During our work we discovered the advantages in the presentation of the results of our research. Nevertheless the project would not desert the way of scientific historical work, which is reading and interpreting sources of all origin. The big memory capacity gives the possibility to show all the originals at the place where they are needed together with the transcription and with the interpretation. Here we are able to show the original historical research.

The foundation of this work was a very large amount of data which made it impossible to rely on written sources alone. Especially the analysis of pictures and the architectural her-

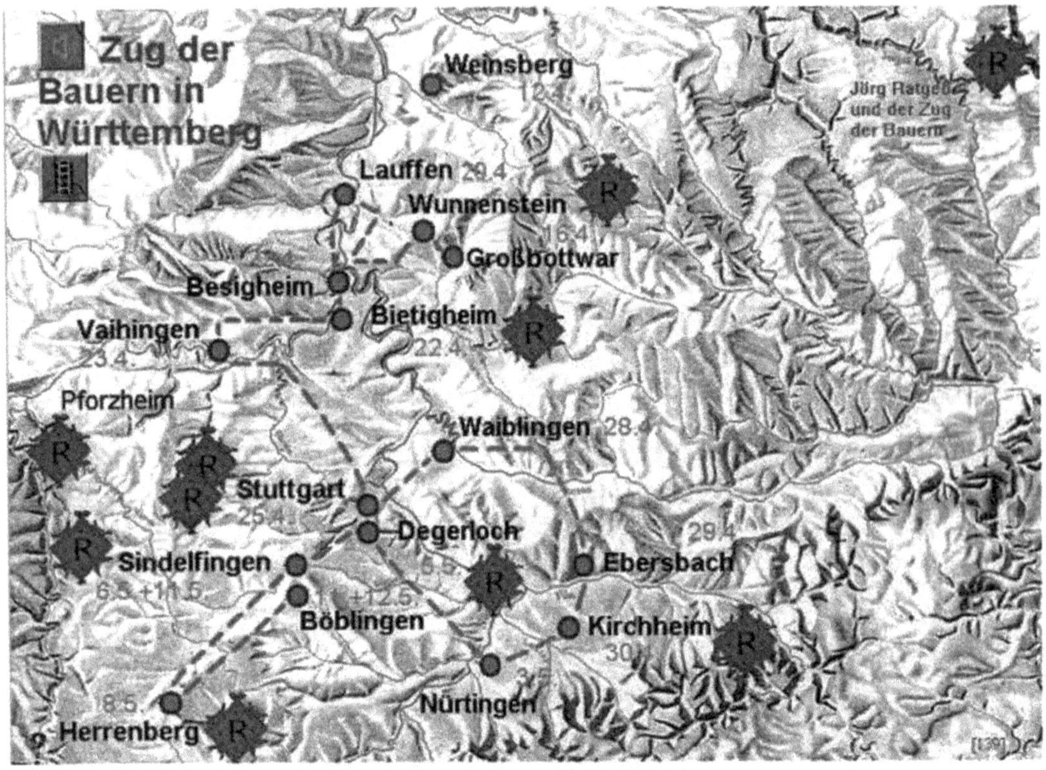

Figure 3 Interactive map of the peasants campaign through the duchy of Württemberg

itage as well as on site visits were important for our research. The cooperation between the various scientific disciplines, such as history of art, history, architecture, geography and IT was essential. The representation had to conform to high standards of precision, especially in regard to the publication on CD.

Naturally the making of these CDs required a lot of thought. Not all of the required detail is provided by historic sources. Even with a maximum of effort not all factors and their consequences could be included. Furthermore is the labour need for a project of this magnitude prohibitively costly. 16 people worked 3 years to produce these CDs. This makes projects like this unrealistic in a commercial environment. It is also imperative to solve the question of the copyright of pictures in a scientific and multi media publishing context. The project would not have been possible if it had not been for the waiving of the license fees by the government of Baden-Württemberg.

In summary it may honestly be said that the aim of presenting the historic material for didactic use was reached. The three dimensional reconstructions as well as the topographical model of Württemberg may prove of value to research, as they enable the user to see the perspective of the protagonists, this in turn will raise new questions.

It is this aspects of the perspective of the protagonists that is of most importance to the historian, because it makes some actions more accessible in its course and result.

In future we hope to embed the model into the CAVE system supplanting it with further data and ultimately link it to a database.

This project in a whole is an experiment to display the variety of our cultural heritage across the borders of scientific disciplines.

[1]For further information about the "Artentity" foundation pleas contact: http://www.transimagine.de/ratgeb.

REFERENCES

BURGBACHER, K. and FAIX, G., 2000. In den Wirren des Bauernkriegs - Jerg Ratgeb und der Herrenberger Altar. Stuttgart.

BLICKLE, P., 1993. Die Revolution von 1525. München:289-320.

AUGMENTED CULTURE: HISTORICAL ENVIRONMENTS FOR CAPTIVATING EDUCATIONAL TV PROGRAMS

ANTONELLA GUIDAZZOLI
CINECA INTERUNIVERSITY SUPERCOMPUTING CENTRE

MASSIMO ALESSIO MAURI
CINECA INTERUNIVERSITY SUPERCOMPUTING CENTRE

RENZO SALVI
RAI EDUCATIONAL - RAI - RADIO TELEVISIONE ITALIANA

MARIA CHIARA LIGUORI
BOLOGNA UNIVERSITY

ABSTRACT

New visual information technologies are increasingly effective in the field of Cultural Heritage mainly for educational purposes and communication. Three dimensional reconstructions of historical or archaeological environments constitute an interesting and flexible tool. It is well known that 3D models can be used to completely construct realities and objects which either no longer exist or exists only as ruins or fragments. A reconstruction which is historically and technically validated can become a good starting point for different educational products. Why should we be limited in showing just the three dimensional models when it is possible to make them alive in a Virtual Set? The researchers can "live" inside the results of their studies explaining and communicating them in the most effective way. This was achieved in a project recently developed by Cineca (Interuniversity supercomputing centre) and Rai educational (RAI - Radio Televisione Italiana - Milano). The reconstruction of a Roman House in Pompeii has become a stimulus for exploring new frontiers in TV narration in the age of digital reproducibility of environments and locations.

INTRODUCTION

Virtual Reality is increasingly perceived as a useful medium. It's a tool capable of simplifying the performance and the execution of various tasks, enabling people to get in touch with the world thanks to more and more realistic simulations. It is also an effective way for improving knowledge. New visual information technologies are increasingly effective in the field of Cultural Heritage. In particular, Virtual Cultural Heritage refers to the usage of Virtual Reality systems in order to generate, navigate, explore and inquire into reconstructed historical environments.

Three dimensional reconstructions of historical or archaeological environments constitute an interesting and flexible tool that enables for example to:

- access monuments that could be damaged and are, therefore closed to the public;
- reproduce, in the best possible way, realities and objects that no longer exist or that have left only some fragments or ruins;
- offer a fruition without limits, take reconstructed objects to viewers and gather objects scattered in the landscape, displaying them in a new context;
- reach a dynamic vision, without spatial and temporal limits, offering points of view not easily reachable in real life (i.e. birds eye view point), and travelling along the time line in order to show different phases of the analyzed reality .

THE VIS.I.T. LAB: METHODOLOGIES AND PROJECTS

Since 1988, Cineca has been promoting visualization techniques to support the research through a dedicated laboratory: VIS.I.T. (VISual Information Technology). The VIS.I.T. Lab is an inter-disciplinary laboratory dedicated to developing human-centred, powerful, interactive 3D graphics tools in a variety of fields, including Scientific Visualization, Bioengineering and Cultural Heritage. The mission of the lab is to present concepts and techniques, as well as massive or complex numerical data, through Interactive Computer Graphics, Virtual Reality and Human Computer Interaction technologies.

The working experience in the field of Virtual Cultural Heritage models has highlighted the importance of multidisciplinary environment among engineers, computer scientists, architects, modellers, historians and archaeologists. The team has to work together in every phase for designing and implementing an immersive emotional and historical experience. By means of this kind of collaboration it is possible to rely on a deep scientific knowledge of historical data regarding their selection and their representation that can be made visible through the interactive application. Therefore, it is mandatory to highlight: 1) the references to the authors of the models; 2) the sources used for the reconstructions; 3) the procedures for validating the reconstructed models.

During the last few years, several projects have been developed: from Ancient to Contemporary History, passing through Modern History, paying a high degree of attention to educational purposes:

- The Nu.M.E. Project Interface has been developed in order to allow visitors to witness the evolution of the city from the end of the first millennium to today (http://www.storiaeinformatica.it/nume/italiano/npresent_a.html).

- 'Bologna delle Acque' concerns the importance of water in the history and in the environmental evolution of the city of Bologna and its surroundings. The aim of the project is remind and renew a relationship with this natural element that has constantly been the base of great part of activities, settlements and commerce (http://www.cineca.it/HPSystems/Vis.I.T/Researches/ibc.html).

- In the MUVI project - the Virtual Museum of Daily Life in 20th century (http://www.cineca.it/muvi), starting from historical researches, a domestic interior of the Fifties has been reconstructed, furnished in any detail. The objects in the virtual world have been linked to a DB with historical info and explanations.

- The Roman Bologna is an archaeological project that unifies cartography, GIS and landscape perception and awareness.

- The Certosa Museum - The Certosa Cemetery in Bologna is an important but neglected monument and the Museum is born in order to valorise it. This kind of urban environment can be enhanced using tools similar to those used for cities: the 3D reconstruction can be integrated with information about the landscape and the archaeological excavations and can become a graphic interface for accessing data archived in several databases of multimedia sources (http://www.progettocertosa.it/).

Applications are being developed with a particular care in their flexibility, quickness of use and improving the context awareness. There are several multidisciplinary undergoing projects that use various computer technology functionalities and attain flexible multi platform realizations. Applications can range from a medium powered PC to High Performance systems, such as Virtual Theatres and Virtual Sets. 3D virtual worlds are starting points for a straightforward interaction with complex environments that can be navigated, modified and queried through the link to more than one relational Data Base at the same time (GIS, Multimedia data, etc.), becoming an interface for accessing different information and developing different educational products as well.

THE VIRTUAL RECONSTRUCTION OF CASA DEL CENTENARIO AND THE MANIFOLD POSSIBILITIES OF ITS USE

Another relevant project is the Virtual Reconstruction of Casa del Centenario in Pompeii. Thanks to this project we will see how a historically and technically validated reconstruction can become a good starting point for a series of educational products.

The "Pompei - Insula del Centenario IX 8" Project was started in 1999, thanks to the efforts of Daniela Scagliarini Corlàita of the Department of Archaeology - Bologna University. It has been realized in the framework of the MUSE Project (funded by MURST within the National Research Program on Cultural Heritage Parnaso) conducted by Boconsult, a Ducati sistemi Division, in cooperation with Bologna University, CINECA and Sinet. Its aims are the documentation, study and valorisation of the Insula IX of Pompeii, called "Insula del Centenario" from the large domus, excavated in 1879-1880, which occupies its main part. Peculiar to the Virtual Archaeology project applied to Insula del Centenario, and even to the whole "Pompei - Insula del Centenario IX 8" Project, is the educational character: from data acquisition to their elaboration and processing, from 3D modelling to texturing and dissemination. The main features of the digital application are:

- Use of digital photoplanes
- Archaeological texture painting
- Time varying models
- Communicative and philological models
- Porting to low resolution environments

The Virtual Insula del Centenario consists of two immersive virtual environments: as it is nowadays - in order to stimulate comparisons and to make possible the access in an area where tourists are not admitted, and as it was at the time of the eruption that, in AD79 A.C., destroyed Pompeii - in order to visualize archaeological hypotheses and theories based on the study of ruins and findings. The project "Casa del Centenario in Pompeii" has reconstructed part of this Pompeian house starting from the present structures: frescos, taken away and stored in museum, have been artificially put again in their original position; frescos left on the walls and damaged due to weather impact have been "replaced" with less spoiled versions worked out using photos taken in the Thirties; ruins have been integrated with reconstructed lakking sections (e.g. the higher part of walls, coverings, fixtures).

As before said, this reconstruction has been used in several ways:

Figure 1 A render of the peristilium

1 - with PDAs, for helping the context awareness on the archaeological site. Portable devices are easily handled and enable a friendly interaction to the majority of users, offering the opportunity of visualizing in a non invasive way, sources - i.e. photos, drawings, plans, texts, etc. - necessary to validate virtual reconstructions and to better understand it.

2 - with a desktop fruition, for both navigating the model and accessing the Data Base. Thanks to the collaboration with the VISMAN project (supported by the Spinner Consortium with Regione Emilia-Romagna and EU funds), 3D models are being linked to different relational Databases. Users, depending on the kind of DB chosen, can inquire GIS or multimedia sources. In this case, selecting objects in the scene, users obtain explanations on the selected object and see original photos. In other applications, as the one we are currently developing for the Certosa of Bologna, 3D models are a graphic interface for accessing the land register, GIS and the historical sources archived in a multimedia database with audio, video and texts about the partisans' sacrarium.

Figure 2 Some experimentations for linking the ninfeum of the reconstructed Casa del Centenario in Pompeii to a database

3 - with the Virtual Theatre, as the one in CINECA, for experiencing a sense of physical immersion in the virtual environment. Virtual Theatres can show models in a 1:1 scale, fostering a direct and natural perception of space and proportions. The large screen amplifies the sense of presence in the reconstructed environment that, of course, can be simulated even better if enjoyed stereoscopically. Moreover, it enables a collective view and, therefore, an immediate comparison of impressions among the audience.

The application for the Virtual Theatre was performed by a "picture in picture" solution so, while navigating the model reproducing the building of the 79 a. C, in a corner of the screen there is the other model following the same perspective, and vice versa.

4 - models have been used also during a conference that was broadcasted via streaming video and that it is still available on the Web (http://www.cineca.it/convegni/ut_natura_ars/).

5 - and then, 3D models have been used in a Virtual Set for creating a TV documentary in collaboration with RAI (Radio Televisione Italiana - Milano). A Virtual Set system merges in real-time shootings taken with cameras on real people and objects, along with three dimension digital scenes.

In a Virtual Set 2D or 3D digital models are synchronized with the movements of cameras and become virtual scenographies. Optometric sensors, infrared LED, radiofrequency pointers and optical grids for pattern recognition are used in order to transmit to the cameras the co-ordinates of the movements. Except for the actors and some real objects, the Virtual Set is empty and covered with a blue-screen. The various elements of the virtual scenography can be seen only on a video.

Now, once the documentary has been shown on TV a national channel, it has entered an archive created by RAI: Mosaico, a TV mediateque for educational support (http://www.mosaico.rai.it/). In a framework called Mosaico, people, teachers in particular, find a list of educational TV programs and can ask for their programmed broadcast on the RAI Educational satellite channel. People need just to be registered and then they will be notified about the date of transmission.

With a simple registration people have the possibility of asking for the broadcasting of the documentary and then they are informed about the date of transmission.

Conclusions

As we have seen, starting from the same 3D model it is possible to obtain several products and reach different goals and users, from tourists to researchers and students, from the on site fruition by PDAs up to Streaming video, Virtual Theatres and Virtual Sets. The prerequisite is to plan and develop applications aiming to a flexible output in order to adapt them easily to whatever system suits the communicational purposes.

Acknowledgements

The authors wish to thank prof. Daniela Scagliarini Corlaita, Antonella Coralini and Erika Vecchietti, of the Department of Archaeology, Bologna University; Giosué Boetto Cohen of RAI - Radio Televisione Italiana and Riccardo Perani - Virtual Set RAI Milano, Maria Elena Bonfigli of CINECA.

References

BONFIGLI, M.E., CALORI, L., GUIDAZZOLI, A., MAURI, M.A. and MELOTTI, M., 2000. Tailored virtual tours in Cultural Heritage worlds. ACM SIGGRAPH2000 Conference Abstracts and Applications, New Orleans, July 2000:264.

BONFIGLI, M.E., GUIDAZZOLI, A. and LIGUORI, M.C., 2002. From Museum to Mouseum. The Virtualisation of Daily Life Museums. In Niccolucci, F. (ed.), Virtual Archaeology Proceedings of the VAST2000 Euroconference, Arezzo, November 2000. Oxford, Archaeopress.

BONFIGLI, M.E., GUIDAZZOLI, A., IMBODEN, S., LIGUORI, M.C., and MAURI M.A., 2001. Virtual Reality Applications for the Didactics of Daily Life History. In EVA2001Scotland, Proceedings, Hunterian Museum, University of Glasgow, 24-28 July 2001.

BONFIGLI, M.E., CORALINI, A., GUIDAZZOLI, A., MAURI, M.A. and SCAGLIARINI CORLAITA, D., 2001. A data description model for Cultural Heritage Hypermedia and Virtual Environments authenticated by archaeologists. In ICHIMInternational Cultural Heritage Informatics Milan, Italy, 3-7 September 2001.

COLUCCIA, P., GAROFALO, F., LIGUORI, M.C., MONFARDINI, S. and SERAFINI, F., 2001. A Survey of Virtual Reality in the VIS.I.T. Theatre: from Research to Divulgation. In Cultivate Interactive, October 2001 (http://www.cultivate-int.org/issue5/cineca/).

SCAGLIARINI CORLAITA, D., CORALINI, A., GUIDAZZOLI, A., CINOTTI SALMON, T., and VECCHIETTI, E., 2002. Virtual Archaeology. La Casa del Centenario. In Scagliarini Corlaita, D. and Coralini, A. (eds.), 2002, L'Alma Mater a Pompei. Il progetto "Insula del Centenario", University Press Bologna, Imola.

Presenting Archaeological Information with the Help of Multimedia Cartography

MARKUS JOBST
JOBSTMEDIA PRÄSENTATION VERLAG,
WELZENEGGERSTRASSE 84, A-9020 KLAGENFURT, AUSTRIA

SEE THE CD FOR THE EXTENDED VERSION

ABSTRACT

Digital media are best suited for the presentation of regional, urban and specialized topics. Especially the field of multimedia-cartography creates new ways of expressing information and transferring discoveries. It is much easier with this technologies to make maps comprehensible to a wide range of persons. Maps are no longer an élite tool of a small number of professionals, but enable the discussion and information in geographical and thematical aspects.
Archaeology as a scientific discipline explores and tries to find and visualize dependencies of non visual connections of findings and building remains. The aim is to receive an understandable general view. For this reason multimedia cartography provides a powerful tool for the communication among scientists and for the presentation to a general public. The consideration that the quality of presentation is set by the perception and the thinking of the user, whose reference throughout the evolution is the 3D space, should define the way of processing and visualizing information.
This article explains the work of multimedia cartography, shows up the main requirements to datasets and mainly describes the role of multimedia cartography in the field of archaeology. With the help of the latest developments in the project Carnuntum 3D the suitability for experts and for a general public should be discussed.

Introduction

In the last two decades new possibilities in the communication with the help of maps have been developed caused by the digital technologies. It is now very common to use interactive maps, combined with pictures, video, sound and three dimensional views. Due to these extended modalities it should be easier to present information with its regional context more effective.

Therefore the major job of cartography is to find an efficient way to present and communicate thematical information in context with geographical one. In this paper and the developed prototype Carnuntum 3D this thematical information, which can be political, economical or cultural, concentrates on archaeological and cultural themes.

Archaeology as a scientific discipline explores and tries to find and visualise dependencies of non visual connections of findings and building remains. The aim is to receive an understandable general view of a culture in a specific epoch and region. For the communication of these spatial related information among scientists and for the presentation to a general public, multimedia cartography provides a powerful tool. During the creation of a multimedia system, the user-interface, the functionality and the visualisation of information is influenced by the perception, knowledge and thinking of the user (Dransch 1999).

There are a lot of examples using digital presentation technologies in archaeology. Products like "Virtueller Rundgang durch das antike Carnuntum" (Schrenk et al. 2000), "Troia - 3000 Jahre Geschichte im Modell" (Zöller 2001), papers like "Digital Archaeology Reconstructions from Egypt on the WWW" (Shiode 2002) or projects like "3D-Murale" (Hynst 2002) use digital reproductions of ancient objects, buildings, districts or towns. These virtual worlds present a realistic impression of objects in a very large, detailed scale. In addition to excavation maps these reproductions and documentations may be used to communicate discoveries of archaeological work and thus to save cultural heritage.

Technical developments in data acquisition and data management extend the way of documentation and enable new possibilities for the implementation of these data in digital cartographic products. An efficient presentation and Geo-communication with the help of 3D multimedia maps, resulting from these implementations, creates the imagination that new discoveries, which were not visible until now, can be made by experts.

The importance of cultural heritage is the reason for this work and article, in which a theoretical explanation of the notion "multimedia" will be given, the main problems with data will be shown and the first steps in designing a prototype of a three dimensional cartographic exploration system will be presented. This paper is based upon the thesis "Der Einsatz multimedialer Technologien in kartenverwandten Darstellungen am Beispiel Carnuntum" (Jobst 2003).

Multimedia Cartography - A Theoretical Explanation

Multimedia cartography may be seen as an extension to the traditional cartography. Instead of dealing with paper based maps, its working place is the digital world. Several ways of communication are combined with the help of computers. The aim is a better communicating presentation of the world to help users cognition.

The use of "multimedia" serves as support to the cartographic communication process (Cartwright et al. 1999) if the task of cartography is the communication of spatial related information (Gartner 2002b). Multimedia is characterised by several attributes (Dransch 2002):

- The manipulation , presentation and communication of information is done on the computer - all actions are digital.
- The different media are not related to one another - they are independent.
- It is able to show different media and processes at the same time - multitasking.
- The media are manipulated and combined by interactivity.
- The information uses different types of coding (or media) - multicoding.
- The media are combined due to their functions and content - combination of media.

The term "media" stands for writings, pictures, sound, video, interactivity and movement. Like in the case of video, where the combination of picture and sound created a new media, we are thoroughly convinced that the independent combination of all media with spatial data creates a new integrative media - multimedia (Dransch 2001).

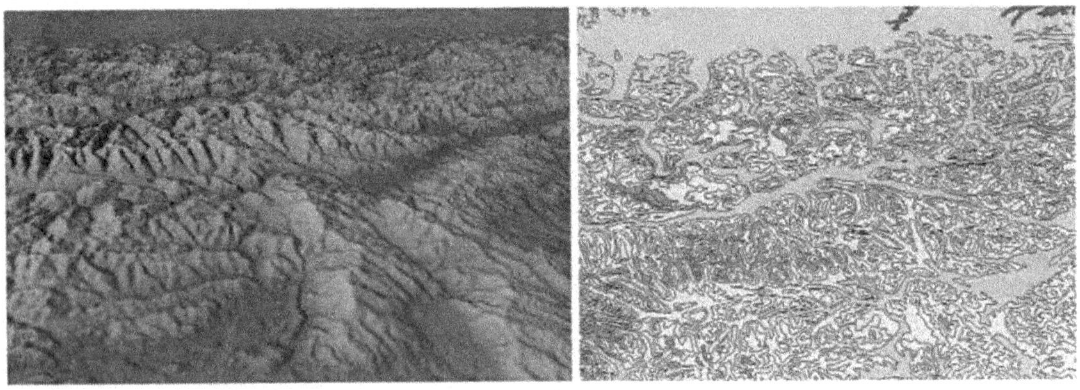

Figure 1 Comparison of a 3D and 2D view of the same area, Jobstmedia

An important part of the communication process is the intuitive recognition of information. In case of multimedia cartography, dynamic 3D views support the intuitive recognition of topography, which is useful for 60% of users of topographic maps, who have problems in retrieving a spatial impression - according to surveys of the university of Dresden (Buchroither 2002).

These results support a further usage of 3D views in multimedia cartography. In this case a 3D view is meant as map related presentation, which is defined by an oblique position of the projection plane in space. The traditional map may be seen as special case, which uses a projection to a horizontal plane. (Hake et al. 2002)

In principle a map related presentation - a 3D multimedia map - is part of the presentation model within the cartographic communication process, which serves for the communication of the information. Objects from this presentation model are coming from the primary model, where measured data like photogrammetric models are administrated. They are modified with techniques of generalisation and symbolisation to enable an "effective" communication (Kelnhofer et al. 2002).

In opposite to these encouraging aspects of an intuitive information transfer, there are three main disadvantages from a cartographic point of view:

- Due to the perspective, it is not possible to compare geometrics.
- One view consists of many different scales, depending from the distance. So it is not possible to read distances between different points without a helping tool.
- Elements of the map may be hidden by other elements. Then some information may not perceived.

It is impossible to solve these problems in static views, but in multimedia cartography an appropriate programming minimises these disadvantages. A camera movement as exploration tool may prevent hidden areas. Measurement tools allow an active distance measurement or orthographic camera projections may support an intuitive distance comparison.

Figure 2 Picture containing all disadvantages of a static 3d view, Jobstmedia

In case of the prototype Carnuntum 3D a map related presentation is combined with objects - photogrammetric models - from the primary model.

The advantage of this technique is the combination of different scales, containing different simplifications, and thus the visualisation of relations between large scale - virtual real - objects, like an excavation object or area, and the smaller scales symbols, e.g. districts of a town, in the background.

The difficulty of this multimedia presentation is the information extraction of the user and the communication form, which are barely considered yet. A useful combination of media and their use in three dimensional views has to be explored in future research projects.

Demands on Data

An enormous amount of data is needed to create this 3D multimedia map. It would be nonsense to build up the data for each project again - this would be too cost effective. Instead a global available data pool should be used.

At the present there are free available datasets. Most of them cannot be used for large scale applications. But they form a useful small scale basis to show an overview before going into detail.

Two classifications of the data can be made. There are height - digital terrain models - and feature - vector - data. The sources and formats are different. Very common are "*.bin" files (readable in GIS systems) or 16bit grayscale heightfields for the height and "shape" or "*.dxf" files for feature. In the step of data-combination mistakes may occur by different scales or map projections of the data. The effect becomes apparent in the resulting map with inhomogeneous point, lines or areas. These mistakes may be avoided by considering metadata, if they are available.

In general the Geo-data-offers are afflicted with some imperfections (Meng 2002):

- Geo-Data of big areas are available, but they are interrupted.
- There are limited areas, where datasets are complete, but there is a lack of detail.
- There are masses of data for limited areas, but without any description of the data quality.

In case of historic archaeology the data of findings or objects were often referred to the area of discovery, the political state, the boundaries of a community or the nearest church. It is very hard, nearly impossible, with this specifications to reference the objects on an exact place on the earth.

It is very important to use geographic coordinates for the positioning of findings. If the objects for a cartographic application cannot be fixed on one place of the earth, a description of a radius around the midpoint coordinates - to describe the area of reference - and an applicable scale for this radius have to be defined. During a political and economical development boundaries change, but geographic coordinates or defined areas stay the same.

The first step to make data available for a multimedia cartographic information system and for further projects is to create a set of metadata - if possible following the international standards of metadata structure.

This description of data due to their quality, accuracy, date of recording etc. makes it possible for cartographers to determine the scales and combinations where and how these data and objects may be used. Only data having the same quality can result in a homogenous presentation.

First Steps of a Prototype - Carnuntum 3D

We decided to take the area of "Carnuntum" for the designing of a prototype. The main intention is to show that this cartographic tool is useful for the visitor of the archaeological site as well as for the expert.

Figure 3 Pfaffenberg in 1900 and in 1980, photos: Werner Jobst, mounting: Markus Jobst

"Carnuntum" was one of the most important Roman cities at the Austrian LIMES. The most interesting district of this Roman town is the arrangement of temples right on the top of a hill, called "Pfaffenberg".

At 1908 the mining of gravel started the destruction of this hill. At the beginning of the 80´s all interesting parts of the temple were brought to depots in Bad Deutsch Altenburg. In the mid 80´s the hill, the regional connections and impressions from and to the temple were destroyed.

The very first step of our prototype was to build up the hill again. It was a challenge to reconstruct the initial heights. Photogrammetric procedures with the help of old photographs could not help - there was only one main view, the west side of the hill, but we needed more for a 3 dimensional model. At least we were able to find some military maps of 1870, from where it was possible to digitise the heights and to put the data into location with reference points.

The next step was the digitising of the archaeological information. Most maps of the temple were only in a local system and not referenceable. With the help of one areal photography done in 1970's we were at least able to reference these data as well.

The reconstruction of the third dimension needed the help of an archaeological architect, who uses the floor plan, the excavation documentation and the knowledge of ancient building construction to design a possible 3D reconstruction. Smaller scales of the prototype Carnuntum 3D used a simple extrusion of the floor plans to present "block"-models. These representations of the objects were interpreted with colour as signature and "mouse overs" to receive some text information.

The three dimensional, large scale models of the temples worked out in 1985 seemed to be scientifically wrong. According to present research, new models are in work since the beginning of 2003.

The only scientific based reconstruction was the "Heidentor". Several interpretations were available. To demonstrate the functionality, some of the interpretations, including a photogrammetric model, were used in the application.

The ongoing development of archaeological work was the origin of the idea that the models shown in the application should be exchangeable - or better: the models should be read from a central working database in case of an internet application and from a local file on a CD-Rom. The problems of topicality in the visualisation could be solved with this demands. In addition different interpreted models could be shown too.

The requirements for a solution like this are the "Geo-referencing" of the models - for the positioning - and their metadata - to use them in the right scale.

The next steps of the project "Carnuntum 3D" are already defined and consist of the designing of a symbol catalogue for the 3rd dimension, the exploring of a powerful usage of sound coding, the creation of a free camera movement with useful parameters. As soon as detailed models will be available the step to a larger scale will be done.

RESUMEE

The present investigations demonstrated that map related 3d views are suitable for the cartographic communication process in archaeological applications. The main technical problems that can be seen now are the changing of scales including the changing of the models detail and the movement from a three dimensional projection view to a topographic map.

The main important issues further work should be concerned with are:

- How can a cartographic multimedia application be developedto become an important tool for the expert? Is there a need for such a exploration tool and if yes, what kind of tools are needed?

- How can a dynamic generalisation be established in a three dimensional world? First steps are made with the Level of Detail (LoD) techniques. But the influence on how the simplification of geometry is done is limited. The possibilities of aggregation and displacement have to be explored.

- The role of the visualisation tool as an interface to a database or content management system. This essential step may enable the techniques of multimedia cartography to communicate the content of a database, which could play an important part in saving cultural heritage.

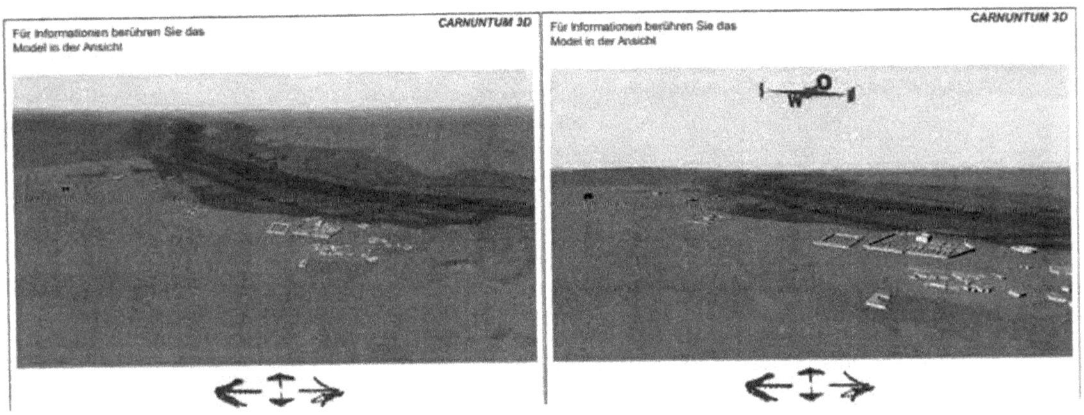

Figure 4 & 5 Screenshots from the application "Carnuntum 3D", Jobstmedia

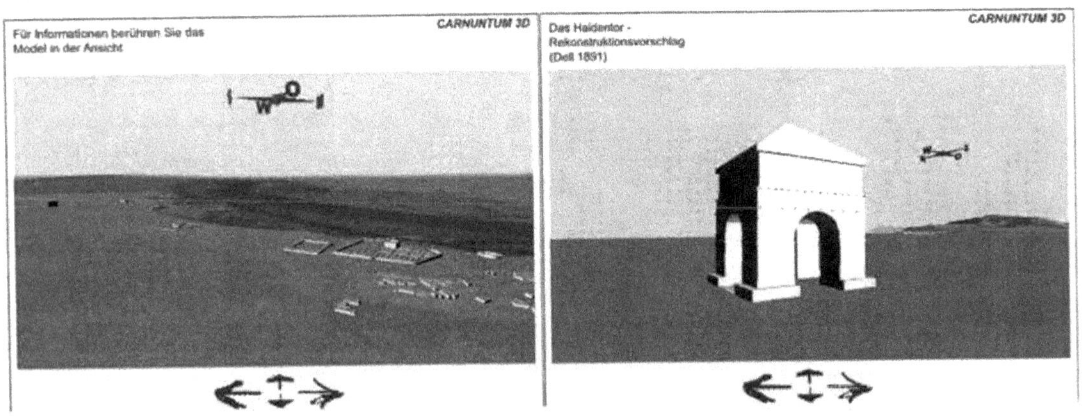

Figure 6 & 7 Screenshots from the application "Carnuntum 3D", Jobstmedia

REFERENCES

BUCHROITHNER, M., 2002. Autostereoskopische kartografische 3D-Visualisierung. In Deutsche Gesellschaft für Kartographie e.V. (ed.), Kartographische Schriften Vol. 6, Kirschbaum Verlag, Bonn

CAMMACK, G.R., 2002. Cartography, Vitual Reality and the Internet. In Maps and the Internet. Geowissenschaftliche Mitteilungen 60, Technische Universität Wien.

CARTWRIGHT, E.W., 2002. New Media Visualizations and Visualising geography, in Maps and the Internet. Geowissenschaftliche Mitteilungen 60, Technische Universität Wien.

CARTWRIGHT, W., PETERSON, M.P. and GARTNER G., 1999. Multimedia Cartography. Springer Verlag, Berlin/Heidelberg/NewYork.

DRANSCH, D., 2002. Handlungsorientierte Mensch-Computer-Interaktion für die kartografische Informationsverarbeitung in Geo-Informationssystemen, Diss. Fachbereich Geowissenschaften der freien Universität Berlin, Berlin.

DRANSCH, D., 2001. Stichwörter zur Multimedia Kartographie. In Bollmann, J. and Koch, W. (eds.), Lexikon der Kartographie und Geomatik.

DRANSCH, D., 1999. Theoretical Issues in Multimedia Cartography. In Multimedia Cartography, Springer Verlag, Berlin/Heidelberg/New York.

GARTNER, G. and UHLIRZ, S., 2002. Maps, Multimedia and the Mobile Internet. In Maps and the Internet, Geowissenschaftliche Mitteilungen 60, Technische Universität Wien.

GARTNER, G., 2002b. Multimedia und Telekartographie. In Deutsche Gesellschaft für Kartographie e.V. (ed.), Kartographische Schriften Vol. 6, Kirschbaum Verlag, Bonn.

HAKE, G., GRÜNREICH, D. and MENG, L., 2002. Kartographie - Visualisierung raum-zeitlicher Information. Walter de Gruyter, Berlin/NewYork.

HYNST, St. and GERVAUTZ, M., 2002. 3D-Murale - Archaeology and Virtual Reality. In Harl, O. (ed.), Workshop7 - Archäologie und Computer (CD), Vienna.

JOBST, M., 2003. Der Einsatz multimedialer Technologien in kartenverwandten Darstellungen am Beispiel Carnuntum. Diss. Institut für Kartographie und Geo-Medientechnik, Technische Universität Wien, 2003.

KELNHOFER, F., LECHTHALER, M. and BRUNNER (eds.), 2002. Kartographie und Telekommunikation. In Telekartographie und Location Based Services, Geowissenschaftliche Mitteilungen 58, Technische Universität Wien.

MENG, L., 2002. Personalisierung der Kartenherstellung und Mobilität der Kartennutzung. In Deutschen Gesellschaft für Kartographie e.V. (ed.), Kartographische Schriften Vol. 6, Kirschbaum Verlag, Bonn.

REISCHER J., 1996. Flugsimulation auf Basis eines digital hergestellten 3D-Panoramas. In Wiener Schriften zur Geographie und Kartographie Vol. 8, Institut für Geographie der Univ., Ordinariat für Geographie und Kartographie, Vienna.

SCHRENK, M., VOIGT, A., HUMER, F. and BOHUSLAV, P., 2000. Virtueller Rundgang durch das antike Carnuntum. In Harl, O. (ed.), Workshop5 - Archäologie und Computer (CD), Vienna.

SHIODE, N., 2002. Digital Archaeology: Reconstructions from Egypt on the World Wide Web. In Harl, O. (ed.), Workshop7 - Archäologie und Computer (CD), Vienna.

TAHIRI, D. and BÉTHUNE, S. de, 1997. DEM production from topographic maps: digitizing or scanning? In Proceedings Vol. 2 of the 18th International Cartographic Conference, Gävle Offset AB.

ZÖLLER, W. and ACKERMANN, R., 2001. TROIA - 3000 Jahre Geschichte im Modell. Konrad Theiss Verlag GmbH.

COMPUTER... WHAT FOR? VITUALITY VS. REALITY IN ARCHAEOLOGY

PAULA CAMPO
ENDESA SCHOLARSHIP -
INSTITUTO DEL PATRIMONIO HISTÓRICO ESPAÑOL
(SPANISH HERITAGE INSTITUTE), MADRID, SPAIN

ABSTRACT

During the last decades, computers and the programs that give them "virtual life" have been used as powerful research tools, even in developing countries. Moreover, the narrow availability of some devices, such as neutron activation analysis, or the high investment they involve do not represent an excuse for avoiding their application in archaeological issues, mainly thanks to encouragement of multidisciplinary research teams.
Constraints of finance, manpower staff and time all conspire to limit what may be achieved, and compromises are inevitable. In this sense, computers and the Internet offer the easiest and most economic means to communicate our research results, both, within the academic community and to the society. However, their "virtual nature" is still argued against "reality" when archaeological interpretations are evaluated by a scientific committee.
It is the wide spread of computers in archaeology and the conservative attitude of academic institutions in South American centres that are of most interest here since they show the "biases of integration" of this facility and the possibilities of avoiding them in the future.

The strength of this paper lies upon the concepts and issues improving the relationship between Heritage and Public through Internet. Over the past few years, Computers and the programs that give them "virtual life" have being brought to the attention of the archaeological community as powerful research tools, even in developing countries. No scientific fieldwork can be imagined without the aid of GPS or GIS, no laboratory results can be managed without the use of statistical tools or computerized typologies or classifications, and no information can be shared without the existence of specific archaeological databases. Furthermore, some analytical methods applied to pottery for its mineralogical characterisation, for instance, the neutron activation analysis widely used for provenance studies, require very sophisticated software and equipment. This situation is solved with the emergence of multidisciplinary research teams, providing a place where chemists, physicians, architects, anthropologists and archaeologists work all together for common interests. So, the narrow availability and/or expensive costs of some analyses are set aside if only by prompting the scholarly investigation. Anyway, these new technologies have been well accepted as research tools and ignored as scientific communicators.

Since the developments in science and technology introduce more comfort to the modern life, Society invests on those branches of knowledge rendering evident benefits. Among Social Sciences, disciplines related to Heritage have recently acquired social relevance, however, the benefits obtained for the ordinary life are not so immediately evident and the government investments are not so generous. Therefore, constraints of finance, manpower staff and time all conspire to restrain what may be achieved in Humanities, and further compromises are inevitable in pursuit of scientific aims. Here, the problem refers partly to the social role of Archaeology. As a discipline interpreting material culture of past societies in order to understand present social diversity, Archaeology should assume its role as a nexus between the Public and their Heritage making more evident the benefits of knowing our past for a present identity building. In this sense, Computers and the Internet offer the easiest and most economic means to communicate the insights resulting from long-term research work, both within the academic community and to the society.

In spite of the advantages of using Information Technologies for communicating archaeological insights about Heritage, main Argentinean government scientific committees are still reluctant to accept the "virtual nature" of these devises as a suitable reference of the "more real" publication in archaeology. Moreover, electronic publication via Internet is regarded as a non-scientific press, and therefore few archaeologists spend their efforts in this media. As a result of this, Archaeology is not able to show the social benefits of fulfilling the gap between Society and its Heritage, and the archaeologists are not allowed to go beyond the academic "bubble".

WHAT IS ALREADY KNOWN?

We order the world in terms of the things it represents and we order things surrounding us in terms of our points of view. Our points of view depend on our previous training and education, and so they are as diverse and changeable as the identities built up by distinct human groups. Thus, Heritage can be an object, custom, or quality, which is passed down over many years within a nation, social group, or family, and is thought of as something valuable and important which belongs to all its members. This feeling of belonging is the very root of the resulting identity. In other words, Heritage is the meaning traditionally borne by any object instead of the object itself, hence the sciences dealing with the Past should assist in the construction of the present identity of any particular group.

Our career in Humanities teaches us that Archaeology is the study of the past as evident in the material remains available to us. But, does it matter who we are and how we see ourselves? Yes, because we can not assist in the construction of any identity if we are not able to identify ourselves. As Fagan (1996:vii) points out "Archaeology in the popular imagination is the stuff of which dreams are made, a world of adventure, and romance, of golden pharaohs, and long-vanished civilisations. Many people still believe archaeologists are tough, pith-helmet-clad men and women slashing their way through clinging jungle or penetrating the secrets of ancient pyramids... The transformation of archaeology from an amateur pastime into a sophisticated, multidisciplinary area of study is one of the great triumphs of twentieth-century science". Fortunately, we are scientists in Humanities but it is still remain to define our social role for the Society whose citizens we are.

Then, what is the place of archaeology in the contemporary world? On the surface, its small place is clear enough. Archaeology consists of noticing, preserving and rescuing buried fragments from ancient and not-so-ancient times, and then striving to understand earlier peoples and societies from those fragments. However, we think that its scientific process include three compounds. The mechanics of archaeology - finding old sites, preserving them, digging and recording and interpreting what is found - are the routine procedures of a field science. The people who perform these tasks - academics in universities, bureaucrats in historic preservation offices, field workers out in the mud - are engaged in the routine of work in our society. The means of communicating their findings - archives, database records, specialised books, magazine articles, and television programs - are the routine ways of dealing with knowledge. And all these three exist in order to reveal the real place of archaeology in the contemporary world (Fagan 1996). Although, as professionals in Humanities, we are able to manage these mechanics and to define these people, little interest is paid on establishing these means of communication (cf. Wolle 2002). This fact leads to the current situation figuring the scientists as people dealing with odd topics completely useless for the society.

Computers and Programs as Research Tools

As Archaeology is about the physical side of our history, it suppose the use of Natural Sciences techniques to process, order and transmit data with regard to the distinction quoted above. For example, multivariate statistic analyses, GIS applying statistical tools and spatial analyses, CAD simulation, tessellations and triangulations procedures, computerized classifications and typologies, and specific database are widely used for accomplishing current archaeological praxis. Sometimes, this implies the emergency of multidisciplinary research teams in order to achieve the appropriate data.

In the level of data processing, multidisciplinary teams offer the possibility of applying high technology to archaeological material with well-trained staff and low costs. Neutron activation analysis, for instance, would be impossible without the laboratory equipment and the computer program (Jones 1991). In the level of data ordering, the task of computerising the collections of a museum or a public institution is a long-term solution for a good housekeeping, i.e. DOMUS is an ambitious program co-ordinating collections from all main Spanish museums (MARQ 2003). Moreover, some projects using virtual infography or 3D reconstruction such as ORION or PEREGRINUS (MARQ 2003) provide the means of avoiding the handle of inaccessible, fragile or minute objects by researchers or visitors.

Now, Computers and programs are confined to serve only as research tools? No. In the level of data transmission, nowadays, computers and the Internet are vital even as teaching tools for long-distance training and updating courses hosted by Universities or Academic Institutions all over the world, including developing countries (cf. Grupo NAyA 1996-2003, Martín Guglielmino 1996).

What is always forgotten?

At the beginnings of Archaeology as a Human Science, the romantic school had established scholars should be in complete isolation to understand the general laws ruling the world. This requirement was suggested to guarantee the scientist's objective interpretations. What was then forgotten referred to our condition of citizens. Since the early '80s, the critical theory has promoted the social consciousness of actors in the present and their relations to the past, fostering important changes into the scientific ontology. After defining our discipline as a Social Science, what is always forgotten is to establish our role in the society. In this direction, archaeologists should be regarded as social technicians assuring the link between Society and Heritage. That is, Archaeology plays a leading role in defining what the Society would like to know through its past, in order to understand its present, giving so the categories to deal with and communicate Heritage (cf. Martín Guglielmino 1996).

The scientific isolation produces the illusion that a few bits of history are possessed as "heritage", because the rest is just something that happened a long time ago and has not much to do with the real and contemporary world. The twenty-first century society is largely persuaded that the world is - or should be - a rational place directed by the logic of scientific reasoning and based on the tested facts of gathered data. So, we are in charge of ourselves, not determined by where we come from, nor driven by the beliefs we have carried along. Therefore, defining Heritage may give place to problems related to its possession, signification, management and preservation (Fagan 1996). This definition varies from people to people and time to time. To this sense we have to know how homogeneous is the society whose heritage is under study because heritage is a social construction based on the meanings people gives to things, facts or feelings in order to build up its own identities. And this is what we should never forget.

THE LINK BETWEEN SOCIETY AND HERITAGE

Thinking on Archaeology, as one of the sciences dealing with Heritage capable of shedding light upon certain past meanings and so assisting in the construction of present identities, it is obvious that it should facilitate the link between Society and its Traditions. That is the reason why archaeological knowledge would not be restricted to the academic community. Nevertheless, government scientific committees will have an important part in this enterprise, since they own the criteria to discriminate rigorous interpretations from fiction re-creations of what had happened long time ago. Therefore, Society would acquire appropriate information upon which rely its memories. In this respect, George Henri Riviere (quoted in Martín Guglielmino 1996:16) asserted in particular that Heritage is "all those material and immaterial goods on which, as a mirror, society looks at in order to recognise itself, where it searches for the reason of the landscape where its roots are and where the ancestors lived before it. A mirror that society offers to its guests to make sense, on respect to labour, ways of behaviour and its intimate". Why don't we help to clear up this mirror?

The current social, cultural and political trends encourage society to search for the ways of knowing more and more about its past and present. In addition, Information Technologies improve their performance and increase the number of users daily. And the Internet seems to be a low cost-effective media to communicate whatever it should be known, because the information can be download freely from the World Wide Web. Since archaeologists are not computer specialists a new kind of multidisciplinary team would arise; but this time to solve problems related to the third compound of the scientific process mentioned above, that is, Internet as a scientific knowledge communicator. It must be understood that this new kind of publication in Archaeology would not necessary replace the traditional one but complement it in order to achieve the social role of archaeologists as technicians assuring fluid dialogue between Heritage and Society (who would become the Public of the new development).

STATE OF AFFAIRS IN SOUTHAMERICAN ARCHAEOLOGY. THE CASE OF ARGENTINA

Nowadays, Argentinean Archaeology follows the precepts of American and British Archaeology. However, it faces major finance constraints, less manpower staff and minor time to achieve the ordinary praxis due to its situation of developing country. So, initiatives such as ECAI, or programs as ArchEd, the S-Plus package, TotalStation devices, or facilities as Orthophotomosaics are not w idely spread in Argentina. Anyway, the more traditional analytical methods, such as NAA or DXR, are accomplished by the emergence of diverse multidisciplinary research teams. Hence, Computer and programs have shown to serve as effective research tools even when they are a still a non-so-common equipment.

As the cultural process of globalisation goes on, the reality of societies in developing countries are not so different as those in developed ones. The main distinction between developing and developed countries refers to the purposes of the inquiries prompted by the Society to the Science about the Past. Nowadays, the Past has become part of the Argentinean tourist government projects due to the cultural and landscape diversity of the country-side, rendering a vast tourist offer. Anyway, society is eager for learning about its pre-Columbian or Historic times promoting the advancement of new ways of attaining it.

Citing examples of the social need of a fluid dialogue between Society and Heritage, CICOP or the International Centre for Heritage Conservation in Argentina appeared as a result of the interests of a group of professionals concerned with architectural heritage of modern American cities. Argentinean society suffers a traditional scarcity of archaeological knowledge because it is always kept away from the massive media in order to maintain its "scientific character". As we mentioned above, this fact isolates the scientists from their society and their social role are then completely erased from their academic agenda. Consequently, although heritage projects such as CICOP have good reasons for their existence, the not-yet-resolved lack of Humanities scientists taking part in these endeavours constrains their concept of Heritage to the materials themselves; and so the present identities building introduces some inevitable biases.

Meanwhile, ethics in Archaeology gained great attention improving the process of scientific knowledge and giving place to the rise of specific professional associations. In 1997, APRA or the Professional Archaeologists Association in Argentina established the Ethics Code and the Professional Praxis Standards as its main goals. Although this current trend in determining the ways of doing Archaeology, the main government scientific institutions and research foundations still show a little conservatism about communicating the archaeological knowledge to Society, therefore the debate about Heritage is restricted to academic meetings. For instance, CONICET or the National Committee for Scientific and Technological Investigations is the main Argentinean institution that controls and promotes Science improvement. Its conception of academic excellence increasingly makes null and avoid the social role of heritage technicians that archaeologists should soon start to practice. Thus, the absence of stimuli for social commitment that funds the projects produces a difficult gap to be filled in for an individual scientist. And then the social need of archaeological knowledge is wrongly satisfied by other related disciplines. Frequently, government archaeology departments are headed by architects or engineers because the social benefits of their disciplines are evident for the society whilst archaeology looks like a hobby. Of course, our isolation is the only responsible of this situation.

It is possible to think that the reluctance of the academic community to spread the scientific knowledge to society bases on the high costs of the traditional publication and of the propaganda needed to attract the public. But this is not the case due to the widespread of Internet throughout the academic domain. For example, since middle '90s University of Quilmes offers long-distance curses for those students that can not attend normal classes. Anyone even can complete the inscription on-line, and some tutorial chats are also foreseeable. By the same time, University of Rosario launched a place in Internet where to meet old students and inform themselves about the social and cultural activities promoted by this institution. Moreover, projects such as the Virtual Anthropological Museum of San Luis shows the availability and well-acceptance of Internet as an useful educational tool. In this sense, NAYA or the Anthropology and Archaeology Virtual City has proved to be vital to the development of certain issues such as Cultural Tourism offering the second Virtual Congress for this topic next October (Grupo NAyA 1996-2003). Unfortunately, the government scientific committees do not support these results and the possibility to evaluate the knowledge communicated to the Society is lost.

As a conclusion, Computer and Internet have shown to be important research and training tools in Argentinean Archaeology; even their communicator capability is well accepted. The problem remains in what we think we are and what we must do as social actors. Heritage is vital for the Society identity building and maintenance, so without a thorough knowledge about the past our present is somehow blurred. Our academic committees should get out of the "bubble" they are immerse, and the only mean to achieve this change is showing the benefits of it. We are aware about "how to do scientific archaeology" and now we should learn about "how to communicate this scientific archaeology to the Society". As Information Technologies are versatile tools there is no necessity of searching for new software or equipment, we only have to use them properly. Therefore, the dialogue between Heritage and Society through Internet is the possible way to do so, mainly in developing countries.

Final Words about Virtuality and Reality

The virtual character of most research and teaching tools, specialised websites and cyber-meetings is a challenge to the reality of specialised books, academic papers and oral presentations. The Internet has shown to be a low-cost fast device to spread knowledge inside the academic community, in developing countries such as Argentina.

As the means of communication are available, the scientific community in Humanities needs to take into account their social consciousness in order to preserve the Heritage constituting its study case. Archaeologists want to study past human lives, but those lives are gone. We see them only by proxy; they are shadows we strain to see from the artefacts, so often worn and broken, they chanced to leave behind. The archaeologists' is a strange view of the world, with its odd basis looking at artefacts in quest of the essence of human lives. It is a view congruent with our society - so cluttered by artefacts, yet so uncertain and so ignorant of how those artefacts may direct our lives and how we define what is Heritage (Fagan 1996).

If Heritage lays the basis of social memory, the ways by which the social group recognises itself and apprehends its social values should be under scientific study. Education in Heritage is a cultural management strategy bridging Heritage and Society (cf. Martín Guglielmino 1996, Wolle 2002), in such a way, that it will help to fill the gap between Heritage Associations and the academic community.

Thus, what should be done? In a word, to assume our social consciousness of cultural heritage technicians and our scientific commitment to the society, encouraging the routine way of dealing with knowledge. This is not a hard task. And Internet offers the tools to achieve these aims.

Acknowledgements

I should like to thank CAA 2003 Organisation Committee for the financial support to present this paper. I am indebted to Ignacio Jimenez, Scholarship Co-ordinator at Instituto del Patrimonio Histórico Español, for his opinions on the importance of internet in Cultural Heritage Research and Management which gave place to the discussion presented in these pages. I acknowledge this topic deserves a deeper treatment than the brief one introduced here, and wish further contributions to the issues considered.

References

FAGAN, B. (ed.), 1996. The Oxford Companion to Archaeology. Oxford University Press, New York.

GRUPO NAYA, 1996-2003. Ciudad Virtual de Antropología y Arqueología (www.naya.org.ar).

JONES, L., 1991. Computerising the collections: the art of successful flea handing. In Bowman, S. (ed.), Science and the Past, British Museum Press:172-181.

MARQ, 2003. I Jornadas Museos, Arqueología y Nuevas tecnologías. Museo Arqueológico Nacional - Museo Arqueológico Provincial de Alicante, Madrid.

MARTÍN GUGLIELMINO, M., 1996. Reflexiones en torno a la difusión del patrimonio histórico. Cuadernos Difusión del Patrimonio Histórico, Instituto Andaluz del Patrimonio Histórico, Junta de Andalucía, Consejería de Cultura:14-27.

WOLLE, A., 2002. Electronic Publication in Archaeology. In Weathley, D. (ed.), Contemporary themes in archaeological computing, Graeme Earl and Sarah Poppy, University of Southampton, Department of Archaeology Oxbow Books Monograph 3:65-72.

INTERWEAVING DIGITAL NARRATIVES WITH DYNAMIC ARCHAEOLOGICAL DATABASES FOR THE PUBLIC PRESENTATION OF CULTURAL HERITAGE

RUTH TRINGHAM
UNIVERSITY OF CALIFORNIA, BERKELEY, USA

SEE THE CD FOR THE EXTENDED VERSION

ABSTRACT

This paper describes a project that is currently being carried out in collaboration with the Interactive University of UC Berkeley in which a relational database comprising the contexts, materials, observations, and interpretations in visual, numerical and textual format from Neolithic excavations in Turkey (Çatalhöyük) and Yugoslavia (Opovo) provides the basic architecture for a series of hyperlinked "vignettes". The vignettes comprise an interface on the Internet through which a multi-generational and multivocal public can explore and respect and contribute to cultural heritage. They are guided through the exploration of real data and can contribute dynamically with their own interpretations. The emphasis here is on multiscalar interpretations including those at an intimate scale, which are followed through and evaluated through the fictional narrative genre. The aim of the project is to create an interface, that seamlessly guides, and provides access to, primary archaeological data in a way which is neither intimidating nor mystifying, but which engages the user and encourages critical enquiry, responsibility and the use of the imagination in the creation of heritage places by lifelong learners.

The Dig OpChats project is a collaborative enterprise at the University of California, Berkeley between the Archaeological Research Facility, the Interactive University (Outreach initiative) and the Multimedia Authoring Center for Teaching in Anthropology (MACTiA).[1] The project began in 2002, but was the outcome of a previous collaboration in 1999 in which we had designed teaching modules for 6th graders around primary research data at Çatalhöyük for 6th graders (12 year-olds) and an after-school project in digital storytelling and archaeology.

THE CHALLENGE

The project focuses on the creation of interactive digital learning objects from two archaeological sites that are both sites where I have conducted excavations: Çatalhöyük and Opovo. Both are Neolithic, that is, early agricultural settlements. Opovo is a 6,000-year-old village in Yugoslavia distinguished by burned wattle-and daub houses. I directed archaeological excavations here in the 1980s (Tringham et al. 1992). During the early 1990s the data were digitized post-excavation, from which was created the Chimera Web (Wolle and Tringham 2000).

The Chimera Web is a hypermedia web that was designed to be a public interface and guide to the primary data but - until the current project - never succeeded in this aim. The problem was in the underlying architecture, which was in a variety of formats, including 4th Dimension. At that time these formats were difficult for an amateur like myself to integrate with a dynamic web interface. And I did not have the resources - or even desire -to farm the problem out to specialists in Silicon Valley.

The second site, - Çatalhöyük - is 9,000 years old and characterized by mud-brick houses, some with painted plaster walls, with burials under the floor (Hodder 2000). The houses were rebuilt one on top of the other so that their debris accumulated to form a mound. Since 1997, I have been directing the BACH team (Berkeley Archaeologists at Çatalhöyük) in the Çatalhöyük project (http://www.mactia.berkeley.edu/catal/default. html), whose overall director is Ian Hodder (http://catal.arch. cam.ac.uk/catal/). Along with the reflexive methodology, we have advocated open access to the primary data of the excavations through the Internet (ref). In practice "open access" has been conceived of as sharing data with other archaeologists. If you look at the Çatalhöyük database on the web-page (http://catal.arch.cam.ac.uk/catal/database/main.html), this is not exactly user-friendly. The user needs to know what to search for in the database in terms of Building, Feature or Unit number, There is no guiding interface on line that links to this database. A more public and popular interface for Çatalhöyük was developed through the Science Museum of Minnesota Mysteries of Çatalhöyük web site (http://www.smm.org/visitorinfo/nowshowing/Mysteries. html), but this is not directly linked to the archaeological database.

The challenge in both cases, therefore, has been to give the public access to the raw data of archaeological excavations in a way that neither overwhelms with the amount of data nor mystifies with the hidden meaning of the data, yet does not simplify the complexities of the data and their interpretation.

THE DIG OPCHATS PROJECT

The goals of the Interactive University in the collaborative Dig OpChats Project were very different - and in many ways much more ambitious - than ours had been in the previous public interfaces of the Çatalhöyük and Opovo projects. The Interactive University were interested in taking our[2] experience with public web-based interfaces of archaeological research to another level of dissemination and to use it as a pilot for the dissemination and archiving of university faculty research to the public. Their primary interest was K-12 and higher education audiences - hence the heavy use of education language in the project. We have since become more inclusive to reach public audiences in museums, continuing (lifelong) education, and other contexts, because it is politically a wise thing to do (the aging of the American public) and (for us as archaeologists) because we hope it encourages the larger public to take responsibility for cultural heritage.

The initial goals of the project were:
- To explore the process of transforming faculty research into learning materials for K-12 and other public audiences
- To record the process so that it seems less intimidating and mystified to the content creators (faculty researchers) who should participate actively in creating the digital archive and its interfaces
- To safeguard the data for sustainable long-term re-use and archiving.

In the last goal - the safeguarding the data for sustainable long-term re-use and archiving - I have been especially inspired by a couple of recent articles that stress strategies to ensure the long-term sustainability and survival of the databases. The first is by Julian Richards (Richards 2003), who stressed the importance of open source access to ensure the use and re-use of digitized primary archaeological data for archiving and for keeping the data alive. Thinking about dissemination is as important as thinking about digitizing. The second is by Stephen Downes (Downes 2003), who has advocated for a distributed network of learning objects and their metadata - open access rather than the "silo model". Both these authors recognize the importance of a flexible networking dissemination of digitized data rather than a centralized silo that stores the data. They also both advocate the interoperability of different systems of database architecture, rather than reliance on proprietary architecture created for specific sets of content.

With this in mind, the technical goals of the Dig OpChats project included

- Embracing the technical dimensions of database interoperability (ensuring the framework is ODBC compliant)
- Striving for open architecture in the project framework
- Exploring the conversion from FilemakerPro-XML (the Dig OpChats database system) to additional learning standards in XML formats (e.g. METS/IMS) enabled by the - Scholar's Box (see below).

In our first understanding of the project, this meant in practice that we would build web-ready interfaces based on archaeological data from my research at Opovo and Çatalhöyük. The interfaces that we created would be designed to give multiple audiences access to real-world archaeological databases, and they would be open-ended to the extent of encouraging an expandable web of investigation taking the learner in many different directions as well as provide opportunities for public feedback.

In order to make this project feasible, we have focused on a couple of the enormous array of possible themes suggested by the data. These are themes that could bring both sites together and are both of current interest in my own research - an important motivation if you are to get faculty researchers to do this kind of work (Tringham 2000). Most researchers - like myself - are not enamored of the kind of work that enters into the realm of textbook writing. And I believe that specific research themes can be of interest to public audiences and do not have to be shared only with professional audiences. In addition, it is a pleasure to think of ways to express your current research to non-professional audiences. And the results and feedback can be extraordinarily constructive. The two themes that were chosen are

- How history and historical places are constructed in the modern and post-modern world
- How the material world and ritual performance (especially around domestic architecture) is used - and was used in the past - to create a sense of place and its continuity

Much of the research at Çatalhöyük focuses on the burial of the dead under the houses as a way of creating a continuous place. One interpretation of the Opovo burned houses suggests that this was done deliberately as a means to creating a continuity of place.

The progress of the project starting in April 2002 may seem to have been straightforward but it was very much a non-linear process. We started with the problem of the public interfaces, because that had always seemed to me the primary importance - how to present what we do to the public. We focused on creating the interface as vignettes as our initial "deliverables". A "vignette" was defined as a "web-based presentation, similar to "learning artifacts" or "learning objects" comprising media plus text to illustrate a point, a concept, a lesson - a 1-2 minute non-linear narrative, linked to other vignettes around a theme and/or site. You could think of it as a kind of postcard. Vignettes have metadata associated with them (for indexing, searching, and transferability across digital environments). But the same problem that I had faced in the Chimera Project started to emerge: how in practice were we going to link the interface to the database? It was halfway through the year that it became clear we had this the wrong way round. Rather than developing interfaces that would be "linked" to the databases, the interfaces needed to grow out of a database architecture that was designed for them, that would act as a bridge to the archaeological project databases that were designed for professional "in-house" exploration of the data.

The Vignette Guides

Thus from August 2002 we started to focus on creating the Dig OpChats database system that would form this bridge and make the archaeological databases more accessible. Using FilemakerPro 6, there have been several iterations of the logical and technical structure of the interoperable Dig OpChats database system. The vignettes comprise an interpretive expression that is built from assets (various forms of media), texts, and numerical data that reside within the Dig OpChats database system. Keywords and other metadata (including the source of the media and texts) are used to categorize the vignettes and link them to their related data, and to other Dig OpChats vignettes.

The Scholar's Box that has been developed by Raymond Yee of the Interactive University has become the pivot for multiple users to search not only the Dig OpChats database but any accessible interoperable database (such as the California Digital Library). The search starts with the vignette (which becomes essentially a guide to the search), and ends with the re-use of the components in a contribution that is fed back into the database by the user (http://interactiveu.berkeley.edu:8000/ IU/News#continued).
The vignettes illustrated in the CD version of this paper focus on the Neolithic site of Opovo. Each vignette has a contextual multiscalar timeline and a contextual menu that leads the user to the vignette-guides. The CD version of the paper illustrates the relationship of one vignette to another, as well as the database components that create the vignettes.

Each vignette also has a fixed menu:

- "Things" leads to different categories of materials we exca vate, how they are recorded and examined: images, VR movies, numerical data, reports on material analyses, links to materials, use-life of things, ethnographic analogies.
- "Places" leads to large places, the sites and their landscape context; and about more intimate places from the individual houses whose histories we investigate to pieces of furniture (features) inside the house. It leads to images and movies, Quicktime VR movies, landscape reconstructions, GIS data, provenience records, Harris matrices.
- "People" leads to the archaeologists, to tourists, to imagined prehistoric people, including the individuals we excavate as burials and many others. It leads to audio and video inter views with archaeologists, images and movies about excavating, recording, interpreting and analyzing, team organization, special jobs, diaries, how we get to where we work, where we come from.
- "Narratives" are the macroscale narratives that archaeologists write about (pre)history as well as more intimate interpretations of objects and places, and fictional narratives; narratives take the form of published texts, diaries, stories, reports, syntheses, quotes, and thoughts.
- "Activities" are exercises, classroom assessment techniques, assignments, quizzes, role-playing games related to the vignette.

The prompt "Feedback" leads to a significant aspect of the Scholar's Box/Dig Opchats Project - the idea of vignettes inspired or related to a Dig OpChats vignette can be created by the user and added to the Dig OpChats database The model for such user contributions to a database was recently piloted at University of California, Berkeley in a course on European Archaeology taught in Spring 2003 in the Multimedia Authoring at the Multimedia Authoring Center for Teaching in Anthropology (MACTiA) (http://www.mactia.berkeley.edu).

In this course many of the ideas of re-usability of data and feedback and user-contribution that will become part of the Dig OpChats interface has been piloted. A Filemaker database of archaeological sites with associated images is the starting point for student contributions. They do research from suggested sources in the library and Internet and then select images from the database and write text that draws together the images into a specific theme (such as "cultural heritage"). Their contribution is created using a variety of possible templates, media-rich, text-rich, equal media and text). The student contributions are similar to the vignettes of the Dig OpChats project. Links are made to other student contributions about the same theme, or site, or by the same author. This is how we envisage the "Feedback" part of the Dig OpChats Project.

It is this aspect of the Dig OpChats Project will ensure the re-usability of archaeological data and the engagement of users in its re-interpretation and re-configuration that, we believe, will keep archaeological places, things and people alive in a global and local society.

[1]In using the word "our" and "we" in this paper, I am not referring to myself as the professorial "we", but mean essentially myself in collaboration with my archaeological colleagues at UC Berkeley, including Margaret Conkey and Rosemary Joyce, and especially Michael Ashley-Lopez.

[2]The project team comprises Professor Ruth Tringham (ARF): the creative director, archaeological content provider, Isaac Mankita (IU): the producer and administrator, Michael Ashley-Lopez (MACTiA): archaeological database designer and media director, Raymond Yee (IU): technical director and developer of the Scholar's Box, with David Greenbaum (IU director).

REFERENCES

DOWNES, S., 2003. Design and Reusability of Learning Objects in an Academic Context: a new economy of education? USDLA (US Distance Learning Association) Journal 17 (http://www.usdla.org/html/journal/JAN03_Issue/article01.html).

HODDER, I. (ed.), 2000. Towards reflexive method in archaeology: the example at Çatalhöyük by members of the Çatalhöyük teams. McDonald Institute for Archaeological Research, Cambridge.

RICHARDS, J., 2003. Digital Preservation and Access. European Journal of Archaeology 5(3):343-366.

TRINGHAM, R., 2000. The Continuous House: A View from the Deep Past. In Gillespie, S. and Joyce, R. (eds.), Beyond Kinship: Social and Material Reproduction in House Societies. University of Pennsylvania Press, Philadelphia, PA.

TRINGHAM, R., BRUKNER, B., KAISER, T., BOROJEVIC, K., RUSSELL, N., STELI, P., STEVANOVIC, M. and VOYTEK, B., 1992. The Opovo Project: a study of socio-economic change in the Balkan Neolithic - 2nd preliminary report. Journal of Field Archaeology 19(3):351-386.

WOLLE, A. and TRINGHAM, R., 2000. Multiple Çatalhöyüks on the World Wide Web. In Hodder, I. (ed.), Towards reflexive method in archaeology: the example at Çatalhöyük by members of the Çatalhöyük teams, McDonald Institute for Archaeological Research, Cambridge:207-218.

A Methodology for Evaluating Archaeological Web Sites

Nicoletta Di Blas
Politecnico di Milano, Milan, Italy

Franca Garzotto
Politecnico di Milano, Milan, Italy

Maria Pia Guermandi
IBC, Bologna, Italy

Franco Niccolucci
Università di Firenze, Florence, Italy

Abstract

Even if nowadays almost every cultural institution, including archaeological museums and sites, owns an institutional web site, there are few investigations on evaluation criteria. MiLE, a methodology for web sites evaluation, was recently specialized for cultural web sites by a team of experts from Politecnico di Milano, original developers of the general methodology, and from cultural institutions, coordinated by IBC. It precisely defines scenarios, i.e. abstract tasks performed by abstract users: evaluators assign marks to specific attributes, as clarity or accessibility, acting as the supposed user. Thus evaluation may take into account the target user population and the intended web site goals. A group of students in Architecture from the Università di Firenze was selected to perform an extensive application of this methodology to a large sample of web sites of archaeological institutions, including practically all such Italian sites, a number of Spanish and Polish ones and a sample of sites from other European and extra-European countries. The results give an insight into the effectiveness of such Internet presentations and test the validity of the methodology, suggesting an extensive application of it and the adoption of widely accepted and objective guidelines for multimedia cultural communication.

Introduction

Evaluating the effectiveness of multimedia for cultural communication is a task which as yet has been performed basing only on the intuitive appreciation of their global appearance. Objectives, intended audience and services to be offered are often unspoken and evaluation is carried on, when it is, with an empiric, subjective and qualitative approach. The quantification of multimedia quality remains therefore somewhat fuzzy, with no guidelines to make it as objective as possible, and comparing different applications or evaluating the attainment of a quality threshold, for instance, rely on a totally individual and perhaps unreliable judgment. Multimedia play, on the contrary, an increasingly important role in communicating Cultural Heritage and for some authors they are in fact the best, or only, solution to store and communicate the intrinsic reflexivity, contextuality, interactivity and multivocality of the archaeological record (Hodder 1999, Wolle, Tringham 2000, Biehl 2002, Tringham 2003); others have evidenced that Cultural Virtual Reality applications need to improve substantial aspects as validation, annotation and philological approach (Frisher et al. 2002), which also possibly impact on the archaeological method and theory, so far as valid 3-D models can provide new tools to archaeological investigation (Niccolucci 2002); finally, the widespread presence on the Web of cultural institutions, including museums and archaeological sites, does not clearly correspond to a well-tested business model, i.e. a set of criteria that guide the planning and enable to evaluate the effectiveness of such cultural web sites. So a specialized and effective methodology for the evaluation of multimedia is an urgent need not only for the valid communication of culture but also for the definitive acknowledgement of the theoretic importance of such technology in the realm of archaeological theory.

The present paper is a first step in that direction. It starts from the application of the MiLE evaluation method, developed at Politecnico di Milano, to cultural web sites as proposed in (Di Blas et al. 2002, Bolchini et al. 2003, Triacca et al. 2003) and it tests the proposed methodology on a large set of archaeologically related web sites. The evaluation was carried on by students in Architecture as part of their yearly assignments for the course of "Urban Models" held by Prof. Niccolucci at the University of Florence in Autumn 2002. The students were concluding their studies (5th year, in practice corresponding to what is called elsewhere "graduate students") and their skills were appropriate for the job, which consisted in filling on-line forms; moreover, they exercised in performing the task during a seminar and were supervised and received remote assistance by the teacher while completing it. For each evaluation record they also had to prepare a short comment to justify the scores, an useful feedback to tune the methodology. All the scores and report were reviewed by the supervisor and in a few cases the evaluator was asked to correct mistakes.

The sites under evaluation were chosen by the teacher to represent a wide panorama of Italian web sites with a good number of Spanish and Polish ones; a selection of other European sites, mainly from UK with a few German and French ones, were also taken into account; other sites dealt with non-European institutions, mostly from USA. The reason of the selection were language skills: students were asked to evaluate sites using their mother tongue, with a few more in a foreign language they declared to understand. A group of Erasmus Spanish and Polish students, attending the course, enriched the international flavour of the experiment.

The experiment had multiple goals: to test the method on a large group of evaluators; to evaluate the archaeological web sites; and to analyze sites usually less considered, for language reasons, than those in English. For space reasons, it is impossible to describe here all the sites under evaluation or to appropriately cluster them (e.g. those related to large institutions, those related to museum collections versus those related to archaeological sites, etc.) and analyze the results in terms of such clusters. However, the original data (i.e. individual marks and scoring) will be available for some time on the web and a more detailed analysis will be performed in future work.

SUMMARY OF MILE

MILE is based on two main concepts, the Abstract Task (AT) and the User Profile (UP).

An Abstract Task (AT) is a general type of action that can be performed by visiting a web site only, e.g. "Get practical information for a visit as opening time, ticket cost, etc." or "Get the data necessary for writing a student's report". An AT needs to be defined in a unique and clear way. ATs may be classified according to scope as specific, complex or general, and according to concern as practical, operational and cognitive.

A User Profile (UP) is a general, but detailed, description of the user visiting the web, e.g. "A university student in Humanities, female, with good knowledge of English and a good (wide bandwidth) Internet connection" or "An adult (male) person, with average cultural background but no knowledge of foreign languages, with curiosity for archaeological treasures". The combination of an AT + a UP makes a Scenario.

A Web site has attributes, that is relevant properties, which are scored by the evaluator. Attributes may depend on the scenario, but it is preferable to give an overall list and then choose only the relevant ones. The current overall list of attributes includes the following:

1. Efficiency: the action can be performed successfully and quickly.
2. Authority: the author is competent in relation to the subject.
3. Currency: the time scope of the content's validity is clearly stated. Information is updated.
4. Consistency: similar pieces of information are dealt with in similar fashions.
5. Structure effectiveness: the organization of the content pieces is not disorienting.
6. Accessibility: the information is easily and intuitively - accessible.
7. Completeness: the user can find all the information required by the AT.
8. Richness: the information required is rich (many examples, data, etc.).
9. Clarity: the information is easy to understand.
10. Conciseness: the basic pieces of information are given; texts are not too long and redundant.
11. Multilevel: different levels are available according to user's profile.
12. Multimediality: different media are used to convey the information.
13. Multilinguism: the information is given in more than one language.

As stated before, attribute relevance may depend on scenario, which therefore will also include a relevance coefficient for each of them.

A web site is then evaluated:

- Considering all possible scenarios (or those considered relevant or defined as such by the site mission)
- Giving marks to attributes by direct inspection, using an agreed scale as 1 = very poor to 10 = excellent (0 = N.A.).

For the evaluation it is therefore necessary beforehand:

- To list all ATs
- To list attributes
- To list all UPs
- To create all scenarios by matching every AT with every UP, discarding incoherent couples
- To decide which attributes are relevant, and how much, for each scenario, that is to assign the attribute relevance coefficients for each scenario
- To decide which scenarios are relevant, and how much, for the site, that is to assign relevance coefficients for each scenario as far as the objectives of the evaluation are concerned.

After this preliminary process there will exist scenarios si, with attributes aj and relevance coefficients wik (0 £ wik £ 1) defined for each attribute ak and scenario si

- The evaluation is performed by assigning marks xik to each ak assuming to be in scenario si
- The overall mark mi for the site regarding scenario si is then given by the weighted average

$$m_i = \sum_k w_{ik} x_{ik}$$

- The same for the overall score S, the weighted average of mi using weights pi (0 £ pi £ 1) expressing the relative importance of each scenario in the site's overall goals, or simply averaging them (pi =1 for all i)if the site mission is unclear or unspoken.

$$S = \sum_k p_i m_i$$

This procedure standardizes the evaluation task as far as possible. ATs, UPs and attributes are defined by the evaluation team as well as weights, but subjectivity may be reduced using an agreed set of such features and in any case evaluation transparency is highly improved by the availability of evaluation criteria. Weights may take into account objective factors (e.g. the actual incidence of specific visitors' types) and the mix of scenarios may be precisely tuned to the target audience and the desired objectives of the Web site. The same method may be generalized for other multimedia applications, e.g. 3D or VR models, what will be the object of future work.

The Evaluation Experiment

In order to test the method and to have an extensive evaluation of archaeological museum web sites, the experiment was carried on as described in the Introduct ion. It was decided to privilege coverage (i.e. the number of Web sites under evaluation) versus depth (the number of scenarios taken into account for each Web site), to achieve a better understanding of the method effectiveness and have a first insight into the quality level of archaeological web sites. In other words, we were less interested in thoroughly evaluating individual web sites than in extensively examining of a number of such sites, to test the impression that in most cases these added little, if anything, to the visitors' understanding and satisfaction. Unfortunately, as it will be shown below, this was in fact the case.

After searching the Internet, 164 sites were selected for evaluation by the 18 students, but only 134 were in fact evaluated for different reasons (unavailability of the web sites, fault of the evaluator, excessive complexity of the site for the scope of the task, etc.). Of these, 98 related to museums, 23 to archaeological sites, 4 to complexes including a museum and a site, 1 to a temporary exhibition, and 8 concerned networks of cultural institutions.

The geographic (and linguistic) distribution of the sites was the following:

State	IT	PL	ES	UK	DE	FR	Other EU	Total EU	US	Other non-EU	Total
Number	72	18	9	8	4	2	8	121	9	4	134
Percentage	53.7%	13.4%	6.7%	6.0%	3,0%	1,5%	6.0%	90.3%	6,7%	3,0%	100,0%

IT, PL, ES, DE and FR sites were examined in their home language; the others in the English version, which of course corresponds to the home language for UK and US sites. Other European include sites from Belgium (2), and one each for Denmark, The Netherlands, Sweden, Croatia, Switzerland and Greece. Other non-European include sites from United Arab Emirates, Malaysia and Israel (2).

The test was performed on scenarios deriving from the following User Profiles and Abstract Tasks:

UP	Name	Description
1	Student	M/F High school student; good general culture; language knowledge at a school level
2	Educated tourist	Adult tourist (M/F) age 30-40; good general culture; cultural interests over the average; fair language knowledge

AT	Name	Description	Concern	Scope
1	Historic knowledge	To get general information about the historic context	Cognitive	General
2	Tourist evaluation	To evaluate the potential interest of a visit	Cognitive	General
3	Data collection	To get information about the holdings of the institution	Operational	Complex
4	Information collection	To get operational information for a visit	Operational	Complex

The scenarios under evaluation did not correspond to all possible combinations (UP, AT) but only to the significant ones: s1 = (1, 1); s2 = (1, 3); s3 = (2, 2); and s4 = (2, 4).

Results and Conclusions

As one of the goals of the experiment was testing the method on a rather large sample of persons, the scoring has been examined to verify the behaviour of the evaluators. Scores concentrate in the range 4 to 7, partly for statistical reasons and partly, perhaps, as it is typical of questionnaires, in which response tends to concentrate on central values. They show that variability is a personal factor and scores should perhaps be normalized before comparing the evaluation by different evaluators. However, the difference among scenarios was rather well perceived by students since scores for different scenarios about the same site show appreciable differences.

Of the 72 Italian sites, 40 (56%) have been evaluated below sufficiency, i.e. score below 6 over 10, and 24 (33%) are barely sufficient, between 6 and 7. Only 8 (11%) are "good", with no excellent ones. This bad evaluation is mainly due to very low scores for attributes related to multilevel, multimediality and multilinguism, all receiving an average score between 3 and 4 ("very poor"), while other attributes generally receive an average score between 6 and 7 ("sufficient"). The situation of the 9 Spanish sites (perhaps too small a sample to be significant of the Spanish presence on the Web in this field) is similar. Polish sites, a larger sample with 18 cases, do not behave much better, with 4 evaluated as "very poor" and a majority of "insufficient" ones. Also in this case the fault is caused by poor multimediality and multilinguism, with also some lack of consistence in the presentation of pages. All evaluated sites are in fact very weak as far as multilevel presentation is concerned, that is the possibility of graduated approaches for different users.

In conclusion, the experiment has shown that the method is feasible and requires little training. The results confirm that the use of the Internet as a communication tool for archaeological heritage is far from optimal and in most cases still unsatisfactory: the pages do not avail of the multimedia potential of the web and in general add little to printed text, which perhaps is still the reference communication model for most curators.

It is the intention of the authors to report in greater detail the aggregate results of the evaluation, here summarized for space reasons. Moreover, they intend to complete the investigation on archaeological web sites in order to obtain a larger sample and report comparable evaluations for the archaeological presence on the web. A first attempt aiming at evaluating virtual reality archaeological recreations with a similar methodology is also on the way, and some preliminary outlines were discussed during a seminar (2003) at the Cultural Virtual Reality Lab at UCLA.

REFERENCES

BIEHL, P.F., 2002. Hypermedia and Archaeology: Towards a Methodological and Theoretical Framework. In Niccolucci, F. and Hermon, S. (eds.), Multimedia Communication for Cultural Heritage, Archaeolingua, Budapest:147-153.

BOLCHINI, D., SPERONI, M. and TRIACCA, L., 2003. MiLE: a reuse-oriented usability evaluation method for the web. In Jacko, J. and Stephanidis, C. (eds.), Human - Computer Interaction: Theory and Practice (Part I), Vol. I of the Proceedings of HCI International 2003, LEA Lawrence Erlbaum Associates, Publishers Mahwah, New Jersey.

DI BLAS, N., GUERMANDI, M.P., ORSINI, C. and PAOLINI, P., 2002. Evaluating the Features of Museum Web Sites. In Bearman, D. and Trant, J. (eds.), Museums and the Web 2002, Selected Papers from an International Conference, Archives & Museum Informatics, Pittsburgh, U.S.A.

FRISHER, B., NICCOLUCCI, F., RYAN, N.S. and BARCELÒ, J.A., 2002. From CVR to CVRO: The Past, Present, and Future of Cultural Virtual Reality. In Niccolucci, F., Virtual Archaeology, Archaeopress, Oxford:7-18.

HODDER, I., 1999. The Archeological Process - An Introduction. Blackwell, Oxford.

NICCOLUCCI, F., 2002. Virtual Archaeology: an introduction.. In Niccolucci, F., Virtual Archaeology. Archaeopress, Oxford:3-6.

TRIACCA, L., BOLCHINI, D., DI BLAS, N. and PAOLINI, P., 2003. Wish you were usable! How to improve the quality of a Museum Web site. In Cappellini, V., Hemsley, J. and Stanke, G. (eds.), EVA 2003 Florence, Proceedings, Pitagora Editrice, Bologna.

TRINGHAM, R., 2003. Interweaving Digital Narratives with Dynamic Archaeological Databases for the Public Presentation of Cultural Heritage. This volume.

WOLLE, A-C. and TRINGHAM, R., 2000. Multiple Çatalhöyüks on the World Wide Web. In Hodder, I. (ed.), Towards reflexive methods in archaeology: the example at Çatalhöyük, McDonald Institute Monographs, Cambridge:207-217.

Virtual Reconstruction of Viennese Synagogues: Sustainable 3D Models

Bob Martens
Vienna University of Technology (Austria)

Herbert Peter
Academy of Fine Arts (Austria)

Abstract

Computer-aided reconstruction of no-longer existent synagogues amounts to a "virtual comeback". Irreversible destruction having removed identity-establishing buildings from the urban surface forever is the principal reason for recreating them by imagination. Following the destruction of the so-called "Reichskristall-Night" of November 1938 the synagogues of the Jewish community in Vienna will only survive by means of virtual reconstruction. 60 years later, in the commemorative year of 1998 the first synagogue reconstruction was initiated. The medium-range goal, however, aims at the reconstruction of at least ten further synagogues within a project to be carried out in stages over a period of several years. Changes in personnel also call for a structure to be tracked down later on. This contribution deals with handling of modelling in a systematic manner taking into consideration software fluctuations aiming at a traceable data structure being of importance for subsequent use and following-up work.

Introduction

The basis of reconstruction work proves to be well-founded archive material determining the validity of virtual reconstruction to a high extent. The majority of the present reconstructions concerns synagogues dating from the second half of the 19th century. Due to the "thoroughness" of the building authorities in Vienna and their resulting requirements governing submission plans and alteration planning (1:100 scale) this source is to be considered highly reliable. The technical journals dating from the time of completion of building, however, issue relevant building descriptions for some cases. Photographs also represent an essential information source as they depict an actually realized situation. Any pictures stored in the various archives mostly are black-and-white shots showing the exterior. The major part is picture postcards not to be regarded as unique specimens, whereas the number of interior shots is pretty low.

Modelling Conditions and Setup

Continuing developments in the field of computer-assisted modelling techniques as well as the implementation of knowledge acquired in cooperation with art historians led to advancements concerning modelling procedures. Modelling, moreover, is tackled by different people (individuals and teams) and in differing frame conditions (workshops, diploma theses and commissioned projects, etc.), very easily resulting into considerable confusion regarding data organization. Generally speaking, the structure principle of separation of building members within a simple floor structure will not prove adequate for keeping "track". In order to provide sufficient vision to a reconstruction it is essential to develop structuring relying on the available sources prior to commencement of producing the virtual model to be based on the CAD-program used.

The aspect of usability at a later time is an essential prerequisite for subsequent modelling procedures. First of all, planning documents of the reconstruction object to be dealt with are structured according to constructive criteria. Wall structures in their differing functions (interior and exterior walls) are to be identified and supplemented by supporting pillars, ceilings, intermediate ceilings, staircases, roof constructions, roof covering, framework (roof truss), facade elements, ornaments, furnishings etc., in order to ensure the required overview within the data organization for the changing user community at all times. In any case coherent (geometric) elements within the respective layer are to be documented in form of an individual three-dimensional representation. Specification of layers and the graphic representation of the specific contents should determine the building components in a suitable manner. Representation of the layer contents should preferably be shaded. A wire mesh model does not lend itself to this form of documentation as all clarity is done away with as soon as a great number of building elements within one layer are involved.

A nomenclature has not been provided for the layers, however a documentation clearly explaining the layers used regarding their name specification and content is to be available. A pre-defined layer management limits the possibilities for this project series (reconstruction of synagogues) regarding modelling and does not prove meaningful considering the maximum of approx. 20-30 layers. Moreover, the building structure has already been defined and will not be resulting from a design process to be performed subsequently.

The implemented software package ArchiCAD® as basic set already provides a number of functions meeting the demands as to required geometry modelling, storey administration and layer allocation. The subsequent steps of material identification and texture verification can be performed not involving any complicated additional efforts. Furthermore, compatibility with other CAD-programs makes for data transfer to other software applications without major information loss. Development of the ArchiCAD-intern language (GDL - Geometric Description Language) doubtlessly represents a real achievement. One of the major advantages of GDL surely is the compact file size next to the open programming language (open source).

The main term of the ArchiCAD-product philosophy is the concept of the Virtual Building not primarily focusing on the production of technical drafts and pictorial rendering, but rather on the information production these are based on in form of a structured project database. Regarding virtual reconstruction the object-oriented working mode makes ArchiCAD so interesting. The possibility of subsequent re-using once produced GDL-objects (object-oriented modeling) over and over again seems meaningful. The option of parametrification not only includes the geometric constellation but also the definition of varying material and colour properties and selectable options. Additionally, further non-geometric particulars and functional properties can be programmed by means of the GDL-script language matching the Basic-computer language in its essential features.

tion". Particularly when several individuals are involved in the course of project work the storey structure helps figuring out correct positioning of building parts within the three-dimensional space.

- Determining a layer structure:

The number of layers to be associated with the matching building parts is to be specified and the criteria for allocation of building elements are to be selected according to constructive aspects. It might be possible that building elements of one layer might fall below or exceed the imaginary horizontal boundaries of the storey management.

Figure 2a-b Layer "exterior walls" per storey at the Synagogue Leopoldsgasse

Example of Implementation: Synagogue Leopoldsgasse

In the course of the project work regarding this synagogue the frame conditions for the structured data model generation were determined. It not only is the geometry defining this structure but also any related (object-specific) parameter, such as material and texture. What is assumed is that the storey management makes up the "horizontal structure", the layer management, however, the "vertical structure" of the building. The terms "storey" and "layer" in this context refer to the software package ArchiCAD®. Below the systematics are described step-by step:

- Research work concerning plan documents, picture material
  and descriptions
- Comparing plans with photographs
- Definition of a storey structure:

Every building element within a virtual reconstruction is to be assigned to a storey. As many stories as desired can be produced which do not have to be identical with the storey structure of the plan documents. It might even prove wise to work with intermediary storeys if a great number of ornaments or ceiling elements occur above the ideational "one-meter-sec-

After completing the reconstruction all layer contents is to be documented. Therefore, first all layers are to be faded out and subsequently the contents of each single layer are to be rendered. Regarding the example Kluckygasse the layers are specified as follows (selection):

- Compiling materials used

Each element on one layer and in a storey provides of a specific surface colour (corresponding to the material). This colour is related to the single geometric faces and thus can differ within one object. The element as such, however, is not split up between the various layers.

- Determining textures

A further step regarding classification is the material texture assigned to the specific surface colour. A texture is the graphic rendering of a building material, which is projected to the geometry of the building element when a photorealistic rendering is being generated. Special effects of photorealistic rendering possibilities of a CAD-software may also issue additional light, gloss or reflecting effects, influenced by various light sources within and/or outside the building model.

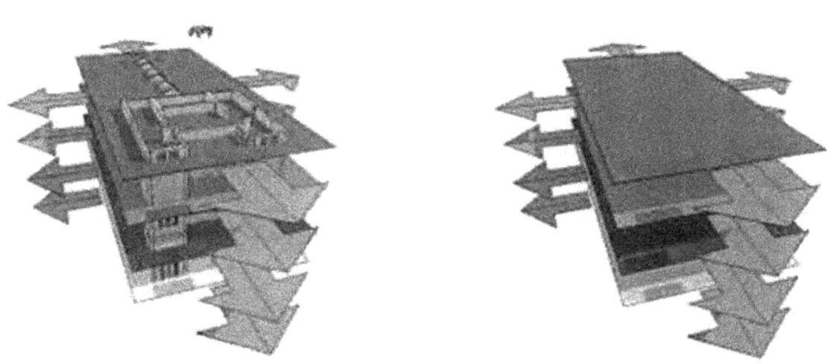

Figure 1a-b Storey management in the Synagogue Leopoldsgasse

- Compiling library elements and modules

The construction of project-related building elements is achieved by implementing all ArchiCAD-standard tools available. Building parts stored as library elements are to be stored as so-called "modules" in a specific directory. This procedure enables any following alterations of individual library elements.

- Archiving project files

Finally, all project data are to be stored in a clearly laid-out directory structure. An expansion by individual directories can be furnished whenever required (e.g. regarding textures). The so far unrecorded interior configuration of synagogues can be comprehensively visualized. By means of comparative studies of related buildings dating from the same time and typical material and surface effects a material- and colour-related reconstruction can be developed. Individual materials are to be specified by means of data organization of the building elements in order to avoid the already mentioned tiresome subsequent editing. Compilation of materials used and specification of textures can be carried out at a later stage, if required information is missing. An impression of the reconstruction of the Synagogue Leopoldgasse is issued below.

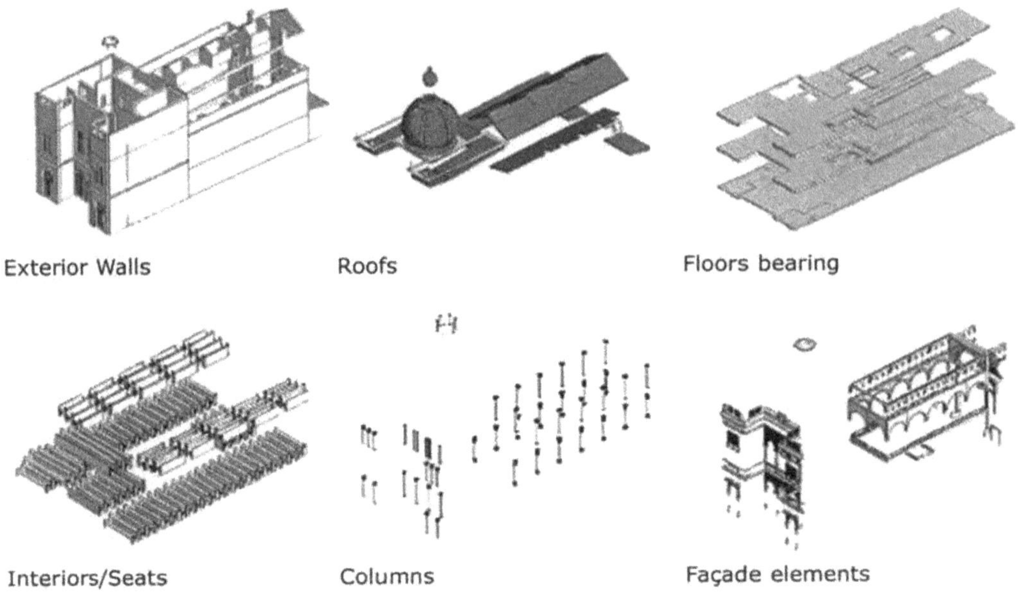

Figure 3 Layer documentation Synagogue Leopoldsgasse

CONCLUSIONS

This paper provides a contribution establishing a relationship between a technique increasingly used in architectural history with a more rigorous process of data analysis and authenticity. CAD is often employed with less care than the authors propose. What is to be considered is that reconstruction work furnishes the specific levels of detail, i.e. any further art-historical findings are to be included subsequently. We might be dealing with longer periods of time where the initial team is no longer involved in modelling work. Moreover, it is difficult to specify if a certain CAD-software package will still be available a few years later. These reasons might severely affect advancement of work and therefore investing in systematic maintenance will be beneficial and pave the way for further utilization - possibly also in a different CAD-environment.

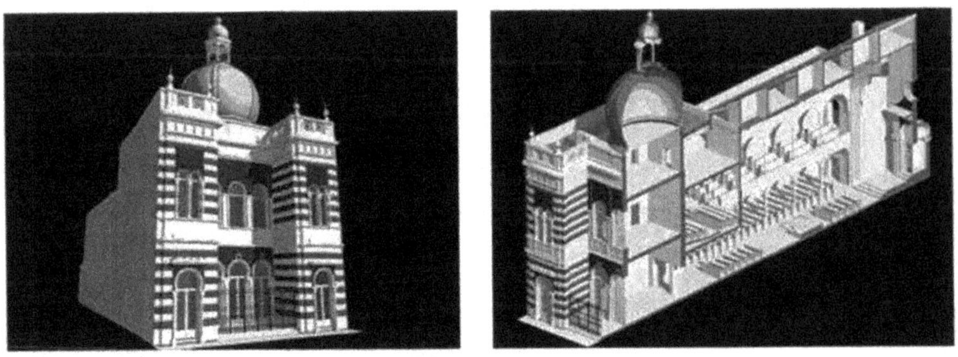

Figure 4a-d Findings of reconstruction work

REFERENCES

GENÉE, P., 1987. Wiener Synagogen 1825-1938. Löcker Verlag, Vienna.

MARTENS, B., UHL, M., TSCHUPPIK, W.-M. and VOIGT, A., 2000. Synagogue Neudeggergasse: A Virtual Reconstruction in Vienna. In Kos, R. (ed.), Constructing the Digital Space, Proceedings IVth Sigradi-Conference Rio de Janeiro:165-170.

PETER, H., 2001, unpubl. Die Entwicklung einer Systematik zur virtuellen Rekonstruktion von Synagogen. Diploma-Thesis, Vienna University of Technology, Vienna.

WALLMÜLLER, D., 2001, unpubl. Computergestützte Rekonstruktion des Leopoldstädter Tempels. Diploma-Thesis, Vienna University of Technology, Vienna.

Cultural Heritage, Public and Architecture in the Time of new Information and Communication Technologies

SEE THE CD FOR THE EXTENDED VERSION

Marc Grellert
Technical University of Darmstadt,
Department CAD in Architecture

Abstract

Architecture has at all times belonged to the important cultural achievements of mankind. From ordinary buildings to prestige buildings, architecture portrays a substantial part of the cultural heritage of human societies. Through the centuries much architecture worldwide was lost. Digital technology, such as three-dimensional computer simulation, has revolutionized the possibility of the presentation of such lost architecture and brought it back to mind for a greater public. First I would like to briefly outline the great variety of these digital possibilities on the basis of selected virtual buildings and projects. These virtual projects were developed in the digital labs of Prof. Manfred Koob from the Technical University of Darmstadt, Department CAD in Architecture - projects that fascinated me years ago and prompted me to go on these digital travels to other times and cultures. Thus I was involved in most of these projects. However the focus will be on a project, which I initiated in 1994 at Darmstadt Technical University. It is the computer reconstruction of the destroyed German synagogues.
Second I would like to cover the potentials of the Internet in context to cultural heritage, using again the synagogues as an example.

CAD and Cultural Heritage

With the 3D-CAD reconstruction of Cluny in the end of the 1980s and the showing of the simulation on TV, it was suddenly evident that a powerful fascination had been generated. This had never been achieved by drawings and architectural models before. Architecture that no longer existed could now be presented as moving pictures to a large public through the mass media television. In the 1990s this insight also reached museums and exhibition halls.

Now there was also the chance to present ideas that had never been built. An example is the so-called "Wolkenbügel" by El Lissitzky. (Fig.1)

Figure 1 Wolkenbuegel by El Lissitzky , 3D-CAD simulation © Darmstadt Technical University, CAD in Architecture, Prof. M. Koob

Figure2 Cathedral of Speyer, 3D-CAD simulation © Architectura Virtualis GmbH,Darmstadt

Figure 3 Schlüter Hof Berlin Castle, 3D-CAD simulation © Architectura Virtualis GmbH, Darmstadt

Also technical processes, for example, construction work in the Middle Ages, can be excellently illustrated. Machines can be shown in action. The Romanesque cathedral of Speyer documents this. (Fig.2)

The destroyed castle in Berlin is for me an example that a virtual model could give a more honest answer, suitable to the times, than an ideological overloaded facade built on a former location.(Fig.3)

A computer model, such as one of the Vatican Palace, also provides the opportunity to show more clearly the spatial function and meaning of still existing historical buildings in cases where these can no longer be comprehended. (Fig.4)

Buildings like the Boyana church near Sophia, which can only be visited by a few people each day in order to protect the important Middle Age frescoes, can thus, with the help of the computer, be accessible for many people. (Fig.5)

The reconstruction on the computer, which shows all distortions, and the precise photographic portrayal of the entire church have made it also possible to simulate first on the computer the urgently necessary restoration work on the frescoes.

The virtual reconstruction of the destroyed German synagogues transports the cultural heritage of a social minority, which was discriminated, expelled and murdered. The visualization of the destroyed architecture is thus part of a general discussion of remembrance, the remembrance of the holocaust.

The destruction of the synagogues denotes in this a beacon light. Those who destroyed the cultural buildings of a minority, want to exclude it from the society - with all consequences. (Fig.6)

Figure 4 Vatican Palace - loggia of Raphael, 3D-CAD simulation
© Darmstadt Technical University, CAD in Architecture, Prof. M. Koob

Figure 5 Boyana Church, 3D-CAD simulation
© Darmstadt Technical University, CAD in Architecture, Prof. M. Koob

Figure 6 Synagogue Nuremberg, 3D-CAD simulation
© Darmstadt Technical University, CAD in Architecture, Prof. M. Koob

Thus the intention of this project was manifold:

- To show the cultural lost
- To bring back to the public the historical importance of the synagogue especially as a characteristic of the cityscape
- To convey to the public spatial impressions of an almost unknown architecture
- To give a representative overview of the architecture of synagogues in Germany
- To make aware that the synagogues are a part of a shared cul- ture of Jews and non Jews
- To remember the circumstances of the destruction
- To remember the desecration and destruction of over 1100 synagogues just in November 1938
- To ask what led to this and what happened afterwards
- To make a contribution to educational and commemorative work for the use in schools, universities and museums in connection with the Nazi period by using the new media.

The idea for this project stems from the impression of an arson attack of a synagogue in Lübeck in 1994 by neo-Nazis, Thus the project should be also seen as a signal against anti-Semitism and racism, which were again gaining strength.

We used an architectural selective criteria to select the synagogues. We chose large cities with the expectation that our intention would reach more people. And we selected the synagogues from the 19th and 20th century, because this was the time of the birth of the great synagogues in Germany.

One can classify the synagogues in the 19th and 20th centuries into three different groups:

First, the Neo-Islamic synagogues which emphasized the independence of the Jews by using an extraordinary architectural style. For example, the synagogue of Kaiserslautern. (Fig.7)

Figure 7 Synagogue Kaiserslautern, 3D-CAD simulation
© Darmstadt Technical University, CAD in Architecture, Prof. M. Koob

Figure 8 Main Synagogue Munich, 3D-CAD simulation
© Darmstadt Technical University, CAD in Architecture, Prof. M. Koob

Figure 9 Synagogue Plauen, 3D-CAD simulation
© Darmstadt Technical University, CAD in Architecture, Prof. M. Koob

Second, the Neo Romanesque synagogues with an emphasis on the affiliation of the Jews with the German nation. For example, the synagogue of Munich. The Romanesque style was used for the old German cathedrals like Worms or Speyer. (Fig.8)

Third, synagogues which follow contemporary or regional styles. For example, the Bauhaus synagogue of Plauen. (Fig.9)

These virtual synagogues were shown in the year 2000 in an exhibition at the Federal Art and Exhibition Hall of Germany in Bonn. Institutions and museums from all over the world showed great interest and there will be an exhibition tour, starting this year on the 9th of November in Tel Aviv which then travel to the United States in 2004.

Culture Heritage and Internet

The new information and communication technologies, especially the Internet, have changed and are changing many spheres of society. Using the Internet will be self-evident for the generations of the 21st century. This will have an influence on the culture of remembrance. I ask myself what are the potentials and possibilities of the new media for cultural remembrance. Which deficits of the traditional forms of remembrance could be compensated? What are the limits of the new media?

I believe
- the Internet can become a new form of cultural remembrance
- the Internet can develop new forms of sharing the process of remembrance
- the Internet can allow people all over the world to participate - independent of time or space
- the Internet can offer a framework for personal involvement
- on the Internet information can always be updated

In my research work I have developed various concepts, which include the remembrance of the synagogues using the new media. One of these concepts is the Synagogue Internet Archive: http://www.synagogen.info.

Here one can
- call up information about more than 2000 German synagogues
- make external contributions and add eyewitness reports, images, comments and links to the individual synagogues, which re directly online.

With this Internet archive the German synagogues that were closed, desecrated or destroyed during the Nazi regime are to be remembered. At the same time, information is given about what happened to the buildings still remaining and the former locations after 1945. Starting point of this concept is the opinion that remembrance is very significant when people take an active part in it. As opposed to traditional forms of remembrance like monuments, where visitors normally only are passive, the Internet user can contribute to the expansion of the Synagogue Internet Archive.

This archive allows for the first time an overview about synagogues in all of Germany. And the Internet archive makes it easily possible to actualize information and to add information, which is a problem for other media, such as books. With this archive it is also now possible to obtain more exact quantitative information. For example that an unbelievable number of more than 350 synagogues and prayer rooms were torn down after 1945.

The Synagogue Internet Archive has tree main menus:

- Searching for synagogues
- Looking for new external contributions
- Background information

[1] Searching for Synagogues

Here either the synagogues of a federal German state are displayed alphabetically, for example, all synagogues in Bavaria or one can search for a particular synagogue in a city or on a street. The selected synagogue will be first presented by so-called base information, which is listed below:

- Information about the duration of the synagogue
- The date and the extent of the damage during the Nazi regime
- Details about a possible demolition
- Information as to what remains of the synagogue today
- Information about in what way the synagogue is remembered - what is now on the former location.

This information is based on bibliographical sources, which can also be found in the archive. This is the scientific part. First sources were used which give an overview about synagogues in a certain region or in a federal state. Thus a very large number of synagogues and prayer rooms could be included, while keeping the necessary amount of work within reasonable limits.

In the menu of the base information one can also see if there are any external contributions. For Bad Kissingen, for example, there are 35 images, 12 comments and 2 eyewitness reports registered. If one wants to call these up, one can choose the images, the comments or the eyewitness reports and a short list is displayed, from which one can select respectively. From this menu the user also can add information to the archive. For example, images that are called up over a personal file manager can be entered into the archive through the Internet. All entered external information is immediately online.

[2] New Entries

A second main menu point is the display of newly entered contributions. These are shown sorted according to the date. The last contribution is always listed first. As an example, one can find an eyewitness report to the synagogue in Munich from a user from England or a link to the synagogue in Essen or an image from the burning synagogue in Darmstadt, which was part of an estate and has never been published before. As far as the links are concerned, first a short description is displayed and then a new navigation window with the respective site can be clicked.

[3] Background information

The third main menu contains background information as a timetable to the Nazi regime, a short summary about the history of Jewish sacred buildings, a glossary and impressions from the virtual synagogues. There is also a guest book on this internet site.

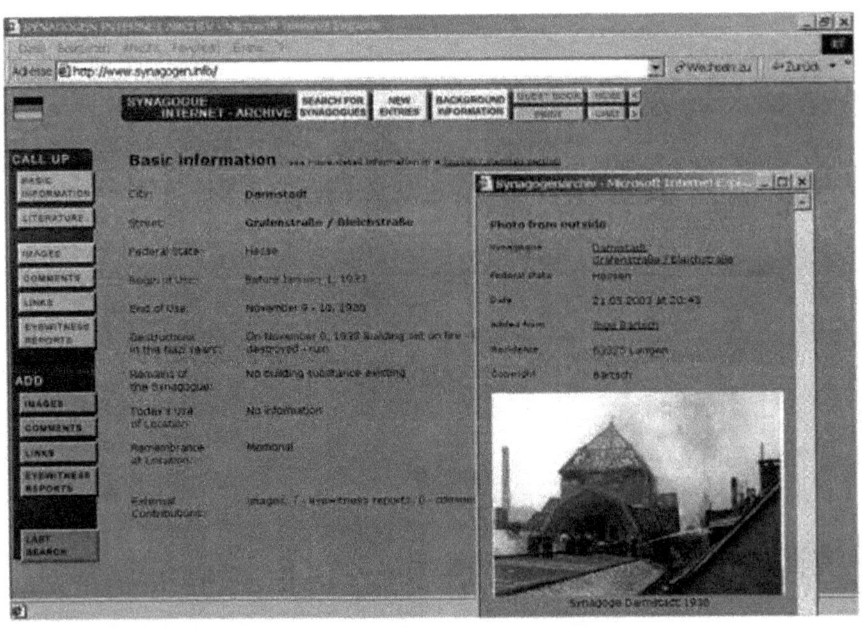

Figure 10 Synagogue Internet Archive, base information with external contribution, http://www.synagogen.info

A first analysis about the archive, which is now online for 6 months, has shown very satisfactory results. Until now there have been over 35,000 accesses of the starting page and about 560 external contributions have been received. The concept of the direct online publication of the external contributions has proved to be feasible because of high quality contributions. There were only a few technical errors on the part of the user and the small number of insulting and unqualified external contributions have been limited to the guest book and were able to be kept from publication without any problem. After the Congress in Vienna, a team was formed by Austrian participants of the congress who will expand the archive to include Austrian synagogues.

In closing one could say that the possibilities of cultural remembrance using the new media does not make it an unnecessary issue in the real world. There must be a parallelism of the sites. Historical authenticity at physical sites and immaterial remembrance in the immaterial digital sites. Both must play out their specific qualities.

One future possibility could be the combination of virtual buildings and authentic location. New forms of cultural remembrance do not lessen the obligation to think about what is to be remembered. Remembrance in context to society and the conveyance of cultural heritage always has something to do with a message for the next generation. In the case of the synagogues, this means the visualization of the loss as an incentive to reflect upon German history. I hope that this also strengthens the sensitivity pertaining to the exclusion of minorities today. (Fig.10)

REALONLINE - IMAREAL'S DIGITAL IMAGE-SERVER

INGRID MATSCHINEGG
INSTITUT FÜR REALIENKUNDE DES MITTELALTERS UND DER FRÜHEN NEUZEIT, KREMS

ABSTRACT

REALonline was launched as a web server in spring 2001 by the Institute of the History of Material Culture of the Middle Ages and the Early Modern Period (IMAREAL), which is affiliated with the Austrian Academy of Sciences. This image-server has evolved from a computer-supported project dealing with the collection and analysis of medieval pictorial sources, mainly of central European provenance, with regard to their information on everyday life and material culture (Thaller 1984:219-228). A continuously growing photographic archive of thousands of elder black-and-white photographs and newer colour slides represents the basis for these investigations. The digitization of this photo-collection has been completed; any new records are immediately stored as digital images. In its present state (June 2003) REALonline contains approximately 16,000 objects from the Middle European area, covering a time span from 1200 to 1600. Furthermore, the systematic photographic documentation activities of the institute yield an annual increase of about 1,000 documents (Vavra 1999:156). The database includes panel paintings, mural paintings, manuscript illuminations, sculptures and preserved original objects (objects of every day life, arms and armours, all kinds of instruments etc.).

TEXT AND IMAGE

REALonline is a searchable image and text database. The verbal description of the images is based on a standardized data-model: The first and essential part of every description contains the formal documentation of the image, i.e. all of the available documentative information on the work form, the type, the title, the artist and the date of the creation as well as the place, country and name of the institution where the work of art is kept and bibliographical references. The second part comprises the iconographical description of the contents of the image. This section consists of a definition of the scene or scenes - one image may consist of a number of different scenes - with subordinate groups for the depicted actions, people and objects (Jaritz 1993:23-39).
Departing from these specifications, we developed a model which arranges and displays the information in hierarchical relations. Our data-model is comparable to the new XML-document-type-definitions with nested structures of repeatable elements and attributes. The images are described in German language and in standardised terminology. Using verbal descriptions of pictorial sources (or of original objects), we are creating our own texts. We thus have to provide the means to make the documentation and description as standardised and as close to the source as possible. Moreover, the terminology we use should be accessible to other researchers in history or cultural studies. At the same time we have to decide to what extent we can use contemporary language in order to classify codes or a terminology derived from the language of medieval written sources. Finally, we must be aware of the fact that our descriptions of the images will never be complete but that we are working with a "living database" to be modified, corrected and extended, whenever new information is available (Jaritz 1993:17).

THE WEB-INTERFACE

The input data is organised according to the rules of our database-software-system KLEIO. Developed by M. Thaller at the Max-Planck-Institute for History in Göttingen/Germany in the 1970s, this program caters for the specific demands of research performed in history and cultural studies (Thaller 1989). The web-interface for REALonline was constructed by Thaller (now at the Institute for Computer Sciences in Humanities University of Cologne/Germany) (Thaller 1998).

In order to search the database, you have two options (see illustration 1). You can choose the main-criterion for your search on a dropdown-list which includes terms such as "topics" or "scenes" and "personal names" as well as "social ranks", "actions" or "gestures". Apart from selecting a main criterion, you can also restrict your search to a specific time in between the years 1200 and 1600. Finally you can determine whether you would like the display of the results to contain basic or extensive information, documentation and details of the respective images by ticking the relevant boxes. In addition to the browsable list of search terms, the program also offers a field for free search.

If, for example, you need material on medieval castles, you can either select the catalogue containing "material objects" and proceed to find the term castle on the browsable list of terms in this specific catalogue. Or you can go to the end of the navigation bar and enter the term "castle" into the search field which also allows right truncation using an asterisk. If you first want to find out whether the database contains any material on the topic in which you are interested, you can use the button "Show terms". The program will then select the closest match on the list of available search terms. If, however, you prefer to use the button "show images", this particular search will process a number of minimized pictures of

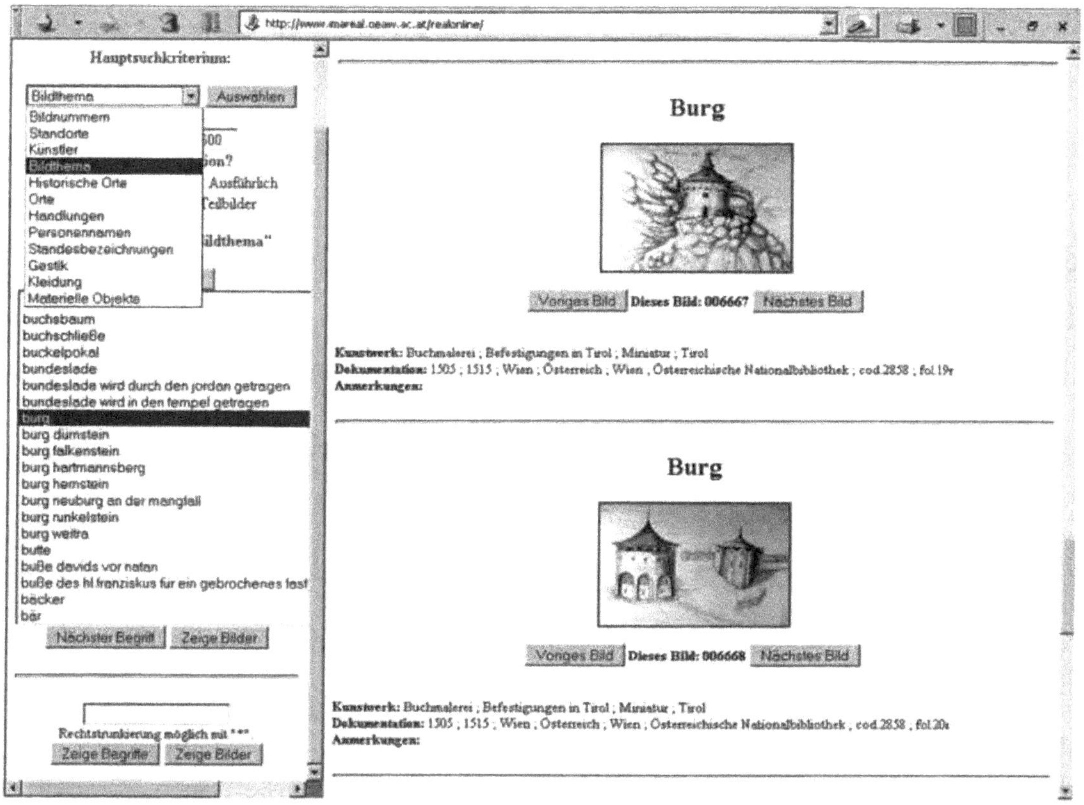

Figure 1

castles in the main frame. In order to view larger versions, you can click on the thumbnails. Some of the thumbnails will not load a larger version of the respective image but inform you that the owner of copyright did not give permission for the institute to provide a higher resolution version on the internet. The thumbnails include extensive descriptions which also list details of the depicted objects. Specific terminology used for the descriptions of medieval clothing can be looked up on a linked clothing dictionary. Other hyperlinks provide information on artists, bibliographical references and text from the bible.

Images of Medieval Everyday Life?

Images contain information on people, objects, situations etc. and their connections or relations respectively. They have to be analysed with an awareness of their role as media for communication between humans, either as their producers, artists, patrons or the theological authority on the one hand, or their recipients of different status, knowledge and needs on the other. Images consist of signs and symbols to make this communication possible. These are based on patterns of intention (Baxandall 1985:1-11). Hence, the interpretation of images certainly does not end with the description of their contents, but must aim to analyse these communicative elements. In trying to come closer to the everyday life and the material culture of the Middle Ages, we are interested in the signs, symbols and patterns of intention. A very important role, for example, is played by medieval symbolism. An image was intended to motivate its spectators, above all the illiterates, in a moral or didactic way. It had to contain such signs and patterns as made those motivations possible and easily applicable. These signs were often elements drawn from the everyday knowledge of the people. They were based on things which were very well known or even used by the beholders. They did not stand for themselves, but were to lead the recipients to the (theological) message of the image. When investigating the image's role as a source for everyday life in medieval society, we must always be aware of this fact. The main emphasis in using the database REALonline, therefore, has been placed on finding signs, on analysing their relevance and on interpreting the 'reality' of such sign-bearing objects in everyday life (Jaritz 1993:17-18).

Homepage of REALonline (http://www.imareal.oeaw.ac.at/realonline).

References

BAXANDALL, M., 1985. Patterns of Intention. On the Historical Explanation of Pictures. New Haven and London.

JARITZ, G., 1993. Images. A Primer of Computer-Supported Research. Halbgraue Reihe zur historischen Fachinformatik A 22, St. Katharinen (http://www.imareal.oeaw.ac.at/ias_primer).

THALLER, M., 1998. Bilddatenbanken im Internet. Kolloquiumsbericht, München (http://www.bayern.de/HDBG/t97thall.htm).

THALLER, M., 1989. Kleio. Ein Datenbanksystem Halbgraue Reihe zur Historischen Fachinformatik B 1, St. Katharinen (http://www.hki.uni-koeln.de/kleio).

THALLER, M., 1984. Mittelalterliche Realienkunde und EDV. In Die Erforschung von Alltag und Sachkultur des Mittelalters, Veröffentlichungen des Instituts für Mittelalterliche Realienkunde Österreichs 6 = Sitzungsberichte der Österreichischen Akademie der Wissenschaften, phil.-hist. Kl. 433, Wien:219-228.

VAVRA, E., 1999. REAL - eine interdisziplinäre Datenbank zur Alltagsgeschichte. In Das Mittelalter 4/2:156-159.

Replication in Archaeological Information Systems

Nisha Joseph

Damian Green

John Cosmas

Take Itegaki

Department of Electronic and Computer Engineering
Brunel University, Uxbridge. UB8 3PH. UK
nisha.joseph@brunel.ac.uk

Abstract

This paper describes the 3DMURALE database, which follows a generic database structure, for the storage of a wide range of data, applicable on a diversity of excavations. The database stores 3D models of buildings, stratigraphy, statues and artifacts and is accessed through a newly developed GIS entitled Strat. This research database is accessible via the Internet 24/7 and can store and retrieve multiple campaign data concurrently. This paper discusses the problem of remote replication, on sites without high-speed access and proposes a viable Replication Tool.

Introduction

This paper presents a new database system, 3DMURALE, which seeks to fulfil the needs of not just one archaeological campaign but may be of use on any archaeological campaign. The database follows a system design, which allows replication and is designed to efficiently accommodate various types of archaeological data. Therefore, this database would be ideal in a modern archaeological campaign, which not only deals with textual data but also multimedia data such as images, 3D data, photogrammetric models and video. The paper also discusses various important issues in the development of sophisticated multimedia databases.

Terminology

In an increasingly inter-connected world moving towards greater integration of information systems, one finds a plethora of databases following different systems and standards in the field of archaeology. Archaeologists are yet to agree on the rules which could ensure that archaeological database systems are standardized which would make it possible to inter-link them and make them accessible globally.

In 1995, Arroyo-Bishop stated, "We cannot allow a myriad of databases to develop independently, each with their own different themes, structures, indexes and vocabularies" (Arroyo-Bishop et al. 1995). Contrary to Arroyo-Bishop's aspirations, examples of individual database systems appear regularly, for example Akasheh (2002) recently presented another archaeological database specific to their excavations, at the Amman Cultural Heritage conference.

Before development of digital storage system for archaeological data can commence, terminology needs to be unambiguous and clearly defined.

A number of confusing and potentially misleading terms have come into use to define the elements that make up archaeological sites (Barker 2001). Words such as feature, artefact, find, stratigraphic unit, context and layer are terminology commonly used in archaeology. Some of these terms are interchangeable, for example layer and stratigraphic unit, whereas others are not so clearly defined, for example, feature and context. A 'context' is defined by Barker as an omnibus term for all stratigraphic units (layers, features, strata and so on) found in an excavation.

Arroyo-Bishop defines features as being part of stratigraphic units. Since features are often used to describe immovable artefacts, i.e. buildings on site, Arroyo-Bishop's method does not apply in all cases.

In order to define the structure for a generic database model, the terminology which archaeologists employ needs to be agreed upon. Efforts are being made in this direction with the development of the CIDOC conceptual reference model (Crofts, Nick et al. 2001). But the development of the Strat GIS required database structure to be established before any recording or storage took place (Green 2002).

The 3DMURALE database adopts the following hierarchy:

Project - Site - Excavation Unit - Stratum - Find

A Project consists of Sites, which in turn consist of Excavation Units, which are composed of Strata, which would contain Finds. Features are being described as external immovable artefacts in the 3DMURALE database.

The Database System

From a requirement analysis of archaeological users, it was established that the 3DMURALE Database should follow a replicated system design where the data is maintained in a central database and during an excavation season, the new data is entered into the local databases in the laptops or personal computers of the archaeologists. Once the season ends, the newly entered data is uploaded from the local databases to the central database.

The 3DMURALE database handles storage and retrieval of text, 2D and VRML image information of archaeological content such as excavation units in the site, specifications of the strata, buildings, artefacts, parts of artefacts, photographs, stratigraphic drawings and documents.

Thus the main design issues are concerned with what types of underlying database systems should be used for the central and local databases, how large the units data should be represented and managed in each of the central and local databases and what functionality is required of the replication tool that is over and above the functionality that is provided by current commercially available replication tool.

Likely candidates for database platforms included MySQL, PostgreSQL, MS-Access, MS-SQL Server 2000. A database system intended for archaeologists must be as low-cost as possible. Using the prevalent open source databases can bring down the cost of the archaeological information system.

Figure 1 The 3DMURALE Database Servers

Database	Cost
Alpha Five Version 5	$349
askSam Professional 5	$395
Microsoft Access 2002	$339
Paradox 10	$489
PostgreSQL	Free
MySQL	Free
SQL Server 2000 Standard Edition	$4.999 per processor
Oracle9*i* Standard Edition	$15,000 per processor

Table 1 Database Cost Comparison

The 3DMURALE database makes its content available by remote Internet access for other archaeological researchers and members of the public. Three physically separate database servers are used as shown in Figure 1. Similar databases investi-gated include "ABCD", a relational database containing informa-tion about records of macrofossil plant remains from archaeologi-cal deposits throughout the British Isles. In ABCD, data are stored in a series of 14 tables linked by common fields. ABCD uses Paradox software for interrogating and manipulating the data (Tomlinson 1996). The details about the archaeological sites, excavation units in the site, the specifications of the strata, the photographs and facts about the different features and finds are entered into the database during the course of excavation.

The database chosen for the 3DMURALE is Post-greSQL, which has the advantage of being free and open-source, thus making this system as low cost as possible. It is one of the most advanced and well-supported open source and advanced object relational databases.

Replication

Archaeological sites rarely have the ideal network capabilities for high-speed remote data transmission. Due to the nature of sites, they are often in remote locations often without any connec-tion to the Internet. The uploading of this data to a central data-base, which consists of large database sets which result from advanced photogrammetric surveying and photographing is a major issue. It may be argued that the immediate upload of this data is not of great importance. Data is often analysed after the excavation allowing time to be spent on the actual excavation, which can often only be carried out during a limited time-period. Burning of the updated database to a CD-ROM, which is then sent by traditional mail methods or transported back with the excavation crew may often be the only plausible method.

Some archaeological sites may have access to a modem connection, albeit sporadic. 3D models of stratigraphy or buildings are often in the realms of Megabytes or hundreds of Megabytes. Using a low bandwidth line to transmit this data is impractical. If laptops are used during the excavation season, once returned to the academic or professional institution, replication becomes more reliable owing to a less intermittent connection and higher bandwidth. The ideal would be to have a low-cost high-bandwidth wireless data communication device such as a satellite device, which could be used instantly or on a daily basis for the uploading of the database.

The Stratigraphic Visualization Tool ("STRAT Tool") is used for the purpose of recording and visualizing data on archaeologi-cal stratigraphy. This tool is used by the archaeologists to insert, query and visualise information from the local Microsoft Access database. The STRAT Tool is developed in Microsoft Visual C++ and connects to the database using ODBC.

For migration of both database schema and data from Microsoft Access to a PostgreSQL database, certain software such as pgAdmin, Access2PgConverter, exportSQL, the MDB Tools exist. However, none supports replication of data on a selective basis. That is to say, the user cannot specify a particular record in a table and any related information regarding that row to be repli-cated into another database.

ExportSQL is a Microsoft Access module that exports an Access Database into a PostgreSQL database (Pavlinusc 2001). It exports all tables in a MS-Access database file into two text files: one containing SQL instructions to delete the new tables to be created, and the other with SQL instructions to create and insert data into the new tables. It is useful in exporting the schema from Access to other databases but does not include the capability of migrating data.

PgAdmin is a general-purpose tool for designing, maintaining, and administering PostgreSQL databases. It runs under Windows 95/98/ME and NT/2K. PgAdmin, installed along with pgMigra-tion v1.4.12 (the Database Migration Wizard Plugin for pgAd-min) is capable of transporting both schema and data. But pgAd-min is not capable of handling the data types such as the 'OLEObject' type used in the Access database. Binary data, such as photographs and 3D models of scanned buildings, artifacts and stratigraphy are stored using this data type. PgAdmin is not a promising option, as it would restrict the application from utiliz-ing the binary object data type, provided in the Access database.

The Access2PgConverter is an application that converts Ac-cess database to conform to the PostgreSQL definition for table names and fieldnames (de Groot 2003). It translates tables, que-ries, forms, reports and modules in Access applications. After translation the PgAdmin Migration Wizard should be employed to replicate the data to PostgreSQL.

MDB Tools is an open source suite of libraries and utilities used to read MDB database files making the data available on other platforms. Microsoft Access stores data in MDB files. Specifically, MDB Tools includes programs to export schema and data to other databases such as Oracle, Sybase, PostgreSQL, others, and MySQL. MDB Tools includes an SQL engine for performing simple SQL queries. A sparse but functional ODBC driver is also included. MDB Tools currently has read-only sup-port for Access 97 (Jet 3) and Access 2000/2002 (Jet 4) formats. Access 2000 support is a recent addition and may not be as com-plete as Jet 3 support.

A replication tool entitled the "Replicator" was investigated for its applicability, since it was described as a tool capable of supporting replication of virtually any database (Davies 1998). Replicator is designed to work with any database, and therefore has to be told which databases and which tables it is supposed to manage, what parent/child relationships exist between tables, and with what frequency it should check for changes. These instruc-tions are contained in a configuration file. What is required is a system, which uploads recent data regardless of the contents of the destination database. Since the database is accessed by multi-ple users, a system, which merely duplicates data, is inappropri-ate.

Consequently, for the replication of data from the local data-base, (stored in the Microsoft Access Database) to the central database, (the postgreSQL database), a replication software tool is being developed. The replication tool selectively accesses the archaeological tables in the local database and replicates to the central database without loss of data-integrity.

Similarly, it is capable of downloading table data such as Project and Sites to the local database.

The replication tool under development appears as shown in Figure 2. On selecting the specific rows in the Projects table and selecting the export option from the menu, it is possible to trans-fer all the related information about the projects, sites and so on from the local Access database to the central postgreSQL data-base.

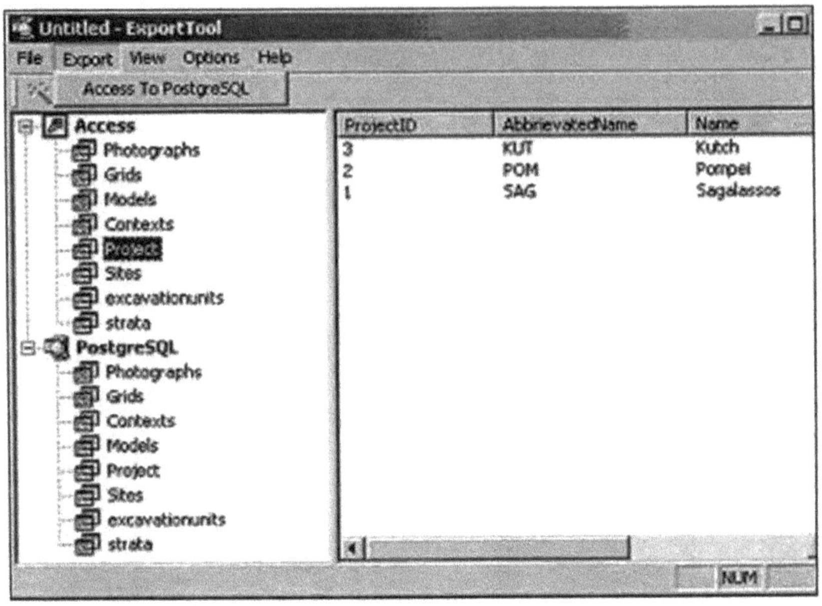

Figure 2 The STRAT Replication Tool

Similarly a user of the tool is able to selectively download information about a specific site or project from the central data-base to the local database.

Future enhancements sought for the replication tool include

- Generic upload. The access database need not be of predefined structure. The tool has the intelligence to handle any table
- Upload based of search criteria. E.g. Projects done between dates x and y or with Project IDs > 2000 can be uploaded.

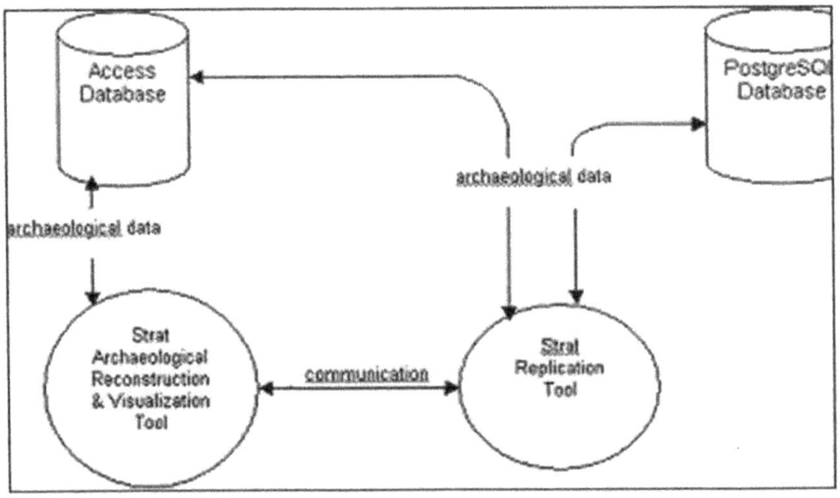

Figure 3 Replication of archaeological data

- Configuration of the ODBC settings from the applica-tion - auto configuration -for PostgreSQL database.
- File Open option for Access Databases rather than specifying them in the ODBC settings which would make it a handy tool for the archaeologists to handle.

Storing Binary Data

An archaeological information system, from which the visuali-sation tools retrieves data needs the Polaroid photographs, 3d models and images to be made available to the front end tools which will go on to process this data. There are two ways of storing this type of data, which is binary data.

- It can be stored in the file system and a path of the lo-cation of the file can be provided in the database
- It can be stored in the database itself

The choice on whether to store binary data in a database is completely a question of the requirements for each specific pro-ject. So the question is not whether to choose the database which provides the additional functionality, but whether the needs of the project demand the increased functionality of the database. If the application requires with absolute certainty that the data cannot be modified without the appropriate constraints, then it makes sense to place it in the database.

So, it comes down to the continual trade-off of functionality/requirements vs. performance/simplicity. By storing images in the DB, the application has better control over them. The chances of the files getting deleted accidentally are not there if this method is opted. Additionally, backups and restores are less complicated. Backing up a site would be easier if the second method is adopted as it would mean that the database does not have to open and close the files each time and this makes the backup process faster. On the other hand, if these binary files are large and are being updated frequently, putting them in the database can create a real nightmare of a storage problem. The system can be slowed down even to crawling speeds when the hits are very high. The per-formance of the database when either of these methods is adopted needs to be investigated to a greater extent.

PostgreSQL database provides two distinct ways to store binary data. Binary data can be stored in a table using PostgreSQL's binary data type bytea, or by using the Large Object feature which stores the binary data in a separate table in a special for-mat, and refers to that table by storing a value of type OID in the table. So as to determine which method is appropriate the limita-tions of each method should be considered. The bytea data type is not well suited for storing very large amounts of binary data. While a column of type bytea can hold up to 1 GB of binary data, it would require a huge amount of memory (RAM) to process such a large value. The Large Object method for storing binary data is better suited to storing very large values, but it has its own limitations. Specifically deleting a row that contains a Large Ob-ject does not delete the Large Object. Deleting the Large Object is a separate operation that needs to be performed. Large Objects also have some security issues since anyone connected to the database can view and/or modify any Large Object, even if they do not have permissions to view/update the row containing the Large Object (PostgreSQL 2003).

Shapiro etal. explore the performance issues while working with Binary Large Objects in various databases. Binary Large Object or BLOB as it is commonly referred to, is a collection of binary data stored as a single entity in a database management system (DBMS). BLOBs are used primarily to hold multimedia objects such as images, videos and sound, though they can also be used to store programs or even fragments of code. Not all Data-base Management Systems support BLOBs. A BLOB has no structure, which can be interpreted by the database management system but is known only by its size and location.

Large data objects can be stored in a field with the OLE Object data type in a Microsoft Access table. Some large binary data objects cannot be represented, however, if they do not have an OLE server that understands the data being stored.

Conclusion

The database system described in this paper can be used for the storage of all forms of archaeological objects. It is generic enough that its usage need not be restricted to 3DMURALE, but can be extended to various archaeological campaigns.

Acknowledgements

This work was carried out as part of 3DMURALE, a European Union funded project (Proposal number: IST-1999-20273).

REFERENCES

AKASHEH, T., 2002. Archaeological Information System and Databas-ing for Petra Monuments. Proceedings Multimedia and Cultural Heri-tage Amman, Royal Scientific Society, Amman, Jordan, 29 September - 2 October 2002.

ARROYO-BISHOP, D. and LANTADA ZARZOSA, M.T., 1995. To be or not to be: will an object-space-time GIS/AIS become a scientific real-ity or end up an archaeological entity? Archaeology and Geographical Information Systems. Taylor & Francis, London:43-53.

BARKER, P., 2001. Techniques of Archaeological Excavation, Third Edition, Fully Revised. Routledge, London (EC4P 4EE, ISBN 0-415-15152).

CROFTS, N., DIONISSIADOU, I., DOERR, M. and STIFF, M., 2001. Definition of the CIDOC object-oriented Conceptual Reference Model v3.2.1 (http://cidoc.ics.forth.gr).

DAVIES, S., 1998. Replicator (http://www.sdc.com.au).

DE GROOT, D., 2003. Access2pgconverter (http://www.talon.nl).

GREEN, D. et al., 2001. A Real Time 3D Stratigraphic Visual Simula-tion System For Archaeological Analysis And Hypothesis Testing. Pro-ceedings VAST 2001.

GREEN, D., 2001. Moving Towards the 3D Visualisation and Automatic Correlation of Stratigraphic layering. Harl, O. (ed.), Proceedings Workshop6 - Archäologie und Computer (CD).

HARRIS, E., 1989. Principles of Archaeological Stratigraphy. Academic Press Ltd.

MDB, 2003. MDB Tools Unlocking your data. (http://mdbtools.sourceforge.net).

PAVLINUSIC, D., 2003. ExportSQL (http://www.rot13.org/~dpavlin/sql.html).

POSTGRESQL, 2003. Storing Binary Data. (http://developer.postgresql.org/docs/postgres/jdbc-binary-data.html).

SHAPIRO, M., 1999. Managing Databases with Binary Large Objects. (http://storageconference.org/1999/1999/papers/18shapir.pdf).

TOMLINSON, P., 1996. A review of the archaeological evidence for food plants from the British Isles: an example of the use of the Ar-chaeobotanical Computer Database (ABCD). (http://intarch.ac.uk/journal/issue1/tomlinson_index.html).

Documenting two Histories at once: Digging into archaeology

Jon Holmen

Christian-Emil Ore

Øyvind Eide

Department of Electronic and Computer Engineering
Brunel University, Uxbridge. UB8 3PH. UK
nisha.joseph@brunel.ac.uk

See the CD for the extended version

Abstract

In this paper we will give an overview of the ways in which the principles of edition philology have been applied to archaeological texts and archives. We will also discuss what way we may encode the semantic content of a text in order to make it possible to get useful answers to queries as well as to map the information into a formal standard ontology (the CIDOC-CRM). We have used SGML to express the semantic as well as the structural information of archaeological texts.

Introduction

In 1992 the National Documentation Project was launched. A major part of the project was to create an information system for the archaeological museums in Norway. For almost 170 years the archaeological museums in Norway ave published information on a yearly basis on their acquired artefacts in specially prepared acquisition catalogues. The descriptions of finds in these catalogues are quite longwinded, including extensive information on the finds, the find contexts, their place and time, the finder or excavator, as well as detailed descriptions and classifications. The series of catalogues served for practical purposes as the main artefact inventory of each museum.

Over the years, computer applications in archaeology have primarily been used to generate a wide variety of statistics as well as to develop inventory databases. Pattern recognition and simulation/AI were at their peaks around 1980 and 1990 respectively. In the 1990s GIS, 3D modelling and the Web were in focus (see Scollar 1999). Most applications were designed to analyse information collected during fieldwork. Archaeologists use texts in the same way as other scholars, but text philology as such is not central central part of their discipline. Thus the task of creating a database on the basis of old reports is normally done through reading the text and keying into a database the information considered to be essential in a normalised form.

The Documentation Project covered several disciplines including electronic text collection. Thus it was quite natural to try to apply principles from the text encoding community and to use SGML to mark-up the catalogue texts. In 1992 the text encoding community had little to offer to archaeology. Archaeological reports seem not to be the focus of interest for text scholars. As a consequence we had to write our own DTDs covering the classification practice over the past 170 years. In 1992-2000 almost 30 000 printed pages of text were converted and SGML tagged (see Holmen and Uleberg 1996, Holmen and Ore 1996 and Ore 1998 for more details on the project).

```
<CATYEAR>
<INTRO> <MNAME>UNIVERSITETETS OLDSAKSAMLING TILVEKST
</MNAME><YEAR>1989 </YEAR></INTRO>

....

<NRPAR><CATNR nrid="37267">C37267.
<ARTEFDATA><ARTEFACT>Axe</ARTEFACT> of <MAT>iron</MAT>
<SHARED>from <PERIODE>Late Medieval time</PERIODE>.
<ARTEFDATA><MEAS>L: 141mm</MEAS>, <MEAS>edge W:109mm</MEAS>.
Carpenter's axe with <FORM>specially shaped blade to accommodate the fingers
</FORM> when it is held just "behind" the edge. Particularly necessary for fine work,
when used at an angle to the edge, or as a gouge. <SHARED>Found<FINDLOC>on
the hill, about 300m above <LOC>ÅROS KAPELL</LOC> </FINDLOC>,
<FARM>SJØGLØTT</FARM>, <PARISHD><PARISH>ÅROS </PARISH>
</PARISHD>, <MUNICIPALITY>RØYKEN</MUNICIPALITY>,
<COUNTY>BUSKERUD</COUNTY></FINDLOC>, in <FINDYEAR>1959
</FINDYEAR> by <FINDER>Berge Narvik</FINDER>, Tjernsrudveien 24, Jar,
Oslo. </SHARED></NRPAR>
```

Figure 1 The SGML mark-up pf an acquisition catalogue entry (translated into English)

Over the last couple of years there has been an increasing awareness of the need to include the information and content found in older archaeological and cultural historical oriented documents into archaeological and cultural heritage systems. Our method of encoding and extracting information from electronic versions of old archaeological reports has been taken up by others (e.g. Crescioli, D'Andrea and Niccolucci 2002). Schloen 2001 suggests an XML formalism for storing and interchanging archaeological information. Meckseper 2001 describes the situation in the UK and points out the usefulness of XML. Both seem to address the question of how to use XML for writing and storing new archaeological documents.

A Methodological Note: Scholarly Text Editions

In traditional text philology the aim is to create a "best text" based on a set of manuscripts (text witnesses). A scholarly text edition is usually a printed version of the "best text" accompanied by a critical note apparatus documenting variation in the text witnesses and possibly a set of facsimiles of (some parts of) the manuscripts. During the last 10-20 years computers have been introduced into text philology. Today a modern electronic scholarly text edition does not necessarily contain a "best text". A "text edition" consists of a bibliographical database, electronic facsimiles and transcripts of the text witnesses combined with a few extra search tools and explanatory texts. The transcripts are usually given a mark-up (e.g. XML) that enables hyper-linking as well as a presentation of different views on the texts.

To the extent that text scholars are interested in texts formed at museums, this does usually not extend beyond texts that are first class museum objects. They do not so often consider newer texts describing the normal activity of the museum such as archives, excavation reports etc. to be of interest. We have chosen a wider approach. The internal archives, reports, acquisition catalogues and other documents are the main sources in the development of scientific/scholarly databases for students, scholars and curators at a museum. As a result of our work during the last ten years, we have refined upon a four-step work procedure for the museum texts developed from and inspired by our close contact with other text based disciplines in the humanities. The first three steps correspond to the steps normally applied to an ordinary text edition. The final step, in which we interpret the semantic content of the text according to an ontology, is not so commonly used by text editors.

Methods of Digitization

The methodology described in this paper covers the digitization of three main types of archaeological: Acquisition catalogues, a series of printed site and monument records (SMR) and the so-called topographical archives. As the latter of these have not been published as printed books, but comprises large heterogeneous document collections, it was simply too expensive to convert them into electronic text. These archives were digitised as facsimiles and indexed according to a predefined data model (step 1, 2 and 4 below), see Holmen and Innselset 2003. The acquisition catalogues and the SMR, on the other hand, were OCR read and SGML tagged.

The digitization was performed according to the following four steps:

1 We created a bibliographic record of the original documents including metadata such as Title, Author, Year of publication, Edition number, etc.

2 According to good practise one should make an electronic facsimile by digitising an image of every single page of documents. Unfortunately this was not done due to high storage costs in the early 1990s.

3 We then created an SGML text by applying OCR to the facsimiles or, in the case of handwriting, by manually transcribing the text. The transcript was then given a mark-up indicating pages, special layout, typefaces, illustrations, inserts etc.

4 Finally, we carried out a step which extended the basic descriptive encoding, as we introduced a semantic encoding of the content. This meant that we looked at the information elements in the text and coded them in relation to a given set of concepts or ontology. In our case an ontology is the set of entities and relations that we consider to would best describe the archaeological finds and objects. (Fig.3)

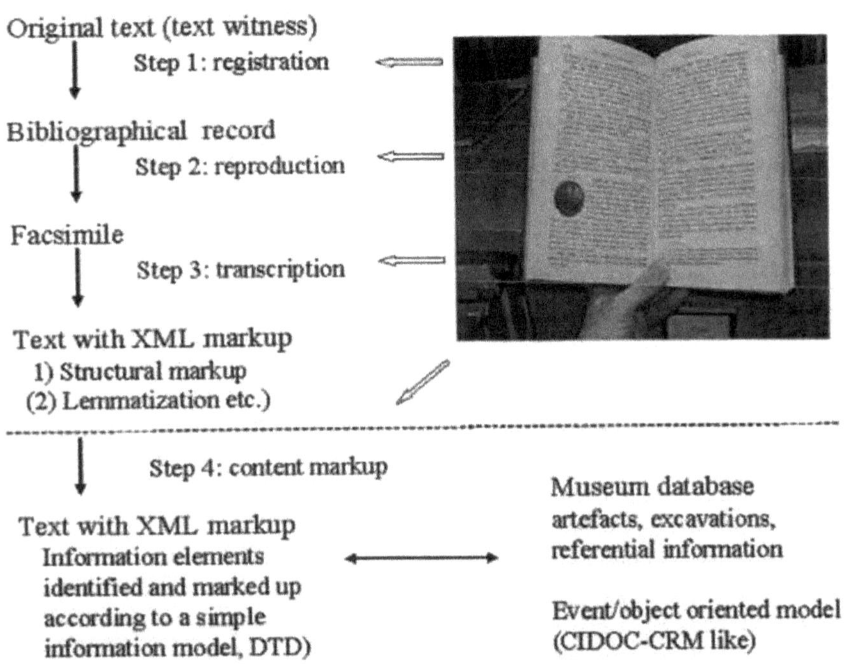

Figure 2 Digitization of museums documents - ideal versus reality

Ontology

The term 'Ontology' is a philosophical term denoting the study of being. The Artificial Intelligence community tends to let this term denote a specification of conceptualizations or simply a model of some part of the reality. In computer science the term 'ontology' is often synonymous with 'data model'.

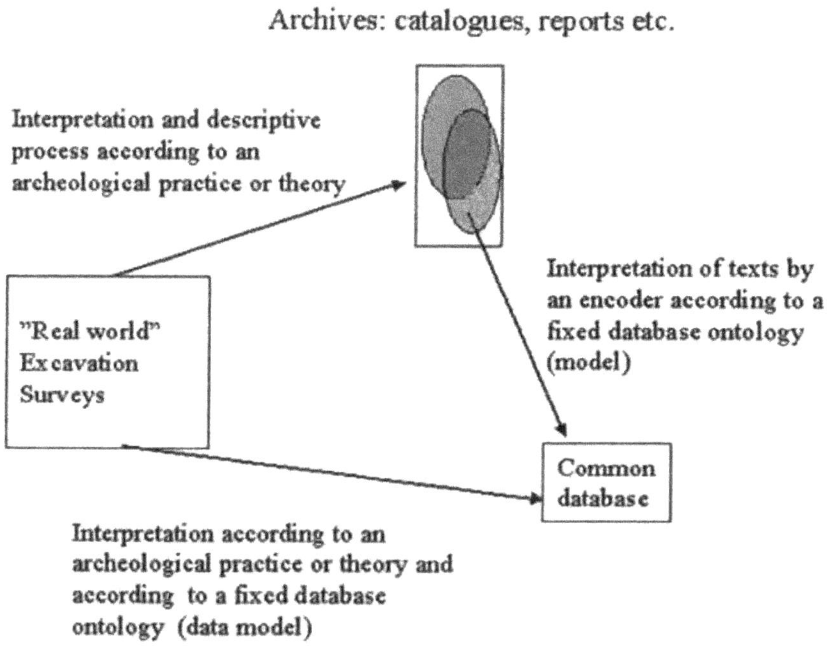

Figure 3 Information extraction - a multiple interpretation problem

The archive material accumulated during the last 180 years illustrates several ontological shifts in the purely philosophical meaning of the word 'ontology'. The scholars' own view on the present world and their implied view on the near and distant past have not been constant over the years. As a result the reading of old reports may change. This is a well known fact and one that is usually taken care of by the note apparatus in scientific papers. However, in most databases one usually does not state or comment on the source of each information unit. A not uncommon complaint among non-computerized scholars has been "that information looks so true when seen on a computer screen (but this is not the case)". Even in the CIDOC-CRM model reification of data is not a part of the model.

The methods found in scholarly text editions and the mark-up process described above are used to trace the source and the interpretation of the original information sources. The encoding of the semantic content must necessarily reflect our view on what constitutes important archaeological information and does not necessarily coincide with the conception of the 19th century author of the original material. The original author included information according to his conceptions, from which we encode the subset that match our conception or ontology. Thus it is important to include at least facsimiles of the original text in order to give the future reader a chance to judge for him- or herself.

As the old material was produced over a period of more than 180 years, it varies significantly with respect to the type of information that is emphasized as well as to style. The methodology used in converting the material gives us additional information. Registers of persons who played an important role in the history of the museums have been produced on the basis of the mark-up. We can also create sequences of the events in the museums' history, thus giving us the opportunity to realize to what extent each individual's "style" influences the content. By connecting all this we can reconstruct parts of the history of the institutions.

The information about new material and new activities at the museums is treated according to a much more unified standard. This standard is necessarily based on the present archaeological school(s) and will probably be changed in the future when this is considered necessary. Thus we are developing a system in which standards are represented as form definitions/DTDs and preserved.

New material included at a later stage will thus conform with one specific standard that is stored in the database, together with information such as author and date. One implication of this is that when upgrading the standard, there will be no need for upgrading older material that was entered according to an old standard. We will only need to perform a mapping of the old and the new standard.

This will ensure the preservation of data in its original form, and assure that no information is lost during upgrades. It will also give us and future users of the systems an overview of the development of archaeology as a science and reveal according to which school the various reports were written.

Conclusion

Through the process of converting the original paper-based catalogues and archives at the archaeological museums into digital form, and as far as possible observing our defined principles for this kind of conversion, we have created reliable relational databases of archaeological archives and artefact information. We have also developed an opportunity of working directly with the source material by making an electronic text archive with powerful indexes for both the catalogues and the documents. This makes it possible to study them as individual historical objects.

Through the coding of the source material in a comprehensive ontology we have also, in addition to providing "normal" database functionality, laid the ground for enabling us to describe the history of the museums and the history of Norwegian archaeology. We have also established a system in which the principles used by archaeologists in their documentation work of today may be continually recorded, thus creating what in the future will be history.

REFERENCES

CIDOC-CRM. CIDOC Conceptual Reference Model. Proposed ISO 21127 (http://cidoc.ics.forth.gr/).

CRESCIOLI, M., D'ANDREA, A. and NICCOLUCCI, F., 2002. XML Encoding of Archaeological Unstructured Data. In Archaeological Informatics: Pushing the Envelope. Proceedings of CAA 2001, Oxford:267-275.

HOLMEN, J. and ORE, C.-E., 1996. New life for old reports - The Archaeological Part of the National Documentation Project of Norway. ALLC-ACH '96, Book of Abstracts: Conference abstracts, posters and demonstration, no. 70, Report Series of the Norwegian Computing Centre, Bergen.

HOLMEN, J. and ULEBERG, E. Getting the most out of it - SGML-encoding of archaeological texts. Paper at the IAAC'96 Iasi, Romania (http://www.dokpro.uio.no/engelsk/text/getting_most_out_of_it.html).

INNSELSET, S. and HOLMEN, J., in print. Digital Archaeological Resources at the University of Bergen: An Efficient Tool in Research and Heritage Management? Paper at the IAAC'99, Dublin, Ireland.

MECKSEPER, C., 2001. XML and the publication of archaeological field reports. Master's dissertation, University of Sheffield.

ORE, C-E., 1998. Making multidisciplinary resources. In Burnard, L., Deegan, M. and Short, H., The Digital Demotic, A selection of papers from Digital Resources in the Humanities 1997, Publication 10, Office for Humanities Communication, King's College, London.

SCHLOEN, D., 2001. Archaeological Data Models and Web Publication Using XML. Computers and the Humanities 35:123-152.

SCOLLAR, I., 1999. Twenty five years of Computer Applicatins to Archaeology. In Archaeology in the Age of the Internet (BAR International Series 750).

The Thesaurus of Archaeological Toponymy[1]

Giulia Pardi
Library of the National Archaeological Museum of Florence

Abstract

This paper describes a data-base created for the geographical and administrative identification of archaeological place-names: the Thesaurus of Archaeological Toponymy (Thesaurus di Toponomastica Archeologica, TTA). The compilation of this data-base began in 1987 and now contains about 8,000 place-names. The TTA is a structured and controlled vocabulary mostly of the archaeological toponymy and micro-toponymy of Tuscany. The data-base is organized in two hierarchical structures arranged from broader to narrower context (one: nations, states, provinces, communes, cities, inhabited or deserted archaeological sites; and two: catchment basins, cities, inhabited or deserted archaeological sites). The TTA provides both the vernacular and Italian names of places, variant in other languages and historical names; the record for a place may also contain the descriptive note, geographical coordinates and bibliographic source of the names.

Introduction

The Thesaurus of Archaeological Toponymy is a result of the cataloguing operations of the books kept in the Libraries of the Museo Archeologico Nazionale of Florence and of the Istituto di Studi Etruschi ed Italici.

The Museum's collection was instituted in 1871. Beginning in 1899 Luigi Adriano Milani curator of the Museum, saw to the transfer of the archaeological heritage as well as the volumes dealing with Egyptology, Etruscology and Graeco-Roman archaeology that had been housed at the Uffizi until that moment; to this first, ancient nucleus many further acquisitions have been added and the present heritage consists of approximately 40,000 volumes. The Library of the Istituto di Studi Etruschi ed Italici was placed beside it since 1930 and the book heritage owned by the Institute stands at approximately 10,000 volumes. Together they form a centre of documentation for the preservation of the Tuscan archaeological heritage.

In the second half of 1987 a program of bibliographical data management[2] led to the beginning of compilation of a Dictionary of Subjects, dedicated to this field. The Dictionary of Subjects of the bibliographical resources is carried out by the adoption of key-words drawn from titles, texts, forewords, abstracts, indexes, as well as from the language of archaeologists. The key-words consist of a single word (lemmas as 'temple') or a compound word (syntagms as 'Hera temple'). In this way we have registered a considerable quantity of associated concepts, agents, activities, materials, archaeological items, and place-names.

The context: The archaeological Bibliography

It is quite simple to reach the origin of the bibliographical information as each subject is available in any of the informatic catalogues which can be found by using a key word or free text. However, this does not hold for the subject 'place-names', since they have been in the archaeological bibliography for only 15 years, and not in every case has it been deemed necessary to perfect the publications with maps, including those connected to excavations or the recognition of such: almost all previous publications were lacking in them. Thus, until 15 to 20 years ago, the sites, on the whole, were not placed in their geographical environment. Furthermore, both the variant place-names (that several sites have been subjected to over the years), and the vast quantity of homonymous names has only added to the confusion. The very same archaeological publications that are supposed to assist in locating the sites geographically, such as atlases and maps, are organized according to a cartographic subdivision or to the chronology, morphology and the illustrations of the objects discovered.

From the very beginning of this work, since place-names and micro-place-names have represented the majority of subjects recorded, it has been necessary to identify a means that will allow us not only to trace the source of the bibliographical information, but also to identify the geographical and administrative location of the site.

Place-Names of Archaeological Sites: Quantity and Characteristics

As described in the Dictionary of Subjects sixty per cent of the items consists of place-names (slightly more than 8,000 but expanding), because they are related to archaeology in a wide sense, including places of discovery, location, restoration, study and events. The most recorded places are the Tuscan ones, even if the Thesaurus is open to the entire Mediterranean area. Obviously, the further we get, the rarer and more casual the places become.

It is common knowledge that archaeological sites have a great deal of variable features: we can have places very rich in buildings, remains and finds, or simply places where only one object was found such as a statue, or a coin or a shard. They can have the same name of one single building or of one street or of one river, etc. They can cover wide chronological periods, or witness only short fractions of time. In spite of such variety, the archaeological sites have two common features: all of them have a place-name and most have a geographical placement.

CHOICE OF METHOD

According to the Guidelines of International Standard Organisation, the definition of a thesaurus in structural terms is a controlled, dynamic of semantically and hierarchically dictionary correlated terms that covers a specific field of knowledge (ISO 2788/1974), followed in 1986 by a further definition: 'Thesaurus: a dictionary to be used as a reference book in which words with similar concepts are grouped together' (ISO 2788/1986). The formes definition is clear enough not to require further explanations, while in the latter it is sufficient to substitute 'concepts' for 'place-names', thus highlighting the usefulness of applying this method to archaeological topography and place-names, indeed it makes it possible to give explicit connections a priori… in a relatively simple hierarchical context, namely the semantic connection defined as 'part - the whole', in which the subordinate defines a part compared to the whole as represented by the dominating part. It is accepted only for those cases where it is considered that the connection 'part - the whole' contains strong, mutual implications, and more precisely in the following taxonomies: the human body, geographical location, and social hierarchical disciplines and structures (Trigari 1992).

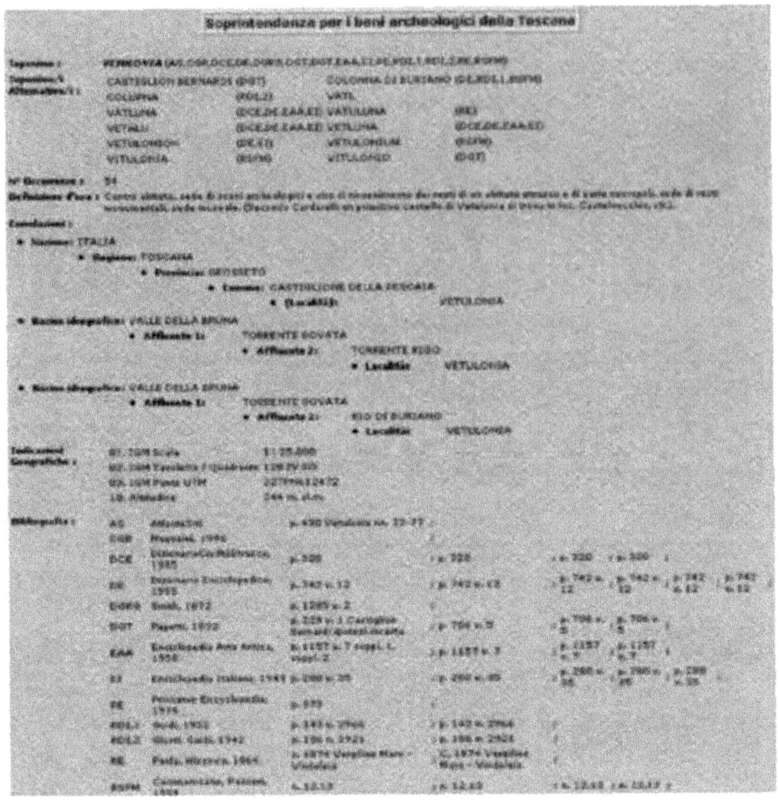

Figure 2 The file of Vetulonia.

We wish to point out that our choice of using a controlled, structured dictionary which in our opinion has the merit of clarifying a priori the connection between the place-names (Fig.4 and Fig.5), has not been adopted by the curators of the Alexandria Digital Library Gazetteer or by other similar prestigious instruments found on the internet[3], because it is likely they were able to utilize other systems of registration from the very beginning. On the contrary the Thesaurus of Archaeological Toponymy was born on paper, with all the limitations this can bring, and on the basis of example available, like local works by Cammarosano and Passeri (Cammarosano and Passeri 1984), and of Papaldo and Signore (Papaldo and Signore 1989) or to international publications such as the Art & Architecture Thesaurus of the Getty Information Program (Petersen 1990). Much has been written about thesauri since the end of the 19th century, but the bibliography of geographical and place-names thesauri goes back only to the last decade of the 20th century.[4]

Symbols Getty's Thesauri	Symbols of TTA	
USE (use)	as preferred	place-name
UF (used for)	as alternative	place-name
TT (top term)	as national	place-name
BT (broader term)	as regional	place-name
BTP (broader term partitive)	as district	place-name
NT (narrower term)	as municipality	place-name
NTP (narrower term partitive)	as locality	place-name

Figure 1

THE FILES FOR THE REGISTRATION OF THE PLACE-NAMES

[1] Our place-names are arranged in records, according to the pattern adopted by the Art and Architecture Thesaurus, and also Getty Thesaurus of Geographic Names, whose hierarchical relations between terms are borrowed for the toponymy. In this way the symbols, that in the Getty Thesaurus express the relations between terms, become as those shown in the Figure1.

[2] The standards of the cataloguing can be well understood from the place-name's file of Vetulonia (see also Fig.2).

[2.1] In the field of preferred place-name is recorded the place name used in the present cartography.

[2.2] In the field of alternative place-name are recorded historical and linguistic variants of the place-name. Each variant provides an approach to information, i.e. the same result can be obtained looking under the Etruscan place-name 'Vatl' or under the late medieval one of uncertain origin 'Castiglion Bernardi'.

[2.3] The number of the occurrences shows the quantity of publications held by the library in which the place-name is present.

[2.4] The scope note shows the essential features of the site indicated by the place-name, thus Vetulonia being a particularly famous site in Italy is described briefly as: an inhabited place, a site of archaeological excavations and an Etruscan settlement and necropolis.

[2.5] In the field of correlations the preferred place-name is listed in a hierarchy according to the present administrative subdivision of the land; so 'Vetulonia' is in the municipality of 'Castiglione della Pescaia', in the district of 'Grosseto', in the region of 'Tuscany' and finally in 'Italy'.

[2.6] Furthermore, the preferred place-name is listed also in a second hierarchy of the names of the catchment basin that embrace the site, that allows us to show the settlements and the remains and to reconstruct the topography of the ancient peopling; so 'Vetulonia' is situated between the 'Torrente Rigo' (North) and the 'Rio di Buriano' (South), affluents of 'Torrente Sovata' in the 'Bruna's' catchment basin.

[2.7] In the field of geographical information we record the scale, the number and the data of the topographical map, the co-ordinates of Merca-tor's Projection and / or Gauss Boaga, the cadastral data, the elevation.[5]

[2.8] Under these data it is possible to insert one or more pictures.

[2.9] In the field of sources we record the titles concerning the place-names' variants.

[2.10] In the last field, not visible in figure 2, the name of the person filling in the registration, has been recorded.

[3] Further information on the file

[3.1] The place-names of uncertain location are tied to the most probable hierarchies. If the effective location is verified, it is possible to cross them out from the wrong hierarchy. In the same manner the place-names of archaeological sites, that are localized on boundary-lines and belong to different municipalities, can be assigned to different hierarchies.

[3.2] We have introduced a further level too[6]: the sub-locality. This has no administrative value, but it allows us to insert at a correct level the place-name pertaining to sites included in a broader one. For example, this device avoids an urban area like 'Rome' being on the same level as 'Largo Argentina', or 'Firenze' being on the same level as 'Piazza della Signoria'.

[3.3] Obviously the free-text research can be applied also to the field of the scope note, allowing us to locate all the inhabited places, settlements, monumental complexes, necropoles, occasional finds and their relative features.

INDEXES AND HIERARCHIES

[1] This type of cataloguing allows us to obtain the sequence of file containing place-names, that may be ordered according to a chosen rule.

[1.1] The alphabetical index of preferred place-names and their variants, perfected by the definition of the use which provides essential features.

[1.2] The index of place-names with respective historical, linguistic and synonymic variants.

[1.3] Another which shows the geographic and administrative location of the preferred place-names (Fig.3).

[1.4] A hierarchical index in which every preferred place-name is arranged in alphabetical order in the catchment basin that embraces the sites (Fig.4).

[1.5] And finally a hierarchical index of the preferred place-names arranged in alphabetical order, level for level, with the place-names belonging to the same administrative area (Fig.5).

[2] Moreover, the structure of the thesaurus foresees the registration of the other hierarchies bound to further territorial subdivisions, like the historical and ecclesiastical ones, in which we will be able to insert the place-names, even if these further data-bases are not yet implemented.

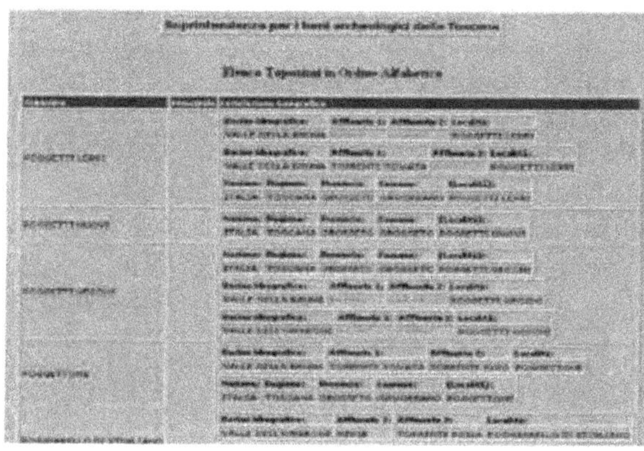

Figure 3 : Index of the preferred place names

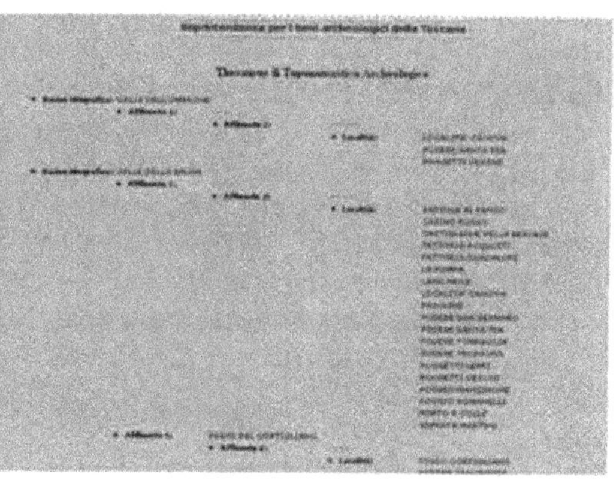

Figure 4 Hierarchical index of preferred place-names in the catchment basin

CONCLUSIONS

The intention of this work is to contribute to the knowledge regarding the distribution of ancient peopling connected with the geomorphological structure and the present administrative division of the territory that reflects the one used in Italian Institutes responsible for the preservation, conservation and fruition of the cultural heritage.

As the implementation of the database is fortuitous (as already explained, it is tied to the Library acquisitions) and therefore not even the territory of our region can be considered complete, however, it should be kept in mind that the list a theoretical contribution to the analysis of the problem has been supplied; as far as our work in concerned, we feel that

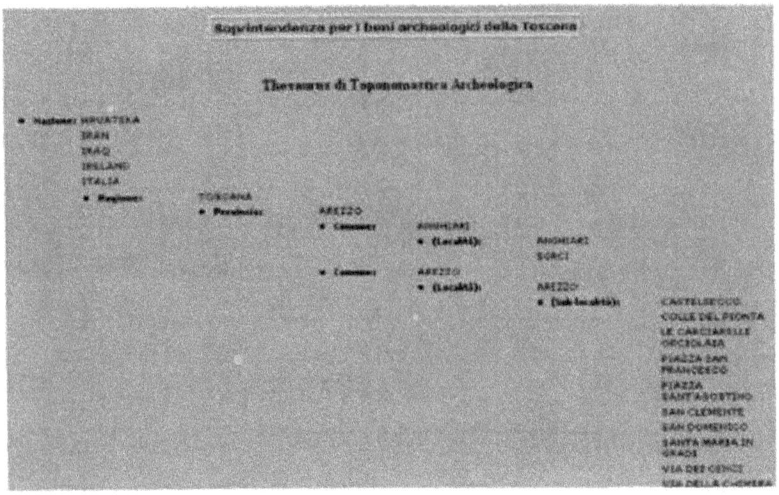

Figure 5 Hierarchical index of preferred place names-in the administrative subdivisions

the initial objective will have been reached when the implementation of all the place-names of archaeological sites in Tuscany has been finished.

The Thesaurus of Archaeological Toponymy is presently available for consultation only in the libraries of the Archaeological Museum and the Istituto di Studi Etruschi[7] that are visited by erudites, professors, archaeologists and university students made up mainly of Europeans and Americans (about 600 persons a year). The following developments, such as the expansion of the geographical area and the neighbouring regions the increase in software adaptability and the capacity to interact with other databases, depend to a large extent on funding that will have to be found. However, the first partial result will be the opportunity to consult, from the beginning of 2004 the Thesaurus of Archaeological Toponymy on the Biblioteca Nazionale di Firenze site (www.bncf. sbn.it), in the space entitled 'Segnalibri'. Only after an initial running-in stage will we be able to assess its merits and demerits based on the answers that visitors to the site will be asked to give to specific questions.

1 The Thesaurus of Archaeological Toponymy is a data-base elaborated in collaboration with Dr. Donatella Venturi. The revision of the Italian text is the work of Dr. M. Iozzo. The translation of this paper is of Mrs. Doris Gage.

2 At first we used an software in DB3 created on purpose, until the end of the year 1999; in the 2000 we have acquired a new software (elaborated by Logica s.r.l. of Florence in collaboration with G. Pardi and D. Venturi and taking advantage of the fundamental suggestions by Laura Corti of the Scuola Normale Superiore di Pisa and Toni Petersen and Marisa Trigari of the International Federation of Library Association and Institutions, IFLA) in which the data is formatted in SQL, it is elaborated in Visual Basic and Access and the results of the queries are formatted in HTML.

3 In Google to the query: Mediterranean Archaeological Gazetteer turn out to be over 600 pages.

4 The Clearinghouse, Instytut Informacji Naukowej, Technicznej, i Economicznej (IINTE) of Warszawa, whom we consulted on this subject at the end of 1988. They replied that nothing had been published.

5 During the debate following the exposition of this work at the convention CAA 2003, a thoughtful colleague kindly pointed out the usefulness of inserting both the latitude and the longitude of the site.

6 A symbol for this level does not exist in the specialized literature.

8 The two Libraries are in the same building and have the same address: (via della Pergola, 65-50121 Firenze, Italia).

REFERENCES

Atlante Aerofotografico delle sedi umane in Italia. Istituto Geografico Militare, Firenze 1964-1989.

BOSIO, L., 1988. Carta archeologica del Veneto.

CAFFARELLO, N., 1971. Dizionario archeologico di antichità classiche.

CALDERINI, A. and CALDERINI, M., 1960. Dizionario di antichità greche e romane.

CAMMAROSANO, P. and PASSERI, V., 1984. Città, borghi e castelli dell'area senese-grossetana. Repertorio delle strutture fortificate dal Medioevo alla caduta della Repubblica Senese.

Carta archeologica della provincia di Firenze, 1995. 1. Firenze.

Carta archeologica della provincia di Reggio Emilia, 1990. Comune di Poviglio, Poviglio.

CASINI, S., 1994. Carta archeologica della Lombardia 4. La provincia di Lecco.

I centri archeologici della provincia di Siena, 1986. Siena.

CORTI, L., 1992. Beni culturali: Standards di rappresentazione, descrizione e vocabolario.

CRISTOFANI, M., 1985. Dizionario della civiltà etrusca.

CURRI, C.B., 1978. Forma Italiae, Regio VII. Vetulonia, 1.

DAREMBERG, Ch. and SAGLIO, E., 1877-1919. Dictionnaire des antiquités grecques et romaines.

Enciclopedia dell'Arte Antica, 1958. Roma.

Enciclopedia dell'Arte Medievale, 1991-2000. Roma.

Enciclopedia italiana di Scienze Lettere ed Arti, 1949. Roma.

GARGIULLO, S. and OKELY, E., 1993. Atlante archeologico dei mari d'Italia.

GASCA QUEIRAZZA, G., 1990. Dizionario di Toponomastica.

Getty Thesaurus of Geographic Names, 1998. Los Angeles.

Guidelines for the establishment and development of monolingual thesauri, 1981. Paris.

ISO 2788. Guidelines for the establishment and development of monolingual thesauri, 1986. Genève 1974, 2nd ed. Genève 1986.

LAVELL, C., 1989. British Archaeological Thesaurus.

Lexicon Iconographicum Mythologiae Classicae, 1981-1999. Zürich/München.

Lexicon of the Greek and Roman Cities and Places, 1992. Amsterdam.

Multilingual Egyptological Thesaurus, 1996. Utrecht/Paris.

NENCI, G. and VALLET, G., 1977. Bibliografia topografica della colonizzazione greca in Italia e nelle isole tirreniche.

PAPALDO, S. and SIGNORE, O., 1989. Un approccio metodologico per la realizzazione di una banca dati storico geografica.

Pauly's Real-Encyclopädie der classischen Altertums Wissenschaft, 1914-1978. Stuttgart.

PETERSEN, T. (ed.), 1990. Art & Architecture Thesaurus of the Getty Information Programm.

PIERI, S., 1969. Toponomastica della Toscana meridionale.

PIERI, S., 1919. Toponomastica della Valle dell'Arno.

Popolazione e abitazioni, 1991, no. 13°. Censimento generale Istat della popolazione e delle abitazioni.

The Princeton Encyclopedia of Classical Sites, 1976. Princeton.

Rationes Decimarum Italiae nei secoli XIII-XIV, 1932. Tuscia, 1, Città del Vaticano.

Rationes Decimarum Italiae nei secoli XIII-XIV, 1986. Tuscia, 2, Città del Vaticano.

Reallexikon zur Byzantinischen Kunst, 1963. Stuttgart.

REPETTI, E., 1833-1845. Dizionario geografico fisico storico della Toscana.

SMITH, W., 1872-1873. A Dictionary of Greek and Roman Geography.

Thesaurus of Archaeological Site Types, 1992. London.

Topographical Bibliography of Ancient Egyptian Hieroglyphic Texts, 1960. Oxford.

TORELLI, M., 1992. Atlante dei siti archeologici della Toscana.

TRIGARI, M., 1992. Come costruire un thesaurus.

VALENTI, M., 1995. Carta archeologica della provincia di Siena.

WHITEHOUSE, R.D.W., 1983. Dizionario di Archeologia.

The Cederberg Rock Art Survey Project: A Co-ordinated Field Record and Database Structure

Bastian Asmus
University of Cape Town
bastian@science.uct.ac.za

Conny Meister
University of Cape Town
meicon@web.de

Abstract

The Cederberg Rock Art Database is the first fully digital database introduced into Southern African rock art research. This paper deals with rock art methodology for recording, storing, manipulating, exploring and preparing collected information for subsequent analyses. The Cederberg Rock Art Database was designed to store the entirety of not only the rock art, but also the topographic and archaeological records.

To demonstrate the potential of this tool Asmus (2003) examined the issue of the existence of different site types. He furthermore analysed the sites based on the evidence of the painted images, the topographical position, and the archaeological surface material. The recorded features were portrayed by descriptive statistics and compared with similar studies throughout southern Africa. Paintings, topography and archaeological surface remains provide the respective foundations for a classification of similar sites into groups. The groups of classifications were compared in order to identify the possible parameters responsible for site choice. The results of the analyses did not demonstrate the existence of sets of characteristics, which seemed responsible for site choice. In fact the outcome was quite the opposite: apart from the fact that shelters and overhangs in closer proximity to constantly flowing water were preferred, the sites seemed to have been chosen rather indiscriminately.

The nature of the database enabled Asmus (2003) to conduct detailed analyses. It also assisted him in critically assessing the capacity of the recorded information in order to answer the previously posed questions. These included: Do different types of sites exist in terms of the nature of painted imagery; how is site choice influenced by topography and accessibility; how does geomorphology influence site location; what is the association between painted sites and other archaeological remains? The database, in conjunction with a novel model of energy expenditure to characterise site accessibility, addressed these questions successfully.

Aims

The Cederberg Rock Art Database is a result of several research interests and activities in the Cederberg. Initially we commenced with three years extensive design and development. Moreover, we advanced our personal research experience during three rock art recording field seasons, to finally conclude with a careful evaluation and comparison of traditional recording methods in Southern African rock art studies (Asmus and Meister 2001, Asmus 2003).

The goals for the development of the Cederberg Rock Art Database were to combine the traditional methods of recording rock art with up-to-date digital technology resulting in a valuable system for recording rock art in the Cederberg. The database is multi-functional; fulfilling the needs of our research interests, and furthermore acting as an archive facility for rock art sites of the Cederberg and beyond. Elsewhere Asmus (2003) proposed a methodology to provide the minimum requirement for recording rock art sites in the Western Cape as a broad and powerful base for comparison.

Our interests included the development of an appropriate rock art recording tool, of which the database is but one part, thus complementing the coordinated survey methods described later on. The realisation of this work encountered numerous challenges, including the creation of a user-friendly interface of the database, the coordination of field observations, the co-ordination of the digitising process of the collected data, and accessibility to this information to a large group of people. Overcoming these obstacles proved as an opportunity to introduce this database method to southern African archaeology.

Concept and Structure behind the Cederberg Rock Art Database

The database consists of eight data files, which were developed during the course of the several field seasons, each providing the framework for data entry. In addition several parts of the database deal with visualizing and displaying issues. These files are programmed to return and display general descriptive statistics of the recorded information, e.g. the display of the frequencies of depicted animals per site in a systematic tree diagram. The eight main files cover the following aspects:

(1) general site information; (2) topographical information; (3) paintings of human beings; (4) paintings of animals; (5) paintings of inanimate objects; (6) occurrence of handprints; (7) digital images of rock paintings; (8) archaeological remains.

The power of the database facility particularly lies in the possibility of generating related files at any necessary moment without changing the information in the master files. The definitions of the relationships between the discrete files are: the value for the site name relates all eight main files to each other. The image cross-reference uses additionally the site unique painting ID number. Files dealing with images of humans and animals have one-to-many relationships. The relationship between the general site information file relates one entry per site to a unlimited number of entries that deal with the description of individual rock art images.

Negative Effects on Databases

Since queries and subsequent data analyses are dependent on the comparison of search strings with the contents of the database, data consistency is a key element to the scientific value of the application. The following aspects have been very important during the design of the database.

(1) All entries carrying the same meaning have to have exactly the same string/value
(2) Cells with freely formulated narrative or descriptive contents cannot be queried meaningfully
(3) Multiple values entries per variable have to be avoided.

The less complicated the strings or values in cells are, the higher the likelihood of consistent data. Instead of leaving the user free to formulate strings freely to describe a feature, he has to choose from defined values for each variable in pull-down menus. More detailed descriptions are broken down into multiple variables, rather than to record these with multiple value entries per variable Table 1 presents three alternate ways to describe the same rock art image.

Narrative description	There is a painted red human figure with breasts and with a bag with tassles. The head is missing. faces left.
Description as an outline	Red female without head; with a tasslebag; oriented to the left.
Description broken down into single descriptive variables	sex: female - head: missing - attribute: tasslebag orientation: left - colour: red

Table 1 Three different examples to describe the same rock art image

The first two options are prone to a variety of interpretations because they are descriptions of an image in ones own words and, therefore, do not allow meaningful queries of the collected information. It is quite clear that the third option maximizes the potential for later analytical work. It is imperative that the values for every variable are clearly defined and comply with the database and recording guidelines. We wish to reiterate that these value definitions are open to modification. New definitions can be introduced during any stage of recording, given that they are included in the pool of value definitions. In some cases, though, it is indeed desirable to create variables that store freely formulated descriptions, such as to explain the directions to a site.

Recording of Rock Art Images

The idea behind descriptive variables is to define and standardize the vocabulary for the description of rock art images, resulting in a very clear, structured way of recording and was first successfully applied to rock art recording in Southern Africa by Vinnicombe (1967) and Pager (1971a, 1971b). A very good introduction to this matter can also be obtained from Evans (1971), who proposed a universal scheme for recording rock art of South Africa, which prepared for a subsequent analysis with computers. All four researchers worked on the basis of separating single images into smaller units, which are described by standardized variables. The idea of describing distinctive sections of a specimen is widely used in the archaeological science, because the level of precision can be adjusted easily depending on the research interest. The main problem is to account for the almost infinite variability of rock art images, as basically every image is unique. It is necessary to reduce the infinite amount of data to a manageable extent of information. To group painted images with each individual sets of characteristics into separate groups of similar paintings the development of categories is mandatory. These are themselves abstractions, and thus are inevitably biased by the research interest. Apart from this, they reflect the researchers way of thinking, rather than prehistoric realities. The study is comparative and thus constrains individual manifestations of the creative human mind to categories consisting of sets of predetermined variables, which can be analysed archaeologically.

To keep pace with the changing needs of recording, this set of variables is open to modification. New variables, or values thereof, can be defined at any time, as long as they are documented and their definitions are added to recording system and database. Data credibility and lucidity are very important issues in this study and each single description is cross-referenced with a digital image.

These descriptive variables are a result of earlier fieldwork conducted with a different recording system, which did not prove feasible for our needs. The flexibility and ease with which the variables can keep pace are an advantage above recording systems, where static categories of "painting types" were defined beforehand. The use of a digital audio recording device replaced the majority of the hand-written paper record. This most recent refinement of the recording system allowed us to collect and digitise a large quantity of information in a short period of time. The time consumption was reduced both, in the field during recording and in the lab during digitising the information. The use of headphones and the co-ordinated structure of the database contributed to data consistency because it minimized mistakes during the digitising process of deciphering hand-written field notes.

Every human figure, animal, handprint or inanimate object is treated as an individual painting, which requires a description. Basic units, which in synthesis yield the description of

Human figure	Animal
biological sex	animal (field observation)
head	
torso	
motion	
arm motion	
attributes	
clothes	therianthrope
context	completeness
orientation	orientation
colour	colour
visibility	visibility
state of painting	state of painting

Table 2 Overview of descriptive variables, as they describe an image of human being

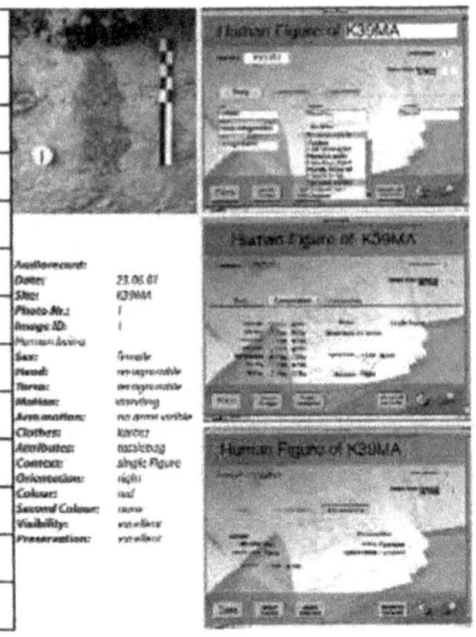

Figure 1 The field record is carefully co-ordinated with the database. Every value of every descriptive variable is available in pull-down menus of the Cederberg Rock Art Database

the image, are related to the research question. We emphasize that an objective basic unit does not exist, and therefore the archaeologist is required to classify the images in an archaeological meaningful way. Our study (Asmus and Meister 2001, Asmus 2003) primarily concentrates on paintings of humans and animals, therefore we dealt with these two classes in more detail than with handprints or inanimate objects. In fact the inanimate objects form such a heterogeneous group, that they were only counted. Handprints were documented photographically and counted. Table 2 lists all descriptive values of the image classes of human beings and animals. The values of these variables are available in pull down menus in the Cederberg Rock Art Database.

The recording archaeologist describes the painting by choosing the appropriate value for each variable (see Fig.1). It describes a particular variable best and records this characterization with a digital audio device. If no satisfactory match could be found, a new value was defined and added to the recording system. The data set for one image was complemented with a site unique painting ID number, the number of the digital photo, date and site name.

Conclusions

The use of low cost up-to-date digital equipment in conjunction with a relational database for the task of recording rock art is a success. Several questions concerning the existence and distribution of rock art sites in the study area could be successfully addressed with assistance of the Cederberg Rock art Database: A relationship between relativ e proximity in distance and altitude could be found. The total amount of 967 paintings of humans and 346 of animals could be analysed. A relation between unusual imagery, such as the paintings of spread-legged females, the so-called ´Mythic Women´ (Goodall 1962, Solomon 1994), or images of elephants, or both, and topographically more isolated sites could be discovered (Asmus 2003).

The time effective way of recording and digitising captured information requires a minimum of personnel. This relatively low cost approach could contribute to the task of protecting the fragile heritage of rock art in Southern Africa.

Acknowledgements

The project was financially supported by the Herbert-Quandt Stiftung der Varta AG, the Bund der Freunde der Universität Tübingen and the University of Cape Town. We are endeared to John Parkington, and the Clanwilliam Living Landscape, who gave us the opportunity to realize the Multimedia application of the Clanwilliam Rock Art Trail (Parkington, Asmus and Meister 2003). Furthermore we would like to acknowledge the tremendous work of Tobias Tonner and Tony Manhire in surveying and recording, and especially Tobias Tonner for making invaluable comments and contributing ideas towards the projects aims. We also acknowledge Joellyn Palomaki for proof-reading this paper.

REFERENCES

ASMUS, B., 2003, unpubl. Painted images, landscape position and site characteristics in the Rondegatvalley, Western Cape, South Africa. MSc. thesis, Department of Archaeology, University of Cape Town, Cape Town.

ASMUS, B. and MEISTER, C., 2001. Projekt zur Aufnahme von Felskunst in den Cederbergen, Südafrika. Archäologische Informationen 24, 1:93-96.

EVANS, R.J., 1971. A draft scheme for computer processing of rock art data. In Schoonrad, M. (ed.), Rock Paintings of Southern Africa, Supplement to the South African Journal of Science, Special Publication Number 2, South African Association for the Advancement of Science, Johannesburg:73-79.

GOODALL, E., 1962. A Distinctive Mythical Figure appearing in the Rock Paintings of Southern Rhodesia. In Nenquin, J. and Mortelsmann, G. (eds.), Actes du IV congrès panafrican de préhistoire, Musée Royale, Tervuren:399-405.

PAGER, H., 1971a. Ndedema. Akademische Druck- und Verlagsgesellschaft, Graz.

PAGER, H., 1971b. The Rock Art of the Ndedema Gorge and neighbouring valleys, Natal Drakensberg. In Schoonrad, M. (ed.), Rock Paintings of Southern Africa, Supplement to the South African Journal of Science, Special Publication Number 2, South African Association for the Advancement of Science, Johannesburg:27-33.

PARKINGTON, ASMUS and MEISTER, 2003. The Mantis, the Eland and the Hunter - A virtual journey through rock art sites in the Warmhoek valley. CD-Rom.

VINNICOMBE, P., 1967. Rock Painting Analysis. South African Archaeological Bulletin 22:129-141.

SOLOMON, A., 1994. 'Mythic Women'- A Study in Variability in San Rock Art and Narrative. In Lewis-Williams, J.D. and Dowson, T.A. (eds.), Contested Images - Diversity in Southern African Rock Art, Witwatersrand University Press, Johannesburg:331-372.

Causality and Cross-Purposes in Archaeological Predictive Modeling

Thomas G. Whitley
Brockington and Associates, Inc., USA

See the CD for the extended version

Abstract

In recent years numerous archaeological approaches to predictive modeling have been presented in the literature. Most of these have taken the "inductive" perspective of applying known site locations to an analysis that estimates probable site location based on a mathematical equation and presents predictive surfaces in a GIS. Conversely, "deductive" models have also been used in which "expert systems" or site selection variables have been quantified as probability surfaces. There has been little discussion, though, of the differences between CRM and academic-based predictive modeling and how it has influenced the state of the "science" today. Generating more refined inductive predictive models either through the use of higher quality site location data or through more complex statistical techniques, runs counter to the implicit goals of CRM-based predictive modeling. A simple deductive GIS approach which assumes a causal explanatory relationship creates comparable or better results (especially in homogenous areas) with no negative effects on these limited goals. Ultimately, the dichotomy between inductive and deductive approaches is not in theoretical orientation, rather it is embodied in our understanding (or failure to understand) that predictive modeling is really a tool useful only for land management, not interpretive archaeology.

The primary use of predictive models to date is almost invariably for large scale Cultural Resource Management (CRM) applications, and they typically occur in North America. The second primary purpose for which predictive models have been developed is that of understanding past land use or what can be called site selection processes (Whitley 2000). It was recognized early on that predictive models provided a quantitative aspect to what was well understood as the qualitative realm of settlement pattern analysis (e.g. Allen et al. 1990). Clearly, predictive modeling is seen as a means by which we might address some of the complex issues of human/landscape interaction, typically in a GIS framework.

Though Bayesian statistics is not invoked in all predictive models, the underlying structure of Bayesian probability is assumed for all of them. The first fully articulated assumption of all archaeological predictive modeling is a form of the Bayesian rule of "total probability" (Pearl 2000:3). This assumption can be stated as: The probability of any land unit being a site (or a portion of a site) is the sum of the probabilities of all exhaustive and mutually exclusive variables that cause a land unit to be chosen as a site or to unintentionally be made one. Put simply, this means that the presence of a site in any area is a factor of all possible influencing variables (intentional or unintentional). This invokes a direct causal relationship between archaeological sites and a host of possible variables; the underlying principle of all predictive models. This is further elaborated by the "conditional rule of Bayesian probability" (Pearl 2000:3-4). This can be stated in archaeological terms as: The probability of any land unit being a site is the sum of the probabilities of each causal variable multiplied by the conditional probability of that variable. Put simply again, this means that not every factor is as influential as every other factor; that they each respond to other conditions, or in effect provide an individual weight to the final probability of any land unit being part of an archaeological site. The diversity in influence of different variables allows predictive modelers to statistically assess, or make expert decisions about, what they find to be the most influential factors and build their models accordingly.

So, what are the theoretical implications of the foregoing characterizations? The most obvious is that the site must be a valid and useful concept to explain human activity. Also, that sites are definitive and absolute in their ability to be identified, and that the characteristics used to identify sites are strictly correlative with human behavior, and not other phenomena. That correlation is assumed to mean that the most common or abundant behavior(s) are the most significant with respect to the purposes of predictive modeling, and that the success of predictive models is a factor of their ability to identify the most common site-selection behaviors.

Are these implications necessarily true, though, and what is not implied by these assumptions, but may be true? First of all, sites may not be easily distinguished by empirical evidence, and they may not be considered strictly representative of human activity (Dunnell 1992). Different activities result in different kinds of sites, and different kinds of sites may have many different kinds of manifestations. Strictly speaking, significance of archaeological sites cannot be equated with abundance, and models geared toward identifying the most common elements of site selection may, in fact, overlook very significant sites in supposedly low potential zones. This becomes manifested in the distinction between so-called inductive and deductive models.

Interestingly, archaeologists employ the terms inductive and deductive in ways that are quite limited with respect to their meaning in the realm of scientific explanation. In philosophy of science, the terms inductive and deductive refer to the difference between logical arguments based on universal laws and statistical tendencies. Deductive explanations are based on mathematically consistent and provable laws, while inductive ones are based on observing statistical trends (Hempel 1965, Salmon 1998). In archaeological predictive modeling, however, inductive and deductive do not refer to methods of explanation, rather to practical means by which probability values are calculated. In the end, all predictive models are trend-based, or inductively explanatory; even so-called deductive ones. Instead, the distinction should be made between models as being either correlative or cognitive in nature.

Correlative models are those which use existing site data and currently measurable environmental variables to build statistical relationships which can then be generalized from previously surveyed areas to those which have not been surveyed (such as through regression analysis). As such, correlative models are strictly empirical by nature and make the assumptions of determinism. Deterministic explanation comes in several forms (Hempel 1965 Salmon 1998) but in general the Bayesian rule of conditional probability, in a deterministic framework, would embrace the idea that the inaccuracies of any predictive model are the result of not having enough information. All probabilities could be identified, in principle, if we only had the ability to do so.

Cognitive models are those which are not limited by existing archaeological data. Hypotheses of site placement are built on understanding more complex issues involved in the cognitive selection of suitable areas. Then, presumed important variables are measured and classified in a way hypothetically similar to how prehistoric populations may have done so, and probability surfaces are projected across the entire landscape. A single cognitive model may, in fact, produce many different permutations which can be tested. Once complete, the known dataset of archaeological sites is then compared to the projected probability zones and accuracy and precision estimates are made. Cognitive models are not necessarily deterministic, though they can be. But they do have the potential to embrace indeterminism, since they are not built from correlative evaluations.

A good way to understand the difference between deterministic and indeterministic explanation can be illustrated by this example (adapted from Mackie 1974:40-41 and Salmon 1998:145-147). Imagine three candy machines. The first is purely deterministic; if you put in a Euro, a candy bar is always ejected. No other coin or object will cause the candy bar to be released, and no candy bar will ever be inadvertently released without the insertion of a Euro. We know that the function of inserting a Euro is both necessary and sufficient to cause a candy bar to be ejected. We can therefore deductively explain the presence of a candy bar with the deterministic rule that a Euro must have been inserted.

A second machine, though, is somewhat different. With it the insertion of a Euro will always produce a candy bar, but sometimes insertion of other items will have the same effect. Thus, a Euro is a sufficient cause of the presence of a candy bar, but it is not the only possible (or necessary) cause. The presence of a candy bar, therefore cannot be deductively explained by a rule stating a Euro must have been inserted, rather it becomes an issue of inductive probability. There is a certain likelihood that a Euro may have caused the appearance of the candy bar (which can be calculated by assessing the state of the system over a period of time or a set number of observations). The machine is still deterministic, though, because we assume that if complete information were available, we could always explain the presence of a candy bar.

A third machine differs in that inserting a Euro or other items triggers an instantaneous analysis of the spin of a quantum particle trapped in the machine. If the spin is in one direction it will result in the release of a candy bar, but in the other direction it will not. Similarly, sometimes with no insertion of any coins, the particle's spin will be measured and a candy bar will be ejected. Here, the insertion of a Euro is neither sufficient nor necessary to explain the presence of a candy bar. We do not have any explicable deterministic way of addressing the presence of a candy bar from this machine. We could assume that more information is needed, but clearly even when we have a suspected causal factor, it does not always result in the release of a candy bar (hence prediction is impossible). In order to fully understand the nature of the machine we need to address all of its constituent parts and the mechanistic effects of all possible causal factors, even if some of those may be reliant on fundamentally indeterministic variables (such as quantum mechanics or human free will).

When dealing with archaeological sites, we are faced with a similar circumstance. Correlative models assume that all variables are either present and have been measured, or that they could be if we only had the information. Ironically, this is termed deductive chauvinism, (Salmon 1998:142-163) and it implies that given all variables and all parameters, all probabilities could be determined and prediction would be 100 % accurate and 100 % precise. Correlative modelers, though, do not deny the influence of human behavior and cognition in making site placement decisions. Yet, the acceptance of this perspective assumes that all human cognition is a deterministic system and implies an absence of free will (Salmon 1998:28). Thus, correlative models are only capable of identifying necessary factors to produce archaeological sites, and only those which are both frequently necessary and commonly observable. Correlative models are not capable of providing insight into the sufficiency of site placement factors to explain the presence of a site, nor the mechanisms of how site selection processes are determined. This is in direct opposition to the second purpose of predictive modeling.

Cognitive models assume, at least, the limitations of the second machine, but allow the possibility of the third (where some aspects of site placement decisions may be, in principle, inexplicable). This assumes that some aspects of human systems are dynamical. Dynamical systems do not imply the absence of predictability for all aspects of the system, merely that they range on a scale between entirely predictable to entirely unpredictable. A correlative analysis, though, is by its very nature limited to only those aspects which are highly predictable, and cognitive explanation must focus on causal-mechanistic issues instead (Salmon 1998). This requires the adoption of explanatory frameworks which deal with indeterministic phenomena.

So, this brings us to the point at which we might ask; why do so many applications of predictive modeling tend to be strictly correlative? The first reason is probably one of convenience. It is relatively easy to take an existing dataset of archaeological sites and environmental values, perform some standard statistical analyses (such as multiple nonlinear regression), and produce a handy formula. The formula can then be turned around and applied in a GIS and a fairly accurate and precise model sometimes results. Second, correlative models are often presumed to have some level of objectivity. With respect to predictive models, I think it is often assumed that more complex statistics means greater objectivity. Such methods have their uses, but the old maxim of "bad data in, bad data out" applies regardless of the methods of mathematical manipulation in between. This was illustrated quite succinctly by Cowgill's warnings against placing too much emphasis on mathematical models at the expense of theoretical profundity (Cowgill 1986:387).

But, ultimately, why is it that correlative models seem to work, at least on occasion? The main reason a correlative model does achieve success is based on what might be called the lowest common denominators for site selection. These are several variables that correlate highly with archaeological sites simply because they are limiting factors on all human behavior; primarily slope and distance to water. Invariably, every successful correlative predictive model uses slope and distance to water, in some form, as key factors in developing correlative formulas, and they almost all occur in semi-arid or highly dissected areas. Causality, though, has not been addressed in such correlative models. It is assumed that because of the correlation between known archaeological sites and particular variables, that they were actively used as suitability indicators for sites. In reality, though, I argue that most environmentally correlative variables (especially slope and distance to water) act primarily as auto-correlations and were probably rarely cognized as variables of choice. In that sense, they may be considered necessary factors for site selection, but their importance is auto-correlative with all human activity not causally conditioned by it. Thus, they cannot be seen as sufficient cause for site placement.

So the upshot of this discussion is; how do correlative models fail to meet the goals of predictive modeling? Several things would be required for successful application of a correlative model. First, the project would need to be located in a region that is sufficiently arid and dissected enough that slope and distance to water are meaningful limiting factors, or suitable alternative limiting factors can be identified. Second, it would require a large dataset of accurate and well described archaeological sites from all temporal periods of similar function, and which do not cross multiple cultural boundaries. Third, it would require an additional dataset which could be used to test the accuracy and precision of the model, or alternately a large enough initial dataset of sites that a "jack-knife" sample can be held back and used for testing. Fourth, it would require a well developed environmental dataset from which measurable variables can be extracted.

The problem is that there are few locations within which all of these criteria are met. Almost invariably the project areas do not have such ideal conditions or large and accurate datasets, and those conditions can only be met by vast amounts of data gathering at a very high cost. It therefore becomes too costly or even impossible to do a correlative predictive model in many cases, and ultimately the resulting model does not provide better insight into site placement processes than intuition. As the cost of doing the predictive model increases, the reasons it was initiated become increasingly irrelevant from a land management perspective. Since slope and distance to water are the primary important limiting factors in most correlative models, they contribute the greatest percentage to a model's success. Any variables beyond the primary ones add proportionally less to the gain statistic and therefore are of smaller and smaller consequence. Conversely, though, the means of extracting the influence of those additional variables adds increasingly to the cost of the model.

Ultimately, this begs the question of; how should we create and apply quantitative archaeological predictive models? What purposes are appropriate, and who should pursue them? I argue that we need to recognize that correlative predictive models (regardless of their methods and area of application) have severe theoretical and explanatory limitations. They can be used in some situations to give insight into land management activities; specifically alternatives analysis. But they should not be mistaken as tools of interpretative archaeology. Likewise, it would be inappropriate to consider correlative models as a means to protect significant archaeological sites or high potential areas, as they would be limited to only abundant kinds of sites and a few, probably auto-correlative, environmental factors. Without causality, predictive modeling may be a useful land management tool in some severely limited settings and with nearly unlimited funding, but it provides no explanatory power and forces a deterministic and facile understanding of human cognition.

REFERENCES

ALLEN, K., GREEN, S. and ZUBROW, E., 1990. Interpreting Space: GIS and Archaeology. Taylor and Francis, New York.

COWGILL, G., 1986. Archaeological Applications of Mathematical and Formal Methods. American Archaeology, Past and Future; A Celebration of the Society for American Archaeology 1935-1985. Meltzer, D., Fowler, D. and Sabloff, J. (eds.), Smithsonian Institution Press, Washington DC.

DUNNELL, R., 1992. The Notion Site. Space, Time, and Archaeological Landscapes. Rossignol, J. and Wandsnider, L. (eds.), Plenum, New York.

HEMPEL, C., 1965. Aspects of Scientific Explanation. Free Press, New York.

MACKIE, J., 1974. The Cement of the Universe. Clarendon Press, Oxford.

PEARL, J., 2000. Causality: Models, Reasoning and Inference. Cambridge University Press, Cambridge, UK.

SALMON, W., 1998. Causality and Explanation. Oxford University Press, Oxford.

WHITLEY, T., 2000. Dynamical Systems Modeling in Archaeology: A GIS Approach to Site Selection Processes in the Greater Yellowstone Region. PhD Dissertation, Anthropology, University of Pittsburgh, Pittsburgh, PA.

Use and Abuse of Digital Terrain/Elevation Models

Willem Beex
ARC-bv and National Museum for Antiquities, Netherlands

Abstract

DTM's and DEM's are widely used for landscape reconstruction and GIS analysis. Very accurate models are possible to construct on a small scale with the aid of a 3d-scanner, or on a larger scale with the aid of photogrammetic processing of aerial images. But far more often archaeologists will have to rely on software to process point-data and contour lines.
The different kinds of software have many algorithms and parameters. They have to be used with care, as changing an algorithm or the parameters for an algorithm, will dramatically alter the resulting DTM or DEM. Especially if the model is used for further GIS analysis, any lack of knowledge on these differences may have serious repercussions on the conclusions of for instance site-locations and ways of communication.
Ken Kvamme already demonstrated these possible dangers almost a decade ago. Since then, many new and more elaborate algorithms came available, capable of both creating a better model as well as an even more distorted image of reality. This suggests that archaeology may benefit by an update on this subject.

Introduction

Letting a computer process a series of co-ordinates, representing a research-area and values within that area, seems pretty straightforward. In this paper, however, I will try to explain that there are many things to be considered, before the end-result may be used or presented, as a valid part of, or in future research.

Awareness of this started formally about ten years ago and has led to my workshop of this CAA in 2003.[1] Between these years, many well-known authors in the proceedings of the conference kept these ideas alive (f.i. Barceló 2000, Lock 1995, Niccolucci 2001, Stancic 1995) but far too often only in front of a "specialized audience".[2]

So in many aspects, this paper is a both a summary of these past efforts and warnings, as well as an enhancement towards the status quo of this moment. Certainly as I have added some calculable items, especially on the grid size of collecting data, and some examples, which will make the understanding of creating a functional DTM/DEM more useful to scientists in archaeology.

Using DTM/DEM's, from Simple to very Complex

In Archaeology, the very basic usage of using a digital model representing z-values within an area is to create a map with contour lines. This is very straightforward, and usually taken from existing maps. In many cases contour lines can be digitised directly form any topographical map, or easily obtained with an application able to convert available point-data into such a map.

More often, archaeologists want to present a "good-looking" 3d-model of the values. This requires a program with an algorithm to process point-data or digitised contour lines. If not used for further scientific research, please use the basic options of the available application, and see if it looks "reasonable". Be sure to provide meta-data, including the origin of the data, the program and the algorithm used.

The last phrase is very important. In the pre-computer era, detailed elevation maps were produced by highly skilled surveyors, using as many accurate measurements as could be taken, and a clear view on the landscape (they were there!). They could enhance their measurements on the map with any visible anomaly in the landscape. No computer has this advantage.

Available Types of Data Sets

All measurements needed for a DTM/DEM can be grouped into two kinds of data sets. They are either gathered within a grid (or the best equivalent to a grid), or they are irregular, as for instance digitising the original point-data maps of surveyors in the Netherlands or contour lines.

The Minimum Distance needed between two Data-Points (Nyquist-Limit)

Whether or not the data comes from point-data measured in an almost regular grid, or from an irregular distribution, the key-factor for a correct creation of a DTM/DEM is the maximum distance between two points in the survey-area. That distance will determine the minimum size of features in the landscape to be detected.

From an empirical test, it was easily discovered, that this distance should be at least one-third of the smallest anomaly in an area that has to be detected, in order to be sure that it reflects itself on a map. Using the knowledge of other sciences however, this idea can be scientifically enhanced and supported by the "Nyquist Limit".[3] This law, also known as "Shannon's Sampling Theorem' states: "A signal must be sampled with a frequency at least twice the frequency of the signal itself." Applied to the creation of a DTM/DEM in a two-dimensional environment, this will have to be: "The maximum distance between two points should be 0.5 times 0.5 equals 0.25 the size of the area/shape of a feature to be detected."

So the "Nyquist Limit", can be used as the best way to determine the lower-limit of any sampling in Archaeology. Please note, that this will also apply to any other spatial sampling.

The Main Algorithms and their Use

Perhaps the most common known algorithm is not actually an algorithm, but just a connection between 3d-points. It is called "Triangulated Interpolation Network", or short TIN. TIN, in it's originally use, does not change any of the given data-points, it just creates a triangular mesh between them. The advantage is of course a full representation of each measurement. The disadvantage is an "edgy" looking DTM/DEM.

The pure TIN would be almost perfect, if data would be evenly spaced within a grid, and with a maximum distance between points not exceeding 1/10th of the necessary "Nyquist Limit" and smoothed. In most cases however, the measurements will be quite close to the "Nyquist Limit", or they have other irregularities, and there is an actual need to "interpolate" the mesh on the basis of the distribution of measurements.

If this is needed, it is important to understand that every single measurement is still fully recognised, but the resulting DTM/DEM may be lower or higher on that exact spot. This can be best explained in describing measurements of a ploughed field on a slope. In that case the field is a mess of small "ditches" and "walls", with no measurement giving the exact slope. Each point, whether too high, or too low, is however part of the general overall slope. And we want to "average" all existing data, in order to filter out the noise. Or in other words: "Too Smooth and Curve."

Most algorithms for DTM/DEM's perform exactly this operation. And therefore they are both very useful and very dangerous. Hence this title "Use and Abuse of Digital Terrain/Elevation models". And there are a dozen of them available. Within archaeology, this amount can easily be reduced to five: "Inverse Distance to a Power; Kriging; Minimum Curvature; Natural Neighbour; Nearest Neighbour." All of these five perform a special operation with the data set, and will result in a less or more accurate 3d-model, giving the needs.

The differences are quite visible, but often not understood. This is the point where the archaeologist should want to have a statistician with knowledge of the subject, as there is no formal "Nyquist-limit" to calculate this effect.

Basic Parameters

Except for "pure TIN", almost every algorithm should have the ability to adjust "Search Radius". This is used to disallow points from beyond a certain distance to flatten mountains like the Alps, or (on the other hand) to allow them to smooth a ploughed field on a slope.

Most common applications for creating DTM/DEM's will also allow a special parameter. This parameter is called "anisotropy", which is best explained as the "knowledge of the most common shapes in a landscape." The Norwegian coastline, for instance, is on many places characterized by fjords.

These fjords show an existing overall tendency in the landscape, which is normally disregarded in an algorithm. Anisotropy as a parameter can use this tendency and change the normal "circular" way in which data-points are allowed to influence each other. It can narrow down the "Search Radius" in one direction. By this, measurements on steep edges could be limited in the "Search Radius" as far as the top of their (average) ridge. This will prevent all algorithms to "down-average" prominent features like fjords, which are very steep and narrow.

Mapmaking

Despite all other warnings, even a DTM/DEM is a map. The "Nyquist Limit" may give an indication for the smallest detail, but it cannot prevent the creation of a terrible representation. As it is no use of making a map with every house in Amsterdam, as part of a roadmap for the entire Netherlands, as it is to use a map with highways as a guide for the centre of Amsterdam. Scale and detail have to be in relation to the size of the documented area.

The "Idrisi" Factor

In the twelfth century the cartographer Abu Abdallah Idrisi[4] presented a silver sphere to the Norman King Roger of Sicily, representing the earth. This may seem an anachronism, given the date, but the king, who apparently fully appreciated the true nature of our planet, kindly accepted it.

Anyone working with DTM/DEM's has to appreciate this too. A DTM/DEM normally is a plan and therefore only a projection of a part of the earth's surface. In general this not so important, but it becomes critical if the research area is over ten kilometres wide, and the model will be used for certain GIS analyses like "viewshed".

Any further analysis that is not entirely map-based, like "viewshed" (the panorama from a certain point in the area), will have to take in account this "Idrisi"-factor. It will need the precise knowledge of the earth's curvature, in order to adjust the DTM/DEM accordingly.

Conclusions

DTM/DEM's are often easily created, but they tend to be just "nice images" instead of a true scientific basis for (further) research. And even the 'most appreciated' examples of the past decade, are presented without any metadata regarding the algorithm, or the parameters used.

This omission does not respect any of these models (and the results based on them), as it is impossible to validate the DTM/DEM for it's specific use.

The data sets themselves are also a main point of consideration. Detection of a feature on the basis of measurements within a 2d-environment is only feasible if the largest distance between two 2d-points follows the "Nyquist-Limit". This scientific limit, based on "Shannon's Sampling Theorem", should be at least 0.25 of the size (smallest diameter) of the least feature to be detected.

And finally, every DTM/DEM is an extended map. It has to follow the rules of a map and it has to know about the "Idrisi"-factor.

[1] The workshop used the program "Surfer 8.0". "Golden Software" allowed the use of their program for CAA2003.

[2] These papers were published in far more specialized publications (very restricted to subject or country, or both) than the CAA-proceedings. The best example is from Franco Niccolucci. A very clear "status quo" of this subject dating from 2001, but only available in the series "Archeologia e Calcolatori", on computer applications in Italy and in Italian.

[3] Many thanks to Sven Haveman, of the "Institut für ComputerGraphik", Braunschweig. He made me aware of this existing mathematical law considering 'sampling', during the CAA2003 conference. We both determined at that time, that this law had to be applicable in this case. Sven Haveman has agreed with the fact, that I have enhanced the paper with this important item.

[4] Abu Abdallah Muhammad Ibn Muhammad Ibn Abdallah Ibn Idrisi al-Qurtubi al-Hasani, as his full name is, was born in Ceuta, Spain; in 1099 A.D. Sources differ on the date of his death, either in 1166 or 1180. His name is nowadays widely known within CAA as the person who gave his name to Idrisi, the GIS-program for the PC.

REFERENCES

BARCELÓ, J.A., 2000. Visualizing what might be: An introduction to Virtual Reality techniques in archaeology. In Barceló, J.A., Forte, M. and Sanders, D.H. (eds.), Virtual Reality in Archaeology, Archaeopress, Oxford:9-35.

BEEX, W.F.M., 1999. Syllabus voor het programma Surfer 7.0, Universiteit Leiden, Leiden.

BEEX, W.F.M. and PETERSON, J. W.M., 2003 in press. The Arminghall Henge in space and time: how Virtual Reality contributes to research on its orientation.

D'ANDREA, A. and NICCOLUCCI, F., 2001. L'informatica dell'archeologo: alcune istruzioni per l''uso. In Archeologia e Calcolatori 12:199-220.

LOCK, G. and STANCIC, Z. (eds.), 1995. Archaeology and Geographical Information Systems. London.

RYAN, N.S., 1996 Computer based visualisation of the past: Technical 'realism' and historical credibility. In Higgins, P., Main, P. and Lang, J. (eds.), Imaging the Past, British Museum Press., London:95-108.

The Application of GIS in Analysing Human Risk to Shipwrecks. Port Phillip Bay, Victoria, Australia

SEE THE CD FOR THE EXTENDED VERSION

Lisa M. Cornish
Department of Geomatics,
The University of Melbourne,
Australia

Abstract

Within Australia, the protection and maintenance of historic shipwrecks is becoming an important part of heritage management. This paper explains how Geographic Information Systems (GIS) can contribute to the management process.
100 shipwrecks in Port Phillip Bay, Victoria, were analysed for potential threats from human interference. The impact of the general population and tourists using the coastal area, foreshore and underwater developments, commercial and recreational fishing, the activities of ships and boats and scuba diving and looting were all studied as part of this research GIS was used to determine the threat from each of the human activities analysed, and to determine an overall threat for each of the 100 shipwrecks. Various methods of combining data were used to determine the overall threat. This found that methods can be used to exaggerate or minimise the level of threat, depending on the aim of the analysis.

This paper summarises research conducted to investigate potential risk to shipwrecks in and around Port Phillip Bay, Victoria, Australia. Geographical Information Systems (GIS) were used to aid in this analysis. The aim of the research was to determine risk to shipwrecks from human activities using a simple method that could be duplicated for similar research conducted elsewhere. Methods of combining data using GIS was also tested to determine how results can alter depending on the approach selected.

Port Phillip Bay was chosen as the study area because it has a large number of shipwrecks currently protected by Australian legislation, there is a high level of shipping, coastal and water activities and research previously conducted has been inadequate (Duncan 2002). It was also selected because its size allowed an assortment of human related activities that occur in the region to be incorporated (Fig.1).
Vessels go through four stages of destruction after being wrecked, but only one of these was relevant to the research. A vessel is initially destroyed and sunk. It is then damaged by natural biological disintegration processes which are then followed by natural chemical disintegration. Finally, a vessel is further destroyed by human activities after its rediscovery (Muckelroy 1978, Ward et al. 1998). As the final stage of the destruction process focuses on the human impact, it was the only section of the destruction process investigated. Risk from human activities was investigated through three groups:

1. Scuba Diving
2. Fishing and Shipping
3. General Population, Tourism and Development.

Scuba divers have played an important role in Maritime Archaeology through discovering shipwrecks and participating in research, however they are also extremely destructive (Bascom 1976). Divers cause damage by creating new entrances to shipwrecks, removing materials that provide natural protection, collecting souvenirs, dropping heavy equipment such as weights and tanks, and colliding with shipwrecks because of buoyancy control problems (Gibbins 1995, Green 1990, Harvey 1996). Scuba divers can access shipwrecks in a number of ways, such as by private boat, charter operators and accessing shipwrecks from the shoreline. They also use a variety of sources to locate them, such as the media and word-of-mouth. To incorporate these into the analysis and get a general idea of the risk scuba divers and scuba diving poses, the proximity of a shipwreck to major beaches and piers, the number of divers visiting shipwrecks, shipwrecks accessed by charter operators and shipwrecks promoted in various media sources were

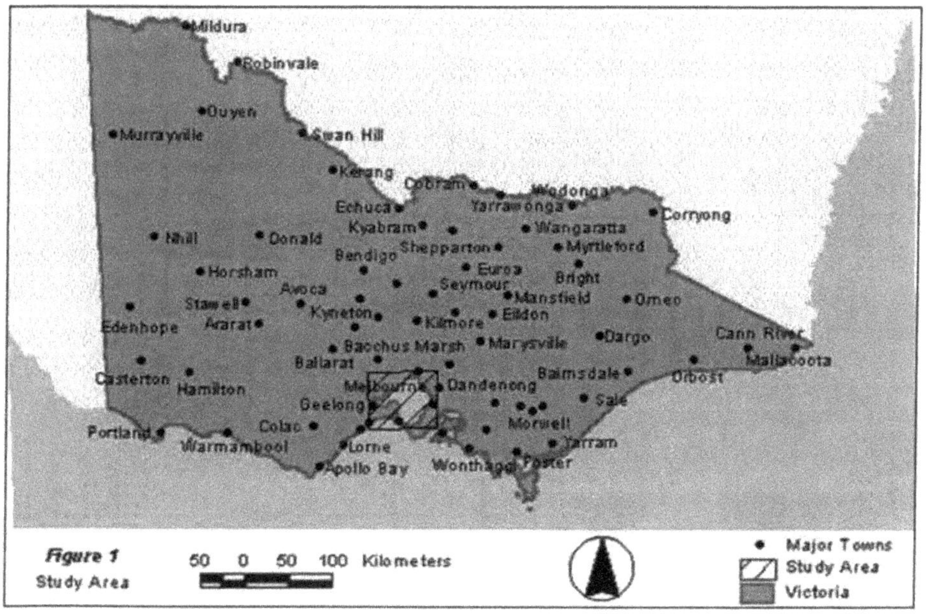

Figure 1 Study Area

analysed as well as the number of shipwrecks previously looted and what type of shipwrecks looters favour.

Fishing and shipping activities are a large part of the human related activities that occur within the study area and have the potential to cause a great deal of damage. Damage is caused by equipment being caught on shipwrecks, removing natural protection, altering the surrounding marine environment, dropping anchors on or near shipwrecks, and producing waves which cause damage to the shipwrecks and the surrounding marine environment (Tomkin 1998, NSW Heritage Office 2000). To get a general idea of the potential damage caused through fishing and shipping activities, locations used for commercial and recreational fishing, the types of boats and ships using the study area and the proximity of both to shipwrecks was incorporated into the study.

The area surrounding Port Phillip Bay is popular for residential housing and popular to tourists. This has led to a large amount of development in the area. The general population and tourists cause damage through water sports and other water activities in the vicinity of shipwrecks. Development causes damage differently depending on the type being analysed - foreshore or underwater. Foreshore developments have the potential to alter the coastal environment. This has been shown to cause damage to the underwater marine environment (Bird 1993). Underwater developments can have a direct impact on shipwrecks or the underwater marine environment (Joint Nautical Archaeology Policy Committee 1990, Breen and Forsythe 2001, Heritage Victoria 2001). The impact of the general population and tourism was incorporated into the analysis by determining the amount of people using the area and the activities they are involved in. The impact of developments was analysed by looking at the number and location of foreshore and underwater developments and the land and property values for local government areas. Proposed developments were also taken into consideration to incorporate future risk. Again, this analysis was not designed to produce an accurate risk result, but rather a result that would be simple, but easy to recreate in other analyses.

For each of the three groups analysed, ratings were produced showing the degree of risk from human activities. The ratings given were between 0 and 3 representing no, low, moderate and high risk. While only one result was created to show risk from Scuba Diving and Fishing and Shipping, two results were produced for the risk from General Population, Tourism and Development. This was done so one result would overlook potential developments and the other would incorporate it.

The three ratings produced showing the risk from human activities needed to be combined to produce a final result and determine an overall level of risk to shipwrecks within the study area. This was also an opportunity to test how results can differ depending on the combination method used. Three combination methods were chosen for analysis: Manual Combination, Rating and Weighting and Adding. For each of these, two results were produced. One overlooked potential developments while the other incorporated them. The final rating system used for each gave values between 0 to 5 rather than the 0 to 3 used previously. This was chosen to allow the extreme high and low values to be distinguished better. For this new rating the values represent no, very low, low, moderate, high and very high risk.

Manual Combination looked at all possible values achieved by combining the three risk ratings and reclassified them with an appropriate value. It allowed for certain risks to be given priority over others and selected combinations to be given higher or lower overall risk ratings if it was considered necessary.

Rating and Weighting involves placing categories in order of importance and multiplying them by values that will increase or reduce their significance before combining. Combining the three risk categories using a rating and weighting method was impossible for this analysis. All three categories were equally important in terms of the risk they posed. They were instead subdivided into nine groups and these were placed in order of importance and given appropriate weightings.

Adding simply involves adding the three risk categories together and reclassifying them into values from 0 to 5. This method was used to test the accuracy of the combination methods used above. Both Manual Combination and Rating and Weighting can be influenced by personal biases on what the results should be.

Manual Combination exaggerated the level of risk within the study area. When using this method, ratings can be increased or decreased depending on the desired result. Adding allowed results to be given quickly, but they were inaccurate. Any results obtained using this method should only be considered a guide for a basic analysis. Rating and Weighting was the preferred method of combining risks (Fig. 2). Important cate-

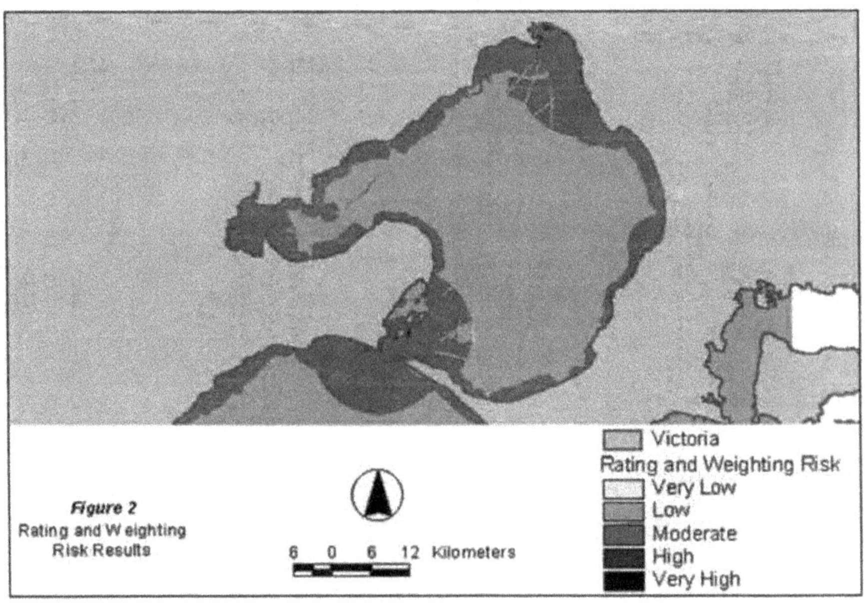

Figure 2 Rating and Weighting Risk Results

gories could be emphasised and changes in the level of risk could be updated or incorporated by simply changing the ratings and weightings assigned to the categories being combined.

In the results of the analysis conducted, moderate and high risk dominated the results. This is because most shipwrecks lie in or near Port Phillip Heads or along the coast where the greatest amount of human related activities occur in the study area. Potential developments were most evident with the results of the Rating and Weighting combination method (Fig.3). Rating and Weighting allowed foreshore developments and underwater construction to play a more significant role in the overall results. Other combination methods masked their impact when they were merged with general population and tourism risk.

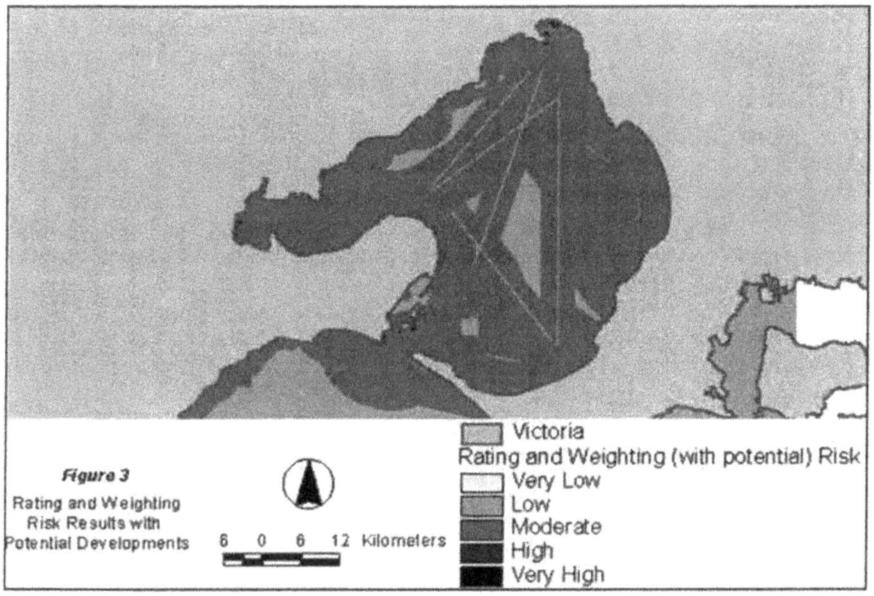

Figure 3 Rating and Weighting Risk Results with Potential Developments

The results of this analysis give a simple and basic understanding of risk to shipwrecks within the study area. To keep the analysis updated, the study needs to be adjusted, revised and conducted regularly to test the impact of changing risk. When using the risk model to determine the protective needs of shipwrecks, it is recommended that shipwrecks with a high or very high level of risk be concentrated on first. However it is necessary to conduct more research to determine the accuracy of ratings and what may need to be done to protect shipwrecks before the risk rating is used for conservation within this study area or other areas.

The complexity in combining and reclassifying data using GIS was downgraded in this study in favour to providing a method of analysing risk that could easily be recreated. Care was taken to use data that could easily be found and replaced by data relevant to the new study area. It was not intended to produce an accurate risk result, but rather a general understanding that could be used as the basis for further research.

References

BASCOM, W., 1976. Deep Water, Ancient Ships - The Treasure Vault of the Mediterranean. David & Charles, England.

BIRD, E.C., 1993. The Coast of Victoria - The Shaping of Scenery. Melbourne University Press, Melbourne.

BREEN, C. and FORSYTHE, W., 2001. Management and Protection of the Maritime Cultural Resource in Ireland. Coastal Management 29:41-51.

DUNCAN, B., 2002, in preparation. The Identification of Potential Archaeological Sites in Port Phillip Bay Based on the Use of Historic Charts and GIS Interpretation. The Globe.

GIBBINS, D., 1995. What shipwrecks can tell us. Antiquity 69:408-411.

GREEN, J., 1990. Maritime archaeology in Southeast and East Asia. Antiquity 64:347-363.

HARVEY, P., 1996. A review of stabilisation works on the wreck of the William Salthouse in Port Phillip Bay. Bulletin of the Australian Institute for Maritime Archaeology 20(2):1-8.

HERITAGE VICTORIA, 2001. Protecting our historic shipwrecks. Heritage Victoria, Victoria.

JOINT NAUTICAL ARCHAEOLOGY POLICY COMMITTEE, 1990. Heritage at sea: proposals for the better protection of British archaeological sites underwater. Antiquity 64:390-400.

MUCKELROY, K., 1978. Maritime Archaeology. Cambridge University Press, UK.

NSW HERITAGE OFFICE, 2000. Guideline note: snagged objects. NSW Heritage Office, New South Wales.

TOMKIN, J., 1998. Commercial Fishing Methods in Victoria. Natural Resources and Environment, Victoria.

WARD, I.A.K., LARCOMBE, P. and VETH, P., 1998. Towards new process-orientated models for describing wreck disintegration - an example using the Pandora wreck. Bulletin of the Australian Institute of Maritime Archaeology 22:109-114.

GIS AND THE EXCAVATION OF THE EARLY MEDIEVAL CENTRE IN POHANSKO, CZECH REPUBLIC

JIRÍ MACHÁCEK
INSTITUTE OF ARCHAEOLOGY AND MUSEOLOGY,
MASARYK UNIVERSITY OF BRNO, CZECH REPUBLIC

MICHAL KUCERA
INSTITUTE OF ARCHAEOLOGY AND MUSEOLOGY,
MASARYK UNIVERSITY OF BRNO, CZECH REPUBLIC

SEE THE CD FOR THE EXTENDED VERSION

ABSTRACT

The Department of Archaeology at University in Brno has fully incorporated GIS into the processing of major excavation at Pohansko, a fortress dating to the 9th century AD. Archaeological research at the site has been underway since 1958 and has uncovered a wealth of artefacts providing insight into Early Medieval society and culture.
In 1995, Department of Archaeology first began applying GIS technology at the Pohansko site. Since then, the researchers have digitised the archived excavation data, incorporated new information and initiated complex analysis projects with the GIS. In 2002 was published the fully digital catalogue, based on GIS technology.
At the present day processing of the intra site spatial analysis by the using of the combination of GIS and multivariate statistics (PCA) continues. Geospatial technology enables not only to store and retrieve a variety of data types instantly, but also to spatially correlate relationships between artefacts and their discovered locations in the fort to better understand how tools and ceramics were used in daily life or rituals.
Application of GIS technology at the Pohansko excavation was awarded by the "2002 GeoMedia Best Practices Award" for GIS implementation within researches and education from Intergraph Mapping and GIS Solutions.

POHANSKO NEAR BRECLAV - EARLY MEDIEVAL CENTER IN CENTRAL EUROPE

The archaeological complex of Pohansko south of the town of Breclav is one of the most important and thoroughly investigated early medieval centers in Central Europe. Built in the ninth century, Pohansko is a sprawling fortress in the southeast corner of the Czech Republic. Archaeological research at the site has been underway since 1958 and has uncovered a wealth of artifacts providing insight into early medieval society and culture. In that time, 140,731 square meters have been excavated, uncovering 1,346 settlement features, 872 graves with skeletal remains, 55 cremations and thousands of post holes. More than 200,000 artifacts have been inventoried from the excavations here. The interdisciplinary research carried out here has been closely associated with a number of academic institutions, first and foremost the Institute of Archaeology and Museology of the Faculty of Arts of Masaryk University in Brno. The results of the scholarly research of three generations of scholars have been published in seven monographs and more than one hundred scholarly articles (a summary of work up to 1995 is in Vigniatová 1996:264-266). For example both Czech and foreign scholars have documented processes at Pohansko, a model early medieval centre that occurred at the historical moment of the birth of a new social order and the creation of the first state-level political organizations. The results of research at Pohansko are also often used as the basis for analogy in the interpretation of other important early medieval centers (among the most recent Fagan 1996).

The fort's impressive ramparts were constructed of stone, dirt and wood to a depth of 6 meters and to a similar height. The total length of the rampart is two kilometers and delimits an area of 28 hectares. Among the most exciting and significant recent discoveries is the court of a local magnate comprising palace, church and cemetery. Extending over a one-hectare rectangular area, the court served as a meeting place for the elite of the Great Moravian society, many of whom were buried in the cemetery. The latter has yielded 407 graves containing swords, axes, spurs, gold and silver pieces and Byzantine-Oriental jewellery. An analogy to this settlement structure can be seen in the Carolingian-Ottonian palatine, which German researchers have called a "palatium" (palace). The feudal court at Pohansko was not isolated. Located nearby was a settlement of craftsmen, which is known as "Lesni skola" [tree nursery] portion of the site, whose tools and grave artifacts offer an intriguing insight into the activities of the various classes of this mediaeval society. A residential area with an entirely different character was found in the outer ward areas.

HISTORY OF GIS PROJECT POHAN

Upon its introduction in 1995, GIS was initially employed to digitize the numerous sketches, plans and drawings resulting from over 45 years of onsite work at Pohansko (Kucera and Machácek 2001, 2003). In addition, researchers applied database techniques to organize and catalogue information relating to tons of artifacts and thousands of photographs. The original GIS laboratory included MicroStation for digitizing and GIS software MGE PC for analysis. Attributes were maintained in a Dbase IV database running under Microsoft DOS. By 1997, this system was upgraded to an Intergraph TD 30 Windows NT workstation, with data migrated to an Intergraph MGE 5.0 GIS and Microsoft SQL Server 6.0. In

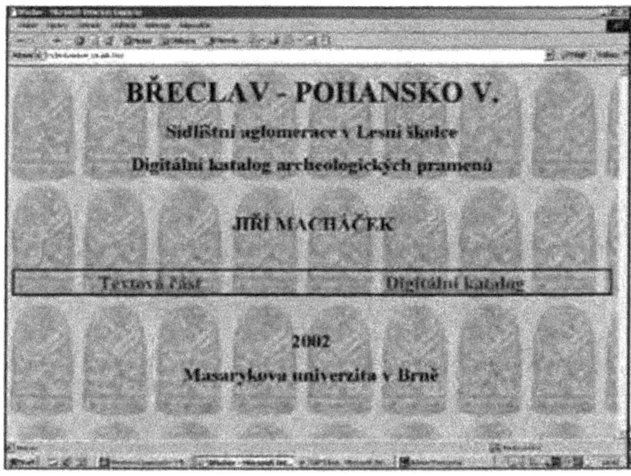

Figure 1 Breclav-Pohansko V. Digital Catalog of Archaeological Sources. Title page

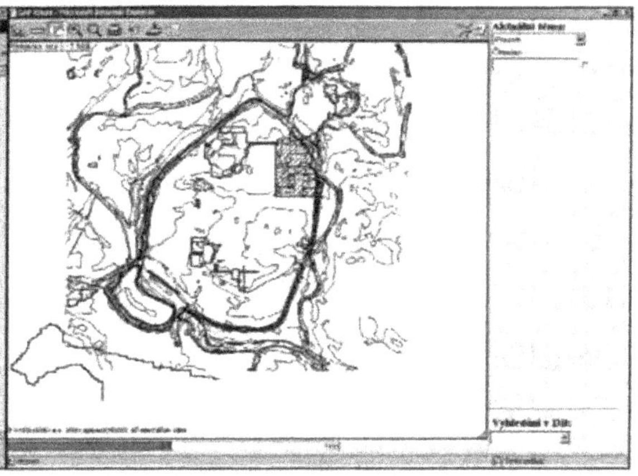

Figure 2 Breclav-Pohansko V. Digital Catalog of Archaeological Sources. Overall map

2002, the Institute of Archaeology became a Team GeoMedia Registered Research Laboratory and, under Intergraph's guidance, once again upgraded its resources.

GeoMedia Professional 5.0 now serves as the main tool for data integration and analysis. Microsoft Access is the primary repository for graphical and non-graphical data, while MicroStation remains the digitizing platform for vector maps and plans.

Researchers now incorporate new information and initiate complex analysis projects with the GIS solution. They use GeoMedia Professional during this phase of data collection to establish relationships between attributes and graphical data files and to perform quality control on the data inputs. GeoMedia also validates the connectivity and geometry of the spatial maps, sketches, and drawings. A critical advantage of the GeoMedia software is that relationships between data sets can be maintained despite the fact that data is stored in a variety of raster and vector formats in different databases. At any given time, Pohansko researchers are managing attributes in Access, vectors in MicroStation, and archived data in MGE. All of these data sets are accessible for analysis and integration in GeoMedia. The geospatial data management solution at the Pohansko site enables researchers not only to store, query and retrieve a variety of data types instantly, but also to spatially correlate relationships between artifacts and their discovery locations to better understand how they were used in daily life or special rituals.

Digital Catalogue of Excavation in Pohansko

One of the most important applications of GIS at Pohansko is the digitalization of the catalogue for the Lesni skolka (Machácek 2002), where excavations were carried out over seventeen seasons between 1961 and 1991. The total area uncovered at the Lesni skola is approximately 1.9 ha. The primary excavation units, which were documented with drawings, photographs and written records, were individual settlement features, graves and postholes. A new phase of comprehensive analysis and publication of the large residential agglomeration at the Lesni skola began in 1995. The phase of research involves the use of the most modern technology including geographic information systems (GIS), which were used in the project with the working title of POHAN.

Information used in the project was in the form of existing drawings and written and photographic documentation. Drawings were primarily in a 1:20 scale on millimeter-grid paper. The documentation presents the substratum, or more precisely the excavated artifact-bearing strata, structures in the fill of semi-subterranean settlement features and graves and their negative remaining (cuts) after excavation. Written documentation is in the form of handwritten research journals and artifact reports. Detailed photographic documentation, generally on black-and-white film, was also prepared in the field. The first step in the creating a GIS for the Pohansko fortification was the digitizing of these analog data. This occurred at three levels:

- transfer of field documentation into a database format using the MS Access program,
- manual vector digitizing of field maps using the MicroStation program,
- scanning the large photographic archive.

This heterogeneous data was then integrated into a unified data model, which is now maintained using the GeoMedia Professional program. This product of the firm Intergraph is based on an object-oriented approach to both graphic and non-graphic data saved in a unified data warehouse using a standard database format (Microsoft Access). Graphic and spatial data in the POHAN project is primarily based on vector models, where areas (polygons) and lines are represented. It is possible to add to these aerial views representing transformed and localized screens. Non-graphic data are saved in the database in the form of alphanumeric or numeric records or hypertext links to external files (screened pictures). The reasoning behind this data model was to present the most complete presentation of the complicated reality ascertained by archaeological research at Pohansko. All other applications are derived from this data model.

The basic properties of archaeological features are their geographic location and topology. For the POHAN project, features are located in terms of the Czech Republic's S-JTSK system of coordinates. Each type of features is depicted with its own color and line type. Significant features contain a record of its non-graphic attributes. Features are divided into several categories that have been derived from the existing Lesni skolka research documentation and logical subdivisions:

Overall site map: a detailed contour map depicting streams and buildings. Also depicted are areas that have been archaeologically explored, with the field season also being indicated.

Surface: depicts the situation prior to the beginning of research following the removal of the overburden. Within the Surface category, non-graphic database information is described in 5-by-5-meter grids that form the base of archaeological research at Pohansko.

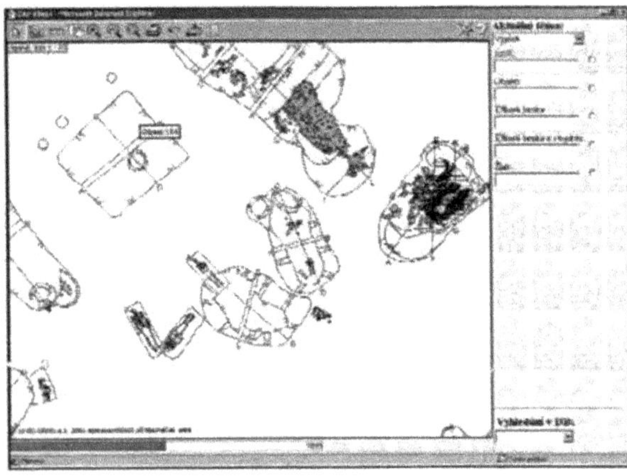

Figure 3 Breclav-Pohansko V. Digital Catalog of Archaeological Sources. Section of excavation, scale 1:75

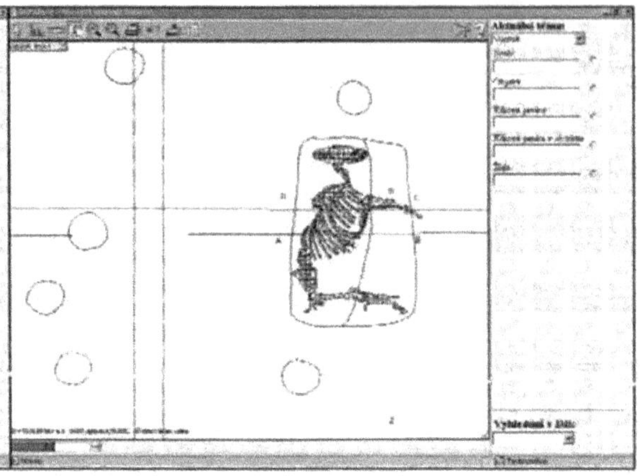

Figure 4 Breclav-Pohansko V. Digital Catalog of Archaeological Sources. Section of excavation, scale 1:25

Fill: contains documentation of the archaeological situation (generally on a 1:20 scale), recording structures and artifacts (deposits) in the fill of settlement features and graves, including profiles left for checking fill artifacts. Non-graphic database records exist for objects that represent individual categories of archaeological features such as settlement features, postholes, graves and palisade ditches.

Cuts: presents visualizations of the archaeological situation in the final stages of field research after the complete removal of all fills and control profiles. This documents the negative form (cuts) remaining after the excavation of settlement features, graves, postholes and palisade ditches. The database records contain the same features as in the Fill category.

Cut-graves, Cut-postholes, Cut- settlement features, Cut-palisade ditch: are logical categories logically derived from the category Cuts. This subdivided category is designed to aid the user in orienting in the complicated archaeological situation as it exists at Lesni skola.

Recent: intrusions into the substrate that has been interpreted by the authors as modern anthropogenic features.

Non-graphic data, which are presented in a relational format accompanying features, are derived from a variety of sources. These include database tables, which were derived from the original handwritten documentation and descriptions of the terrain, and formalized descriptions of artifacts from Lesni skola that were made by several authors during the second half of the 1990s. These also include drawings and recent digital photographs.

Presentation of Archaeological Data in Digital Catalogue

The digital catalog for Lesni skola contains e.q. 7052 vector polygons for stones, 9372 vector polygons for bones, 847 vector polygons for clay daubs, 954 vector polygons with non-graphic attributes for postholes, 285 vector polygons with non-graphic attributes for settlement features, 80 vector polygons with non-graphic attributes for graves or 2419 terrain monochrome photos, 1502 color digital photos of non-ceramic artifacts, 642 drawings of ceramic fragments in relational databases and so on.

A significant advantage of this approach is new possibilities for publishing scientific information. For these purposes, GeoMedia was used to generate raster graphics from the data Warehouse, HTML pages and the Microsoft Access database. The transformation (data migration) was carried out using VARS GAP technology in 2001 and 2002. Data in this new format allows the use of GIS applications in a Web browser environment without being permanently connected to the Web and map servers by using CDs, local hard disks or local computer networks. The digitized catalog of archaeological research at Lesni skola was published on CD (Machácek 2002), making it available to the widest possible range of experts (Fig.1). The program allows the drawing of graphic

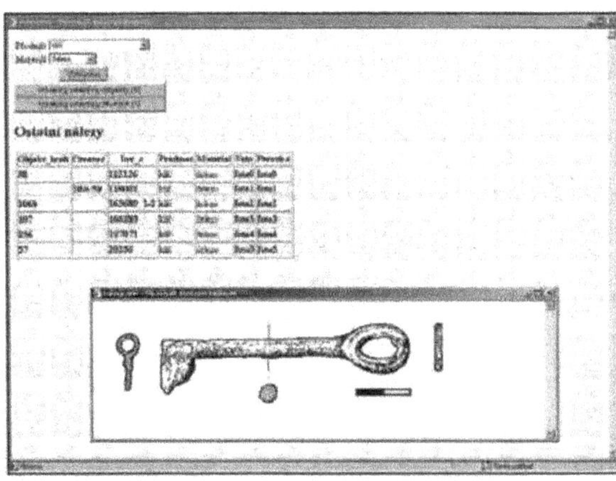

Figure 5 Breclav-Pohansko V. Digital Catalog of Archaeological Sources. Printout of database table resulting from a selection of all keys found at Lesni skola

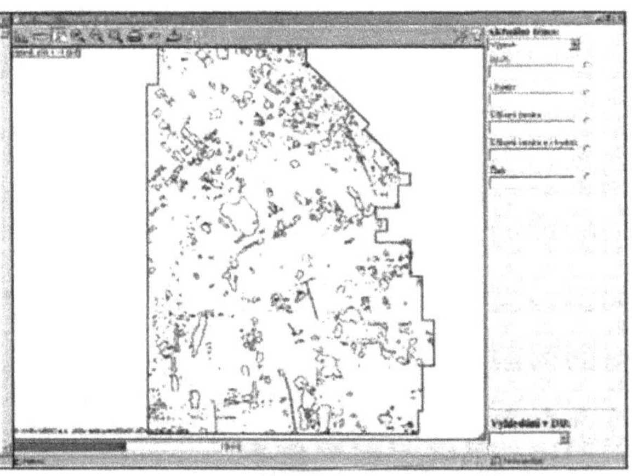

Figure 6 Breclav-Pohansko V. Digital Catalog of Archaeological Sources. Localization of selected features in the map window using arrows

objects in scales ranging from 1:7,000 to 1:25 (Fig.2 - 4). Information from a number of interconnected database tables can be called up interactively for important features by clikking on them. On some tables, hypertext links are marked in bold that allow the calling up of additional relevant information. These are primarily raster drawings (ink drawings, scanned and digitized drawings, etc.) and detailed object descriptions. This data appears in a new window. It is also possible in this application to search for information in the database tables of non-graphic attributes using a variety of criteria (Fig.5). Features that match the search criteria can be depicted on the map using an arrow (Fig.6).

The application of GIS technology at Pohansko earned Masaryk University the 2002 GeoMedia Best Practices Award for GIS implementation within education from Intergraph Mapping and Geospatial Solutions. This annual award is presented to the educational institution that has implemented a groundbreaking application of GeoMedia technology in the classroom and research.

References

FAGAN, B.M., 1996. The Oxford Companion to Archaeology. Oxford.

KUCERA, M. and MACHÁCEK, J., 2001. Räumliche Analysen der archäologischen Daten von Pohansko bei Breclav/Lundenburg mit Hilfe des GIS. In Archäologie und Computer 2000, Workshop 5. Wien. FORSCHUNGSGESELLSCHAFT WIENER STADTARCHÄOLOGIE:44-61.

KUCERA, M. and MACHÁCEK, J., 2003. Back to the Future. Geo-Connexion, volume 2, issue 1:48-50.

MACHÁCEK, J., 2002. Breclav - Pohansko V. Sídlištní aglomerace v Lesní školce. Digitální katalog archeologických pramenu. Brno.

VIGNATIOVÁ, J., 1996. Zum Stand der Erforschung des Burgwalls Breclav-Pohansko. In Stana, C. and Polácek, L. (eds.), Frühmittelalterliche Machtzentren in Mitteleuropa - mehrjährige Grabungen und ihre Auswertung, Brno:261-266.

Efficiency of Chipped-Stone Techniques: An Analytical Model Based on GIS Data

Tsoni Tsonev
Inter-Disciplinary Department, Institute of Archaeology and Museum, 1000 Sofia, 2 Saborna st. Bulgaria
tsts@bas.bg, tsts1113@yahoo.com

Emanuela Montagnari Kokelj
Department of Antiquity Sciences, University of Trieste, via Lazzaretto Vecchio 6, 34100 Trieste, Italy
montagna@univ.trieste.it

Chiara Piano
Department of Geological, Environmental and Marine Sciences, University of Trieste, via E. Weiss 2, 34100 Trieste, Italy
piano@univ.trieste.it

Abstract

The paper proposes an analytical model based on GIS-built terrain models for assessment of efficiency of chipped-stone techniques. The model includes four successive steps. The first one gives an approximate estimate of initial sizes of the nodules, out of which successive stages of core reduction occur. The second one includes GIS data that permit assessment of traits based mostly on digital terrain models. The third step assesses the type of relationship: linear or not between the amount of waste and the GIS defined features. The fourth one allows to build up various strategies based on the dependence between the platform position (optimal value), the amount of waste, the number of secondary platforms, the number of the long blanks detached, and the relative amount of core exploitation failures.
GIS analysis mainly starts from characteristics of the present-day landscape but the aim is to put greater emphasis on the palaeo-geographic context of the sites.
The combined analytical model with GIS data makes it flexible to assess independently different strategies of past human behaviour.

§1 The present approach creates new analytical methods and GIS applications for addressing the question of measuring the efficiency of chipped-stone technological cycles and aims at developing wider theoretical approaches in defining the factors for raw material procurement and technical choices for tool manufacturing of prehistoric communities. So far we came to the conclusion that the degree of technological complexity largely depends on natural and human factors. The first ones concern the geological situation of flint and other raw material outcrops within the terrain where prehistoric people moved and settled. Human factors also contribute to the degree of technological complexity and raw material procurement. So far we developed two models: upper Palaeolithic risk-avoidance strategies and the choices of Mesolithic/early Neolithic people made for raw material procurement and chipped -stone techniques relative to other constant resources. We confine, however, our presentation only to natural factors that contribute to the complexity of chipped-stone techniques. As a starting point we propose a reconstruction of theoretical ellipse that encircles symmetrically a Levallois core. The present approach can be equally well applied to other major core types. Brantingham and Kuhn (2001) suggested ellipse as a method for mathematical modeling of core reduction. We apply their idea of working within an ellipse for reconstruction of the maximal dimension of a nodule from which a given core (leftover) was derived through successive stages of core reduction.

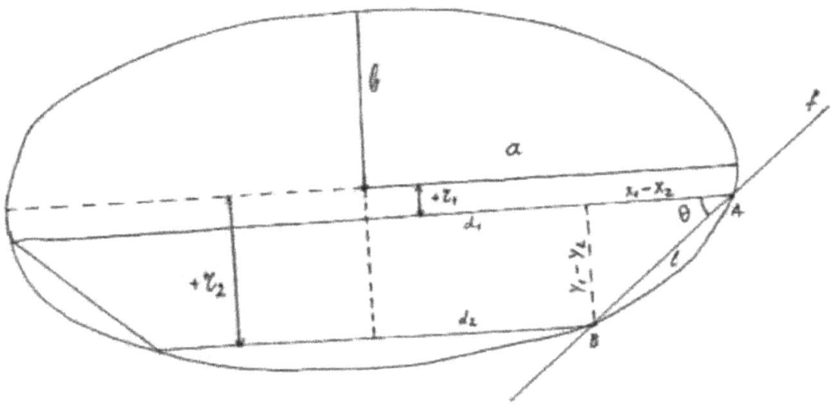

Figure 1 Calculation of the theoretical ellipse that encircles a Levallois core leftover. The same geometry can easily be applied for calculation of the theoretical ellipses of the other major chipped-stone core types.

The calculation has three stages. The first one defines the coordinates of points A and B in terms of x1 =d1/2, x2 = d2/2 and y1 and y2. These are also points of intersection of the straight line f defined as y = mx + n, where m = tgq. It establishes a relation between d1/2 and m, n, a, b within an ellipse equation.

The second group of equations is based on the fact that the straight lines g1 and g2 cross the ellipse at points A and B. The third stage defines a system of two equations for both axes of the theoretical ellipse a and b which are derived from the first two stages of calculation. It is proven that the system has only one solution (Fig.1).

By calculating the theoretical ellipse on different in size and in type cores at a given archaeological site we are able to reconstruct more precisely the original dimensions of the nodules selected by prehistoric people. These data can easily be compared with the size, shape, quality, etc. of the natural nodules (concretions) from the outcrops used by prehistoric communities. This allows us to reconstruct and describe the reduction stages relative to different technical cycles. We can expect a strong correlation between technical cycles and reduction stages that define in two independent ways the amount of waste at each archaeological site. While the assessment of waste of the first one can be controlled through the characteristics of the debitage, the assessment of waste of the second one can be defined directly by the difference between the number of reduction stages and cores (leftovers) and the natural sizes of the nodules (concretions) that occur at raw material outcrops. Each of these technical lines of reconstruction of waste can be defined by polynomial expressions whose coefficients would strongly depend on features such as availability of raw materials, distance, quality, etc.

Using only a minimal set of formal procedures that can easily be adapted to GIS applications, the present approach provides a number of analytical tools for assessment of optimal exploitation of raw material sources. It provides a set of automatic applications that can define the overall technical efficiency in minimizing preparation waste and maximizing tool manufacturing. However, technological complexity and optimal exploitation of resources cannot be confined only to mini/max solutions of core reduction geometry (Tsonev 1999, Tsonev in press). There are other human factors that also contribute to the overall complexity of chipped-stone techniques. I shall present a simple case study that can show the difficulty in defining the right approach for assessment of these technical cycles.

In Temnata Cave Gravettian sequence, Bulgaria, there is a marked difference in the contribution of meso-local flint varieties and the extra-local flint varieties to the efficiency of lithic techniques (Tsonev, in press). Starting from bottom level X the significant increase of the imported high-quality flint varieties (both meso-local and extra-local) is positively correlated with a significant increase of the slenderness of the blades. From this point of view, the increased supply with high-quality flint varieties contributes to the efficiency of chipped-stone techniques.

There is another view, if we consider the distribution only of the extra-local flints from the Gravettian levels. Their percentages fluctuate considerably: from 1.4% at level X to 18.7% at level Vb. In most cases these percentages do not contribute to the overall efficiency of chipped-stone techniques and have no economic value. They just mark the directions of "communication corridors" the Palaeolithic people used.

Thus within the frame of GIS database, it is possible to adapt different approaches for verifying various working hypotheses about the efficiency of chipped-stone techniques.

Tsoni Tsonev

§2 The study of the cultural choices reflected by the lithic techniques adopted by individuals and groups in prehistory should be based on the knowledge of the geo-environmental conditions in which they lived and operated. Geo-referenced maps of both chert outcrops and archaeological sites with lithic assemblages, related by means of GIS software to databases of the chert regionally available and the flint actually used at each site, can be valid tools, and can allow further cultural inferences through the application of specific statistical analyses.

This approach has been recently tested in the easternmost region of northern Italy - Friuli Venezia Giulia - where a flint dedicated database has been created starting from the case study of non-systematic surface collections of lithic artifacts belonging to different prehistoric periods (Montagnari Kokelj et al. 2001, Montagnari Kokelj et al. in press 1). Particularly in such cases - but the observation can have a more general value - a traditional chrono-typological approach must be combined with the study of technology and lithology in order to go beyond the pure determination of when the site was used, and try to make hypotheses on the role that it might have played within one, or more, subsistence-settlement systems (Binford 1983; Binford 1989).

What was evident from the very beginning of our study is the absolute necessity to have specific geological maps assessing the presence and location of regional chert outcrops. Due to the fact that in our case these did not exist, vector thematic maps in .dwg and .dxf format have been elaborated by using AutoCAD Map: the topographical basis is given by the Regional Numerical Technical Map at 1:25.000 and the graphical elements are related to categories of spatial data (points, lines and areas).

The result is shown in Figure 2 This image combines separate representations: the map of the primary formations, i.e. of those lithologies where chert is syngenetic to the enclosing sediment; that of the secondary deposits, including both chrono-stratigraphical units of coherent lithologies and loose deposits, where chert is present as re-deposited material; the map of the main tectonic structures (see Fig.2).

These maps are related to specific database tables of deposits containing their geological description, with parameters such as lithology, color, lithofacies, sedimentary structures, main fossils and age.

Each table is completed by a photo of the sample taken from the deposit under study during preliminary field surveys aimed at checking the geo-morphological and geological features of the area. Accurate macro-analyses of the deposits and thin sections of each sample have been carried out contextually.

The more geological part of our study would not be complete without database tables for the chert contained in all these deposits, where a general description is followed by a more detailed one of cortex, mineralogical, morphometric and morphological characteristics of the chert and of its distribution.

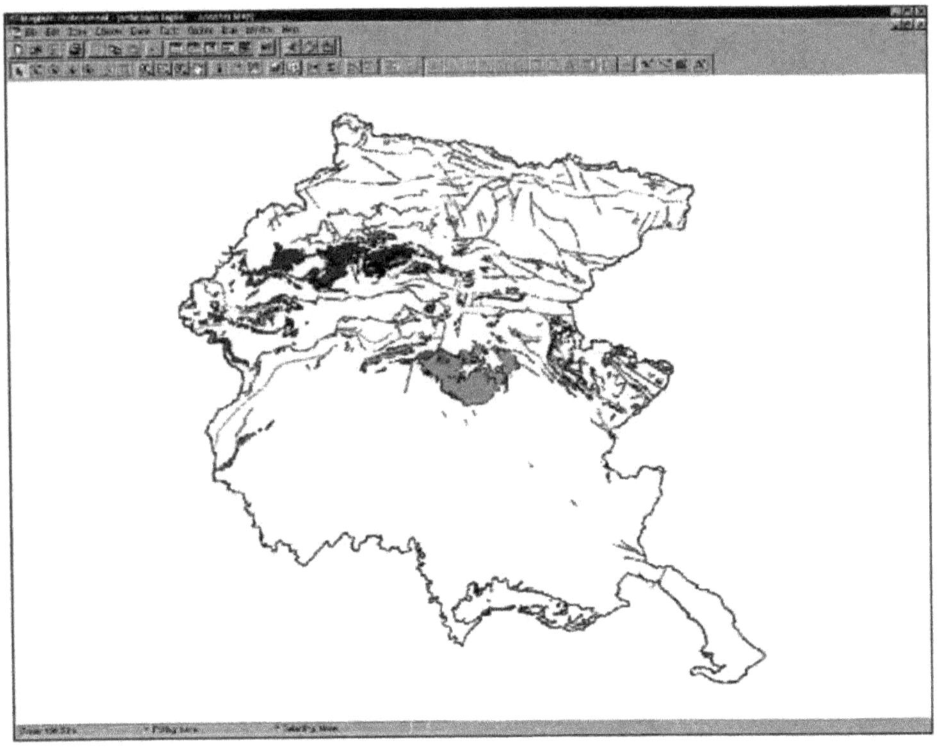

Figure 2 Friuli Venezia Giulia - northeastern Italy: overlay of primary formations, secondary deposits and main tectonic structures maps

- the code tables of the thematic maps: a pre-defined code allows in fact to link them to textual and image data stored into a relational database Access, importable into GIS MapInfo;
- 4 technical tables about deposits and chert description;
- the artefacts table, that contains some of the geological parameters already included in the other tables, plus others that belong to a more traditional archaeological approach.

The last table has been prepared for the first case study concerning lithic surface scatters, but it can clearly be used in any other case, and certainly more profitably for tools coming from stratigraphical contexts.

Moreover, when looking at the structure of our GIS, one can see that it has been created as a flexible system, linkable to other databases: in particular, we refer to one already implemented - ArcheoGIS, the archaeological map of the Isonzo Valley (Montagnari Kokelj 2001) - and to another - CRIGA, the GIS supported database of the archaeological caves of the Karst region (Montagnari Kokelj et al. in press 2) - in course of implementation by the same interdisciplinary team of the University of Trieste.

Certain parameters included in the tables have a more archaeological character, as they focus specifically on the knapping qualities of a particular rock, which depend not only on the factors indicated so far, but also on another important one, i.e. the tectonic situation of the area. The map of the main tectonic structures allows in fact to identify, already on the surface, those areas where the chert is of bad quality due to fracture.

At this point it is important to underline that the geo-lithological part of our research, briefly dealt with so far, actually stems from continuous feedbacks from the archaeological area, and at the same time is very tightly connected with it at informatic level.

The geo-lithological maps and tables are in fact only some of the components of the whole system, the management of which has been facilitated by the use of a GIS software. The structure of the GIS contains:

- 5 thematic maps, 4 of which - geological sample points, primary formations, secondary deposits, structural map and archaeological sites - have been elaborated at regional level and one - single surface collection area - at local level;

Within the CRIGA project, for instance, we have already planned to test the artefacts table on lithic assemblages from stratigraphical levels.

We believe that the integration of the different research lines indicated so far can be useful to analyse the relationship between chert, as natural rock potentially usable for making tools, and flint, as raw material actually used by man in specific sites in the past, in order to make hypotheses on the dynamics of procurement strategies and territorial exploitation. These strategies are likely to change through time, but the knowledge of the geo-lithological characteristics of a specific area is one of the elements indispensable to evaluate the cultural choices involved in long/short distance transport of raw materials, as the first test studies show already very clearly.

More information on the Italian interdisciplinary project can be found in the web: http://www.units.it/lithics.

Emanuela Montagnari Kokelj, Chiara Piano

REFERENCES §1

BRANTINGHAM, P.G. and KUHN, S.L., 2001. Constraints on Levallois Core Technology: A mathematical model. Journal of archaeological science, 28:747-761.

TSONEV, Ts., forthcoming. Adaptation strategies and raw material supply networks of the Evolved Gravettian in Central and South-Eastern Europe on the example of Temnata Cave (Bulgaria), Willendorf II (Austria) and La Cala Cave (Italy). Proceedings of the conference The Gravettian along the Danube, Mikulov, Institute of Archaeology, Czech Academy of Sciences, Brno, Czech Republic, November 2002.

TSONEV, Ts., 1999. Application of a mathematical optimization procedure on data for raw materials supply of the Gravettian culture in the Balkans. BAR S750. Archaeology in the Age of Internet - CAA97, 25th Anniversary Conference. University of Burmingham, Archaeopress, Oxford, 1997.

REFERENCES §2

BINFORD, L.R., 1983. Working at archaeology. Academic Press, New York.

BINFORD, L.R., 1989. Debating archaeology. Academic Press, New York.

MONTAGNARI KOKELJ, E. (ed.), 2001. Gorizia e la valle dell'Isonzo: dalla preistoria al medioevo. Gorizia: Comune di Gorizia.

MONTAGNARI KOKELJ, E., PATRIZI, C., CUCCHI, F., PIANO, C. and BERTOLA, S., 2001. Geo-archeologia nella valle dell'Isonzo (Italia nord-orientale) e oltre [Italian and English text]. Aquileia Nostra 72:29-58.

MONTAGNARI KOKELJ, E., CUCCHI, F., BERTOLA, S., PATRIZI, C. and PIANO, C., in press 1. Surface lithic scatters: interpreting a north-eastern Italian site. Proceedings of the 14th International congress of prehistoric and protohistoric sciences, 2-8 September 2001 Liège - Belgium.

MONTAGNARI KOKELJ, E., CUCCHI, F., MAZZOLI, T., MEREU, A. and ZINI, L., in press 2. GIS and caves: an example from the Trieste Karst (north-eastern Italy). Proceedings of the 14th International congress of prehistoric and protohistoric sciences, 2-8 September 2001 Liège, Belgium.

Iterative Approach to Ancient Paths Modelling in the Iron Age Study of the Dolenjska Region (Slovenia)

Tomaž Podobnikar
Spatial Information Centre
Scientific Research Centre of the Slovenian Academy of Sciences and Arts
SI-1000 Ljubljana, Slovenia
tomaz@zrc-sazu.si

Sneža Tecco Hvala
Institute of Archaeology
Scientific Research Centre of the Slovenian Academy of Sciences and Arts
SI-1000 Ljubljana, Slovenia
tecco@zrc-sazu.si

Janez Dular
Institute of Archaeology
Scientific Research Centre of the Slovenian Academy of Sciences and Arts
SI-1000 Ljubljana, Slovenia
janez.dular@zrc-sazu.si

See the CD for the extended version

Abstract

In this paper we analysed the potential routes between Iron Age settlements. In order to obtain optimal paths, classical GIS-domain algorithms based on the least-cost distance grid analyses, involving isotropic and anisotropic friction and applying different weights based on cognitive perception and common knowledge of the landscape were employed. Our model employed the digital elevation model (DEM) only as spatial data. Our case study area was the central-southern part of Slovenia, named the Dolenjska region, which is limited by natural boundaries of the river Krka on the south-west side and the river Sava on the north-east. Taking archaeology into account, this is the most explored territory of Slovenia. More than 400 late prehistoric sites have been registered with surveys, excavations and digging trenches. Within the framework of the project "Late prehistoric settlement pattern in Dolenjska", conducted by the Institute of archaeology, records have been gathered for the last 20 years. It can be seen that paths in the hilly terrain generally follow the ridges, however on the flat plains numerous potential routes can be noticed. Alternative paths have been statistically and visually evaluated and related to the known late prehistoric remains and evidences.

Introduction

Our research focussed on the potential relationships between Iron Age sites, based on the anticipation that prehistoric people used natural paths (Bellavia 2002) in order to move through the landscape, and the general idea that the natural topography has not changed significantly since the prehistoric period. Furthermore, it might be expected that the paths of movement run between the sites and monuments placed near to them, and that the most basic personal spatial experiences are shared with prehistoric populations in our common biological humanity (Tilley 1994).

The study area is the central-southern region of Slovenia, named Dolenjska, which is limited by natural boundaries of the river Krka on the south-west side and the river Sava on the north-east (Fig.1). 14 Iron Age settlements from a total of 38 documented in this area were selected and we studied 18 paths that run between them.

Geographic Features

The major part of the Dolenjska region is represented by a hilly terrain with altitudes ranging between 200 m and 600 m, except for the north-west part, where the altitude exceeds 600 m above sea level, and the floodplain, extended along the lower course of the river Krka, where the altitude does not reach 200 m. The typical difference between the valleys and ridges is less than 300 m. The surface of this region is predominantly covered by limestone formations and dolomite karst with deposits of lead, zinc and iron ore. The territory is rich with water with the exception of the western part, named Suha Krajina, which means dry country (region). The watercourses are short and passable and do not represent a serious natural barrier for movement, except for the rivers Krka and Sava, which are navigable. The region has a moderate continental climate. The precipitation regime has its peak in the summer and the lowest point in the winter. The threat of avalanches is barely likely. Floods seasonally occur along the river Sava and in the plain of Krka, which is partially marshy. Today forests cover two thirds of the surface, while the rest consists of vineyards and cultivated fields or meadows. It can be seen that the variations of the local environmental conditions and climate regime are not drastic, so we presume they have a minute significance for the analyses of paths.

Archaeological Records

Concerning archaeology, this is the most explored territory of Slovenia. More than 400 late prehistoric sites have been registered with surveys, excavations and digging trenches. Within the framework of the project Late prehistoric settlement pattern in Dolenjska, conducted by the Institute of Archaeology (Dular 1994) records have been collected for the last 20 years.

Archaeological records consist almost entirely of settlements, tumulus cemeteries, single barrows and deposits of artefacts. Until now no remains of prehistoric roads were discovered, except for a probable road, which is still under discussion.

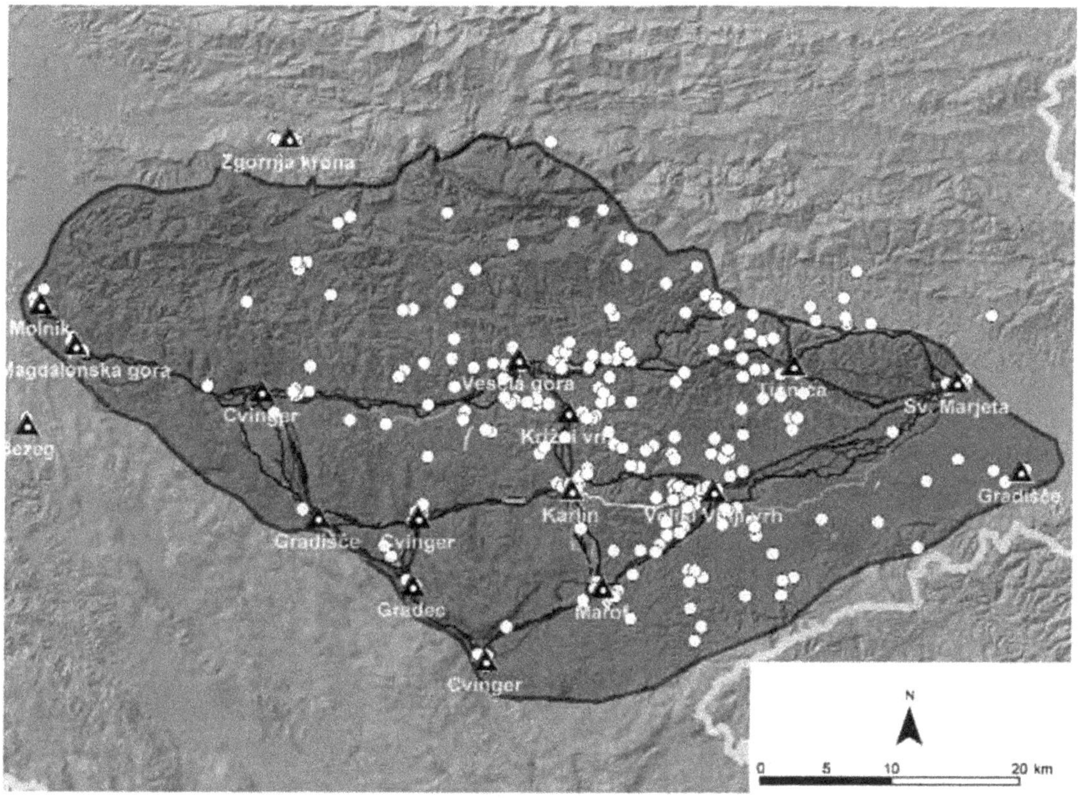

Figure 1 Case study area: Dolenjska region (100 x 90 km)

This probable road has been revealed last year during a preventive excavation for the purpose of motorway construction and so far it is not clearly confirmed as a prehistoric road.

The predominant type of the late prehistoric settlement is a hill fort, situated on top of the hill (Stancic et al. 1995). The majority of burial barrows are located on the ridges of hills (Dular 2003). This choice of localities on landmarks that must have been dominant within the local environment seems highly significant, and this fact is also considered in our study.

Assumptions of the Model

On the basis of the similarities of archaeological remains, we assume strong contacts and exchanges between the settled Iron Age centres. Our assumptions of the optimal paths modelling were the following (see also Harris 2000):

- on the predominantly hilly terrain in our case study geomorphology is assumed as the most important factor,
- climate regime and other environmental conditions such as marshy soil, size of the rivers, vegetation, geology could also be of importance; however, we suggest that these conditions have not been changed drastically to the present day,
- evidence of boats for river movement and wagons are not available, so we assume the settlers used to walk between the chosen settlements within the landscape,
- the impact of cultural and socio-economic factors on the decision making process.

For our case study we modelled ancient paths using only a digital elevation model (DEM) as a first approximation.

The Model of the Ancient Paths Simulation

Our task was to find an appropriate algorithm for predicting the optimal natural path between Iron Age settlements using the best walking strategy. We used classical GIS-domain algorithms and tools based on the least-cost distance grid analyses that are in most cases useful for determining optimal paths from one or more points of origin to one or more destination points (Lee and Stucky 1998). The optimal path may be based on numerous criteria. The basic friction surface could be the Euclidian distance that produces a straight path. More complex paths with different applied frictions are scenic, strategic, hidden, withdrawn, etc. (Lee and Stucky 1998, Bellavia 2002). Our friction surface was based on the slope of DEM and a path modelled under such conditions is considered to be the optimal economical path. We involved isotropic and anisotropic frictions. The anisotropic approach allowed different friction regarding both possible directions between two settlements.

As we assumed at the beginning, the most important factor that influenced the path modelling was geomorphology. For this purpose we used a high accuracy InSAR DEM 25 - digital elevation model with the resolution of 25 m, produced with radar interferometry and enhanced with other DEMs (Podobnikar et al. 2000). Its main advantages are found in geomorphologic homogeneity, high height accuracy of approximately 5 m and low appearance of gross errors. For path modelling it is important that DEMs resolution, height accuracy, expanse of gross errors, smoothness and other attributes that approximate a real surface, are appropriate as regards the strategy of movement (walking, driving…) and the possible intervention into the landscape (paths, roads and even bridges, tunnels…). For example, it is preferable to use a smoother DEM to model road lines instead of paths.

It is well known that even with a slightly different quality of DEMs we can obtain completely different results of path simulation. A possible way to test such conditions is to apply the Monte Carlo simulation of DEMs heights. However, we decided to change some other model parameters:

- friction surface regarding different slope classification,
- applying isotropic and anisotropic friction,
- biasing the modelled path direction by applying an additio-

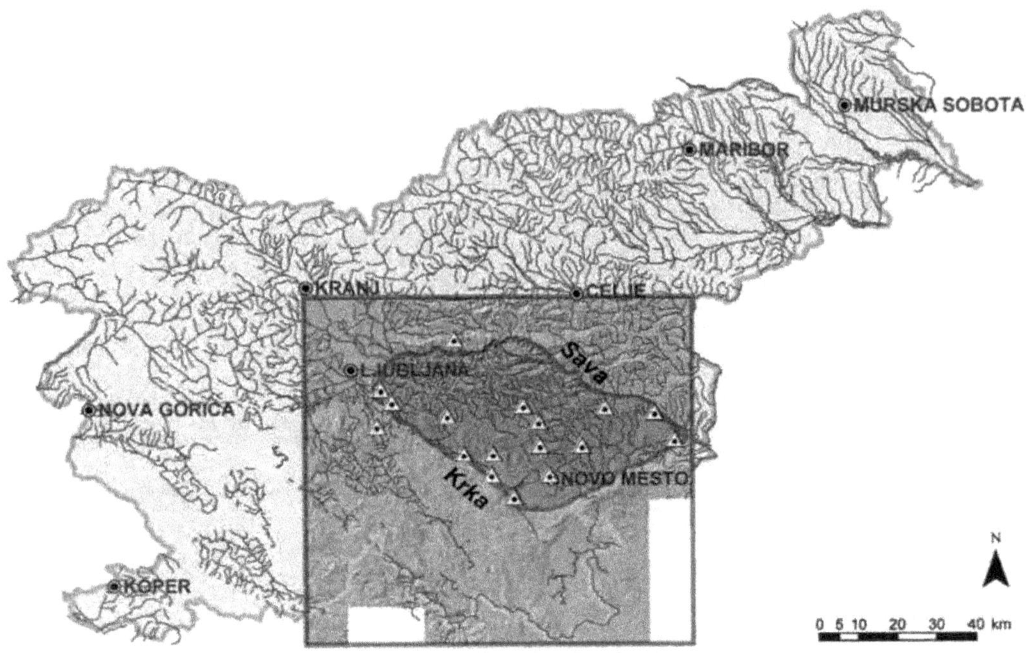

Figure 2 Path distribution using different algorithms and other parameters between 18 settlements and barrow distribution (circles)

nal transversal slope to DEM (linear and parabolic) towards the shortest line between two settlements.

With the listed empirical adjustments of the model parameters we produced different routs (Fig.2). With intermediate control by an experienced field surveyor and considering a greater number of assumptions for the behaviour of walkers, we gradually produced more realistic paths with less spatial variations (step by step the paths were more similar to the previous ones). To illustrate the described procedure: at the beginning we produced friction surface using legends of hiking maps, than we comprised more sophisticated knowledge regarding Llobera (2000:71), Bell and Lock (2000), Pandolf and Givoni (1977)... According to Minetti (1995) we assumed that a longer path means higher energy consumption, we can not obtain energy on the way down, human muscles are less effective than man made machines, metabolically the minimum is reached by 0.15-0.17 m/s uphill and 0.20-0.50 m/s downhill, metabolic and speed optimisation is at slopes ±25%.

We should also suggest (but we did not apply) that a slightly different position of the starting point (settlement, approximated with a point) in some geomorphology configurations could cause simulated paths, which would be considerably different.

Testing the Path Models

After producing nine different algorithms, all with iso- and anisotropic frictions, on 18 links between settlements, using various parameters to produce path models, we decided to extract the best ones. For this purpose we suggested to use or produce the following reference data:

- (1) empirical and deductive: a field surveyor with the knowledge of the archaeological sites drew paths using digital maps in the background (using hill-shaded DEM, the hydrological network, springs and lakes),
- (2) empirical and subjective (cognitive mapping): four non-archaeologists with not a lot of knowledge about ancient paths and topography drew paths using digital maps in the background (hill-shaded DEM, the hydrological network, springs and lakes),
- (3) mathematically: generating Euclidean distances between pairs of settlements,
- (4) using a database: selecting barrows that are near to the paths.

In all the cases we calculated the deviation between the described paths (or barrows) and the modelled paths. Thus we calculated the following parameters: minimal distance, maximal distance, mean distance and standard deviation of divergence for both groups of lines. As the most important we considered the minimal and mean distance. Empirically we also suggested a relevance list of the described four groups of data, ranges from (1) to (4) with the highest relevance of the view of the experienced field surveyor and with the lowest relevance of the significance of barrow positions. It was also not really reasonable that the cognitive mapping (2) in our case was not as relevant as we expected. The problem lies in the different previous knowledge of the landscape, the level of interpretation of topographic maps (hill-shaded DEM), imagination of ancient paths, different perception of the environment, etc.

However, after all the set criteria the best considered algorithms are those with numbers 6, 9 and 7. Within all of them we implemented frictions according to Minetti (1995) with the condition that slopes over ± 30 ° were more weighted. In algorithm 6 we employed an additional moderate transversal linear slope of 5% to DEM, in algorithm 9 - an additional moderate transversal parabolic slope, and in 7 - an additional high transversal linear slope of 12%. The best considered algorithm was that with number 6.

Conclusion

In our research we introduced an iterative and empirical methodology for modelling paths. Hypothetical ancient paths were simulated with variations of parameters using the least-cost algorithm and DEM (as environmental variable) only. With the use of different algorithms we eliminated the possible gross errors of DEM and iteratively produced the most probable paths.

On a hilly terrain the simulated paths generally follow the ridges (or valleys), however, in plains they are more unpredictable.

The geomorphology has a significant impact on the predicted paths on a rough and hilly landscape, while in plains it would be necessary to consider also other factors, such as for instance the local environmental conditions i.e. marshy terrain, larger rivers, vegetation, etc. It is important to note that the concept of natural paths does not explicitly imply their human use as a path, but only their potential to be used as such. The applied path model has to be further verified or estimated with the following additional investigations:

- observation simulated paths towards the entrances of settlements,
- highlighting visibility and inter-visibility from different - archeologically significant points along the routes,
- comparison with the traces of Roman roads and with the courses of modern roads,
- physical walk along the hypothetical roads generated by algorithms,
- evaluation proximity of archaeological sites to the simulated paths, and explanation of certain monumental cultural features from the perspective of movement along the optimal path,
- exploration of the ability to navigate the rivers, etc.

At this stage of the study the most relevant result for archaeological interpretation is that some locations of isolated barrows and artefacts became understandable within the context of movement.

References

BELL, T. and LOCK, G., 2000. Topographic and cultural influences on walking the Ridgeway in later prehistoric times. In Lock, G. (ed.), Beyond the map:85-100.

BELLAVIA, G., 2002. Extracting "Natural Paths" from a DTM Applications to Landscape Archaeological Studies. In Burenhult, G. (ed.), Archaeological Informatics: Pushing the Envelope, CAA 2001:5-12.

DULAR, J., 1994. Befestigte Prähistorische Siedlungen in Slowenien. In Ciugudean, H. and Borofka, N. (eds.), The early Hallstatt period (1200-700 B.C.) in South-Eastern Europe: proceedings of the International Symposium from Alba Iulia, 10-12 June 1993 (Bibliotheca Musei Apulensis, 1). Alba Iulia, Muzeul National al Unirii Alba Iulia:41-47.

DULAR, J., 2003. Halštatske nekropole Dolenjske = Die hallstattzeitlichen Nekropolen in Dolenjska. Opera Instituti archaeologici Sloveniae 6.

HARRIS, T., 2000. Session2 discussion: Moving GIS: exploring within prehistoric cultural landscapes using GIS. In Lock, G. (ed.), Beyond the map:116-123.

LEE, J. and STUCKY, D., 1998. On applying viewshed analysis for determining least-cost paths on Digital Elevation Models. International Journal of Geographical Information Science, vol. 12, No. 8:891-905.

LLOBERA, M., 2000. Understanding movement: a pilot model towards the sociology of movement. In Lock, G. (ed.), Beyond the map:65-84.

MINETTI, A.E., 1995. Optimum gradient of mountain paths. American Physiological Society:1698-1703.

PANDOLF, K.B., GIVONI, B. and GOLDMAN, R.F., 1977. Predicting energy expenditure landscape: incorporating sacred landscapes into cultural traditions. World Archaeology 30:159-176.

PODOBNIKAR, T., OŠTIR, K. and STANCIC, Z., 2000. From DEM to DTM with advanced modelling and integration of existing data. In 3rd AGILE Conference on Geographic Information Science, Helsinki/Espoo 25-27 May 2000. Helsinki/Espoo:103.

STANCIC, Z., DULAR, J., GAFFNEY, V. and TECCO HVALA, S., 1995. A GIS-based analysis of later prehistoric settlement patterns in Dolenjska, Slovenia. In Wilcock, J. and Lockyear, K. (eds.), Computer applications and quantitative methods in archaeology, 1993, BAR International series, 598:161-164.

TILLEY, C., 1994. A Phenomenology of Landscape. Places, Paths and Monuments. Berg.

The Interconnectivity of Cultural Sites: Sights and Sounds across a Landscape

Dora Constantinidis
University of Athens, Greece
Thyatiron 28, Nea Smirni, 17121
dorac@postmaster.co.uk

SEE THE CD FOR THE EXTENDED VERSION

Abstract

It may be assumed that sites rarely developed in a vacuum but maintained contact with surrounding areas. Therefore sites of a common cultural heritage were connected and communicated with each other across a given landscape. But exactly how were culturally related sites connected with each other and on what basis were they dispersed across a region for the purposes of communication? Site catchment analysis has been an important research area. However beyond resource and socio-political influences this paper investigates how to go about determining possible communication-network patterns that presumably helped maintain a dominant culture in a region. A GIS can be used to evaluate the possibility that distances between sites were established in order to facilitate easy communication with each other. Mycenaean sites in Central Greece were chosen to demonstrate how an evaluation of site interconnectivity might reflect inter-site communication patterns in the past. Apart from having couriers travel with news from site to site, sights and sounds may also have been used to relay messages across the land. This paper proposes how parameters that may have influenced communication patterns can be defined and subsequently investigated.

Introduction

Time takes its toll both on ancient landscapes and artefacts at an archaeological site. Retrieving meaningful and clearly understood data from the remains of past human activity is difficult enough. Even more elusive are remnants that could be associated with means of communication that people used across landscapes in the past. Communication has assumedly always been important to people and was perhaps another factor, apart from reasons such as trade, that may have influenced site interconnectivity. Some of the most illustrious examples of early means of long distance communication come from images of people blowing on shells or animal horns. The remains of possibly the earliest trumpets, one bronze or copper with gold overlay and another silver, were found in the tomb of King Tutankhamen (Reeves 1990). Though such instruments may have been used solely for the purposes of ritual at just one site (Gardiner 1966), it is possible that they could have been used to relay messages to other sites as well. References to additional means of communication are available from the Old Testament in the Bible: "...along the rivers of Cush, 2which sends envoys by sea in papyrus boats over the water. Go, swift messengers, ... to a people ... whose land is divided by rivers. 3All you people of the world, you who live on the earth, when a banner is raised on the mountains, you will see it, and when a trumpet sounds, you will hear it." (NIV:Isaiah 18:1-4). It is possible that the location of sites across ancient landscapes provided the communication medium by which couriers, sights or sounds could have been transmitted.

In the past sights and sounds would have moved across the land as messages were relayed. This vivid means of communication contrasts to the invisible and inaudible signals of communication satellites that now relay messages across our relatively "silent" landscapes. This paper presents some ideas for the development of a GIS that can incorporate variable buffer zones for ranges of both sights and sounds that could have been used to send messages across an ancient landscape. The Central Greek landscape of the Mycenaean era (Tsountas et al. 1969) is used to illustrate how a GIS can integrate and analyse distances in relation to parameters that impacted the three major means of communication: 1. couriers, 2. sights, and 3. sounds. When applied to data from "intensive" field surveys, such analyses may reveal sites that were established solely for the purposes of providing effective communication across a landscape. The results of any analyses would of course have to take into consideration other reasons as to why sites were located where they are. Reasons for site location would have to be proposed in conjunction with the possibility that some sites were maintained as "satellite sites" for the purposes of communication.

Settlement Foundation Theories

A fundamental question underlying any landscape analysis is "Why are archaeological sites located where they are?" Registering sites across a landscape allows for insights as to preferred site locations. A site's position in relation to mountains, rivers, coastlines, soil types and other sites may reflect decisions that were based on strategic influences (Hillier et al. 1984). Settlement foundation theories provide theoretical reasons as to why a group of people chose a location where a site was to be established. An interplay between environmental and cultural factors influenced site location to a certain extent. The necessity for close proximity to basic subsistence resources such as water and arable land are perhaps the major factors that determined site location in the past. Certainly socio-economic and religious factors may also have influenced the location of sites to some degree. However the latter reasons are difficult to ascertain and determine with any certainty, unless, for example, definite sites of worship are identified. During certain times in the past it may have been necessary for a site to be located in a relatively safe area which could be easily defended from enemies if needed. Another aspect of site location is the possibility that their position allowed for easy communication across a large area. Maintaining communication networks is a rarely considered reason for the establishment of culturally linked sites across a given area. However it is possible that relaying messages across a landscape may have determined how sites were interconnected, and their position may reflect how they may have relied on the available means of communication.

It is generally agreed that geography and climate were major factors in establishing a site during the Mycenaean era and that this is reflected by settlement patterns (Hope Simpson 1981). Given the fact that agricultural tools made of wood or wood sheathed in bronze were more basic than today (Hope Simpson 1965), then proximity to easily cultivatable and fertile soil was necessary to sustain a community's subsistence requirements. Most important of all was the necessity for a perennial water supply. In many cases a spring or well water may have provided the water requirements at smaller sites. However it is difficult to ascertain this since many of these sources are no longer evident. Besides these factors there was also the requirement for a site to be well drained and easily defended. This would therefore explain why many of the sites during the Mycenaean era are on hills or slopes, which are also relatively close to fertile plains. At such sites both subsistence resources would be provided for and the location could also be defended. To these reasons Bintliff (1977) has added the possibility of the existence of a hierarchy of Mycenaean settlement sites which was based on the size of the site and the power that it had presumably attained.

Depending on how large a settlement site was and the power that it held over a region, distances varied from site to site, thus allowing appropriate farming areas to sustain a community according to its size (Fossey 1988). Major dominant settlements or primary sites had approximately a one hour distance from other major settlements (Bintliff 1977:135-141). This one hour distance translates to approximately a 5km radius around each site, without taking into consideration the varying terrain which may vary the amount of time it takes to travel from one point to another. Major or primary settlement sites were large in size, generally over 40,000 m2, and consisted of at least one megaron, rural domestic buildings, and at least one 'royal' tholos tomb. Secondary settlements usually developed in areas surrounding primary sites at approximately a 2.5 km distance from them. These settlements were smaller in size and had chamber tombs instead of tholos tombs, the differences between these tombs are described by Mylonas (1966:111-130). In between these secondary sites there were smaller or tertiary sites which consisted of only one or two houses and few or no tombs at all. Linear B texts indicate that "palatial" centres had control over the collection and redistribution of goods which entails that communication with surrounding areas would have been enforced. Since most of the Linear B texts have been found within palatial contexts they provide limited knowledge about the smaller sites (Ventris et al. 1956). It is possible that this hierarchy of sites may have been an indirect outcome of the requirement for smaller sites to provide a link that would allow for messages to travel relatively quickly across the land.

Communication Methods and Parameters

Assumedly people communicated by travelling with news from site to site, or in times of emergency used visual cues with fires or flags if distances allowed for that. It is also possible that audible messages were relayed between sites by beating on drums or blowing on animal horns, trumpets or other instruments that would amplify sounds. Unfortunately since minimal tangible archaeological evidence exists for communication, research is limited to analysing distances between sites and the possible site hierarchy in an area that may help reveal possible connections that allowed for any of these means of communication to have been used. So far only these three communication methods have been identified as the most likely means for transmitting messages across ancient landscapes. The effectiveness of these methods depends to a large degree on the distances that the message was required to travel. All of the three means are inadvertently influenced by parameters that would either facilitate or hamper their effective transmission. Ultimately it is the landscape and the location of sites that may reflect the ease or difficulty of utilising any one of these means of communication. In order to explore any underlying communication patterns in the past it is necessary to coordinate and integrate a number of investigations that are currently facilitated by most GIS programs. Some improvements and new methods need to be introduced in order to investigate seasonal visibility and audibility ranges across the land respectively. The already existing facilities that provide Digital Elevation Models (DEM) and have been used in many case studies for cost surface analyses and viewshed or line of sight analyses (Lock et al. 1995) will be discussed in relation to each of the relevant means of communication. Factors or parameters that affected the viability of transmitting messages between sites also have to be considered.

The major factors or parameters that would have affected transmission of messages in the past are the natural environment or topography, the climatic conditions throughout the year, and the culture of an area. Parameters that relate to the surrounding environment are whether there are any natural barriers that would slow down travel or prevent visual or audible messages from being transmitted, and whether sites are close enough to necessary resources that would ensure their survival. Mountains can effectively exclude some sites from communicating with each other unless mountain passes can be identified as alternate routes that couriers could have taken. Rivers on the other hand could have been exploited as a means of travel and if this were not the case then likely crossings would have to be proposed. Additionally, the topography of an area should include soil maps so that the availability of arable land can be determined. Seasonal climatic variations during the year could inhibit travel during the winter months or decrease the visibility range between sites (Zamora 2003). For such observations to be included in results there is a requirement to incorporate data from experimental archaeology and fieldwork throughout the year that can reveal further details about the landscape under investigation. What is likely to be influenced by year round fieldwork are the establishment of possible travel routes that are reliant on the weather, "climatic" or seasonal viewsheds, and how far sound travels across the land with various instruments during calm or windy conditions. Finally the extent of sites of a common culture would have to be identified because assumedly it would have been difficult to traverse across "enemy" territories. A dominant culture may have flourished because an effective communication network was available to culturally linked sites dispersed across a landscape. Smaller sites should be considered for the possibility that they were located where they were in order to maintain communication in the region. In theory small sites that are not close to natural resources may have existed as satellite sites, much like "lighthouses", for the purposes of relaying messages across a landscape.

Communicating across the Mycenaean Landscape in Central Greece

The digitised Mycenaean sites in the areas of Boeotia, Phocis, Malis, Northern Euboia and Eastern Locris are from the survey in Central Greece that was undertaken by Hope Simpson (1981). Global Positioning Systems (GPS) can now be used to incorporate the position of new sites, and additional environmental, geological, geomorphologic and archaeological data into a GIS database for more detailed analyses. For the purposes of this paper the limited data that were available were used to examine methods for investigating possible communication between sites. Hope Simpson (1981) classified the sites by types. The site categories are: 1. Major settlement, 2. Settlement, 3. Settlement with tombs, 4. Major fortified settlement, 5. Fortified settlement, and 6. Tombs, with a further subdivision by size in area (m2) as defined by this paper for the 2. Settlement category: 2a. Small - up to 10,000 m2, 2b. Medium from 10,000 to 20,000 m2, 2c. Large 20,000 to 40,000 m2, and 2d. Undetermined size. The various site types are depicted with different symbols in Figures 1.- 3. that follow. Patterns of site interconnectivity may be revealed showing how the different site types were able to communicate with each other. All of the distances between the various sites have to be examined for the feasibility of using any of the three identified means of communication.

For couriers travelling with news, relay stations may have been provided where replacement runners accepted the message and took it onto the following site. Possible communication routes (CR) may be plotted and then analysed to determine distances and therefore the likely time it would take for a message to get from one site type to another. The region of Central Greece seen in the digitised map (Fig.1) consists of extensive mountainous regions, in particular those of Parnassos and Elikona. These mountain ranges effectively separate the coastal regions from the valleys and fields of the interior. It appears that Central Greek Mycenaean sites of the inland regions are located within close proximity to a central communication route (Hooker 1977). There was possibly a major communication route that ran from Thebes, along the edges of the former lake Copais and then towards the Kifissos valley and finally towards the Sperchios river delta. Lines of communication, especially through mountainous regions, need to be analysed in association with DEMs that allow for gradients or slopes of the countryside to be taken into consideration for the determination of Cost Surfaces or the amount of energy it would require to travel over the land. It can be expected that routes requiring the least energy would have been selected. Van Leusen (2002) discusses in detail the implications of using cost surface analyses. The distances of possible communication routes (CR), plotted in Figure 1., that couriers may have used across the land can be analysed statistically.

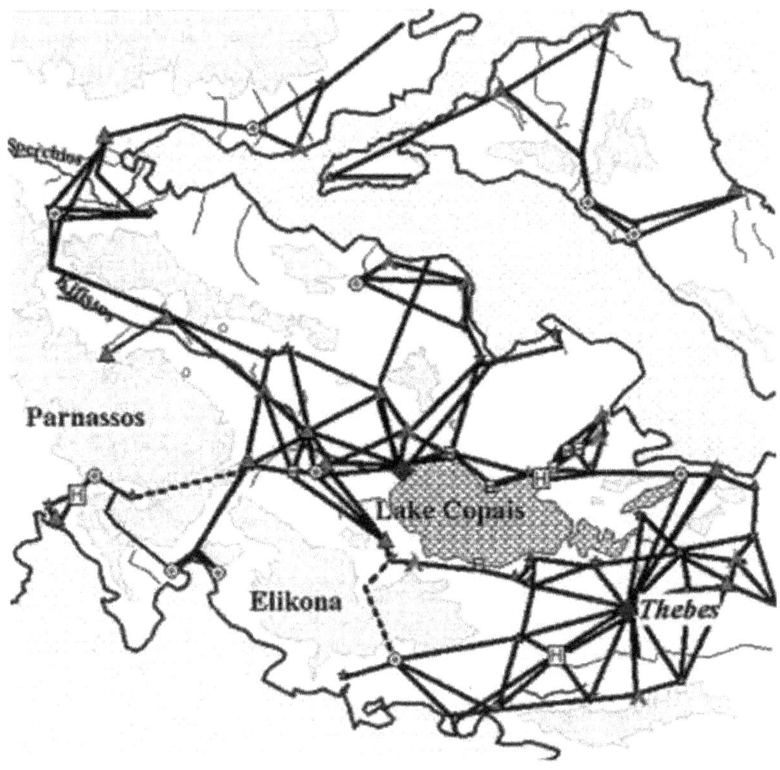

Figure 1 Possible Communication Routes across the Mycenaean landscape

A generalised communication network (CN) where the topography is ignored and all the possible permutations and combinations of different site interconnectivity are plotted can be analysed to determine "average" distances between various sites. A more feasible method would be to use Cluster analysis. The squared Euclidean distance, where distance (x,y) = Si (xi - yi)2 , can be used to determine how objects, in this case sites, that are further apart are distributed. These results could be examined in conjunction with the Chebychev distance where distance (x,y) = Maximum½xi - yi½ would take into account the different site types. This may determine if patterns of distances existed between the different sites. Combined statistical and cluster analyses of both CR and CN respectively may reveal further patterns. (Fig.2)

As mentioned above, sending messages overland would require the results of a DEM. In the case of visual messages Line of Sight analyses or viewsheds also rely on DEMs (Van Leusen 2002). Factors influencing the accuracy of viewsheds

Figure 2a-b Examples of a generalised Mycenaean Communication Network:
a) Major sites to settlements, and b) Forts to settlements

are the amount of available light, the visual cue that was used- fires or smoke are more visible than flags- and the season or prevailing climatic conditions. As Zamora (2003) proposes natural variations of visibility would have to be incorporated into GIS programs. The introduction of "soundsheds" to determine the extent of audible messages across the landscape entails further development to GIS programs. It is proposed that viewsheds (Vsheds) and soundsheds (Asheds) based on the Visibility and Audibility across a landscape be represented as variable buffer zones that determine the viewing and hearing potential at each site under varying conditions. If small relatively isolated sites appear in the landscape within these zones it is possible that they existed so that visual and audible messages could be transmitted.

Future Developments: Incorporating Sights and Sounds into a GIS Database

Sound and improved sight algorithms will have to be developed and incorporated into the spatial analytical capabilities of a GIS. Sound algorithms may reflect differing analytical requirements. For instance urban planners could use GIS soundsheds to analyse the impact of varying decibel levels on neighbourhoods located close to sources of noise pollution. Whereas archaeologists could use soundsheds to determine the range of audible communication across the landscape. This would entail experimental archaeology and the results of research dealing with the transmission of sound. The integration of GIS with multimedia technology (Scholten et al. 1997) may allow for interactive analyses of the varying intensity of "sound" zones across the landscape. This will require incorporating the necessary sound algorithms that take into account the varying terrain and prevailing wind conditions that would either reflect or absorb sound by varying degrees. Subsequently variable buffer zones would allow for the depiction of variations in how sound is propagated. The same principle would apply to viewsheds with any variations, such as climate, taken into consideration. Incorporating both vsheds and asheds as buffer zones allows for the ability to analyse the visual and audible contact between sites. Figure 3 proposes that both sheds are superimposed along with any possible variations (e.g. seasonal viewsheds) that may affect the range of message transmission.

Average buffer zones could be determined for year round communication patterns. Otherwise variable zones should be defined and plotted according to the results of fieldwork or theoretical interpolations of existing data. These developments would enhance future research into the various parameters that affected the viability of communication and hence the interconnectivity of cultural sites across ancient landscapes.

Acknowledgements

Thank you to the CAA conference organisers for financial support to attend the CAA2003 meeting in Vienna, and Professor T. Sellis for providing computing facilities at the National Technical University of Athens.

Figure 3. Proposed variable ViewSheds and SoundSheds, superimposed

REFERENCES

BINTLIFF, J.L., 1977. Natural Environment and Human Settlement in Prehistoric Greece based on original fieldwork. BAR Sup. 28, London.

FOSSEY, J.M., 1988. Topography and Population of Ancient Boiotia. Ares Publishers, Chicago.

GARDINER, A., 1966. Eygpt of the Pharaohs. Oxford University Press, Oxford.

HILLIER, B. and HANSON, J., 1984. The Social Logic of Space. Cambridge University Press, Cambridge.

HOOKER, J.T., 1977. States and Cities of Ancient Greece: Mycenaean Greece. Routledge and Kegan, London.

HOPE SIMPSON, R., 1965. A Gazetteer and Atlas of Mycenaean Sites. University of London.

HOPE SIMPSON, R., 1981. Mycenaean Greece. Noyes Press, New Jersey.

LOCK, G. and STANCIC, Z. (eds.), 1995. Archaeology and GIS: A European Perspective. Taylor and Francis, London.

MYLONAS, G., 1966. Mycenae and the Mycenaean Age. Princeton, New Jersey.

NIV New International Version of the Bible. Excerpts from Isaiah 18:1-4.

REEVES, N., 1990. The Complete Tutankhamun - The King. The Tomb. The Royal Treasure. Thames and Hudson Ltd., London.

SCHOLTEN, H.J. and LOCASCIO, A., 1997. GIS Application Research: History, Trends and Developments. Proceedings of Geographic Information Research at the Millennium GIS DATA Final Conference Le Bischenberg, France, 13-17 September 1997.

TSOUNTAS, C. and MANATT, J.I., 1969. The Mycenaean Age: A study of the monuments and culture of Prehistoric Greece. Argonaut, Chicago.

VAN LEUSEN, M., 2002. Pattern to Process: Methodological Investigations into the Formation and Interpretation of Spatial Patterns in Archaeological Landscapes. PhD. Thesis, University of Groningen.

VENTRIS, M. and CHADWICK, J., 1956. Documents in Myceneaean Greece. Cambridge University Press, Cambridge.

ZAMORA, M., 2003. Viewshed Variations throughout the year at the Alhonoz Site (Southern Spain, Iron Age) related to solar route and its implications on the local communication network. Proceedings CD of the CAA conference in Vienna, 8-12 April 2003.

GIS Spatial Analysis in the Etruscan Trading Centre of Bagnolo S.Vito (Mantova, Italy)

Cristina Longhi
Civico Museo Arcgeologico di Bergamo

Claudia Mangani
Museo "G. Rambotti" Desenzano del Garda

See the CD for the extended version

Abstract

Forcello (Bagnolo S. Vito, Mantova) settlement, one of the most important Etruscan trading centres, has been investigated on a 600 sq.m. surface.
In area 'R18' the presence of 7 main phases has been recorded. These are marked with letters (A-G) and dated from the beginning of the 4th century BC until the second half of the 6th Century BC.
The porpoise of the present paper is 1. computerisation of paper records and data pertinent to phases F and G finds; 2. their management through a GIS software to test the advantages of a data processing analysis compared with the traditional method. During F phase a fire destroyed 2 houses separated by a canal; the GIS analysis of placing of elements, allowed us to confirm and improve the former hypothesis regarding use of the identified room collapse and placing of archaeological finds dynamics and their subsequent events.
Moreover, computerisation allows us to quickly handle a 20 digging-years data collection and to set up a starting point for the forthcoming researches adopting the new methodologies.

The Site

The Etruscan site of Forcello is located in Northern Italy, about 6 km south of Mantua, in Bagnolo S. Vito municipal district. The site, which was already known from some discoveries made there in the XIXth century, was recognised as being an important Etruscan site only at the beginning of the 1980's, when, owing to the intensive farming, distinctive pottery and inscription were found.

Probably the settlement was set up in the VI century B.C. (around 550 B.C.), during the Etruscan colonization of the Po plain, and dropped at the beginning of the IVth century B.C.

The topographical and morphological survey of the site and the surrounding areas has been carried out by means of an electronic apparatus Electric Total Station Topcon GTS 212 and the topographic software Meridiana Geotop. The result is a map, which records the existing modern constructions (such as ditches, farmsteads etc.) as well, and points out that the settlement lies on a slight morphologic culmination, rising above the paleovalley of the river Mincio, close to the slope of the paleoterrace on the hydrographical right hand side, was triangular in shape, about 13 hectares in size, oriented NW-SE. On the north-western side was enclosed by an earthen rampart, at least 320 meters long. On the basis of the present morphology of the terrain, the rampart was probably bordering the other two sides of the settlement as well.

Presumably the settlement had an orthogonal layout, like other Etruscan colonial establishment, such as Marzabotto. One of the main street, 15 meters wide, perpendicular to the rampart, passed through the whole of the settlement; in the middle of it two channels ran, which would act as main sawers for the drainage small canals that bounded the houses.

These were rectangular in plan and shared a common orientation (NE-SW) - with the short sides perpendicular to the main street - and had wooden posts for walling (of wattle and daub construction) and beaten earthen floors; only during the latest building phase roof tiles replaced thatches.

In the southern part there was an area which probably had a sacred function, since the majority of the inscriptions on pottery were found here, as well as the broken top of a small pillar altar made of trachyte.

The excavation area, of about 600 m², is located in the middle of the settlement, in this point the archaeological deposit has appeared to be 1.8 m thick.

- phase A. Remains of negative evidences pertinent to a rectangular room, dating back to the end of the Vth -beginning of the IVth century B.C.;
- phase B. One house rectangular in shape. Only the hearth and some outer negative evidences are preserved. This phase is dated to the second half of the Vth century B.C.;
- phase C. One house rectangular in shape (12,3 x 5,2 m) with a perimeter of post-pits. It was destroyed by a fire during the second quarter of the Vth century B.C.;
- phase D. Badly preserved as the house of phase C was built directly on it. We recognized only some hearths and negative evidences like pits and small canals;
- phase E. The investigated area was devoted to metallurgy. Fire-ditches and hearths mark out this phase, dating back to the beginning of the Vth century B.C.;
- phase F: two houses bounded by a canal were destroyed by fire. The plenitude of pottery sherds from Greece indicate that they belong to the end of the VIth century B.C.;
- phase G. One house with perimeter and rooms' division corresponding to the NW house of phase F. A black figure lekythos of the Cock Group found, let us date this phase around 525-520 B.C.

Nowhere the virgin soil has been reached.

The Spatial Analysis

The paper lists (for contexts, recorded finds, plans, sections, etc.) have been inserted in Excel tables. In our opinion, this format allows a quick management of this kind of data and, moreover, it is perfectly consistent with Access and ArcView (the GIS software in use). Besides, using the software Arched, we produced a new matrix, that gave aid to rectify some inner discrepancies, such as impossible circles or relationships, included both in lists and context record sheets. Last, but not least, a relational data base, regarding contexts and their finds, was processed last year.

All the paper maps have been converted into digital images, brought to the same scale and georeferred. This process forced us to convert the alphanumeric coordinates, which indicate each square mater on the dig, into the corresponding Cartesian coordinates. Plans have been added to an ArcView project, to gain a mosaic. This mosaic allows a global picture of the structures unearthed in sectors R 18 and R19 of the dig.

The first outcome was the implementation of thematic map about finds layout. These maps confirmed or even increased our knowledge about the role of the structures recognized during the dig.

Phase E: The first case here analyzed concerns the layers of the beginning of the Vth century. As already mentioned, the area was devoted to metallurgy. We identified hearths, burnt areas, pits (whose use remains unknown) and small oval ditches, marked by burnt inner side and coal fill. These ditches have been explained as small furnaces to forge metal objects. They were ranged in groups and had often a small hearth or a burnt area nearby.

To corroborate this interpretation, as small furnaces to forge, we decided to work out a map. In this map, we represented the layout of slags, bronze and iron bars etc. on the basis of their weight: the metal wider concentration is by the small furnaces and hearths (Fig.1). We have got the same result when considering the selected finds layout (in this case, we have been able to georefer each single finds, because we had all the coordinates).

Phase F: At the end of the VIth century, a big fire destroyed at least two houses, whose perimeter has not been yet completely brought to light. One of them (house F I) has been more intensely investigated. It has an extension of about 78-79 m2 and it is divided into eight rooms. Thanks to the sudden fire, most of the objects in the house remained on the paleosurface, even the carpological and wooden finds; usually this kind of finding are not preserved in the site.

Having no doubts about practical interpretation and stratigraphic position of these structures, the purpose of the analysis was to test efficacy and benefits of the data processing systems.

To better understand the rooms' functions and the collapse's trends, we analyzed the layout of different type of finds, on the base of the number of sherds presents in each square meter.

The plan of burnt areas and the layout of the doub and wattle gave us more information about the house's structure. The big doub and wattle sherds by the south side of the hearth in house F I show that the wall behind it had been insulating with a thick layer of clay; on the other hand, the few sherds in the rest of the house seem to point out that no coating was on the others walls.

Around the hearth, in house F II, we haven't found big quantity of burnt clay, probably it was in the middle of the room. The presence of red colored burnt areas shows that in house F I-room 4 the heat was very intense; it is possible that the fire propagated right from this area. Besides that, a stove, totally warped by the fire, has been found.

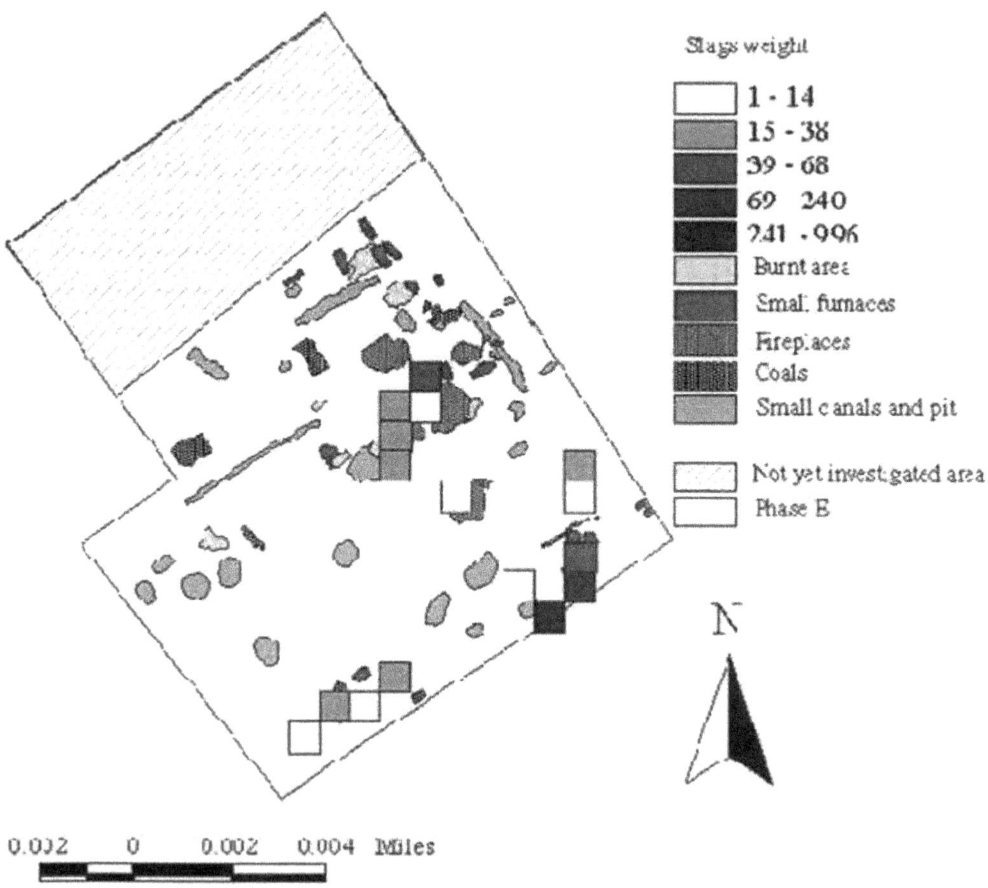

Figure 1 Layout of slags, laminas and iron bars based on weight (grams)

Thanks to the layout maps, we received confirmation that room 2, where carbonized seeds and pulses heaps but no pottery sherds have been found, was used to storage dry supplies in vegetal cases (such as baskets). This room was used for textile works as well, as here we found carbonized remains of the loom's bars and a great concentration of loom's weights.

The data processing analysis makes it clear that there was a second area devoted to textile works, by the wall facing the hearth; probably in this area there was another loom.

House F I-room 4 was the kitchen, as the presence of mortar and stove sherds points out. They are totally warped by the. The greek amphorae sherds layout gave us excellent results to understand their original location in the two houses. The two amphorae from Taso, almost completely rebuilt, were one by the NE corner in house F I-room 3 and the other by the NE corner in house F II-room 2. Two or three amphorae, from Corinth and Milieth, were by the SW corner of house F I-room 7: in fact most of the fragments lied on the slope and the bottom end of the canal.

Figure 2 Attic pottery sherds' layout

The amphorae, almost certainly containing wine, were kept in rooms separated from the kitchen, may be in a room devoted to the banquet.This supposition is backed up by the attic pottery sherds layout (Fig.2). The attic pottery, in house F II, gathers especially, both in the big room with the heath and in the contiguous small one, in house F I by room 3 SW wall.

Information about social customs of the residents can be deduced from fibulas layout. The male ones (Schlangenfibeln and Bandbogenfibeln) and the female ones (Sangusugafibeln, Navicellafibeln, Certosa fibeln) have been found in different rooms. The female fibulas gather in house F I-room 4, that is the kitchen: The thematic maps were useful also to point out the dispersions' trends:

In some cases pottery "exploded" because of the great heath or felt from wooden shelves, while they were burning. In this circumstance the sherds remained in situ, as the bowl in room 3: all the sherds gather by the wall.

The sherds' displacement can be dued often to the leveling and renewal of the collapse surface after the fire, as well as to the need of fill the canal difference of level. As a vase at source probably in house F I kitchen, whose sherds have been found, for the most part, in the canal; also the scattering of the fragments of big vase for provisions can be linked up to the renewal activities.

In others cases the pottery located by the wall by the canal, when the wall felt down, "slipped" in the canal. Most of the sherds of the eye cup have been found in the canal, but a few of them were still laying on the canal's edge.

Conclusions

The processing of the previous documentation helped us in corroborating and reviewing in details the information we've got with traditional research methods.

For example, about phase E structures, we already knew the use of the oval ditches, but we did not notice that the metals findings gathered by these structures. For phase F, we have been able to focus and distinguish in details the dispersions trends after the collapse thanks to the analysis of the different layouts.

Anyway, in our opinion, the most important benefits must be singled out within the documentation's management and query. We have more than 1,500 context and more than 700 plans and sections, as well as thousands of recorded finds.

The high expenses, to talk of the time spent digital recording the paper documentation, have been fully rewarded by the quickness in data searches.

REFERENCES

AA.VV., 1989. Gli Etruschi a Nord del Po. Atti del Convegno, Mantova 4-5 October 1986.

CATTANEO, A.C. and DE MARINIS, R.C., 1996. Le anfore greche da trasporto di Adria, S. Basilio e del Forcello di Bagnolo S. Vito e il commercio del vino nell'Italia Padana. In 2500 anni di cultura della vite nell'ambito alpino e cisalpino, Trento:317-348.

CASINI, S. and ODONE, S., 1996. C'era una volta un abitato etrusco, un percorso didattico tra documentazione archeologica e ricostruzione storica. Quaderni Didattici 6, Bergamo.

DE MARINIS, R.C., 1982. Bagnolo S. Vito (MN). In NSAL 1981, Milano:39-45.

DE MARINIS, R.C., 1982. Bagnolo S. Vito (MN), Scavi e scoperte. In Studi Etruschi L:495-502.

DE MARINIS, R.C., 1983. Manta. In Rivista di epigrafia etrusca, Studi Etruschi vol. LI:196-213.

DE MARINIS, R.C., 1983. Bagnolo S. Vito (MN), Località Forcello, Scavo nell'abitato di V sec. a.C. In Notiziario della Soprintendenza archeologica della Lombardia 1982:31-33.

DE MARINIS, R.C., 1984. Bagnolo S. Vito (MN), Abitato del V sec. a.C. In NSAL 1983:30-34.

DE MARINIS, R.C., 1986. Bagnolo S. Vito (MN), Località Forcello. Scavi nell'abitato etrusco. In Notiziario della Soprintendenza archeologica della Lombardia 1985:43-47.

DE MARINIS, R.C. (ed.), 1987. Gli Etruschi a Nord del Po. Catalogo della Mostra, Palazzo Ducale, Mantova, September 1986 - April 1987, vol. I-II, Mantova, 1986-1987 (2nd ed. Udine 1988, Campanotto ed.).

DE MARINIS, R.C., 1987. Fibule tardo-hallstattiane occidentali dell'abitato del Forcello (Bagnolo S. Vito). In Celti ed Etruschi nell'Italia centro-settentrionale dal V sec. alla romanizzazione, Atti del Congresso internazionale, Bologna, 12-14 April 1985:89-99.

DE MARINIS, R.C., 1988. Nouvelles données sur le commerce entre le monde mediterraneen et l'Italie septentrionale du VII au V siécle av. J.C. In Les Princes Celtes et le Mediterranée, Colloque de l'École du Louvre, Paris:45-56.

DE MARINIS, R.C., 1990. Bagnolo S. Vito, loc. Forcello. In NSAL 1988-1989:38-40.

DE MARINIS, R.C., 1991. La stratigrafia dell'abitato del Forcello di Bagnolo S. Vito e i rapporti cronologici con le culture dell'area circumalpina. In Miscellanea di studi etruschi-italici in onore di Massimo Pallottino, Archeologia Classica 43:237-259.

DE MARINIS R.C., 1991. L'abitato etrusco del Forcello: opere di difesa e di drenaggio e importanza delle vie di comunicazione fluviale. In Gli Etruschi maestri di idraulica, Perugia:75-85.

DE MARINIS, R.C., 1996. L'abitato etrusco del Forcello. In Italia padana e centro-alpina, Guide Archeologiche 5, XIII Congresso Int. Scienze Preistoriche e Protostoriche, Forlì:74-81.

DE MARINIS, R.C., 1999. Rapporti culturali tra Reti, Etruria Padana e Celti Golasecchiani. In I Reti/Die Räter, ArcheoAlp 5:603-649.

DE MARINIS, R., ATTENE FRANCHINI, S., BOSCATO, P., CASINI, S., FRONTINI, P. and TUZZATO, S., 1988. Bagnolo S. Vito (MN), loc. Forcello. Scavi nell'abitato etrusco. In NSAL 1987:34-36.

DE MARINIS, R.C., CASINI, S., CATTANEO, A.C., DE GASPERI, N. and FRONTINI, P., 1994. Forcello (Comune di Bagnolo S. Vito, Mantova). In Studi Etruschi LX:534-549.

DE MARINIS, R.C., LONGHI, C. and MANGANI, C., 1999. Bagnolo S. Vito - Loc. Forcello (MN). In Notes (http://www.unife.it/notes).

DE MARINIS, R.C., BARATTI, G., LONGHI, C. and MANGANI, C., 2001. L'insediamento etrusco del Forcello di Bagnolo S. Vito (Mantova). In Analisi informatizzata e trattamento dei dati delle strutture dell'abitato di età preistorica e protostorica in Italia, Atti del convegno IIPP, Ferrara, 26-27 November 2001.

GIS-Analysis in the Reconstruction of an Early Medieval Landscape. The Upper Lusatian Case-study

George Indruszewski
Geisteswissenschaftliches Zentrum Geschichte und Kultur Ostmitteleuropas e.V. Leipzig and Nationalmuseets Marinarkæologiske Forskningscenter, Roskilde

See the CD for the extended version

Abstract

The GIS-software has been used as a main tool in reconstructing the early medieval landscape in the Upper Lausitz (Saxony, Germany) territory. Quality cartographic data (Landesvermessungsamt Sachsen) was used to produce GIS-based (ArcView) 2D maps of natural and man-made features tied afterwards to a relational database of non-spatial attributes containing archaeological, historical, and linguistic data. The preliminary work included also the elaboration of a DEM (ArcView, ArcInfo, ArcGIS) used to further the reconstruction and the analysis of landscape themes. GIS was used as a working platform onto which spatial statistical analysis and more common GIS-based spatial analysis (viewshed, line-of-sight, distance, proximity, etc.) were integrated with satellite imagery and other photo documentation. The results, so far are encouraging us to plead for the scientific benefits and potentials of GIS-based analysis and reconstruction of early medieval landscapes.

Introduction

Bautzen/Budyšin is considered the de facto cultural capital of the Sorbian minority in the larger region of the Sorbian/German Upper Lusatia, this perception being grounded mostly on decades of intermittent historic and archaeological research that underlines the central importance of this place for the region. The reconstruction of an early medieval landscape for this region focuses therefore, on the reconstitution of the multiple relationships between Bautzen/Budyšin as central place and the peripheral settlements during the early and the High Middle Ages. It takes in consideration the settlements around Bautzen/Budyšin and those on the Upper Spree Valley, those from the entire Upper Lusatia region, and those from the larger area including Northern Bohemia and Lower Silesia.

Settlement Pattern: The Budissin Land

The temporal variation of settlement pattern constitutes an important element in the process of landscape reconstruction, inasmuch as it emphasizes matching of information from various sources: archaeological, historical, and linguistic. The digitized place-names, representing either the site of the local church or the core of the village or hamlet, constituted therefore not only the geo-referenced place-name but also the presumed core-location of the earliest settlement or a core-settlement area for street (Strassen-, Gassendorf) or forest hamlet (Waldhufendorf) village type. Slavic place-names seem to cluster in an area along the Lusatian Land (Upper Lusatian Gefilde), the Upper Neisse Valley, and the Upper Elbe Valley. Historically, these clusters are also the earliest that appear in the written documents. Archaeological evidence about medieval settlement traces and the oldest core of existing villages seem also to refer to the same clusters, except the one on the Upper Elbe Valley. From a physical point of view, the core-region known in German as the 'Lusatian Gefilde' (Lusatian Land) is bordered by the heights of the Lusatian Highlands in the South while a wide stretch of inhospitable flatland covered by marshy areas and forest lies to the North. The two major physical elements of the landscape, the mountain range and the river and creek network, running on two different axes, create invariably a natural network of landscaping ‚nodes' where earliest habitation sites are concentrated. The aspect mapping shows most of the Upper Lusatia divided in two halves with the southern half with hill faces oriented alternately to north and south and a northern half with flatlands oriented northward. Onto this northern half most of the earliest settlements are located. On

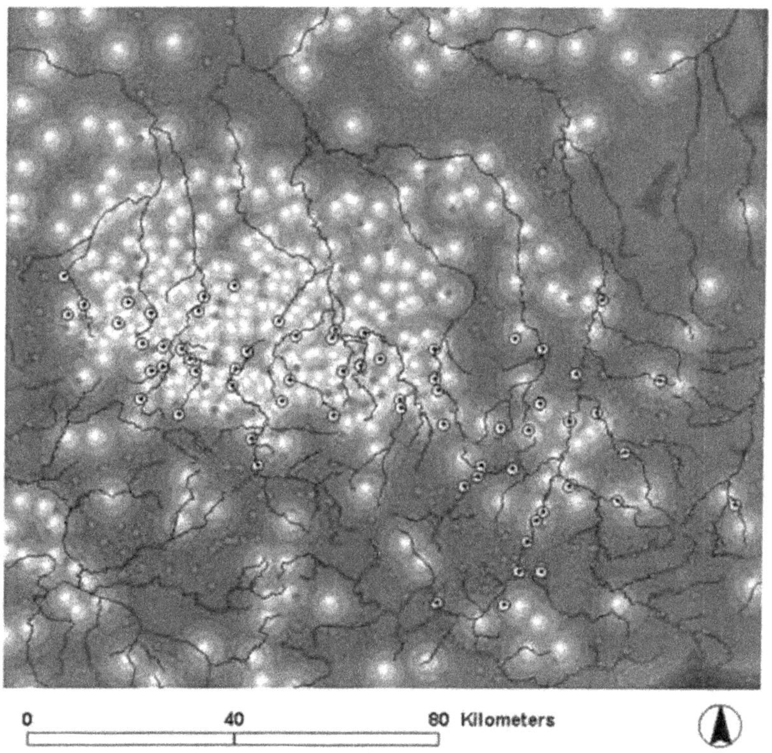

Figure 1 Distance mapping for Slavic (light dots) and German (dark dots) settlements. Black dots on white circles represent early medieval forts

the other hand, the eastern part of the region shows more randomly patterned terrain around the Upper Neisse Valley, acting as a north-south divider. The central part of the Upper Neisse Valley was occupied by earliest settlements in the area, and this cluster is separated from the main one by an uninhabited ridge located between the Black and the White Schöps Creeks. The overall settlement theme laid onto a DEM-model shows Slavic settlements concentrating at elevations lower than 300 m. One could conclude at this stage that Slavic settlers have avoided the Highlands at all times and that the German colonists did not have other choice but to settle on what was a free land zone in the 12th c. However, after place-names classes are summarized and plotted in relation to the mean elevation of the elaborated DEM, it seems obvious that both Slavic and German patronyms tend towards the 200 m elevation level, while possessives differ to some extent: the Slavic stop at the 220 m elevation, while the German ones reach the 280 m elevation. Assuming that patronymic place-names are the earliest settlements in the region, then both Slavs and Germans regardless of their arrival date in Upper Lusatia chose to settle at lower elevations expanding afterwards in the highlands. This is confirmed also by the convergence between Slavic place-names and the Bauernweiler/Rundweiler (peasant/round hamlets) village type distribution.[1] Distance mapping of Slavic place-names shows settlements clustered in the Lusatian Valley within a distance ranging from 2 to 4 km (Fig.1). This seems to come close to Timoščuk's results from Ukraine where intra-cluster distances between Slavic settlements were calculated at less than 2 km (Kobylinski 1997:108). These aggregate in what can be called a settlement unit or cell (Kobylinski 1997:109). Density mapping, on the other hand, identified three distinctive and different types of clusters: the westernmost one appears as an agglomeration around Coblenz, Dahren, and Passditz, the middle one being in fact an encircling band around Bautzen/Budyšin, while the easternmost cluster is linear on a east-west axis. Less dense agglomerations are spotted on the Upper Neisse Valley between Jauernick, Tylice, Reczyn, and Tauchritz, between Kamenz and Wittichenau, around Litschen, and on the Upper Elbe Valley between Krippen and Dorf Wehlen. These clusters could correspond with a higher hierarchical territorial unit, the so-called opole, pogost, zupa (Kobylinski 1997:109). Proximity mapping with ethnic affiliation as identity field for cells shows the degree of nearness between Slavic place-names and also that the direction of penetration of German place-names is from the southwest for both Slavic clusters. Spatial statistics such as Quadrat Analysis are applied to detect how settlement density changes over space. In order to discern this, the test compares the given point distribution with a theoretical random pattern (Quadrat analysis results with qsize = ½ n: 458, Lambda - 1.94805, Variance - 1.94805, K-S Dstat - 0.370782, Alpha level at 0.05 - QA critical value is 0.04533). This value differs significantly from the calculated K-S absolute difference of 0.370782 between the dispersed and clustered patterns, and we may therefore conclude that Upper Lusatian settlements do not distribute in a dispersed manner. The Nearest Neighbour Analysis (NNA) is applied to the same dataset in order to check the result of the first performed test. The NNA test gave (Nearest neighbour analysis: Observed Neighbour dist: 2.06481, Expected n dist: 2.09557, N. neighbourhood R Statistic: 0.985324, StD Zr Score: 0.842362) R = 0.985324 which confirms the earlier result, by indicating that the settlement pattern has a tendency to cluster and is farther from dispersing. The preliminary results in the elaboration of a Predictive Model for Slavic Settlements in Upper Lusatia show that not all variables are sensitive to settlement location. The slope degree is indifferent to settlement location, these being located on both flat and steep areas. Closeness to watercourse does not influence settlement selection, many settlements being spread randomly between two riparian zones. The soil type variable seems to be related to settlement location with Slavic toponyms located on loess and clayey soils.

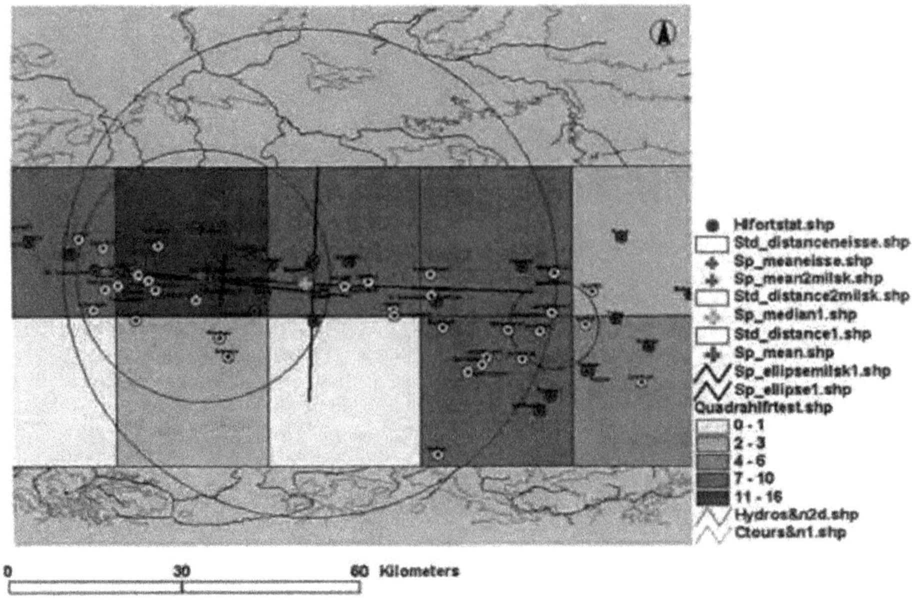

Figure 2 Quadrat Analysis on forts divided into earlier (dark dots) and later (light dots) than 10th c. AD

Bautzen - Zentralort?

Historical information, coupled since the beginning of the 20th century with archaeological evidence, affirmed the role of Bautzen/Budyšin as the most important political, cultural, and economic town in Upper Lusatia. During the course of history, the town knew several phases of urban development, expanding eastward from a core settlement area located somewhere between the St. Peter & Paul Cathedral and the Ortenburg Castle. Situated at a natural crossing/ford over the Spree River, and commanding the via regia trading route that connected Western Germany with Silesia and Eastern Europe the town was first mentioned by Thietmar of Merseburg as 'civitas Budusin' in 1002 and in 1004 as 'urbs Budusin' as being of strategic importance for both German and Polish military interests in the region (Thietmar Chron. 230:9 and

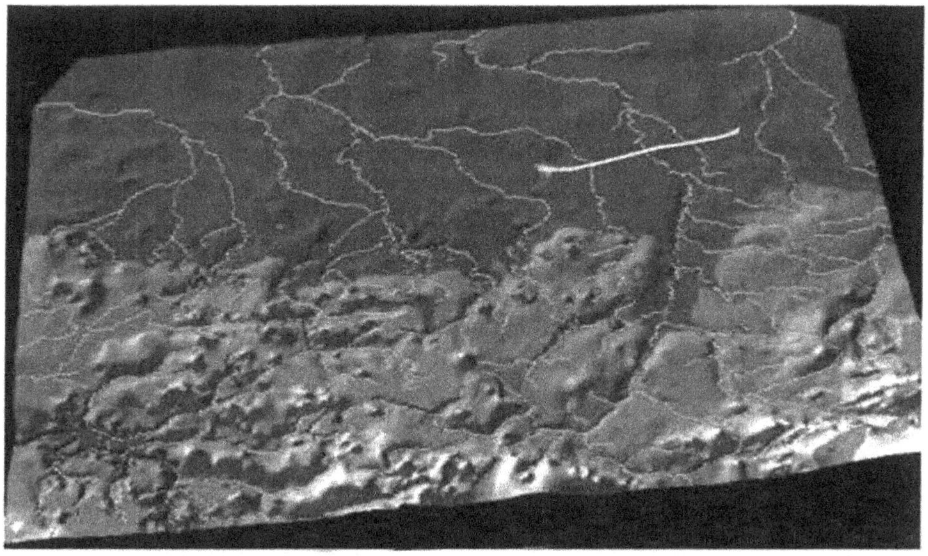

Figure 3 3D DEM of Upper Lusatia (red dots represent locations of early medieval forts)

formed the initial random distribution into a cluster distribution: one located in the Lusatian Land and the other in the Neisse Valley and the 11th -12th -century fortifications accentuated the clustered pattern showing this time a clear tendency to fortify towards southwest in both clusters. The location of the latter in the Lusatian cluster between Bischofswerda and Göda corresponds also with the axis of one of the main trading routes through the region. Under Czech suzerainty forts were built not to defend the southwest corner anymore, but to reinforce the eastern and the northern flank of the Lusatian Land. Simple statistical measures of centrality such as spatial means and the corresponding standard deviations were calculated for the fort group as a whole and for each of the fort clusters. Both the spatial mean and the spatial median for the entire group are located at Belgern, this being the central point of the entire region. This characteristic seems related to the political conditions in the High Middle Ages, when the Czechs controlled the entire region. If the whole region is split into the Lusatian and the Neisse Valleys, then the spatial mean for the Neisse cluster falls between Jauernick and Landskrone, while the mean for the Lusatian cluster falls west of Bautzen/Budyšin. A Quadrat Analysis carried out on fort distribution shows a linear agglomeration for the Lusatian Valley with a tendency to cluster in the Bautzen/Budyšin quadrat (11-16 forts), while radial-like dispersion is to be seen on both sides of the Neisse Valley (Fig.2). On the other hand, line-of-sight calculations indicate that observers based at almost all of the forts from Lusatian cluster would have no visual contact with each other. The opposite holds true for the Neisse Valley where observers at Landskrone can visually contact most of the surrounding forts. The analysis can be carried further through the elaboration of a DEM for the entire region (Fig.3). On the 3D model one can readily visualize which forts are commanding the view, and which can be visually connected from a specific location. With patronymic settlements plotted, the Lusatian Land emerges as a cluster of settlements surrounded by a stronghold ring, whereas the Neisse Valley shows radially dispersed strongholds intermingled with settlements that cluster towards Landskrone. The 1-km buffering of the core-area settlements coupled with the 3-km buffering of the earliest strongholds shows that settlements aggregate in smaller units that aggregate in turn around a stronghold to form a stronghold territorial unit with a radius of ca. 5 km (Burgwallbereich, okreg grodowy, pole, laukas, campus). Subsequently, these units aggregate further into stronghold clusters (opole, pogost, zupa), covering areas up to 20 km while the distance between strongholds units in the Milsca Valley is well under the average estimate of 10 km (Kobylinski 1997:109).

292:14). These premises created the idea of Bautzen as Zentralort, a place destined to be the seat of the feudal power and the converging point of the entire region. D. Scholze stated that before its first mention in the written sources, a stronghold was erected on the high ground surrounded by the meander of the Spree (Scholze 2002:30). This stronghold would have been, according to K. Blaschke, the tribal seat of power of the Milzeni, transformed subsequently into a ducal stronghold of the entire Lusatian Land (Blaschke 2002:46).

Archaeological investigations started in Bautzen/Budyšin inner-city as early as 1906 when W. Frenzel found on the Castle perimeter (Ortenburg) Bronze Age, Slavic, and German ceramic material and subsequently carried out in conjunction with construction and renovation works (CAQ 1985:123, Sczech 1999:98). The results indicate that under the present Baroque Castle the ruins of 12th-century medieval fortifications overlay a Bronze Age earthen rampart while a 10th-century AD Slavic settlement was found extending south and east of the Ortenburg (Sczech 1999:98). The dilemma arising from this situation is not only practical but also theoretical. Was the landscape of the Milzeni really structured around a central place of power, such as Bautzen/Budyšin, or is it a wishful theoretical construct propagated by modern research based heavily on historical tradition (text-driven archaeology case)?

Strongholds: The Visible Landscape of the Past

A total of 74 strongholds were plotted on the GIS map, 68 being mapped with the help of a GPS receiver during the 2001/2002 field work. With very few exceptions most of them belong to the category of earthen ramparts with one or two fortified precincts and one entrance usually on the right side of the compound. From the ensuing mapping it can be clearly observed that early medieval strongholds appeared sometime during the 8th century AD in several places between Kamenz and the Spree, north from Bautzen/Budyšin and Löbau, and east of Neisse Valley between Görlitz/Zgorzelec and Luban. 9th-century strongholds trans-

Conclusions

The spatial coverage of the earliest 152 patronymic settlements corresponds with that of strongholds erected earlier than the 10th c. AD. The buffering of 3 km radius of hillforts theme shows that forts are networked according to local hydrography and that the Milsca Valley took shape out of numerous valley settlements (family nests or settlement units/cells) aggregated around each of the several strongholds located on the valley's watercourse (pole-laukas-campus). An extended valley-wide community consisting of stronghold groups (opole-pogost-zupa) was established in the Milsca Valley (Gau Milska), while the Besunzani Valley (Zagost) continued to forge its radial pattern initially without Landskrone as the centre place (Zentralort). No central places (Zentralort) are envisaged for this period. Forts served, in this stage, as places of social prestige and exchange with the outsiders and less as defensive works. As a direct result of the German-Polish Wars in the region during the 10th-11th centuries hilltop strongholds appeared[2] and forts were built towards southwest as an attempt to stop not the invading armies but the colonization waves started with the process of land gratification to the Bishopric of Meissen (Knebel 1965:23, Cod. Dipl. Lus. Sup. I:25). Bautzen/Budyšin and Landskrone[3] emerged now as strategic locations for both German and Polish warring parties, transforming the region in a centre-and-periphery structure. From now on Bautzen/Budyšin was to become the centre of the medieval Upper Lusatia, while Landskrone receded and eventually disappeared before the newly founded trading centre Goreliz (Görlitz/Zgorzelec).

1 Small villages consisting of few households spread around a central point or at random (cf. Blaschke, HAS, 2002).
2 Elstra, Schmölln, Naundorf, Hochstein, Binnewitz (exception), Rothstein, perhaps Löbauerberg, Schönau a. E.
3 Details about Landskrone and its role in the German-Polish conflict see Von Richthofen 2002:5.

References

BLASCHKE, K.H., 2002. Die Entstehung der Stadt Bautzen. In Von Budissin nach Bautzen, Bautzen:46-60.

BLASCHKE, K.H., 2002. Ortsformen. In Der Atlas zur Geschichte und Landeskunde von Sachsen, Bl. II 2.

HERRMANN, J. and DONAT, P. (eds.), 1979: Corpus für Archäologischer Quellen zur Frühgeschichte auf dem Gebiet der Deutschen Demokratischen Republik (7. bis 12. Jahrhundert). CAQ 2, Berlin:182-234.

KNEBEL, J., 1965. Betrachtungen zum Landschaftsbild und zur Herausbildung frühgeschichtlicher Siedlungskomplexe in der Oberlausitz. Letopis Jahresschrift des Instituts für sorbische Volksforschung, B. 12, Bautzen:5-37.

KOBYLINSKI, Z., 1997. Settlement Structures in Central Europe at the Beginning of the Middle Ages. In Origins of Central Europe, Warsaw:97-115.

SCHOLZE, D., 2002. Bautzen als politisches und kulturelles Zentrum der Sorben. In Von Budissin nach Bautzen, Bautzen:30-46.

SCZECH, K., 1999. Stadtkernarchäologie im Dreiländereck - Bautzen in der Oberlausitz. In Archäologie als Sozialgeschichte - Studien zu Siedlung, Wirtschaft und Gesellschaft im frühgeschichtlichen Mitteleuropa - Festschrift für Heiko Steuer zum 60. Geburtstag, Rahden/Westf.:97-102.

THIETMAR VON MERSEBURG, 1955. Thietmari Merseburgensis episcopi Chronicon. MGH IX, Berlin.

VON RICHTHOFEN, J., 2002. Die Landeskrone bei Görlitz. Eine bedeutende slawische Befestigung in der östlichen Oberlausitz.

Fields of Wheat Back to the Land. A GIS Environment for the Study of Medieval Village History in Central Greece

SEE THE CD FOR THE EXTENDED VERSION

EMERI FARINETTI
UNIVERSITY OF LEIDEN - FACULTY OF ARCHAEOLOGY, THE NETHERLANDS
EMERIF@TIN.IT

KOSTAS SBONIAS
IONIAN UNIVERSITY - DEPARTMENT OF HISTORY, GREECE
KSBONIAS@ATH.FORTHNET.GR

ABSTRACT

The paper aims to broaden the discussion about the integrated use of archaeology and computer applications, bringing into focus the potential of historical data in the GIS based studies. We particularly explore possible ways of processing historical information (such as population and agricultural production data) bringing them into the GIS system as constitutive factors of analysis, in addition to the commonly used environmental data.
As a case study we examine historical sources (Ottoman Imperial Archives) available for the history of Boeotia (central Greece). The paper addresses methodological questions about ways of structuring and managing those historical sets of data. At the alphanumeric level the diverse data-sets are structured and gathered into a database. Accordingly, at the graphic level all the recorded information are given a spatial character and spatial features are attached to them. The medieval and post-medieval data on population and economic production of the province of Boeotia, which come from the Ottoman Imperial Archives, are linked with villages located in the landscape and sites recorded by the archaeological survey. An attempt is made to model the demographic and economic development of the region.

In the archaeological discourse GIS is usually referred to as an environment within which to process information concerning environmental and physical factors, with the aim to study and analyse broader historical and cultural phenomena. In few cases though, historical data have been employed as meaningful variables in the creation of GIS-based geographical and historical models (see for instance Doorn 1993). The present paper aims to broaden the discussion about the integrated use of archaeology and computer applications, bringing into focus the potential of historical archive data processed and analysed within a GIS environment.

As a case study we examine the Ottoman Imperial Archives of the province of Boeotia in Central Greece, which were studied by Machiel Kiel for the Boeotia survey project (Kiel 1997, 1999). Such Tahrirs, which registered for tax reasons the population and the agricultural production of the region, were made as a regular practice of the Ottomans both in synoptic and detailed forms after every accession of a new Sultan or, in the case of long reigns, every 20-30 years (Kiel 1992, 1997:319). They offer us information on the total tax revenue per village, the number of households and, in the case of the detailed registers, the tax share of each product. To these sources we should add travellers accounts of the late 17th and 18th centuries as well as proper census accounts of the newly founded Greek state in the 19th century.

In the present study these diverse data sets were structured in a relational database that contains information on the village names and their location, on the periods in which the villages are recorded, data from the synoptic and detailed Ottoman registers on the population and agricultural production, as well as information from travellers and modern census statistics. The economic data from the registers come in an indirect way, that is tax revenues. Therefore within the alphanumeric database they had to be converted applying specific algorithms in order to get data in a format suitable for processing within the system. The number of tax paying units, i.e. households, was corrected by a factor of five people per household to get an impression on the total population per village. The transformation of the tax share of each product into the metric system was facilitated by the description in the Mufassal tahrir defters of the local lows (Kanunname) and customs about taxation practices as well as the way in which the local measures are related to the standard measure of Istanbul (Alexander 1985, Kiel 1997:320-321). In the Kaza of Thebes the local 'load' for cereals was 6.5 Istanbul bushels (keyl) or 167 kgr, in the Kaza of Levadia it was 6 Istanbul bushels or 154 kgr. The Medre, the measure for wine, equals to 71 litres. Sheep were taxed 1 akce (small silver coin) for two sheep, and this allows to calculate from the total revenue derived from sheep the total number of sheep per village. On the basis of this information given for Boeotia by Kiel (1997:318) and furthermore converting the 13 %, that was the tithe in the Sandjak of Euripos, to which Boeotia belonged, into 100 % we came to an average estimation of the production of each crop in the metric system (see also Balta 1989).

GIS gives us the opportunity to study these parameters in the landscape. As base-layers we used the 1:200,000 and 1:50,000 topographical and geological maps of Boeotia which were digitized according to different info-layers and put into the GIS system as part of the larger work of one of the authors on the ancient settlement patterns of Boeotia (Farinetti, forthcoming). Subsequently GIS layers were produced illustrating the virtual continuous representation of the surface morphology (a DEM), as a TIN and subsequently as a GRID to allow further analyses, the slope, and the geology and arable soils of the region. Next step was to give a spatial character to the information recorded in the registers and collected into the database and attach spatial features to them. At this stage we used the results of an ongoing research by

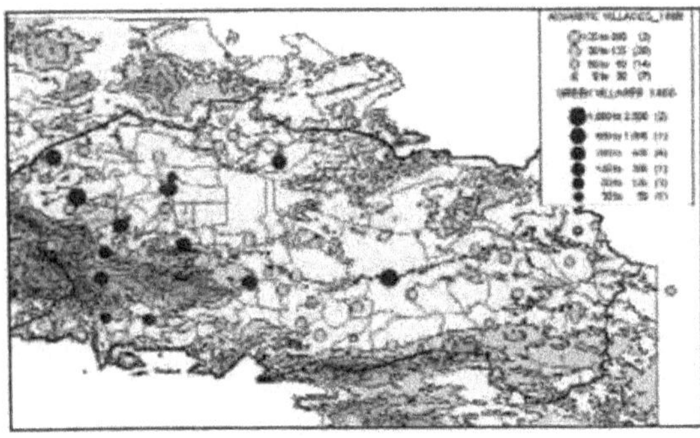

Figure 1 Villages of Boeotia in 1466 according to population and ethnicity

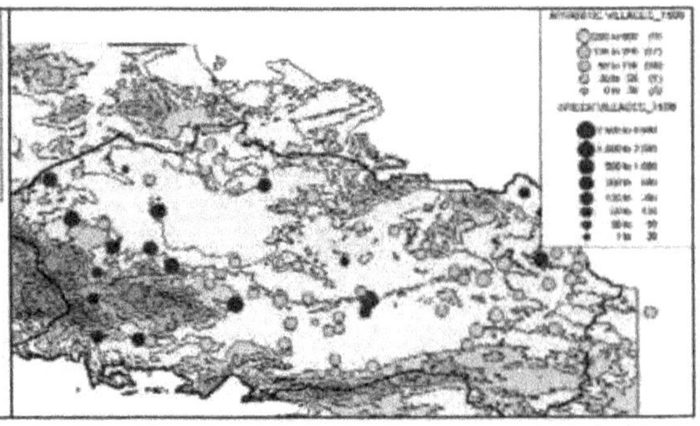

Figure 2 Villages of Boeotia in 1506 according to population and ethnicity

Bintliff and Kiel (forthcoming) who managed to locate a large number of the villages appearing in the documents. A large percentage of them is identical with present day villages, whose name history and location change can be followed through oral history and existing sources, such as travel journals and maps of the 19th or earlier centuries. Problematic are villages that ceased to exist and for which there is no indication of their location. Problematic are also cases of locations bearing the same name or of present day villages for which there are more candidates of the same clan in the archives. In such cases indications can be obtained from the known location of the previous or subsequent village recorded in the defters and this proximity factor has been used to define the approximate broader geographical setting (see also Balta 1992:59-64). Additional information comes from the sites recorded by the Boeotia archaeological surface survey. The project recorded the abandonment of numerous Byzantine village sites as well as the foundation of new ones, connected with a new pottery style and marking the relocation of villages (Bintliff 1995:115).

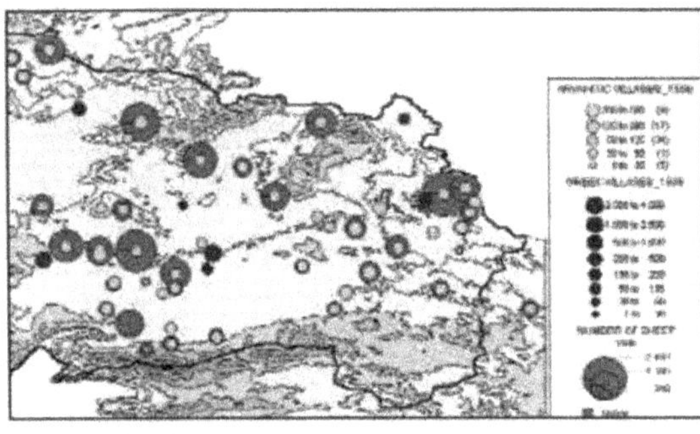

Figure 3 Sheep razing in the district of Thebes in 1506

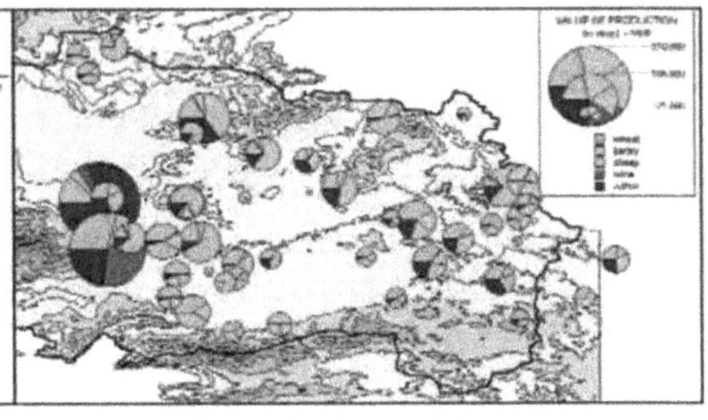

Figure 4 Monetary value of the agricultural activity in 1506

Our aim in this occasion is to present a preliminary discussion of the village history of the late 15th and mainly 16th century, giving emphasis to the spatial component that GIS allows us to include in our study. The oldest registration for Boeotia is a synoptic one and comes from the year 1466/67, few years after the annexation of Boeotia by the Ottomans (Fig.1). The archives make a distinction between arvanitic (ARNAUDHAN) villages, i.e. Albanian speaking, and Greek speaking (RUM) villages, recording in that way the discontinuity in settlement and new population influx which took place in the Late Frankish period (Bintliff 1995). In 1466 the Greek villages were large stable villages of 200-500 people clustering on or beside ancient foci (such as Chaironeia, Chostiai, Thisbe, Koroneia, Copai, Haliartos, Askra) and were concentrated mainly in western Boeotia. To these we should add the two almost entirely Greek towns of the region: Levadia, which was rather a large village of 1,000 people and the town of Thebes with 2,400 inhabitants. On the other hand, the arvanitic villages were tiny hamlets of 30 to 130 people, mobile around several village locations.

Figure1, which also displays the distribution of modern village boundaries, shows that most of the medieval villages stand in the middle of present village territories. In some cases several small arvanitic hamlets exist within a present village community area, the large territory of modern Thebes corresponds to that of the medieval town, medieval Levadia has a smaller hinterland and is surrounded by several other villages, whilst the blank areas are occupied by the sites in the next census of 1506 as a result of population growth and gradual infilling of the landscape. The majority of the divisions displays a comparable territory for both Greek and Arvanitic villages, which does not take into account the different size of the villages, creating therefore an even distribution in the landscape, which reflects probably the organized character of the colonization that filled the deserted settlement niches of the region. Yet the small village size and lower population of the arvanitic Eastern Boeotia created more favourable conditions in comparison to the crowded and confined Greek villages. The implications of these differences for the economic and demographic development become visible in the registers of 1506 and especially of 1570, for which also detailed economic data are preserved.

In 1506 a general demographic growth can be noticed in both Greek and Arvanitic settlements (Fig.2). In the lowlands the arvanitic hamlets multiplied sharing the existing communities areas, a phenomenon related probably to their smaller size as well as to the fertility of these lowland areas. Their size remained small as most of them continued to fall in the range of 30 to 135 people. Yet already in 1506 many of the arvanitic hamlets were developing into substantial villages as a large number of them falls in the range of 135-200 people and a few reached the 300 inhabitants. Despite that, in this period the main distinction between smaller arvanitic hamlets and large Greek villages remained (five of the latter were in range of 200-500, other five had 500-1,000 people, Levadia doubled its population coming up to 2,000 and Thebes the same reaching 4,500 inhabitants).

Differences can be observed also in the economic structure of the villages, which presumably go back to the C15th and are related to the different ethnicity of the villages. In Figure 3, which illustrates sheep razing in the Kaza of Thebes, we observe that all the arvanitic settlements had a strong pastoralist component and were evenly distributed having comparable size of flocks. Their economy was supplemented by cereal and cotton production. On the contrary the Greek villages were not involved in stock raising but were practicing Mediterranean poly-culture.

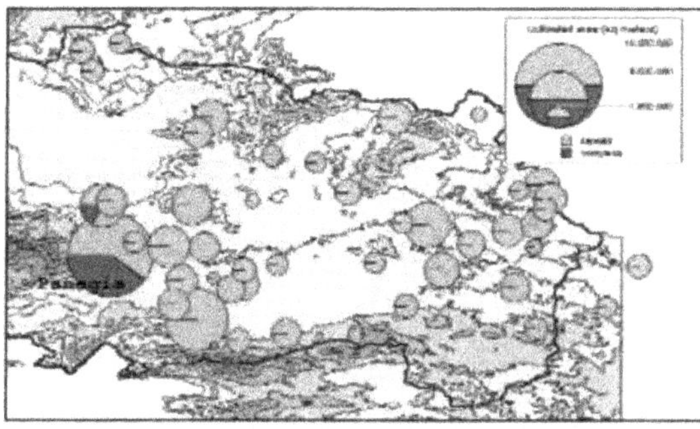

Figure 5 Calculated area devoted to grain cultivation and viniculture in 1506

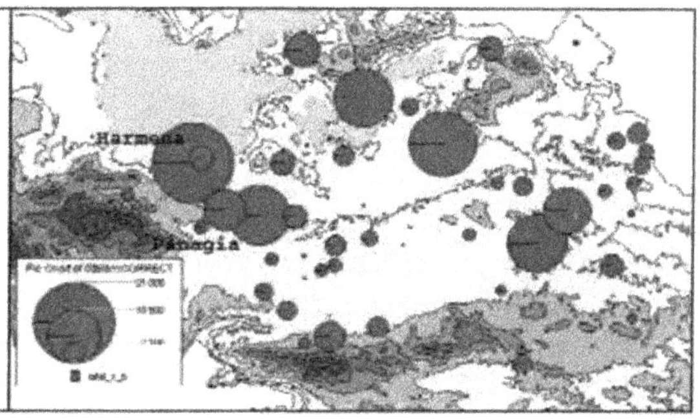

Figure 6 Average income per household in 1506

Figure 4 which shows the monetary value in akce of the different types of agricultural activity, calculated on the basis of the official price (not the fluctuating market price) of each article mentioned in the registers (Kiel 1997:321), lets us evaluate the importance of the different crops for the economy of the villages. The Greek villages, which are very few in the Kaza of Thebes, had an economy based on a considerable grain production for subsistence and in the meanwhile they were orientated towards labour intensive activities. Panagia (the successor of ancient Askra) was specializing in viniculture, while Harmena (the successor of ancient Haliartos) in cotton and the textile industry. We should notice the total absence of olives, a crop that was introduced in other parts of the Aegean area during the 17th century, in connection with an export oriented economy. The inhabitants of the arvanitic villages, alongside the systematic sheep raising, were devoted to supplementary activities such as cereal and cotton production. We notice also differences among them, as some had a higher income from the exploitation of grazing land while others near the plain from the grain production. Therefore in 1506 they were moving already away from a subsistence model based on tiny pastoral hamlets, which might have been the case immediately after the colonization.

In Figure 5 we attempted to assess the production recorded in the registers in terms of the territories of the different villages, calculating the area devoted to grain cultivation and viniculture. We used here the figures that Allbough (1953) gives as the yield of 1 ha of land for different kinds of crops and we took also into account the fallow areas, which practically double the area devoted to cereals. As we should expect, the smaller Arvanitic villages had plenty arable land available, which together with the exploitation of the grazing land created more favourable circumstances. Panaya on the other hand, the largest Greek village after Thebes, with a considerable rise in population seems to have exceeded the potential of its territory.

What these different economic strategies meant for the peasant can be seen in Figure 6 which estimates the average income per household generated by the agricultural production. Small population numbers and larger territories created the circumstances for intensified production, higher surpluses and a better income per household. Striking, on the other hand, is the case of Panagia which in our previous figures appears always as a large village with large production and high population number. Yet these were not a sign of prosperity as the very low average income per household illustrates, indicating the crisis that was facing this village economy. Better was doing Harmena which was specializing in cotton production.

As a result of this deficit but also because of the continuing population growth in both Greek and Arvanitic villages in the course of the C16th an essential transformation, which is reflected in the 1570 register, took place in the Boeotian landscape. First of all a transformation of scale is observed in the Arvanitic villages, which doubled their population and fell in the range of 200-500 people, a size similar to the majority of the Greek villages. In terms of village size there is therefore a growing homogeneity in the rural landscape. On the other hand certain Greek villages and towns grew further and kept dominating the urban landscape: three villages were in the range of 600-1,000 people, another two in the range of 1,000-2,500 and the two towns of Levadia and Thebes had 1,500 and 7,500 people respectively.

The observed homogeneity becomes more evident when we move to the economic structure of the villages (Fig.7). With demographic increase the arvanitic villages turned to supplementary activities and viniculture appears along with cultivation of cereals, cotton and stock raising. They resemble therefore from now on to a typical Greek village in respect to economy and village size. Yet this transformation was not one sided. The Greek villages pressed by their large population and limited availability of land were forced to embark to other activities. In Panagia for example viniculture was decreased drastically and cereal production was increased to feed the growing population. The higher income though derived by the introduction of stock raising, which changed essentially the character of this village economy. This reorientation should be seen as an answer to the crisis of the previous 70 years and not as an investment of surplus money from wine or the textiles into sheep breeding as it has been argued (Kiel 1997).

Our point is illustrated better in figure 8, where we can see the area devoted to cereal cultivation and viniculture in two up today surviving villages, Panagia and Mavromati. Displayed are also the boundaries of the present day villages and a classification of the soils of the region in a scale from 1 (bare rock) through 2 (grazing land) up to 5 (best arable

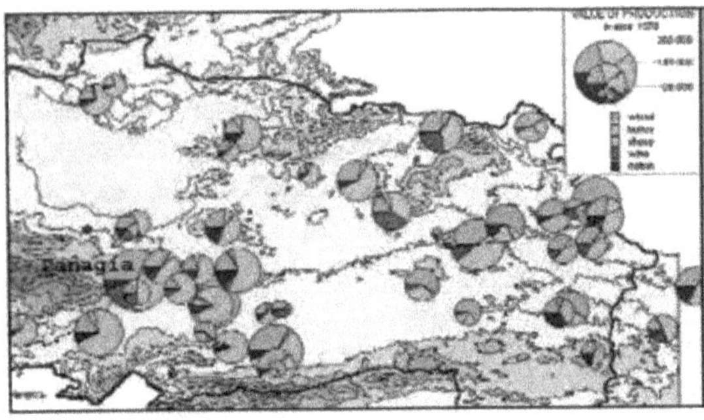

Figure 7 Monetary value of the agricultural activity in 1570

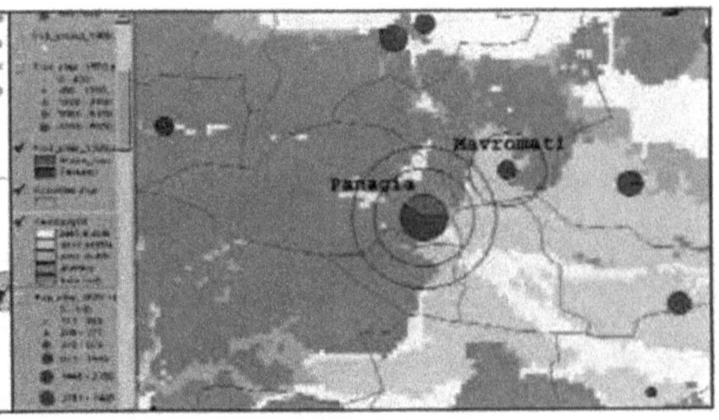

Figure 8 Calculated area devoted to grain cultivation and viniculture in the villages of Panagia and Mavromati

soil). In the case of Panagia we can notice the limited area for cultivation which had been surpassed by the agricultural production as well as the existence of plenty grazing land in the interior mountainous area. The embarkation upon pastoralism was exactly what this community did to ensure its viability and carry on its demographic expansion.

Furthermore, looking at the size of floks and their distribution in Figure 9, we can notice clusters of sites with systematic sheep raising indicating a more specialized landscape (in comparison to 1506 in Fig.3), which was determined by topography (lowland-upland) and soil potential rather than by ethnicity. Finally, the value of production per household and therefore the per capita income shows, as was the case also in the 1506 register, an inverse proportion of the highest income to the village size. This affects mainly the Greek villages, which were larger in size. As a working hypothesis we could connect this deficit with migratory patterns and the population influx which is observed in certain Greek towns such as Thebes, and which can not be explained with natural demographic growth (see also discussion in Balta 1992:97-98 for similar trends in Euboea).

To sum up, with this paper, as part of a research which is just at the beginning, we made an attempt to make a meaningful use of an historical source, the Ottoman registers, combining them with the potentials for spatial analysis that Geographical Information Systems offer. We considered the territories of the villages and the agricultural potential of their hinterlands, the economic structure and importance of the different products for the village economy, as well as the demographic development and changes that took place in time, trying to illustrate the transformation of the Boeotian landscape during the 16th century.

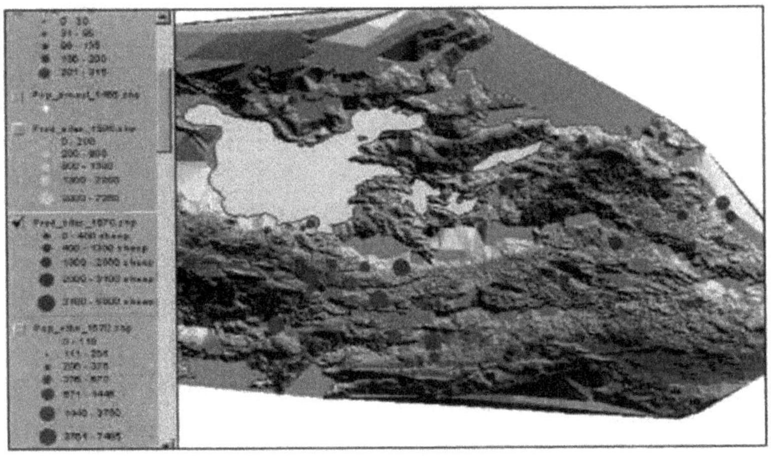

Figure 9 Sheep razing in the district of Thebes in 1570

ACKNOWLEDGEMENTS

We would like to thank Prof John Bintliff, director of the Boeotia Survey Project, for his encouragement. The data used for the analysis are based on the research on the Ottoman archives carried out for the Boeotia Project by Machiel Kiel, to whom we are grateful.

REFERENCES

ALBAUGH, L.G., 1953. Crete - A Case Study of an Underdeveloped Area. New Jersey.

ALEXANDER, J., 1985. Towards a History of Post-Byzantine Greece: the Ottoman Kanunnames for the Greek Lands circa 1500-circa 1600. Athens.

BALTA, E., 1989. L'Eubée à la fin du Xve siècle; économie et population. Les registres de l'anné 1474. Athènes.

BALTA, E., 1990-91. Rural and Urban Population in the Sancak of Euripos in the Early 16th century. Archeion Evoikon Meleton 29:55-185.

BINTLIFF, J., 1995. The two transitions: current research on the origins of the traditional village in Central Greece. In Bintliff, J.L. and Hamerow, H. (eds.), Europe Between Late Antiquity and the Middle Ages. Recent Archaeological and Historical Research in Western and Southern Europe, Oxford (BAR IS 617):111-130.

BINTLIFF, J. and KIEL, M., forthcoming. A Gazetteer of Ottoman Boeotia.

DOORN, P., 1993. Geographical Location and Interaction Models and the Reconstruction of Historical Settlement and Communication: The Example of Aetolia, Central Greece. Historical Social Research 18:35-71.

FARINETTI, E., forthcoming. A GIS based Analysis of the Settlement and Landscape History of Ancient Boeotia (Central Greece). PhD Thesis, University of Leiden.

KIEL, M., 1992. Central Greece in the Süleymanic Age - Preliminary Notes on Population Growth, Economic Expansion and its Influence on the Spread of Greek Christian Culture. In Veinstein, G. (ed.), Soliman le Magnifique et son Temps, Ecole de Louvre, Paris:399-424.

KIEL, M., 1997. The Rise and Decline of Turkish Boeotia, 15th-19th Century. In Bintliff, J.L. (ed.), Recent Developments in the History and Archaeology of Central Greece, Oxford (BAR IS 66):315-339.

KIEL, M., 1999. The Ottoman Imperial Registers: Central Greece and Northern Bulgaria in the 15th-19th century; the Demographic Development of Two Areas Compared. In Bintliff, J. and Sbonias, K. (eds.), Reconstructing Past Population Trends in Mediterranean Europe, Oxbow Books, Oxford:195-218.

THE "CONTEMPORARY MIND". 3D GIS AS A CHALLENGE IN EXCAVATION PRACTICE

STEFANIA MERLO
DEPARTMENT OF ARCHAEOLOGY
UNIVERSITY OF CAMBRIDGE NEWNHAM COLLEGE
CAMBRIDGE, CB3 9DF
SM399@CAM.AC.UK

ABSTRACT

The recent publication of a number of manuals dedicated to excavation highlights a renewed interest in this subject. Nevertheless reference literature fails to address one of the major contributive elements of the last decades excavation practice: the use of IT. On the other hand, dedicated contributions (see CAA papers) tend to focus more on technical issues than on the implications of using IT in humanities.
This paper explores three-dimensional GIS for excavation as a framework within which the integration of theory and practice has the potential of solving practical problems and revising strategic choices during and after fieldwork. A cognitive approach is used to test the response of 3D GIS technology to a more rounded concept of excavation. Conversely, the influence of such technology on our perception of the archaeological record is discussed.

INTRODUCTION

What do archaeologists do? They dig. Or at least this is what the general public still thinks of contemporary "Indiana Jones". Contrary to this widespread and general belief, archaeology has developed into several branches of research that have little to do with excavation. In particular, in the field of computational research, the study of landscapes and the use of the so-called non-invasive methods (remote sensing) has surpassed what was and is still believed to be the most invasive but also information rich archaeological practice. As a consequence, a discussion about excavation methodology and its implications for knowledge production has been somewhat neglected since the early 1980s, when it was decided that the way forward was the stratigraphic method and the recording had to be carried out using single context planning (Harris 1978).

The recent publication of a number of manuals dedicated to excavation itself (Drewett 1999, Collis 2001, Roskams 2001) highlights the renewed interest in this subject after a period of silence. Following the postprocessual claims of Ian Hodder, a discussion on related issues in monographs and papers (Hodder 1997, Chadwick 1997, Hassan 1997, Hodder 1998, 2000, Andrews et al. 2000) has fuelled the academic debate. Nevertheless, the attention focused mainly on statements of the respective theoretical positions on the topics of scientific versus interpretative, subjective versus objective methods. As a consequence, considerations on the practicality and applicability of the different approaches became a secondary issue.

Reference literature fails (an exception being Ian Hodder 1999 and Steve Roskams 2001) to address one of the major contributive elements of the last decade's excavation practice: the use of IT. An interesting and animated debate followed the application of IT to field archaeology (in particular the use of GIS for predictive modeling, view-shed analysis and so on) and it focused on "the ancient mind". The same cannot be stated of computing and in particular GIS at intra-site level. Although some causes of this disparity have been identified and discussed (Biswell et al. 1995, Huggett 2000), a further debate on the implications of using IT during all phases of the excavation process has been ignored. In order to overcome this problem, the use of IT in excavation, should, in my opinion, consider the "contemporary mind" of the archaeologist.

Starting from these premises, I aim to shift the attention from current concern with data and accuracy at the core of data collection and management to a broader discussion about world views and metaphors that are involved in the process of using the world of computation for spatial representation of archaeological contexts, in particular that of excavation.

Within a cognitive science framework, the focus shifts from what we perceive to how we perceive it and more importantly how we represent it. The need to understand mental models of excavation and, in parallel, of GIS cannot therefore be understated.

COGNITIVE ARCHAEOLOGY AND THE CONTEMPORARY MIND

Cognitive archaeology, as defined by Renfrew in 1994 is "the study of past ways of thought as inferred from material remains" (Renfrew 1994:3). Although the focus of cognitive archaeology has always been the "ancient mind" both Renfrew and other contributors underlined that "the analysis has to enfold both 'us' and 'them'. Only by understanding our own perceptions can we recognize the particular way in which we engage with the past and thereby accept that the way in which we listen to them is subjective" (Bender 1993:257). Moreover Renfrew states: "Perhaps we shall soon see some convergence between such fields as cognitive psychology, studies in artificial intelligence, computer simulation and cognitive archaeology. The time may be ripe for a great leap forward. But I don't see this happening until those archaeologists interested in the symbolic and cognitive dimensions devote more attention to the formation of a coherent, explicit and in that sense scientific methodology by which that dimensions can systematically be explored through the examination and analysis of the archaeological record" (Renfrew 1993:250).

Bearing in mind that "cognitive science is the study of intelligence and intelligent systems, with particular reference to intelligent behavior as computation" (Simon and Kaplan 1989:1), Renfrew's call for archaeologist's interest in an integrated methodology of exploration does not come as a surprise. What surprises nonetheless is the fact that cognitive archaeology has been concentrating on trying to understand how the mind of ancient people worked, disregarding other aspects and implications of using cognitive science in archaeology.

The call for an alternative program in cognition studies in archaeology is clear in Gardin (1992), who underlines the difference between an approach focused on people of the past and one considering researchers in the present. The two approaches are not mutually exclusive, but it seems to me challenging and fundamental to confront the second to better contextualize and therefore understand the first.

As a consequence, when we consider cognition and the use of GIS for archaeological excavation, we must shift the attention from the ancient mind to that of the contemporary mind: our ways of thought, now, today and as archaeologists.

The spatial nature of archaeology does not need further discussion and it is the very reason why GIS entered archaeology at a very early stage in the history of its application. What needs development is a discussion of spatial knowledge in archaeology, within the framework of cognitive science, therefore in terms of interaction with computation when translating concepts such as vision, touch, experience.

Spatial Cognition

"Using a GIS requires the use of spatial knowledge. But what kind of spatial knowledge?" (Nyergers 1993:38). A variety of ways can be used to define spatial knowledge. In this paper a contribution by Mark (1993) is used as a general reference for its classification, which can consider the nature of spatial knowledge, the sources of spatial knowledge for cognition, or experiential interaction and linguistic uses. In the following section try and explore how the first two classifications can help in giving insights into spatial cognition and archaeological research, in particular in excavation practice.

A classification by the nature of spatial knowledge was formulated by Golledge (1990) and it divided spatial knowledge into declarative knowledge of geographic facts, procedural knowledge in terms of way-finding and navigation (the so-called sensi-motor knowledge) and configurational knowledge intended as map-like Euclidean geometry which, in its lowest form consists of topology and in its fully developed form would allow a person to estimate absolute distances and directions between known points as accurately as they could while looking at a geometrically correct map.

In connection with the importance of the link between knowledge of space and the use of computer interfaces, Golledge underlines that declarative and procedural knowledge are quite well developed in everyone, whereas configurational knowledge is not developed by everyone to the same extent.

If it is true that different people have different levels of configurational knowledge, how do we deal with it? How can we be sure that any archaeologist in front of a computer is going to be able to explore and understand data with well-developed survey and overview skills? This is an issue when we are designing or we are asking computer scientists to design user computer interfaces for us. Moreover, this is an issue at the very core of excavation planning.

The other classification taken into consideration is the one presented by Mark in 1993. It consists of "three fundamental and distinct concepts of space used in human spatial cognition, differentiated according to the perceptual or cognitive source of that information" (Mark 1993).

Mark divides space into haptic, pictorial and transperceptual, in a hierarchical arrangement (see table 1 for the definition). This classification helps to explain and build the metaphor of excavation, in terms of abstraction and transformation processes from a space that we can directly experience (the haptic space of excavation, where we see and touch and experience) to a superior abstraction that allows us to picture and put together in our mind physical spaces otherwise incompatible (transperceptual space). The transperceptual space does not have any kind of materiality and it can be associated with our mind or the computer (in computational archaeology). Through the metaphor "investigations of space (and excavation, nda) are limited only by the power of intellect and mind" (Golledge 1990:147).

Excavation as a Metaphor

The concept of archaeological excavation as pure destruction has been recently challenged by new ideas such as explosion (Jones 2002) and displacement (Lucas 2001b). Pushing the boundaries into new conceptualisations of the practice of archaeology, I would here argue that the process of excavation is the creation of a metaphor. The concept of metaphor as a purely rhetoric trope has been challenged and applied to the world of knowledge and comprehension. Metaphor is used by humans to understand concepts otherwise incomprehensible, to put together fragments of unknown information into a known unity (Lakoff and Johnson 1980). Here is the parallel with archaeological excavation. The levels of spatial knowledge listed by Mark are comparable to the different phases of the archaeological excavation from the field to the computer.

The GIS model of an excavation is a metaphor of materiality. If in one way we loose the materiality of haptic space, during the abstraction process, we have also gained new insights through the creation of a metaphor in the transperceptual space. What is interesting and intriguing in using the concept of metaphor for the process of excavation, is that it does not really matter which new structure we use to represent the original concept. What is important is to maintain the characteristics that Aristotle assigns to metaphors: to be pertinent and vivid. To fulfill the first requirement the need for questioning and making explicit the metaphorical process and its implications in the production of knowledge becomes self-explanatory. The second requirement underlines how, even when we decide to use mathematical models to conceptualize perceptions of the "real" world in synthetic geometric representations, the value is not given by the detail richness, but by the ability to explain reality (Forte 2000).

The preoccupation of the archaeologist with spatial cognition should occur at different levels (at least two). The first level is the excavation. How do we interact with space when we are in the excavation? How does the understanding of space during excavation allow us to record in different ways? This is further discussed in the section dedicated to 3D GIS. The second level is post-excavation. How do we interact with the computer? How much does our knowledge of computer spatial structures and computer environment influences our way of conducting an excavation? GIS is very much embedded in GIS users' minds. When I go out in the field I cannot help thinking of ways of representing what I am seeing and experiencing in the excavation in the computer environment, whereas years ago I was thinking of reproducing this knowledge into papers, context sheets and notebooks. This is a paradigm shift in how we approach excavation, and it should stimulate new ways of thinking.

3D GIS and Excavation: An Idea

When comparing the computer data structures used in GIS and the archaeological data forms used to represent archaeological excavation it is interesting to notice that a parallel can be created between vector model and stratigraphic method, raster model and spit method.

The stratigraphic method, which stems from the principle of superimposition and from the concept that people lived on surfaces, is concerned with recording surface entities. The data structure that geometrically supports representation of surfaces is the vector one, not only in a computer environment, but also on traditional paper support. Euclidean space is used to contain points, lines and surfaces. In terms of volume organization, there is still a problem of its representation and manipulation, both in terms of paper records and computer environment. Raster data models, which are the structures reflecting space in terms of continuous Cartesian coordinates in two or three dimensions, are discretised in terms of grid cells (pixels). They intriguingly reflect excavations carried out using the spit methodology, where the explored area is divided up in a regular grid and the material is lifted using discrete units. This latter method, which is usually employed in sites "without features" (Drewett 1999:118, is generally considered a refuge (Roskams 2001:112), when the stratigraphic approach fails. Moreover it is considered simply as an alternative find recording system.

Category	Geographical definition	Archaeological metaphorical stage
Haptic	Sensi-motor and haptic perception is the most important early form of spatial information that reaches the mind, and in many ways are the most basic. Defined by touching and bodily interaction Solid-body motion is central to haptic space. With Newtonian physics this leads to Euclidean geometry. Euclidean geometry applies to haptic space	Space of excavation at a human scale (small-scale)
Pictorial	Based primarily on visual perception and indirect sensing. People talk about the visual scene in part using the language of touch and manipulation, which is one form of evidence that pictorial space is metaphorically grounded in haptic space	Space of site excavation at a superhuman scale (large-scale) Landscape archaeology
Transperceptual	Transperceptual It is composed or assembled in the mind from a number of independent haptic and pictorial spaces or objects experienced over time. Ability to walk through mental maps Spaces not perceived all at once	In archaeology associated with our mental space and models. In the field of computational archaeology it is the computer. Space where the metaphor becomes explicit.)

Table 1 A summary of Mark's classification of knowledge of space and, in parallel, the creation process of the archaeological excavation metaphor

As a matter of fact, archaeological excavation methods tend to underestimate the importance of deposits. Raster models have developed into voxel models, which are particularly suited for representing and analyzing volume data. We are therefore provided with a way of representing volumes that could be a challenge in excavation methodology. Obviously this is not a straightforward process and there are implications in this way of approaching excavation, the challenge this time being to learn something new about deposits. And "to learn a new field, according to the cognitive science approach, is to build appropriate cognitive structures and to learn to perform computations that will transform what is known into what is yet not known" (Posner 1989:xi).

Everyday life makes us experience surfaces. The volume within which we travel is empty and is usually the air. Although the stratigraphic method's aim is to record and study surfaces, under the assumptions that people lived on surfaces, the deposit in its volumetric status has been subjected to transformations through time. Excavation is a unique opportunity of experiencing volume in terms of materiality in the haptic space and transform it into the metaphor of the transperceptual space, in order to infer new information about the present and the past.

Research on GIS: Setting an Agenda

Archaeologists have been accused of using GIS "as little more than a mapping system" (Goodchild 1995:46) and geographers seem to have monopolized areas of research on GIS. The goals of 3D GIS should therefore be not only to explore excavation data with GIS as just another tool with which to collect and analyze data. The archaeologist must understand how GIS (in particular multidimensional GIS[1]) interacts with the way we do research and produce knowledge.

A preliminary set of question may be posed, such as:

- What are our ways of engaging with the space of excavation?
- Can multiple ways of knowing be integrated in a GIS? How?
- What types of knowledge and forms of reasoning are not well represented within GIS?
- What are the consequences of their exclusion?

Another priority in the agenda should be the study of the dynamics of metaphors. How do they interact with each other and how do they influence our way of producing knowledge? The engagement with the space of excavation has varied through time, from the era in which the archaeologist was in his study and the material was collected by fieldworkers under his[2] sporadic supervision, to contemporary practice which seeks interactivity and reflexiveness of participants (Lucas 2001a). Today, the use of computers allows for a completely different engagement. One of the priorities of meetings such as the CAA Conference should be to create discussion groups in order to address research on GIS tailored to archaeological research.

[1] For a thorough treatment of theoretical aspects of multidimensional GIS, see Raper 2000.
[2] and I leave his deliberately, as it is almost always male figures that represent the "Golden age" of archaeology.

REFERENCES

ANDREWS, G., BARRETT, J.C. and LEWIS, J.S.C., 2000. Interpretation not record: the practice of Archaeology. Antiquity 74:525-530.

BENDER, B., 1993. Cognitive archaeology and cultural materialism. Cambridge Archaeological Journal 3(2):257-260.

BISWELL, S., CROPPER, L., EVANS, J., GAFFNEY, V. and LEACH, P., 1995. GIS and excavation: a cautionary tale from Shepton Mallet, Somerset, England. In Stancic, Z. (ed.), Archaeology and Geographical Information Systems: a European perspective, Taylor and Francis, London:269-285.

CHADWICK, A., 1997. Archaeology at the edge of chaos: further towards reflexive excavation methodologies (http://www.shef.ac.uk/assem/3/3chad.htm).

COLLIS, J., 2001. Digging up the past. An introduction to archaeological excavation. Sutton Stroud, Thrupp-Publishing.

DREWETT, P.L., 1999. Field archaeology: an introduction. UCL Press, London.

FORTE, M., 2000. About virtual archaeology: disorders, cognitive interactions and virtuality. In Sanders, D.H. (ed.), Virtual reality in archaeology, Archaeopress, Oxford, BAR International Series 843:247-259.

GARDIN, J.-C., 1992. Semiotic trends in archaeology. In Peebles, C.S. (ed.), Representations in archaeology, Indiana University Press, Bloomington and Indianapolis:87-104.

GOLLEDGE, R.G., 1990. The conceptual and empirical basis of a general theory of spatial knowledge. In Papageorgiou, Y.Y. (ed.), Spatial choices and processes, Elsevier Science Publishers B.V., Amsterdam:147-168.

GOODCHILD, M.F., 1995. Geographic Information Systems and geographic research. In Pickels, J. (ed.), Ground truth, The Guilford Press, New York:31-50.

HARRIS, E.C., 1978. Principles of archaeological stratigraphy. Academic Press, London.

HASSAN, F.A., 1997. Beyond the surface: comments on Hodder's "reflexive excavation methodology". Antiquity 71:1020-1025.

HODDER, I., 1997. Always momentary, fluid and flexible: towards a reflexive excavation methodology. Antiquity 71:691-700.

HODDER, I., 1998. Whose rationality? A response to Fekri Hassan. Antiquity 72:213-217.

HODDER, I., 1999. The archaeological process - An introduction. Blackwell Publishers Ltd., Oxford.

HODDER, I. (ed.), 2000. Towards reflexive method in archaeology: the example of Çatalhöyük. McDonald Institute of Archaeological Research, Cambridge.

HUGGETT, J., 2000. Looking at intra-site GIS. In Birliba, V.M. (ed.), CAA96. Computer applications and quantitative methods in archaeology, Archaeopress, Oxford, BAR International Series 845:117-122.

JONES, A., 2002. Archaeological theory and scientific practice. Cambridge University Press, Cambridge.

LAKOFF, G., and JOHNSON, M., 1980. Metaphors we live by. The University of Chicago Press, Chicago/ London.

LUCAS, G., 2001a. Critical approaches to fieldwork - Contemporary and historical archaeological practice. Routledge, London.

LUCAS, G., 2001b. Destruction and the rhetoric of excavation. Norwegian Archaeological Review 34(1):35-46.

MARK, D.M., 1993. Human spatial cognition. In Hearnshaw, H.M. (ed.), Human factors in Geographic Information Systems, Belhaven Press, London:51-60.

NYERGERS, T.L., 1993. How do people use Geographical Information Systems? In Hearnshaw, H.M. (ed.), Human factors in Geographic Information Systems, Belhaven Press, London:35-50.

POSNER, M.I., 1989. Preface: learning cognitive science. In Poster, M.I. (ed.), Foundations of cognitive science, Massachusetts Institute of Technology, Boston:xi-xiv.

RENFREW, C., 1993. Cognitive archaeology: some thoughts on the archaeology of thought. Cambridge Archaeological Journal 3(2):248-250.

RENFREW, C., 1994. Towards a cognitive archaeology. In Zubrow, B.W. (ed.), The ancient mind. Elements of cognitive archaeology, Cambridge University Press, Cambridge:3-12.

ROSKAMS, S., 2001. Excavation. Cambridge University Press, Cambridge.

SIMON, H.A. and KAPLAN, C.A., 1989. Foundations of cognitive science. In Poster, M.I. (ed.), Foundations of cognitive science, Massachusetts Institute of Technology, Boston:1-47.

Reconstructing Sites and Archives: Information and Presentation Systems at Troy

Peter Jablonka
Eberhard-Karls-Universität Tübingen, Institut für Ur- und Frühgeschichte und Archäologie des Mittelalters, Troia-Projekt, Schloss, D-72070 Tübingen, Germany
peter.jablonka@uni-tuebingen.de

See the CD for the extended version

Abstract

As a result of more than a hundred years of archaeology at the legendary site of Troy and in the surrounding landscape (north-west Turkey) an immense amount of information has been excavated, recorded, and re-buried in a huge archive consisting of a variety of documents (notebooks, plans, drawings, photographs), computerised data (databases, CAD plans, digital images), and a library of printed publications. Any use of Information Technology (IT) in this context has to serve the practical needs of a large-scale, still ongoing, archaeological field project: documentation of excavations, archiving information, research and analysis, publication and presentation of results for audiences ranging from archaeologists and scientists to the wider public.

Presentation of results has been accomplished by means of 3-d reconstructions built into a Virtual reality (VR) presentation system. Research is now being supported by an information system integrating existing and new databases with Geographic Information System (GIS) applications. This is linked with other software where needed. Two examples briefly described in this paper are the analysis of stratigraphy combining GIS, databases, and a Harris matrix program, and GIS mapping of archaeological sites and palaeolandscapes with the help of high-resolution georeferenced satellite images. GIS is also used in the planning of a national park which will protect at least some of the archaeological sites from further destruction.

From the experience of the author, technology does not seem the limiting factor in archaeological computing. Two other issues are much more important. First, concepts and contents of presentation and information systems: the translation of archaeological methods and work processes into data structures and software procedures. Second, social constraints: the acceptance of IT by decision-makers and within organisations, the level of computing skills, and the availability of funding.

Presentation System

Faced with the complexity of the tasks briefly outlined in the abstract, it seems tempting to take a shortcut into a fanciful world of 3-d reconstructions and Virtual Reality (VR) to save oneself from despair, which I admit we did on a grand scale during the past two years (Project "Virtual Archaeology-TroiaVR" supported by the German Federal Ministry of Education and Research; consortium under the direction of ART+COM AG, Berlin; detailed descriptions in Jablonka et al. 2003 and at http://www.uni-tuebingen.de/troia/vr/). For use at museums and exhibitions reconstructions of three main phases of Troy and the surrounding landscapes were put into a VR presentation system that can be used to give visitors real-time tours of Troy at three different points in time. Reconstructions are accompanied by context information (plans, texts, images of finds and excavations to show the difference between reconstructions and reality, time line). The system was part of an exhibition on Troy in Germany (Bundeskunsthalle, Bonn).

As archaeologists we need to communicate the results of our work to the wider public. After all, most of us are being paid from taxpayers' money, and it will certainly be appreciated if we can offer an interesting and beautiful experience especially appealing to younger people in return. Virtual reality systems can communicate and present the results of archaeological work. At least archaeologists working for museums or exhibitions should therefore make themselves familiar with content and technology of presentation systems.

A by-product of our work on a presentation system was a still growing library of 3-d reconstruction models which can and have been used to generate images, animations, multimedia content and TV footage (Figure 1 shows an example). Assembling 3-d reconstructions also helped archaeologists to understand excavation results. Students working on the project could acquire skills which might improve their chances on a tight job market.

Towards an Archaeological Information System

Even the non-specialist will ask at some point on what actual information our 3-d reconstruction models are based. Reconstructions are visual interpretations of excavation results. Obviously, these results have to be brought together in some kind of information system. Such a system can be defined as a non-redundant collection of all information pertaining to a certain subject, with procedures to search, sort, filter, analyse, combine, and retrieve this information.

Figure 1 3D reconstruction example for VR presentation (interior of Early Bronze Age building)

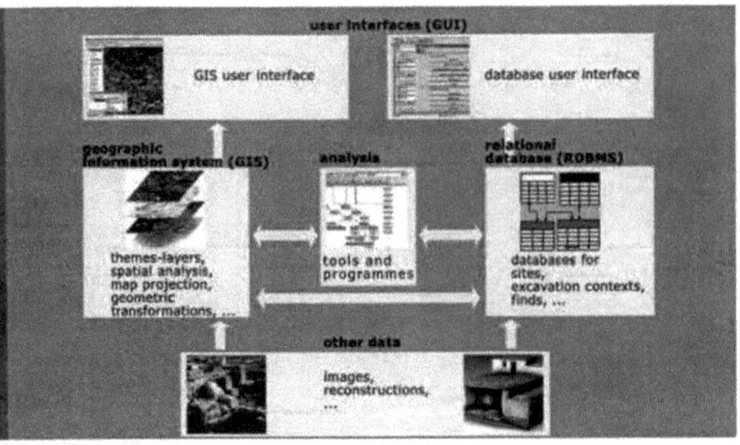

Figure 2 Information system for Troy, system components

On the computing level, the architecture of an archaeological information system seems fairly straightforward. Looking at archaeology, we see maps or plans, and catalogues or lists, with illustrations. We note that almost everything from sites to finds has a spatial aspect. In fact, archaeology mainly deals with matter distributed in space and time by past human activities.

We can thus design a system with a spatial and a conceptual point of entry, or GIS map and database form user interfaces (Fig.2). The data are stored in a GIS and a relational database (RDBMS). To this tools for analysis (finds statistics, stratigraphy) are added. Non-spatial and non-(alpha)numeric data (images, reconstructions) can be stored separately and hyperlinked to the GIS or database. In this way, all kinds of data can be combined to produce different kinds of output (for printed or digital reports, maps and plans, visualisations).

The system can be implemented using standard software, in our case ArcGIS 8 and MS Access, combined with existing archaeology software (e.g. WINBASP for finds statistics and stratigraphy). Computing skills needed are at the level of simple application programming (SQL, Visual Basic for Applications) to customise applications and transfer data between different software and file formats. The use of standard software should make it likely to find archaeologists who have the technical skills to manage and develop the system, or are willing and able to learn them. Several examples of similar systems have been described in the literature. Lock (2003), Wheatley and Gillings (2002), and of course numerous case studies published yearly in the proceedings of the CAA conference are good starting points. Burrough and McDonnell (1998) is an excellent source on GIS.

Modelling Archaeological Information

Publications on archaeological computing tend to focus on technical aspects. The conceptual work before or behind a given technical solution is rarely made explicit. Many systems which may be termed "archaeological information systems" have been and are being developed mainly for practical use in rescue and contract archaeology (examples in Lock 2003:78-123, CAA conference proceedings; an excellent German example can be found at http://www.arc-tron.de/).

From the point of view of contents such systems should at least cover what has been termed "the research archive" (English Heritage 1991, Phase 4), with added capabilities for the acquisition and analysis of data. Since existing examples are carefully studied by those who design new systems we can observe an evolution of best, or at least, common, practice. The key step in the development of such systems is the translation of archaeological concepts into data models. To bridge the gap between excavations, finds, pits and postholes on one hand, data and software on the other, seems bold enough a step to deserve more than the few words it is usually given.

As a first stage in this modelling process, we can identify the following semantic categories in archaeology:

Representations of real-world objects: landscapes, sites, excavation contexts (layers, pits, ruins, ...), and finds.

Conceptual abstractions: time, chronology, stratigraphy, classifications and typology.

Products of work processes: excavation recording systems, catalogues, reports, plans, archives, bibliographies.

The properties of each of these objects can then be described in terms of different data representing it (measurements, textual descriptions, coded values, images, geometry, ...), the connections (interfaces) it has with other objects, and methods used on it (order, query, write report, calculate statistics, draw map, ...). For example, landscapes - sites - excavation contexts - finds are obviously connected in a hierarchy. Different classifications (or typologies, as archaeologists may prefer to call them) can be a property of finds or excavation contexts. Chronology will consist of calendar dates, procedures to calibrate C-14 measurements, and lists of periods or phases. These descriptions can be formalised using tables to note the properties of each object, and diagrams showing connections between objects. The process resembles concepts used in object oriented programming (classes, interfaces, class hierarchies).

The resulting data model should consist of few large objects that could be implemented as separate (but connected) database or GIS applications, thus reducing the complexity of the system by a "divide and conquer" strategy. In the case of Troy, these main objects are: landscapes, sites, excavation contexts, finds, chronology, bibliography, archive, administration. Classifications (typologies) are treated as properties of other objects (e.g. sites can be burials, settlements, ...; finds can be pots of type C21, and so on). As work continues and the system grows, different or new parts and combinations of these objects or modules can be developed and used.

Details of different excavation recording systems, as well as particular technical implementations, can be separated from the data model. Starting with real-world objects or archaeologically meaningful concepts should ensure that we arrive at a meaningful representations of our data and the tasks we wish to accomplish. Ideally, analysis of similar problems should lead to similar solutions, and information systems designed for different projects should in the long run converge into common standards. It seems worthwhile for archaeologists to devote time and effort on this conceptual side of computing rather than hunt for the latest technology, or concentrate on technical details. A comprehensive description of an archaeological information system is beyond the scope of this paper. I should therefore like to give two short examples putting the above considerations into practice.

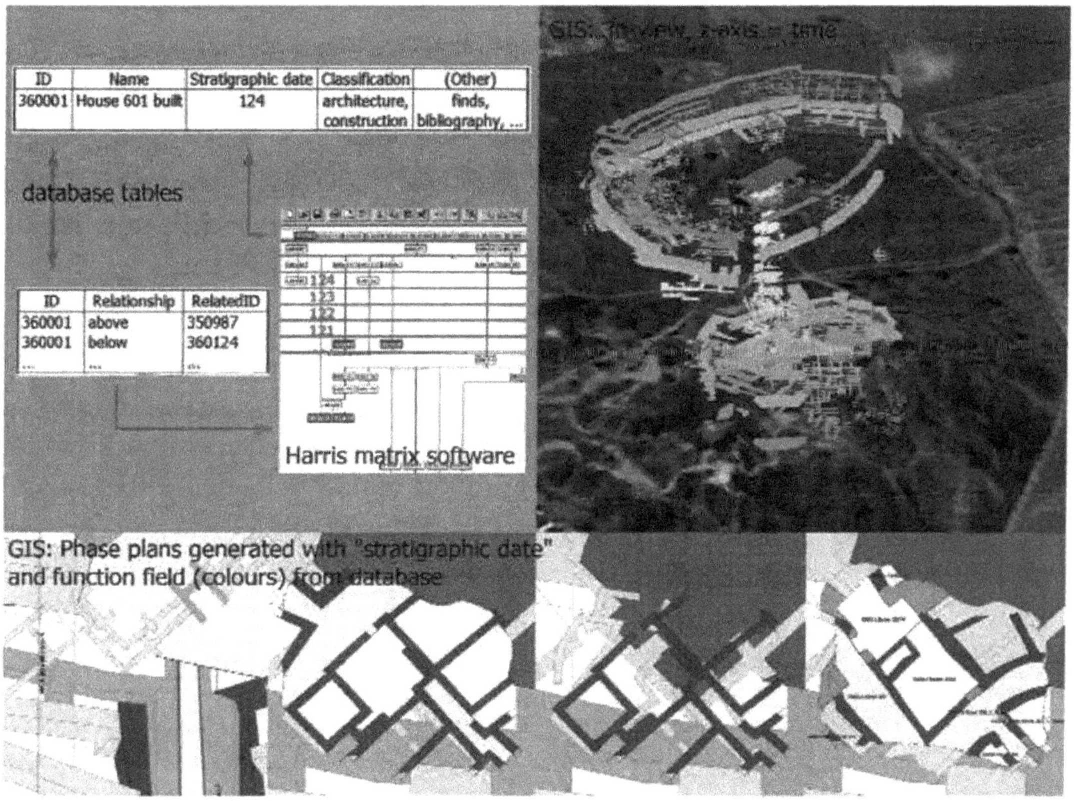

Figure 3 Stratigraphy example: Database tables and stratigraphy program, pseudo 3D display of layers (old=bottom to young=top), automatically generated GIS-phase plans (colour: functional classification from database)

EXAMPLE 1: STRATIGRAPHY

Troy consists of up to 15 meters of settlement deposits. Thousands of structures (layers, pits, surfaces, buildings and so on) can be identified and must be brought into a sequence based on the relationship of neighbouring structures. In other words, a stratigraphy has to be derived from observations recorded during excavation or from the study of published or unpublished information (plans, section drawings, textual descriptions). Since the theory of archaeological stratigraphy is well understood (Harris 1989) and has been implemented in several computer programmes (e.g. Gnet, HARRIS, ArchEd; now also Stratify, see Herzog, this volume), stratigraphy can easily be modelled (Fig.3). In fact, the method applied here has already been described by Alvey (1993).

Two database tables hold data on excavation contexts and their stratigraphic relationships. Contexts are represented as plan or section drawings in a GIS linked with the database table via a unique identifier. A procedure (written in Visual Basic in the MS-Access database environment and run by clicking a button on a database form) then performs the following steps to calculate the stratigraphy: 1.) export data on context and their stratigraphic relations into the format needed in step 2; 2.) run a Harris-Matrix-programme (Stratify); 3.) get the vertical position (line number) of each context from the graph ("Harris Matrix") generated in step 2; 4.) for each context, write this number into a field called "stratigraphic date". With this number an ordered sequence of all contexts can be produced, finds linked to excavation contexts can be assigned a date, and the GIS can draw phase plans, sections with layers numbered according to their position in the sequence, or pseudo 3d-representations of time mapping contexts from low=early to high=late. By visual inspection of phase plans or phased section drawings errors or gaps in the stratigraphy are easily detected. The stratigraphy will be updated, new phase plans, or catalogue descriptions of excavation contexts will be produced automatically as new contexts or stratigraphic relations are entered (note that the stratigraphic position of all contexts, and all finds associated with them, will change whenever new information is entered or existing data are corrected or refined). At a very complicated site like Troy with an excavation history going back to the days of Heinrich Schliemann (1822-1890), one consistent, dynamically growing stratigraphic sequence can thus be constructed and maintained as work continues.

A disadvantage of using GIS to work with stratigraphy is their poorly developed capacity to handle three-dimensional data. Most existing excavation data are also two-dimensional (plans and section drawings) - archaeologists are therefore used to work within the "2.5-dimensional" space typically offered by GIS systems.

EXAMPLE 2: ARCHAEOLOGICAL SITES IN THE TROAD

During the ongoing fieldwork at Troy the landscape surrounding the site (the Troad), has been extensively studied. To model a whole landscape turned out to be easier than expected. All that was needed was a rather straightforward GIS application connected to a sites database and auxiliary databases that are also used by other parts of the information system (bibliography, lists of and links to archive material - mostly scanned images, chronology).

Space, or a landscape, is conveniently represented as a set of thematic layers in a GIS system that will also provide procedures needed to manipulate these data (map projections, transformations, georeferencing, spatial analysis). Currently this system contains the following themes: surveying points, GPS locations of archaeological sites, scanned topographic maps, scanned historic maps, digital elevation models, vector data base maps, IKONOS and Landsat satellite data, geological maps, locations of palaeogeographic sediment cores, reconstructions of palaeogeography for different points in time, and detailed representations of archaeological sites (see below).

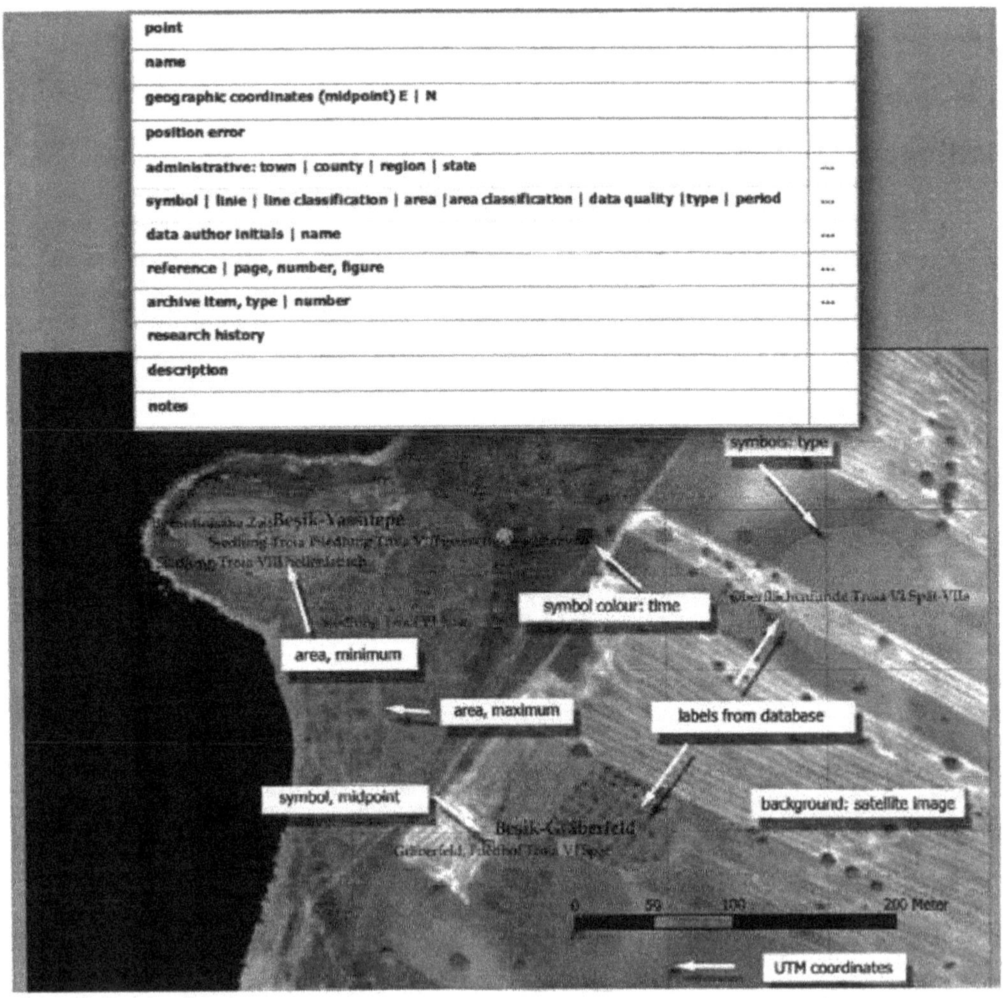

Figure 4 Data model for archaeological site as table; GIS representation of site data (IKONOS satellite image courtesy Compton J. Tucker, NASA and Space Imaging, Inc.)

Time can be represented in GIS or database systems in several ways. Any spatial or other data can carry time information as an attribute field (periods or phases, calendar years, time spans as years from-to). Periods, phases, or whole systems of relative chronology, can be calibrated against each other simply by adding "from" and "to" fields (in years). This information is localised in a chronology database shared by all parts of the information system (e.g. finds and excavation contexts databases). Palaeolandscape reconstructions and historic maps can be displayed as "time slices" by the GIS, if desired, with other data carrying time attributes (e.g. a selection of archaeological sites).

Information on archaeological sites is kept in a database: location of site midpoint (from GIS mapping) and accuracy, modern administrative units the site falls into, research history, description, notes, contributor(s) of data, reference(s) (link to bibliography database), archive material (plans, photos: link to database of archive materials and images), type (e.g. settlement, burial, ...) and period (e.g. Bronze Age, Roman, ...). Since sites can consist of remains of different types belonging to any number of different periods, an unlimited number of such type-period pairs can be associated with one site. By linking this database to a GIS, sites can be mapped in several ways: As site midpoint (with site name); as a row of symbols next to the site midpoint (different symbols for types, different colours for periods present); as multipoint (scattered remains), line (e.g. roads), or area features (different areas for certain or reconstructed site extents) for each type and period with known extent present at a site. In the field, sites can be mapped on printouts - fairly easy with 1m-resolution satellite data showing every tree. GPS readings can be taken in the field and plotted. It is easy to re-visit sites even in difficult terrain by loading their coordinates into a GPS and using a "go to" function to find them.

Figure 4 shows a data model of an archaeological site as table and GIS representation of sites with a georeferenced satellite image as background map.

Conclusion

A century of excavation has reduced the settlement mound of Troy to a heap of rubble resembling a stone quarry more than anything else. Research results have been re-buried in a huge archive that has become as difficult to access as excavation at the actual, multi-layered and badly preserved site can sometimes be. From our own, still ongoing, fieldwork, we have several thousands of hand-drawn plans, hundreds of notebooks, a collection of more than 40,000 photographs, and data that may be anything from texts, a maze of databases, scanned images, CAD-plans, to satellite data - not to speak of three sets of information from earlier excavations under the direction of Heinrich Schliemann, Wilhelm Dörpfeld, and Carl W. Blegen. Traditional, comprehensive post-excavation analysis and publication of research results at the site and in the surrounding landscape (including descriptions, plans and stratigraphy of all excavation contexts and complete finds catalogues and statistics) will be very difficult, if not impossible due to the size and complexity of the task.

Information on landscape, archaeological sites, excavation contexts, and finds, is therefore being integrated in an information system. This is possible using customised standard software (GIS and databases). Structure and content of such a system deserve more consideration than the technology used. It is important to translate concepts and methods of archaeology into formal data models and procedures to create systems that serve the needs of archaeologists, and will be understood, accepted and used by them. Such decisions about the ways archaeology is represented in computer systems - often tacitly and implicitly made by specialists - will certainly shape the future of the discipline. Once an information system has been set up, it should be possible to publish its contents either on CD/DVD or online (for examples see Clarke et al. 2003, the Electronic Cultural Atlas Initiative, this volume and http://www.ecai.org/).

Apart from being an object of academic research, Troy, and archaeology in general, is very popular. To serve the wider public, a VR presentation system showing reconstructions of the site with accompanying context information has been created. Ironically, the spectacular results obtained within a short time in our VR reconstruction project helped to create an awareness of the possibilities of archaeological computing even with colleagues who had sceptical views both on the use of computers in archaeology and 3-d reconstructions. The project also made it possible to obtain resources needed to make progress with the information system. Cultural, sociological, and economic factors - the acceptance of IT within organisations, the level of computing skills, the importance ascribed to computing in teaching and research, and the availability of funding for the acquisition of equipment and the recruiting of staff - will be at least as important for the success of IT projects as the technologies used (Lock 2003:263-268).

References

ALVEY, B.A.P., 1993. Interpreting archaeology with Hindsight. The use of three dimensions in graphic recording and site analysis. In Harris, E.C., Brown, M.R. and Brown, G.J., Practices of Archaeological Stratigraphy, Academic Press, London/San Diego:218-228.

BURROUGH, P.A. and MCDONNELL, R.A., 1998. Principles of Geographic Information Systems. Oxford University Press, Oxford.

CLARKE, A., FULDORD, M. and RAINS, M., 2003. Nothing to Hide - Online Database Publication and the Silchester Town Life Project. In Doerr, M. and Sarris, A., CAA 2002. The Digital Heritage in Archaeology. Computer Applications and Quantitative Methods in Archaeology. Proceedings of the 30th Conference, Heraklion, Crete, April 2002, Hellenic Ministry of Culture, Athens:401-404.

ENGLISH HERITAGE, 1991. The Management of Archaeological Projects. HBMC, London (2nd ed.).

HARRIS, E.C., 1989. Principles of Archaeological Stratigraphy. Academic Press, London/San Diego (2nd ed.).

JABLONKA, P., KIRCHNER, S. and SERANGELI, J., 2003. TroiaVR: a virtual reality model of Troy and the Troad. In Doerr, M. and Sarris, A., CAA 2002. The Digital Heritage in Archaeology. Computer Applications and Quantitative Methods in Archaeology. Proceedings of the 30th Conference, Heraklion, Crete, April 2002, Hellenic Ministry of Culture, Athens:13-18.

LOCK, G., 2003. Using Computers in Archaeology - Towards virtual pasts. Routledge, London/New York.

WHEATLEY, D. and GILLINGS, M., 2002. Spatial Technology and Archaeology - The archaeological applications of GIS. Taylor & Francis, London/New York.

Understanding the Protohistorical Territorial Heritage by Means of Iron Age Settlement System Analysis in GIS: A Case Study in the Eastern Languedoc (France)

SEE THE CD FOR THE EXTENDED VERSION

LAURE NUNINGER
POST-DOCTORATE
LABORATORY OF CHRONO-ECOLOGY (NATIONAL SCIENTIFIC RESEARCH CENTRE, UNIVERSITY OF FRANCHE-COMTÉ, BESANÇON, FRANCE)
CENTRE FOR SPATIAL STUDIES (SCIENTIFIC RESEARCH CENTRE OF THE SLOVENIAN ACADEMY OF SCIENCES AND ARTS, LJUBLJANA, SLOVENIA)

ABSTRACT

We often tackle the territorial problem only with a regressive analysis. However this reasoning tends to enhance the status of settlements still seen nowadays as villages or towns. Otherwise, we think that protohistoric agglomerations were much less significant if they were deserted during gallo-roman period. That's why our reasoning is influenced by our visible cultural heritage. Yet a spatial approach about the dynamic of settlement systems during the Iron Age underlined a non-linear development with relative "recession" and "expansion" periods for each agglomeration. Starting with the fifth century BC, these compete in establishing their (extractive) settlement networks, and in developing their land use. We can observe in a relatively detailed way the intensification of this competition in the form of symbolic expressions of the networks' growth through the visibility control, the increasingly massive fortifications and the increasingly high towers. Extern stimuli, as the Mediterranean exchanges or the roman colonisation for example, seem to accentuate this competition and contribute to develop some agglomerations or to pressurize others. We propose an analysis model which cross both reasonings, regressive and spatial that would give a better understanding of the evolution of territorial patterns during Iron Age and what it brought in our territorial heritage.

Introduction

In classical archaeology, the study of proto-historical territories is often approached by means of regressive analysis. This type of analysis is based on the assumption that "the memory of societies takes care of the transmission of the inherited data, making it possible to go back as far as the first settlements" (Leveau 2002:9). This principle can be questioned as it tends to emphasize the status of settlements still seen nowadays as villages or towns, and deserted agglomerations are considered to be less significant. This type of analysis is highly influenced by the visible cultural heritage. However, even though the Greek and Latin sources written during and after the Roman conquest can tell us something about the situation of the pre-Roman peoples, it is also true that they can deform reality or erase some aspects of pre-Roman territorial dynamics. For this reason, we must apply a constructive and systematic approach to studying these territories by using archaeological evidence that can shed new light on the processes of territorial evolution. By combining data sources, I may be shown how the protohistorical territorial pattern has contributed to our own territorial heritage. After a presentation of the geographical and historical context, the spatial models and the methodology applied, an applied example is given from the Nîmes region in the south of France. Using this case study as an example, this paper will focus on specific aspects of the results to emphasize how a different methodological approach can influence our overall perception of the protohistorical territorial system.

The Volcae Arecomici Area and its "Capital": Nîmes

The data set used in this case study is from the Languedoc area in the south of France, more precisely between the two towns of Nîmes and Montpellier (see Fig.1). According to ancient sources (Christol and Goudineau 1987-1988), it was occupied by the pre-Roman people of the Volcae Arecomici, whose capital was located in Nîmes. The territorial system controlled by Nîmes, however, is an important topic of discussion. It seems that during the protohistorical period (6th c. BC - 1st c. BC), the status of Nîmes was not yet clearly outlined. The assumption of the central role of Nîmes rests on the interpretation of texts handed down from the second Iron Age[1], as well as archaeological evidence which has proven to be rather weak. Furthermore, these data seem to indicate that Nîmes did not have the same status throughout the protohistorical period; rather, it was the object of progressive construction. Under these conditions, it is necessary to situate the evolution of this community within the more general process of a regional settlement pattern, in order to verify if Nîmes actually played a dominant role and controlled a broad territory from its beginnings. If not, it is a matter of identifying the factors of its success in relation to other agglomerations like hillforts in the region. Unfortunately, there is no exhaustive archaeological information available on the entire Volcae Arecomici area. It is accepted that it could have covered the Eastern zone of the Hérault department and the whole of the Gard department with some extensions into Lozère and Ardèche (Fiches 2002:119). But for practical reasons I have decided to study only the area around Nîmes which covers nearly 900 km2. The settlement analysis carried out is based on the hypothesis that each settlement should not be studied in isolation in order to define its territorial influence. Instead, every settlement interacts with its neighbours, forming part of a dynamic system controlled by a centre. We can only understand the geographical position of each settlement, its hierarchical status and its capacity to last, that is to say, to actively intervene in the territorial development, when it is studied together with the other settlements surrounding it. Taking this into account, the analysis aims at emphasizing the links, which the various habitats could maintain between them and with the space exploited. In this way we can try to rebuild the local settlement networks, which correspond to the territorial influence of a rural community. On a different scale, it is a question of finding clues for the connections between the various centres of these rural communities. It is believed that in this way we can better understand the territorial organisation of the Volcae Arecomici people as well as the real role of Nîmes in it.

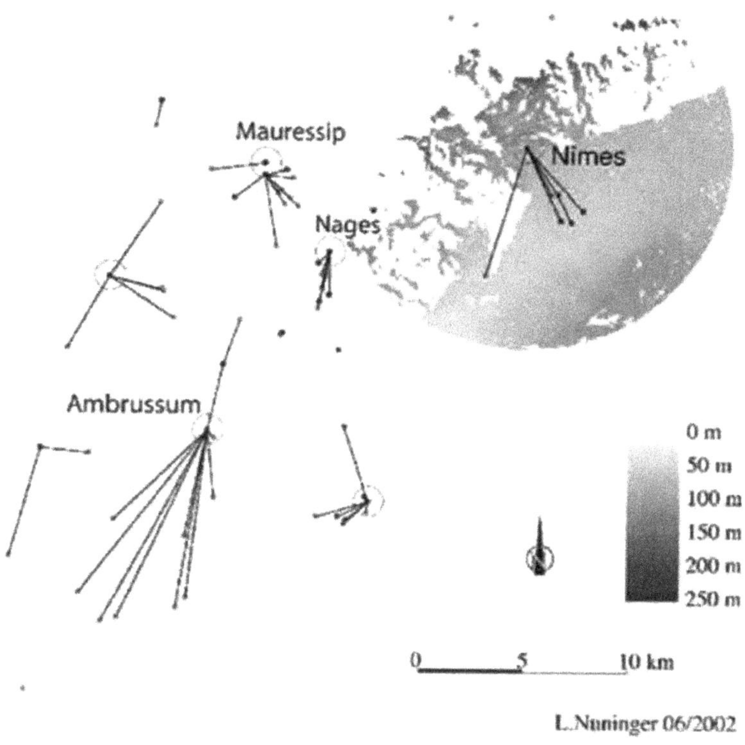

Figure 1 Visibility and calibration of the networks model: the example of Nîmes during the 2nd C. BC

Analysis

We present briefly the global methodology which is more detailed on the CD. First I tried to devise a classification system for the archaeological settlements in order to rank them in a hierarchical typology. With this intention, each settlement was described in a systematic and homogeneous way, using a set of variables that was elaborated and tested within the framework of the Archaeomedes Programme (van der Leeuw 1995, Durand-Dastès et al. 1998, Favory et al. 1999). The most relevant variables-especially surface area, the nature of building materials, and the symbolic status-were processed by means of factorial correspondence analysis, followed by an ascending hierarchical classification (Sanders 1989, Nuninger 2002a:100-129). These statistical analyses made it possible to distinguish seven classes of settlements: starting from the most modest ones, such as temporary constructions, shelters or farm annexes, toward the most significant settlements, such as the agglomerations which constitute the major centres of settlement (Nuninger 2002a:117-125). The projection of the settlements on the first axis of the factorial analysis, which is strongly structuring from the point of view of hierarchisation, enables me to rank all the settlements. In this way I can locate and evaluate the hierarchical weight H(i) of each one on a relative scale.

In order to define the links between the various settlements, I applied a gravity model already experimented with in the formalization of the networks of Gallo-Roman settlements (Durand-Dastès et al. 1998:203-217). This model allows us to attach each settlement to a central place according to distance and offers the possibility to construct theoretical settlement networks, per century (Pumain and St Julien 2001:184).

We integrated to this basic model a cost surface of movement by foot as a function of the topography. This weighted distance between each settlement and its contemporaries was calculated with GIS. The gravity model was then calibrated: on the one hand, according to a critical distance at 10 km (because on average this corresponds to the distance of the nearest neighbour between centres); on the other hand according to the visual from the centres, with the assumption that each settlement connected to a centre must be able to be controlled by that centre (Nuninger 2002a:137-149) (see Fig.1).

In order to generate "territorial maps" from the information generated by the settlement network analysis I used the 'off-site' finds. These are scattered sherds that are generally rather fragmented with blunted edges due to repeated agricultural activities. These scattered sherds are interpreted to be the result of spreading by domestic manuring. Their spatial distribution can therefore allow us to evaluate the cultivated area by protohistorical communities. By reason of the imprecise chronology, I developed a chronological model based on the outcome of all the settlement excavations in the studied area. Then, I developed a spatial modelling integrating distance from settlement using ethnological references. The entities selected and defined according to the model can constitute the theoretical community lands[2] or catchment area (the agricultural territory exploited by a rural community) of the communities whose evolution we can follow by century (see Fig.2).

Lastly, in an attempt to understand the basis of occupation of each community, I worked out a final variable, which corresponds to the visual competition between contemporary centres. This layer is simply obtained by a boolean overlay of zones with visual control over each centre.

The confrontation between these various archaeological or derived findings within a dynamic analysis of the settlement in the region then enabled me to shed new light on the problem of protohistorical territories, supplementing the classical regressive reasoning.

Towards New Observations of Protohistorical Territorial Dynamics in the Area of Nîmes

To illustrate this, I present an example from the Nîmes area, a case study centred around three hillforts situated in the

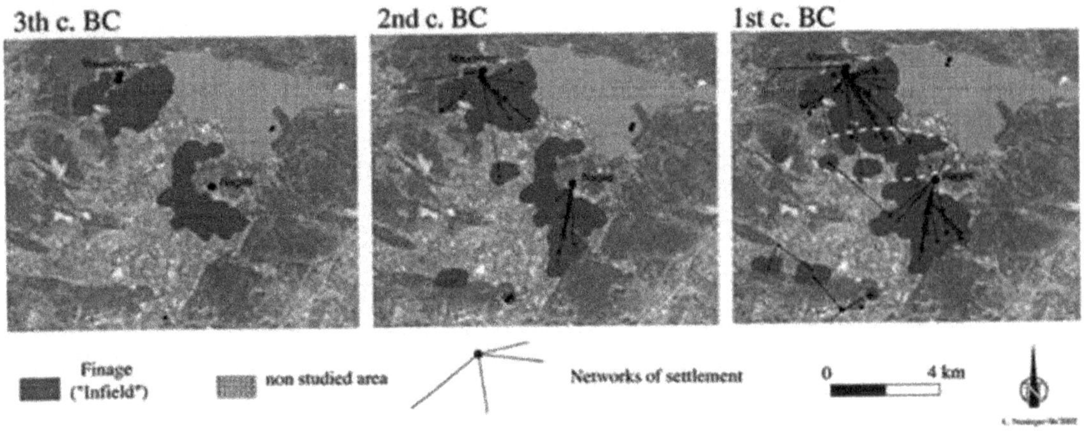

Figure 2 Networks of settlement and theoretical community lands or catchment areas in the Vaunage region (Gard department, France)

micro-region of the Vaunage: Mauressip, Roques de Viou and Nages. The Volcae Arecomici occupied this area during the protohistorical period, and the goal of the analysis was to understand how the territory of this people was organized. Besides, we were interested in knowing when the hillfort of Nîmes started to play the role of capital, or at least, of central place.

The first settlements of the highest hierarchical level, Nîmes and Mauressip, were created in the region at the end of the 6th or the beginning of the 5th c. BC. Both agglomerations seem to show that their influence is similar, at least until the beginning of the 2nd c. BC. In this period, no centre seems to have clearly dominated the others and we could envisage a mosaic of communities whose emergent centres have full control of their territory. Even if we consider a system of alliances between these communities in the form of a confederation, it is highly probable that this has been sufficiently flexible to ensure a relationship that, if not conflictual, was at least competitive. This impression is reinforced by the evolution of ostentatious and prestigious constructions. They follow one another and continue to increase in qualitative and quantitative terms between the fourth and the 1st c. BC, with the construction of more massive fortifications and higher and higher towers on each regional hillfort (Nuninger 2002b). Moreover, a spatial anomaly in the generally regular settlement pattern, found around the hillfort Mauressip, attracts the attention and raises a number of questions about this competitive spatial control.

Until the beginning of the 4th c. BC, the hillfort of Mauressip was undoubtedly the only nucleated settlement within the Vaunage micro-region. Its strategic position, on a butte, enabled it to visually control all of the inlier.[3] In addition, it controls all the entrances to the inlier to the south, the northwest and to the northeast towards Nîmes. However, from 375 BC, the situation changes due to a new habitat, Roque de Viou that takes residence on a hill which faces the Mauressip mound. The ostentatious/prestigious character of the surrounding elevated wall facing Mauressip[4], and the vast surface area enclosing Roque de Viou, makes it implausible that this settlement was established without a conflict with the Mauressip community or without its tacit agreement obtained by external support of the newly installed community. Thus, the development of this site could signify a territorial dispute between these communities of the Vaunage and may be indirectly linked with Nîmes. At the end of the 4th c. BC, the community of Roque de Viou moved a few hundred meters to the neighbouring hill in Nages without any archaeological explanation: the site was not destroyed, and it happened rather rapidly. Even though that in itself does not constitute an explanation, it should be stressed that the settlement was relocated from a place within a zone of visual control of the hillfort of Mauressip to a protected site. The hill which dominates the present village of Nages, is oriented toward the south, at the foot of the spur of Roque de Viou, and is not visible from Mauressip. When we cross the various zones of visibility from each of these hillforts, we can see that the zone of visibility of Roque de Viou makes contact with almost all the space controlled by Mauressip (see Fig.3), whereas the visual influence of Nages covers only on half of the zone seen from Mauressip (see Fig.3).

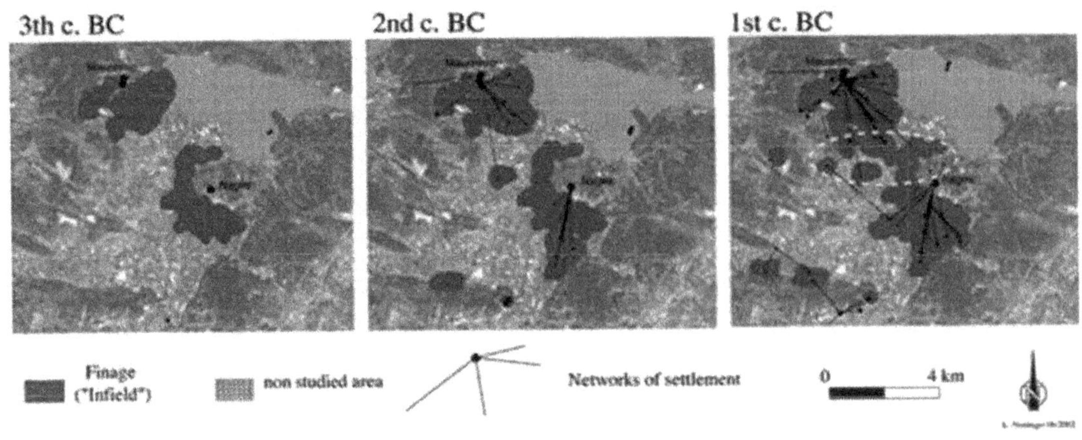

Figure 3 Visibility from the hillforts of Mauressip, Roque de Viou and Nages

The chronological succession of these settlements then enables us to suggest the following hypothesis: the site development of the Roque de Viou community, during the 4th c. BC, could have become a threat, a rival in the face of Mauressip, whoscreation in the fifth century BC would ensure a certainly older and better anchored control of the land. In addition, from the end of the 5th and 4th c. BC, we can observe the distribution of agricultural spread indices. This points to the extension of the cultivated space and a decrease of the amount of new land available for agricultural activities. This phenomenon could have led to a relatively equal division of the zones of influence of the two communities of Mauressip and Roque de Viou/Nages, especially from the 3rd c. BC onwards.

The absence of violent destruction in the agglomeration of Roque de Viou could indicate the existence of a relatively peaceful agreement with the community of Mauressip, but highly unfavourable for Roque de Viou. Because of this, the community of Roque de Viou then sees itself obliged to invest in the construction of a new agglomeration and probably in the development of new land for the construction of their territory. This interpretation of the event would mean that Mauressip is still able to oppose a strong force of other communities to preserve its territorial base. Even so, it appears to be weakened by the division of the zones of influences in Vaunage. Consequently, if we accept this hypothesis, the example of Mauressip - Roque de Viou/Nages, implies the existence of politico-economic relations between the communities, which one can attribute to the tribes. A question remains, however, concerning this type of territorial conflict: could it be settled between the two communities? Did it presuppose the intervention of an external body, in the form of an "inter-tribal council" or of a "dominant leadership"?

The state of the archaeological knowledge unfortunately does not permit us to resolve this question, but it does point to a rather limited regional political organization. Under these conditions, we can ask ourselves which agglomerations in this area could profit from such a situation. Of course, we think of Nîmes, which in the same period keeps all the necessary space to develop itself. Then, it's interesting to follow the extension of each centre in the settlement networks and cultivated areas. In the beginning of the 2nd and especially in the 1st c. BC, competition between the centres seems to have reached its climax, and while certain communities mutually contribute to their own suffocation, others are able to continue their development. From the 2nd c. BC, one can observe the creation of small sites, classified as farm annexes or small habitats. They don't last for very long, but they are found within the settlement networks of the centres present in the 3rd c. BC. This diffusion of small settlements can be interpreted as the expansion of the centres. All in all, each centre appears to be able to control the space needed for their expansion. Only those of Mauressip and Nages have seen their progression rapidly limited by one another, all the more so because they are the most "productive" centres as far as the small sites are concerned. Moreover, for these two communities, it is necessary to emphasize the extent of their cultivated area which ends up meeting some time in the 1st c. BC, pointing out new signs of possible conflicts (see Fig.2).

Conclusion

With the help of joining different developed models and by analysing the visibility area from each centre, we can observe some interesting anomalies which definitely reveal conflict areas. Moreover, it is possible to analyse their evolution, suggesting that the territorial protohistorical history is not a continual process of development until the Roman conquest. On the contrary, it seems to be a more complex and dynamic system. Finally, my conclusion is that the protohistorical territorial system is too difficult to be understood solely by way of regressive analysis, because this system probably did not follow a strict hierarchical pattern with territories fitted together. On the contrary, we can think about a territorial dynamics pattern, based on a settlement network and different types of alliances between the centres. According to this model, the status of leader of Nîmes seems rather late and it is probably linked with the Mauressip's decline.

Thus, applying a reasoning that also includes the approaches presented in this paper, we are far from the continuity principle, which is often used, in regressive analysis, because of a lack of archaeological evidence. In this way, spatial archaeology and modelling can completement our perception of territorial dynamics.

Acknowledgements

Many thanks to Philip Verhagen for rereading and correcting the paper. I would like to thank the eastern Languedoc team and the participants in the Archaeomedes programme for the development of the approach which constitutes the fundamental basis of this study. I wish to thank the French Ministry of Research and Education, the laboratory of Chrono-Ecology and the MTI@SHS centre (University of Franche-Comté, France), the Regional Archaeological Service of Languedoc-Roussillon (France) and the Archaeomedes Project (EU) for their funding of this work. Finally, I would like to thank the French Ministry of Foreign Affairs for their funding of my post-doctoral training within the framework of the Lavoisier Programme.

1 Notably those by Caesar, Strabo and the Pliny the Elder, between the middle of the 1st c. BC and the middle of the 1st c. AD.

2 According to the definition from Leonard R. and Longbottom J. 2000, see "Terroir":55.

3 "An area or group of rocks surrounded by rocks of younger age", according to Le Grand Dictionnaire Terminologique: http://w3.granddictionnaire.com/btml/fra/r_motclef/index102_1.asp.

4 Since the creation of the Roque de Viou site.

REFERENCES

CHRISTOL, M. and GOUDINEAU, C., 1987-1988. Nîmes et les Volques Arécomiques au Ier s. av. J.-C. Gallia 45:87-103.

DURAND-DASTÈS, F. et al., 1998. Des oppida aux métropoles. Coll. Villes, Anthropos, Paris.

FAVORY, F. et al., 1999. Archaeomedes II: une étude de la dynamique de l'habitat rural en France méridionale, dans la longue durée (800 av. J.-C. - 1600 ap. J.-C.). AGER 9:15-35.

FICHES, J.-L., 2002. Volques Arécomiques et cité de Nîmes: évolution des idées, évolution des territoires. Territoires celtiques: Espaces ethniques et territoires des agglomérations protohistoriques d'Europe occidentale, Proceedings of the XXIVth International Conference, Martigues, France, 7-11 June 2000, Errance, Paris:119-128.

FFRP, 1994. Des Vosges au Jura, topo-guide des sentiers de Grande Randonnée (GR59). Topo-Guides, French federation of hiking, Paris, 1994.

LEEUW, S. van der (ed.), 1995. Volume 3: dégradation et impact humain dans la moyenne et basse vallée du Rhône dans l'Antiquité, partie II: la mobilité de l'habitat rural dans la moyenne et basse vallée du Rhône dans l'Antiquité. In The Archaeomedes project: understanding the natural and anthropogenic causes of soil degradation in the mediterranean basin, European Union, inedited.

LEONARD, R. and LONGBOTTOM, J., 2000. Land Tenure Lexicon: a glossary of terms from English and French speaking West Africa. International Institute for Environment and Development, London (http://www.iied.org/docs/drylands/lexicon.pdf).

LEVEAU, P., 2002. Les territoires. Un bilan des méthodes d'étude. Territoires celtiques: Espaces ethniques et territoires des agglomérations protohistoriques d'Europe occidentale, Proceedings of the XXIVth International Conference, Martigues, France, 7-11 June 2000. Errance, Paris, 2002:9-17.

NUNINGER, L., 2002a. Peuplement et territoires protohistoriques du VIIIe au Ier s. av. J.-C. en Languedoc oriental (Gard-Hérault). Ph.D. thesis, defended in December 2002, University of Franche-Comté, France (http://tel.ccsd.cnrs.fr/documents/archives0/00/00/29/81/index_fr.html).

NUNINGER, L., 2002b. Pôles de peuplement et organisation territoriale au second âge du fer en Vaunage, Vidourlenque et Vistrenque (Gard). Territoires celtiques: Espaces ethniques et territoires des agglomérations protohistoriques d'Europe occidentale, Proceedings of the XXIVth International Conference, Martigues, France, 7-11 June 2000. Errance, Paris:139-149.

NUNINGER, L., 2003. Approche méthodologique d'un espace agraire protohistorique: la Vaunage (Gard). Actualité de la recherche en Histoire et Archéologie agraires, Proceedings of the 5th International Conference of AGER, Besançon, France, 19-20 September 2000. University Press of Franche-Comté, Besançon.

NUNINGER, L. and RAYNAUD, R., 2000. Dynamique et système de peuplement en Vaunage (Gard) du VIIIe au IVe avant J.-C. Mailhac et le premier âge du fer en Europe occidentale, Proceedings of the International Conference, Carcassone, France, september 17-20, 1997, Monographie d'Archéologie Méditerranéenne 7, Lattes:35-52.

PUMAIN, D. and SAINT-JULIEN, T., 2001. Les interactions spatiales. Cursus géographie, Armand Colin, Paris.

PY, M., 1990. Culture, économie et société protohistoriques dans la région nîmoise. French School of Rome, Paris/Rome.

SANDERS, L., 1989. L'analyse des données appliqué à la géographie. Alidade, RECLUS, Montpellier.

VAN LEUSEN, P.M., 1999. Viewshed and Cost Surface Analysis Using GIS (Cartographic Modelling in a Cell-Based GIS II). Computer Applications and Quantitative Methods in Archaeology, Proceedings of the 26th CAA International Conference, Barcelona, Spain, March 1998. BAR international, Archaeopress, Oxford.

WHEATLEY, D. and GILLINGS, M., 2000. Vision, perception and GIS: developing enriched approaches to the study of archaeological visibility. Beyond the map, Proceedings of the NATO Advanced Research Workshop, Ravello, Italy, 1-2 October, 1999. Nato science series, IOS press, Oxford:1-27.

WHEATLEY, D. and GILLINGS, M., 2001. Seeing is not believing: Unresolved issues in archaeological visibility analysis. In Cost Action G2, On the good use or geographic information systems in archaeological landscape studies. European Communities edition:25-36.

Managing Different Scales in Intra Site and Micro Regional Analyses using GIS

Stefano Rossi
Scuola di Specializzazione di Archeologia, Università degli Studi di Milano, Milan, Italy

Roberto Maggi
Soprintendenza Regionale per i Beni e le Attività Culturali della Liguria, Genoa, Italy

SEE THE CD FOR THE EXTENDED VERSION

Abstract

The main objective of the paper is to present micro regional and intra site analyses of the mainly Bronze Age site of Pianaccia di Suvero, La Spezia (Eastern Liguria, Italy). During ten years of field work on the site and its vicinity, large and complex data have proved evidence of a settlement, ranging from the Early Neolithic to Bronze Age. Most significant is the discovery of a workshop for the production of steatite ornaments. An intense occupation, probably related to a Middle Bronze Age anthropic activity and recent agricultural practices, have caused erosion of lower layers and a deep accumulation of modern colluvium.
During the field work a variety of different techniques was used, ranging from non destructive techniques like artefact surface collection and geophysical surveys to pedological investigations and excavations. All relevant documentation on site and its micro region was integrated into a geographic information system. The main objective of the analyses performed was to correlate intra site data and micro regional data enabling insight into the prehistoric frequentation of the site and elaboration of the archaeological sequence. During the work innovative solutions on handling inconsistent spatial information, especially coming from varying scales were used.

Introduction

One of the reasons of the outstanding success of GIS technology in archaeology, is the possibility, as it is offered by such systems, to put together maps with different scales in a same context.

But while planning a GIS based on a multi-scale cartography, we cannot ignore the different precisions of the embedded geographical data. This inconsistence can be solved using multi-precision as it happens in the modern numeric cartography.

Pianaccia di Suvero: a Case Study

Some of the methodological approaches the Authors would like to suggest in this paper have been applied to the elaboration of a GIS of the prehistoric site of Pianaccia di Suvero (near La Spezia, Eastern Liguria) (Rossi 2002).

Figure 1 The site during the 1984 campaign

The site area has been studied for over 30 years since the beginning of 70s (Maggi 1984, 1987, 1990, Giardi and Maggi 1980). In particular, during ten years of field work on the site and its surroundings, carried out by Soprintendenza Archeologica della Liguria in the 80s, large and complex data have proved evidence of a settlement, ranging from the Early Neolithic to Iron Age.

The site area comprises a generally north south trending gently sloping interfluves sharply bounded on the north and west sides by precipitous slopes.

In the southern part of the site, characterized by Neolithic findings, artefacts occur in the deeply weathered soil apparently, most of sites and upper archaeological levels may have been lost by erosion. Otherwise, in the north-eastern area, because of terracing on the southeast and southwest sides, the archaeological sequence has been preserved from the modern erosional phase. The main Copper Age occupation involved the construction of a stone platform or some other structure. Intense occupation took place. Most significant is the discovery of a workshop for the production of steatite ornaments in the Middle Bronze Age Layers (Gernone 1994, Gernone and Maggi 1998).

During field works a variety of different techniques was used, ranging from non-destructive techniques like artefact surface collection and geophysical surveys to pedological investigations and excavations.

Almost 30 years of research produced a great amount of data and a complex multi-scale cartography

MULTI-PRECISION

The data available for an information system of the archaeological area of the Pianaccia di Suvero were embedded in a wide range of map of different precision:

- 1:25,000 cartography (IGMI, Italian Military Geographic Institute);
- 1:25,000 geological map;
- 1:10,000/1:5,000 cartography (CTR, Italian Regional Technical Cartography);
- 1:10,000 orthophoto, reporting erosion and vegetation cove rage;
- Geophysical survey data drawn on CTR 1:10,000;
- 80s Survey map free-hand drawn on CTR 1:10,000;
- 1:2,000 cartography, reporting the positioning of archaeological excavations;
- Free-hand - no scale - drawings reporting the positioning of 70s surveys;
- Archaeological planimetric cartography 1:10.

Of course these data are not directly comparable (Voorips 1998:256). On handling the informative contents of this cartography we must carefully consider the representation of the geographical data in different scale. In fact the presence of different scale involves a different precision of the relative spatial data.

For instance, it is obvious that, if we consider the draw of a street on a little scale map (i.e. 1:25,000) transposed into a smaller scale (i.e. 1:1,000) the street's dimensions will not correspond to the dimensions it would have had if drawn in this scale.

The same thing happens considering the perimeter of a building. If we draw it on a 1:25,000 scale map it will be 2 millimetres thick (Fig.2a) while the same outline, in a 1:100 scale map would correspond to 5 meters on the ground (Fig.2b). At this scale, we could draw a more precise profile of the building inside that line even rotating it according to real data (Fig.2c).[1]

On the other hand we cannot give up to use more precise data from a great scale map (as, for instance, the 1:10 archaeological documentation) into a wider context (i.e. 1:2,000 cartography) only for the difficulty connected to their dissimilar representation.

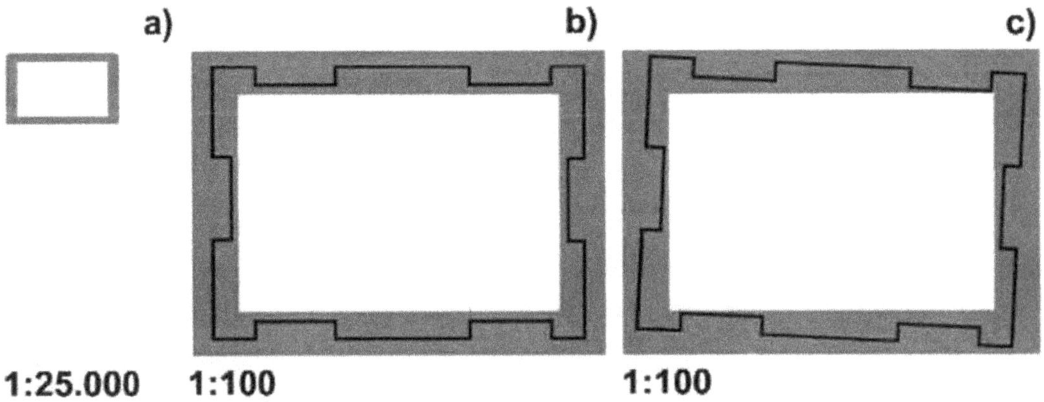

Figure 2a The perimeter of a building drawn (in grey) on a 1:25.000 scale b-c) the same object in a greater scale (1:100) with a more precise drawing of the building (in black).

The inconsistence of different spatial data can be solved using multi-precision as it happens in the modern numeric cartography where a little scale is used for acquiring urban areas (1:1,000) and a medium scale (1:5,000) for all the other area. This method requires to resolve the consistence of the great scale details into minor scale representation (Jones et al. 1996, Egenhofer, Clementine and Di Felice 1994).

In this case, all the spatial data were acquired into the GIS, without affecting their precision, and georeferred using minor scale cartography. For Instance, the 1:10 archaeological plan were georeferred using the 1:2,000 map and in a similar way, this map, reporting the positioning of stratigraphical excavations, was integrate into the 1:5,000 CTR Cartography in a Raster Tiff format that was already georeferred according to the Italian Geographical System, Roma 40 and Gauss-Boaga coordinates.[2]

All the subsequent operation, regarding the different maps described above, were operated applying rototranslation, projective dilating and other similar transformations trying to superimpose points well recognizable on the both maps considered.

For the archaeological plan were used conformal transformations which preserve the consistency of the internal proportions of the image.[3]

This process, trying to compensate the differences instead of completely cancelling, consents to maintain the geometrical consistency inside the major scale cartography.

Once completed the vectorialization of the raster images, inside the GIS, the attributes of the relational database were fully connect to the relative spatial data represented with the different precisions of the vectorial cartography.

As cleared above, we chose to discard simply formal collimations to guarantee the internal consistency of the major scale data without using rotations and other transformations that we cannot surely verify.

The organization of data with a multi-precision method was possible since a Geographical Information System does not share the limitations of the traditional drawing on paper and may consequently represent objects of different detail with a progressive precision.

Conclusion

All the step described above have determined the organization of all the data in a unique Information System that allowed to carry out archaeological analyses on both an intra-site and a micro-regional scale (see, for example, Fig.3).

The main objective of the analyses performed was to correlate intra site data and micro regional data enabling insight into the prehistoric frequentation of the site and elaboration of the archaeological sequence.

Pointing out the existence of a specific "cartographical threshold" does not mean to forget the relative precision required by the context and your own goals. Of course, you have always to apply a logical discrimination (a "logical" threshold") which depends on the different uses of your spatial data.

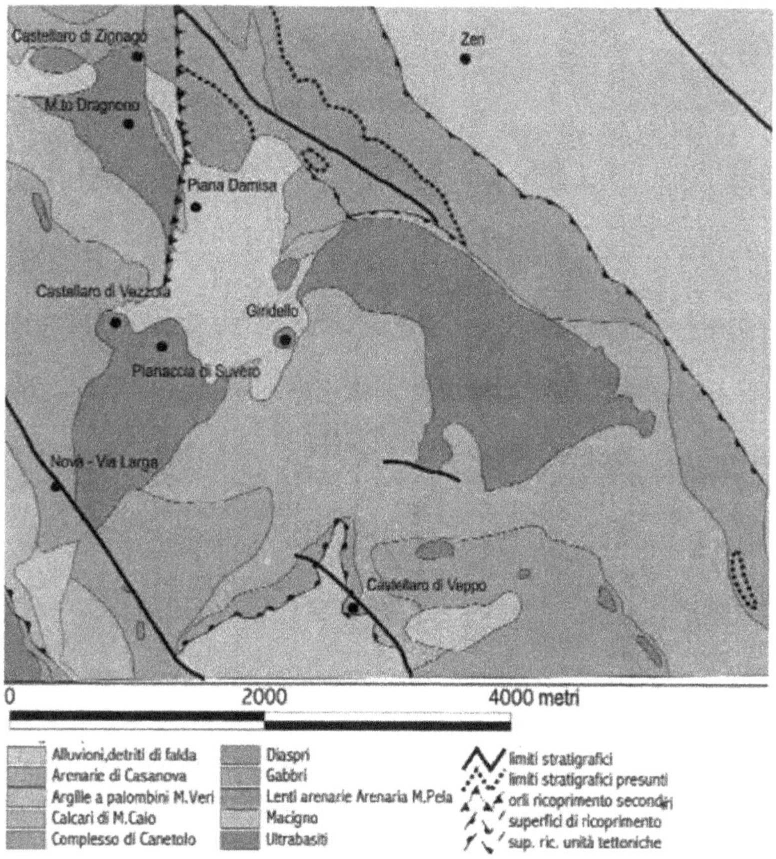

Figure 3 The positioning of prehistoric site around Pianaccia di Suvero on the geological map. It's possible to find a strategic choice connected to the geological parent material

1 This would result more striking if we consider a 1:10 archaeological plan.
2 In this case, rather than a bilinear warping method, that apparently will have guarantee a better precision, it has been preferred a method that allowed to maintain the measures of the major scale cartography in a minor scale.
3 In fact, it has been applied geometrical corrections based on the relative distance between known points of the image using projective transformations rather than deformations like rubber sheeting.

References

EGENHOFER, M., CLEMENTINE, E. and DI FELICE, P., 1994. Evaluating Inconsistencies Among Multiple Representations. In Waugh, T.C., Healey, R.G. (eds.), Sixth International Symposium in Spatial Data Handling, Edinburg, Scotland:901-920.

GERNONE, G. and MAGGI, R., 1998. Lavorazione della Steatite alla Pianaccia di Suvero. In Nicolis, F. and Mottes, E. (eds.), Simbolo ed enigma - Il bicchiere campaniforme e l'Italia nella preistoria europea del III millennio a.C., catalogo della mostra tenuta a Riva del Garda, 12 May - 30 September 1998, Provincia Autonoma di Trento, Trento:95-97.

GERNONE, G., 1994, unpubl. La produzione della steatite della Pianaccia di Suvero (SP). Tesi di specializzazione, Universitá di Pisa, 1993-94.

GIARDI, M. and MAGGI, R., 1980. Pianaccia di Suvero - Rocchetta di Vara (La Spezia). In Preistoria Alpina 16:150.

JONES, C.B., KIDNER, D.B., LUO, L.Q., BUNDY, G.Li. and WAVE, J.M., 1996. Database Design for a Multiscale Spatial Information System. International Journal of Geographical Information Systems Vol. 10, no. 8:901-920.

MAGGI, R., 1984. Pianaccia di Suvero. In Melli, P. (ed.), Archeologia in Liguria, II, Scavi e scoperte 1976-81, Soprintendenza Archeologica della Liguria, Genova:37-58.

MAGGI, R., 1987. Pianaccia di Suvero. Il sito campaniforme. In Melli, P. and Del Lucchese, A. (eds.), 1987. Archeologia in Liguria, III, Scavi e scoperte 1982-1986 1, Preistoria e Protostoria, Soprintendenza Archeologica della Liguria, Genova:23-24.

MAGGI, R., 1990. Officina per la produzione di oggetti in steatite dell'età del vaso campaniforme. In Bollettino di Archeologia 1-2:119-122.

ROSSI, St., 2002, unpubl. L'area preistorica della Piamaccia di Suvero (La Spezia): Analisi Informatizzata dei dati di scavo e ricostruzione della storia del sito. Tesi di Laurea, Università degli Studi di Milano, 2001-2002. Relatore Prof. De Marinis, R.C., Correlatore Maggi, R.

VOORIPS, A., 1998. Electronic Information Systems in archaeology. Some notes and comments. In Archeologia e Calcolatori 9:251-267.

GIS Analysis of Historical Cadastral Maps as a Contribution in Landscape Archaeology

Michele De Silva
Dipartimento di Archeologia e Storia delle Arti,
Università di Siena, Italy

Giovanna Pizziolo
Dipartimento di Archeologia e Storia delle Arti,
Università di Siena, Italy

See the CD for the extended version

Abstract

The use of ancient cartography is presented by the authors as a useful source of information in landscape archaeological studies, performed throughout GIS. In particular the paper is focused on historical cadastral maps to enhance the understanding of landscape transformation and in pointing out evidence of landscape perceptions and territory organisation of the past. Moreover historical geographical research carried out via GIS take great advantage by the integration of different kind of sources. The elaboration of different type of data can provide useful contribution in understanding archaeological contexts. Case studies concerning ancient tuscany maps will be presented by the authors.

Introduction

The aim of this paper is to present some considerations on the role of historical cartography as a source of information in landscape archaeology studies and to explore the expansion of its potential through integration with other sources. This theoretical contribution is presented with the help of two case studies in which the Reverse Research methodological approach is undertaken within a GIS environment.

First, we would like to focus our attention on the nature of historical cartography, analysed in terms of an historical manuscript, a document, representing the archaeological map at the time of its creation (Azzari et al. 2002). In fact, an ancient map represents the territorial settings of a landscape related to the historical context of its execution, providing a portrait of that historical landscape; moreover an historical map may contain traces and signs of previous settings still recognizable at that time. In some fortuitous instances, maps have provided a record of the position of ruins and/or archaeological evidence from which we have gleaned useful information about their existence, extent and position. Furthermore, it is necessary to interpret historical cartography, when viewed in this way, i.e. from the perspective of historical document and as a product of human invention, in terms of direct and indirect information that can be accessible to scholars today. It is especially important to systematically verify the data illustrated in a map in relation to its characteristics and position. In fact, maps signify, in a symbolic and iconographic way, the reality of the landscape of the past as perceived by the ancient cartographers who drew them. They are an intrinsic interpretation of the reality represented.

Historical Cartography and the Reverse Research Approach

We can approach the study of historical maps by following two parallel and complementary directions of analysis: by interpreting the historical document from a semiotic and anthropological perspective, or by developing research from an historical and geographical point of view (Cuccoli et al. 1985). We are strongly convinced of the importance and efficacy of complementary studies but in this paper we are going to predominantly focus on the second type analysis.

Historical cartography offers a wide range of map typologies classifiable in terms of scale, thematic contents, techniques of representation and so on. In particular, a significant difference in cartographic production may be due to the adoption of various drawing practices such as a view in perspective, or a bird's-eye view, or other pictorial methods, which, from their complex results, can be very stimulating and exciting as regards the perception of the landscape and its representation, but which radically compromise, (when compared to planimetric illustrations, for example), the possibility of using ancient maps in a georeferenced environment (Azzari et al. 2002). For this reason, in the case of GIS analysis, it is important to subdivide historical cartography in terms of the geometrical characteristics essential to reliable GIS layer input: historical maps need to be transformed and warped in order to be adapted into a present-day geographical reference system. Consequently, the geometric quality of maps is a critical element to bear in mind. In this sense, planimetric maps, and in particular the cadastral maps of the 18th and 19th centuries, are optimal historical sources to be used in landscape archaeology reconstructions performed with GIS. In fact some of these are excellent examples of geometric accuracy and, moreover, they offer detailed information and a wealth of thematic data. In order to carry on the research using historical cartography, it is important to assess and contextualise ancient maps and to enhance information in a diachronic and synchronic framework. It is thus important to integrate into the GIS several kinds of sources in order to obtain different thematic and temporal layers, taking advantage of the potential of thematic and spatial overlay. It thus is necessary to correlate the study of ancient maps with other information derived from recent maps and aerial photographs in order to investigate the changes in landscape during different historical phases. The methodological approach proposed in this paper is based on reverse historical geographical analysis. Research is developed starting from the analysis of today's landscape settings, going back in time to the interpretation of the settings of the past throughout the individuation of previous phases. In this way landscape settings are sliced in their historical stratification, proceeding in reverse through time, as in an archaeological excavation, from the most recent to the most ancient.

Case Studies

[1] The sources

In this paper we are presenting some examples extracted from two work-in-progress case studies in which we have utilized the catasto lorenese, a geometric cadastre carried out for the whole territory of "Granducato di Toscana" between 1817 and 1835. The catasto lorenese is particularly rich in detailed information documented in the form of maps, in the form of records in tables or inventories, and in a descriptive form as related texts. Maps are truly full of data and allow good and accurate georeferencing, while avoiding too much unwanted deformation of the features represented in the document. The subject matter particularly purposeful to our studies are the "Quadri d'insieme", i.e. the general framework topographic maps varying in scale from 1:25,000 to 1:80,000, and the "Fogli", detailed maps generally at a scale of 1:2,500 or 1:5,000. The historical maps have been acquired into the system and the main features (such as hydrology, settlements, road networks and place names) have been digitised in thematic layers. Besides the acquisition of these ancient maps into the GIS and the subsequent analysis of the historical information which they offer, it is important to contextualise them within the dynamic transformation of the landscape. For this reason it seems fundamental to integrate the research with other sources to input into the GIS. In this case the comparison with aerial photographs produces very exciting results. In this paper we begin by analysing two particular photo series that are portraits of two different phases of 20th century landscapes in Tuscany: the most recent one, which refers to 1997, at a nominal scale of 1:10,000, provides information for contemporary landscape settings. In terms of the previous settings, that is to say how the landscape appeared before the extensive transformation in the post sixties period, we have chosen the VOLO GAI, shot in Tuscany in 1954 at an approximate scale of 1:33,000. The quantitative and qualitative comparisons between these sources performed throughout GIS allow us to highlight the transformation trends occurring from the beginning of the 19th century until today. The analysis was developed focusing on two zones: the Sesto Fiorentino area and the Grosseto plain.

Figure 1 The vectorialisation of cadastral map overlaid on the aerial photo of 1954. A detail of Sesto Fiorentino Plane

Figure 2 The road (interrupted line) and water network (black line) derived by cadastral map overlaid on the probable centuriation grid (pale grey line) in Sesto Fiorentino Plane

[2] Case study 1: Sesto Fiorentino

Sesto Fiorentino is located on the northwest side of Florence's surroundings, partially occupying the Florentine plain and partially extending to the gentle slope of Mount Morello. Currently, the Municipality of Sesto is a satellite urban centre of the large metropolitan area of Florence and, in the last twenty years, its territory has been subjected to intense urbanisation, involving an increase in communication infrastructures. It was during this urban expansion that a large amount of archaeological evidence was uncovered, and the ongoing systematic rescue excavations have exposed several periods of occupation from Neolithic to medieval times (Martini et al. 1993). Recent development has strongly affected the landscape, which appears significantly transformed when compared with the 1954 settings. The comparative analysis between aerial photos (1997 - 1954) reveals general changes in land use, which have deeply overwritten the previous land texture. Dramatic changes occurred not only on the level of urban expansion. In fact even where the agricultural zone has been preserved, changes have taken place in terms of the type of cultivation, and we can observe a shifting from the present single crop system to the traditional mixed agricultural system of the past. Proceeding thus in a reverse direction in this matter, we have assessed what was the situation at the beginning of the 19th century. The cadastral maps[1] testify to a general continuity in landscape settings between the beginning of the 19th century and the first half of the 20th century. In order to better perform a field system comparison, we have elaborated the historical maps by means of the Erdas Imagine 8.5 software. Using a series of filter applications we have highlighted the orientation presented by field patterns and subsequently vectorised them in a semiautomatic way. This elaboration allows us to easily compare different phases. The comparison between the cadastral maps and the 1954 aerial photos shows that the general settings, in the rural plain areas, remain almost the same, as can be seen by observing the field patterns (Fig.1). Road and water networks show the same trend. Moreover the analysis of the 19th century landscape, as it is represented in the maps, offers evidence of the territorial organisation in more ancient periods, recording the continuity and changes that have occurred in the territory. In order to distinguish traces and signs of previous territorial settings we have focused our attention on the Roman period, which had a huge impact on the landscape and continued through time, the permanence of the centuriation network, to represent a very significant historical and landscape marker. Moving forward from this perspective we have built up a grid with a 355 m side module that in the Sesto Fiorentino Plain corresponds to the module of Roman centuriation and subsequently we have "anchored" and oriented that grid, based on the archaeological field evidence attested to in literature (Cresci et al. 1999). In this way we have built up a layer reproducing the highly probable centuriation grid (Peterson et al. 1995). We have consequently overlaid the grid onto the thematic layer extracted from the 19th century cadastral maps (Fig.2). The comparison between these frameworks confirms the continuity in field system and territorial settings. In fact the position or orientation of a lot of segments of the centuriation grid matches with the road and water network of the 19th century, providing evidence of the strong markers imposed on the landscape in the Roman period. Moreover, the strength of the Roman pattern is observable in the landscape subdivisions executed by the cadastral cartographers. In fact the borders between the different maps, the "fogli", follow partially the centuriation limites. The strength of the Roman landscape organisation is also exemplified by the continuity in orientation in the field pattern both in the northeast-southwest direction or the northwest-southeast direction.

[3] Case study 2: Grosseto plain

The Grosseto coastal plain is located in southern Tuscany. A large part of this wetland area was occupied during the Etruscan and Roman periods by Lake Prile. In the northern part of the area the ancient cities of Vetulonia and Roselle were facing the lake from an elevated position. After the Roman period the lake progressively turned into a swamp. From Medici times through the last century, the area underwent a series of important reclamation and drainage activities.

Throughout the case studies presented we would like to underline the potential of the Reverse Research analytical approach undertaken in conjunction with the help of other sources. In particular, by means of the integrated analysis of aerial photos (especially from the past) and historical cartography, we can obtain a great deal of precise and varied information, both explicit and implicit. This methodology allows us to highlight the occurrence of interesting features from a landscape archaeology perspective. In addition, this type of analysis sometimes provides chronological information and, above all, it offers - contrary to other analysis techniques - the exact localisation of these features. The Reverse Research analysis on the Grosseto plain, presented in this paper focuses on landscape transformation with regard to hydrology, coastline and traces of ancient roads.

First we will examine the Fosso Razzo, a small right side tributary of the Ombrone River in the vicinity of its mouth. Today the small watercourse is constricted into a regular channel and the observation of the recent aerial photo doesn't allow us to individuate any trace of its natural course. By contrast, in the 1954 aerial photo, we can see, in the vicinity of the artificial channel, clear signs probably denoting the ancient natural riverbed. The overlay with historical cadastral maps[2] confirms the interpretation of the anomalies seen in the aerial photo of 1954 (Fig.3). In fact, in the 19th century map we can observe a sinuous watercourse, which exactly match the traces previously individuated, giving evidence that in 1823 the fosso was not yet rectified.

The Reverse Research analysis technique may also be applied to the study of coastal shoreline change. In fact, the well-executed geometrical characteristics of the Lorenese cadastre permit us to perform an accurate analysis of the extension of the coastline in the proximity of the Ombrone River mouth. The overlay of aerial photos and historical cadastral maps of different periods permits us to follow the steps in erosion of the coastline, which has retreated about 1km since 1823 (Fig.4).

Furthermore the integration of different sources may help in understanding archaeological evidence. Particularly in the case of the Grosseto plain, the historical cadastral maps testify to the presence of remains of the Aurelia Roman road. The interrupted line symbol, adopted by historical cartographers to portray the remains of the ancient Via Aurelia, and the text line on the map "traccia dell'antica via Aurelia" indicates that the road was not in use in the early 19th century. The Lorenese cadastre also indicates the presence of ruins of a bridge ("rovine del ponte del diavolo") located on the edge of the Ombrone River's final bend, which the ancient maps called the

Figure 3 The 1954 aerial photo shows traces of the ancient natural riverbed. In detail the feature shows the fosso as it appears in Lorenese cadastral map (1823)

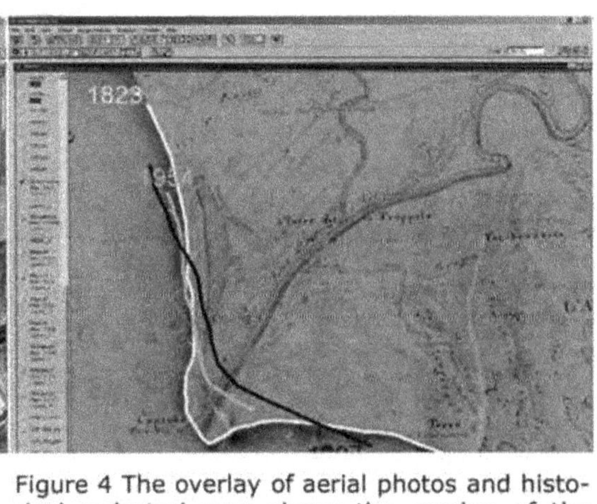

Figure 4 The overlay of aerial photos and historical cadastral map shows the erosion of the coastline in the proximity of the River Ombrone mouth (lines represent the coastline: white 1823, grey 1954, black 1997)

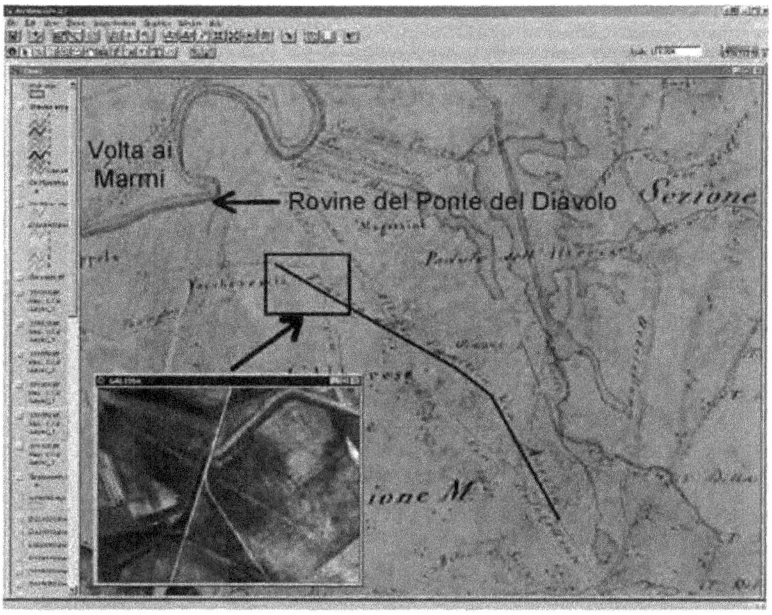

Figure 5 "Trace of the ancient Aurelia" roman road in the Lorenese cadastre (black line). In detail the road trace in the aerial photo of 1954

"volta ai marmi" ("river bend of marbles"), a clear reference to the presence of a roman bridge. In the 1997 aerial photos, the traces of the Roman road have almost disappeared. By contrast, in the 1954 aerial photos, these traces are still present and very sharp. It is exciting to realize that by comparing the aerial photos with historical cartography we are now able to interpret these photo anomalies as the Aurelia Roman road (Fig.5).

The effectiveness of data integration has been widely demonstrated (Vermeulen et al. 2001) and it becomes especially relevant when we are dealing with landscape features with complex stratification. Also pertinent to this mode of research is the integration of data derived from different kinds of analysis. In particular, the results obtained in the last example have been correlated to data gathered from a geometrical analysis applied to the Lorenese road network. This analysis (Azzari et al. 2000) highlighted anomalies in network connectivity, perhaps bearing witness to road segments belonging to a higher level, or having an older origin than the local inter-village road network. If we compare these anomalies with the results previously presented, we can recognize some of these segments as a part of the Roman ways, thus establishing a more comprehensive view of the Roman viability in this area.

Conclusions

The case studies presented above are focused on the advantages of data integration involved in our Reverse Research method of analysis. Following this approach, Landscape has to be considered not in a synchronic way, but as the dynamic result of an historical sedimentation process, in which different events have left traces one on top of the other, becoming a complex whole. The Reverse Research method of analysis together with the integration of different sources allows us to display separate layers of historical sedimentation. From this vantage point, the use of georeferenced historical cartography and its integration with data derived from other sources leads us to explore new complementary interpretations and opens stimulating research possibilities in landscape archaeology studies.

Acknoledgement

This work has been carried out in full collaboration by the authors. In particular paragraph "Case Study [2]" has to be attributed to Giovanna Pizziolo and paragraph "Case Study [3]" has to be attributed to Michele De Silva.

Aerial photos reproduced in this paper derived from photographic archive of Istituto Geografico Militare, (Autorizzazione n. 5763, in data 1.07.2003).
The images of ancient cartography of Lorenese Cadastral map related to Grosseto Plain (ASG Antico Catasto Toscano-Comunità di Grosseto- Quadro d'Unione) have the following authorization: Archivio di Stato di Grosseto, n° prot. 2572/X.1-1 del 22/09/2003

[1] These maps of the "Catasto Lorenese" are placed in the Archivio di Stato di Firenze. We are really grateful to Dr. Francesco Martelli for his kind collaboration and precious suggestions.

[2] These maps of the "Catasto Lorenese" are placed in the Archivio di Stato di Grosseto. We are really grateful to the Director of the Archive Dr. Maddalena Corti for her kind collaboration.

REFERENCES

AZZARI, M., DE SILVA, M. and MARCACCINI, P., 2000. Road network and cultural places in Tuscany. Atti del II congresso internazionale su 'Science and technology for the safeguard of cultural heritage in the Mediterranean Basin', Paris, 5-9 July 1999, Elsevier, Paris :1189-1192.

AZZARI, M., DE SILVA, M. and PIZZIOLO, G., 2002. Cartografia del passato e GIS per l'analisi della trasformazione del paesaggio - Cartography of the past and GIS for the analysis of landscape transformations. Geostorie, April-August 10, 1-2:30-31 and CD Rom.

BACCI, M. and GIACHETTI, M., 1995. Insediamenti romani nella piana fiorentina e il loro rapporto con la centuriazione di Florentia. L'Universo: 546-561.

CRESCI, M. and ZANNONI, M. (eds.), 1999. Itinerari nella storia del territorio sestese. ArcheoProgettiEdizioni. Firenze.

CUCCOLI, L. and TORRESANI, S., 1985. Introduzione alla cartografia e alle rappresentazioni grafiche. Bologna CLUEB.

DE SILVA, M., 2002. Lo studio dei tessuti viari storici in ambiente GIS. Fonti, software, oggetti geografici e database relazionali - The study of historical road networks in a GIS environment. Sources, software geographical objects and relational databases. Geostorie, April-August 10, 1-2:54-55 and CD Rom.

DE SILVA, M., PIZZIOLO, G. and SARTI, L., 2002. Dallo Scavo al territorio: un GIS per la gestione integrata del patrimonio archeologico dell'area di Sesto Fiorentino - From excavation to territory: a GIS for an integrated management of the archaeological heritage in the Sesto Fiorentino area. Geostorie, April-August 10, 1-2:25-27 and CD Rom.

MARTINI, F. and SARTI, L., 1993. Costruire la memoria,archeologia preistorica a Sesto Fiorentino. Garlatti e Razzai, Firenze.

PETERSON, J.W.M. and RAYWARD-SMITH, V.J., 1995. GIS study of potential traces of a Roman cadastre and soil types in Romney Marsh. In Lockyear, K. and Wilcock, J. (eds.), Computer Applications and Quantitative Methods in Archaeology 1993, Tempus Reperatum, Oxford (BAR Int. Ser. 598):155-160.

VERMEULEN, F., ANTROP, M., HAGEMAN, B. and WIEDEMANN, T., 2001. Ancient Roads and Fields in Northwestern Gaul - A GIS Based Analysis. Stancic, Z. and Veljanovski, T. (eds.), Computing Archaeology for understanding the past, CAA 2000, Computer Applications and Quantitative Methods in Archaeology, Proceedings of the 28th Conference, Ljubljana, April 2000 (BAR Int. Ser. 931):187- 196.

Computer Applications for a Reconstruction of Archaeological Stratigraphy as a Predictive Model in Urban and Territorial Contexts

M. Cattani
Department of Archaeology, University of Bologna

A. Fiorini
Department of Archaeology, University of Bologna

B. Rondelli
Department of Archaeology, University of Bologna

SEE THE CD FOR THE EXTENDED VERSION

ABSTRACT

The developments of new software suggest new trends in computer applications supporting the organization of the archaeological record. Archaeological stratigraphy is one of the best frames for the knowledge of the past and many attempts have been made to control each value recognized on the field (excavations, geognostic research) both in urban sites and in the landscape.
With the help of software used in geology it is possible now to relate stratigraphical data to build predictive maps for the management and preservation of archaeological heritage. Other software used for a three-dimensional visualization can add information of architectural elements to integrate the reconstruction of stratigraphy. With the support of GIS and statistical software we can complete the path toward the analysis of archaeological content.

This paper is based on three main assumptions:

1) The archaeological deposit is a complex stratigraphy with layers, features and interfaces that have surfaces and volumes information. Main deal and duty of each archaeologist is to find, describe and record in the best way each of this information.

2) To proceed in this direction it has been more and more evident that the 3d data are significant and can add results in the interpretive step. It is necessary therefore to support stratigraphic documentation with three dimensional single information (survey points) and drawings or CAD interpolations (DEM, DTM, solids).

3) GIS is the best tool for archaeological data management, especially if supported by techniques of survey (Total Station or 3d laser scanner) and digital image photomosaics.

The following step of this paper is to present the results of different trial methods and techniques able to acquire a specific documentation of archaeological deposit in order to reproduce in any moment the original stratigraphy destroyed by the excavation. This procedure, far from a simple virtual didactic reconstruction, is intended as a reasoning tool useful to understand what information we need for a more sophisticated stratigraphic analysis. In this condition we need to be aware of our scientific limits and to present interpretive or predictive models about all the elements that took part to the formation or transformation of the archaeological deposit.

Attempts of reconstruction of archaeological stratigraphy become fundamental for predictive model in urban and territorial contexts, where they are used for town-planning development, scientific research and archaeological evaluation (Berger and Verhagen 2001). At the base of these predictive models is the quality of each information and the capability in managing and interrelating different types of data (Cardarelli et al. 2001). Since each value and description of 3d graphic entity is not sufficient for a proper evaluation of archaeological record it is necessary to obtain and build 3d features inside a more sophisticated system. To control spatial and volumetric information is the best solution for a true management of stratigraphy.

When we apply models to analyse stratigraphy we need accurateness of field information, facility in 3d data management, capability of 3d analysis through the use of a Geographic Information System that allow us to link and join several kinds of data.

Since the data management of excavations or archaeological maps has been well experienced with ArcView and now with ArcGIS (Esri), we try to test the software respecting fixed requirements. We are aware yet that ArcGIS software does not allow to model correctly 3d volumes and it is not a complete 3d analyst tool:

1) A TIN is an object to represent a surface with a set of contiguous non-overlapping triangles (ArcGIS 3D Analyst 2002). It means the interpolation methods manage only a bi-dimensional plan.
2) ArcGIS software does not include as entity the solid; there fore it does not include analyst tools for solids.

Waiting for an upgrading of software in this direction we propose some examples to stress 3d data management and relative problems applied to the excavation context. The data used for this application belong to an excavated site of Bronze Age (Montale, MO), where the whole documentation has been recorded since 1996 within a GIS through digital equipment supported by an Electronic Total Station and a digital camera with zenithal photographs (Candelato et al. 2001). A latter example regards an architectural feature to emphasize problems with 3d management. It has been chosen a part of a ruined wall from Middle Age context.

For each of these archaeological contexts we tested two different ways to manage 3d data surveyed with the Total Station: a voxel modelling and a vector modelling.

First evidence to stress regards the acquisition of data. For a reconstruction of any stratigraphic layer we tried to respect the following requirements:

- To get available a high amount of surveyed points (usually from 500 up to 3,500 points in 40 m square).
- To respect a regularity in data acquisition for flat surfaces and more density in morphological variability.
- To avoid excessive loss of time and resources on the field research (experimentation is often tolerated but not always accepted).
- To repeat the same methodology for all the recognized strati graphic units.

One result is to determine the regularity of data acquisition and control the media of distance between the needed points. The results here presented regard only some of the best-surveyed layers recognized in the excavation, but they have been sufficient for an analysis and discussion toward a three-dimensional reconstruction.

The first attempt concerns the voxel modelling. Creation of raster structure through algebraic and geostatistical algorithms is widely used solution to discrete the real world, including both digital image representation and numerical values modelling. The advantage of raster matrix lies in the easier use of filters or Boolean operators applied to single pixel or groups of pixels through sophisticated image processing or modelling. With voxel dimension (a cube pixel with three dimension) the filter capability support volumetric and informative added value (usually colour).

Voxel methodology has been largely applied on Physics, Tomography and Geology. In this last environment several software have been developed for the management of stratigraphic boreholes and slices and for these software we tried to find a similar application in the archaeological context.

In particular we have tested RockWorks2002 (© RockWare, Inc.), largely used software for geological applications. With Rockworks we can visualize and manage data from boreholes to obtain stratigraphic sections, profiles, models and interpolated reconstruction combining stratigraphic points with litologies. It allows to build solid models made by voxel starting from points with x, y, z values, characterized by an added numeric value qualifying properties of volumes (colour or type). Good examples have been tested for geophysics or geochemistry models, where the use of filters of image processing allows producing sophisticated analysis and variations. Barcelo showed in a paper in the previous meeting of CAA 2002 an example of application in archaeological context (site of Shamakush VIII, Argentina) (Barcelo 2003). The modelization proposed is based on the interpolation of overlapping points with different z value belonging to different layers, following the method largely applied in Rockworks of boreholes. From these points they have obtained interpolated surfaces (GRID Upper and Down) closed by each side. This method is strongly limited by the choice of fixed surveyed points, in contrast with the complexity of the archaeological stratigraphy and it appears not to exploit the full capability of the software with the interpolation of voxel solids.

Rockworks is based on a main difference between stratigraphy model and solid model: it can obtain thickness geometry model from boreholes, but also it can build solids from whole excavated deposit. We applied the use of filtering Two Bounding Grids to obtain a model shaped by two surfaces Upper and Lower. Rockworks import the text files with coordinates from a Total Station survey and it has two possibilities to create a TIN or a GRID with points related to the top surface of a solid, filtered by the polygon that is defining the limits of the stratigraphic unit. The same can be done with the lower surface (Fig.1).

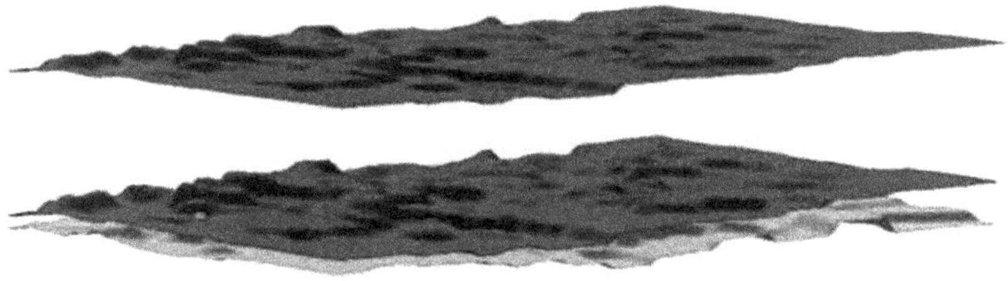

Figure 1 Voxel modelling of upper surface and lower surface of a stratigraphic unit

If we use the datasheet with all the surveyed points characterized by US number as Grade value Rockworks creates a solid model of each stratigraphic unit with the Closest Point option, under the Create New Solid Model / Algorithm settings. This is the most basic solid modelling method, in which the value of a voxel node is set to be equal to the value of the nearest data point, regardless of its distance from the point or the value of its other neighbours. Between the advantages we recognised that the model's value range will be equal to the data point range, with maximum and minimum model node and data point values corresponding exactly. Solid model nodes will honour the control points. This method is useful when generating models in which the values are not gradational. Between the disadvantages, this algorithm produces a grid model with abrupt changes between grid nodes. For us it was not successful because of the poor quantity of coordinates.

In our test we tried to use two different voxel modelization of the same set of selected points:

- a unique model with joined together all the information of several layers
- a single model for each layer.

The total voxel model is a result of interpolation of all the points in tabular shape with coordinates x, y, z and a numeric value (value G) added for the identification of stratigraphic unit (US). To avoid interpolation faults created inevitably by the algorithm, it is necessary to create an offset at the limit of each US. For all the points that belong to the bottom surface or to the side surface have been reduplicated to a specific distance (1 cm) and ascribed to the neighbouring US.

In this type of modelization we can obtain a good result through the algorithm "Closest Point (largely used in raster 3D model) or through the algorithm Inverse Distance Anisotropic. For the top surface or the side surface we can apply filters for polygons or surfaces that allow isolating and removing from the "camera-voxel" the "air" parts, particularly important to obtain a good result in the voxel modelization (Nigro et al. 2002). This first type of modelization offers the advantage of a total model, that is possible to filter (by selection) or to analyse in a cumulative system obtaining slices, profiles in every direction beside the capability to select or exclude desired stratigraphic units.

The modelization through voxel does not solve some problems: if the number of points is limited there are serious errors of interpolation, especially if we want to build a cumulative solid model; the resolution of models and management of models depend strongly from hardware and graphic cards. Nevertheless the wide aptitude of this model type is limited by the high amount of memory required, necessary to maintain the high resolution. The number of nodes of the volume determines Voxel resolution. Quite as larger is the volume of the stratigraphic units as higher will be the needed resolution in order to maintain the reliability in the stratigraphical analysis. The high resolution will imply long time for the computation and regeneration of the model. With an ordinary personal computer it has not been possible to exceed the voxel resolution of 5 cm. This measure is anyway lower or equal to the density of points surveyed by the Total Station.

Figure 2 Voxel models of isosurfaces imported in ArcGIS with exploded representation

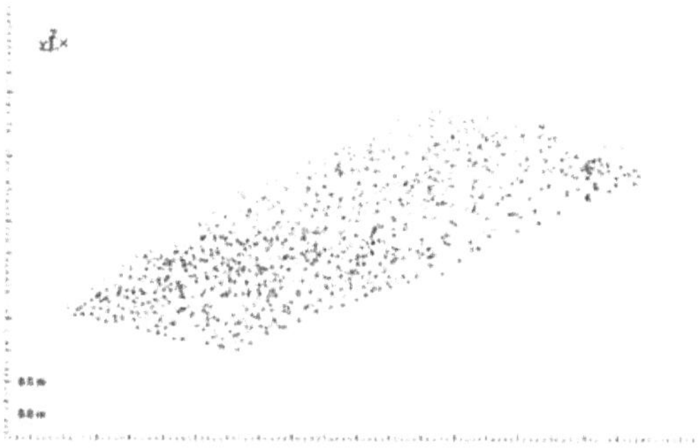

Figure 3 Cloud of points surveyed by a Total station representing top and bottom of a stratigraphic unit imported for vector modelling

The second type of modelization has been applied especially to overcome these problems. For each US it has been created a model filtering the top and bottom surface (UPPER and DOWN GRID FILTER) through the algorithm Kriging. With this system it has been possible to achieve the resolution of 1 cm. The entire visualization of all the stratigraphic units is possible with the command APPEND.

Once created the solid we can proceed to realize profiles, slices, regularly spaced sections, isolate volumes and layers, apply filters or Boolean operations, density analysis satisfying a large part of the requirements of the archaeological research and enhancing the three dimensional informative capability. More relevant and useful for the archaeological analysis is the peculiarity of voxel model to create three-dimensional density buffers. Rockworks allows to visualize colour-scaled three-dimensional volumes pertinent to single points. We tested this procedure to the distribution of metallurgical slag found in the excavation and recorded with each coordinate. This visualization is largely more representative and explicative respect to the traditional two-dimensional density analysis.

At the end of the voxel modelization we can affirm that it represent a powerful tool for a three dimensional analysis of stratigraphy. The reliability is based to the high number of surveyed points and this is contrary to the disadvantage of the already mentioned high cost computation (Nigro et al. 2002). Moreover important and significative are the export possibilities of Rockworks2002 both compatible with CAD and GIS software: fences, vertical or horizontal slices can be exported raster to vector as two-dimensional GRID (x,y; x,z; y,z), sections and profiles as dxf or shapefile, besides to format Voxel Analyst for Intergraph. The export of voxel models in shapefile format still needs an intermediate step that transforms the model as isosurfaces. Then we can import in ArcGis models of isosurfaces (Fig.2). In the GIS environment we can proceed with exploded display or create sections of each isosurface.

In spite of this capability, since ArcGIS does not include solid primitive, the exported models from Rockworks as isosurfaces become a set of contiguous polygons. A single US can contain thousands of records and the capability of GIS of topological analysis is reduced if not invalidated. A solution of this problem is to merge the contiguous polygons in one entity.

The second attempt of identifying a modelization system of stratigraphic data concerns the use of a vector modeller.

SDRC ® Imageware Surfacer is a specific software for three-dimensional modelization, for systems of rapid prototyping and therefore appropriate to create, verify, and analyse solid models. A three dimensional model satisfies the required features for a material reproduction when it is verified the continuity of the surface both mathematical and visual. The vector modeller allows to create a closed surface with double zed value and therefore obtain a solid able to calculate volume, surface, curvature analysis or to obtain profiles or sections along any line of representation, even cumulative, to subtract, merge or to extract the contact surface with other solids. The model considers topology but we need to develop controlling and culling systems, not yet available in all GIS software.

Starting from the same stratigraphical context used for voxel model, we import as text files the coordinates of a layer divided in upper surface and lower surface (Fig.3). Then we can create a TIN for each surface through a Polygonization. The software creates a triangular mesh by connecting points to neighbouring points of a selected cloud. After that we merge together the two surfaces and obtain a single object that can consequently closed as a solid. The final process is to export this object in ArcGIS.

The current archaeological research employs the three dimensional survey of US by top and bottom surfaces, but we propose as more correct, the use of tools of modelization able to describe the shape as regions of volume explicated by a closed surface. A Triangulated Irregular Network, based on criteria of triangulation proposed by Delaunay and available in most of GIS software cannot generate this kind of models and therefore does not provide elaborations of solid modelization, like Booleans operations and cumulative slices. Imageware Surfacer has a data model for geometric vector entity (Closed Polygonized Point Cloud) able to describe the shape of objects as a closed collection of bi-dimensional triangles, oriented in the three-dimensional space.

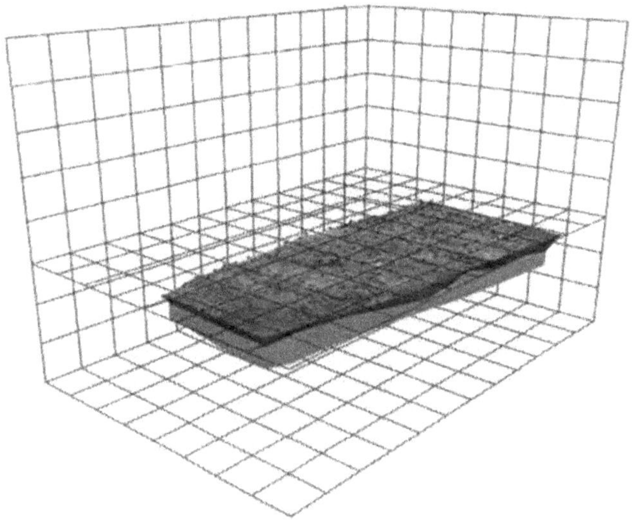

Figure 4 Closed polygonal entity with points and digital photograph applied on the surface and vector models

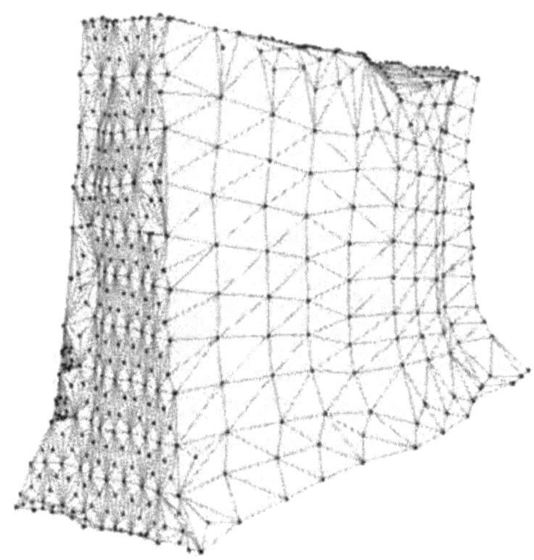

Figure 5 Vector model of an architectural remain

We propose a process of modelization with three steps: optimisation and polygonization of points, evaluation and repairing of surfaces, analysis and final output of the model. The data-entry of two main stratigraphic units is represented by four lists of coordinates of the spatial location of points of surfaces top and bottom. These points are organised on different layers and submitted to the process of polygonization. The algorithm creates a triangular mesh by connecting points to neighbouring points of a selected cloud with the result that volume of each US is completely described by a process of automathised lofting of edges of the top and bottom surfaces. This method as the advantage to create a mesh of triangles directly from original data but it can produce errors of mathematical continuity of the surface, especially in zones of high disorder of points (holes or imperfections in the connection between polygons). More the original surveyed points are organized more efficient will be the software ability. This phenomenon has been remarked processing points in a pseudo-regular grid with increment more or less high respecting the complexity of the shape.

In the process of modelization we have tools to verify and correct the mathematical continuity of surfaces so to check the accurateness of all the information and eventually to reproduce many times the modelization. To put to test a formal description more symbolic and more immediately understandable we can apply algorithms for automatic triangles decimation through the reduction of vertex with same approximate value. The removal of vertex is based on the combination of three parameters with different weight to assign in the process of compression of the shape. First parameter guarantees that all removed vertex leave a surface of polygons with same size. Second parameter assures that surface will be maintained, as it was where the curvature is higher. Third parameter minimizes the error between original and newer surface through the analysis of vectorial difference. Through the command Reduce Polygons to Count the software will remove points from flat surfaces, avoiding changing edges or other basic points of the geometry of model.
The result will be saved in output format with coherent geometry to be imported in the GIS. Since ArcView 8.2 does not include the closed polygonal entity, it is necessary to export in DXF format, and then explode the geometrical continuity. The final output will be readable in ArcScene as collection of single bi-dimensional polygons oriented in three-dimensional space (Fig.4).

One of the best features of modelling software is the possibility of Verifying models. It has tools to control the congruence of the model to the entry points with the possibility to avoid anomalous points or to address us to repeat measurements in some areas. SDRC ® Surfacer provides a full set of tools for verifying whether the polygonal models meet the requirements for a valid STL (Stereolithography) model. STL is the de facto standard data format used to drive rapid prototyping systems. STL is a standard file format for representing shapes as a closed collection of triangles. The software can display the volume, can obtain sections and profiles, union, add or subtract different features and propose other geometrical analysis (e.g. curvature).

We obtained best results with vector modeller is the application of this software to architectural features (Fig.5). The software ArcGIS can transform two TIN surfaces in a single graphic entity called MultiPatch through the lofting process. This multipatch cannot be considered as real solid entity since there is no direct method to take a collection of individual 3D polygons and construct a single multipatch geometry out of it. The alternative is to import from different software and it has been released that the forthcoming ArcGIS 9.0 has some advanced functionality for importing as a form of "multipatch", 3D Objects from 3D modelling software such as 3D Studio Max.

Our experimentation finds more confirmation about the path to follow from several examples of 3d GIS applied by urbanist scholars in collaboration with main software houses. In such cases new trends are directing to the explication of spatial relations between three-dimensional object (Arens 2003, Arens, Stoter and van Oosterom 2003).

[1] TIN represents surfaces as functional surfaces. Functional surfaces have the caracteristic that they store a single z value, as opposed to multiple z values, for any given x,y location."..."Functional surface models can be contrasted with solid models which are true 3D models capable of storing multiple z values for any given x,y location." (Surface Modeling with TIN. ArcInfo Users Guide; Version 7.0, ArcDoc, 1994. Environmental Systems Research Institute, Inc., 380 New York Street, Redlands, CA 92373).

[2] Thanks to the courtesy of Prof. A. Cardarelli, director of the excavation carried out by the Archaeological Museum of Modena.

[3] Imageware V10, 2000. Command Reference.

References

ARCGIS 3D Analyst, 2002. Three-Dimensional Visualizazion, Topographic Analysis, and Surface Creation. An ESRI White Paper, January.

ARENS, C.A., 2003. Maintaining reality: modelling 3d spatial objects in a Geo-DBMS using a 3d primitive. MSc thesis, March.

ARENS, C.A., STOTER, J. and VAN OOSTEROM, P.J.M., 2003. Modelling 3d spatial objects in a Geo-DBMS using a 3d primitive. AGILE conference, Lyon, April.

BARCELO, J.A., DE CASTRO, O., TRAVET, D. and VICENTE, O., 2003. A 3D Model of an Archaeological Excavation. In Doerr, M. and Sarris, A. (eds.), Computer Applications and Quantitative Methods in Archaeology, CAA 2002:85-87.

BERGER, F. and VERHAGEN, Ph., 2001. Predictive modelling of buried archaeological sites in the Tricastin-Valdaine region (Middle Rhone Valley, France). In CAA 2000: Computing archaeology for undestanding the past, 18-21 April 2000, Ljubljana.

BERGSTROM, A. Techniques for Implementing Three-dimensional Computer Modeling in Archaeological Research and Education (http://atl.ndsu.edu/resources/scanning/).

CANDELATO, F., CARDARELLI, A., CATTANI, M., LABATE, D. and PELLACANI, G., 2001. Il sistema informativo dello scavo di Montale. In Preatti del workshop Soluzioni GIS nell'informatizzazione dello scavo archeologico, Siena, 9 June 2001 (http://192.167.112.135/NewPages/TESTIWORK/06.rtf).

CANDELATO, F., CARDARELLI, A., CATTANI, M., LABATE, D. and PELLACANI, G., 2003. Il sistema informativo dello scavo della terramara di Montale (Castelnuovo Rangone - MO). In Analisi informatizzata e trattamento dati delle strutture di età preistorica e protostorica in Italia, Ferrara, 26-27 November 2001, Firenze.

CARDARELLI, A., CATTANI, M., GIORDANI, N., LABATE, D. and PELLEGRINI, S., 2001. Valutazione del rischio archeologico e programmazione degli interventi di trasformazione urbana e territoriale: l'esperienza di Modena. In Gelichi, S. (ed.), Dalla carta del rischio archeologico di Cesena alla tutela preventiva in Europa, Atti della giornata di studio, Cesena 1999, Firenze:31-40.

NIGRO, J., LIMP, F., KVAMME, K., DE RUITER, D. and BERGER, L., 2002. The Creation and Potential Applications of a 3-Dimensional GIS for the Early Homin Site of Swartkrans, South Africa. In Computer Applications and Quantitative Methods in Archaeology, CAA 2001:113-123.

STOTER, J. and ZLATANOVA, S., 2002. Visualisation and editing of 3D objects organised in a DBMS. Proceedings of the EuroSDR Com V Workshop of Visualisation and rendering, 22-24 January 2002, Enschede, The Netherlands.

ZLATANOVA, S., RAHMAN, A. and PILOUK, M., 2002. 3D GIS: current status and perspectives. In Proceedings of ISPRS, CD-ROM, 8-12 July, Ottawa, Canada:8p.

Stratigraphic Modelling of Multi-period Sites Using GIS: The Case of Neolithic and Early Bronze Age Knossos

M. Katsianis
Institute of Archaeology,
University College London, UK
markoskatsianis@hotmail.com

SEE THE CD FOR THE EXTENDED VERSION

Abstract

Knossos in the island of Crete, Greece, is a multi-period tell site with a complex depositional and excavation history. Efforts to preserve the Late Bronze Age 'Palace' at the site and make it accessible to the public have necessarily impeded the study of the earlier levels. Indeed, most of the data gathered from these lower levels is fragmentary and, in its current form, resists any coherent overall synthesis.
The present paper summarises the results of an attempt to model the depositional record of the Neolithic and Early Bronze Age of Knossos on the basis of the recovered evidence using Geographic Information Systems (GIS). The methodological problems inherent in digital stratigraphic reconstruction of complex multi-period tell sites are discussed and the ensuing picture assessed.
Finally, the analytical value of such reconstructions is explored further, in relation to demographic modelling of multiphase sites via the concept of packed house volume (PHV).

Introduction to the Project

This paper describes an attempted application of GIS to the stratigraphic modelling of multi-period tell sites. The case study is the site of Knossos, which is situated in the north-central part of Crete, Greece. It is a multi-period tell site known for the large LBA "Palace" that was excavated by Sir Arthur Evans at the beginning of the 20th century (Evans 1921, 1928). Evidence for earlier periods dating back to the beginning of the Neolithic was uncovered in many areas, but the superimposed structures restrained large scale excavation of these levels and has prevented the coherent diachronic synthesis of the site's early history (Winder 1991, Broodbank 1992, Whitelaw 1992, Evans 1994).

The idea was that using GIS, the surviving extent and thikkness of the archaeological deposits that underlie the palace would be modelled, on the grounds of the existing stratigraphic and ceramic evidence (see also Richards 1990, Catani et al., this volume). In this sense, a picture of the depositional state for the whole of the tell would be created.

Methodology

In GIS terms, the project is about Digital Terrain Modelling. The software packages used were ArcView and ArcInfo.

Initially, a DTM of the present surface was produced in order to provide the basis of the analysis. A total area of 600 x 700 m centred on the hill of Knossos was selected, since it provided a visual extent that enabled the comprehension of the surrounding landscape. Contour data (at 1 and 0.5 m interval) and individual spot heights (3,780 in total) were recorded from topographic maps, plans and sections of the palace (Hood and Smyth 1981, Papanis 1982). The interpolation was performed using the Topogrid function of ArcInfo (Hutchinson 1988, 1989). The resulting DTM had a resolution of 2 m (Fig.1).[1]

Next, a DTM of the pre-excavation tell topography was generated in order to establish the original surface from which early test-pits were dug. In this way, it was possible to use the recorded depths from trenches, where surface measurements were absent (Fig.2).

All stratigraphic data were organised in a binary database that recorded the presence or not of a period and a numerical that provided the absolute heights of the upper surface of each period. They were inserted as points (399 in total) having their exact position taken from record plans, excavation reports and descriptive accounts (see Katsianis 2002:Appendix).

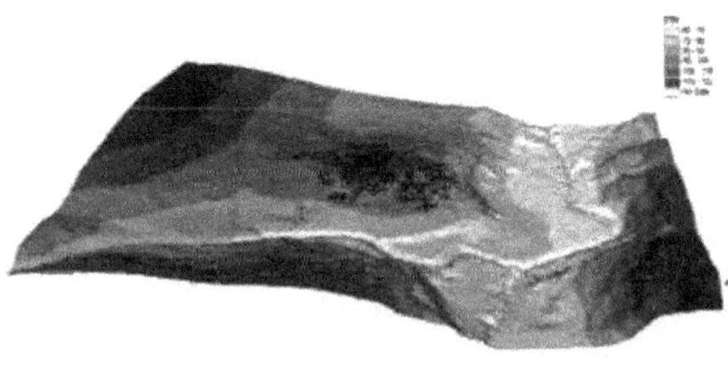

Figure 1 Present DTM (res. 2m)

Figure 2 Pre-excavation DTM (res. 2m)

The study area was further localised around the palace, where there was the bulk of the relevant archaeological evidence. The reconstruction of the stratigraphy begun from the bottom upwards, since the bedrock was the only surface that certainly covered the entire area (Fig.3).

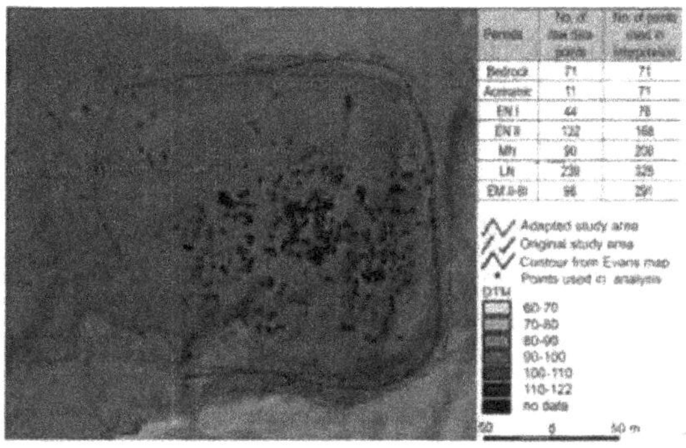

Figure 3 Study area and distribution of points used in the interpolation of the surfaces

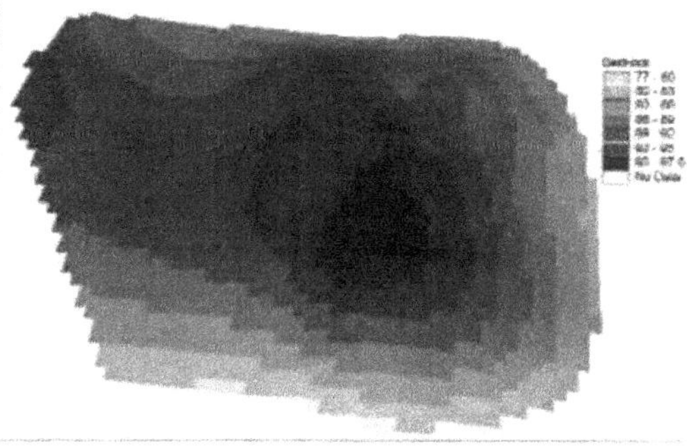

Figure 4 Bedrock DTM (res. 5m)

They provide insights into the slope of the surviving deposits as well as the depositional relationships (Fig.6).

Additionally, six DTMs were produced from the respective surfaces giving a visual and measurable account of both the extent and thickness of the deposits (Fig.7). They included measurements of the total area and average depth in every case. From those it was possible to calculate the total surviving volume of each deposit and suggest new estimations of their probable original volume and extent.

Lack of data at the edges prevented point based interpolation procedures (IDW 12) to perform well with the creation of the bedrock surface, even in low resolution (30 m!). Only the incorporation of a boundary could constrain the significant edge effect. In his excavation accounts from Knossos, J. D. Evans produced a probable contour map of the bedrock, based on its absolute heights and inclination, where exposed (Evans 1971:pl.VI, 1994:Fig.1). One of these outer contours was used in order to control the noise from the edge-effect. Its inclusion was enough for Topogrid to interpolate successfully over the whole area at a 5 m resolution and minimise all previous edge effects. The probable introduced error from the inclusion of the contour line was reduced by limiting the study area after the interpolation. In this way, the bedrock DTM was created (Fig.3, 4).

The rest of the DTMs were even more troublesome, because they had to account not only for the depth of the deposit, but also its extent. The binary database provided the areas of presence of the surviving deposits in each period. The points covering this area were then supplemented by the points of the underlying surface that lied outside it. The new point theme was the basis for each interpolation. In effect, each DTM produced, had an increase in height only in the places, where the deposits were present, and kept the previous surface in areas, where it was absent (Fig.3 - table). In this way, six DTMs representing the major chronological phases were produced at a resolution of 5m (Fig.5).

Assessment of the Results

The archaeological value of such attempts is largely dependent on the accuracy of the result. In the present application, these surfaces are not precise representations of the deposits. They should be viewed as a rough approximation of the stratigraphy based on the ceramic sequence. Their accuracy is related to two factors: the nature of the primary data and the interpolation performance.

The performed interpolation was based on a set of scarce and irregularly distributed points. The interpolation scale of 5 m was the highest possible supported by the current dataset. This scale managed to provide a comprehensive account of the shape of each deposit and allowed quantifiable observations of archaeological interest. The comparison of the surfaces with the primary data reveals that the shape and the depth of the resulting deposits are in principle compatible (Fig.7), mainly because the interpolation procedure maintained the values of the original data.

However, significant problems were also indicated. By far the biggest methodological fallacy is the edge effect. The extents of the site were impossible to approximate by the interpolation procedure alone, especially to the west of the palace. This suggests that future applications are largely dependent on adequate data around the edges of the study area.

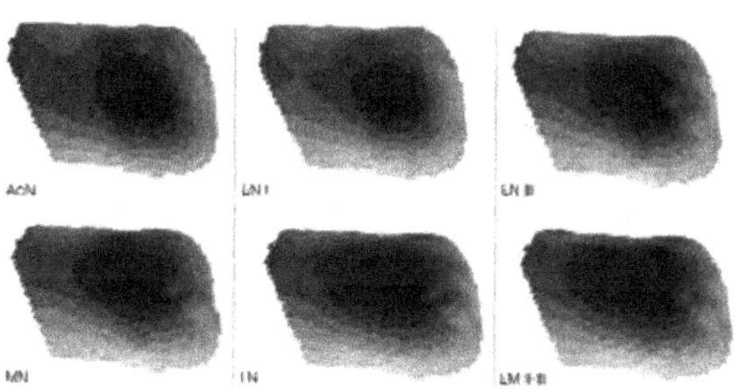

Figure 5 DTM's showing the topography of the surviving deposits in each period (res. 5m)

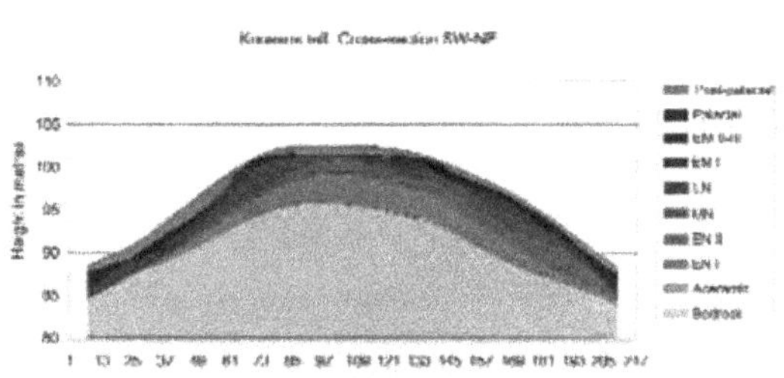

Figure 6 Cross-section of Knossos tell (SW-NE) according to the modelled deposits

A number of inaccuracies are also present in places where there is lack of data of the earlier periods (not all trenches reached the bedrock). This probably resulted in later periods appearing disproportionally deeper in those areas. This is a problem of the interpolation procedure that is related to the spatial distribution of the data.

The assessment of the resulting surfaces in defining the actual shape of the deposits was facilitated by two ways. By extracting profiles from the DTMs (Profile Extractor 6.0) the modelled stratigraphic sequence was displayed as sections.

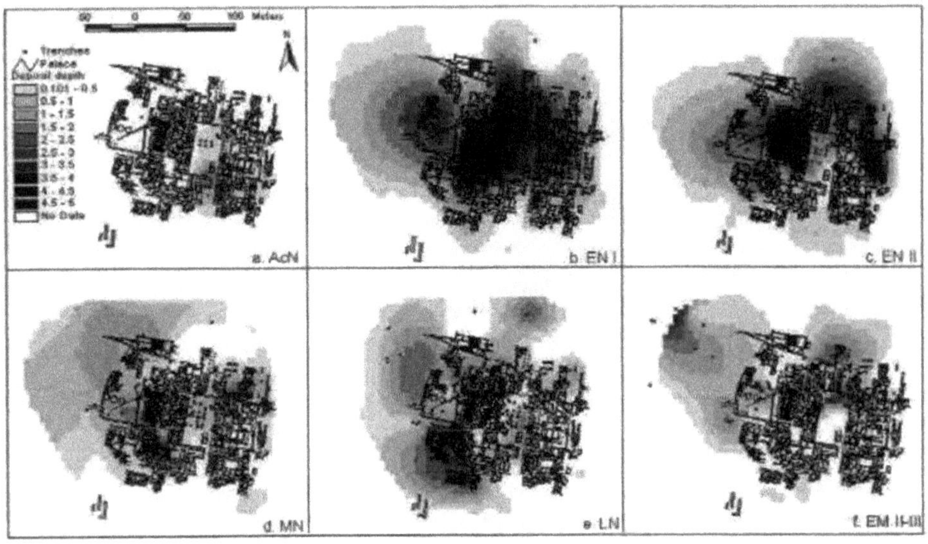

Figure 7 Extent and depth of the surviving deposits in each period according to the interpolation

Another deficiency is the misrepresentation of minor local truncations (i.e. local absence of the deposit). The basic convention of creating one area encompassing all points that showed the presence of a period, and the resolution employed caused the interpolation to ignore secondar y truncations. Binary interpolation procedures could have helped in the better definition of these areas.

On the whole, the accuracy of the surfaces can be considered at least informative of the actual depositional situation. The tracing of the original extents of each deposit was helped by the interpolation, but had to be correlated with the archaeological evidence and is open to interpretation (see Katsianis 2002). Future refinements that will incorporate more local variation are possible, but they are largely dependent on the collection of more data.

Population Estimates Based on Depositional Measurements

The modelled deposits were employed in the exploration of issues regarding the diachronic growth pattern of the site. Using a different approach to population estimates advanced by Ammerman et al. (1976), the accumulated deposits of a site can be used as referents of its population, thereby incorporating an idea of occupation intensity as well as extent.

The concept of Packed House Volume (PHV) signifies the volume of debris deposited on a site following the collapse of a house. It can be calculated by multiplying the total wall length of a house with its average wall width and its average wall height. In this respect, the quantity of houses standing together at an individual horizon within a period (N) can be found by multiplying the total deposit volume (V) with the average life expectancy of a house (T) and dividing the result by the packed house volume (H) and the length of the period (P)[2]:

$$N = V \times T / H \times P$$

For the house life expectancy variable (T) two values were used, 30 and 40 years, since most pisè house structures are claimed to fall within that range (Ammerman et al. 1976:44, note 7). The resulting number of houses was then multiplied by the number of residents. This can vary according to the household type. Depending on whether a household is equated with a nuclear or an extended family the calculation used 5 and 9 people respectively. The derived estimates represent the average population any time during that period. In reality, demographic variation should be expected within a period. If an idea of population growth rate can be established, then the population number can be distributed within a period accordingly (Table 1).

For this paper the surviving deposits' volume was used in the calculation. The resulting figures, therefore, represent the minimum population number according to the used variables during a certain period. Due to the loss of the original deposit volumes (i.e. LN) in later periods and the noted interpolation problems, these figures deviate from the numbers one would actually expect. However, they set a minimum of expectations. Should a reconstruction of the original deposits be made in an interpretative manner, more realistic results can be obtained (for such an example see Katsianis 2002).

Phases	Vol. of surviving deposit	House use-life	Length of period	PHV	No. of houses	Population number (household of 5)	Population number (household of 9)
	(m^3)	(y)	(y)	(m^3)	(n)	(n)	(n)
AcN	2756	30	500	36	4.6	23	41
EN I	51444	30	1500	48	19.8	99	179
EN II	44652	30	400	60.3	55.5	278	500
MN	24037	30	200	73.5	49	245	441
LN	32266	30	400	84	28.8	144	240
EM II-III	15477	30	800	73.2	8	40	72
AcN	2756	40	500	36	6	31	55
EN I	51444	40	1500	48	28.6	143	257
EN II	44652	40	400	60.3	74	370	665
MN	24037	40	200	73.5	65.4	327	589
LN	32266	40	400	84	38.4	192	346
EM II-III	15477	40	800	73.2	10.6	53	96

Table 1 Population estimates based on Packed House Volume.

Concluding Remarks

The above example shows that GIS applications of this kind are not mere predictive models of the archaeological record. They also have analytical value, as they can be used to monitor the processes of the past that might have led to the modelled picture and address complex issues.

The main problems lie in the fact that such modelling attempts and the consequent interpretations are very difficult to verify and their success is largely dependent on the primary data and the capacities of GIS interpolation methods. In Knossos, the modelled outcome was an advance in the evaluation of the whole site, but still remains under critical evaluation. Verification and data problems can only be assessed by further excavation, while software performance by more applications and special GIS functions that take into account the problems present in such studies.

Acknowledgements

This project is summarising the results of an MSc thesis conducted at the Institute of Archaeology, UCL. I am indebted to Todd Whitelaw for entrusting me with this project. Much of the data used were assembled by him, drawing on published and unpublished records of the excavation at Knossos. His guidance and suggestions during the writing of my thesis and this article was invaluable. Also, to James Conolly for his help and encouragement. I would also like to say thanks to Spiros Tsipidis, Andy Bevan, Kostas Venetis, Tasoula Dimoula, Rena Veropoulidou and Nikos Valasiadis for their help and presence during the presentation and the preparation of the final version of this paper.

[1] All DTMs are shown with an exaggeration factor of 1.5 for the better comprehension of the modelled landforms.

[2] Ammerman et al. (1976) note four factors that may affect the outcome of the calculation: reuse of material in later phases, erosion of the tell, use of non-durable material and contribution of total volume from other sources (i.e. ceramic, refuse disposal).

References

AMMERMAN, A.J., CAVALLI-SFORZA, L.L. and WAGENER, D.K., 1976. Towards the Estimation of Population Growth in Old World Prehistory. In Zubrow, E.B.W. (ed.), Demographic Archaeology, University of New Mexico Press, Albuquerque:27-61.

BROODBANK, C., 1992. The Neolithic Labyrinth: Social Change at Knossos Before the Bronze Age. Journal of Mediterranean Archaeology 5/1:39-75.

CATANI, M., FIORINI, A. and RONDELLI, B., this volume. Computer Applications for a Reconstruction of Archaeological Stratigraphy as a Predictive Model in urban and Territorial Contexts.

EVANS, A.J., 1921. The Palace of Minos I. London.

EVANS, A.J., 1928. The Palace of Minos II. London.

EVANS, J., 1971. Neolithic Knossos - the Growth of a Settlement. Proceedings of the Prehistoric Society 37:95-117.

EVANS, J., 1994. The Early Millennia: Continuity and Change in a Farming Settlement. In Evely, D., Hughes-Brook, H. and Momigliano, N. (eds.), Knossos: A Labyrinth of History: Papers in Honour of Sinclair Hood, Oxbow Books, Oxford:1-20.

HOOD, M.S.F. and SMYTH, D., 1981. Archaeological Survey of the Knossos Area. 2nd ed., The British School at Athens, London.

A Volumetric Assessment of Ancient Maya Architecture: A GIS Approach to Settlement Patterns

Kathryn Sorensen
University of California, USA

Jeffrey B. Glover
University of California, USA

Scott L. Fedick
University of California, USA

Abstract

This paper will discuss the general applications of GIS technology to our research in the Yalahau Region of northern Quintana Roo, Mexico. In particular we will address the use of a volumetric analysis as a means of developing an architectural comparative framework at both the intrasite and regional scales. The comparative framework is a powerful tool that allows us to investigate and visualize the distribution of social power both within the site of T'isil and across the region. The direct relationship between social power and architectural volume is predicated on the assumption that actors who utilized the largest dwellings were able to coerce (or force) the greatest number of people to aid in their construction.

Introduction

The Yalahau region, where the current research is being conducted, is located in the north-eastern corner of Mexico's Yucatán Peninsula and represents a unique physiographic landscape dominated by extensive freshwater wetlands. This water rich environment stands in sharp contrast to the dry karstic plane that characterizes the rest of the northern Yucatán Peninsula and is perhaps the most biotically rich and diverse ecosystem in the northern Maya Lowlands (Lazcano-Barrero 1995).

Limited archaeological reconnaissance had been conducted in the region prior to the 1993 initiation of the Yalahau Regional Human Ecology Project. More pertinent for our discussion here, however, is the project's implementation of Geographic Information System (GIS) technology, which began in 1999. Initially used as a data management tool, the GIS has more recently been applied as an analytical tool to the settlement pattern data from the ancient Maya sites in our study area. This paper investigates the application of GIS for the analysis of volumetric calculations that can be used to address questions concerning the socio-political, cosmological, and economic organization of the sites in the region at both the individual site and regional levels.

Why Volume?

It is our assumption that the relationships that create social networks within a society are predicated on "power" relations. Therefore, to begin understanding social relationships, the allocation of "power" within those relationships must be determined. "Power" here is defined as "the control of resources and the productive means to exploit and distribute these resources" (Hendon 1989:894). Using this definition, how is power displayed in the archaeological record, recognizing that it is not absolute power that we are after but differences in relative power?

As Eliot Abrahms (1989:53) noted "architecture, by virtue of its capacity to absorb relatively large amounts of energy during production, can hypothetically reflect a significant range of organizational behaviors requisite for such construction, an important index of cultural complexity." Following this statement, our interest in architectural volume stems from its ability to represent the labor force that was organized by administrators at a particular site and thereby a proxy measure of administrative power. The volumetric mass of the largest architecture in the site center will be the main factor in measuring the size (administrative power) of a site.

Site Level Analysis

Volumetric analysis has different applications at the site level and at the regional scale. At the site level the questions being addressed by the volumetric analysis correspond to a model of Maya community organization first proposed in the archaeological literature by Michael Coe (1965). Coe proposed a concentric zone model for community structure based on ethnohistoric evidence of Yucatec Maya communities. The communities tended to be situated around cenotes (limestone sinkholes). Cenotes are interesting features on the landscape that blur the line between sacred and profane spaces. In the dry northern Maya Lowlands they were important sources for drinking water, yet at the same time cenotes represent portals of communication between the "lived" world and the watery underworld. Because of their conduit properties between the levels of existence, they were conceptualized as the axis mundi of communities.

Using Coe's model, we would expect to find the cenote bounded by the most impressive architecture, marking the residences of the elite and temples of worship. This pattern can be seen as a reflection of the unequal distribution of social power within the community where "(h)ouses...are 'structuring structures' - culturally loaded spaces that socialize by encouraging practices consistent with the meanings they encode" (Johnston and Gonlin 1998:145). Additionally, Hirth notes "residential architecture is perhaps the strongest and most consistent expression of wealth and rank in agrarian societies and is a product of both a household's social functions and energetic capabilities" (Hirth 1993:123), and at the site level will be the object of study. Therefore, the relative grandeur of architecture is a symbolic form of communication stating the importance of those individuals to the rest of the community. In Coe's model, social differentiation within the community is expressed through peoples proximity to the site center, and the access to the supernatural realm, the temples and cenote, which that proximity brought. This implies that the largest and most elaborate architecture will be located closest to the community center.

At the site level, Coe's model has been applied to the 2,000 year old site of T'isil, located about an hour drive west of Cancun. The site appears to be centered around a cenote, and over the past four field seasons, the northern half of the site has been almost completely mapped. Mapping the site of T'isil has been achieved by the use of optical transits, Brunton compasses, metric tapes, range finders and other conventional equipment. The site covers approximately 3.5 square kilometers, and further mapping of T'isil is continuing. In addition to the mapping, extensive surface collections have been made along with preliminary test excavations that have revealed a short major occupational period dating to the late part of the Late Preclassic period to the early part of the Early Classic period (c. 100 B.C - A.D. 350).

Regional Survey

At the regional scale, however, the questions change due to the intensity of the study being conducted. The major research question being addressed at this scale of analysis concerns the socio-political organization of the region. Specifically, the research will test four models of the spatial distribution of administrative centers within the Yalahau region in order to address the complexity of the socio-political organization. The focus at the regional scale is not on the distribution of domestic architecture but on "monumental" architecture at the site center, which includes domestic and public buildings.

The "size" of each center is to be measured by mapping the largest structures at the site, while the basic elements of the site plan will be recorded by mapping all of the structures within the defined site core. The mapping strategy is different on the regional project than at T'isil. The sites are being geographically positioned using a handheld GPS and mapped with a Topcon total station. As the number of structures mapped at the site core will be judgmentally determined and will certainly vary between sites, a standard of the five largest structures will be used to calculate the architectural volume of each site and create a comparative volumetric database. While the number five is primarily arbitrary, it does have significance in Maya cosmology (center and four cardinal divisions), and may represent a principal of site planning as discussed above (see Ashmore 1991, Coe 1965). The creation of a comparative volumetric database will be the main tool used when creating a hierarchy of sites within the region.

GIS Methodology

ArcView 3.3 with the 3D Analyst extension was used to calculate the volume for each structure at both the site and regional scales. Points corresponding to each structure were highlighted in the view using the select button. Then using the Surface pull-down menu, we selected the Create Tin from Features option and created a Tin (Triangulated irregular network) surface. At that time, the Tin was highlighted in the ArcView Table of Contents. Using the Surface pull-down menu again, the Area and Volume Statistics option was chosen. This results in planimetric and surface areas, as well as volume calculations for each structure. A table created from the data generated by the Area and Volume Statistics option was then used to generate a polygon theme based on volume calculations at T'isil, allowing us to visually display the distribution of volumes across the site.

Other Methods

A search of the Environmental Systems Research Institute's (ESRI) webpage produced an ArcScript extension, written by wildlife biologist Jeff Jenness, that allows the user to calculate various surface and topographic characteristics for Point, Line or Polygon themes that overlay a GRID or TIN surface. This script allowed us to automate the volume calculations for each structure quickly and easily and directly placed the volumetric calculations in the structure's table.

It was suggested to the authors at the 2003 CAA meeting that the use of the Cut and Fill tool, part of 3D Analyst, would provide a way to more accurately calculate structure volumes. To use the Cut and Fill tool two TIN surfaces must be created that are bounded by the same polygon. The bottom surface would be created using the points defining the basal platform and the upper surface would be the TIN of the whole structure. This method seems to be the most accurate but we have had little time to fully apply it to our data.

Pros and Cons

There are pros and cons associated with both of these methods. The Point, Line and Polygon ArcScript extension from the ESRI web page was written to calculate all volumes from a single base elevation and this proved to be problematic because our sites, like the vast majority around the world, do not occur on perfectly level planes. Even though the topographic variation across T'isil is no greater than a meter and a half, volumes generated using Jenness' script proved to have a 3-6 % error rate in comparison to the more manual method described previously. The advantage of Jenness' method is that it allows volume calculations to be automated, saving many hours of labor. This method would be acceptable when working with a relatively flat site, where accuracy is not critical.

Volume calculations obtained by using the 3D Analyst extension also had a slight error associated with the data produced. The Area and Volume Statistics option calculates volume using an elevation of the researcher's choosing, typically the lowest elevation point of the structure, generating a horizontal plane to represent the bottom of the structure. Again, this calculation would only be accurate if each structure was on a perfectly level plane. However, the error from this method was minimal in our calculations because there are no major slope changes associated with the individual structures at the site. The measurements from this method are skewed high, and the Cut and Fill method would alleviate this error. The Area and Volume Statistics method did turn out to be very time consuming at the site level, as there are approximately 600 structures currently mapped at T'isil, and a TIN had to be individually generated for each structure. This problem is compounded with the Cut and Fill method because 2 TINs must be generated for each structure. At the regional scale, however, this was not a problem due to the limited number of structures mapped at each site.

Results

At this point in time any discussion of results on the proposed power of this tool in the elucidation of a regional hierarchy of administrative nodes across the landscape would be premature due to lack of data. However, two of the larger sites in the region, Site 7 and Kimin Yuk, provide an example of the possible strength of the volumetric tool. The sites are located approximately 7 km from each other and have very different organizational properties in terms of the way in which they were laid-out and the manner of building construction used. Glover's initial impression after spending some time at each site was that they were approximately the same size. After calculating the volume of the structures at both sites, however, his initial impression proved to be very wrong. The 3 platforms mapped to date at Kimin Yuk measure 83,206 cubic meters while the 5 largest structures at Site 7 measure 13,704 cubic meters.

More concrete results came at the site level. Volume calculations at T'isil appear to support the concentric zone model proposed by Michael Coe. Volumetric assessment of structures at T'isil revealed a natural break in the size distribution, with the eleven largest structures at the site measuring more than 450 cubic meters. As would be expected when applying Coe's model, the vast majority of these large structures (9) clustered closely (within 180 m) around the cenote that marks the geographic center of the site. The eleven largest structures at this site all had a volume greater than 450 cubic meters, and as would be expected when applying Coe's model, nine of these structures are situated within 180 meters of the cenote at the site center. The site of T'isil has a maximum radius of approximately 1,100 meters, so it is significant that the majority of the largest structures are located in such close proximity to the cenote, supporting the concept of a site center. Site planning principles, such as those described by Coe's model, have been documented at Maya sites during the Classic, Postclassic and Contact periods, but have not been widely researched at Preclassic sites such as T'isil. This volumetric study is significant because it supports the idea that these same site planning principles have their origins in the Preclassic period and have deep temporal roots.

Conclusion

In conclusion, this paper presents the results of our initial foray into the realm of GIS driven volumetric calculations for the explicit purpose of developing a proxy measure for social power. It appears that our preliminary attempt has produced interesting and useful results. Further applications of GIS technology to volumetric calculations may become a useful tool for archaeologists in various areas of future research. We are still hoping to create an automated script that will be nuanced enough to incorporate the topographic variation across a site.

Acknowledgements

The current research of the Yalahau Regional Human Ecology Project is conducted with permission and support of Mexico's National Institute of Anthropology and History, and with generous funding from Michael Baker and the Baker Family Foundation, and the UC MEXUS program of the University of California.

References

ABRAMS, E., 1989. Architecture and Energy: An Evolutionary Perspective. In Schiffer, M.B. (ed.), Archaeological Method and Theory Vol.1, The University of Arizona Press, Tucson, AZ:47-87.

ADAMS, R.E.W., 1981. Settlement Patterns of the Central Yucatan and Southern Campeche Regions. In Ashmore, W., Lowland Maya Settlement Patterns, University of New Mexico Press, Albuquerque, NM:211-258.

ADAMS, R.E.W. and JONES, R.C., 1981. Spatial Patterns and Regional Growth among Classic Maya Cities. American Antiquity 46(2):301-322.

ASHMORE, W., 1991. Site-planning Principles and Concepts of Directionality among the Ancient Maya. Latin American Antiquity 2:199-226.

COE, M.D., 1965. A Model of Ancient Community Structure in the Maya Lowlands. Southwest Journal of Anthropology 21(2):97-114.

DUNNING, N.P., 1992. Lords of the Hills: Ancient Maya Settlement in the Puuc Region, Yucatan, Mexico. Prehistory Press, Madison, WI.

FORD, A., 1986. Population Growth and Social Complexity: An Examination of Settlement and Environment in the Central Maya Lowlands. Anthropological Research Papers no. 35, Arizona State University, Tempe, AZ.

GUDERJAN, T.H., 1991. Maya Settlement in Northwestern Belize: The 1988 and 1990 Seasons of the Rio Bravo Archaeological Project. Maya Research Program and Labrinthos, San Antonio, TX.

HENDON, J., 1989. Status and Power in Classic Maya Society: an Archaeological Study. In American Anthropologist 93(4):894-918.

HIRTH, K., 1993. Identifying Rank and Socioeconomic Status in Domestic Contexts: an Example from Central Mexico. In Santley, R. and Hirth, K. (ed.), Prehispanic Domestic Units in Western Mesoamerica, Boca Raton, CRC Press:121-146.

JENNESS, J., 2003. Surface Tools for Points, Lines and Polygons. ESRI (http://arcscripts.esri.com/details.asp?dbid=11665,23 February 2003).

LAZCANO-BARRERO, M.A., 1995. Ecological Diversity in El Eden Reserve, Northern Quintana Roo. Paper Presented at the Conference on Biochemical Disease Diversity in the Mexican Tropical, University of California, Irvine, CA.

TURNER, E., TURNER, N.I. and ADAMS, R.E.W., 1981. Volumetric Assessment, Rank Ordering, and Maya Civic Centers. In Ashmore, W. (ed.), Lowland Maya Settlement Patterns, University of New Mexico Press, Albuquerque, NM:71-88.

The Potenza Valley Survey: Towards an Explanation of the Settlement Patterns Through the Combined Use of GIS and Different Survey Techniques

Geert Verhoeven

Ghent University - Department of Archaeology and Ancient History of Europe, Blandijnberg 2, 9000 Ghent, Belgium

Frank Vermeulen

Ghent University - Department of Archaeology and Ancient History of Europe, Blandijnberg 2, 9000 Ghent, Belgium

Abstract

In 2000, Ghent University started an international geo-archaeological survey project in the valley of the Potenza River, which links the Apennine hills to the Adriatic coastline of Central Italy (Marche region). Under the direction of Prof. Frank Vermeulen, the Departments of Archaeology and Geography aim to measure the evolution of social complexity within this valley and evaluate the phenomena of acculturation, in time as well as in space.

In addition to systematic fieldwalking and geomorphological research, aerial photography is also undertaken in an exhaustive way. The implementation of the data achieved by these survey techniques into a GIS will be outlined, as well as the difficulties of attaining this in an accurate way.

Besides this function of GIS as a spatial data management and visualising tool, it also offers possibilities in analysing settlement patterns, as some results of locational analysis will make clear. Furthermore, the results of spatial analysis will permit incorporation into the archaeological heritage management of the region in the near future.

Introduction

Since Ward-Perkins started his South Etruria Survey in the early 1950s, archaeological Mediterranean field surveys sprang up like mushrooms. Especially in Italy, systematic prospections really expanded in the form of many small-scale and large-scale projects. In these fifty years, the Marches in Central Italy as well witnessed a lot of smaller and bigger survey projects, almost -if not all- executed by Italian researchers. Nevertheless, an area that still had not been subjected to a major archaeological survey was the valley of the Potenza. This river flows over its ca. 80 km long course from the Umbria-Marches Apennines through a wide and fertile Apennine foothill landscape before it ultimately runs into the generally flat Adriatic coastline zone. Already important in Prehistory (Percossi Serenelli 1981:135-144), the Potenza river valley became one of the most commercial routes of the Central Italian Protohistory and even in later periods the valley remained an important corridor for political, economical and cultural contacts between the Thyrrenian and the Adriatic coast. In Roman times, several cities developed in or near the valley floor (Potentia, Helvia Recina, Trea, Septempeda & Prolaqueum) and even a southern branch of the Via Flaminia passed through it. In the subsequent period, the area remained of importance as it formed the contact zone between Longobards and Byzantines.

These are only a few of the reasons why a geo-archaeological survey called the 'Potenza Valley Survey' -PVS- was initiated in January 2000. This interdisciplinary project was set up by Ghent University (Belgium), more in particular by the Departments of Geography and Archaeology. Under the direction of Prof. Frank Vermeulen, the major aim of the PVS was - and still is, as the project will last till at least 2006 - to measure the evolution of social complexity within this valley and evaluate the phenomena of acculturation, in time as well as in space. Therefore, the project got the subtitle 'From Acculturation to Social Complexity in Antiquity: A Regional Geo-Archaeological and Historical Approach'. To forefill the initially outlined goals[1], three specific sample zones, systematically spaced around the Potenza river, are focused on.

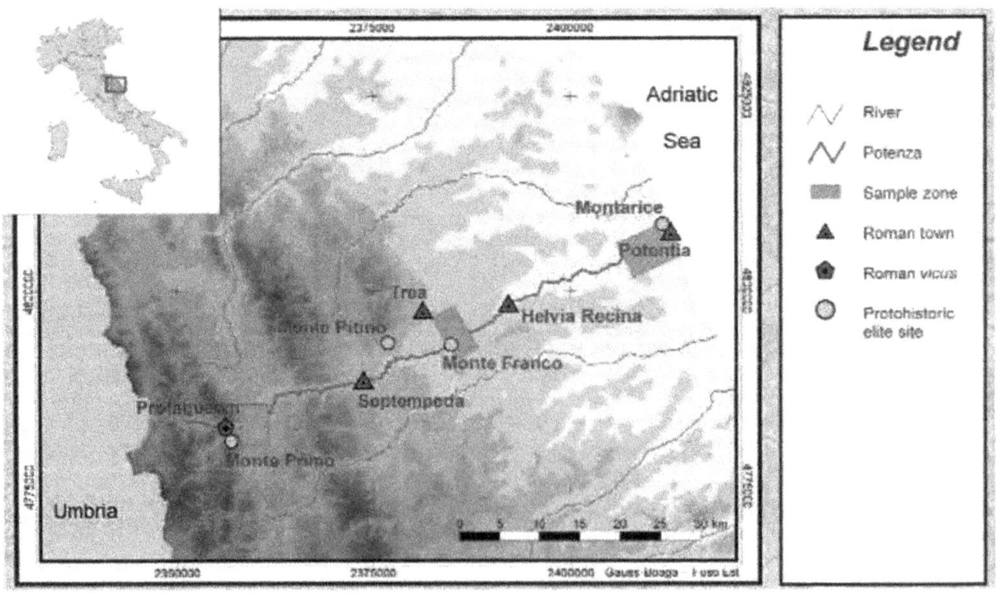

Figure 1 Location of the three sample zones

(Fig.1) The geo-archaeological research consists of four important cornerstones: aerial survey, field survey, geomorphological survey and a historical survey. The aerial survey comprises the collecting of as many vertical photographs and satellite images as possible. Besides this passive part, this survey also has an active part: taking oblique aerial photographs from a low flying aircraft. Till now, this has been done in a very thorough way: on the one hand, different flights were made in different seasons; on the other hand, pictures were taken of the whole valley with a specific interest in the three sample zones. Systematic prospection of the ploughed fields, always carried out in the month of September, is the method used in the field survey. In search for archaeological evidence, intensive linewalking was chosen as prospection method. Therefore, an interval of 10 to 15 meter between different walkers is aimed for. Scholars of the Department of Geography conduct the geomorphological part of the project by making as many field observations as possible. Additionally, some augering is also performed to reconstruct the evolution of the coastline and study the problem of alluviation and colluviation.[2] Finally, an intensive historical survey to study known archaeological sites of all periods into detail is implemented as well. Moreover, some research of toponymical and historical written information is on-going.

Since the PVS started more than three years ago, one can be convinced that the amount of gathered information is substantial.[3] Managing such quantities of -almost all geographically linked- data has become much easier with the introduction of Geographic(al) Information Systems. From the beginning of the project, it was the aim of director Frank Vermeulen to incorporate all gathered survey data into a GIS-context.

Building the GIS

In fact, before data could be managed and analyses computed, the building of the GIS had to occur. Working with different researchers in different disciplines, each of them with own particular data, a centralization of all information seemed to be necessary. The creation and storage of data was for a long time performed on different computers, but in January 2003, one central PC was purchased and configured to contain all possible data in an orderly and easily accessible way. In consequence, this PC also functions as the centre of all GIS-analyses.[4] The data stored in the GIS consists of access-databases (although they should lead to one, all-embracing database which contains all data of the four surveys) and a lot of raster and vector files. The greater part of the vector files are digitized topographical maps (scale 1/10,000, CAD-format), obtained thanks to the Regione Marche. Where needed, manual digitizing (with digitizer tablet or on-screen) completed the vector data. However, spatial data can also be stored as raster, a fundamentally different format. In ArcView, a distinction is made between raster images and raster maps, also called grids. Till now, the number of grids is very limited (some DEMs and distance grids), in contrast with the image data. Besides the huge collection of aerial and on-ground photographs (mostly of artefacts), some scanned maps (topographical, orthophoto and cadastral), a few Landsat satellite images and a -fast expanding- number of rectified images complete the GIS-data at present.

After outlining the project, the structure of the GIS and the data that are available, just one more step has to be taken before one can start visualising and analysing: be(come) aware of the data quality. After all, it is important to understand that it is possible to have 'error free' attribute data (in terms of the method employed), but impossible to have 'error free' spatial data, because our multi-dimensional reality will always be different from the digital representation of it. Therefore, the entering of spatial information can rightly be seen as the Achilles' tendon of almost every GIS (Voorrips 1998:255).

Quality affecting factors that can be mentioned are the accuracy and precision of the source data, the interpolation methods employed, the scale of the data, the georeferencing system used, the data collection technique and the sampling strategy that are applied, the process of scanning and digitizing, etc. In this respect, a lot of attention is paid to metadata and -a limited amount of- quality control.

Metadata and Quality Control

Metadata can be seen as 'data about the data' as it describes the content, quality, condition and other characteristics of data. It provides us with information about the who, what, when, where, why and how of a data set. In consequence, proper metadata are critical in preserving the usefulness of data over time and it is crucial to supply all kinds of GIS-data with this information. Different metadata standards exist. In the PVS, the geospatial data is documented using the Federal Geographic Data Committee's (FGDC) Content Standard for Digital Geospatial Metadata, which the Federal Geographic Data Committee approved in June 1998 (Federal Geographic Data Committee 1998). To build FGDC-compliant metadata, an easy-to-use application called the ArcView Metadata Collector was achieved, allowing to create metadata for any data type supported by ArcView.[5]

Besides metadata, quality control is another important topic in the PVS-GIS. Using spatial data in a GIS-operation implies that error (defined as the discrepancy between a given - digital - value and its value in reality) in the input will propagate to the output of the operation. In this respect, the resulting output is a function of the input values and inaccurate input values will automatically affect the computed results (Heuvelink 1993:23-5). Therefore, it is significant to test the accuracy of some digital maps. One example is the Digital Elevation Model or DEM. In the majority of GIS-data sets, the DEM is the most important layer. Except information about the height, it is also used to generate slope and aspect and it serves as well as basis for cost path analysis, distance calculations etc. As a consequence, it is the fundamental basis to study the topographical features of archaeological sites.

One always needs an interpolation algorithm to construct a usable DEM. Some digitized high points or contour lines form the basis of the DEM, but they do not form a continuous surface with height values yet. To retrieve this, an interpolation method is ran and will attribute the places without data a figure that represents the height of that specific place. However, different interpolation methods produce different terrain values. We must surely accept that the final DEM will be an approximation of a continuous phenomenon, but how closely this approximation reflects reality can be calculated (Hageman and Bennett 2000:114-5).

Presently, the accuracy of the DEM applied for the first sample zone was calculated. This control occurred on the basis of two tests. Initially, the digitized contour lines were checked visually by generating a TIN (Triangulated Irregular Network).[6] A closer look at the characteristic points and lines as well as the hillshaded relief revealed some profound errors, owed to the manual labelling. Afterwards, a quantitative test was ran. Eleven different interpolation methods[7] were executed. The predicted elevations for a given DEM were compared to some reference points (none of them was located on a contour line) and the discrepancy between both was calculated. With these figures, a global RMSE (Root Mean Square Error, which is mathematically the same as standard deviation) could be calculated for a given DEM. This procedure was repeated for each interpolation. Subsequently, different errors could be compared and the lowest RMSE revealed the preferable DEM. In the near future, homogeneous computations will be used to create the DEMs for the second and the third sample zone. Furthermore, it is also hoped some more digital files can be tested on their accuracy as these tests really give a good idea of the quality of the files one is working with.

First Analytical Results

Setting up the GIS on the one hand and still on-going analysis from the different researchers on the other, caused the analyses to be very limited and temporary till now. The executed computations presented here are all applicable to the first sample zone.[8] The research was divided into two parts: analyses of data linked with the cadastral maps and analyses of the environmental characteristics. The first part covered items as the calculation of the precise prospected area (3.2 km²), the determination of the proportional ratio of the visibility classes (19.6% bad; 3.4% moderate; 77% good) and the computation of the proportional ratio of the fields with erosion (29.9%). Furthermore, a chi-squared test of goodness-of-fit revealed for a significance level a = 0.05 that sites are equally distributed across fields with and without erosion. A determination of the off-site proportions (0 = 5.7%; 1 = 56.5%; 2 = 19.2%; 3 = 18.6%)[9] as well as a calculation and visualisation of the total and periodical site density (total = 17.19 sites/km²; pre- and protohistoric = 3.13 sites/km²; Roman = 13.75 sites/km²; medieval and post-medieval = 1.12 sites/km²) complete this first part. One can notice the high density of Roman sites. However, a further subdivision per period is needed and was on-going during the writing of this article.

The second part dealt with the relation between the sites and the environmental characteristics as height, slope, distance to the nearest water source, etc. The analyses of these environmental characteristics are based on a study by Hodder and Orton (1976:226-229). They examined the distribution of 173 Iron Age coin finds in relation to Roman road locations in central and southern England. Using manual methods of pre-GIS days, they performed a statistical analysis which revealed a significant association between the coin distribution and the Roman road network. With GIS, such analyses can be executed a lot faster and with results that are even more pre-

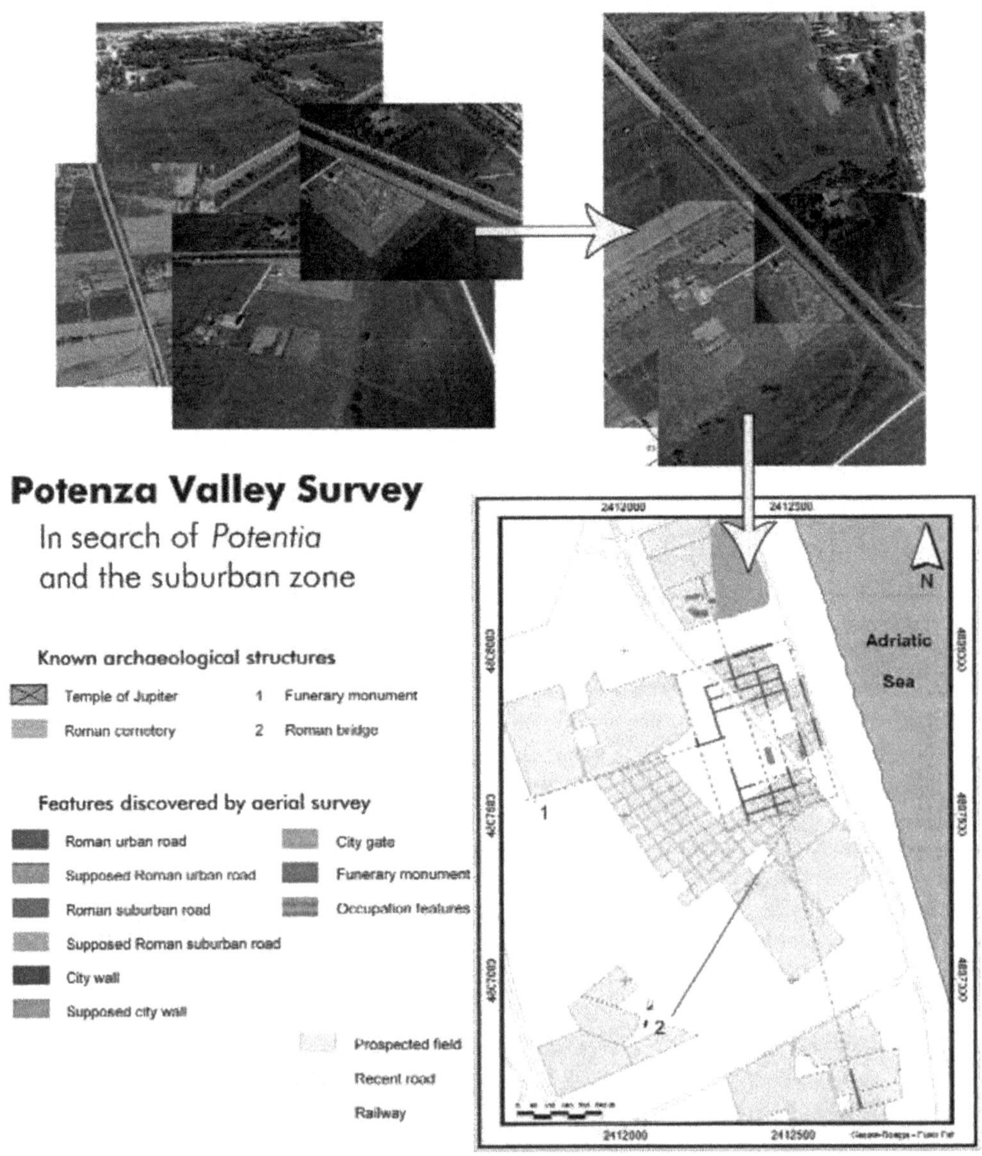

Figure 2 Mapping Potentia and the suburban zone

cise and accurate. Using the Kolmogorov-Smirnov one-sample test, the relation of the sites (per main period) and isolated finds to the following environmental characteristics was determined: elevation, slope and aspect; distance to the waterways, to the Potenza, the springs and the total hydrographical network; distance to the nearest road, to the nearest flint; view on the Potenza. Some relations did occur (e.g. the isolated Stone Age artefacts were obviously linked to the springs and the Roman sites to the contemporary rural roads), but generally spoken, no specific connection between the environmental characteristics and the sites could be deduced for the moment, partly due to -as underlined before- the fact that the periods were too broad delineated for detailed analyses.

PROSPECTS

In the near future, more and new analyses should replace these preliminary results, although the small number of sites for some periods will make analysis -even in the future- problematical.

For the moment, a new kind of research is in progress: rectifying oblique aerial photographs to map all crop, soil and shadow marks. Of course, a thorough interpretation of the aerial images will follow, but we want to take the research one step further by comparing the field data with the oblique photographs. Do features of both data sources match perfectly? What is the amount of artefact scattering due to ploughing or erosion? Are these relations constant in one sample zone and -of course- how big are the (dis)similarities between the zones? These are just some questions that could be asked. Furthermore, the combination of both data sources can also give a surplus value to each data source, leading to a better interpretation of the archaeological features and a greater understanding of the site's structure and its occupation phases. The Roman site of Potentia can function as a nice example. Using AirPhoto 2.20[10] and a digital topographical map, the most interesting photographs of the former Roman city were rectified. On the photomosaic, a lot of different archaeological features could be spotted and interpreted. (Fig. 2) Besides the city, also some images in the suburban area were rectified. Here, some Roman roads and several funerary monuments could be distinguished. In the near future -i.e. when the material processing has finished- these results will be compared with the data from the field survey. On the one hand, the oblique images will give a better idea of the 'real' proportions of the city, on the other hand the sherds and pieces of marble, glass and tesserae can tell something about the intra-site spatial organisation and temporal delineation of the features.

Besides, this highly significant data can -and will- be combined with the geomorphological and historical information in the three survey zones, just to make the whole settlement process easier to understand and the interpretations more reliable. In this way, the capabilities and advantages GIS offers will be completely expressed in the PVS-project. Apart from the amount of work that still needs to be done, it can already be stated that the combination of GIS and different survey techniques is answering a lot of questions posed at the start of the project.

1 For a detailed overview of the main goals, the reader is kindly asked to consult Vermeulen 2002a.

2 An overview of this work as well as the first results can be read in Vermeulen, De Dapper, Boullart, De Vliegher and Goethals 2002.

3 For more detailed information about the project and the work performed till now in the three sample zones, the following articles are recommended: 'Vermeulen 2002b', 'Vermeulen and Boullart 2001' and 'Vermeulen, Monsieur and Boullart 2002'.

4 Hardware: 2.4 Ghz processor; 512 MB RAM; 75 GB hard disk; 1,44 MB floppy drive, CD-R/RW 48/16/48 and several printers. Software: ArcView3.2 as GIS-software, Minitab 13 to perform statistical analyses, AirPhoto 2.20 to rectify images and Photoshop 7.0 to edit images. All programs are working on a Microsoft Windows XP operating system.
It is also worth to underline that, besides this central PC, every member of the PVS-team has his own PC, sometimes with some important peripheral instruments as a CalComp 9500 digitizer tablet, an A3-scanner (EPSON 1640 XL) and a slide scanner (Canoscan 2700F).

5 See http://www.csc.noaa.gov/metadata/text/download.html.

6 This is a vector-based structure, composed of a set of triangular facets.

7 Exponential, Circular, Spherical, Gaussian and Linear Ordinary Kriging, Universal Kriging -with linear and quadratic drift-; Spline, Inverse Distance Weighting; Trend and TIN. All of them were run with different values for the variable parameters. As a result, 35 grids were calculated.

8 A detailed report of all analyses, including the computation of the DEM, can be read in Verhoeven 2002.

9 Off-site 0: 0 artefacts /50m; 1: <=5 artefacts/50m; 2: <15 artefacts/50m and off-site 3: >=15 artefacts/50m

10 Designed by Irwin Scollar, this software is specifically developed to rectify archaeological images made with hand-held uncalibrated cameras (Scollar 2002, 167).The quality of the program has to do with the amount of interesting features. It's one of the only low-cost programs that offers different transformations, with and without height values. Furthermore, it offers three interpolation methods, a calculation of the mismatch between the source and the target image (control of the accuracy of the rectification!), has a lot of image processing techniques and supports different GIS-formats.

REFERENCES

FEDERAL GEOGRAPHIC DATA COMMITTEE, 1998. Content Standard for Digital Geospatial Metadata. FGDC-STD-001-1998, revised June 1998, Washington, D.C.

GILLINGS, M. and WISE, A. (eds.), 1999. GIS Guide to Good Practice. In Archaeology Data Service, Oxford.

HAGEMAN, J. and BENNETT, D., 2000. Construction of Digital Elevation Models for Archaeological Applications. In Wescott, K. and Brandon, R. (eds.), Practical Applications of GIS for Archaeologists. A Predictive Modeling Toolkit, London:113-127.

HEUVELINK, G., 1993. Error Propagation in Quantitative Spatial Modelling - Applications in Geographical Information Systems. In Nederlandse Geografische studies 163, Utrecht.

HODDER, I. and ORTON, C., 1976. Spatial Analysis in Archaeology. In New Studies in Archaeology, Camebridge.

MILLER, P., 1999. The Importance of Metadata to Archaeology - One View from within the Archaeology Data Service. In Dingwall, L., Exon, S., Gaffney, V., Laflin, S. and van Leusen, M. (eds.), Archaeology in the Age of the Internet - Computer Applications and Quantitative Methods in Archaeology, CAA 97, Proceedings of the 25th Anniversary CAA Conference Held at the University of Birmingham, UK, 10-13 April 1997. Oxford (= BAR Int. Ser. 750):133-136.

NOAA Coastal Services Center, 07 01 2003. ArcView Metadata Collector v2.0 Extension (http://www.csc.noaa.gov/metadata/text/download.html).

PERCOSSI SERENELLI, E., 1981. Le vie di penetrazione commerciale nel Piceno in età protostorica:Nota preliminare. PICUS. Studi e ricerche sulle Marche nell'antichità 1:135-144.

SCOLLAR, I., 2002. Making Things Look Vertical. In Bewley, R. and Raczkowski, W. (eds.), Aerial Archaeology - Developing Future Practice, NATO Science Series, Series I: Life and Behavioural Sciences 337, Amsterdam:166-172.

VERHOEVEN, G., 2002, unpubl. The Potenza Valley Survey. Een GIS-analyse van de eerste veldcampagne en een duiding van de begrippen "Geografisch InformatieSysteem" en "field survey". Dissertation Ugent, Gent.

VERMEULEN, F., 2002a. The Potenza Valley Survey (Marche). In Attema, P. et al., New Developments in Italian Landscape Archaeology, Oxford, BAR Int. Ser. 1091 :104-106.

VERMEULEN, F., 2002b. Uit de lucht gegrepen: de bijdrage van archeologische luchtfotografie in het "Potenza Survey" project. In Tijdschrift voor Mediterrane Archeologie 26:33-42.

VERMEULEN, F. and BOULLART, C., 2001. The Potenza Valley Survey: Preliminary Report on Field Campaign 2000. In Babesch 76:1-18.

VERMEULEN, F., DE DAPPER, M., BOULLART, C., DE VLIEGHER, B-M. and GOETHALS, T., 2002, forthcoming. Geo-archaeological Approaches in the Valley of the Potenza (Marches, Central-Italy). Proceedings of the congress of Paris.

VERMEULEN, F., MONSIEUR, P. and BOULLART, C., 2002. The Potenza Valley Survey: Preliminary Report on Field Campaign 2001. In Babesch 77:49-71.

VOORRIPS, A., 1998. Electronic Information Systems in Archaeology. Some Notes and Comments. In Archeologia e Calcolatori 9:251-267.

Ancient Tanagra (Grimadha): Geophysical Prospections and Modelling for Understanding of Urban Plan and On-site Activity Areas.

Branko Mušic
University of Ljubljana

Božidar Slapšak
University of Ljubljana

Emeri Farinetti
University of Leiden

Abstract

Research in Tanagra in the framework of the Boiotia Project, co-directed by John Bintliff and Božidar Slapšak, is since 2002 part of the Late Roman Cities and Countryside project, an Interuniversity Attraction Poles programme (IUAP V/09). It is a survey oriented research of a complex urban site for which no foreseeable excavation can be envisaged. It applies a battery of non-desctructive methods, including archaeological, geophysical, surface morphology, and architectural survey.
The paper will concentrate on the potential of combined use of raster based GIS; 2D and 3D magnetic modelling; DEM, and of unsupervised classification for interpreting the results of geophysical prospection. 15 ha, about one half of the intra-muros area have been surveyed. Data sets acquired by geophysics are complementary in many details. Multivariate statistics is used to build up a site map in a single image, where archaeologically relevant units are discernable by spatially defined classes, based on the level of correlation between input data sets. The output is a composite image of differing, but normalised input data, and that could lead to partial automation of the interpretation procedure. 3D visualization is used for presentation of measured anomalies in physical fields, and of the final interpretation of the urban plan.

For the third consecutive year, Tanagra is the focus of research within the Boeotia Project, now co-directed between Leiden and Ljubljana Universities, and since 2002 also part of the Urban and Rural Transformation in the Western and Eastern Roman Empire Project, an Interuniversity Attraction Poles programme (IUAP V/09) funded by the Belgian federal government and co-ordinated by Marc Waelkens of the Catholic University of Leuven (Sagalassos).[1] A number of city territories in Boeotia have been studied during the 20 years or so since the project was initiated by John Bintliff and Anthony Snodgrass back in 1978 (Bintliff and Snodgrass 1985, 1988, Bintliff, Howard and Snodgrass 1999). All research within the project is survey-oriented, no excavation is envisaged as part of the scheme, and at Tanagra, no other institution is contemplating excavation at this point.

Figure 1 Results of magnetic prospection (Geometrics G-858). Dynamics: - 12 nT to +12 nT. View from the South

Neither has the urban area of Tanagra been the object of any substantial archaeological excavation in the past. While the cemeteries have been extensively excavated and robbed for the lovely Tanagrinas since the 70s of the 19th century, the only excavation within the city walls has been that of a Greek team of an early Christian church and possible temple structures on the acropolis, published in 1890 (Konstas 1890). Surface architectural remains are scarce, and have been largely removed at the end of the 19th century for construction in the nearby village of Ag. Thomas. An attempt was made in the mid-eighties by a Canadian team directed by Douane Roller to identify the urban plan based on architectural survey. His map published in the BSA in 1987 (Roller 1987) has the status of academic authority among classical archaeologists, and was naturally taken as the starting point of our investigations. The intra-muros area is now protected and is grazed by a flock of sheep.

The novelty Tanagra brings to the Boeotia project, which has by now accumulated an impressive database from systematic archaeological surface survey, both off-site and on-site, including urban, is on the one hand, the GIS support developed by Emeri Farinetti, and on the other, the new interest in integrated urban survey, applying a battery of techniques, including surface morphology (Farinetti and Sigalos 2002) and architectural survey (tested in 2002, see Bintliff et al. 2003, and launched fully in 2003), as well as geological and geophysical prospections (Slapšak 2001). This is where the Project has developed a most fruitful synergy of the Leiden and the Ljubljana teams (Bintliff et al. 2000, 2001, 2002). The goal here is to explore the potential of non-destructive methods in the study of classical urban sites, and the rather favourable conditions in

Tanagra make it a good case in point, which permits such exploration to be pushed to the limit.

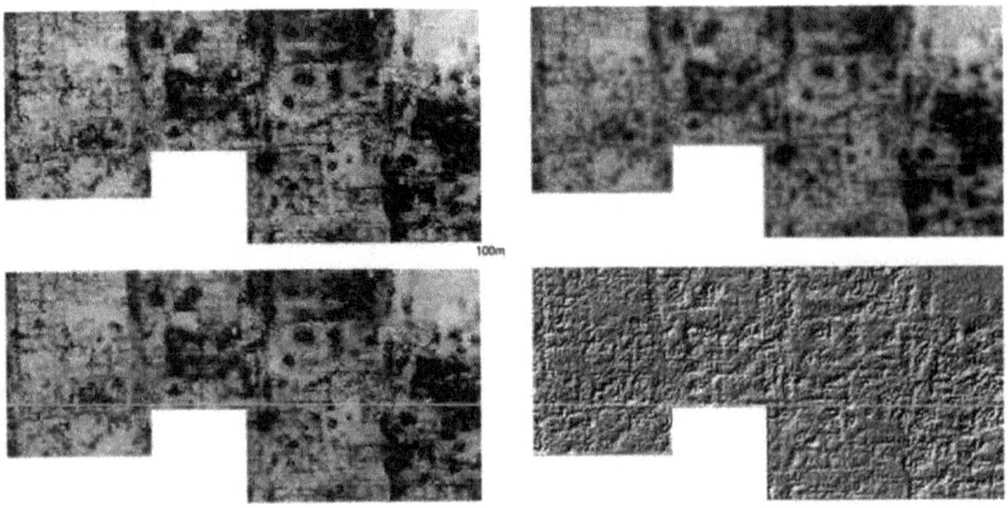

Figure 2: Resistivity. Raw data (top left) and processed images

An area of 1 ha was chosen in 2000 in the north-eastern corner of the walled city, to test the equipment and techniques of geophysical prospection for best performance in this geological and archaeological setting. Based on this testing, two methods were chosen to map the totality of the site: measurements of total magnetic field by caesium magnetometer Geometrics G-858, and geoelectric mapping by resistance meter Geoscan RM15. Such choice is justified because of the best labour input / result ratio for total mapping, and because of the complementarity of the two techniques in detecting types of archaeological features present. Further methods can then be applied for targeted survey, to address specific problems identified after completion of mapping (Bintliff et al. 2000:109-126, Slapšak 2001:10,19-23).

The geological survey identified types of bedrock and building materials at the site, as well as their magnetic susceptibility values. For the purpose of our geophysical prospection, it was important to note that both the bedrock and the prevalent building material (liassic and jurassic limestones) have very low magnetic susceptibility. Building material made of igneous rock is also found, but the frequency is low, so this material was obviously used only exceptionally, and in a limited number of buildings. Such material is not found in the vicinity of the site (Slapšak 2001:11-14, Bintliff et al. 2001:73).

On the other hand, magnetic susceptibility of the soil shows relatively high values on the site, so a good magnetic contrast can be expected between built structures and the soils in which they are embedded. Naturally, the highest values are those of brick and burnt clay. Thus the contrast in magnetic susceptibility of the typical materials found on the site should enable us to identify both objects with induced magnetization, and those with thermoremanent magnetization. The magnetogram shows clearly that that is indeed the case here. Linear anomalies in induced magnetization point to walls and streets, and discrete areas of thermoremanent magnetization witness to activities connected with intensive use of fire.

In contrast to magnetometry, the results of geoelectric mapping seem less informative at first sight. An additional problem here is that of variable surveying conditions due to changing moisture of the soil. By filtering the results by areas surveyed under comparable moisture conditions, we can nevertheless extract important information from geoelectric maps. It is important to note here that, while the major features such as streets are less clearly discernible than by magnetometry, this methods adds important detail when it comes to identification of walls within the building blocks. The Early Christian church in the eastern part of the city is a good case in point: such crucial detail as the apse is in fact only visible on the resistivity map.

As the result of geophysical prospection, the Roller map can now be corrected in many parts. The position of the North-South axes of the urban structure are confirmed by geophysics, and so are many other details. On the other hand, the

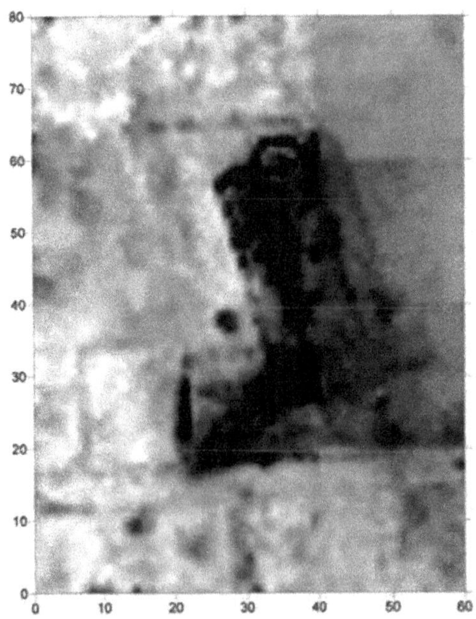

Figure 3: Results of resistivity measurements (Geoscan RM15) at the area of the Early Christian church

Figure 4: Results of magnetometry vs. Roller map (1987) (top right)

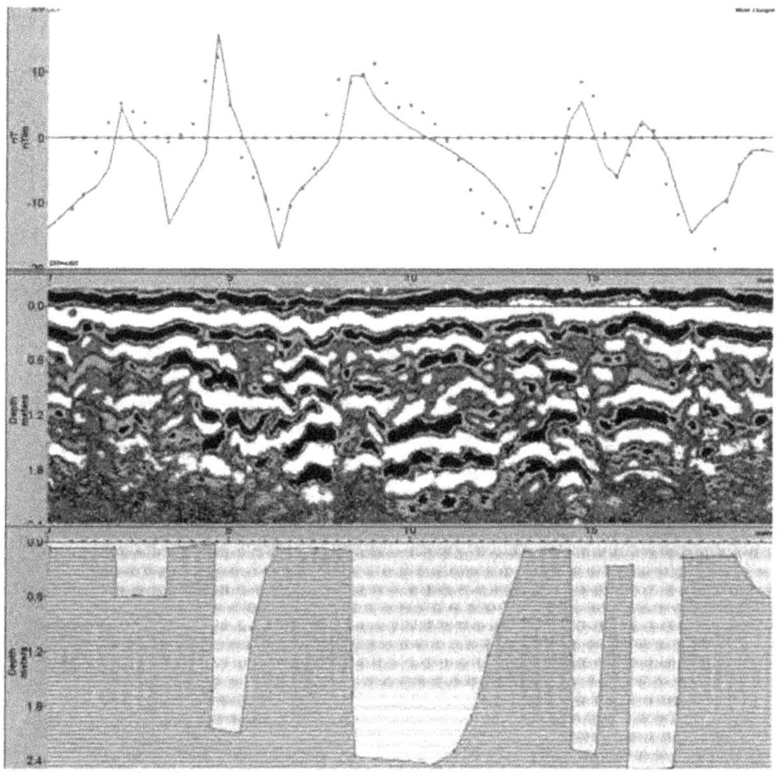

Figure 5 Magnetic modeling of architectural remains. Georadar and magnetic profiles were used for more realistic models of architectural remains. Crossection across the street is represented

East-West axes are positioned more to the South, and the Agora is not where Roller would have it, so the main structuring axis, which crosses the central-southern part of the city looks different on our magnetogram, and consequently the South Gate must be sought in another place. The visible architectural remains, which Roller relied upon exclusively for his reconstruction of the urban grid, were for him frustratingly scarce, so we should be impressed by the many features he identified correctly (however the current Project with a larger team and more time have recorded a large number of in situ architectural features in all parts of the city).

A high priority topic in archaeological investigations in Tanagra is certainly industrial activity, and most notably its ceramic production. While the area within the city walls is not the obvious place to search for potters' quarters, geophysics did revealed foci of high thermoremanent magnetization there, which are indicative of some kind of (industrial) activity using high temperatures. If the objects with induced magnetisation are best identified by mapping the gradient of magnetic values, those with thermoremanent magnetization show best on the readings of the upper sensor of the magnetometer, which is not affected by the iron objects on the surface and in the topsoil.

Discrete areas of thermoremanent magnetisation will appear as magnetic dipoles. Ideally, their orientation will be that of the North magnetic pole at the time of the magnetization event. In practice, the orientations will be affected by the state of preservation of the objects, by the vicinity of other magnetic objects at the time of the magnetization event, and by the actual archaeological context. Regularities observed in orientations of dipoles may permit dating of the structures concerned. Interestingly enough, analysis of surface finds in the area where such features appear, point to some medieval / early modern ceramic production at the site (Poblom, J. in: Bintliff 2002).

The geometry of the structures with thermoremanent magnetization can be analysed by modelling. Again, the state of preservation will be a crucial factor determining the results. Geometry can be assumed to be based on the shape of the anomaly on the magnetogram. An hypothetical model can then be constructed, and then adapted to the measured values to develop a more realistic model of the archaeological feature. Both simple and complex objects can be analysed in this way.

The geometry of objects with induced magnetization will in most cases be rather complex, so radar profiles can be used to facilitate modelling. In Tanagra, radar profiles were taken as part of the initial experimentation over interpretable anomalies read on the magnetogram, and chosen for testing such an analysis. One profile was taken across a street, another over the city wall, and yet another over walls of a building within a city block. The profiles were taken to further refine initial hypothetical models of the structures. The degree of fit between the model and the empirical values was then observed, and the hypothetical model adapted to the measured figures. Thereby, a more realistic model of the structure could be developed.

This is only part of the experimentation, which we are engaging in, with a view to explore ways to extract the most of the archaeological information from the exciting images produced by geophysics in Tanagra. At future conferences, we hope to be able to present more of the same, and soon we hope to have also some consistent applications on architectural complexes chosen for detailed analysis.

[1] The Tanagra Project is jointly directed by John Bintliff and Bozidar Slapsak, with Kostas Sbonias as Assistant Director; for the IUAP Project, see Waelkens, M. e.a. 2003.

REFERENCES

BINTLIFF, J. and SNODGRASS, A.M., 1985. The Cambridge/Bradford Boeotian Expedition: the First Four Years. Journal of Field Archaeology 12:123-161.

BINTLIFF, J.L. and SNODGRASS, A.M., 1988. Mediterranean survey and the city. Antiquity 62:57-71.

BINTLIFF, J.L., HOWARD, P. and SNODGRASS, A.M., 1999. The hidden landscape of prehistoric Greece. Journal of Mediterranean Archaeology 12-2:139-168.

BINTLIFF, J.L., FARINETTI, E., SBONIAS, K., SIGALOS, L., SLAPŠAK, B. and MUŠIC, B., 2000. The Tanagra survey. Report on the 2000 season. Pharos 8:93-128.

BINTLIFF, J.L., EVELPIDOU, N., FARINETTI, E., MUŠIC, B. and SLAPŠAK, B., 2001. The Tanagra survey. Report on the 2001 season. Pharos 9:33-74.

BINTLIFF, J., FARINETTI, E., MUŠIC, B., POBLOME, J., SARRI, K., SBONIAS, K., SCHACHTER, A., STISSI, V., SLAPŠAK, B. and VIONIS, A., 2002, forthcoming. The Tanagra project: Third preliminary report, The 2002 season.

FARINETTI, E. and SIGALOS, L., 2002. Detailed Topography and Surface Survey. What is the point? Tanagra City Survey 2000. In Burenhult, G. (ed.), Archaeological Informatics: Pushing the Envelope CAA 2001 - Proceedings of the 29th Conference, Gotland, April 2001, BAR Int. Ser. 1016:385-391.

KONSTAS, D.D., 1890. Anaskafai Tanagras. PAE:33-35.

ROLLER, D.W., 1987. Tanagra survey project 1985, The site of Grimadha. BSA 82:213-232.

SLAPŠAK, B. (ed.), 2001. Tanagra 2000: Report on preliminary geoarchaeological and geophysical investigation. University of Ljubljana, Ljubljana.

WAELKENS, M. et al., 2003. IUAP-PIA V/09 Urban and rural transformation in the western and eastern Roman Empire. Interdisciplinary archaeology of the late antique and early medieval times, Report on the activities of the network during 2002, Leuven.

A Study on the Cognitive Structure of the Historical European Quarter in Cairo

SEE THE CD FOR THE EXTENDED VERSION

ALI ESSAM EL-SHAZLY
ASSISTANT PROFESSOR, ENVIRONMENT AND TECHNOLOGY DIVISION, GRADUATE SCHOOL OF ECONOMICS, HITOTSUBASHI UNIVERSITY, 2-1 NAKA, KUNITACHI-SHI, TOKYO 186-8601, JAPAN.
ELSHAZLY@ECON.HIT-U.AC.JP

ABSTRACT

In this cognitive study, the incorporation of digitized map-overlay clarifies the chronology of urban change for the European Quarter in Cairo. Layers of shape files with attributed data were generated for the quarter's detailed 'Insurance Plan' in 1905, and overlaid by similar surveys in 2003. The urban tissue in 1905 superimposed a network of boulevards to link major open spaces, as well as integrating with Old Cairo through transitional spaces. Meanwhile, the node of 'Ezbekieh Plaza' filtered the function of buildings in correlation with the urban tissue, thus defined as the composer of the quarter. When overlaying the urban information in 2003, the squat or retreat of micro spaces had weakened the network of boulevards. This together with the lack of enforcing the function of buildings and open spaces had facilitated the dissolution of node structure in present Cairo. The neglect of the core 'Ezbekieh Plaza' is obvious to affect the assemblage of historical urban structure in Cairo.

INTRODUCTION

The foundation of the European Quarter in Cairo dates back to the reign of the Egyptian ruler, Ismail (1863-1879), who adored the European civilization. The French engineer, Haussmann, in 1867 first planned the quarter with, Le Grand, in 1874 further realizing the plan (Ali 1998:32). The British occupation in 1882 continued to develop the colony, and later in 1905 the European Municipality set the 'Insurance Plan' for the enlarged colonial urban structure (Caneri 1905). This plan resembled a GIS scheme, manually though, of linking the design plan to the detailed street-building information, function, and insured building items as a precaution measure against the risk of damage by the colonized (El-Shazly 2003:175-182). However, in 1952 the revolution for nationalization resulted in the mass exodus of Europeans from Egypt, and followed to the present by urban transformation. Meanwhile, in 2002 Cairo Governorate addressed the architectural conservation of the historical quarter (The Urban Planning Association 2002). The description of physical characteristics, nevertheless, lacked the strategies of conservation. In this study the GIS technique of overlaying maps attempts to compare the cognitive structure of the colony between the formation (1867-1951) and the nationalization (1952-to date) according to: 1) urban tissue, 2) nodes and landmarks, and 3) function (among the readings this paper is based on are: Kostof 1999, 1991, Krier 1979, Rossi 1986, Lynch 1981, 1960, Cohen 1999). Through this analysis the key issues of Cairo's European Quarter are pointed-out for conservation.

THE FORMATION OF THE EUROPEAN QUARTER (1867-1951)

The status quo of Cairo in 1867 secluded a walled town eastwards of the Nile bank. The Haussmannization plan in 1867 of Cairo formed the plaza named 'Ezbekieh' to border the Islamic Town. In the same year the French engineer, J. P. De Schamps, landscaped the open space of the plaza on the precedence of Parisian gardens (Mubarak 1889:211-213). From Ezbekieh, in 1867 'Rue Neuve' crossed the town straightforward to the eastern wall-gate, while in 1872 'Boulevard M. Ali' split the town diagonally towards the 'Manshieh Plaza' located at the town's southeast edge. The latter boulevard was extended through 'Rue Claude Bey' to the Central Station of railway terminal (Scharabi 1989). These major boulevards were superimposed on the irregular Islamic urban tissue to structure the axial orientation of Cairo. The total buildable area of blocks webbed interior courtyards and narrow passages into the Islamic urban tissue. Later in 1874 the French engineer, Le Grand, planned the network of radial streets to spread over the alluvial green field bound between the town wall and the Nile course (Ali 1998:27-29). The state ownership not only eased the urbanization of this field, but also the ruler, Ismail, gifted the land parcels to any citizen (regardless of nationality) on condition to develop European-style buildings with gardens (Owen 1969:337). The real-estate business adopted the free-market of ownership and rents, with taxes imposed to reinvest in public facilities (Owen 1969:337-339). Despite the geometric radial pattern of the European Quarter, another gridiron pattern was overlaid following the Nile stream. In 1874 the streets occupied 30% of the total land area, while the buildings formed only 13 % leaving the rest of the land as gardens of approximately 63 hectares (Ali 1998:40). In the 1900's the European Quarter experienced a building boom of average 3,344 permissions of construction per year compared to 1,477 during the 1880's and 90's (Owen 1969:338). This resulted in the partial replacement of previous gardens by buildings along the perimeter of blocks.

On the macro level of Cairo, the projected streets from the 'Ezbekieh Plaza' terminated in major nodes of the city (Fig.1.) In this regard, the peripheral railway plazas of 'Manshieh' and 'Central Station' were linked to the central 'Ezbekieh' for commercialization. Meanwhile, the traditional Islamic node of 'Azhar Mosque' was integrated through 'Rue Neuve' to the central plaza. The radial network of the European Quarter, however, nodded in round-points that distributed allover the area. The pole of 'Ismailia' round-point on the Nile radiated the axial boulevards of 'Kasr El-Nile' and 'El-Bustan,' respectively, towards the other poles of 'Ezbekieh plaza' and the royal 'Abdin Palace' with its front plaza. The latter two plazas were linked by boulevard 'Abd El-Aziz,' of dense pedestrian flow. Another node of special character developed all through the Nile to form a recreational promenade of continuous gardens. The landmarks culminated in the focal point of 'Ezbekieh' with The Opera in 1869 designed by the Italian engineers Scala & P. Avoscani (Ali 1998:70), in addition to Hotels Continental & Shepherds, The Mixed Tribunal and the Central Post Office. Other landmark buildings developed in the various nodes of the European Quarter such as the Museum of Egyptian antiquities located in 'Ismailia' round-point. Furthermore, vistas of sculptures were adorned to the round-points, which together with towers and minarets of religious buildings formed vertical landmark features.

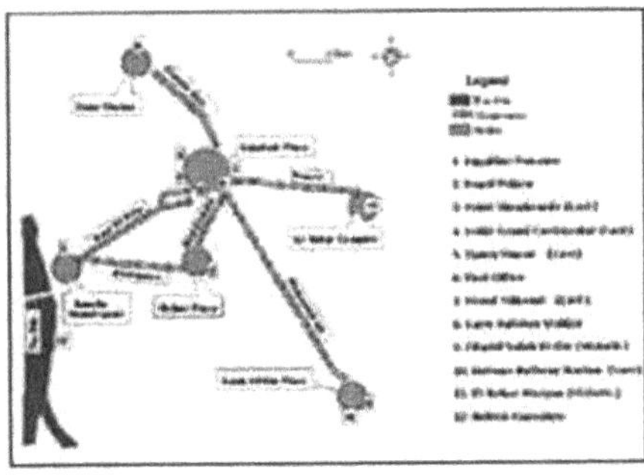

Figure 1 The Nodes & Landmarks of Cairo Between 1905 & 2003

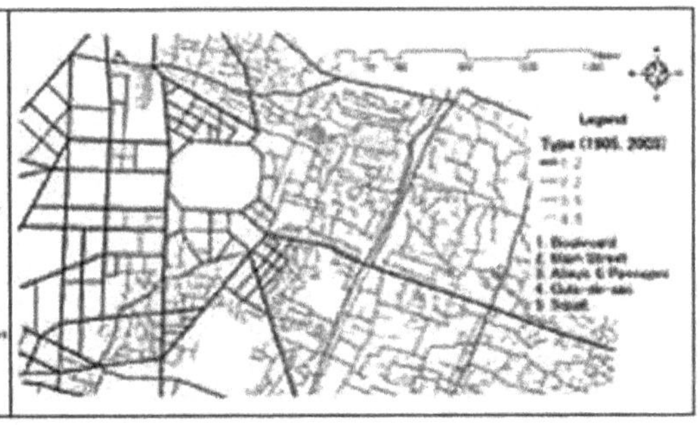

Figure 2 Overlaying the Street-Networks of 1905 and 2003

The rise of Cairo's European population from 35,385 in 1895 to 55,987 in 1905 demanded new policies of development (Owen 1969:337). The European Municipality of Cairo was founded in 1905 to enforce the colonial urban structure. The Municipal policy of 'Insurance Plan' in 1905 not only correlated the urban form to the function of the colony, but also the 'insurance' of building items in risk of damage (Insurance Plan of Cairo 1905). In this plan, the commercial center of 'Ezbekieh Plaza' was regulated a 27 m building-height of Neo-renaissance style façades. Commercial 'galleria' buildings, however, developed on both sides of 'Rue Neuve' with characteristic interior passages of average three-stories (ibid). Arcades were built along boulevards 'Claude Bey' and 'Mohamed Ali' for retailing activities with residential function on the upper floors (ibid). Further east and south of 'Ezbekieh' prevailed the preexisting housing of the Islamic Town (ibid). The zone to the west of 'Ezbekieh' replaced the single-family houses of surrounding gardens by maximum 36 m building-height of Victorian and Neo-Gothic façade styles. In this zone flourished the European business offices, commercial banks, insurance companies, prestigious shops and restaurants (Ali 1998:59-65). The villa-type of housing, meanwhile, shifted to the new residential area of 'Garden City' in the southwest zone of the colony. This developed Baroque style houses and palaces of flower gardens along characteristic curving street-pattern on the British conception (Ali 1998:80-83, 184). Later during the 1920's and 30's emerged apartment buildings and governmental buildings of Art-Deco styles that prevailed in Egyptian cities.

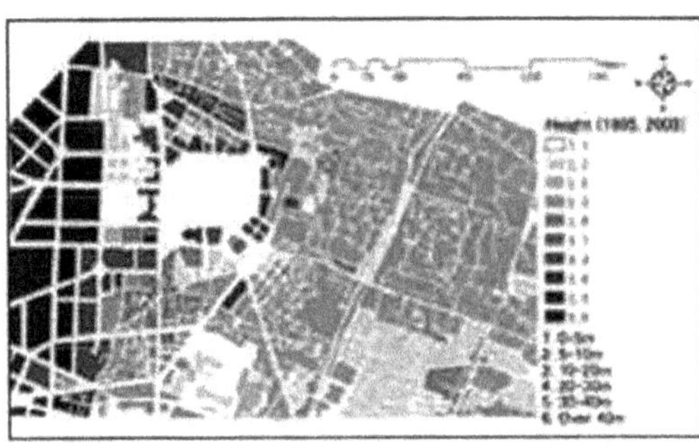

Figure 3 Overlaying the Building-Heights of 1905 and 2003

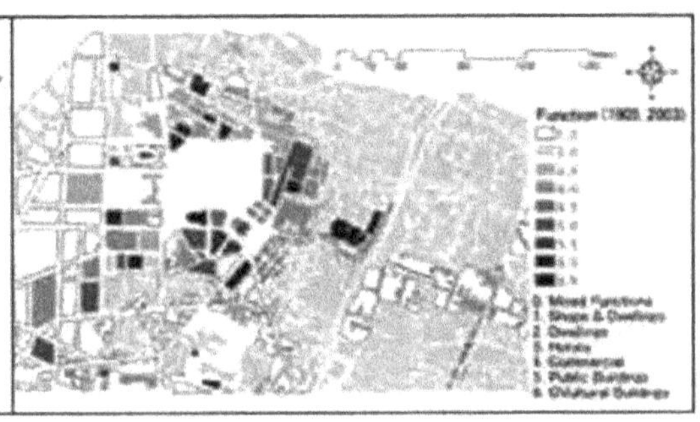

Figure 4 Overlaying the Building-Functions of 1905 and 2003

THE NATIONALIZATION OF THE EUROPEAN QUARTER (1952 - TO DATE)

The revolution in 1952 for nationalization occupied the European Quarter of major transformations in progress. In order to compare the situations of the quarter before and after the turning point in 1952, the method first digitizes the 'Insurance Plan' in 1905 that structured the cognition of urban tissue, nodes and landmarks, and function of the colony until the nationalization. Second, the present situation in 2003 as investigated by the author is digitized for the same criteria of the 'Insurance Plan' in 1905 to make the possibility of comparing the two situations. Through this access to urban information, the two values in 1905 and 2003 of the same field are overlaid to categorize their transformation.

The bylaws related to urbanism changed from the closed Municipality to the National Government of different policies. The concept of round-points linked by major boulevards is not classified by the code of urban design or public streets to transform into ordinary vehicular streets[1] (Fig. 2.) Meanwhile, the occupation by street vendors is not planned in the urban tissue in contradiction to the regulation specifying: 'The squat of public spaces without authorization is prohibited for selling goods, shop-windows, seats or tables, kiosks, bars and the like.'[2] Besides, the characteristic passages of building interiors are being squatted for workshop extensions without pedestrian functioning.

Landscaping the outdoor environment lacks the public amenities, which suggests the law stating: 'The Local Authority should plan the elements of pavements, lighting & landscaping for the pedestrian amenity'[3] as too general to be enforced. Furthermore, the Government considers the historical

European Quarter through: 'special bylaws of the Central District (European Quarter) should consider the capacity of streets in terms of vehicular circulation, parking lots, loading and unloading of goods, public transportation and the pedestrian circulation.'[4] In reality, however, the ever-increasing vehicular circulation overloads the capacity of streets and plazas in the historical quarter without realizing the law's expression 'should consider' into practice. The redevelopment of buildings, nevertheless, is regulated by setbacks for the purpose of widening the narrow streets as: 'a minimum setback ratio of one-third the street width up to 12m, and a one-fourth the street width up to 15m'.[5] This regulation is adopted without considering the conservation of the historical urban tissue, in addition to different building- height of: 'a maximum building-height of one and a half times the widest street of the building's site.'[6] Accordingly, the urban tissue experiences duality of figure-ground relationship between the original buildings and the redevelopment (Fig.3).

Parallel to the chaotic urban tissue, the nodal structuring of the European Quarter ceased to develop at Cairo's different levels of resolution. The central 'Ezbekieh Plaza' lost the enclosure by landmark buildings, while deprived from the recreational activities of its gardened open space. Other nodes had been cancelled such as the railway station of 'Manshieh Plaza' to affect the city-level structuring of railway nodes linked to the 'Ezbekieh' center. The round-points followed suit without nodal importance as for the 'Ismailia' one limited to vehicular through passing, while the Nile promenade transformed to a mere vehicular street. It is certain that the loss of 'Ezbekieh' node had affected, in consequence, the whole nodal structure of Cairo. Besides, the modern high-rise buildings overwhelm the original vertical features of religious buildings and sculptures. These high-rises, however, are scattered without criteria of structuring the spatial experience in the European Quarter. Similar process of high-rise development is taking place along the Nile, which destroyed the uniformity of original palaces and villa-type housing.

The governmental designation of the European Quarter as the Central Business District of Cairo specifies the functions of 'commercial and monetary establishments, business offices, hotels, recreational and cultural facilities, main administrative buildings, small-scale industry, and also the residential func tion'[7] to take place. The actual functions in the European Quarter, however, mix workshops, private clinics, residential flats, shops of different items, business offices, governmental offices and others in a laissez-faire manner (Fig.4). The law proves too general to control the functions in the historical quarter. The mixture of functions in any one building decorates the building façade by different materials, colors and signboards that spoil the characteristic historical façades. Furthermore, the conversion of building interiors to suit the various functions causes technical problems such as the removal of structural walls for space extension and improper fittings of building equipment.[9] Besides, the ownership of real-estate properties transferred from the European nationals to the National Government or sold to natives. The Government in 1952 changed the real-estate business from the free market to the controlled economies without revenues.[9] The low-income rent constrains the deteriorated condition of buildings in the European Quarter without maintenance or insurance. Meanwhile, the real-estate taxation law exempts the old buildings before 1944 that applies to the European Quarter's building stock without evaluation.[10]

Conclusion

The historical formation of Cairo's European Quarter in comparison to its nationalized situation at present concludes the following:

1. The locus of 'Ezbekieh Plaza' in 1905 projected radial boulevards to synthesize Islamic Cairo with a 'New Paris' by the Nile. The three thoroughfares of 'Rue Neuve,' 'M. Ali' and 'Claude Bey' welded in the Islamic urban tissue through building galleries and arcades in transition to the irregular alleys of the traditional Islamic blocks. The European network of round-points linked by boulevards, meanwhile, fused into another gridiron one after the Nile orientation. New blocks in 1905 maintained the amalgamation of 'Garden City' doted by buildings with prismatic blocks of specific depth/height ratios. At present, however, the historical core of 'Ezbekieh' lost a definite role on the city level, with major boulevards transforming into ordinary vehicular streets. The historical urban tissue, nevertheless, is changing under the present regulations of building setbacks and heights that vary from the original.

2. The common-theme in 1905 of boulevards terminating in major open spaces structured the nodes of the European Quarter. The foci of 'Ezbekieh Plaza' overlapped the three-sided structuring of: 1) railway nodes at the two poles of the northwest-southeast axis, 2) the Islamic 'Azhar' node along the east-west axis, and 3) the triangular flow nodding in 'Ismailia Round-point,' 'Abdin Plaza,' and 'Ezbekieh Plaza' itself of concentrated landmark buildings. The longitudinal node of the Nile promenade, nevertheless, formed the special urban landscape of the European development. On the contrary, the situation in 2003 of outdoor nodes dissolved into the vehicular circulation to replace gardens. In specific, the dismantlement of socio-economic activities from the core 'Ezbekieh' node reflects the catastrophe of shaping nodes in present Cairo. In chain to the lost nodes, the landmarks develop inharmonious high-rises that are scattered without concept.

3. The Municipal 'Insurance Plan' in 1905 of Cairo planned the function of the European colony through the process of invasion and succession. This developed the western zone of 'Ezbekieh Plaza' into business establishments instead of the single-family houses that shifted further southwest. Eastwards of 'Ezbekieh' developed commercial galleria buildings to stretch along 'Rue Neuve' in the residential Islamic town. It is obvious for 'Ezbekieh Plaza' to have structured the functions, while itself represented the socio-economic hub of grand hotels, The Opera, Mixed Tribunal and recreational open space. Although the present regulation classifies the building-function, the actual situation in 2003 proves a chaotic mixture of functions, which together with the controlled economies of real-estate without revenues cause the building deterioration.

[1]Publishing House, 1995.

[2] The Law No. 140 in 1956 and No. 84 in 1968, The Law of Public Streets and Advertisement, Third Edition, El-Amiriyah Publishing House, 1992:2.

[3] The Law No. 3 in 1982, op. cit.:15.

[4] Ibid:67.

[5] The Ministry of Housing and Reconstruction, The Law No. 106 in 1977, The Law of Building Works, Tenth Edition, El-Amiriyah Publishing House, 1995:75.

[6] Ibid.

[7] The Law No. 3 in 1982, op. cit.:66-67.

[8] Field surveys in 2003 by the author.

[9] The Law No.49 in 1977, The Law of Renting Building Spaces and Organizing the Relationship Between the Owner and the Tenant, Thirteenth Edition, El-Amiriyah Publishing House, 1994:55.

[10] The Law No.49 in 1977, op. cit.:25.

REFERENCES

ALI, A., 1998. Cairo in the Era of Ismail. The Egyptian Lebanese Publishing House, Cairo.

CANERI, T., 1905. La Ville Du Caire Son Present Et Son Avenir: Essai Sur La Creation D'Une Municipalite. L'Institut Francais D'Archeologie Orientale, Cairo.

COHEN, N., 1999. Urban Conservation. The MIT Press, Cambridge/Massachusetts.

EL-SHAZLY, A., 2003. The Prospects of the European Quarter in Cairo. Journal of Asian Architecture and Building Engineering, The Architectural Institute of Japan Vol. 2, no. 1, May.

KOSTOF, S., 1999. The City Assembled: The Elements of Urban Form Through History. New Edition, Thames & Hudson Ltd., London.

KOSTOF, S., 1991. The City Shaped. Thames & Hudson Ltd., London.

KRIER, R., 1979. Urban Space. International Publications, New York.

LYNCH, K., 1960. The Image of the City. The MIT Press, Cambridge/Massachusetts.

LYNCH, K., 1981. Good City Form. The MIT Press, Cambridge/Massachusetts.

MUBARAK, A., 1889. Al-Khitat: The New Plans of Khedieve Tewfik in Cairo. Vol. I, Bulaq Press, Cairo.

OWEN, R., 1969. The Cairo Building Industry and the Building Boom of 1897 to 1907. The International Conference of Commemorating the Millennium Foundation of Cairo City, Amyrieh Press, Cairo.

ROSSI, A., 1986. The Architecture of the City. The MIT Press, Cambridge/Massachusetts.

SCHARABI, M., 1989. Kiro: Stadt und Architektur im Zeitalter des Europaischen Kolonialismus. Ernst Wasmuth Verlag Tubingen, Germany.

THE URBAN PLANNING ASSOCIATION, 2002. Historical Architecture of Cairo, Final Report, The Town Planning Center, Cairo.

Effective - Easy to Use - Economical: PGIS, the Database for Archaeological Monuments in Speyer (Palatinate), Germany [1]

Andrea Zeeb-Lanz
Landesamt für Denkmalpflege Rheinland-Pfalz
Archäologische Denkmalpflege Amt Speyer

Abstract

The monumental heritage department in Speyer/Palatinate has developped a GIS-supported inventarisation for archaeological monuments which guarantees the user an easy handling and a very effective procedure not only thought for the needs of the monumental heritage but as well for scientific research. Besides this data bank offers minimal costs concerning purchase and support. The following article gives a short insight into the possibilities of the system.

Introduction

In the last decade, the use of GIS-supported applications has multiplied in the archaeological world and is beginning to become a standard-tool for the monumental heritage of Germany as well (see e.g. Zeeb 1999, Wilbertz 2002, Schwarz 2002, Hess 2003). Whereas some of the bigger state-departments for monumental heritage in Germany were able to provide the necessary funds for the development and later support of GIS-based data banks, the smaller and financially less amply furnished departments are still forced to manage with analogue data and maps.

Out of the need to accelerate statements on building plans and the like and for more effective procedures in the daily work, the department for archaeological monuments in Speyer took advantage of the fact that a PhD-student working on his thesis in Speyer offered his skill in programming data banks. The authors experience in working with an archaeological GIS gained during her years of work at the monumental heritage department of Saxony offered additional help in creating a GIS-supported data bank for the archaeological monuments of the Palatinate (southern part of Rhineland-Palatinate). The goal was to achieve a system which would be modest in costs, easy to handle but nevertheless effectively working for the necessities of the monumental heritage administration.

The need for more effective tools becomes more and more demanding in the preservation of archaeological monuments. The amount of all kinds of building projects is steadily rising. Each one concerned with any changes in the ground requires a statement from the archaeological side. The way these statements are traditionally achieved shows that the time required and the output of the analogous work is not up to date any more.

For working on a plan for land utilization for instance it was necessary to examine all the registered sites in the communities concerned. Every known archaeological site had to be looked up in the analogous record, maps had to be looked at and the parish-files with detailed descriptions and extended pieces of information had to be read.

The result of this extended research was a copy of the original plan into which were entered by hand the sites concerned. A list of the sites, their significance and potential endangering had to be written down separately. This proceeding is neither effective nor in keeping with the times. It does not add to the reputation of the archaeological departments as equal partners for investors and building administrations in the communities.

A second important fact is the endangering of the analogous data - parish-records are often not kept fire-proof; one major fire and the archaeological record as a whole would be lost without any chance of recovery.

The summary beneath lists a number of advantages that make a GIS-supported data bank an outstanding preference to the conventional course of the administration of archaeological sites:

- lists of selected sites can be created instantly
- specific formulations of a question can be followed up easily
- if details to a site are needed they can be provided with a single mouse-click
- detailed or generalized maps can be drawn up
- parish-records are available in standardized digital form
- safeguarding of analogue data is guaranteed

A second main subject on the agenda of the departments for monumental heritage must be mentioned in connection with PGIS: archaeological research. Specific scientific questions can be answered more easily with the help of a GIS. Quantitative and qualitative analyses are performed within the frame of a data bank. Culture-historical maps quickly answer questions concerning settlement history or changes of settlement patterns.

Technical Details and Possibilities

The data-bank U. Himmelmann, responsible for programming PGIS, created, is a product fulfilling a multitude of tasks.[2] It is usable without lengthy introduction, it is compatible with all Microsoft products and must not be supported by a specific firm as access-programmers are widely available.

Extended discussions about structure, contents and meaning of the sources led to a concentration on the archaeological site which is the focus of the archaeological work in monumental heritage administration. A multitude of primary relations and connections with other data were defined (Fig.1). The arrows with broken lines indicate that many secondary relations had to be defined as well before definitely working out the structure of the data bank.

PGIS consists of three central parts: PGIS (the data bank), the GIS-Software/Digital map and the parish-records (in different formats under HTML-surface). In a second step the integration of the items shown on Figure 2 is planned; then the entire record of an archaeological site will be provided by PGIS.

PGIS can be used on a single station as well as on an internal network or via internet. The programme was realized on the basis of MS access with extended SQL-orders; the next stage will be programmed in Visual Basic. A small programme allows the transfer of the data bank including all of its functions to an unlimited number of other computers. No MS access-licence is needed, PGIS works with an integrated runtime-version free of charge.

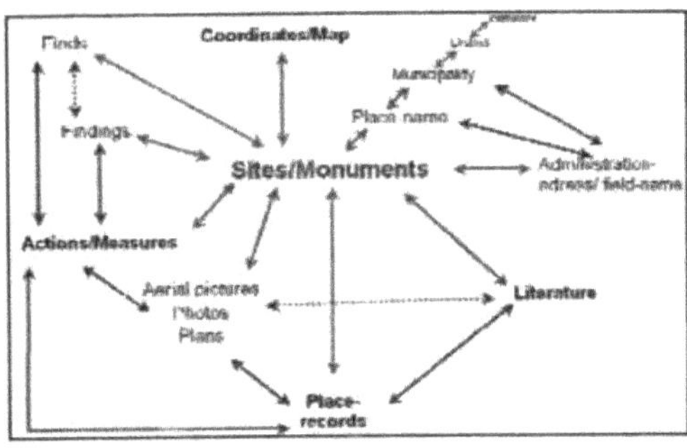

Figure 1 Relations and connections of the archaeological record

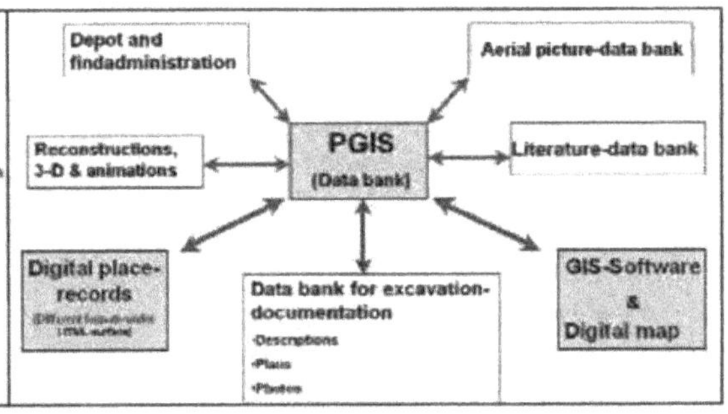

Figure 2 PGIS. Relations in the databank

As explained further below ("Visualisation"), our access-based data bank is attached to the GIS-programme "Hill-GIS", but a combination with standard-GIS-programmes like ArcView or MapInfo could as well be realized without any problems at all.

OUTLINE AND FUNCTIONS

One of the main functions is the enquiry-tool (Fig.3). The parishes containing archaeological sites are arranged alphabetically. Specific place names can be filtered as well as single site-numbers, all sites of a bigger administration unit or all sites of a specific period like Roman for instance. Besides it is possible to run several requests at the same time. Detailed information about the archaeological background of a certain site is provided like the grid reference, field name of the location or details concerning the finds from a site and their chronological classification. It is also possible to get access to the original parish-folders containing a lot of details including old maps, drawings and site-descriptions. These are formatted as html-pages, thus needing only a small amount of memory capacity so that even lengthy documents can be integrated.

The tools for input, corrections and additions to the information-pool of each site form an important part of the system. Data can easily be changed, new information can be added as well as new objects. An unlimited amount of data can be filled into the data-line of a single site which is important as a lot of different archaeological objects are possible on one site.

Thesauri for the fields "archaeological object" (e.g. "settlement site", "burial mound") as well as for chronology and archaeological culture (e.g. "Neolithic", "bell-beaker culture") guarantee the avoidance of writing errors. The information in the thesauri is structured hierarchically so that general assignments like "graveyard" are as well possible as exact descriptions like "barrow with skeleton-burials". Each thesaurus can be changed and new terms can be added.

Every action like "excavation 1987" or "aerial photo 1992" on a site is recorded, new actions can be added in unlimited number. For each kind of action a thesaurus is used also. A short inventory of finds which are assigned to each single action is provided. This inventory is only thought as a general overview - for detailed information entering the finds data bank is asked; this data bank will be integrated in PGIS in 2004.

Naturally, as PGIS is a data bank with GIS-application, a map for every research can be seen and printed out. When making an enquiry, it is important not only to create a map but also to receive a list of the monuments selected. PGIS offers the choice between a short list providing just basic information and an extended inventory containing more details for each site.

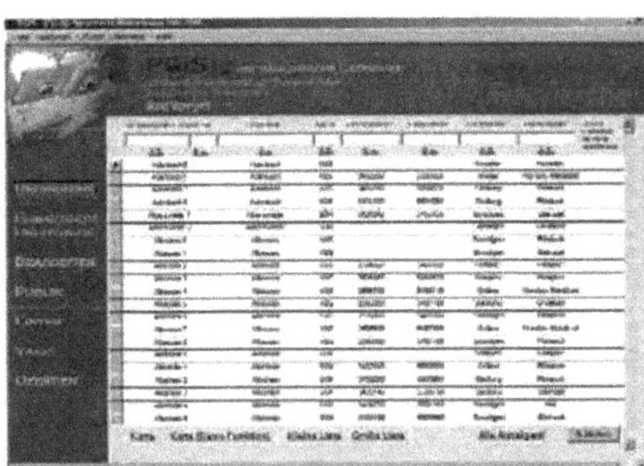

Figure 3 PGIS. Main window for enquiry

Figure 4 PGIS. Integrated ortophoto of the restaurated Roman villa from Ungstein (Palatinate)

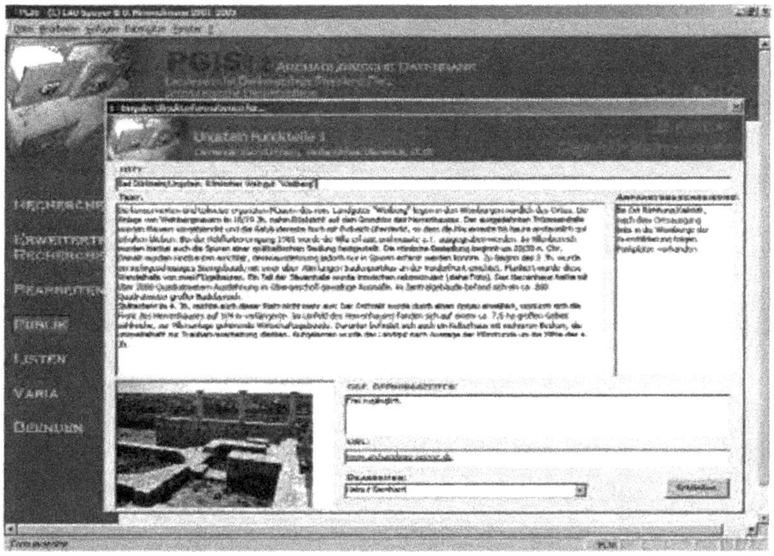

Figure 5 PGIS. Flyer-text for the restaurated Roman villa from Ungstein (Palatinate)

These lists can be printed directly out of the programme or be transferred into any MS office-format like Word documents ore Excel data banks for further use.

VISUALISATION

The location of the chosen sites on digital maps is realized by a cooperation with the Department for Land Survey and Geological Basis Information of Rhineland-Palatinate. By providing the afforded maps and a programme (Hill-GIS) for integrating our site-symbols the department is in return offered information on unendangered, visible monuments for integration into various maps produced by this department.

Concerning the maps there is a choice between a scale of 1:200,000, 1: 25,000 and various zoom-scales below the latter. Also provided by the Land Survey Office are orthophotos covering the whole of the Palatinate. These are very helpful especially for the location of visible sites like burial mounds or restaurated roman villas (Fig.4).

Short texts for public relations work concerning selected monuments, added by pictures, opening times and journey descriptions are an additional possibility when using PGIS (Fig.5). All such pages can be transformed easily into html-documents for further use in the internet, for the creation of flyers or other items.

ENDNOTE

PGIS, as described above, is a financially attractive, easy-to-use and effective alternative for the procedures of daily work in state-departments for monumental heritage as well as in archaeological science-work. Such tools, mostly still very cost-intensive at the time, become more and more demanding in the process of preserving archaeological monuments as well as facilitating scientific research.

1 Sligthly altered and extended version of my lecture given at the Conference "Enter the Past", April 8-12th in Vienna. PGIS stands for "Pfalz-("Palatinate)GIS"; I give my thanks to Dr. U. Himmelmann, who is responsible for the programming of the system and took part in setting up the PowerPoint-presentation presented in Vienna. Thanks also to Dr. Christoph Steinmann for looking over my english manuscript for the lecture in Vienna.

2 Introduction to PGIS see Himmelmann and Zeeb-Lanz 2004, Himmelmann and Zeeb-Lanz, in preparation, www.archaeologie-speyer.de.

REFERENCES

HESS, S., 2003. Webmapping und verteilte Datenbanken - Entwicklungen und Perspektiven moderner Denkmalpflege und Forschung - Dargestellt am Beispiel des Fachinformationssystems ADABweb. Arch. Nachrichtenbl. 8:46-54.

HIMMELMANN, U. and ZEEB-LANZ, A., 2004, in preparation. PGIS - Datenbank mit GIS-Anbindung für die archäologischen Denkmale der Pfalz (Arbeitstitel). Archäologie in der Pfalz, Jahresbericht 2002.

HIMMELMANN, U. and ZEEB-LANZ, A., 2004, in preparation. PGIS - eine effektive und kostengünstige Alternative zu bestehenden archäologischen GIS-Systemen (Arbeitstitel). Arch. Nachrichtenblatt 9.

SCHWARZ, R., 2002. Luftbildarchäologie und Denkmalinventarisation in Sachsen-Anhalt. Arch. Nachrichtenbl. 7:313-324.

WILBERTZ, O.M., 2002. Archivierung und Erschließung von Luftbilddaten im Archäologischen Archiv des Niedersächsischen Landesamtes für Denkmalpflege (NLD). Arch. Nachrichtenbl. 7:309-312.

ZEEB, A.,1999. Erste Schritte zu einem Flächenmanagement - Ein GIS in der sächsischen Archäologie. In Harl, O. (ed.), Workshop3 - Archäologie und Computer 1998 (CD), Vienna:85-106.

Site Reconstruction of Ancient Vindobona

R. Gietl
Inst. for Classical Archaeology,
University of Vienna, Austria

M. Kronberger
Vienna, Austria

M. Mosser
Stadtarchäologie Wien, Austria

Abstract

Archaeological and geological researches in Vienna come to the conclusion that the surface of Roman Vindobona even in the restricted area of the nowadays inner district was quite different in contrast to present Vienna. Beside the cognition of a mutation of the Danube course in a northward direction over the centuries, there are indications in changing the river beds of tributary brooks. A flood of the Danube in ancient times and a mudflow resulting from this flood in the northern area of the settlement entails a chronological division of the site reconstruction. This implicates completely new perceptions on terrain and settlement structures in Roman times. A three dimensional reconstruction of that surface should include the current developments in researching the urban area of Vindobona, that means river courses, elevations and altitudes, routes of streets, buildings of the legionary fortress and the surrounding canabae legionis and the river docksides in reconstruction. Implements in ensuring this intentions are a GIS-application, 3D-studio, altitude and geological plans of present day Vienna and new two dimensional plans of the legionary fortress and of the dimensions of the surrounding canabae legionis. In future the project should be enlarged around the whole area of the Vienna basin.

The taking place of the project is based on the cooperation of three branches of research:

Archaeology: with the results of Roman settlement research in the inner city of Vienna.

Geology: with the evaluation of drilling logs and a new digital geological map of Vienna resulting from it.

Surveying and mapping: with the establishment of a digital elevation model of Vienna.

Historical and Topographical Remarks

Vienna was a thoroughly important base on the border of the Roman empire. In ancient times it was called Vindobona and existed as an Roman legionary fortress from the 1st to the 5th century AD in the province of Pannonia. In flourishing times there lived 5 to 6000 soldiers within the walls of the fortress (Fig.1).

Figure 1 Imagination of landscape around the legionary fortress of Vindobona in Roman times (© Wien Museum Karlsplatz)

The outlines of this fortress within today's inner city of Vienna are well known, although archaeological excavations could have been made in very different qualities and only in very restricted areas during the last 150 years. It was located at a very favourable site, surrounded by the river Danube in the north, by a brook, which does not exist today, in the west and a today canalised river in the east. These watercourses will play an important role for the appearance of the model. Collecting and mapping the results of elder and new excavations led in 2002 to a provisional new reconstruction plan of the legionary fortress approximately of mid 2nd century AD (Fig.2). These mappings enabled a full-coverage acquirement of the ancient Roman altitude level inside the inner city of Vienna.

Reconstruction

First computer supported attempts to create buildings of the fortress three-dimensionally began with the reconstruction of the three archaeologically known gate-towers in AutoCAD. It based on archaeological ground plan drawings from the beginning of the 20th century and on comparable gate towers from other Roman forts and fortresses especially in Britain, Germany and northern Africa (Fig.3). In consequence of the gate-tower project a voluntary employee, Mrs. Helen Sarah Lacey from Britain, tried to place the gate-towers into the modern urban features and made little cinematic sequences of her work.

The Ancient Surface

Last year it was R. Gietl, who tried to build an attractive three dimensional model of the legionary fortress of Vindobona in the context of the so called

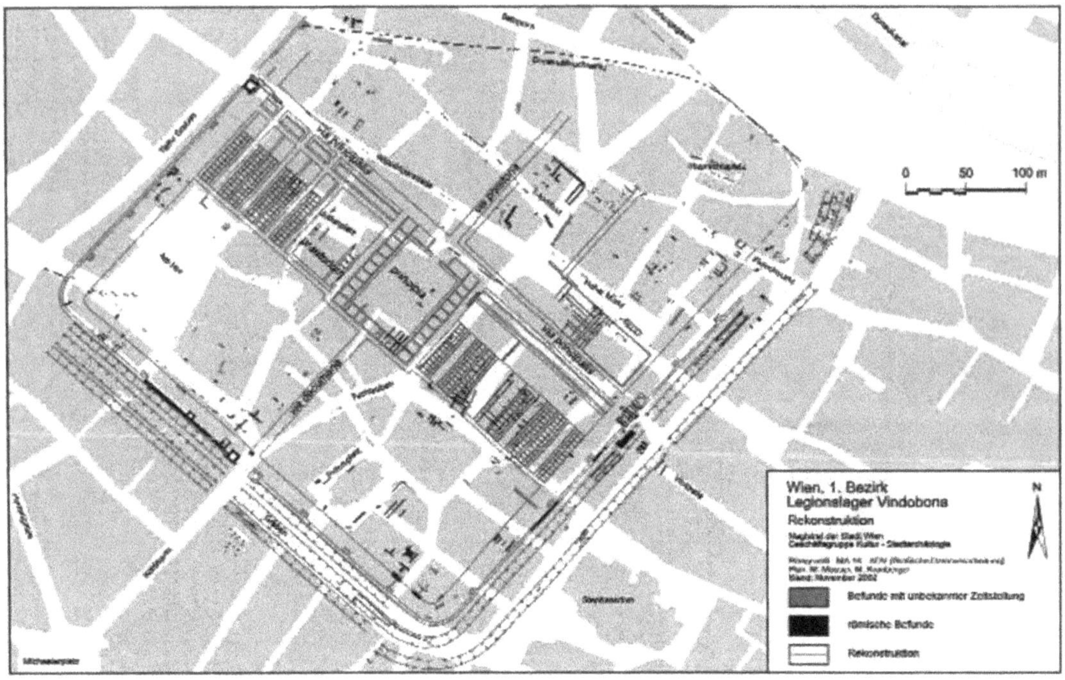

Figure 2 Vindobona - legionary fortress, approximately of the mid 2nd century AD
(Map based on MZK MA14-ADV)

"Science week", a yearly open air event of the Universities to bring science nearer to the people. In this context the cooperation was born, the aim was the reconstruction of the ancient conditions around the Vienna fortress and its three dimensional rebuilding. But soon problems occurred concerning the ancient landscape, available maps and technical possibilities. The archaeological researches and the parallel carried out geological evaluation of drilling logs in Vienna come to the conclusion that the surface of Roman Vindobona was quite different in contrast to present Vienna.

The most outstanding result from the cooperation with the geologists was the recognition of a flood of the Danube in ancient times and a mudflow resulting from this flood in the northern area of the settlement. This fact required at least two stages of the ancient model: The rectangular fortress and its surroundings before the mudflow and the reduced settlement after the flood.

A three dimensional reconstruction of that surface should include the current developments in researching the urban area of Vindobona, that means river courses, elevations and altitudes, routes of streets, buildings of the legionary fortress and the surrounding canabae legionis (military suburb) and the river docksides in reconstruction. For this a suitable elevation model of Vienna as a basis to reconstruct the ancient surface and to build up the architecture was necessary.

Such elevation models have already existed since the 19th century. The advanced measurement systems of the surveying departments of the Habsburg monarchy led, beside an exactly measured land register, also to detailed altitude plans and models (Fig.4). But the technical prerequisites improved over the times, and finally the department of surveying and mapping of Vienna has made a digital elevation model of the city, which they made available for our purpose.

Three main sources have been available to produce the ancient surface:

- altitude level points of the ancient Roman surface from certain excavations
- evaluated drilling logs from the whole area of the inner city defining the water courses
- the digital elevation model of the inner city

The current working was a very educational one, because you pushed on problems you didn´t expect before, you have been forced to logical solutions and finally you have more questions than answers. One of the results of this finding process is, for instance, the fact of three ditches around the fortress, a fact which has not been so clear before. These three ditches also seem to have intentional drainage function in direction to the Danube river in times of inundation of the brook west of the fortress.

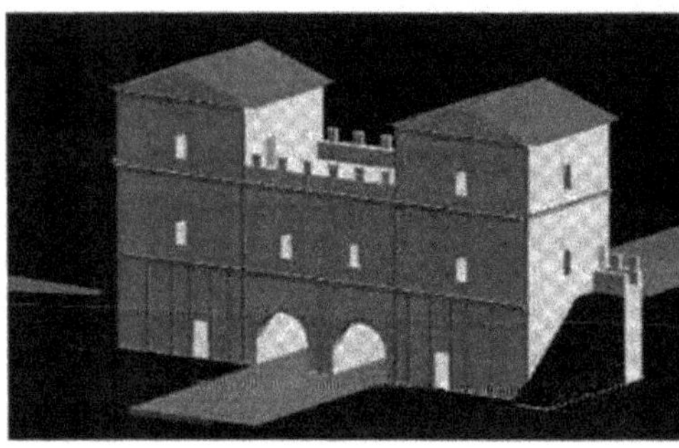

Figure 3 Three-dimensional reconstruction of the gate towers in the legionary fortress of Vindobona

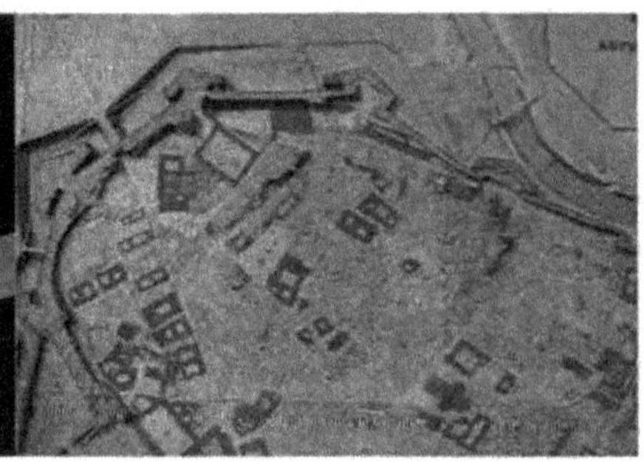

Figure 4 Relief-map Vienna - inner district (Otto von Altvatter 1858)
(© Wien Museum Karlsplatz)

To form the river beds, to reconstruct their ancient courses, to hold in mind also Roman canalisation and docks is one of the most difficult tasks of our work.

Technical Process

To create a site model we had to look for data with z-values and we found an archaeological map showing the contour lines of Vienna's 1st district, used by F. Kenner, the leading archaeologist of Roman Vienna around 1900 (Fig.5).

By digitising the contour lines of this map and triangulating them we generated a first DEM (digital elevation model) that reflects the topographic situation about 100 years ago. In a second step we started to compute and to model this DEM using the archaeological reconstruction of the legionary fortress. The result was a DEM of the fortress area, that was quite exactly in his ground plan, but with z-values of modern Vienna. It wasn't useable for archaeological aims because the present-day surface of the 1st district is totally different compared with the Roman age surface. Today the majority of the fortress is a couple of meters under the modern surface, but some other parts were dug away during the Middle Ages or in the 19th century and now there we walk under the sandals of the Roman soldiers. To solve this problem we tried to include new sources of height data which reflects the third dimension of the terrain surface in Roman age.

Another geological source are the borderlines of the alluvial plains. They show us the range of possible river movements around the fortress since the last glacial period.

The third and last source is still the archaeological reconstruction of the fortress, of the ancient street system and other buildings.

On the base of this data our team composed of geologists, a civil engineer expert for DEM's and archaeologists started to edit the points, to filter accidental error in reference data and to model the ancient surface. We used a software called SCOP, which the department of Geodesy made available to us. SCOP is designed for interpolation, management, application and visualisation of digital terrain data, with special emphasis on accuracy. SCOP aims at high quality of interpolation. It has been developed and continuously improved over the last 30 years from the Institute of Photogrammetry and Remote Sensing of the Technical University of Vienna. With this software it is possible to modify easily the height and the position of every mass point or break line. Additionally we can give different weights to our data, depending on their accuracy. That means for example that the coordinates from recent excavations have more influence on the interpolation process than their older and less exact neighbours.

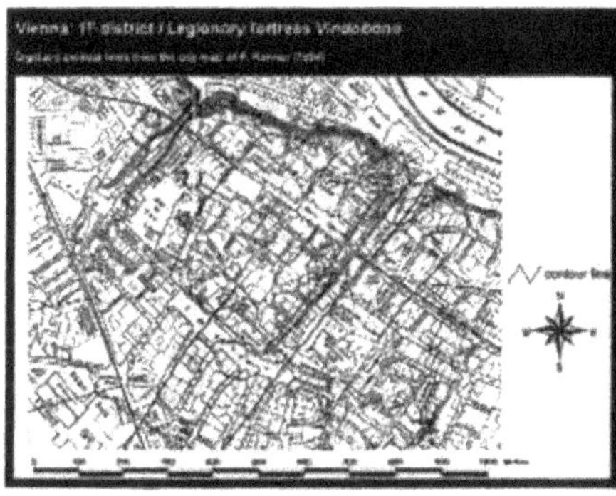

Figure 5 Digitised contour lines from the city map of F. Kenner 1897

Figure 6 Vienna: 1st district - Legionary fortress Vindobona 3D scene: SCOP; DEM superimposed by the archaeological reconstruction, vertical exaggeration factor: 1.5 - View from North

First of all we started to collect every three-dimensional coordinate which was ever recorded during an archaeological excavation in the legionary fortress or in the canabae legionis over the period of the last 150 years. As we have different Roman surface levels between the time of occupation in the 1st half of the 1st century and the late antiquity, we selected as far as possible points from the oldest Roman layers for this DEM. The height data of the single excavations has different precision: from the 19th and the beginning of the 20th century we have only old archaeological diaries with rough sketches and vague values. Over the last 150 years the land surveys became more and more accurate and in contrast to the oldest data we have modern excavations in the fortress area surveying with a precision of a few centimetres.

Our second source of three-dimensional data were about 2100 geological profiles taken from bore drills analysed by the Viennese company "DonauConsult" on authority of the city of Vienna. They are only a small part of over 50.000 profiles which form the public "Baugrundkataster" of the department of Geology of the city of Vienna (MA 29). In every profile the geologists documented the level of the deepest human-caused layer. Normally this indicates more or less the arrival of the Romans. But sometimes the bore drills penetrated deep medieval cellars or pitches, which have destroyed the Roman layers and for this reason one has to discuss about this data from case to case.

During this process we have found out numerous new details about Roman Vienna: For example that inside the fortress the ground fell away from northwest to southeast for about 3,5 m. Further we think that we can locate the harbour of the Roman Danube-fleet. The process is just going on in discussing with river management engineers about the ancient course of the Danube and the brooks surrounding the legionary fortress to reach a higher accuracy in the outlying districts of the DEM. We want to use the results for example to analyse visibility or as a starting point for a GIS of the territory of Roman Vindobona (Fig.6).

Further we will update the DEM in regular intervals with new data from recent excavations to improve its quality and it will be used as surface for a new generated 3D-Studio model of the legionary fortress of Vindobona.

GEOSTATISTICAL APPROACH TO THE TOPOLOGY OF THE PREHISTORIC SETTLEMENT SYSTEM: A CASE STUDY IN MUSASHINO UPLAND, TOKYO, JAPAN

TSUMURA HIRO'OMI
DEPARTMENT OF MUSEUM SCIENCE,
NATIONAL MUSEUM OF JAPANESE HISTORYJONAI-CHO 117,
SAKURA, CHIBA, 285-8502, JAPAN,
TSUMURA@REKIHAKU.AC.JP

ABSTRACT

The purpose of this study is to discuss the prehistoric settlement systems in the view of Geostatistical analytics, and to introduce the theoretical concepts of GIScience into Archaeology. The subject area is the eastern part of the MUSASHINO upland that has an area of approximate 300 square kilometers, with 98 archaeological sites from the middle Jomon period (cal BP. 5400-4500). The population in highest stage of its development is estimated to be 2000. First, database that has a mutual comparability of the space and site's archaeological properties was created. The digital site distribution maps, which can refer to miscellaneous attributes were also created. Then the following three analyses were applied. 1) The Voronoi method (as interpolation) was applied because the geometrical properties of the diagrams symbolize the topological characters. 2) The inter-site visuability was drawn by applying Viewshed Analysis for elucidating the inter-site spatial relations. 3) Trend-surface analysis was enforced using point properties of archaeological cultural elements to reconstruct the spatially concrete images. Finally, the results were lined up along the "unconditional" time-series. Thus, I propose considering this approach as Four-dimensional "4D" GIS.

INTRODUCTION

Until recent times, the prehistoric settlement has been discussed from the multifarious point of view in archaeology. In order to reconstruct the human evolution, "Process or Marxism" archaeologists in the 1970s, eagerly argued about the socioeconomic context. But, after 1980s, Many Western archaeologists thought that this approach is insufficient in understanding the human-nature interactions since the settlement reflects human behavior corresponding to the surrounding environments. "New" and "Post-Process" Archaeology were regarded as much important. Spatial archaeology (Clarke 1977) is the one study area developed in this atmosphere. It was the phenomenal challenge toward the old analytics in which any archaeological information was never grasped as spatial attributes.

Such awareness of the issues was gradually adopted by geographers in Japan, although in "Prehistoric Geography" from the 1940s, the systematic linkage of Archaeology and Geography had not been achieved. But, because most of the Japanese archaeologist blindly believed that "Archaeology as Archaeology" based on materialistic data is the only positivistic approach, their potentialities of collaboration have not grown.

In 1990s, GIS developed rapidly by the expansion of PC and GIScience (Geographic Information Sciences; as epochal interdisciplinary studies) was formed. Its application in Archaeology began in 1990s, nevertheless it had not been acknowledged as an effective method, rather it had been thought as a useless tool, for there was no theoretical foundation for processing the spatial information in archaeological methodology with the exception of a few studies (Tsumura 2002). Thus, in this preliminary study, I would like to practically and theoretically organize some former individual approaches by especially focusing on the interdisciplinary perspectives in order to discuss about the prehistoric settlement in the view of Geostatistical analytics as well as introduce the GIScience theoretical concepts into Archaeology.

MUSASHINO ARCHAEO-ECOLOGICAL PROJECT

[1] Target Area

The target area is the eastern part of the MUSASHINO upland that has an area of about 300 square kilometers with 98 archaeological sites (see Fig.1, 2).

The palaeo-geomorphologic characteristics of this area are 1) there are less topographic undulations which enormously

Figure 1 Subject area

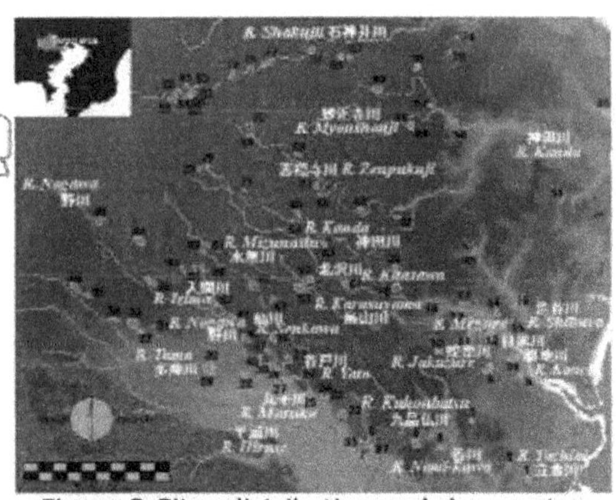
Figure 2 Sites distribution and river system map on DEM

influence the human-walking costs compared to other areas. Its mean altitude: 38.59m and standard deviation: 17.79, 2) the slope angle is not so steep, and its slope direction is none (flat) or SE in the area. Its mean angle: 1.81 degrees and standard deviation:

3.55, mean aspect: 139.72 degrees and standard deviation: 98.49, 3) streams or creeks are abundant, and some springs are along the streams, 4) the contrast between upland and lowland where nestle close to the Tama river is very clear and there are many fountains in this area that we call "HAKE".

[2] Digitizing the Site Distribution Maps

For equipping GIS with reference system based on site's archaeological properties and spatial coordinates, I digitized the site distribution maps as polygon data by tracing the usual paper maps with digitizer. I thought that the address matching, which is common method, is unsuitable method, for the old and usual maps were not made by taking its spatial coordinates in to account. In other words, the data recording ways should follow the paper map's methods as well in transcribing, if the observers' judgments about the site area are regarded as important. Since address matching refers to the text form address, the most important information about the locality may be omitted, for many sites are in an out-of-the-way anonymous place.

It is a big issue in how we divide the site area. Since all the sites are not fully excavated, "The site area" (in the strict sense of the word) must be indistinct permanently (Tsumura 2002). I considered the investigator's assessment is the most important information; there adopted "The potential site area" (estimated by Kobayashi et al. 2002) as "The virtual site area". This approach is very simple and unpolished, but most certain in converting the raw archaeological information into digital form.

[3] Digitizing the other spatial element maps and constructing the database with reference system

I digitized the other spatial elements, such as river system as a line data and soil system as a polygon data. Fundamentally, we should reconstruct the palaeo-environment by some appropriate methods, but it is very difficult with GIS technology at present due to a massive alluvial soil cover (Tsumura 2002).

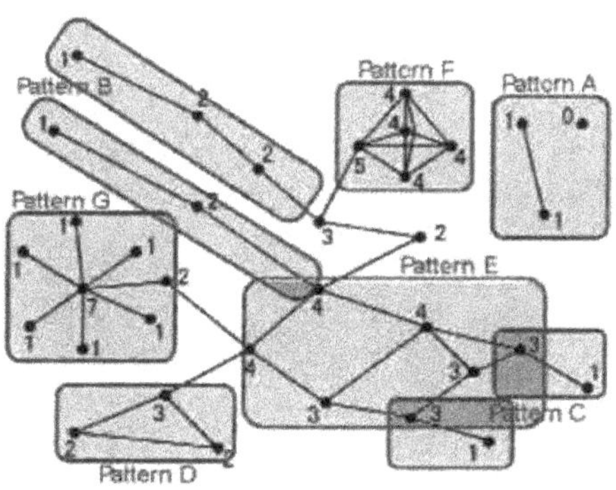

Figure 4 Inter-site visuability patterns

The digital maps of spatial elements, which has each spatial coordinates, or the site digital distribution maps were combined with archaeological properties by GIS technology with the original discrimination IDs. The mutual references, spatial attributes from/to site's properties, were made possible, and this system made many digital site distribution maps.

Large quantity of digital maps were arranged along the calibrated radiocarbon date "cal BC" to clarify the temporal transitions of human activities. This approach will promise sublimation to the next stage of archaeological studies with the biological perspectives from the old style that slipped into the materialistic view of history. For the former relative timescale could only make a static cultural image.

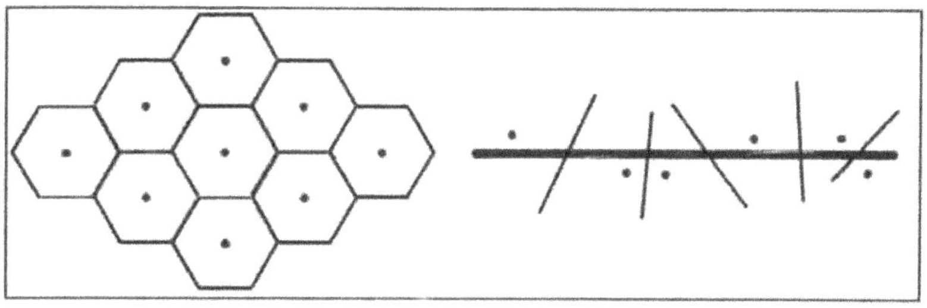
Figure 3 Models of Voronoi Diagrams Left: Hexagonal diagrams, Right: Distorted diagrams

GEOSTATISTICAL APPROACH 1 - VORONOI METHOD

[1] Making the Voronoi Diagrams

The Voronoi method is one of the interpolation techniques in Geometry. Its diagrams can indirectly express the properties about the topology of the spatial objects. For example, a set of hexagonal diagrams indicates that these objects are being uniformly distributed, or a set of distorted diagrams may mean that these objects are spatially influenced by some other dimensional spatial attributes (see Fig.3). Therefore, in order to know the spatial behaviors of prehistoric human societies, the inter-site relations will become clear along the time-series by using many digital site distribution maps with "cal BC", if the Voronoi diagrams that based on each locations of the archaeological site are illustrated.

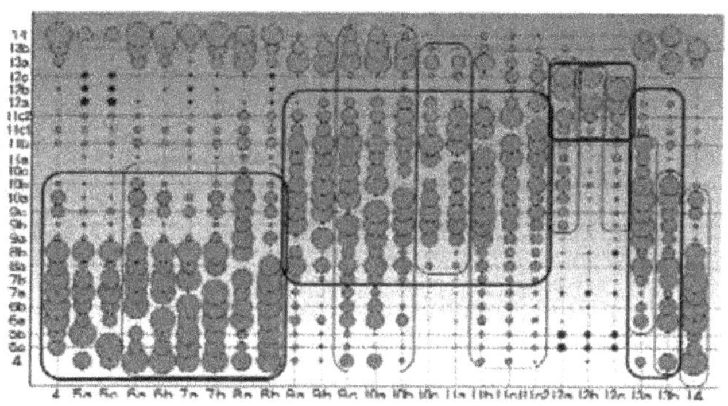
Figure 5 Graphic chart of the visuability patterns determinant

[2] The Voronoi Diagrams

The Voronoi diagrams are extracted by

i. The straight lines are drawn between each object.
ii. Its vertical bisector is drawn as far as their intersections.
iii. The inside area of these is made the connections to the

object as a polygon. This method has been well applied for the analytics of marketing or ecological area because of iii. I analyzed the site distribution maps with this method (Fig.6).

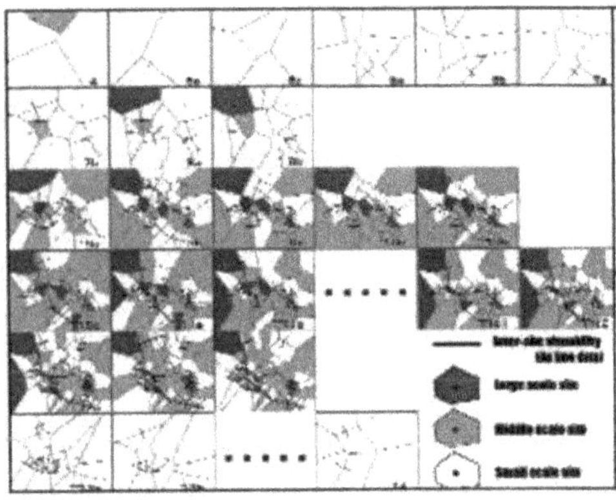

Figure 6 Voronoi Diagrams and inter-site visuability lines

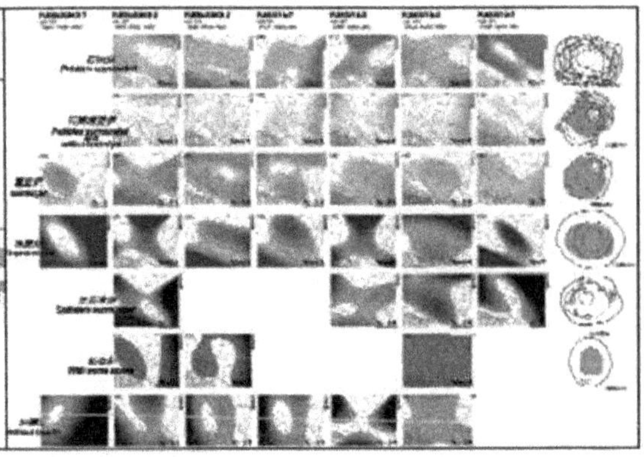

Figure 7 Trend-surfaces of various hearths' composition ratio

[3] The spatial correlations between the Voronoi Diagrams and the inter-site visuability patterns

To examine the spatio-temporal relations between site locations and their own relations, the inter-site visuability lines that were reconstructed in chapter 4.2 were overlaid on the Voronoi diagrams along the time-series (Fig.6). It is clear that the former traditional explanations, for example, these prehistoric dwellings were being arranged with spatial uniformity to avoid intervention, has some problems, for there are many diagrams more distorted than hexagonal in any phases. And we will also become aware the other old explanations about site locations, such as "Its decision makings might be influenced by the existence of rivers or springs" are not necessarily so simple. Instead this analysis made it clear that the site location and area had been connected with their own spatial interactions. This method made possible to create some concrete images about the inter-site relations as a linear situation and elucidated a mechanism that lies behind making decisions on site location with the geometrical character of diagrams that symbolize the topological characters.

Geostatistical Approach 2 - Viewshed Analysis

[1] What is the Viewshed Analysis?

The Viewshed analysis is a method to extract all visible area from the whole target area based on relations between the view point and the angle of elevation. In GIS technology, this process is executed based on DEM. Although this approach may have some problems (e.g. a huge gap of the assessment of data and the interpretation), it illustrates the actual images of the prehistoric human spatial interactions.

[2] Describing the Inter-Site Visuability

In assessing the inter-site visuabilities, histograms on the number of visuable or unvisuable relations were made, for these spatial topologies of visuabilities can be classed by these statistical quantities (see Fig.4). For instance, when its mode is 2 or 3, its topology of inter-site conjunctions will look like a line or a circle (Pattern A, B or D). Similarly, when its mode is 4 or 5, or its standard deviation is smaller, it will look like 2D or 3D web (Pattern E or F). More over, when its maximum and minimum value gap is large, it will look like a spokes of a wheel (Pattern G). We have only to assess its patterns that were modeled from some inter-site visuabilities to reconstruct the prehistoric human-human interactions.

As a result of this approach, it became clear that the inter-site relations of the target area gradually became complicated from pattern A to B, to C/D, to E/F, to G (Tsumura et al. 2002). This pattern show that the prehistoric peoples not only had decided their settlement locations based on natural environmental elements, but also made decisions under the pressure of their own social interactions. If they decided freely, their settlements would have been accumulated on the same place where the natural environmental condition is most suitable.

Finally, I computed the visuability determinant patterns from correlation coefficients, and created the graphic chart (Fig.5). It is clear that several epochs would be extracted for explanation about human social evolution from this result.

Geostatistical Approach 3 - Trend Surface Analysis

[1] The Keys of Trend Surface Analysis

The trend surface analysis is one of the Geostatistical methods that assess the spatial tendency in vector field such as directivity or intensity of subjects from several spatially dotted attributes. If we were to apply this approach to the archaeological spatial analytics, several archaeological cultural elements will play the mouthpieces of their own prehistoric social contexts. Presupposing this idea, pottery, hearths, and lithic types were adopted as archaeological elements. Pottery types' trend surface will indicate an inter-site trading system because pottery were conventionally a movable cultural attributes. Hearth types' trend surface will show matrimony system since conventionally woman had managed hearths. And lithic types' trend surface will reconstruct the spatial expansion of various subsistence. With GIS technology, we would be able to explain several spatial interactions or correlations about material culture elements left by human activities in various contexts in high probability.

[2] The Trend Surface Analysis

The trend surface are computed by

i. LINEAR $z = b0 + b1x + b2y$,
ii. QUADRATIC $z = b0 + b1x + b2y + b3x^2 + b4xy + b5y^2$,
iii. CUBIC $z = b0 + b1x + b2y + b3x^2 + b4xy + b5y^2 + b6x^3 + b7x^2y + b8xy^2 + b9y^3$,

I sorted out a suitable object from these results by the goodness of fit and F ratios. These trend surfaces were arranged along the "unconditional" time-series (e.g. in case of hearth types', see Fig.7).

As a result of the analysis, I could suggest the following; 1) the spatial deviation of trend surfaces of hearth types is larger than pottery, 2) in case of the trend surfaces of hearth types, there are several small centers (peaks) of directivity and/or intensity, 3) these results of 1 and 2 reflect the spatial characteristics of cultural elements, namely, movable (like pottery) or un-movable (like hearths), 4) an appropriateness of the spatio-temporal complex could be reconfirmed as a conceptual premise in archaeology that is the multiple spatial layer of infinite dimensions.

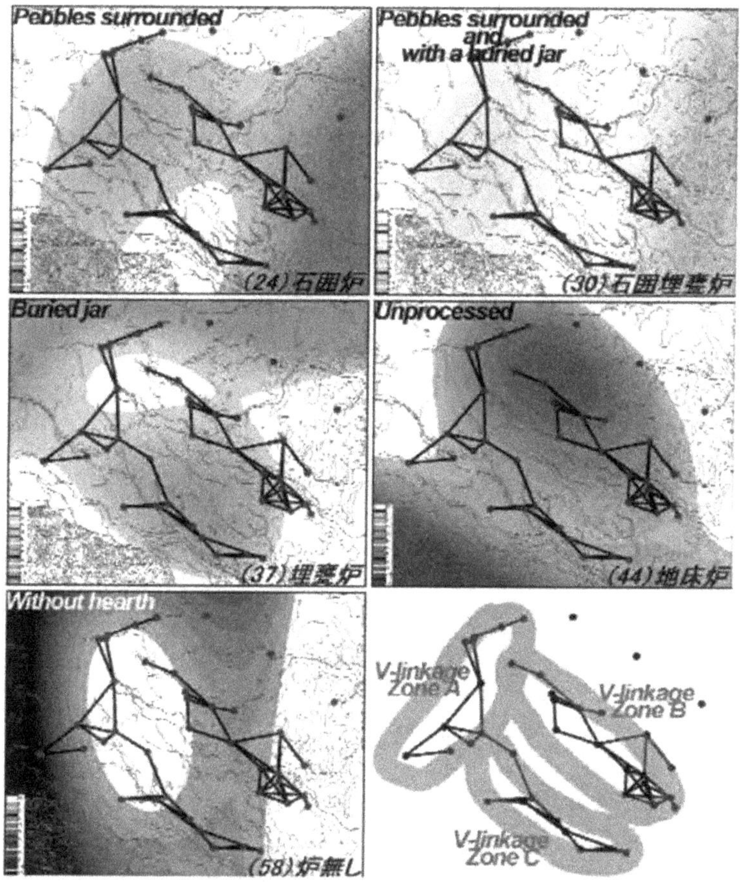

Figure 8 Spatial correlations between trend-surfaces and inter-site visuability lines

[3] Spatial Correlations within the Cultural Elements

Finally, to detect more concrete correlations from these spatiotemporal complexities, I unfolded the inter-site visuability patterns on these trend surfaces (see Fig.8) using the overlay analysis. As a result, there were positive correlations between them. In this case, the visuability linkages that are zoned by its spatial density (: V-linkage zone) were well matched with the hearth types trend surfaces. The V-linkage zone A correlates closely with the pebble-surrounded hearth' trend surface, the V-linkage zone B has a mutual relation to the unprocessed hearth's trend surface, and the V-linkage zone C has an interrelation with the buried-jar hearth's trend surface.

Therefore, we can see that the distribution of the sites might have been decided with the inter-site visuability based on their matrimony system. For example, a system of marriage in which the wife visits the husband at his home, and its spatio-temporal complexities were gradually became complicated. Although this conclusion is very simplified, at least, we would be able to understand that these spatial situations have not occurred by simple one spatial context like a "Water supply".

Conclusion - Real Ability of GIS Technology

Finally, Voronoi diagrams, inter-site visuability lines, and trend surfaces of cultural elements were lined up along the "Unconditional" calBC time-series. The interpolation of temporal gap was morphed by leaping. This approach made clear the dynamism of culture.

By applying the GIScience in archaeological spatial analysis, I was able to approach "eloquently" to the prehistoric human spatial activities through several "silent" material cultures. These are new methodologies that are neither "Geo-Archaeology" nor "Prehistoric Geography". I would like to propose this methodology as "Time-GIS" or "Four-dimensional GIS". The real abilities of GIS are to treat much interdisciplinary information founded on unified systems, to compute much-complicated statistical problems, and to show the process of "eloquent" reconstruction of the human history.

Acknowledgements

I would like to acknowledge that this work was done in the framework of the MUSASHINO Archaeo-ecological project, funded by National Museum of Japanese History. I would like to thank Prof. Toyohiro Nishimoto (Dept. Archaeological studies, National Museum of Japanese History) coordinator of this project, for his support. And I wish to thank my advisor Dr. Takashi Oguchi (Center for spatial information science, The University of Tokyo), and Prof. Izumi Niiro (Dept. Archaeology, The University of Okayama) for their assistance with my research. And I would like to special thank Ken'ichi Kobayashi (The Graduate University for Advanced Studies) for his fundamental supports in my technical researches and thank my project members. And I would like to thank Tomoko Anezaki and Dr. Mark Hudson for their inspection of my clumsy English.

REFERENCES

CLARKE, D. L. 1977. Spatial Information in Archaeology. Spatial Archaeology. London.

KANEDA, A., TSUMURA, H. and NIIRO, I. 2001. Geographical Information System for Archaeology. Kokon, Tokyo.

KOBAYASHI, K., TSUMURA, H. and TATEISHI, T. 2002. A fundamental approach to the archaeological site distributions in eastern part of the Musashino Upland at the middle Jomon period. Settlement studies 4.

NISHIMOTO, T., TSUMURA, H., KOBAYASHI, K., et al. 2001. An Ecological Study of the Jomon Settlement (1). Zoo-archaeology 17: 73-82.

TSUMURA, H. 2000. The assessment of archaeoenvironment by using GIS. Zoo-archaeology 15: 1-19.

TSUMURA, H. 2002. A Methodological Approach to the Assessment of Inter-site Conjunctions. Zoo-archaeology 18: 39-54.

TSUMURA, H. 2002. A review of the "culture" concept in archaeology by application of GIS -Through a case study of the Jomon settlement at the MUSASHINO Upland-. PAPERS and PROCEEDINGS of the GEOGRAPHIC INFORMATION SYSTEMS ASSOCIATION 11: 393-396.

TSUMURA, H. 2003. An Analysis of Prehistoric Settlement Systems in the Perspective of Spatial Information Science. From Decipherment to Application of Site and Urban Information: 9-18. The University of Kyusyu.

TSUMURA, H., KOBAYASHI, K., SAKAGUCHI, T. et al. 2002. An Ecological Study of the Jomon Settlement (2). Zoo-archaeology 18: 1-37.

TSUMURA, H., KOBAYASHI, K., TATEISHI, T. et al. 2002. An Ecological Study of the Jomon Settlement (3-1). Zoo-archaeology 19: 39-72.

TSUMURA, H., KOBAYASHI, K. and NISHIMOTO, T. 2003. A GIS Approach to the Prehistoric Settlement in Japan. New Developments in History and Culture Research using Information Technology: 117-124. National Museum of Japanese History, Chiba.

Modelling the intensity of early Neolithic land use with the help of GIS - an example from the "Mörlener Bucht", Wetterau, Hesse, Germany

R. Ebersbach
Inst. für Ur- und Frühgeschichte,
Universität Basel, Switzerland

C. Schade
Seminar für Vor- und Frühgeschichte,
Universität Frankfurt, Germany

See the CD for the paper

Abstract

The survey project BBM ("Besiedlungsgeschichte der Bandkeramik in der Mörlener Bucht") yielded a great deal of data about early neolithic settlement acitivities. Settlement densities, sizes and durations of villages of the linear pottery culture (5300 - 4900 BC cal) are fully known in the test area. Based on the number of contemporaneous houses a minimal field size can be calculated. With the help of GIS minimal necessary territory sizes are modelled. The diachronic comparison between the supply of arable land and the demand of the Neolithic farmers and their stock leads to a discussion of the intensity of early Neolithic land use (and abuse?).

Active 3D Sensing for Heritage Applications

J-A. Beraldin

F. Blais

L. Cournoyer

G. Godin

M. Rioux

J. Taylor

Institute for Information Technology
National Research Council Canada
Ottawa, Ontario, Canada, K1A 0R6
angelo.beraldin@nrc-cnrc.gc.ca

Abstract

Active three-dimensional vision is concerned with extracting information from the geometry and the texture of the visible surfaces in a scene, reasoning about the data and finally, communicating the results. In order to take full advantage of these vision systems, one must understand not only their advantages but also their limitations. This paper covers some issues that must be addressed before embarking in a 3D scanning project.

Introduction

In the last twenty years, many advances have been made in the field of solid-state electronics, photonics, computer vision and computer graphics. Non-contact three-dimensional (3D) measurement techniques like those based on structured light and passive stereo are examples of fields that have benefited from all of these developments (Jähne, Haußecker and Geißler 1999). These non-contact measurement techniques have found widespread use in heritage, environment modeling, virtual reality, and many industrial fields. Figure 1 shows a classification of non-contact measurement strategies that rely on light waves. In the case of passive techniques (that use ambient light), only visible features with discernable texture gradients like on intensity edges are measured. Laser-based systems are used to structure the environment in order to acquire dense range maps from visible surfaces that are rather featureless to the naked eye or a video camera. Hence, the 3D information becomes relatively insensitive to background illumination and surface texture. This paper reviews the basic principles that underline the majority of active 3D vision systems. Some pointers are presented in order for the reader to become more critical when picking a 3D vision system. The target application is in the context of cultural heritage (Godin 2002).

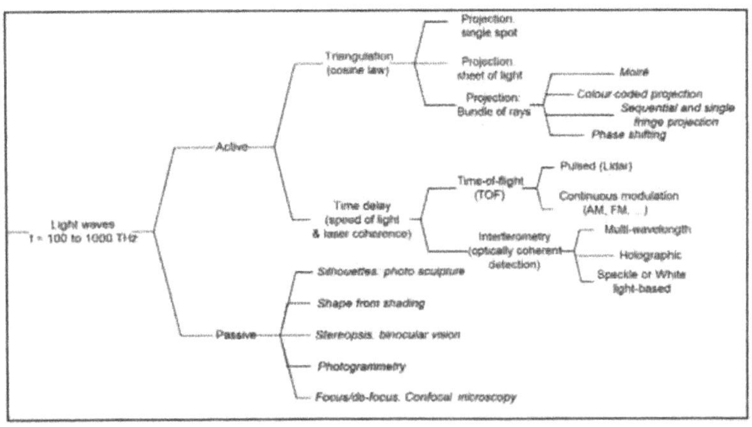

Figure 1 Classification of non-contact 3D surface measurement techniques based on light waves

Concluding remarks appear at the end of the paper.

Representations of Three-Dimensional Surfaces

The simplest way of representing and storing surface coordinates of a scene is with the use of a depth map. A depth map is like a grey scale image generated by a 2D camera except that the z information replaces the intensity information. If the elements of the image are organized so that for a given 3D coordinate all the coordinates surrounding it are physically neighbors of that coordinate, then we have an organized point cloud. If a grey scale value is assigned to each surface element (e.g. using a triangular patch) according to their slope angles, then this method provides a way to artificially shade an image in order to highlight surface details.

THREE-DIMENSIONAL SENSING

[1] Triangulation

Triangles are the basis of many measurement techniques, from basic geodesic measurements performed in ancient Greece to more modern laser-based 3D cameras. The Greek philosopher Thales (6th century BC) has been credited with the discovery of five theorems in geometry. Two of which are used to solve the triangulation equations in a triangulation-based range camera: opposite angles of intersecting straight lines and the cosines law. Triangulation is used in applications requiring an operating range that is less than 10 meters. For ranges between 2 m and 10 m, there are a limited number of triangulation-based sensors available commercially. The geometrical principle of optical triangulation is shown in Figure 2a. Knowing two angles (α and $\Delta\beta$) of a triangle relative to its base (baseline d) determines the shape of the triangle. For an incremental change of distance, Δz, one measures the change $\Delta\beta$ (assuming α constant) using the theorem of opposite angles. The angle sensor shown on that figure is in fact a solid-state image sensor. To measure a profile, one needs to scan the spot.

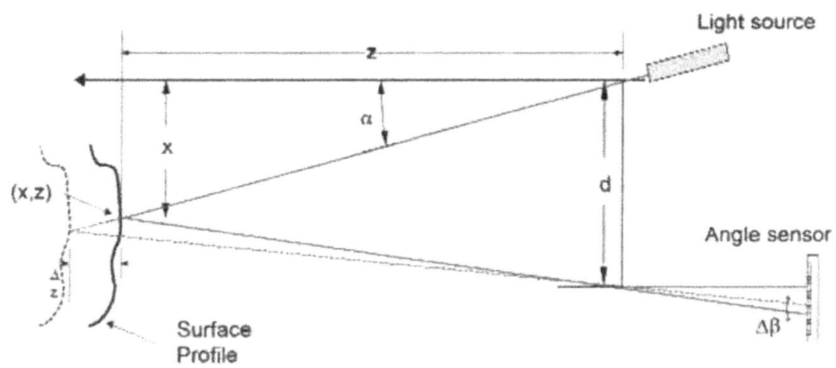

Figure 2a Principle of active triangulation: a triangulation-based laser probe

To show the limitations of triangulation, people approximate the measurement uncertainty of the error in z by the law of propagation of errors. One finds, that the measurement uncertainty in z is inversely proportional to both the camera baseline and the effective focal length of the lens, but directly proportional to the square of the distance. Unfortunately, f_o and d cannot be made as large as desired. The baseline d is limited mainly by the mechanical structure of the optical setup (stability of the whole system decreases as d increases) and by shadow effects (self occlusion problems increase with d). Rioux 1994 presents an innovative approach to triangulation-based range imaging that allows very large fields of view without compromising the performance of the system. In order to acquire a full 3D image, one has to either move the scanner or use another technique that yields full field 3D measurements. Galvanometer-based systems use two mirrors to raster scan a scene. Depending on the application, line scanners can be mounted on rotation, translation stages or coordinate measuring machines (CMM). Finally multi-line laser projection and coded pattern projection systems are also used for full field 3D cameras (Jähne, Haußecker and Geißler, 1999). Moiré effect is another 3D technique.

[2] Time delay

A fundamental property of light waves is their velocity of propagation. In a given medium, light waves travel with a finite and constant velocity. Thus the measurement of time delays created by light traveling in a medium from a source to a reflective target surface and back to the source (round trip) offers a very convenient way to evaluate distance. With an approximated speed of light of 3 108 m/sec, then, 3.33 nanoseconds correspond to about 1 meter. Interesting enough, 3D systems based on time delay measurement are used either for very small fields of view (< 10 mm) or very large ones (> 10 m). For ranges between 0.2 m and 10 m, there is a limited number of this type of sensor available commercially. Long-range sensors (range exceeding 10 meters) are usually based on the time-of-flight (TOF) technology (also known as laser radar or lidar for short). The camera to object distance (z) is measured by sending a relatively short impulse of light on a reflective surface and measuring the round trip,τ , $z=c*\tau/2$. If one wants a resolution of 1 mm then a time delay of about 3.33 picoseconds will have to be measured. Most commercial systems provide a resolution of about 10 mm to 100 mm. Other systems based on continuous wave (CW) modulation get around the measurement of short pulses by modulating the power of the laser beam. Again the modulated signal is projected onto a surface, the scattered light is collected and a circuit measures the phase difference between the two waveforms and hence a time delay. For instance, with a frequency of 10MHz and a phase resolution of 0.01 degree (not too difficult with standard electronics), the resolution in z is about 0.5 mm. Because the returned wave cannot be associated with a specific part of the original signal, it is not possible to derive the absolute distance information from a simple CW method. This is known as the ambiguity interval. To get around the inconvenience of a range ambiguity interval, one can use multiple frequency waveforms.

[3] Interferometric

The high coherence of laser light is very helpful in measurement involving interference of light beams. If a light beam is divided into two parts (reference and measuring) that travel different paths, when the beams are combined together interference fringes are produced. Such devices are called laser interferometers. Very small displacements (in the order of a fraction of wavelengths) can be detected (using coherent detection), and longer distances can be also measured with low measurement uncertainty (Jähne, Haußecker and Geißler 1999).

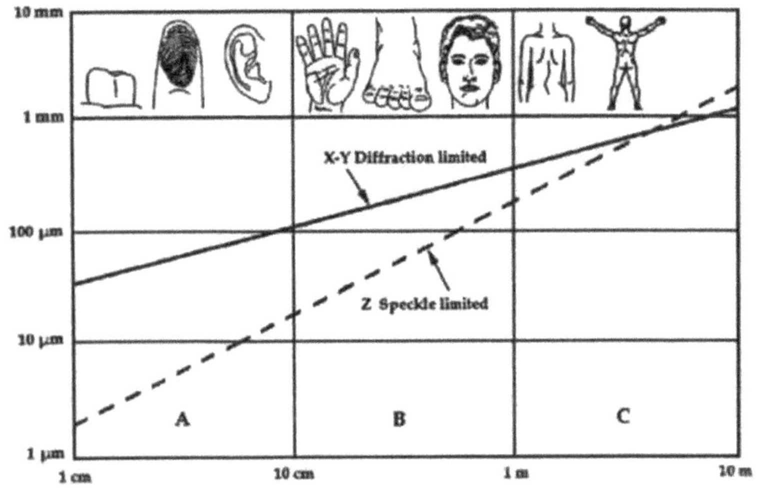

Figure 2b Principle of active triangulation: physical limits of 3D optical measurements based on laser projection and triangulation

PERFORMANCE EVALUATION OF LASER-BASED TRIANGULATION SENSORS

[1] Laser beam propagation

Optical triangulation resolution is limited by the diffraction of the laser light. Even in the best emitting conditions (single mode) the laser light does not maintain collimation with distance. In fact, the smaller the laser beam, the bigger is the divergence produced by diffraction. Fig. 2b illustrates that constraint. The solid line shows the relationship between the X and Y axes resolution (direction perpendicular to the laser projection) and the physical dimensions of the object to be scanned. Figure 3 presents results obtained with our laser scanner.

[2] Measurement uncertainty
Diffraction limits impose a constraint on the resolving power along the X and Y-axes as described above, but along the range axis (Z) one could expect a continuous improvement as the amount of laser power is increased. Unfortunately this is not the case; indeed the coherence of the laser light produces damaging interference effects that limit the resolving power of a position sensor Fig. 2b. This is known as the speckle effect.

[3] Calibration
The calibration of any range camera is concerned with the extraction of the internal parameters of the range camera when a mathematical model exists (focal length, lens distortions, scanning parameters, etc.) or with a mapping of the distortions in a look-up table when some elements cannot be modeled very well. After this phase, the range measurements are available in a rectangular co-ordinate system. After the calibration process, the accuracy, repeatability and measurement uncertainties of the range camera can be determined. This is the verification stage.

CONCLUSION

We reviewed some of the basic principles behind 3D range cameras: from simple trigonometric relations to parameters involved in the digitizing of a 3D object. The purpose of measuring 3D objects and sites could be for documentation, inspection or visualization and possibly all of them. The potential of active 3D vision for applications in heritage or as an input to virtualized reality environments was discussed. Real world acquisition and modeling is now possible. Technological advances are such that difficulties are more of a logistical nature than technology per se. Additional information on commercial 3D digitizing systems is available on the following web site: http://www.vit.iit.nrc.ca/Pages_Html/English/Links.html.

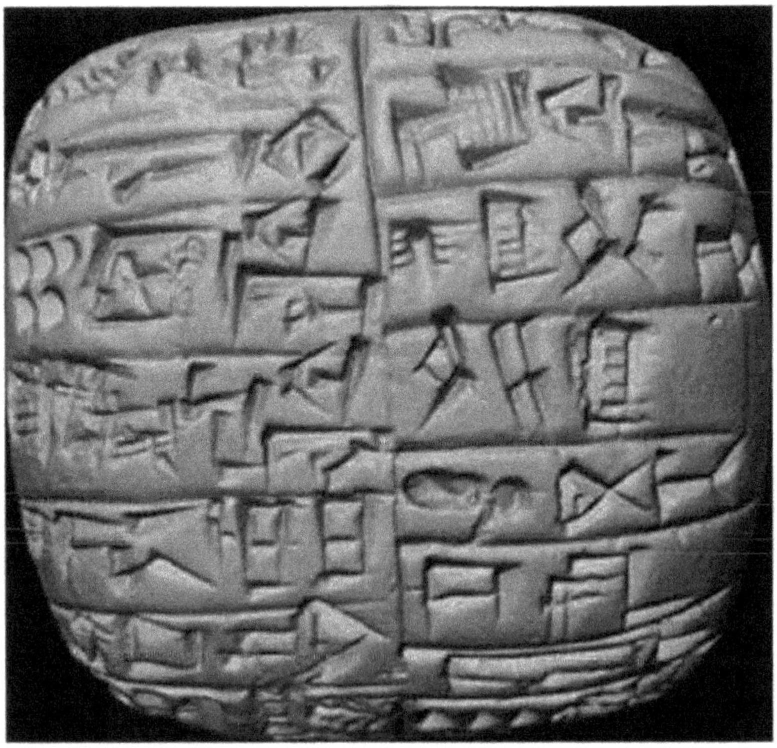

Figure 3 Shaded 3D model of a fragile cuneiform tablets can be interactively examined and the writing "unwrapped" for study. The X-Y resolution is 0.05 mm and Z uncertainty is about 0.01 mm.

REFERENCES

GODIN, G. et al., 2002. Active Optical 3D Imaging for Heritage Applications. IEEE CG&A 22(5):24-36.

JÄHNE, B., HAUSSECKER, H. and GEISSLER, P. (eds.), 1999. Handbook of Computer Vision and Applications. Vol. 1, Sensors and Imaging, Chap. 17-21, Academic Press, San Diego.

RIOUX, M., 1994. Digital 3-D imaging: theory and applications. SPIE Proceedings, Videometrics III, International Symposium on Photonic and Sensors and Controls for Commercial Applications, Boston, MA. 31 October - 4 November, Vol. 2350:2-15.

SOFTWARE ARCHAEOLOGY - AN INTERDISCIPLINARY VIEW

GERHARD CHROUST
KEPLER UNIVERSITY LINZ
SYSTEMS ENGINEERING AND AUTOMATION
GC@SEA.UNI-LINZ.AC.AT

SEE THE CD FOR THE EXTENDED VERSION

ABSTRACT

Maintenance is one of the key problems of software engineering, often nicknamed 'software archaeology'. This paper discusses analogies between software maintenance and archaeology, emphasising similarities and dissimilarities. It shows some surprising parallels and insights concerning what one calls legacy systems or legacy artefacts.

THE LEGACY PROBLEM, MAINTENANCE AND KNOWLEDGE ELICITATION

Maintenance of software products is a key problem today, i.e. repairing and enhancing so-called 'Legacy systems'. These systems have outlived their planned useful life, their programmers, their base technology etc. The general public recognized this in the course of the transition to the year 2000 and (for Europe) in the change to the Euro.

There are several causes why software becomes less usable or erroneous during its life time (Lehman 1985, Basili 1990). These causes should sound familiar also to archaeologists:

- (Lehman's Law) Successful systems have to change in order to remain acceptable to their users (Lehman 85).
- Stable systems which do not need change are 'dead' systems.
- Most of the changes are not caused by programming errors, but are due to external changes (changed legislation, different requirements, unforeseen change to the environment, e.g. changing to the Euro). Industry sources indicate that approx. 40% of all changes (i.e. 'maintenance') are actually changes to adapt a system to changing environments (Sneed 1990).

One has also to admit that legacy systems have several advantages which distinguish them from systems to be newly written and therefore justify maintenance:

- Existing and operational systems often contain considerable hidden domain knowledge not documented or know to the users.
- Old system work, which is not self-evident for newly built systems.
- etc.

Therefore legacy systems are not only old burdens, but also old treasures - like in archaeology. They pose, however, some problems:

- Their developers do not exist any more, one cannot ask them.
- Documentation (for design and operation) is non-existent, is lost, is unreadable, written in nowadays unknown language (Who still knows the programming language IPL-V?).
- Existing documentation is unreliable and often outdated with respect to the current system in operation.
- The requirements, motives, objectives and the environment in which the systems were build and operated do not exist any more or cannot be understood.
- Parts of the system are missing and forgotten.
- These systems contain extra parts which are not useable any more, even not accessible by normal operation (software calls this 'dead code'), but make understanding more difficult.
- Large parts of the system have been changed over and over again.
- The systems were are build in a technology with is outdated and often not safe and reliable any more.
- The original (probably clear design) has been modified over time and was obliterated by various minor modifications.

We recognize that one of the major problems is acquiring enough knowledge about the legacy system using available artefacts. One has to elicit knowledge from the available sources, structure it and preserve it in adequate and hopefully better accessible and understandable form. Archaeologists fight with the same problems (Hunt 2002). Therefore Harry Sneed, a well known German-Canadian software pioneer, coined the term 'software archaeology' (Sneed 1994) for maintenance work in the software industry (Hunt 2002, Dennet 1986). We will develop this idea further.

A major distinguishing characteristic is the aim of these two fields:

- Archaeology puts the emphasis on putting the observer into the historical "original" environment, striving to preserve the past for analysis and contemplation, while
- software maintenance tries 'to bring the legacy system into today users' environments', striving to keep old systems in productive use.

HANDLING LEGACY SYSTEMS

Software engineering provides a range of techniques to handle legacy systems, the so-called 'Re-techniques' because of their common prefix (in italics you find remarks targeted at archaeology).

REcognition: Recognizing and identifying useful information (using data mining and pattern recognition) are important methods in software maintenance. *This turns out to more difficult in archaeology because most of the data are hidden, but software engineers often are also at loss to find a certain critical software module's source code.*

REallocation: Artefacts often are brought into another environment, mostly for safeguarding or protection from destruction, in archaeology often out of pure greed. *Especially in archaeology this is difficult, cumbersome, error-prone and not always successful: artefacts get lost, broken, stolen and confused during transfer.*

REcombination: Related information is not necessarily in one place, legacy software has functions distributed over the code, largely due to maintenance patchwork. *Human intuition can be augmented by massive computer support to identify potentially matching pieces and interfaces dispersed over the world.*

REpair: Artefacts need repair in order to preserve them, a fact well-known in software engineering, where maintenance is important but also difficult due to software's idiosyncratic properties like invisibility and easy changeability (Brooks 1986). When repairing software the original code has only to be preserved if there are old systems still using it and this can easily achieved by copying. *In archaeology preservation of original artefacts and the precise distinction between original and replacement is a key concern, especially in the case where a site has many strata.*

Restoration: *This is one of the major challenges and source of controversy in archaeology (but not an issue in software): to which epoch and status should the artefact be restored? Typically after the fire in the famous Redoutensäle of the Vienna Hofburg, the discussion arose whether to restore these rooms to their last 20th century appearance or to their original appearance (1705).* Software can easily be duplicated to allow both versions to exist in parallel. *For archaeology only Virtual Reality (Billinghurst 2002) offers the chance to see several views.*

RE-documentation: Traditionally (not only in archaeology) documentation of artefacts is rudimentary, often not existing and unreliable or unreadable. Some documentation is not even recognized as such: initially even cuneiform inscriptions were misunderstood as decorations without deeper semantics. *Fortunately archaeologists are trained in documentation and see this as one of their major professional tasks - in contrast to the archaeological adventurers of the 18th century and in contrast to most software engineers.*

RE-structuring: Due to maintenance the structure of a software product is gradually deteriorating.

- It is therefore necessary to re-structure a software product, compatible with the original concepts and the changes made since.
- The structure and organisation might - sometimes even unintentionally - be completely transformed into another structure. *This effect is well known to archaeologists, when different uses of a buildings cause more or less small changes which in sum, however, often completely change the - outlay, the appearance and the usage pattern of a building.*

REverse Engineering: Key issues when confronted with some unknown artefact are: What does it accomplish? How does it function? What was its purpose? Why has it been built like that? *Archaeologists, perhaps more than software engineers, understand that these questions have to be answered on different semantic levels: If we know what the form of a house was (of which we may discern only the foundations), we still do not know what rites or professions were performed in the various rooms, let alone why a certain activity was performed at all. And we know that more than one interpretation is possible.*

REengineering: Reengineering is the rebuilding of a system with different means and/or technology carrying over the information or functionality of the old system but for sustained/improved usage. *Archaeology has the privilege not to have to cater for current usage of most archaeological artefact. Such use would be contra-productive and destructive for scientific research of archaeological sites.*

Nevertheless in some rare instances even archaeology has to use reengineered artefacts. Examples are copies of statues from medieval churches or the duplication of the caves of Lascaux in Paris.

A true blend between archaeology and computer technology is Virtual Reality, which allows us - in a true re-engineering fashion (same 'look and feel' but completely different technology) to 'virtually re-build' every artefact sometimes even indiscernible from the original artefact in order to enable many people to see and to even touch (!) *without leaving their home.*

REuse in different context: In software production, as in every other industry, the reuse of partial products is one of the keys to productivity and quality. *With respect to archaeology this is a highly undesirable human activity since it usually meant carrying away archaeological artefacts for some other unknown, profane use, like using the Pyramids or the Carnuntum site as a cheap source of building material like a stone quarry.*

Cross-Fertilization

We can observe that the maintenance of software products and archaeological work have considerable overlap, especially in the following areas (Hunt 2002):

Preservation Problem: The identification and preservation of artefacts, etc. is of utmost importance. In archaeology - due to the uniqueness of artefacts - it has to be done with utmost precaution, the production of identical copies eases this problem for software engineering.

Understanding Problem: The understanding of the meaning of the available documentation and its validation is a key challenge, misinterpretations often are long-living (cf. interpretations of Knossos) and counter-productive.

Documentation Problem: Reading and understanding ancient documentation is a multi-levelled problem, starting from identifying characters or symbols (Doblhofer 1990) to trying to read and pronounce the utterances etc. Programming languages do not pose similar deciphering problems like some of the ancient languages. In the domain of software lack of proper documentation, lack of visibility of the dynamics of a program causes problems by forcing maintenance engineers to deduct bottom-up the functionality of a program.

Matching Problem: In any complex system a key to understanding is knowing the relation of artefacts to one another. *Archaeology has the disadvantage that many of the artefacts have been removed from their original site (often illegally and secretly),and have gone through many hands (and countries!). Establishing the original relationships needs modern technology and algorithmic approaches to pattern matching and data mining, only possible nowadays.*

Presentation Problem: For different reasons both archaeology and software have a similar problem: How to explain to outsiders (including those who can provide the necessary sponsor money) what actually the underlying structure, concepts, and plans were. For software this is mainly caused by the invisibility of software and the difficulty to show dynamic *behaviour. For archaeology part of the problem is the lack of some important parts of an artefacts. Virtual Reality or Mixed Reality can be very supportive there (Billinghurst 1992, Forte 1997, Tarumi 2000 and Stone 1992).*

SUMMARY

Archaeology can be helpful in providing understandable, obvious examples for the rather abstract, ephemeral observations and problems of software, while software can bring new ideas and technology to the long-established field of archaeology, introducing new approaches and methods. In this paper we have shown some similarities between the field of Software Maintenance (*"Software Archaeology"*) and Archaeology. Some of the problems where software engineers have difficulties to accept them on an intuitive basis, are obvious in archaeology (e.g. the destruction of structure by maintenance, the drifting of architecture by enhancements, the ambiguity of reverse engineering, etc.) On the other hand archaeology can profit from software's ability to process masses of data and supplying new representational means by Virtual and Mixed Realities. Like in many other fields, interdisciplinarity pays off.

REFERENCES

BASILI, V.R, 1990. Viewing Maintenance as Re-use oriented Software Development. IEEE Software January:19-25.

BILLINGHURST, M. and KATO, H., 2002. How the virtual inspires the real - Collaborative augmented reality. CACM Vol. 45, no. 7:64-70.

BROOKS, F.P.Jr., 1986. No Silver Bullet - Essence and Accidents of Software Engineering. In Kugler, H.J. (ed.), Information Processing 86, IFIP Congress:1069-1076.

DENNETT, D., 1986. Julian Jaynes' 1986 Software Archeology. Canadian Psychology April, Vol. 27(2):149-154.

DOBLHOFER, E., 1990. Zeichen und Wunder. Weltbild-Verlag (ISBN 3-89350-X).

FORTE, M. and SILIOTTI, A., 1997. Die neue Archäologie - Virtuelle Reisen in die Vergangenheit. Gustav Lübbe (ISBN 3-7857, 200888-2).

HUNT, A. and THOMAS, D., 2002. Software Archaeology. IEEE Software Vol. 19, no. 2:20-22.

LEHMAN, M.M. and BELADY, L.A., 1985. Program Evolution - Processes of Software Change. APIC Studies in Data Proc. no. 27, Academic Press.

SNEED, H.M., 1990. Software Wiederverwendung. ADV (ed.), EDV in den 90er Jahren, Jahrzehnt der Anwender - Jahrzehnt der Integration ADV:199-214.

SNEED, H., 1994. Claimed to have coined the term 'Software Archyeology. Personal Communication.

STONE, R. Virtual Reality and Telepresence. Robotica no. 10:461-467.

TARUMI, H., MORISHITA, K., ITO, Y. and KAMBAYASHI, Y. Communication through virtual active objects overlaid onto the real world. Proceedings of the third international conference on Collaborative virtual environments, ACM Press:155-164.

Computerised Methods for Palaeolithic Art Studies

Ludmilla Iakovleva
Institute of Archaeology,
National Academy of Sciences of Ukraine
& CNRS UMR 7041 France
L.Iakovleva@wanadoo.fr

Abstract

Since the discovery of portable and cave Palaeolithic art in the second middle of XIX° century, the research methods have made major progress, mainly since the seventies. The purpose of this paper is to show the role of computerized methods for the study of Palaeolithic art, focusing particularly on:

- Recording techniques (various photography, photogramme try, laser 3D) and image processing,
- Style analysis of paintings and sculptures,
- Spatial structuring of cave art figures,
- Chronological and regional studies by data analysis.

It is then possible to discuss the reasons of both the potentialities and the limits of cave art studies in the purpose of explaining the meaning of Palaeolithic art, and to propose new developments to achieve it.

Introduction

Palaeolithic art research is based upon numerous methods and techniques, come from other scientific fields like first Archaeology, but also other Human Sciences (Ethnography, History of Art and History of Religions), from Exact Sciences: (Physics and Chemistry) and from Engineering Sciences (Semiotics, Computer Sciences, Statistics and image processing). In this paper, I would rapidly recall the different methods, with several case examples: firstly, acquisition methods beginning with recording techniques, intrinsic and extrinsic characterization and determination of art objects, secondly with structuring methods like sequential analysis, chronological analysis, stylistic analysis, intrasite and intersite spatial analysis, and finally the methods for the reconstitution of the symbolic art system in its global context (Iakovleva 2002).

First Phase: Acquisition

[1] Acquisition by recording techniques

The records of a cave art painting or a portable art sculpture are the result of an interaction between the art object to be recorded and the specialist who has to record it.

Recording is not else than an interactive and retroactive training process between the art object and the specialist. He is of course helped by appropriate optical, photographical and computerized systems, in order to improve it. But digital recording needs always a visual control nearby the wall or the support of the art objects. The recording method is starting with a recording convention which is a formalized descriptive syntax, like for a key of a map or the colour matrix of a satellite image. Such an example of a recording convention designed for the study of the magdalenian engraved and sculptured frieze of the "Roc-aux-Sorciers" rock-shelter at Angles-sur-l'Anglin (Vienne, France) is given in (Iakovleva and Pinçon 1997).

The 2D recording for cave art is using various and well adapted techniques :

- Freehand drawing on a plastic film, at a very short distance of the wall,
- Drawing directly on a photograph printing (scale 1/1),
- Photography, enhanced by more or less sophisticated techniques like special films, digital records, image processing and photographic fit.

Going back to the wall, with a microscope and macro-photographs, it is then possible to record the details which cannot be visible with long-distance records, needed to limit the negative effects of image distortion. Such techniques allow making conventional records. An example of conventional recording is given in a study of the magdalenian engraved and sculptured frieze of the "Roc-Aux-Sorciers" rock-shelter in Angles-sur-l'Anglin (Vienne, France): the panel of Venus and the panel of Ibex. The conventional records, after validation, may be then transformed into realistic records (Iakovleva and Pinçon 1997).

The analysis using high (binocular) or very high (ESM) magnification on the portable objects, or when it is not possible due to their volume on replicas obtained by casting, are allowing to study art objects at a scale of detail never obtained before.

The 3D recording is using the last digital recording techniques like photogrammetry, and a short while ago 3D scanner. By controlling the reconstitution in the three dimensions, the 3D records are handling the volumetric surface of the wall or of the support. Then 3D recording is different from 2D recording which is realizing de facto a conic projection with the well-known distortions. Digital 3D techniques are allowing differentiated processing of the wall, integrating 2D or 3D various records (photographs, macro-photographs, frames) depending on the needed level of detail. But it is necessary to point out the time involved to do it, then the cost of a high quality 3D reconstitution. 3D records are naturally a way to use virtual reality techniques, offering a dynamic walk into the discovery of the art space, opened then to a new public area without any risk for the originals (fac-simile).

[2] Acquisition par object characterization

The choice and the preparation of a support are a major issue in the relationship between the artist, the art object and its raw material, whatever the support is a stony surface for cave art or a volumetric support of various raw material (ivory, antler, bone, stone, amber, etc.). Numerous studies have discovered the important marks of preparing stony walls by pricking, scraping, polishing and others, and drafts before painting, engraving or sculpturing of a figure (Iakovleva and Pinçon 1997). Concerning portable art, the relationship is stronger because the available volume and the morphology of the support will be often the reason of the choice of the figured animals. And its attitude is integrated in a composition which will be embedded into the support. Studies have also revealed the sound importance of the stony walls inside the caves, by resonance (Dauvois and Boutillon 1994).

Characterization is concerning also the art object itself, by its morphology (outline, surface, volume) and the techniques used for its representation: painting (stencilling, blowing etc.), engraving, incision, pricking, sculpturing (low-relief and high-relief in cave art; in the round, flattened in portable art) and casting. The techniques of production of the figures is handling numerous parameters, like the use of volumes of the wall, the lightening with natural or artificial light, the looking at the figure under different points of view or constrained by the geometry of the place, the choice of the raw material and the preparation of the colour, the tools needed to realize the figurations, determined by experimentation and use-wear analysis. For example, the spectrometric analysis on colour samples of the Lascaux cave (Menu and Walter 1996), have shown the existence of mixtures of colours with organic binders.

The representation of an object is following a complex life cycle from the initial draft, but often modified by repentances, transformations, re-use, integration in a new composition, modification, quite-complete or partial destructions and substitution. The specialist must reconstitute such a story of the art object.

[3] Acquisition by object determination

When the represented art object is an animal, zoology is helping us to determine the species, the sex, the age, the anatomic accuracy, the anatomic ratio exactness. Zoology may explain also the realism of a particular attitude. It owns to the specialist of Palaeolithic art to study the completeness of the animal, the richness of anatomic details, the obvious wish of deformation or exaggeration, in relationship with the realism or the schematization of the representation. It is also necessary to study, which is evident for any art historian, the figuration convention (face, side-face, ¾ face), the perspective rules, the animation and the composition of the scene, involving to reveal a typology of styles of figuration. Data Analysis is a useful statistical technique to determine such styles of figurations and their regional widening. For example, the typological analysis of "Contours découpés" of magdalenian horses by correspondence analysis (Buisson et al. 1993) is showing the large homogeneity of these portable art objects over all the franco-cantabrian area.

Some particularly rare objects are much less easy to determine, for the evident reason that the palaeolithic artist has not intentionally represented an animal. They are called generally anthropomorphs or anthropozoomorphs, these often schematic figurations, mixing human and animal anatomic features.

The signs, at the opposite, are very numerous objects with a so wide morphology, that very few specialists had the audacity to design a systematic description. Only, the use of semiotics allows approaching rigorously the problem, like G. Sauvet (1987) who has proposed a grammar for upper palaeolithic signs.

1.POINTS ET ELEMENTARY LINES
- Points
- Straight line: stickers
- Simple broken line: chevron, hook
- Curve line : simple, spirally, wavy

2.ELEMENTARY SURFACES
- From straight lines to polygons: Triangle, square, rectangle, rhomb, pentagon ("tectiform"), hexagon.
- From curve lines to circles and ovals,
- Mixed of lines and curves: streamlined, escutcheon,

3. COMPOSITIONS
- Clustering
  - Of stickers around a centre: cross, star, y, direction sign, bundle,
  - Of stickers around a line: comb, palm,
  - Of stickers around/inside an elementary surface: grid surface,
- Rhythm (repetition)
  - repetition of points and linear figures
    - duplication,
    - drawing up,
    - concatenation: zigzags, scallops,
  - Repetition of elementary surfaces,
    - encasement, check surface,
  - Organization of space
    - symmetry,
    - division of the surface into registers.

Figure 1 A grammar of palaeolithic signs (Sauvet 1987)

[4] Acquisition: Dating, Location and Association of art objects

Extrinsic information must be handled, beginning with space and first the location of the art objects in their surrounding space (cave, cave entrance, cave depth, rock-shelter, open-air site, open air rocks, portable support), and then inside with the location on a wall for cave art or on a support for portable art.

Another extrinsic information is the association with other art objects, which may be:

- a superposition of paintings or engravings often difficult to decipher, which may be get visible by image processing,
- a composition of several figurations on a wall or a support, which is often interpreted, not as a random association, but as the result of a design of the palaeolithic artist, and then carefully located on a chosen panel, specially prepared for that,
- a naturalist scene, easily understandable from the knowledge acquired from the ethology of the represented animals.

Another extrinsic information is chronology, which may be obtained by several ways:

- association of the object with a dated archaeological layer: portable art object or fragment of a frieze fall down in a dated archaeological layer,
- wall painting or engraving recovered by an archaeological layer,
- cave art with a unique archaeological layer or hearths or isolated bones or charcoal snuffing out of a torch etc.,
- absolute 14C dating of a painting made with charcoal colour (for example, the bison cellar of Altamira cave by Freeman and Echegaray 2001).

Second Phase: Structuring

The structuring phase is processed by correlations between intrinsic information and extrinsic information obtained by the previous acquisition phase.

The sequential analysis is reconstituting the whole set of actions and decisions taken by the Palaeolithic artist : choice of the representation space (cave, cave entrance, cave depth, rock-shelter, open-air site, open air rocks, portable support), use of tools and colours, preparation of the portable or wall support, technique, process and sequence of production of the figures, composition of the scene.

The chronological analysis is making a synthesis of all the elements allowing dating a composition of figures from superposition, absolute 14C dates and archaeological context (for example, the four steps chronology of the superposed engravings by incision and pricking of a rock of the archaeological park of Foz Coa (Portugal) (Baptista and Gomes 1995)).

The stylistic analysis of the figures is concerned by the analysis of the following features: proportions (anatomic ratios), production technique, representation rules, perspective rules, animation, deformations and exaggerations, involving the evidence of chronological phases or production schools (for example, the school of engravers of hypertrophied horses of La Madeleine (Apellaniz 1987)).

The intrasite spatial analysis is a method designed to study the spatial variations of the intrinsic information acquired from the objects of a site. It is the case, for example, concerning the correlation analysis of the represented bestiary and its topographical location. We will develop here the famous study made by A. Leroi-Gourhan (1965) concerning the preferential location of certain association of animals inside the cave.

The correspondence analysis is separating distinctively different areas:

- The central composition areas where are represented the associations horse-bison or horse-aurochs (depending on the respective zoocenose of bison and aurochs),
- The periphery of the central areas where are represented ibex, reindeers, dears and does,
- The entrance areas where are represented deers and does,
- The end areas (crossing, dead end and small lateral dead end) where are represented mainly felines, bears and anthropomorphs,
- Mammoths and rhinoceros are in a special case to be located in the central areas for the early periods (Aurignacian and Gravettian) and in the periphery of central areas for the late period (Magdalenian).

The intersite spatial analysis is a method designed to study the spatial variations at a regional scale. A good example is the study of the assemblage of animal species of the cave art of Western Europe. It is a correlation analysis between the represented bestiary, the location of the art caves, and their chronology. Such a study has been first directed by A. Leroi-Gourhan (1965), and after expanded by Sauvet G. and S. (1979) several times on a more complete assemblage and the use of correspondence analysis. The analysis is showing the regional and chronological variations of the represented bestiaries during the upper Palaeolithic in Western Europe .

Third Phase : The Reconstitution of the Symbolic System

The third phase is corresponding to the highest ambition of the study of Palaeolithic art that is it is the integration in the cultural system and then, the intention to try to decipher the symbolic meaning of the Palaeolithic art.

The reconstitution of the technological and stylistic know-how has been detailed previously, and it is not necessary to come back again on it.

The reconstitution of the value of art objects in the social system has been relatively not deeply studied in Western Europe, for the evident reason that few portable art objects have been found in association with dwelling structures, at the opposite of the discoveries of portable art objects made in the numerous settlements in Eastern Europe.

Those studies allow handling either the identity or differentiation of portable art objects in the intra-group social structure, or the identity and differentiation in the inter-group social structure. For the same reasons, the social role of cave art (initiation, preparation to hunting, communication) is a field of research always to investigate.

Finally, the reconstitution of the symbolic meaning of Palaeolithic art objects is always a virgin field, by a lack of methods to handle it. Several ways are open, sometimes explored, but their results have not convinced the scientific community:

- Correspondence between a hunting territory and a bestiary represented in a symbolic space,
- Correspondence between the intra-group social identity and a system of identification or a system of values, inspired by the environment ("totemism"),
- Correspondence between the events at the origins of tracks found in art cave and the evidence of an intermediation system with the strengths which are believed to control the world ("shamanism"),

Then, it is trivial to say that the scientific studies of art Palaeolithic are only at the very beginning.

CONCLUSIONS

A short description of contemporary methods for studying Palaeolithic art is showing how computerized techniques are embedded in the methodological approach: semiotics for description of art objects, statistics (using exploratory data analysis, and mainly correspondence analysis), image processing (for 2D and 3D acquisition but also for improving the quality of the record), virtual reality. Palaeolithic art studies are then a good example of the interest of use of computerization in Archaeology, and how may be used computerized techniques within a methodological framework to obtain a successful integration.

REFERENCES

APPELLANIZ, J.M., 1987. Modèle d'analyse d'une école dans l'iconographie mobilière paléolithique: l'école des graveurs de chevaux hypertrophiés de la Madeleine. In Clottes, J. (ed.), L'Art des objets au Paléolithique; Tome 2: Les voies de la recherche, Actes des colloques de la Direction du patrimoine, Colloque du Mas d'Azil, November 1987, Paris, Ministère de la Culture:105-138.

BAPTISTA, A.M. and GOMEZ, M.V., 1995. Arte rpestre do Vale do Côa. I. Canada do Inferno - Primeiras impressôes. Trabalhos de Antropologia e Etnologia 35 (4):349-422.

BUISSON, D., FRITZ, C., KANDEL, D., PINÇON, G., SAUVET, G. and TOSELLO, G., 1996. Analyse formelle des contours découpés de têtes de chevaux: implications archéologiques. Congrès National des Sociétés historiques et scientifiques 118° Pau, 1993, C.T.H.S., Paris:327-340.

DAUVOIS and BOUTILLON, 1994. Caractérisation acoustique des grottes ornées paléolithiques et de leurs lithophones naturels. In La pluridisciplinarité en Archéologie musicale, MHS, Paris:209-251.

FREEMAN, L.G. and ECHEGARAY, J.G., 2001. La grotte d'Altamira. La Maison des roches, Paris.

IAKOVLEVA, L., 2002. Un essai de formalisation des études sur l'Art paléolithique. Archeologia E Calcolatori 13:31-40.

IAKOVLEVA, L. and PINÇON, G., 1997. La frise sculptée du Roc-aux-Sorciers à Angles-sur-l'Anglin (Vienne). R.M.N. and C.T.H.S., Paris.

LEROI-GOURHAN, A., 1965. Préhistoire de l'Art occidental. Mazenod, Paris.

SAUVET, G., 1987. Les signes dans l'Art mobilier. In Clottes, J. (ed.), L'Art des objets au Paléolithique, Tome 2: Les voies de la recherche, Acte des colloques de la Direction du patrimoine, Colloque du Mas d'Azil, November 1987, Paris, Ministère de la Culture:83-99.

SAUVET, G. and S., 1979. Fonction sémiologique de l'art pariétal animalier franco-cantabrique. B.S.P.F. t.76, n°10-12 :341-354.

VISIBILITY AND THE LANDSCAPE: AN EXPLORATION OF GIS MODELLING TECHNIQUES?

M. VAN LEUSEN
CLASSICAL AND MEDITERRANEAN ARCHAEOLOGY,
GRONINGEN INSTITUTE OF ARCHAELOGY,
THE NETHERLANDS

SEE THE CD FOR THE PAPER

ABSTRACT

As part of the author's PhD studies, attempts were made to compare and merge the results of several field survey projects which have taken place in the Pontine plain (south Lazio, Italy) since the early 1980s. For most regional studies, field surveys are the only source of diachronic information about settlement and land use. It was found that even basic comparisons between quantitative aspects of field surveys conducted by the same team but during different campaigns are extraordinarily difficult to make. Ensuring the compatibility of field survey methodology and results must therefore become a major international research aim. The paper presents the author's attempts to compare field survey data sets, the reasons for their failure, and potential remedies.

Revealing Collections: Discovery, Access and Interoperability

Kate Fernie
Archaeology Data Service and HEIRNET, UK

Abstract

This paper will look at a practical application of collection level description as a means of improving access to analogue and digital collections. Collection level description services provide a way of discovering the rich collections of physical objects, books, archives, digital resources, digital surrogates and of learning objects that are held by libraries, museums, archives and others. They may perhaps provide a way of bridging collections held across Europe. Participants at CAA will be familiar with the use of resource discovery metadata to offer network access to digital collections - and familiar with the difficulties in finding and locating analogue collections. Collection level descriptions provide a common framework for describing collections of all types (object, physical archive, book, digital resource, learning object) and a means of discovering their location, content and access arrangements. The HEIRNET register provides an example of a collection level description service enabling researchers and cultural heritage managers to discover collections held by multiple content providers across the UK.

In the United Kingdom the phrase 'historic environment' is used to encapsulate the archaeological monuments, historic buildings and other structures that survive in our landscape. Many people and organisations are involved in recording different aspects of the UK's historic environment in databases and in other information systems for conservation management, research, public information services and for a range of other reasons. Few would think of these information bases as being collections, however they form only one element of our knowledge base about the historic environment. Other elements, such as the finds from an archaeological excavation of a monument and the associated excavation archives are more familiarly referred to as 'collections' as are the reference collections and library collections that are so important to research.

The term 'collections' is used in this paper as a generic way of referring to clusters of information about the historic environment whether the cluster comprises of a database, GIS, museum collection, documentary archive, library collection or type series. This paper will suggest that, by taking a step back from the detail of the format and looking at the collection, we may be able to find ways of improving access to these resources for research in the UK and internationally.

The Benefits of Sharing Information

Those who are involved in collecting information about the historic environment will recognise the fact that no collection is ever complete. However well developed the collection, invariably it will be complemented by a collection that is held by another. For example, in the UK the National Monument Records provide an index to known archaeological monuments and historic buildings which is complemented by local authority Sites and Monuments Records and enhanced by specialist thematic databases developed by research projects. Completing the distribution map of the UK's archaeological resource involves gathering together information held in different collections.

There are many reasons why different information about the historic environment is divided between institutions. Some are practical - a museum has the resources and skills to curate collections of archaeological archives but not to provide the services of a research library. Some are political - a local authority will only collect information that relates to sites and monuments within its administrative boundary. Others still are historical - an archaeologist may deposit material with different institutions at different stages during a career or posthumously. The result is fragmentation - with information about specific aspects of the historic environment being held by many different institutions at different geographic locations.

Discovering Collections

The problem that faces us is finding out what information is available and where it is held. Without this most basic of information, undertaking a programme of research can involve a lifetime journey visiting institutions to check their catalogues. It is extremely difficult for the national heritage agencies to build up a picture of the historic environment and to plan conservation strategies. There may also be a risk of duplication of effort, with people and organisations developing collections that overlap rather than complement each other.

HEIRNET and its Register of HEIRs

HEIRNET, the Historic Environment Information Resources Network, is a consortium of organisations that came together in 1998 under the auspices of the Council for British Archaeology. The members of HEIRNET, recognising the difficulty in knowing who holds information about different aspects of the UK's historic environment, set out to create a register of available collections (Baker et al. 2000). This register (online at http://ads.ahds.ac.uk/heirnet/) is being developed to provide a searchable online index to collections.

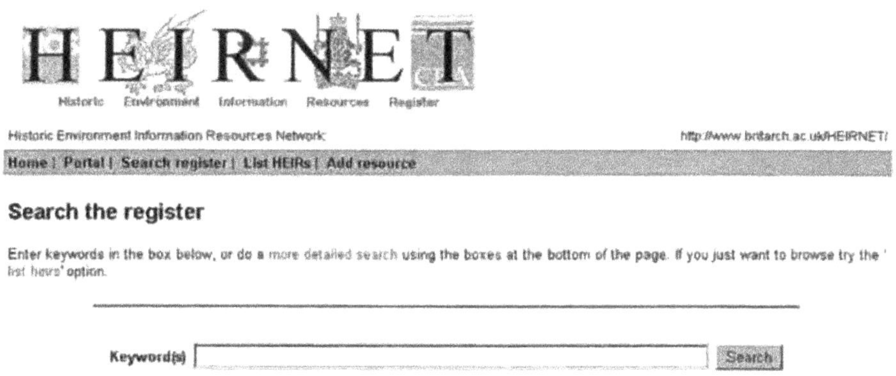

Figure 1 The HEIRNET Register search interface

The information contained within the HEIRNET register provides a collection level description of the resources that have been identified. A register entry provides a general overview of a collection - a brief description of its scope together with an indication of its subject coverage, geographic area and the formats of the material held. Perhaps the most important information of all, are the details of how and when to contact the person or organisation which holds the collection. Where a collection is available online, the Register provides researchers with a direct link to the resource.

HEIRNET's aim in developing its register is to help researchers to find resources (Fernie 2003). The register offers a simple and easy to use mechanism for searching and browsing through the available collections to find out what material is available for a research topic, which organisation holds it and what the access arrangements are.

Other UK Initiatives

HEIRNET is not alone in developing collection level descriptions. In the UK, there are a number of other initiatives that are developing similar indices that are of potential interest to archaeological researchers. For example, Re:source (England's national agency for museums, libraries and archives) is developing Cornucopia, a national database which describes museum collections. In another initiative, the JISC (Joint Information Services Committee) is developing the Archives Hub to provide descriptions of collections of archives and manuscripts held by UK universities and colleges. Further initiatives include AIM 25 (an index to archives held by institutions within the greater London area), Enrich UK (an index to collections developed through the New Opportunities Fund) and SCONE the Scottish Collections Network.

Reasons for Collection Level Description

There are two principle reasons why organisations in the UK have begun to create collection level descriptions. Many institutions have significant cataloguing backlogs. By creating an index to their own collections, these institutions are able to prioritise their cataloguing programmes, identify vulnerable collections that require conservation and plan programmes of work. Another reason, both within and between institutions, relates to the desire to provide users with a finding aid to aid their discovery of collections that are often under-used. Both the HEIRNET register and Cornucopia are examples of collection level description services intended primarily to aid resource discovery.

Much of the work to create collection level descriptions in the UK has, directly or indirectly, been influenced by the Research Libraries Support Programme (RSLP). The RSLP funded a large number of digitisation projects in UK higher education institutions and developed a schema for describing the collections that were created to make sure that these were both consistent and machine readable (Powell et al. 2000).

Describing Collections

The RSLP schema covers both concepts that are familiar to users of library catalogues (title, description, author, date) and concepts that are familiar to collections managers (physical location, deposit conditions, access arrangements). The RSLP schema was developed within the libraries sector with reference to existing schemas. ISAD[G] (the General International Standard Archival Description) and EAD (Encoded Archival Description) are comparable schemas which are in use by archivists (Powell 2003). Although comparable to the RSLP schema, ISAD[G] is much richer and requires archivists to provide more detailed information about an archive collection.

Although it was originally developed for use only by digitisation projects funded by the RSLP, the RSLP schema has had impact across the cultural heritage sector in the UK. This is because, like Dublin Core metadata, the RSLP schema is relatively simple and provides a consistent core of useful information. Mapping between the RSLP and both the ISAD[G] and EAD schemas has provided a framework for searching across the collection level descriptions being created in libraries, museums, archives and the historic environment sector.

Using Collection Level Descriptions

Most researchers who have undertaken a programme of research that required them to find material held in many different institutions will recognise the potential of a service that helps to identify what is available. Such services are invaluable at the start of a research programme. For institutions the benefits are twofold. Firstly, creating an index to collections helps to plan their management and secondly it provides a mechanism for informing potential users about the existence of collections.

Having a sound standards framework for the creation of collection level descriptions by different institutions is important as it enables this information to be shared between institutions. If the collection level descriptions that are created by HEIRNET, Cornucopia and the Archives Hub are sufficiently similar this provides a framework for machine-to-machine communications and for cross-searching. Rather than duplicating work undertaken by Re:source and the JISC, in future HEIRNET might be able to provide a portal (Fernie 2003) to enable users to search both its own register and the Cornucopia and the Archives Hub.

Inter-operability between collection level description services requires a good measure of consistency in the way in which information is recorded and also compatible subject classification. Classification at the collection level requires a broad-brush approach that does eliminate some of the differences between subject disciplines, although terminology remains a hot topic for discussion at Collection Description focus meetings in the UK between librarians, archivists and museum curators (Robinson 2003).

What about Europe?

Work in the UK is developing the RSLP schema to provide a consistent recording framework for museums, archives and libraries and similar work is already underway in Canada (Dunn 2000). The involvement of a wide range of individuals and institutions is important because it is enabling the methodology to be tested against a range of situations and refined to provide a simple and consistent way of recording collections (Miller 2000). If this is successful in providing researchers with a useful finding aid for UK-based collections, collection level description could provide a framework for resource discovery across European collections. After all archaeology does not confine itself to modern political boundaries. The development of a register of collections relating to the historic environment across Europe would have great potential for research.

Challenges for the Future

To some extent the challenge in developing a collection level description service for the historic environment is facing up to the past. There are many reasons why information about the historic environment is divided between institutions that have different professional traditions, funding regimes, responsibilities and emphases. Building a picture of the information resource means working with institutions that have very different backgrounds. Getting the information that we need about collections and making sure that it is consistently recorded is one of the main challenges that we face in developing such a service. In the UK, HEIRNET is meeting this challenge by being pro-active in its development of the HEIRNET register and in seeking opportunities to bring institutions with an interest in the historic environment together.

References

BAKER, D., CHITTY, G., RICHARDS, J. and ROBINSON, D., 2000. Mapping information resources: A report for HEIRNET. Council for British Archaeology.

DUNN, H., 2000. Collection Level Description - the Museum Perspective. D-Lib Magazine, September 2000 (http://www.dlib.org/dlib/september00/dunn/09dunn.html).

FERNIE, K., 2003. Getting it together on-line: HEIRNET and Internet-based resource discovery tools for the Historic Environment. Internet Archaeology issue 13, February 2003 (http://intarch.ac.uk/journal/issue13/fernie_index.html).

POWELL, A., HEANEY, M. and DEMPSEY, L., 2000. RSLP Collection Description. D-Lib Magazine 6 No. 9 (http://mirrored.ukoln.ac.uk/lis-journals/dlib/dlib/dlib/september00/powell/09powell.html).

MILLER, P., 2000. Collected Wisdom: Some Cross-domain Issues of Collection Level Description. D-Lib Magazine, September 2000 (http://www.dlib.org/dlib/september00/miller/09miller.html).

ROBINSON, B., 2003. UKOLN Collection Description Focus organisational website (http://www.ukoln.ac.uk/cd-focus/).

Animation by Computer. A Tool for Understanding the Dynamical Behaviour of Ancient Machines

SEE THE CD FOR THE EXTENDED VERSION

Friedrich Balck
Institut for Physics und Physical Technologies,
Technical University of Clausthal, Germany

Fritz Keller
Institut for Geophysics,
Technical University of Clausthal, Germany

Abstract

Computer animation by Java-Applets can help to study principles and behaviour of machines from the past. In former times the knowledge of "how to do something" was not written down because it was traditionally given from the older to the younger (e.g. craftsmen, machinists). As the simulation of processes in nuclear plants or airplanes is absolutely necessary before starting them, these tools can also help industrial archaeologists to understand for example working of waterwheels or steam-engines. Since 1995 there had been excavated two waterwheels in the mining region of the Harz-Mountains, which were used more than hundred years before as winding machines to bring up rocks and ore from the earth and put down materials and machines.
What are the elements of the construction and which are the important parameters for driving the machine? Using a computer applet for this example we can show the principles of working, restriction by physical laws and tricks to reduce water consumption.

Enter-the-Past! Obviously it is impossible to enter the past in real live. But under one assumption it is possible nevertheless: Let us assume, we are in the future, then the presence may become the past.

Every excavation should or will reveal a lot of finds and also many questions. Some of the most important questions about the excavated object are: How did it look like? How was the structure? How was it built? How was it used? Using the mentioned time transfer above , the arising problems of the fourth questions can be explained by the following example: An archaeologist excavates an complete car in the time 100 years later from now on. After cleaning the inside of the car he will see three strange pedals on the floor. Traces on the rubber surfaces show that they were used frequently. Since the technique for moving persons and objects has changed several times until now (e.g. controlling vehicles by joystick, bio-neural interfaces, transporting masses through energy-conversion and vice versa) the archaeologist from the year 2103 does not know the meaning of these pedals. Possibly he might suggest that both, the left and the middle pedal was used for moving direction ("left" and "right") and the other one for the input of the desired velocity ("down"=fast, "up"=stop).

Today everyone with a driving licence, knows how to use these pedals, but in the future this knowledge might be lost. A transfer from generation to generation might no longer possible when the technique changes. Only in some old books written for driving schools or in museums one could get information about how to drive a car in former times.

Back to the presence. An old water wheel from the end of the 19. century (from the presence 100 years before) was excavated in 1997. This wheel was used as hoisting engine in the mining region Oberharz (northern Germany), where silver, lead and zinc were excavated from depths of several hundreds of meters for several centuries.

Figure 1 Reversible water wheel (G. Agricola, 1556)

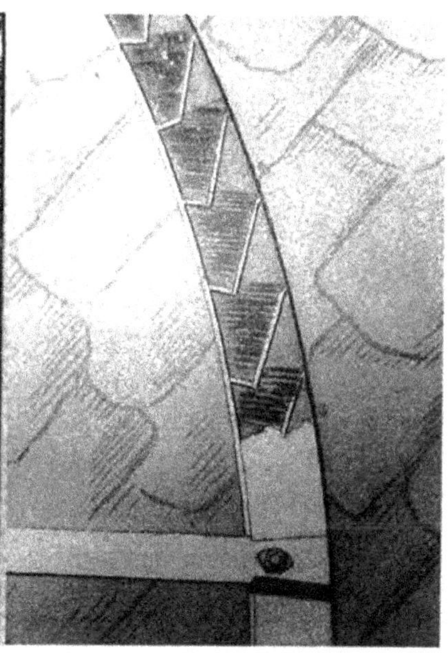

Figure 2 Elbow-like buckets (A. Polle, 1848)

The construction of such wheels is well described and depicted (Agricola 1556, Leupold 1724, Delius 1772) (Fig.1).

Two types of wheels are known (Fig.4):

1. The "Kunstrad" (Kunst = art, technique, Rad = wheel) delivers the energy for the "Kunst", a system of two reciprocating rods which drive the pumps in the shaft to drain the pit. The rods are connected to the motion of the wheel shaft by a crank (Fig.5).

2. The "Kehrrad" (reversible wheel) possesses two drums on its axle where two cable (or chains) are wound in different directions. The upper ends of the

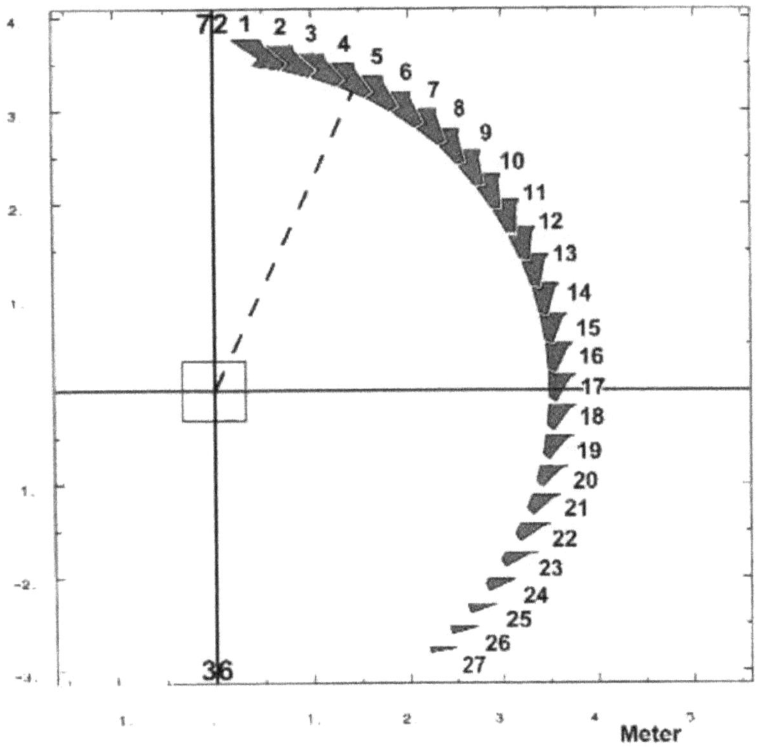

Figure 3 The effective content of the buckets decreases due to the motion of the wheel (FB)

cables are led into the shaft. Their reciprocating movement allows an almost balanced use of the wheel, similar to an elevator with counterweight.

The following will deal with construction and using of such reversible wheels. Driving such a "Kehrrad" requires a lot of experience, similar to the problem stopping a train in a station exactly at a predefined position. For the driver of the train there are a lot of unpredictable parameters which have influence on the dynamical behaviour of the train such as: load and velocity of the train, friction, wind, incline of the lane, temperature of the brakes, humidity and so on. It is not a trivial stopping at the desired point neither for a train nor for a "Kehrrad".

The finds of the excavation revealed an "Kehrrad" (an overshot wheel, the water comes from the top). The construction is more sophisticated in respect to durability in contrary to the sketches of Agricola et al. The shaft consists of a heavy oak block with cross section of 0.6 m x 0.6 m and length of about 8 m. At each of the ends there is a gudgeon made out of cast iron. The two cable drums consist of iron rims with wooden inlays for the cables.

In Figure 2 the construction of the bukkets can be seen. Three wooden boards build the container for the water. The outer (lower) are mounted "elbow like". The inner board, the soal, has been removed, but the wooden rims can show the arrangement.

Since the beginning of the last period of ore mining in the Oberharz in the middle of the 16. century the miners had build a large number of artificial ponds with hundreds of kilometres of drains collecting, storing, transporting and distributing the rain water. This system is preserved till today as cultural heritage and is called "Oberharzer Wasserregal".

As the miners produced a large amount of dump material, which they put down in the direct near of the shaft, the waterwheel had to be protected against this dump by a large ring wall made out of stones. During the long operating time of the mine and the less space in that valley the height of the wall had to be increased for several times to protect the wheel continuously. After closing the shaft in the beginning of the 20th century the inner of the wall including the remains of the water wheel were filled with dump material, which was now excavated in 1997. Wet conditions in the lower part and heavy metals in the water preserved the wooden remains from decay so that they now can still be presented. In an exposition of the Oberharzer Bergwerksmuseum, located in the whim of the former Ottiliae-Shaft a reconstruction of the wheel (scale 1:1) is presented. Most of the finds could be integrated into the construction (Fig.6).

Helpful for the analysis of the finds and their use in former times are drawings and models which are preserved and presented in that museum. A mine named "Dorothea" at Clausthal was visited by a number of famous persons in the past, e.g. Wolfgang von Goethe and James Watt. A precise model of that "Dorothea" mine from 1820 and a perfect drawing from the middle of the 19. century still exists. Both describe the technique in a perfect manner and beyond the model can show and explain the mechanics of several parts of the machines and their construction. Some of these important parts are three wooden levers, with which the driver could control the movement of the machine (two of them can be seen at Agricolas picture too). They were used similar to that proposed for the car before mentioned: "left", "right" and "stop". The first two levers control the function of the two water taps above both halves of the wheel. Each of them can

Figure 4 Two wheels are necessary, pumping water (l.) and transport of material, hoisting engine (r.) (Beyersdorf 1908)

Figure 5 A crank converts the rotational motion of the wheel to a reciprocating movement of the rods (as Fig.4)

put water on the wheel so that it will move in the one or the other direction. The third lever can serve the tongs of the brake, which will touch either the rim of the wheel (drawing) or an additional wheel (model) from the outside.

Figure 6 Remains of the excavated wheel, exposition (FB)

The most important rule of the miners in the Oberharz was to save water during its application on water wheels. Less consumption meant longer use of the stored water in the ponds. In dry periods the miners could not work in the mines, because they were unable to pump the groundwater out of the shafts.
It was the goal to lead the water into the buckets and let it remain there on its way down as long as possible. Splashing or dropping down led to losses in energy transfer.
As the effective volume in a bucket decreases, because the bucket during its going down is turned respectively to the centre of gravity, more than half of the water is lost after half of the way (Fig.3). In consequence of this reason one should fill the buckets not by 100 % but only by 20 % or 25 %. This water will stay about 75 % of the way down in the buckets so that the energy efficiency may reach 75 % at best conditions.

The driver of the engine stands in the near of the shaft and can watch the work at the landing stage, where the load of the cable is moved into and out of the shaft. A hammer is connected to a long line down to the bottom of the shaft, it is used as signalling system. In almost all mines the driver has no visual contact to its engine.

During the moving of the ore drums the driver can read their actual depth at an mechanical instrument like a clock. The hands of the instrument are coupled by a gear to the shaft of the water wheel. In order to transfer the rotation of the shaft to the instrument a pair of two reciprocating rods are used which were connected on both ends by two cranks on each axle respectively. The cranks are oriented in such manner that during the highest velocity of one the other has its turning back point. This device which shows (to show = "weisen", tool = "Werkzeug") the driver the depth is called "Weis-Zeug".

Fortunately there is an ancient film (16 mm) which was produced in 1923. The film shows some scenes from the environment of the driver and its techniques.

In the near of the excavated wheel (80 m) there was another water wheel which worked in the years about 1800. Its construction is slightly different. (see application) Instead of one shaft with wheel and cable drums side by side there are two axles one upon another one for the wheel and the other for the drums. Both axles are coupled by two systems of two rods respectively. Each rod is connected to a crank at both ends similar to that Weis-Zeug mentioned above. The advantage of this construction is to have shorter axles and less space necessary for the wheel. The disadvantage is a more complex construction with more wear.

To study the dynamical behaviour of such system an internet application was created using JAVA. (www.pe.tu-clausthal.de/AGBalck/kehrrad/kehr_app-d.html)

This program takes into consideration:

- the masses of the cables in the shaft depend on their length hanging,
- the mass of a long cable is comparable to the mass of a fully filled ore drum,
- the diameter of the cable drums decreased, while the cable is unrolled,
- the wheel has a moment of inertia,
- high velocity will produce noticeable centrifugal power in the water buckets,
- the buckets are not elbow-like, they are simplified.

For purpose of training the program can run in a special mode without ore drums.

The switches are:

opening the taps (left, right)	L, R
brake (open, close)	AUF, ZU
ore drums, full or empty	VOLL, LEER
cable with /without load	TONNEN

Some surprising observations can be done:

A full ore drum can drive the wheel without water.
Using water from the opposite tap to stop the wheel will not work effectively.

All buckets fully filled with water will override the brake.

REFERENCES

AGRICOLA, G., 1556. De Re Metallica.

BALCK, F., 1999. Wasserkraftmaschinen für den Bergbau im Harz. Papierflieger, ISBN 3-89720-341-3.

DELIUS, C.H., 1772. Anleitung in der Bergbaukunst.

LEUPOLD, J., 1725. Theatrum Machinarum Generale.

THE ARHEOTIM ARCHAEOLOGY PORTAL. PROMOTING ROMANIAN ARCHAEOLOGICAL SCIENCE ON THE INTERNET

DOREL MICLE
WEST UNIVERSITY OF TIMISOARA, ROMANIA

ABSTRACT

The Romanian archaeological patrimony is one of the richest ones in Europe, but it is not sufficiently popularised. The ARHEOTIM archaeology portal provides a way to integrate the Romanian archaeological values in the great European patrimony.

Following the footsteps of important archaeological portals, already acknowledged at an international level, ARHEOTIM wishes to fill an empty page on www Romanian space.

The structure of the portal is a didactic one, addressing primary to college students but also to all those interested in archaeology. ARHEOTIM includes approximately 2000 links grouped in 12 categories: Regions; Techniques/Methods; Themes; Epochs; Theory; Artefacts; Romanian archaeology; Personalized pages; Maps/Images; Portals/Directories; Archaeology for children; On-line magazines.

In order to be a genuine guide in archaeology, ARHEOTIM also includes a section referring to the Methodology of suffering and interrogation in www, a specialized Forum for discussions, an Electronic Magazine and an internal Search motor.

The most interesting chapter however, is the one referring to projects undertaken by History and Archaeology Study Centre (CSIATim), which has as a goal the familiarization, at a medium level, of the archaeological sites in western Romania.

Within this project 2 multimedia products are presented: e-Tibiscum and e-Praetorium.

Both web pages are complex projects that combine written text with sound, image and 3D modelling in an attempt to offer an image as clear and complete as those provided by the best archaeological sites concerned with Greco-Roman civilization, in western Romania.

GENERAL PRESENTATION

Romania is one of the European countries, which are least represented on the Internet in the field of archaeology. This is why the existence of a Romanian archaeology portal has become a necessity, now that different pages that popularize the Romanian archaeological sites are being published. However, there is not yet a modality to group and classify these pages into one accessible shape. Although shy attempts to create a Romanian portal[1] were made in the past, the projects have remained unfinished to this day.

ARHEOTIM (http://www.uvt.ro/csiatim/arheotim/index.htm) is meant to be a complex archaeology portal, realized within a project of the Centre of History and Archaeology Studies Timisoara (CSIATim)[2] by local specialists with the Centre's own resources. The Server that hosts the portal is that of the West University of Timisoara, and is administered by the Department of IT and Communications[3] within the University.

ARHEOTIM promotes any initiative to publish on the Internet pieces of information connected to archaeology from within and outside our countries borders. It is open to any kind of collaboration with Romanian or international institutions in the field.

The main objective of ARHEOTIM is to promote the Archaeological Science at regional, national and international level by publishing specialized scientific articles in its own Electronic Journal "ArheoVest". Other ways of promoting are publishing WebPages with archaeological themes (digging reports, scientific monographs, popularization papers, didactic papers, courses for students etc.), as well as the inauguration of a Specialized Discussions Forum in the field of archaeology. Through this Forum, the Romanian or foreign specialists may exchange opinions, ideas and information connected to their job profile.

ARHEOTIM is addressed to the specialists in archaeology and to the students that study this science. This is why it also presents didactic elements connected to the methodology of navigating and evaluating a webpage, as well as bibliographical and webographical references for those who want to study further this domain.

The structure of the portal was conceived by accumulating several ideas already put into practice by other European portals that belong to this field. There was a desire to achieve a simple and aerated graphic, without complex dynamic presentations that would require special resources for navigation and search.

The fundamental concept underlying this portal was its practical function, in an attempt to offer the specialists (and archaeology amateurs in general), Romanian or foreign, a useful, easy to use instrument, where they may find all types of information connected to archaeology. The portal is created in Romanian language, but in order to facilitate the access of foreign users, the technical terminology was preserved in English, with which the Romanian users are already familiar.

Since ARHEOTIM is not limited to the promotion of research done by the CSIATim specialists, but it is also meant to be a promoter of archaeological science in general, several common inter-institutional research projects have been initiated. In this direction, a first contact was established with the Museum of Banat (from Timisoara), which has a special section in the Home Page of the portal. Therefore, an attempt is made to elaborate common projects, to publish digging reports of the Museum specialists, to publish thematic WebPages with monographic profile.

A special concern of those in charge with ARHEOTIM is the promotion of specialized papers (monographs or other types of documents with a scientific character) of the researchers in the domain of archaeology. There are 3 perspectives in this direction:

a. the publishing of lists of papers edited by certain specialized institutions;
b. promoting scientific papers through the publication of certain abstracts;
c. actual publishing of certain specialized papers in electronic format.

ARHEOTIM PRESENTS THE FOLLOWING SECTIONS:

[1] The Categories section of the ARHEOTIM Archaeology Portal

There are 82 domains from the area of archaeology which are presented hierarchically, and are classified according to the international standards respected by almost all archaeology portals: Regions, Techniques and Methods, Themes, Epochs, Theory, Artefacts, Romanian Archaeology, Personal WebPages, Maps and Images, Portals / Directories, Archaeology for children and Online Journal.

Although there is no strict rule in this sense, 12 thematic domains are however accepted. The sub-domains may vary according to the complexity and dimension of the portal. There are also known and accepted several dozens types of software for webpage making, the usage of any of them being left to the decision of the web master. For the ARHEOTIM Archaeology Portal, the program used was a licensed HTML editing program, with a FrontPage 2002 interface, realized by Microsoft. For the good functioning of the portal, no complex graphic elements or Java or C.S.S. (Cascading Style Sheets) were introduced. Thus any type of browser may quickly open the portal without requiring further resources. All the scientific data concerning a certain webpage are presented under a tabular form so that the user may have the whole image of the recording. The user can easily identify the name of the webpage, the language in which it is written, the classification and the elements of multimedia it includes. The legend of the graphic symbols used in the description of the web document is present at the foot of the page. In order to keep the aesthetic aspect and not to force the user to navigate too much on the vertical axis to obtain the needed information, the table was limited to 20 recordings. The preferred variant is the creation of more pages which are numbered at the heading and the foot of the table. The user can navigate every 20 recordings through the hyperlinks.

Since it is a didactic portal, the only elements considered necessary to be published here were the name of the page, the language in which it is written, the type of document and the elements of multimedia it includes. The URL address is hidden within the name of the page, but in the future we will attempt to publish a short description for every registered webpage.

[2] The Methodology section of the ARHEOTIM Archaeology Portal

The Methodology section of the portal refers mainly to students and to those who do not master the methodology of navigating, searching and identifying the useful information in an Internet directory.

This section makes available for the user the general criteria of identification and selection of a scientific (or useful) webpage, according to the international standards, drawing attention upon the copyrights that protect the text, sound and images included in the electronic document: the accuracy, authority, objectivity, currency and coverage of the web documents.

[3] The News section of the ARHEOTIM Archaeology Portal

The News section presents periodically (once a month) the new publications from the archaeology domain which can be edited in an electronic format or in a classical format.

The pieces of news are short, strictly informative in the form of short articles, along the text (or images), the coming out date of the item of news and the source of documentation are presented.

a. classic documents - an informative summary (in the case of a long article) or an informative note (if it is a piece of news) is presented. The source is also mentioned with the author, title, page, publishing house and publishing year.
b. electronic documents - a summary or an informative note is presented. The source is mentioned with the author, title, web address. The web address is interactive, the user being able to log on directly with a simple click on that particular address.

The ARHEOTIM will constantly keep in touch with the professional institutions and organizations of history and archaeology in the country, to whom it will present in a special section the exhibits organized, the organized debates and meetings, the publications, the scientific activities in general.

The Museums will have a special status, the section destined for them presenting the permanent exhibits and the temporary and itinerant exhibitions.

The political debates, the Symposiums, the Congresses or any other scientific manifestations from the archaeology domain, with regional, national or international character will have a special presentation in a section destined for this particular purpose.

[4] The e-Journal section of the ARHEOTIM Archaeology Portal

The e-Journal section makes available for the users a specialized electronic journal "ArheoVest" edited by the Centre of History and Archaeology Studies Timisoara with a biannual edition, that wishes to promote scientific research through the publishing of articles and studies in the domain of archaeology and that of connected disciplines. Permanent sections: Archaeology and History, Modern Methods and Techniques in Archaeology, Digging reports, Interdisciplinary Studies, Book review, others. Scientific coordinator: Doina Benea, Ph.D.

The articles will have a scientific or didactic character and will be edited in PDF format. The necessary program for visualizing and printing is Acrobat Reader 5.0. The text of the articles will be written in a European language of circulation or in Romanian with a summary in a European language.

Although the team is young and this is the first project of this type, the ARHEOTIM Archaeology Portal is meant to become, in a short period of time, an instrument appreciated both in Romania and abroad. A new graphic and a new way in processing data using PHP (Hypertext Preprocessor) are already in progress. This is also the case of multimedia facilities, which are now being improved.

[5] The Forum section of the ARHEOTIM Archaeology Portal

The Forum section of the portal offers the students, specialists and amateurs in archaeology the possibility to contact and notify each other (in writing) on different themes concerning archaeology, in an environment provided by online services.

The users may create their own account and password and may choose the theme of discussion (about the portal, about archaeology, about archaeological news or didactic element) and then they may send a written text to the portal, just as an e-mail.

The discussion Forum is based on a newsgroup selector that allows the user to choose from the list the theme (s)he is interested in. ARHEOTIM uses a simple discussion Forum, displaying the information in an unrolling list on 3 main fields:

- the title of the message.
- technical data: e-mail address and the forwarding date.
- the counting of the number of readers and of the number of answers.

The actual reading of the text is done by activating the title of the message in a new window that also offers complete data of the author.

[6] The Research section of the ARHEOTIM Archaeology Portal

The Research section has 2 spheres of interest: the projects of the CSIATim team (WebPages to popularize the regional archaeology, digging reports, different projects) and the links to the Romanian and international research institutes.

a) The projects of the Centre of History and Archaeology Studies Timisoara. WebPages meant to popularize the regional archaeology: e-Tibiscum (the Roman camp, the military vicus and the Roman town of Tibiscum - called nowadays Jupa, in the Caras-Severin county, Romania) and e-Praetorim (the Roman camp and military vicus from Praetorium - called nowadays Mehadia, in the Caras-S everin county, Romania). Several other projects are in progress: The Dacians in Banat, The Roman Legions in the Province of Dacia, The villa rustica at Dalboset (in the Caras-Severin county, Romania), and The Neolithic Site at Dudestii Vechi (in the Timis county, Romania).

b) Institutions and organizations of archaeology from Romania and abroad. Here the main Romanian and international institutions and organizations belonging to this profile are presented in alphabetical order.

E-TIBISCUM

The Roman archaeological complex at Tibiscum in the locality Jupa, Caras-Severin county, is one of the largest ones in Western Romania. The area filled with archaeological load spreads on about 27 ha, of which only 5 % have been excavated. Still, the diggings that have been done so far, showed the general aspect, the evolution phase and the economic and spiritual aspects of the Roman camp, the military vicus and the Roman town at Tibiscum.

The webpage of the e-Tibiscum project structured the entire information known so far, in the form of a monograph based on digging reports, the articles and the monograph published by the director of the site and the team of archaeologists. The monograph comprises 6 sections with a historical and archaeological character (Introduction, The Roman Camp, The Vicus, The Town, The Economic Life, The Spiritual Life) and 6 sections with a technical character (Glossary of terms, Bibliography, Images, 3D Graphic Reconstruction, Summary in English, Summary in German).

The webpage is made up of 2 frames: on the left the navigating buttons (Java application) and on the right the actual page that presents the information on 2 columns (text and images). The text makes frequent references to the glossary of terms, to the images, maps and also to bibliography. The entire webpage is an interactive multimedia document, made up of text, images, maps, boards and video-movies.

E-PRAETORIUM

The Roman archaeological complex at Praetorium, in the locality Mehadia, Caras-Severin county, is one of the latest sites in Western Romania. After a first attempt of systematic research, dating back to the Second World War, the diggings were resumed by the team of archaeologists led by Prof. Dr. Doina Benea. As a result, the webpage presents the information as grouped on archaeological campaigns and not as a monograph.

The entire project is an interactive multimedia document made up of text, image and sound, realized in Romanian, with a summary in English. The information is presented on 2 columns (text and image), being grouped again in 6 sections with a historical and archaeological character (General Presentation, Short History, Archaeological Prospections, The Roman Camp, The Vicus, Archaeological Campaigns) and 5 sections with technical character (Content, Glossary of terms, Bibliography, Images, Summary in English).

Although it is a smaller archaeological complex, the Praetorium promises to be, in the light of new discoveries, a point of reference for the Romanian classic archaeology in the region Banat.

1 Institute for Cultural Memory (CIMEC) http://www.cimec.ro/ default_eng.htm and Archaeological Professional Association of Romania (APAR) http://www.archaeology.ro.

2 Centre of History and Archaeology Studies Timisoara http:// www.uvt.ro/csiatim/index.htm.

3 Department of IT and Communications http://www.uvt.ro/ itc.

REFERENCES

BENEA, D. and BONA, P., 1994. Tibiscum. Bucuresti.

BUCKI, L., 1998. Internet 6 în 1. Bucuresti.

JENNINGS, R., 2000. Totul despre Access 2000. Bucuresti.

JOCKEY, Ph., 1999. L'Archéologie. Paris.

KENNEDY, A.S., 1999. The Internet. London.

LAZAROVICI, Ghe. and MICLE, D., 2001. Introducere în arheologia informatizata. Timisoara.

LEHNERT, W.G., 1999. Light on the Internet: essentials of the Internet and the World Wide Web. Harlow.

LEVINE, J.R., BAROUDI, C. and LEVINE YOUNG, M., 2001. Internet. Bucuresti.

MACREA, M., GUDEA, N. and MOTU, I., 1993. Praetorium - castrul si asezarea romana de la Mehadia. Bucuresti.

MULLER, N.J., 2002. Enciclopedia Internet. Bucuresti.

NAVARRO, A., 2001. Effective Web Design. San Francisco.

NORTON, P., 2000. Retele de calculatoare. Bucuresti.

PRICE, G. and SHERMAN, C., 2002. The invisible Web: uncovering information sources search engines can't see. New Jersey.

TYLER, D. and ERICKSON, C.D., 1998. Microsoft FrontPage 98. Bucuresti.

VAUGHAN, T., 2002. Multimedia - Ghid practic. Bucuresti.

An Internet-Geographical Information System for Multidisciplinary Research in Oman

Jutta Häser
Deutsches Archäologisches Institut, Orient-Abteilung
Podbielskiallee 69-71, D-14195 Berlin, Germany

Angelika Schulz
Prenzlauer Allee 32
D-10405 Berlin, Germany

Abstract

Since 1999 a project, entitled "Transformation Processes in Oasis Settlements in Oman", was undertaken by orientalists, ethnologists, architects, urban planners, plant physiologists, and archaeologists from universities in Germany and Oman, as well as from the DAI. The research is focused on processes of transformation throughout history, regarding the architecture and material culture, the economy and ecology of the oasis settlements in a regional and interdisciplinary approach.
An Internet-GIS was developed to provide all involved scientists as well as other researchers in this field with the most recent results of the archaeological investigation. The cartographical data are based on aerial photographs. They were processed with the programs AutoCad Map and AutoCad Overlay to create a georeferenced map. Topographical and archaeological features like tombs, settlements, cultivation terraces, water management systems etc. dated from the third millennium B.C. to sub-recent times were digitised. With the help of the program AutoCad Mapguide these data were transferred to the Internet. Each site is represented by a photograph and a short description. However, the internet users are not only able to look on these data, but they can also evaluate these data with GIS-tools (e.g. distances, puffer) under various aspects without installing a special GIS-System.

Introduction

Co-operation between scientists of different disciplines are standard in modern research projects. When these enterprises take place in a foreign country, usually local scientists are also involved. In many cases special support is granted to those projects, in which various partners from different countries co-operate. This is true not only for the natural sciences but also for the cultural sciences and therefore also for the archaeology.
The work in spatially distributed and specialist groups requires an intensive exchange of information, which is made easier by modern information technology. However, the exchange of documentation, which includes maps, plans, drawings, photographs, reports and databases in their various stages of processing, is still problematic. An Internet-Geographical Information System (Internet-GIS) has been applied to solve this problem in the project "Transformation Processes in Oasis Settlement in Oman", which shall be introduced in the following.

Description of the Project

Against the background of rapid economical and social change in the Sultanate Oman since the 1970s, a German-Oman, interdisciplinary research group has been involved in documenting the traditional way of life in all its material, economical and social aspects and in developing strategies for the preservation of oases in a modern world (Häser 2000).

The research team includes orientalists from the University in Tübingen, architects and urban planners from the University in Stuttgart, agriculturalists from the University in Kassel, archaeologists and agriculturalists from the Sultan Qaboos University of Muscat as well as archaeologists from the Orient-Department of the Deutsches Archäologisches Institut in Berlin.

Initially the project was financed by the Ministry of Research of the State of Baden-Württemberg, the Sultan Qaboos University in Muscat and the Deutsches Archäologisches Institut. Since 2002 the Deutsche Forschungsgemeinschaft has provided financial support for the project.

In the years 1999-2001 a pilot study was conducted in a region of central Oman, which extends along the Wadi Bani Awf, crosses the al-Hajar mountains and ends in the al-Hamra region. The area under investigation is located in the dry belt and encompasses various ecological zones, which are characterized by the northern and southern foothills of the al-Hajar massif and its up to 3,000 m high upland. No archaeological survey had been conducted in the northern foothills of the al-Hajar mountains before; therefore, a broad gap in research could be filled.

The aim of the archaeological project is to determine the chronological dimension of the oasis settlements and to describe the sequence of occupation in the study area. Not only settlements were integrated in the study, but also tombs, irrigation systems, towers, protective walls and rock art.

With reference to historical periods, archaeological findings are a supplement to the knowledge on the history and social structure, which could be gained by the orientalists in the research group from historical reports and documents in archives (e.g. in al-Hamra).

Agriculturalists participating in the project have investigated the agricultural potential and the bio diversity in the region by studying ecological resources, such as the availability of water and the quality of the soils, further the impact of man on the environment in improving natural conditions, for example, with irrigation systems or manure. The question of the interrelationship between agriculture in the oases and the keeping of animals by semi-nomads is of great importance in this special case.

Architects and urban planners have dealt especially with the development of the architectural structure of the oases in their topographical situation during the last 300 years as well as with the comparison of rural and urban settlements.

Methods of Investigation

Different methods were used in the study of archaeological sites. The survey was prepared with the help of aerial photographs scaled 1:20,000, which were examined in respect to archaeological remains. They were also used in addition to a Global Positioning System (GPS) as a basis for mapping, since the existing topographical map scaled 1:100,000 is not exact enough to map the archaeological remains accurately. The aerial photographs were set in a mosaic for this purpose, which is the map base for the Internet-GIS (see below).

Pottery was collected from different archaeological sites during the survey. The sherds were determined in respect to ware, shape and date and statistically processed in a database. Not only prehistoric pottery sherds were registered, but also Islamic material, since this was generally almost neglected in previous research projects.

Architectural remains at selected sites were recorded using large-scale aerial photographs that were taken from a helium-balloon at a height of 300-1,000 m above the ground. The photographs were made with a normal camera. The ground plan of the buildings was drawn with the help of these images, which had been previously rectified. Afterwards the plans were controlled on the ground.

The architecture of the inhabited settlements, al-Hamra and Balad Seet, was measured and drawn by hand to determine the structure of the buildings. The ground plan of these settlements was also mapped and analysed with the help of the rectified photographs. Based on these plans rural and urban structures can now be compared with one another and the development of house types can be observed and defined.

Interviews with the inhabitants of the oasis settlements have provided information on the internal division of the houses, the use of each room as well as the inventory.

Orientalists from Tübingen received permission to study an older archive in al-Hamra, which contains various documents and letters that provide insight into economical and political conditions. The documents were photographed and registered and are now being analysed from different aspects.

The focus of the agriculturalists is the estimation of agricultural production and carrying capacity of oasis settlements and their surroundings. This group produced the large-scale photographs mentioned above and utilized them together with a differential GPS for mapping agriculturally arable land and the radius of grazing grounds for sheep and goats. These spatial data were connected with related attribute data (e.g., the kind of agricultural products) and analysed afterwards.

Results of Archaeological Investigations

Occupation of the region under study started during the early Bronze Age around 3000 B.C. It probably served as grounds for grazing. No settlements were found, but burial tombs were discerned that can be dated to the 3rd and 2nd millennium B.C. (Häser 2003).

The oldest settlement remains are evidenced from the early Iron Age period around 1000 B.C. Settlement activities were probably closely connected with the use of a new irrigation technology. Only this "falaj"-technologie made crop cultivation and planting of date palms possible in that region (Häser, in press).

Settlement occupation was continuous since the early 1st millennium B.C. Diverse ecological conditions led to differences in the regional settlement development. While the settlements in the al-Hamra region shifted - probably due to shifting irrigation systems - the settlement occupation in Wadi Bani Awf remained in one particular place, the Balad Seet, where an abundant and stable outflow of springs ensured a constant water supply until the middle Islamic period between the 11th and 16th century A.D. (Nagieb, Häser, Siebert and Buerkert, in press).

A change in the occupational structure was enforced by investments of the new Ya`ariba dynasty starting in the first half of the 17th century A.D. A new irrigation channel was built in al-Hamra, which made a shift in settlement to an area that was never occupied before possible. Some small oases settlements were also established in the Wadi Bani Awf during the same period, which met the requirements of newly built irrigation channels. It can be assumed, that these investments were also granted by the Ya`ariba dynasty to encourage the tribes in that region.

Only with the oil boom since the 1970s has the settlement structure changed again due to modern roads, different construction techniques of houses, supply of electricity and use of water pumps.

Design and Use of the Internet-GIS

During the investigations outlined above all project partners prepared various material and numerous documents to describe their research results. The primary objective of the Internet-GIS is to support both publication and dissemination of this documentation, not only for the scientists involved in the project but also for the broader public. Using the Internet, new results can be incorporated without delay, and all information can be retrieved world wide. The user only needs access to the Internet, although display of the interactive map requires an additional browser plug-in, which is available free of charge.

The Internet-GIS and its major component - the dynamic map - is accessible via the project homepage http://www.dainst.org/index.php (search for "Oman" and/or "Oasensiedlungen"). Building an overall structuring frame, the website includes different parts of documentation, e. g., a general project outline, a map, an index of archaeological sites or contact information.

The digital map serves as an orientation base and includes the underlying spatial structure (e. g. surface topography, road and path network, land use) as well as archaeological sites. Further explanatory material such as text documents or photographs are referenced via incorporated hyperlinks (visualized as site number and/or cartographic signatures).

In absence of an appropriate map of the survey area, coloured aerial photographs (scale of 1:20,000, taken in 1985) were used for digitization. Following a graphic rectification (using Autodesk® CAD Overlay® 2000), based on the topographical map "Rustaq 1:100,000", the clearly visible settlement areas, the road and path network and the drainage system were digitized (using AutoCAD Map® 2000). A conventional CAD drawing (*.dwg vector file format) with thematically separated layers was the result.

All relevant data associated with archaeological sites, such as geographic coordinates and further information regarding type and date, are stored in a relational data base (Microsoft® Access 97).

The linkage between spatial objects (e. g., archaeological sites) and the related attribute data (e. g., site names) was realized within the CAD drawing file, which had to be connected with the relational database. The linked objects were exported and saved in a specific spatial data file format (*.sdf), which in turn was used as input for the dynamic map.

The software package Autodesk® MapGuide® is a Geographical Information System, which supports the compilation of complex, static or dynamic maps to be used via Internet. The exported spatial files, connected with attribute data, were put together in a so-called map window file (*.mwf). Each source file (*.sdf) already corresponds to a thematic map layer in the map window file and may be configured static or dynamic.

According to user requirements each layer can be switched on or off. If a layer is configured as a dynamic layer, each request will initiate an update based on the current state of the connected data base.

Different zoom levels and navigation tools allow selective views showing either an overview or details. Further tools are available to calculate area sizes or distances. Buffers can be created to define specific catchment areas, for example, to determine the range within tombs are located around settlements. These buffers can be saved automatically on a separate layer for further analysis.

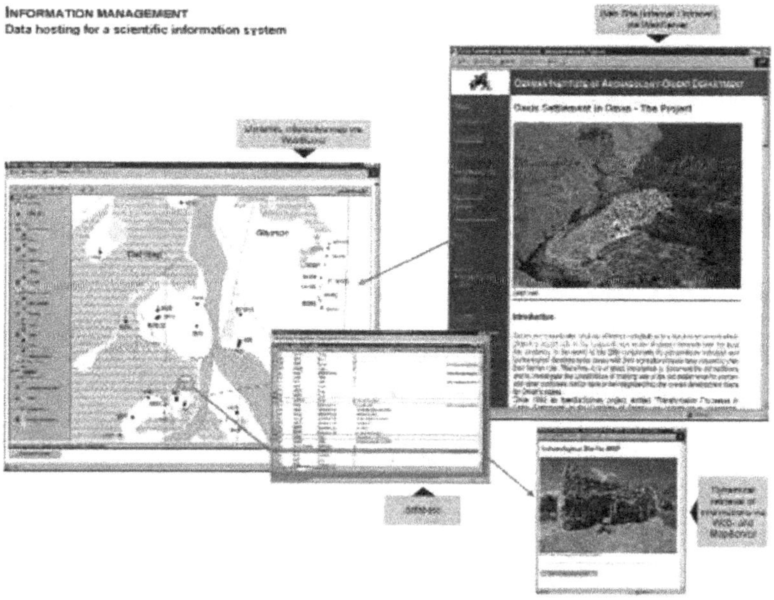

Figure 1 Information Management, Data Hosting for a Scientific Information System

Additional information such as site descriptions, charts and photographs can be prepared as separate HTML documents, which will facilitate easy updating. Photographs were scanned from slides and saved in the graphical *.jpg format. The resolution was optimized for screen display (72 dpi), which also minimized memory capacity. Access to these additional documents is provided via hyperlinks incorporated in the dynamic map.

The development of an Internet-GIS and all its components requires appropriate software. In this case the following software packages were used: the CAD program AutoCAD Map®, the database Microsoft® Access, the GIS system Autodesk® MapGuide® (including MapGuide Server, MapGuide Raster Workshop, MapGuide Author), the web server Microsoft® Internet Information Service, and the image-editing software Adobe® Photoshop. To host all data an adequate server with sufficient working and storage memory capacity is required.

On the client's side a common web browser (e.g., Microsoft® Internet Explorer, Netscape® Navigator) and a browser specific plug-in or ActiveX® control is needed; java script must be enabled.

The Internet-GIS described above is accessible for everyone without any restriction. However, if necessary, single parts of the documentation can be protected so that access is limited to authorized persons, e. g., project participants. Browsing through the Internet-GIS, the user can retrieve the documented results of the whole archaeological survey in terms of type of site and date.

Conclusion

An Internet-Geographical Information System was established for the interdisciplinary project "Transformation processes in Oasis Settlements in Oman", which enables the various working groups to retrieve interactively the results and the latest state of processing of the archaeological survey via Internet. It is planned to integrate the results of the other project partners in a future stage of work.

References

HÄSER, J., 2000. Formation and Transformation Processes of Oasis Settlements in the Sultanate of Oman: Preliminary Report on a new Field Project. Proceedings of the Seminar for Arabian Studies 30:115-118. Belgium.

HÄSER, J., 2003. Archaeological results of the 1999 and 2000 Survey Campaigns in the Wadi Bani 'Awf and in the al-Hamra-Region. Proceedings of the Seminar for Arabian Studies 33:21-30. Belgium.

HÄSER, J., in press. Agricultural Water-Management on the Oman Peninsula. In Bienert, H.-D. and Häser. J. (eds.), Men of Dikes and Canals - The Archaeology of Water in the Middle East, Orient-Archäologie, Rahden/Westf.

NAGIEB, M., HÄSER, J., SIEBERT, S. and BUEKERT, A., in press. Agriculture, hydro-infrastructural conditions and settlement history of a mountain oasis in northern Oman. In Die Erde.

When the People are the Context
Putting the "Minor" Picasso- Arias Collection and its "Major" Context onto the Web

Alfredo Calosci
Universidad Europea de Madrid, Spain

Abstract

This paper discusses the experience promoted by the Directorate General for Museums and Libraries under the Madrid Regional government's Department of Arts when publishing the Works of Pablo Picasso included in the Arias Collection on exhibit in Buitrago del Lozoya's Picasso Museum.
Both the collection and its publication on the Net are merely minor anecdotes considering their respective contexts.
The collection comprises few more than 60 objects bearing testimony to the long, warm friendship between Picasso and his barber Eugenio Arias, born in the village where the museum stands. And it is precisely this anecdotal nature that makes the Picasso Museum a veritable museum of the story of a friendship, an exile and a shared passion for bullfighting, instead of just a modern art museum.
This is precisely why the biggest effort in the creation of the Web site was geared towards documenting the two protagonists' lifestyle, their friends in common and the sites in the south of France serving as the backdrop for this small story. While the objects are exhibited in the museum rooms, it is on the Web site where we aim to document the context.

This paper discusses the experience promoted by the Directorate General for Museums and Libraries under the Madrid Regional government's Department of the Arts when publishing the Works of Pablo Picasso included in the Arias Collection on exhibit in Buitrago del Lozoya's Picasso Museum[1] on the World Wide Web.

Both the collection and its web site are merely minor anecdotes considering their respective contexts.

The collection comprises few more than 60 objects bearing witness to the long, warm friendship between Picasso and his barber Eugenio Arias, born in the village where the museum stands. And it is precisely this anecdotal nature that makes the Picasso Museum more than just a small museum of modern art, but a veritable museum of the story of a friendship, an exile and a shared passion for bullfighting,

Making Sense of a Digital Exhibit Design

The making of the Museum Web site offered the opportunity to develop an alternative exhibit design which complements the way in which the pieces of the physical collection are presented.

Like many other art museums, the Picasso Museum displays unique pieces of work. Far from competing with Madrid's great galleries, the exhibition of Picasso's Works, often merely "minor", takes on its full meaning in the context of a small village in the mountains north of Madrid where it comprises one of the major tourist attractions.
The Web site's "digital exhibit design" is just a few clicks away from the greatest centres of art around the globe, and this is why the anecdotal aspect of the collection's origins become a true point of interest when browsed from a global perspective.
Driven by the desire to complement rather than to replicate the experience of the museum visit, we came to the conclusion that a virtual walk through the exhibition, elegantly set in a limited amount of space with artificial lighting in the basement of the Town Hall, was not appropriate.

While the objects are exhibited in the museum rooms, it is on the Web site where we aim to document the context.

Making the Collection a Part of the Whole

The core section on the site serves to document each of the pieces in a reproduction of the catalogue put together and digitalized according to the standards drawn up at the beginning of the 1990s by the Spanish Ministry of Culture. This enabled its potential inclusion in wider general database.

Figure 1 and 2 The digital library: a record

In the layout of the catalogue cards, a particular accent was placed on the photographic reproductions of the works, doubtlessly one of the site's most enticing facet, while the rest of the graphic and navigational elements took a discrete back seat.

Figure 3 Relationships and friendships: Jacqueline Roque

Figure 4 Itineraries: the exile

ONE PLACE, SEVERAL ITINERARIES

Cataloguing each object with its own card provides a useful tool for both researchers and demanding users. But it also tends to focus attention on each individual object rather than on readings of the collection as a whole.

One of the digital exhibit's advantages is that it can offer several itineraries by "packing" the same collection into alternative, thematic views.

The museum's web site came about with the intention of offering itself as a place where a host of different itineraries can be traced out by topics related to the collection's origin rather than by other criteria usually used by contemporary art museums or for Picasso's work.

MAKING SENSE OF THE COLLECTION

The invention, evocation and documentation on the Web site of the thematic threads was of the keys to provide meaning to the historic and geographical context of the events that gave rise to the collection. Most of the pieces in this collection come from or are related to Vallauris, a village in the south of France where Picasso and Arias met and lived for several years.

Eugenio Arias had fought in the Spanish Civil War on the Republican side and, after the war was lost, he was able to flee to France where he fought in the resistance against the Nazi occupation. In 1946, he set up his barber shop in Vallauris where he remained an anti-Franco activitst in exile.

Picasso's years in Vallauris were characterized by several ceramic works, comprising one the most interesting parts of the collection. Picasso learned various pottery techniques in George and Suzanne Ramié's Madoura pottery shop.

The aspect that we feel is of most interest in documenting this context lies in the evocation of the closely knit personal relationships and friendships that Picasso and Arias had over those years. We thus discover that the two friends met through a common friend of the Ramiés, who in turn were friends of Jacqueline Roque, later to become Picasso's wife, and that all of them were friendly with people in the French Communist Party, such as Louis Aragon and Jean Cocteau.

We feel that the context of this collection is made up to a large extent by a closely knit web of friends and it is therefore easy to relate the works on display with one or several characters in Picasso and Arias's surroundings during those years.

The production of the web site was therefore the occasion to recover and publish documents and images that are not part of the collection but that can play a crucial role in the understanding of the whole.

KEEP IT SIMPLE

In order to present the collection's context on the site, "low technology" based on the very roots of the WWW were used: hypertext links providing the chance to offer minimum references that the user can complement with information on other more specialized sites.

Both the description of the works and the references to places and persons are systematically cross-linked with hypertexts. The brief bibliographical notes describing Picasso and E. Arias' circle of friends underscore the major traits of their relationship with the collection.

Although the quality of the interaction offered on the site is much more modest than what can be found in most interacti-

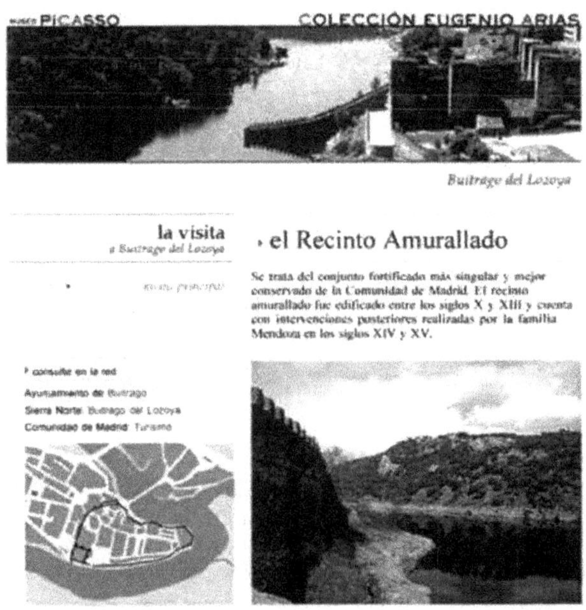

Figure 5 Visiting Buitrago del Lozoya

ve multimedia applications, we feel the overall result meets our initial goal, i.e. to present an interesting testimony of the history of a friendship between two men in exile and their shared passions. On a couple of sections of the site, a greater effort was made in using interactive resources offered by the multimedia publications. We used the shockwave multimedia format to make the biographies of the collection's two protagonists more dynamic and to create a collection browser, an application where users can organize and browse all of the pieces according to several different criteria.

Further efforts could be made in order to more interactively represent the circle of friends that surrounds the collection. Some recent results in conceptual graphic representation such as Smart Money's Map of the Market[2] or the Plumb Design's Visual Thesaurus[3] may serve as a good starting point.

Figure 6 The collection browser

THE WORLD WIDE WEB ON A SMALL SCALE

Speaking of and using the WWW often leads us to make considerations of a global nature: large scale encyclopaedic data bases, a potential audience stretching out across the globe, international standards for information exchange, and other large scale projects.

We hope that this project will be considered among many other examples of how a site with a local focus can be published without loosing sight of certain global considerations.

CREDITS: THE WEB SITE' S TEAM

Servicio Regional de Museos y Exposiciones, Dirección General de Archivos, Museos y Bibliotecas, Consejería de las Artes, Comunidad de Madrid

Cristina Conde de Beroldingen, Maria Dolores Jiménez-Blanco, Julián H. Alonso, Marina Prieto, Andrés Mengs, Tomás Antelo, Ana González Mozo

Arquimática S.L., Francisco Rodríguez de Partearroyo , José Carlos Greciano Merino

1 Museo Picasso - Colección Arias: http://www.madrid.org/museo_picasso/.

2 Smart Money' s Map of the Market: http://www.smartmoney.com/marketmap/.

3 Plumb Design's Visual Thesaurus: http://www.visualthesaurus.com/online/ index.html.

REFERENCES

CZERNIN, M. and MÜLLER, M., 2001. Picassos Friseur - Die Geschichte einer Freundschaft. Verlag Kiepenheuer & Witsch, Köln.

ONLINE: WEB DEVELOPMENTS AT THE ROYAL COMMISSION ON THE ANCIENT AND HISTORICAL MONUMENTS OF SCOTLAND AND THE DEVELOPMENT OF A SHARED HERITAGE PORTAL WITH HISTORIC SCOTLAND

MARK GILLICK
RCAHMS JOHN SINCLAIR HOUSE, 16 BERNARD TERRACE, EDINBURGH, SCOTLAND, UNITED KINGDOM

PETER MCKEAGUE
RCAHMS JOHN SINCLAIR HOUSE, 16 BERNARD TERRACE, EDINBURGH, SCOTLAND, UNITED KINGDOM

RICHARD STRACHAN
HISTORIC SCOTLAND, LONGMORE HOUSE, SALISBURY PLACE, EDINBURGH, SCOTLAND, UNITED KINGDOM

ABSTRACT

In 1998 the Royal Commission on the Ancient and Historical Monuments of Scotland (RCAHMS) launched CANMORE: the on-line version of the National Monuments Record of Scotland database. CANMORE enables the public to search a database of over 200,000 site-based records spanning Scotland's rich heritage from prehistory through to 21st century architectural and engineering achievements. Users can access detailed accounts of many of the monuments, browse through bibliographic references, or see what collections material (drawings, manuscripts or photographs) is held in our archives. With the launch of CANMAP (http://www.rcahms.gov.uk/canmore/ login.show), a web-based GIS, in June 2002, complimenting the database search engine, users now have the option to execute simple geographic searches against a map-background.
In 2003 the service will be enhanced (and re-branded) to include statutory information managed by Historic Scotland. The same software engine has also been adapted to present information about our extensive holdings of vertical aerial photography through

INTRODUCTION

The Royal Commission on the Ancient and Historical Monuments of Scotland (RCAHMS) records and interprets the sites, monuments and buildings of Scotland's past and promotes a greater appreciation of their value through the National Monuments Record of Scotland (NMRS). CANMORE, the database of the NMRS, has been online since 1998 and in July 2002, search facilities were enhanced through the addition of a user-friendly web-GIS interface - CANMAP. Historic Scotland acts on behalf of the Scottish Ministers to safeguard the nation's built heritage and promote its understanding and enjoyment. Historic Scotland and RCAHMS are now collaborating on a shared web-GIS, or 'heritage portal', to provide a 'one stop' point of access to the information each organisation maintains.

THE NATIONAL MONUMENTS RECORD OF SCOTLAND

The RCAHMS holds information on over 220,000 archaeological and architectural sites in Scotland. These records range from prehistoric through to Roman, medieval and later archaeological monuments and landscapes as well as architectural, including vernacular architecture, industrial archaeology and maritime records. Site information is supported by a related archive of over 700,000 items, including 241,000 prints or drawings, 469,000 photographs, including oblique aerial photography, and 21,000 manuscripts. The resource is dynamic with new material catalogued daily.

Before looking at presenting this tremendous wealth of information and knowledge over the Internet in more detail, it is worth pausing to remember that the NMRS database is not a homogeneous resource compiled for the specific purpose of disseminating information to remote users. Until 1983, when the function was transferred to the RCAHMS, the Ordnance Survey Archaeology Division maintained a separate card index, which included summary descriptions of individual sites, for the primary purpose of selecting antiquity information for cartographic representation on their maps. Incidental to that purpose, the card index also formed the backbone of the NMRS database when the record cards were optically scanned in the late 1980's. The Royal Commission's own survey and recording programme, of which detailed accounts were published through the County Inventory Series and successor reports focusing on more specific study areas, also contributes to the database. Themed projects, such as the First Edition Settlement Project have also added over 20,000 new records to the database. Information is also regularly added through bibliographic searches through the literature. Under agreement with the Council for Scottish Archaeology, the annual summary accounts of current archaeological work, which are reported in Discovery and Excavation in Scotland, are routinely added to the database verbatim. The archives are also steadily growing through the deposition of primary records from archaeological projects. Computerisation of the NMRS architectural records collections indexes and Heritage Lottery Funded projects, such as the Scottish Architects Papers Preservation Project, has also seen the database grow considerably in recent years. So information, whether gathered to aid selection of antiquities for depiction on the maps, from the RCAHMS survey programme, the annual summary reporting of archaeological fieldwork in Scotland, or from cataloguing the wealth of architectural archives, is presented seamlessly in an Oracle database.

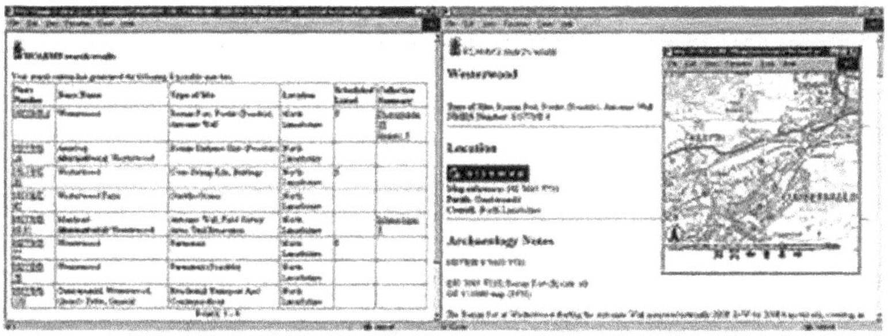

Figure 1 (left) CANMORE, results screen for a search on the place name 'Westerwood'.Figure 6 The collection browser
Figure 2 CANMORE, sample site details for the Roman Fort at Westerwood (NS77NE 8.00)

WWW.RCAHMS.GOV.UK AND CANMORE

In 1998 the RCAHMS launched it's website (http://www.rcahms.gov.uk). As well as containing information about the role of the organisation, current projects and a showcase of recent survey results, the website offers users the opportunity to search the database of the National Monuments Record of Scotland through CANMORE. The acronym stands for Computer Application for National MOnuments Record Enquiries but Canmore also means 'Great head' or 'Chief' the Gaelic name given to Malcolm III who was crowned king of Scots in 1058. On completing a simple registration form, the user is given free access to the entire computerised database of the NMRS through the search screen, although the number of records returned is restricted to 500. The user may search either singly, or through a combination, on building or site name, type of building or site, administrative area such as current Council authorities, the former Regional boundaries or by parish, Ordnance Survey (OS) 1:10,000 quarter sheet and NMRS site number or by keyword. In addition, searches may be filtered to return all records or, through radio buttons, archaeological, architectural or maritime records or any combination of those groups. In the illustrated example (Figs.1 and 2) the user is searching for information on the Roman fort and frontier work, the Antonine Wall at Westerwood, North Lanarkshire. A search on the name 'Westerwood' will return eight possible matches (Fig.1). The results table comprises six fields: the unique number for each site in the NMRS database, the name, including any alternative names or spellings, by which a site is recorded in the database, the type of site, the current administrative area it lies in, any statutory protection (if the site is a Scheduled Ancient Monument or a Listed Building) and a summary of the collections holdings by category with hyperlinks to further indexes. On selecting the hyperlink in the NMRS number field, the user may view more detailed records about a particular site. In the illustrated example (Fig.2), the user has retrieved the site details for the Roman fort and a stretch of the Antonine Wall at Westerwood (NMRS NS77NE 8.00). The screen returns further locational details, including the National Grid Reference (NGR) for the particular site. In a feature, added in 2002, clicking on the 'site map' button opens a scalable location map in a separate window. The screen reports any archaeological or architectural notes associated with the record as well as selected bibliographic references and a repetition, from the results table, of the summary totals for the types of collections held. By following the hyperlink set up in either the site details report or in the results table, a separate screen delivers an index to that particular aspect of the collection. A further hyperlink, embedded in the collections archive, will return the detailed catalogue entry for any given record. Although the collections indexes include digital images associated with a particular record, the images themselves are currently not available on the Internet, though this is anticipated within the next two years.

CANMAP

In June 2002, a map enabled query system, CANMAP, was launched to compliment CANMORE. Accessed through the CANMORE login screen, CANMAP is a map enabled query system for CANMORE using a web-GIS, ArcIMS 3.1 (HTML Viewer), Apache 1.3.19 and Jakarta-Tomcat 3.2.3 on Windows 2000 Server. CANMAP enables users to zoom into any part of Scotland through a click and drag tool and view the NMRS records against an appropriate scale of Ordnance Survey raster map determined by pre-set scale thresholds. Above 1:250,000 scale, the NMRS records are displayed as a series of blue dots, with the level of accuracy for the NGR of an individual record expressed through the intensity of the shading. Lightly coloured dots reflect poorly located sites where the quoted NGR may only be accurate to the nearest 10Km or 1Km. Progressively darker shades reflect the increasing confidence in the accuracy of the site location may be expressed to the nearest 100m, 10m or 1m. The underlying data is a shape file created daily from the NMRS Oracle database and information is therefore rarely more than 24 hours out of date.

Figure 3 (left) CANMAP, distribution of NMRS records in the vicinity of Westerwood Roman Fort (NS77NE 8.00), displayed against OS 1:10,000 raster map, site selection and summary table of results
Figure 4 (middle) CANMAP, sample site details for the Roman Fort at Westerwood
Figure 5 (right) shared web-GIS 'beta version', a combination of Scheduled Ancient Monument area extents and Listed Building locational data provided by Historic Scotland and NMRS data from the RCAHMS may be displayed and searched together. The map extracts shows Scheduled Ancient Monument area extents and NMRS data for the Roman Fort and Antonine Wall at Westerwood

Users may either query NMRS records by selecting an individual blue dot or, through an area search (Fig.3), select and generate a report on one or more NMRS sites. Once selected, the user can view the associated site records which are drawn from CANMORE (Fig.4).

TOWARDS A HERITAGE PORTAL

Historic Scotland (http://www.historic-scotland.gov.uk) is responsible for discharging the functions of the Scottish Ministers in relation to the protection and presentation of Scotland's built heritage, and advises them on built heritage policy. It gives statutory protection to ancient monuments of national importance ("scheduled ancient monuments") and to historic buildings of special architectural or historic interest ("listed buildings") and seeks to ensure that no avoidable damage is done to them.

There are currently over 7,500 Scheduled Ancient Monuments in Scotland. These include an extraordinary range of types of monument, from prehistoric chambered tombs, stone circles, Roman forts and ruined castles, to the first ironworks of the Industrial Revolution and the remnants of the German Grand Fleet scuttled in Scapa Flow in 1919. Others are not decorative or impressive but are still important, from ancient rubbish tips through to medieval cultivation ridges and deserted crofting settlements. Many can only be seen from the air.

Scotland has around 46,000 Listed Buildings. A 'building' can be defined as any man-made structure, and the coverage of lists includes telephone kiosks to tenement blocks and water pumps to cathedrals. They are divided into three categories based on their importance: national or international importance (Category A), regional importance, or major examples of a particular period, style or type (Category B); and local importance (Category C(S)). The lists are compiled and maintained by Historic Scotland on behalf of Scottish Ministers, and are organised in council areas, burghs, parishes and city wards.

All of the information relating to the Scheduled Ancient Monuments and Listed Buildings is held within an Oracle database. In 2001 Historic Scotland digitised the boundaries of all of its Scheduled Ancient Monuments, and more recently a programme of capturing the locations of Listed Building information digitally (as point data) has been underway. As GIS compatible digital datasets, these have become a valuable resource both within the agency and outwith, where the datasets are provided free of charge under a licensing system, to council archaeology services, public sector bodies and utility companies.

Since the launch of CANMAP in 2002, the development of a portal has been underway, to provide a 'one-stop-shop' for heritage information in Scotland. After login, the user will be able to search the whole of Scotland for Scheduled Ancient Monuments and Listed Buildings, as well as the NMRS records from the RCAHMS, using the same functionality as CANMAP. Listed Buildings are displayed as purple stars (with a 1m and 10m accuracy distinguished by depth of colour). Scheduled Ancient Monuments are displayed as red dots at scales of less than 1:50,000 and as red polygons (boundary) at scales of greater than 1:50,000 (Fig.5). Users are made aware from on-screen dialogues that both datasets are digital representations of statutory designations and are seen as indicative and for information purposes only. Further guidance is made available in a 'Help' section and there are links to discovery level metadata as well as meta-information documenting field definitions and sample values. The user may query on individual sites or a report generating function allows the user to make an area selection on either one, two or all three of the datasets and run a report. The report will extract information on the selected records and list them in a separate window, under the heading of the respective originator (Historic Scotland or RCAHMS). The report window also includes any notes, disclaimers and contact details for further information.

The launch of a heritage portal for Scotland forms a platform for users to openly access information on statutory designations and heritage data within a single environment. It represents a major advance in the e-delivery of heritage information in Scotland to both the general public, researcher and commercial contractor. It is anticipated that additional datasets may be added in due course, enhancing the functionality further.

ACKNOWLEDGEMENTS

The authors would like to thank Diana Murray and Rebecca Jones at the RCAHMS and Gordon Barclay, David Breeze and Nick Haynes at Historic Scotland for their comments on this paper. In addition the authors are grateful to Stephen Thomson at the RCAHMS for enhancing the screenshots to publication standard. Development of the CANMORE system was undertaken by John McLeod, CANMAP and the shared web-GIS by Mark Gillick.

A Brief Proposal for an Online Presentation of the Architectural Heritage in the United Arab Emirates (UAE): An Application in Al-Ain City

Amal Abdullah Al Qubaisi
Department of Architectural Engineering,
College of Engineering, United Arab Emirates University
amalq@emirates.net.ae

Ihab Nabil Elkhawas
Department of Architectural Engineering,
College of Engineering, United Arab Emirates University
ielkhawas@link.net

SEE THE CD FOR THE EXTENDED VERSION

Abstract

In this short paper, the authors are proposing and briefly describing an online system for the gradual surveying, recording, documenting and presenting of architectural heritage in the UAE. On one hand, the immediate aim is to present these documented historical sites online for the current and the future generation of Emirates-nationals. This would help in spreading knowledge of the architectural historical icons among students eager to know their own heritage. The online documentation would also help in promoting the cultural awareness on the subject of U.A.E. architectural heritage among the general public.

On the other hand, the long term aim of this study is to accurately document the architectural heritage of the UAE. This documentation can further be used by current and future generations in their effort to conserve their own history and heritage. The brief description of the online system will only focus on the architectural heritage of Al-Ain city, a major city in the Eastern Region of Abu-Dhabi Emirate the capital of the U.A.E., as a pilot model. This model can be expanded and applied in the future to cover more cities and sites in the UAE.

In this proposal and its pilot web-site, the authors adopted a simple WEB interface that works on standard PCs and utilizes standard virtual reality technology. Based on a study presented by Haval in 2001, this system would not compromise the strong need for gathering and linking heritage information, and yet, its' simplicity would save time and produces a user-friendly environment for the general public (Haval 2001).

Introduction

Over the last few decades, western modern-style buildings had completely altered the original Emirates architectural style and the local identity. Fully glazed towers and high-rise modern buildings are now the dominant features of all major cities in the United Arab Emirates (UAE). However, there is a strong feeling among the Emiraties of the urgent need to conserve their own heritage and to preserve their local identity. In the last few years, there have been several successful projects that aimed to conserve some of the main historical icon-buildings in the country. However, these projects are not fully known and represented to the general public. The majority of the Emiraties are not fully aware of the significance of their own architectural heritage except of these located in their immediate vicinity.

Furthermore, the UAE lacks a full inventory and a proper documentation of its architectural heritage. The absence of these elementary data made it very difficult to assess and conserve the historical buildings and sites. Moreover, there is no central place within the UAE where general information on the architectural heritage can be obtained or easily accessed by those requiring information on the subject. From the above discussion, the need for a complete, easy accessible information management system that hosts all information on the architectural heritage of the UAE is urgently needed.

In this short paper, the authors are proposing and briefly describing an online system for the gradual surveying, recording, documenting and presenting of architectural heritage in the UAE. On one hand, the immediate aim is to present these documented historical sites online for the current and the future generation of Emiraties. This site should be made available to all users requiring information on the subject. This would help in spreading knowledge of the architectural historical icons among college and school students thirsty for knowing their own heritage. It should also be accessible to all interested parties, specially, conservation authorities, researchers, educational institutions and tourism promotion activities. The online documentation would also help in promoting the cultural awareness among the general public.

On the other hand, the long term aim of this study is to carefully and accurately document the architectural cultural heritage of the UAE. This documentation can further be used by current and future generations in their effort to conserve their own history. This brief description of the online system will only focus on the architectural heritage of Al-Ain city as a pilot model. This model can be expanded and applied in the future to cover more cities and sites in the UAE.

In this proposal and its pilot web-site, the authors adopted a simple WEB interface that works on standard PCs and utilizes standard virtual reality technology. Based on a study presented by Haval in 2001, this system would not compromise the strong need for gathering and linking heritage information, and yet, its' simplicity would save time and produces a user-friendly environment for the general public (Haval 2001).

In the next Section, a briefly describe the difference between the web-site proposed in this study and similar online web-sites. Section 3 illustrates the different phases that were followed to produce the digital application. A brief description of the proposed web-site is illustrated in Section 4. Section 5 reports on the results of a questionnaire designed to test users' response to the proposed web site. Conclusions and future work is then outlined in the last Section.

Literature Review

The authors reviewed a large number of similar web-sites on the Internet. These sites are listed among references. The main difference between those listed sites and the web site proposed in this study is the global national content of the site. Thus, it covers all the different Emirates and cities in the UAE, and would serve as a reference for all the architectural heritage sites in the country.

For the purpose of this study, Static HTML was the authors' choice. This is because users of Dynamic sites have to know exactly what they are looking for. In a dynamic setting, the web pages are not actually pre-set; instead, they are called upon users' request for information. This would not be the case in a heritage web-site. Users may not have any previous architectural heritage information before using the site. Thus, a Static setting would really help them to gradually browse through the different subjects and increase their information as they go through. Most of the intended users would have an easier time surfing through a Static site with fewer choices and clear hierarchy than a Dynamic site where they have to select specific targets and skip others.

Digital Application

The work on the digital application was divided into eight phases. These phases are briefly described in the next sub-Sections.

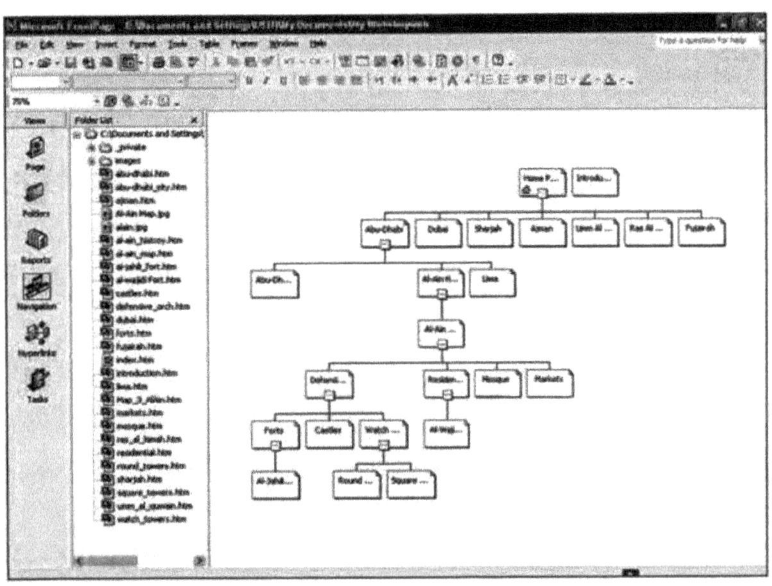

Figure 1 Web-tree and hierarchy

[1] Data Gathering and Field Visits

That included an initial survey of the architectural monuments and sites in Al-Ain city, their classification by type, importance and location. This phase also included some field visits to acquire digital images and the verification of available data.

[2] Web Site Hierarchy and Organization

That included the design of the Web-Tree. This tree is based on a well defined hierarchy that will be explained in full in the Web site description in section 4. In this phase, decisions had to be made on the sequence of web-levels, and possible links that would maintain the homogeneity and the integrity of the Web site. The Web-tree was then executed on FrontPAGE for Windows XP as the base of the work (Fig.1).

[3] Graphic Work

That included the design and the execution of the main graphic theme of the Web site, which included in turns, the design of buttons, text effects, background images, images selection, sizing the Web pages, and finally, integrating all the graphic work in a harmonious design. The graphic work was mainly executed on Adobe PhotoSHOP 7.

[4] Buttons, Special Effects, and Animations

That included the design of buttons "Overlay" and "Down" effects. It also included text and banner animations. This work was mainly executed on Adobe ImageREADY7.

[5] Executing Web Hierarchy and Links

That included putting individual pre-saved Web pages previously constructed on Adobe ImageREADY 7 in their pre-designed location on the Web-tree executed on FrontPAGE for Windows XP. This phase also included assigning each button to its proper URL link.

[6] Adding Special Web Components

That included adding image galleries and database. It also included links with special "Plug Ins" which included Anfy and PixMaker used in managing and displaying special image effects.

[7] Adding Virtual Reality Viewers and Files

A Virtual Reality viewer was embedded into the HTML code of the Web page. Links to VRML files would trigger virtual reality viewer to appear. Users may then view, manipulate, and walkthrough VR models at their own will.

[8] Testing the Web on Different Browsers

During the course of work, the previewing of the various steps and phases on actual Web browsers was essential to validate the work on a step by step base to prevent wrong or dead-end links as well as long loading time.

Web-Site Description

The web-site can be categorized into six interconnected levels. These levels include: (1) The Introduction and the homepage level; (2) Individual Emirates level; (3) The cities level; (4) Architectural heritage buildings categorized based on buildings' function; (5) Building types within each function category; and finally, (6) Within buildings' function and type, buildings are then listed by name.

Although the above levels can be explored in a sequential order, the web-site is designed to allow the user to directly skip from the home page directly to individual Emirates and cities through a clickable map. This map illustrates the geographic relationship between the UAE and neighboring countries. It also includes the position of the seven Emirates of the UAE. Another clickable map is provided within the major cities level. This map can also take the user directly to individual historical buildings without the need to go through all the intermediate levels.

Testing Users' Response

The product of this research work has not been posted on the Internet. It was, however, presented to the students and the staff of Department of Architectural Engineering at the United Arab Emirates University through open sessions. Students from other departments were also introduced to the proposed web site through similar sessions. The aim was to test the output of this study on a number of random users. This test was intended to evaluating the users' responses and to guide us in adjusting our work accordingly.

In this context, the views of a random sample of users were investigated using a simple questionnaire. This questionnaire was designed to measure and evaluate the following: (1) The frequency of using internet among the intended users; (2) The need for information on the cultural heritage of the UAE on the internet; (3) The effectiveness of the proposed web site as a tool utilized to raise the awareness of architectural heritage issues in the UAE; (4) The appropriateness of the information hierarchy proposed in this site; (5) The adequacy of the information posted; (6) The view of the intended users on the practicality of surfing the proposed site; Finally, (7) the enthusiasm among the intended users for possible future expansion of this digital tool.

Two main groups were targeted to participate in the evaluation process. The first group was those with architectural background and a direct interest in architectural heritage. The second group, however, was the rest of the public. Going through the results, it is clear the first group uses the internet more frequent than the second. The results also show that at some point the majority of both groups had searched the internet for heritage information. This proves the internet is an effective medium for conveying heritage information to the intended users. In general, both groups agreed that having this web or similar sites would increase the cultural heritage awareness in the UAE. It is fair to say that except on the subject of the frequency of using the internet mentioned above, both groups showed similar or close responses to all other questions.

With regards to the questions that specifically targeted the participants' opinion on this proposed web site, the majority of both groups strongly agreed that it is a good representation of the cultural heritage of the UAE. They also agreed that the information provided were sufficient and reliable. On the issue of the actual design of web site, most of the participants' graded the information provided as very good to excellent information. They also agreed that the site is user friendly. Finally, the vast majority of the participants are in favor of expanding this site to cover all the architectural heritage of the UAE.

Conclusions and Future Work

In this short paper, the authors introduced a pilot model for a national architectural heritage web-site. The current digital application covers only one city; however, one of the main future aims of this study is to gradually expand the web-site to include all historical architectural icons of the UAE.

Adopting this digital approach in documenting the architectural heritage has changed the work practice of those involved. Originally, the work was done manually in a non-centralized manner for some selected buildings. Even those buildings were not properly documented or easily accessible. Now and in light of this work, the beginning of a central, intensive, properly documented inventory of the architectural heritage is being realized. For those buildings already in the database, there is a direct easy-access way to obtain proper digital information on demand. Although, the digital documentation process is tedious and time consuming, on the long run it should be more reliable and easy to update.

The response we encountered though our open presentations showed that the vast majority of participants have found this web application to be effective, desirable, and reliable. They were also strongly in favor of expanding this application to cover the rest of architectural heritage icons in the UAE. However, this anticipated and recommended expansion will not be an easy task. The project is in need for adequate funds, more experienced working team, and a lot of patient.

REFERENCES

A DIGITAL ARCHIVE OF AMERICAN ARCHITECTURE (http://www.bc.edu/bc_org/avp/cas/fnart/fa267/, January 2003).

A DIGITAL ARCHIVE OF ARCHITECTURE (http://www.bc.edu/bc_org/avp/cas/fnart/arch/, January 2003).

ADDISON, A., 2000. Emerging trends in Virtual Heritage. In IEEE Multimedia, April-June.

BATH TOWN CENTRE VIRTUAL TOUR (www.bath.co.uk, February 2003).

DIGITAL IMAGE PROJECT (http://www.bluffton.edu/~sullivanm/, January 2003).

GREEK ART AND ARCHITECTURE (http://history.evansville.net/greece.html#Art%20and%20Architecture, January 2003).

HAVAL, N., 2000b. Three-Dimensional Documentation of Complex Heritage Structures. In IEEE Multimedia, April-June.

IMAGES OF MEDIEVAL ART AND ARCHITECTURE (http://www.pitt.edu/~medart/index.html, March 2003).

KEMAH BOARDWALK, Austin City Tour, Galveston Island Tour (http://www.virtualtour360.co.uk/, January 2003).

NATIONAL TRUST: Architecture (http://www.nationaltrust.org.uk/historicproperties/index.cfm?page_id=71, January 2003).

ROMAN OPEN AIR MUSEUM - Hechingen-Stein (http://www.villa-rustica.de/indexe.html, January 2003).

THE ENGLISH HERITAGE (http://www.english-heritage.org.uk/, April 2003).

THE GREAT BUILDINGS COLLECTION (http://www.greatbuildings.com/gbc.html, January 2003).

THORNES, R. et al., 1998. Documenting the Cultural Heritage (http://www.object-id.com/heritage/index.html).

A Web Oriented Database for Archaeological Spreading

Marianna Baldi
Vast-lab, PIN - University of Florence, Prato - Italy

Fausto Carminati
Vast-lab, PIN - University of Florence, Prato - Italy

Abstract

We would like to illustrate one work developed inside "Prisma Project" (coordinated by Prof. Franco Niccolucci) presented during last CAA. The aim of the project was to test informatic instruments for archaeological museums. Our case of study was the creation of a web site for the Archaeological Museum of Grosseto - Italy. During this work it was decided to create a "virtual section" of the museum to allow people to look at objects not available in the showcases (in this case a collection of coins).
For this purpose we have utilized a web oriented DB made with Postgre and dynamic PHP pages. Omitting details of the DB plan, we have made an on-line publication, in our opinion interesting:
1. It communicates archaeological contents with a simple language.
2. It allows a guided consultation of the objects organized, in this case, in chronological phases.
3. The research form contains lists of key words to help not specialist users.
4. This work includes some forms (made with PHP) to update, to delete or to insert new data, using a simple internet connection.
This prototype can be utilized for other purposes easily.

Introduction

This contribution reports the experience of one of the case-studies realized during the PRISMA project. [1]

The objective of the present case-study was to define a method to allow people to look at objects, usually not visible, stored in museums warehouses. The presence of "invisible" materials is important specially in small archaeological museums in Tuscany addressed by the Prisma project. We want to underline that this work has been thought for a specific user that we define "general public".[2]

Focusing on this target audience, we planned work as follows:

1. Define needs and limits of small archaeological museums.
2. Chose the best method for web-publishing, determining:

   I. Tools
   II. Techniques
   III. Structures

Needs and Limits of small Archaeological Museums

Small archaeological museums in Tuscany present a series of problems:

- Many artefacts.
- Low budget.
- Few employees.
- Small space for exhibitions.

So often many artefacts are not visible, there are few publications on paper either to help visitors during the tour of the museum or for increasing visitors' knowledge about the history and culture of the civilizations that made the exhibited artefacts. Then, the few employees and the low budgets don't allow an adequate number of guided visits. In sum there are poor services for the "general public".

In the light of these facts the goal is to determine how small museums can give better services and, above all, show all their archaeological collections without increasing their budgets.

Our idea was to create for the museums one or more "virtual" expositive sections. In this case "virtual" does not refer to reconstructions known as "virtual reality", too expensive for these institutions, but it means merely "not real" exhibition, we can also call it cyber- exhibition.

Our aims were, therefore, to let museums show their archaeological collections, to increase visitors' satisfaction and to keep costs low.

The solution could be a web oriented publication, integrating the museum web sites with this "virtual section". A web publication can be consulted either in the museum, creating an internet area is really a low cost solution, or at home.

Web Publishing for Archaeological Collections

We decided to test this web-oriented publication with the small heterogeneous numismatic collection of "Museo archeologico e d'arte della Maremma, Grosseto".[3]

After the decision to use the Internet to communicate archaeological content the main problem was choosing the best way to do it. We also considered that this product should grow with the contribution of the museum employees, usually unable to manage web editors. So we analyzed two ways of web-publishing: static HTML pages and dynamic pages with a database management system in the background.

STATIC PAGES VS. DYNAMIC PAGES

We think that is clear that the biggest problem for small archaeological museums is money, so in the following analysis we considered mainly costs. We divided cost in two classes: development cost and management and expansion cost.

[1] Static pages

[1.1] Development cost

The starting cost for developing static pages is definitely lower than for developing dynamic pages. The following steps are required

1. Data analysis
2. Data structure design
3. Creation of static web pages (one by one the cost is lower, but even the publication of a small corpus needs the creation of a lot of pages), requiring these following sub-steps.
    1. Layout design
    2. Pages construction (editing).

[1.2] Management and expansion cost

To extend the on-line publication it is necessary to know the use of an HTML editor and of an FTP client. Moreover any change in the graphic layout or in the page structure requires re-building all the pages, and this is very expensive. As far as accessibility is concerned static web pages allow only a visit along a pre-built path.

[2] Dynamic pages

[2.1] Developing cost

At the beginning the cost is higher than static HTML. The following steps are required:

1. Data analysis (*)
2. Data structure design (*)
3. Database creation
4. Creation of dynamic web pages, more expensive then static pages, requiring the following two sub-steps
    1. Layout design (*)
    2. Pages construction (programming).

(*) Required also for static HTML

[2.2] Management and expansion Cost

Extending the on-line publication (in our case) is very simple because inserting, updating or removing one record from the database can be done using web interfaces, no computer science skill is not necessary and the task can be performed by museum employees.

Also the cost of changing the graphic layout is lower because just few pages need to be changed. Moreover, and really important for users, dynamic web pages allow a visit along a pre-built path but also using search forms.

AND THE WINNER IS...

Analyzing pros and cons, we decided to develop a database and dynamic pages, in particular because we wanted to give the museum an easy tool to widen the publication of its collections. Following project aims we chose to develop the system using open sources software, that is:

OS = Linux
Web server = Apache
DBMS = Posgree SQL
Server side pages language = Php

SOMETHING TO REMEMBER FOR WEB ORIENTED DATABASE

Using the World Wide Web you open your data to everyone who is connected, so having in mind that the target audience is the "general public", the language used for describing artefacts must be simple and it's better to create one or more guided paths for users, for instance a chronological path, a geographical path, one explaining technology and so on. All this may be easily created using data stored in the database.

Research forms must be friendly, they must show a list of key-words, either because the visitors are not experts or to avoid empty results deriving from inappropriate search terms.

DATA SEARCHING IN THE NUMISMATIC DATABASE OF "MUSEO ARCHEOLOGICO E D'ARTE DELLA MAREMMA, GROSSETO"

In this work we decided to create a complete search form showing all the fields that can be used for data searching. Moreover also the page containing this form is dynamically generated, using the data stored in the database to create lists of keywords to avoid empty results. The user can choose to start a research using only one field or using more fields, in this case he can also change the Boolean operator to AND (default is OR).

The answering page collects the data submitted by the search form, then creates an SQL query depending on the Boolean operator and testing which fields are filled and which are empty. After this a connection to the database is opened and the query is submitted. Query results are used to create a web page following layout instructions listed in a separate php file. The resulting page is then displayed to the user.

Producing the code to generate the dynamic web pages (not only for searching the database but also for inserting new data and images) required two person/months work. Records were already available on paper, but not published by the museum.

This methodology can be applied to any kind of museum collection. It requires only the availability of records, generally produced in museums at least for administrative purposes, and some simple programming. We point out that all server side software used is open source and free,. therefore it can be easily adapted by small museums, regardless how small their budget may be. Our code for queries and display of results is freely available on request.

ACKNOWLEDGEMENTS

We thank Dario Battini for his collaboration in developing the php code used for this project.

[1] One of the aims of this project, coordinated by Professor Franco Niccolucci, was to develop tools for multimedia dissemination of archaeological resources. See F. Niccolucci, M. Baldi et al.(2003).

[2] People who want to get Archaeological information but do not necessarily have a professional know-how.

[3] Archaeological and Art Museum of Maremma in Grosseto.

REFERENCES

BALDI, M., CARMINATI, F., 2002. Realizzare un sito web per una struttura museale, il Museo Archeologico e d'Arte della Maremma - Grosseto. In Niccolucci, F. and Hermon, S. (ed.), Multimedia Communication for Cultural Heritage, Proceedings of a Workshop held in Prato, 1 October 2001:25-34.

NICCOLUCCI, F., BALDI, M. et al., 2003. The PRISMA Project: an Integrated Network for Local Archaeological Museums, In Doerr, M., and Sarris, A., The Digital Heritage of Archaeology. Proceedings of CAA2002, Athens, Archives of Monuments and Publications, Hellenic Ministry of Culture:411-414.

SANDERS, D.H., 2002. Virtual Archaeology and Museums: where are the exhibits? In Niccolucci, F. (ed.), Virtual Archaeology, Archeopress, Oxford (BAR International Series 1075):187-194.

CIPHER (Communities of Interest to Promote Heritage of European Regions) - Internet Cultural Portals and the Development of an Irish Cultural Heritage Forum - (www.cipherweb.org)

Anthony Corns
Discovery Programme
34 Fitzwilliam Place, Dublin, Ireland

Oliver McHugh
Discovery Programme
34 Fitzwilliam Place, Dublin, Ireland

Eoin Kilfeather
Digital Media Centre
Dublin Institute of Technology, Dublin, Ireland

John McAuley
Digital Media Centre
Dublin Institute of Technology, Dublin, Ireland

Abstract

This paper outlines the concept of heritage forums and introduces the CIPHER project, which aims to encourage public participation in heritage through the provision of access to bodies of Internet data. By developing multi level web sites ('Cultural Portals') CIPHER aims to provide meaningful, access tailored information to a wide range of interest levels. The system involves descriptive and interpretive assessments combining elements such as text, mapping, photography, documents, drawings and reconstruction models CIPHER within an ontological narrative structure.

One example of the CIPHER project is the Irish Cultural Heritage Forum (www.explorer.ie), which aims to provide an Internet access point that will help to guide both public and professional interests to a greater understanding of Irish heritage information. The forum will enable users from novice to professional level to explore archaeological narratives and search for relevant artefacts and monuments. The Explorer Cultural Heritage Forum utilises an ontological database to store all narrative and archaeological information.

Introduction

Initiated in April 2002, the CIPHER project is a thirty-month project supported by the European Union. The CIPHER project aims to provide public access to cultural heritage information around the region of Europe utilising new technologies. The principle activity of the project is to investigate and establish a series of Cultural Heritage (CH) Forums. A CH Forum is a portal website which allows all those interested in its content to explore, learn, interact and contribute information to the web site.

Traditionally CH Forums allowed users to gain access to heritage datasets, however, the domain and context within which this data resides in is often neglected. The resulting user experience is acceptable if the user has previous domain knowledge, but if the user is new to heritage concepts the CH Forum does not fulfil its intended purpose. The CIPHER project identified that new methodologies for knowledge modelling were required to satisfy this remit. The development of ontological database and narrative content structures enable users to explore CH domains, with CH information contained within an ontological CH framework. To increase shared user experience content creation and forum tools allow users to build and share heritage stories and information.

The Irish cultural heritage forum explorer.ie is one of the four CH Forums developed as part of the CIPHER project. This forum was developed by The Discovery Programme and The Dublin Institute of Technology, and provides public access to an Irish archaeological domain including: The Discovery Programme archaeological databases containing the Irish National Sites & Monument Record (SMR), and archaeolo-

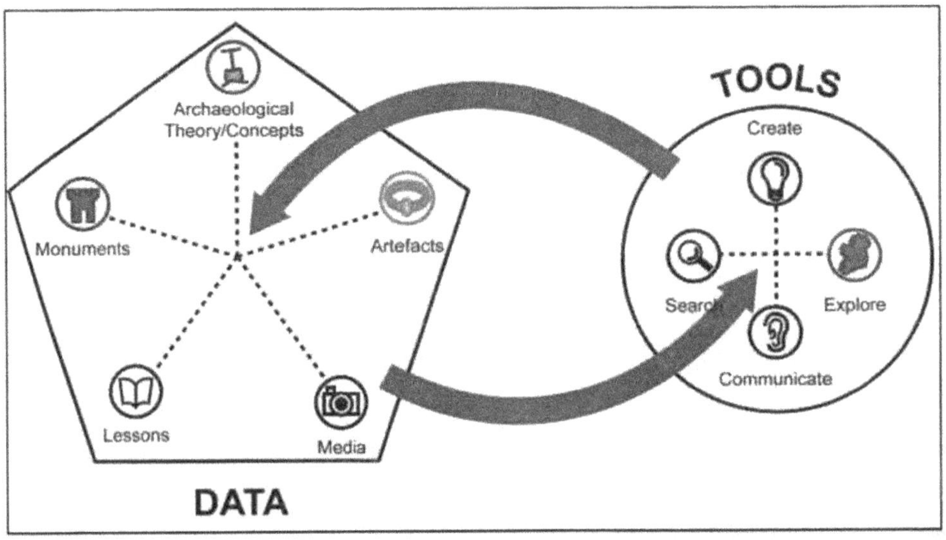

Figure 1 Schematic diagram illustrating the two core components of the Explorer CH Forum

gical artefact data provided by The Discovery Programme and the National Museum of Ireland.

Overview of the Explorer Heritage Portal Structure

The structure of the Explorer CH Forum can be divided neatly into two sections: firstly the data structure and data content that describes and represents the CH Forums heritage domain, and secondly the tools that allow users to explore and interact with the described domain (see Fig.1). The first major component of the Explorer CH Forum is the storage of cultural heritage data. This consists of national cultural data sets and associated media, including: images, text, and video linked together in a hybrid ontological - relational database management system (RDBMS). The second major component is the Explorer toolset, which allows users to access, interact with and contribute to the previously mentioned datasets. Tools that will be available include: semantic search functions, lesson exploration and creation tools, and forum tools for sharing and discussing cultural heritage information.

Domain Content

The domain content for the explorer CH forum is provided by several national and state organizations who have the responsibility to collect and catalogue information relating to archaeological and historical monuments and finds, resulting in over a centuries worth of Irish cultural heritage information. These catalogues at present are still in the process of digitisation and their structure is poorly defined and monitored. Even though there are problems with the totality and accuracy of these data sets, the Explorer CH Forum uses them as the central archaeological data set.

[1] The National Sites and Monument Record (SMR)

The SMR represents the complete list of categorized monument data for the Republic of Ireland that have archaeological or historical interest. The SMR is maintained by Dúchas; the National Heritage Service and consist of 123,504 monuments. Each individual SMR contains positional and dating information and is a classified by within one of 786 non-hierarchical monument classifications types that have evolved over time.

[2] The Discovery Programme Digital Archive and Research Knowledge

The data sets represent the totality of the digital archaeological artefacts and monuments information collected over the previous 10-year period the Discovery Programme has been in existence. Research projects carried out by the Discovery Programme have provided a large amount of well described and catalogued information relating to Iron Age and Bronze Age Ireland, including artefact databases, three dimensional survey plans and illustrations. The second element of data provision by the Discovery Programme consists of the staff's expert knowledge of the archaeological and historical domain, and consists of descriptions of archaeological concepts, theories and relationships.

[3] National Museum of Ireland Artefact Catalogues

The National Museum of Ireland houses and archives the largest collection of archaeological and historical artefacts in Ireland. Of this complete dataset only a small proportion exists in digital form and no standardized classification system has been adopted.

[4] Standardisation of Data Sets

As specified above in the there has been little comprehensive standardisation of Irish archaeological data sets. As Explorer is utilising three disparate data sources, effort has been made to formulate a standardised classification system that both organises and captures archaeological relationships.

The development of a new heritage classification system was beyond the remit of the short research time of the project. It was acknowledged by Explorer domain experts that research into standardised heritage classifications has been carried out by national heritage bodies (Baines 2001) and should be taken into account when developing an Irish classification system. Adoption of a recognised CH terminology would allow an Irish heritage data set to inherit the established standards including the MIDAS (Monument Inventory Data Standard). With the assistance of English Heritage (EH) the Explorer CH Forum has adopted the online EH monument and artefact thesauri (Charles 2001) to create a standardised Irish archaeological thesaurus. The EH thesauri represents a controlled vocabulary that allows archaeological objects to be related by similarity within a hierarchical domain. Using Irish CH experts, the Dúchas SMR has been mapped to corresponding classifications within the EH thesaurus and new preferred terms created for certain Irish monuments and artefacts (e.g. EH Burnt Mound would have a preferred Irish Thesaurus term of Fulacht fiadh) and new subclasses created (e.g. new Torc jewellery class within the artefact thesaurus).

Data Structure

To model archaeological objects and concepts the explorer CH Forum utilises the concept of ontologies to create a data structure in which to place all central CH Forum data. Ontologies represent a way for domain experts to model knowledge by specifying a concept. Gruberd describes ontologies as being "...a description of the concepts and relationships that can exist for an agent or a community of agents" (Grüber 1993). When applied to digital data structures ontologies allow us to model both hierarchical, inherited and associative relationships between objects, concepts and people. Using a series of hierarchical class structure including super-class, sub-class entities, objects can be grouped into conceptual groups.

CIPHER project partners the Knowledge Media Institute (KMi) and the Czech Technical University (CTU) has designed Apollo: a dedicated ontology creation software. Apollo uses the concepts of: "classes" to develop an organisational hierarchy within data structures e.g. castle class, "instances" that allow actual real archaeological objects to exist within the ontology (e.g. Dublin Castle); and "slots" that provide the attributes and descriptive characteristics of the concept/object you are trying to capture. The Apollo software also creates hereditary functions with all attributes (slots) from super-classes being inherited by their children (e.g. a sub-class of daggers would inherit all the applicable attributes of edged weapons).

Four separate ontologies have been designed for use within the Explorer CH Forum, which model its central domain content. They include: monuments, artefacts, archaeological concepts and theory, and people. Each of these ontologies can be related to each other through linking mechanism within the classes and instances structure. By using the adapted EH thesaurus the majority of the monument and artefact thesaurus were already provided. Due to the experimental nature of Apollo, completed ontologies were exported to a standard RDBMS environment (MySQL) to allow for easier data management and Internet implementation.

CURRENT CH FORUM SEARCHING METHODS AND THE EXPLORER AUDIENCE

The Explorer CH Forum uses exploration mechanisms common to most other CH Forums, including search facilities based on semantic methodologies and geographic extents. Many on-line CH Forums are designed in such a way that some previous knowledge of the domain is required e.g. a public user would carry out a search on a CH Forum to find a list of hillforts within their county. However, allowing this user to search through online data using a common semantic search mechanism assumes that they understand both the concepts of hillforts, and where they are located within the complete CH classification structure. These preconditions to getting access to CH Forums content has not been intentional, but more of an over sight by the CH Forum designers who have often used a specialist vocabulary for such a long period of time that it becomes second nature to use it liberally.

As the CIPHER project is a member of the EU "Heritage for All" programme and much of the information held was being directly, or indirectly funded by the general public, it was decided that the Explorer CH Forum should be aimed at a large as possible spectrum of potential user groups. Three major user groups were identified as potential users of Explorer, with a combination of their previous knowledge and age group being the determining factor to which group a user would belong. A brief description of each user group and possible examples is summarised in Table 1.

User Category	Description of Users	Example of Users
Basic	Users have no to little previous knowledge of cultural heritage domain Users are 12 years old or younger Non computer experts	Primary school children Users with learning difficulties
Intermediate	Users have no to little previous knowledge of cultural heritage domain Users are over the age of 12 Non computer experts	Secondary school children Adult users with little or no experience of cultural heritage General members of the public
Expert	Users have differing levels of cultural heritage domain knowledge Previous computer knowledge	University and college students Members of special interest groups Professional users e.g. cultural researchers, national organisations

Table 1 Summary of potential Explorer user groups

The Explorer CH Forum will also have a user authentication system that will record user details such as age and possible previous cultural knowledge, so they are guided automatically towards the correctly targeted content.

NARRATIVES AND LESSONS

Traditionally CH forums would present information methodically using lists of cultural data simply extracted from cultural databases, e.g. monument lists. For the Explorer CH Forum to work effectively for the prospective audiences identified in Table 1, new methodologies of facilitating cultural information exploration were needed.

[1] Narrative and Story/Lesson Concepts

In the tradition of teaching and learning it is taken that people will understand subjects through the concepts of story telling rather than trying to comprehend a subject by reading facts. Examples include the difference between parables and rules, and general conversations where learning information from people often consists of a story containing a series of facts placed in a comprehensible context (Mullholland 2002). This tool of story telling has been applied as a main concept within the Explorer CH Forum. The concept of a story within the Explorer CH Forum can be described as a series related narratives that encapsulate and relate a cultural heritage philosophy to its target audience. An example of a story could be an account of a Bronze Age conflict between two warring groups, with the story being able to explain the theory of Bronze Age weapons and society. Narratives can exist as several formats including: text, images, and video or audio media. Narratives are also aimed at their respective audiences outlined in Table 1, with an example being a story aimed at children that explains food during the medieval period, which consists of a series of pictures that illustrate the process of milling wheat. However, one could explain this same concept to an expert audience by writing a series of narratives based around evidence of medieval cooking artefacts, together with appropriate photographs and illustrations.

A framework matrix of stories, or lessons describing the central concepts of archaeology and early cultural history in Ireland has been created (see Fig.2). Domain experts within the Discovery Programme and other CH institutions have created stories that populate this matrix, and act as the central learning resource within the Explorer CH Forum.

[2] Integration of Narrative and Lessons to Explorer's Central Data Ontology

The concept of the lesson and narrative are intrinsically linked to the central CH data sets and ontologies described in Section 3 and 4. Where a lesson or a narrative is related by its content to a class or instance within the Explorer ontology; such as an artefact, a relationship can be created to link a story with a reference to an actual object. By making this relationship the concepts of lessons, CH classification ontologies and object instances are fundamentally linked. An example of this could be were a lesson describing the use of hillforts in the Bronze Age can be linked to the classification of hillforts within the monuments ontology and possibly and an instance of a hillfort such as Mooghaun.

A lesson can be related to as many classification and instances as possible. This link can now act as a two way connection in exploring cultural heritage, where you can learn about objects and there context and then get access to examples of such objects from the national CH data set, and alternatively one can approach the subject from the other direction and search for a specific object and get access to any lesson or narratives that have been related to it. The concept of lesson and cultural objects being related within the CH Forum allows for more "lateral" exploration of cultural heritage. Users of the CH Forum can move from a lesson to related cultural object and onto another related lesson and so on.

[3] Narrative and Lesson Construction Tools

Using the authentication system described within section 5, users of the CH Forum are allowed to create, edit and publish their own CH lessons and stories. Allowing the user to do this produces three clear benefits to both the user and CH Forum:

1. Increases the amount of online cultural content
2. Creates an empowered experience for user
3. Allows for users to increase their understanding of a subject as the process of organising ideas into a comprehensible lesson forces the user to fully learn and understand the subject.

Using a Java applet the user enters a lesson title and some text that will make up a narrative for an object called a meta-lesson. The author/user is then encouraged through a series of wizard interfaces to relate their lesson narrative to classes and instances within the Explorer monument, artefacts, theory and people ontologies. The author can also upload and relate media to the lesson narrative, and subsequently this media is then related to relevant classes and instances within the Explorer ontology. Once a meta-lesson has been created it can be linked together with a series of other meta-lessons into a larger lesson and published upon the Explorer CH Forum. Users can also group together meta-lessons that have been created by other users into a new lesson that they find interesting.

Access by other forum users to user created lessons may be limited to a specified used defined interest group, created using the forum tools. An example of this would be were the stories created by a primary school teacher would be available to only the students in their class. Outside of this private group access will be restricted. It is envisaged that certain users will gain accreditation to become expert lesson creators, that allows for their lessons to be publicly accessible to all Explorer users. Certain Explorer domain administrators will govern the nomination of the expert lesson creators.

Future Explorer Developments

At present the Explorer CH Forum is only half way towards its full completion. Over the next 16 months increased functionality is being added to Explorer CH Forum.

[1] Forum Communication Tools

The Explorer CH Forum will provide users with the ability to share ideas, lessons and experiences with other Explorer users and Explorer domain experts. Users will have the ability to set-up forum groups that share similar interests. This maybe a local history group, or an individual school class. Within their personal CH Forum space they can publish and share stories, and using communication tools discuss Explorer content. One envisaged use maybe from local interest forums that would construct lessons and stories on local monuments, artefacts, and history. Provision is being made to allow groups to have designated experts related to the, e.g. County cork Local History CH Forum would have a link through the Forum to the local heritage officer.

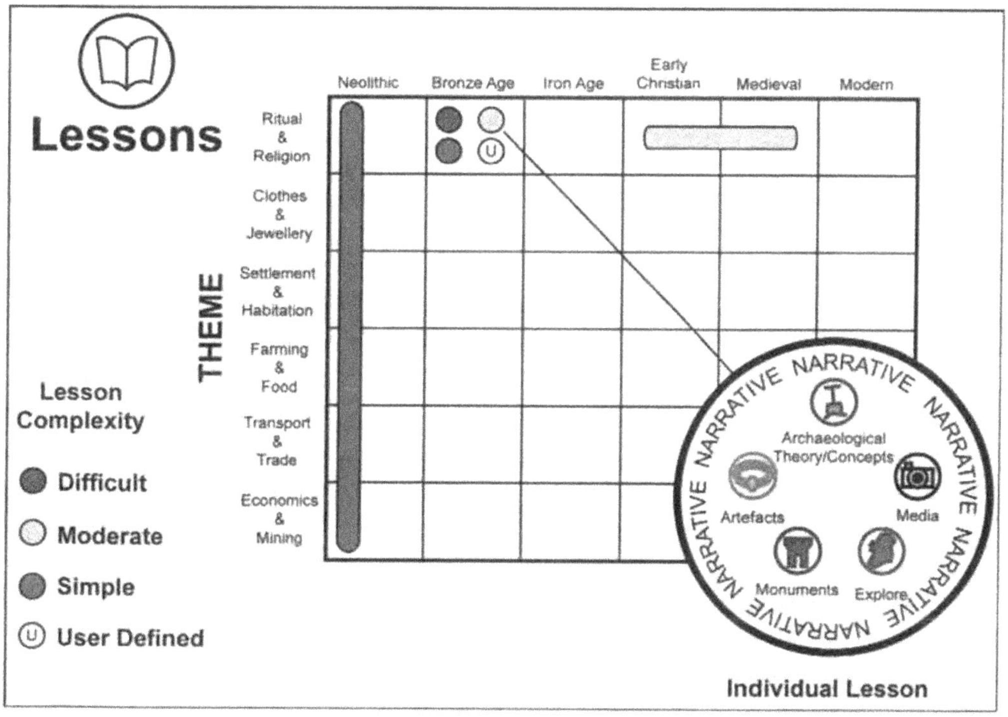

Figure 2 Illustration of the Explorer lesson matrix containing example stories/lessons aimed at the various CH Forum audiences. Included is a closer look at the concept of a story consisting of media and data encapsulated by a narrative context

[2] VRML Narrative Interface

Utilising the increasing amount of three-dimensional information from detailed archaeological surveys of monuments and artefacts, the Explorer CH Forum will have a dedicated VRML browser that will enable users to explorer interactive three-dimensional models. These models will contain embedded narratives similar to those used within the lesson concept, e.g. a three-dimensional model of a monument can be explored, with the user being able to visit defined interest points within the model which through the use of text, audio, image and video media explain the context and the "story" behind the monument.

Conclusion

In this paper, we have presented an original approach to organising, creating and sharing cultural heritage information. The CIPHER project and the Explorer CH Forum demonstrates a coherent way to manage and combine disparate cultural data within an environment that is beneficial to both heritage data standards, classification systems and user comprehension due to the creation of cultural ontologies. The conception and construction of a lesson methodology and the narrative exploration environment allows CH information to be targeted at various audience groups, and allows CH Forum audiences to become empowered by the presence of online lesson creation tools and forum communication tools. (A version of the web application software can be found at http://www.explorer.ie/)

References

BAINES, A., 2001. Scottish Thesaurus of Monument Types: final report. RCHAMS Press, Edinburgh.

CHARLES, D., 2001. In the Beginning Was The Word: Making English Heritage Thesauri Available On-line. Cultivate Interactive Issue 3, 29 January.

GRÜBER, T.A., 1993. Translation approach to portable ontologies. Knowledge Acquisition 5(2):199-220.

MULLHOLLAND, P. and COLLINS, T., 2002. Using Digital Narratives to Support the Collaborative Learning and Exploration of Cultural Heritage. International Workshop on Presenting and Exploring Heritage on the Web, DEXA 2002, IEEE CS Press PR01668.

Peer-to-Peer Ways to Cultural Heritage

Dimitris C. Papadopoulos
Cultural Heritage Management Laboratory (CHMLab)
Department of Cultural Technology and Communication, University of the Aegean, Greece
dimxp@ct.aegean.gr

Efthimios C. Mavrikas
Cultural Informatics Laboratory (CI-Lab)
Department of Cultural Technology and Communication, University of the Aegean, Greece
tim@ct.aegean.gr

Abstract

As acknowledged by most scholars, practitioners and heritage institutions, the digital resources of documented Cultural Heritage form a broad corpus of diverse attributes relating to content, representation and target audiences. Until recently, Web publishing of such resources has been inconsistent in addressing these attributes, by lacking formal structure, coherence, information discovery and retrieval mechanisms, and support for custom points-of-view (individualised, thematic, spatiotemporal, multilingual, multicultural) and collaborative work. Peer-to-peer computing encourages a distributed architecture which could amend such inefficiencies.

This paper explores the emergence of peer-to-peer content networking and the ways in which it meets the requirements arising from specific aspects of the Cultural Heritage field: to encourage exploration and collaboration by identifying universal themes; to maintain contextual information; to provide the interpretation necessary to tell a story; to authenticate knowledge; to seamlessly document and record in a variety of media; to police intellectual property rights.

The Digital Data of Cultural Heritage Dialogue

Scholars, practitioners and heritage institutions are continuously adapting their tools and methods to fit current needs and address emerging problems in the creation, publishing, management and use of the digital data of Cultural Heritage. Not only are the carrying media new, but the content and distribution methods have also changed completely and new stakeholders have emerged among both users and creators of information (Abid and Radoykov 2002:65). The growing use of Internet applications in the Cultural Heritage field favours and encourages the identification of common themes and issues of global scale and interdisciplinary nature. In some cases, standards have already been set on the basis of collective, collaborative work (one such recent achievement being the CIDOC CRM). Furthermore, many heritage projects are community-based or are making attempts to include new, non-expert audiences in their production and sharing of knowledge. However, widespread collaborative work and research still rise as key future challenges rather than everyday realities, as more appropriate communication channels of direct dialogue and real-time collaborative work have yet to be explored. Users and information providers are still unaware of the potential use of the Internet not only as a communication medium, but also as a tool to share and interchange processed and raw data (D'Andrea 2000:318). This paper introduces peer-to-peer computing as one such catalyst of communication and collaboration in the Digital Age of Cultural Heritage.

About Peer-to-Peer

The notion of peer-to-peer in computer science is hardly new: sharing computing resources, integrating disparate networks, and freely cooperating in an academic research environment were the driving forces behind the original Internet of the late Sixties, which followed host-to-host communication, resource and bandwidth equality concepts. These concepts gave way to asymmetric, unequal server/client concepts with the commercialisation of the Internet and the consequent trend to fit activities on this bustling new mass medium into communication patterns used by television, radio and newspapers.

The "renaissance" of peer-to-peer computing can be attributed to a growing need to overcome current barriers to the formation of ad hoc communities, whether of people, of programs, of devices, or of distributed resources (O'Reilly 2001:58) that were put in place during the shaping of the Internet, since the inception of the World-Wide Web. Barriers both technological (Network Address Translation, dynamic IP addressing, firewalls, asymmetric bandwidth connections) and social (unsolicited mail, virus spreading, copytheft, plagiarism) in nature. The return to the original Internet is embodied by the creation of a class of applications that takes advantage of resources -storage, cycles, content, human presence- available at the edges of the Internet (...), tolerating and even working with variable connectivity (Shirky 2001:22). These applications address file sharing and licensed media distribution, messaging, web publishing, ad hoc collaboration (pervasive computing), groupware and distributed computation (grid computing).

Cultural Heritage Shared Spaces

Peer-to-peer computing represents an important social promise to the Cultural Heritage field. The creation of self-operated peer-to-peer communities organised around Cultural Heritage shared spaces will pool the efforts of individual professionals and enhance existing communication and collaboration patterns. It will integrate resources, disseminate knowledge and democratise information.

A Cultural Heritage shared space represents a collection of content and peer-to-peer applications. Shared spaces support applications that:

- fluidly integrate rich text, equations, vectors and raster images, voice and video
- are made of and use standard components (e.g. peer-to-peer integration of ESRI ArcView, Adobe Acrobat Reader, Extensis Portfolio, etc.)
- represent Web data in standard formats (e.g. HTML, XML/RDF, MPEG, SVG, etc.) and communicate using standard protocols (e.g. HTTP, TCP/IP, Bluetooth, IEEE 802.3/802.11b, SOAP, etc.)
- encapsulate collaborative tools (e.g. discussion, document review, event scheduling, project planning, etc.)

There are many existing peer-to-peer applications fulfilling the above criteria. The following are especially noteworthy: Edutella (educational resource sharing, based on the open-source JXTA development platform), Groove ("peerware", peer-to-peer enterprise groupware), Akamai EdgeAdvantage and Digital Parcel Service (content distribution system), Publius (encrypted publishing, named after the 18th century authors of The Federalist Papers), Radio Userland (weblog publishing), Meerkat (syndicated content reader), Rhymbox (integrated chat, based on the open-source Jabber instant messaging platform).

A shared space is very likely to be intermediated. Intermediaries such as a Ministry of Culture, an Archaeological Service, or a professional body can regulate access, safety and content distribution policies for shared spaces within their expertise. Control always remains at the hands of community members. Peer-to-peer community user membership can be drawn de facto from already established communities of special interest (e.g. ICOM) or on a voluntary, personal interest basis, according to the developer community model encountered on the Web (e.g. Java developer forums, JXTA community projects, SourceForge community bug report lists).

The applicability and use of Cultural Heritage shared spaces can be demonstrated in the case of archaeological content.

The Case of Archaeological Content

A peer-to-peer community organised around Cultural Heritage shared spaces provides an environment for real-time collaborative work which could be highly valued in the case of digital archaeological data. In terms of archaeological research, practice and knowledge dissemination, these shared spaces would host a wide range of activities and processes pertaining to the excavation process, academic dialogue and research collaboration, and the dissemination of archaeological knowledge to the public.

Contemporary excavation projects tend to handle a growing bulk of digital data. Field notes and diaries, pictures and drawings, geographical data, all elements of archaeological recording are turning into digital material, sometimes through direct on-site input. Excavations also incorporate a great deal of interdisciplinary work or, in some cases (e.g. rescue archaeology), demand immediate expert consultation. In large-scale national and international projects, a fine-tuned collaboration between various expert groups (e.g. geologists, conservators, architects) has to be established. Collaborative projects between different institutions or between a field research team and a "mother institution" often have to face problems of poor technical support and data exchange, which result in a time (and fund) consuming collaboration. Field archaeologists have to deal -in many cases- with a local area being technologically isolated, which burdens on-site recording and systemisation of digital data.

A community consisting of peers who practice archaeology or any other involved discipline could deal with these issues in a real-time collaborative manner. This community could set up a shared space covering different stages of an excavation project, or connecting distant excavation projects and research teams supporting defined sets of activities within the following:

- identification of common themes and problem solving strategies
- immediate expert consultation and comparative, interdisciplinary work
- collaborative authoring, cooperative interpretation and narration building
- collaboration in transforming raw data into systemised data
- arrangement of work schedules, timetables, whiteboards
- unified on-site documentation and data processing
- enhancement of existing communication and dialogue

These sets of activities would integrate and enhance existing excavation toolboxes: GIS software, portfolio management, email. The future development of advanced methods for remote operation, distributed computation and ad hoc connectivity in archaeological shared spaces would further enable the field archaeologist to use remote sophisticated hardware and software for real-time 3D modelling/rendering, and interconnect all the technical equipment available on-site (including portable 2D/3D/barcode scanners and GPS tracking devices).

Archaeological shared spaces would also enable peers beyond the excavation practice to:

- incorporate fragmented data (e.g. in excavation sites, museums, labs) through shared availability
- set and apply metadata and ontology standards to received data
- set and apply standards for authenticating archaeological knowledge
- set language annotation preferences and use multilingual tools
- co-manage work schedules and data processing
- apply selected media formats, and determine means of documentation, authoring and publishing in order to ensure compatibility and interoperability
- apply security and safeguarding policies, self-manage access and data sharing criteria
- increase the academic awareness and transparency of

archaeological research and knowledge
- e-publish without major labour requirements, sophisticated software, high costs, or time-consuming work (a motivation against the "digital dilemma" and the divide between archival data and publications)
- manage change and adapt to new technologies

In less formally structured and strictly safeguarded applications, a shared space membership could include non-expert group peers and individuals. This membership composition would enhance an expert-to-public dialogue and encourage participation in archaeological interpretation and knowledge building. Many archaeological projects have recently made attempts to allow customised, personalised points-of-view and include non-expert, local or global audiences into excavation practice, archaeological research and interpretation (e.g. Çatalhöyük, Ename 974, PAST, see Hodder 1999, Kotsakis et al. 2002, Pletincx et al. 2001). Such an approach is also justified by "community" or "community-based" archaeology which has an interest in involving local communities into archaeological research.

Thematic shared spaces (e.g. focused on the theme of a Neolithic settlement excavation) encompassing a broad audience ranging from the excavation team itself to high school students, PC-equipped locals and random, geographically-spread Web visitors, would result in original interpretation perspectives and establish direct dialogue between distant communities. For individuals representing "less dominant" views, being able to attend the stages of an ongoing excavation (through shared textual and audiovisual data) or taking part in the archaeological story-telling, gives a sense of belonging to a community that jointly produces knowledge and forms the body of narration. In a community of peers, "authority" (in this case: the archaeological expertise) is no longer the sole content provider since it is a receiver at the same time, allowing a greater initiative space to non-expert peers.

Peer-to-Peer Risks

Peer-to-peer communities and shared spaces demonstrate some inherent risks owing to their reliance on digital communication technologies, to their content and their peers. While the reliance on digital communications is no different than any other Internet application (access being a prerequisite to participation, this is still a highly discriminatory limitation as most of the Earth population is "offline"), content and peer risks should be examined more closely.

Shared spaces apply to that part of digital archaeological data which can be made available on a shared basis. Peers are able to determine the range of content coverage within their community. In the domain of cultural resource management -here taking the case of archaeology- access control to a peer-to-peer community and the regulation of digital rights are crucial issues. It is therefore necessary to set standards for the safeguarding of intellectual/cultural property rights and fine-tune them to national policies.

In a context of data exchange, collaborative authoring and participatory interpretation, where both communication protocol and human errors or mishandlings are involved, the risk of losing primary data is real. Ways to preserve primary data and avoid the loss of "information commonsense", such as data duplication (backup) and encryption, should be taken into consideration.

Even in a peer-to-peer, decentralised environment, the development of dominant peers and regulative authorities is still possible, as is the tendency to homogenise academic, linguistic and cultural diversity. "Elite" expert groups and western "developed" institutions may tend to marginalise non-expert groups and small, developing institutions in a community of peers. English will still tend to be the language exclusively used in international and even non-western projects.

The creation, regulation, and communication function of a peer-to-peer community and shared space depend on the peers that create it. Dominant views and schemes may still survive, but alternative views, diverse communication patterns and multivocality should always be guaranteed. Peer-to-peer communities and shared spaces can be self-managed in terms of setting regulation policies, language preferences, range of participation and dialogue. If the centralised systems are so constructed that they overflow with diversity and alternative perspectives (Hodder 1999), then decentralised systems have the true potential to answer these issues in a more fluent way.

In every case, there has to be a firm will for collaboration, data sharing and exchanging, and dialogue not only between archaeologists and other researchers but also with interest groups and non-expert audiences. Data flow within peer-to-peer communities should remain unburdened by rigid mentalities, bias and hesitation relating to the electronic publishing and exchange of data, academic and research collaboration, and the sharing of knowledge with the public.

Thinking Peer-to-Peer: A Trend Setting

Data sharing, collaborative work and research have been identified as future trends of Internet use in the field of Cultural Heritage and archaeology. As described by Hodder, *'there is a shift from hierarchy to networks and flows'* (Hodder 1999). Such a shift can also have an impact on archaeological theory and practice, since *'the potential for participation is such that it does possible to talk of the erosion of hierarchical systems of archaeological knowledge and the emergence of a different model based on networks and flows'* (Hodder 1999).

Peer-to-peer is a model based on networks, flows and collaboration between communities of peers. The main consequence of the implementation of peer-to-peer content networking is the vertical disintegration of the provider/user or server/client scheme. In the case of the digital data of archaeology, this respectively leads us to a series of transitional effects which signify a shift from authoritative, linear schemes to communities of peers creating and using data on a shared basis. Being a decentralised, "democratised", participatory model of Internet use, peer-to-peer content networking provides spaces for collaboration, dialogue and multivocality.

Shared spaces can only be viable as long as trends favour a shift from central to peripheral, from dominant or "western" views to alternative or local interpretations, from "safekeeping" of academic "certainties" and archaeological knowledge to publishing, exchanging, and communicative and collaborative researching. The decentralised nature of peer-to-peer communities and shared spaces -excluding the notion of a sole content possessor- has the potential to meet modern challenges such as the discovery and quality control of resources or the survival of primary digital data (Richards 1998:347-348).

Peer-to-peer computing is still being re-introduced as the fundamental collaborative basis of the Internet. This paper has attempted to demonstrate a first link between such promising technologies and the field of archaeological theory, practice and the dissemination of knowledge, since no choice of tool should be regarded as incidental (Richards 1998:331) or detached from a discipline's body.

ACKNOWLEDGEMENTS

The authors wish to thank Dr. Gerasimos Pavlogeorgatos at the Cultural Heritage Management Laboratory (Department of Cultural Technology and Communication, University of the Aegean), Dr. Evangelia Kavakli at the Cultural Informatics Laboratory (Department of Cultural Technology and Communication, University of the Aegean), Prof. George Vouros at the Intelligent Cooperative Systems Group (Department Of Information and Communication Systems Engineering, University of the Aegean), and Prof. Nicolas Nicoloyannis at the ERIC Laboratory (Université Lumière Lyon 2) for their comments and support.

The authors also wish to thank CAA 2003 conference attendants and peers for sharing their thoughts on the applicability of peer-to-peer networking in the field of Cultural Heritage and archaeology, and Prof. Franco Niccolucci for initiating a small, ad hoc "community" of dialogue on this issue.

REFERENCES

ABID, A. and RADOYKOV, B., 2002. Access and Preservation in the Information Society. Museum International Vol. 54, no.3.

D'ANDREA, A. et al., 2000. Web Access to an Archaeological GIS. Proceedings of the 2000 Conference on Computer Applications and Quantitative Methods in Archaeology (CAA 2000 - Computing Archaeology for Understanding the Past), Ljubljana, Slovenia, 18-21 April 2000.

HODDER, I., 1999. Archaeology and Global Information Systems. Internet Archaeology issue 6.

KOTSAKIS, K. et al., 2002. A Wireless E-guide for Archaeological Sites. Proceedings of the 2002 Conference on Computer Applications and Quantitative Methods in Archaeology (CAA 2002 - The Digital Heritage of Archaeology), Heraklion, Greece, 2-6 April 2002.

O'REILLY, T., 2001. In Oram, A. (ed.), Peer-to-Peer: Harnessing the Power of Disruptive Technologies. O'Reilly & Associates Inc.

PLETINCX, D. et al., 2001. Presenting a Monument in Restoration: The Saint Laurentius Church in Ename and its Role in the Francia Media Heritage Initiative. Proceedings of the 2001 Conference on Virtual Reality, Archaeology, and Cultural Heritage (VAST 2001), Glyfada, Greece, 28-30 November 2001.

RICHARDS, J., 1998. Recent Trends in Computer Applications in Archaeology. Journal of Archaeological Research Vol. 6, no. 4.

SHIRKY, C., 2001. In Oram, A. (ed.), Peer-to-Peer: Harnessing the Power of Disruptive Technologies. O'Reilly & Associates Inc.

From Vectors to Objects: Modeling the Nasca Lines at Palpa, Peru[1]

Martin Sauerbier
Institute of Geodesy and Photogrammetry,
ETH Hoenggerberg, CH-8093 Zurich, Switzerland,
www.photogrammetry.ethz.ch
martin.sauerbier@geod.baug.ethz.ch

Karsten Lambers
Institute of Geodesy and Photogrammetry,
ETH Hoenggerberg, CH-8093 Zurich, Switzerland,
www.photogrammetry.ethz.ch
karsten.lambers@geod.baug.ethz.ch

Department of Prehistory,
University of Zurich, Karl-Schmid-Str. 4,
CH-8006 Zurich, Switzerland, www.prehist.unizh.ch

SEE THE CD FOR THE EXTENDED VERSION

Abstract

The Nasca Lines around Palpa are currently investigated by archaeologists and geodesists from Switzerland, Germany, and Peru. A photogrammetric analysis of high resolution aerial images allowed the accurate 3D recording of more than 1500 geoglyphs, most of which have also been thoroughly described in the field. A digital 3D model of the area of investigation is the first important result of this effort. However, the vectors recorded during the mapping process that mark the preserved outlines of the geoglyphs do not coincide with the geoglyphs themselves, which contain also an interior surface. Furthermore, they show certain traits important for an archaeological analysis, some of which (geometric attributes like location, size, and orientation) can be derived from the processed vector data, while others (contextual attributes like stratigraphic relations and associated objects) have been recorded during fieldwork. An important prerequisite for the intended GIS-based analysis of our data is to define meaningful archaeological objects (i. e., the geoglyphs) and assign them different kinds of attributes like those mentioned above. In this paper we describe the original data, the process of object definition, and our data model and discuss some preliminary results.

Introduction

If archaeological objects, be they single finds, features such as graves or buildings, or whole sites, are to be analyzed using a GIS, the underlying database in which information concerning the objects is stored has to contain both spatial data (defining the location of the object) and attribute data (describing its characteristics) (cp. Bill 1999:11, Wheatley and Gillings 2002:23). The main purpose of the GIS for the Nasca Lines at Palpa, Peru, which is currently being developed and implemented at the Institute of Geodesy and Photogrammetry at ETH Zurich (IGP) is to enable the analysis of the relationship between the geoglyphs and their surroundings in order to better understand the criteria that guided the creation of the ground drawings (Grün, Sauerbier and Lambers 2003, Sauerbier and Lambers 2003). The recording of spatial as well as attribute data on the more than 1500 geoglyphs in the Palpa area required a two fold approach resulting in two separate data sets:

- Spatial data: On a base provided by a series of large scale aerial images suitable for stereoscopic analysis, all geoglyphs were identified and photogrammetrically mapped (Grün and Lambers 2003). The resulting geometrical data was stored as 3D vector data (lines).

- Attribute data: All geoglyphs were then located, numbered and described in the field using a standardized feature sheet (Reindel, Lambers and Grün 2003). The archaeological data was stored in a relational database.

These two data sets cannot be linked easily due to a peculiarity of the spatial data set resulting from the state of preservation of the Palpa geoglyphs. Unlike objects in a cadastral GIS, where geometrical information is derived from measurements of clearly defined objects (e.g. buildings, real estate; see Bill 1999), the Palpa geoglyphs usually cannot be measured in their entirety (as polygons) since many of them are partly damaged by erosion or modern land use. Another difference to cadastral GIS, where measured objects of the same class usually do not overlap, is that many geoglyphs had already been cut or covered by other geoglyphs in prehispanic times. During the image-based photogrammetric recording of the geoglyphs, only the actually preserved or securely deducible parts of geoglyph borders were mapped in order to maintain this important information. This has resulted in a set of digital 3D vectors that do not represent entire geoglyphs, but rather discontinuous sections of their borders. In other words, the spatial data and the attribute data do not then refer to the same category of archaeological entities. In order to enable a meaningful archaeological analysis, the 3D vectors have to be combined and complemented into polygons representing the most likely original shape of each geoglyph based on additional archaeological information. Such reconstructed objects can then be assigned the geoglyph numbers and attribute data recorded in the field. The aim is to store the data in an object-relational database management system (DBMS), which brings about the benefits of an object-oriented approach (methods, inheritance etc.). For analysis, the combined data sets have to be made accessible via an intuitive graphical user interface that should allow the retrieval of all available information about a given geoglyph by simply clicking on its graphical representation.

In the following sections we describe a preliminary workflow for object definition that represents an intermediate step towards a system that stores both attribute and spatial data in a single database and allows read/write-access from a graphical user interface.

Technical Requirements and Software Solutions

The archaeological research problems to be addressed, which have been described in Grün, Sauerbier and Lambers 2003 and Sauerbier and Lambers 2003, bring about specific requirements concerning the technical design of the GIS. An important specification is that the basic data should be accessible in a standard data format in order to allow the different users (Grün, Sauerbier and Lambers 2003:Fig.5) to access the database using GIS software of their own choice. Furthermore, a standard data format is important for the publication of the data (on CD or web based, e.g. in XML format). Therefore, it was decided that Oracle 9i DBMS would be used as a central database of the GIS. Most GIS software packages have an interface to Oracle DBMS, and the data is also accessible via the SQLPLUS interface, through the use of embedded SQL, or from programs written in C++ or Java.

Three different software packages were evaluated for the GIS being developed for use as a tool for data manipulation, query and analysis, and as graphical user interface that provides access to the data. These were TopoBase 2.03 by C-Plan, GeoMedia 4.0 by Intergraph and ArcGIS 8.1 by ESRI. The most important criteria for evaluation were user friendliness, the efficiency of tools for data manipulation, the querying and editing capability for raster and vector data, and cost. Furthermore, the GIS software should be able to be customized, i.e. it should allow enhancement of its functionality through use of programming language. For fieldwork in Palpa, as well as for publication and presentation of the project, a further important requirement of the GIS is that it provides an efficient tool for the production of 2D paper maps at different scales.

As a result of the evaluation it was decided that ArcGIS 8.1 (and later on release 8.3, which was not yet available for the work described here and was therefore replaced by ArcView 3.2 for some steps) would be adopted as a basis for the GIS, though this did not preclude the use of other applications for specific tasks.

Development of a Workflow for Object Definition

As mentioned earlier, the first requirement for a GIS-based analysis of the Palpa geoglyphs is to combine and integrate the two distinct data sets (spatial and attribute data) in order to obtain meaningful archaeological objects. The workflow for object definition is then as shown in Figure 1.

[1] Spatial data

The preserved borders of the geoglyphs which are visible in the stereopairs of the aerial images were mapped on an analytical plotter Wild S9 using XMAP 4.00 by Aviosoft as illustrated in Figure 2 (Grün and Lambers 2003).

The resulting vectors were stored in proprietary XMAP files (XMP). In order to process the data, these files are converted to DXF format using the XMPDXF converter by Aviosoft. Then, in ArcGIS 8.1 the DXF-file is converted to a coverage using the DXFARC command. It should be noted that although it is also possible to obtain objects of area type directly in XMAP, the conversion of polygonal topologies using XMPDXF causes a loss of information.

The next step is the completion of the line vectors to obtain closed polygons. For this purpose, the high resolution orthoimage (25 cm footprint) is shown as a background layer in the ArcInfo user interface, with the vector layer in the foreground as indicated in Figure 3.

Based on what is visible in the orthoimage and what is known about the geoglyph stratigraphy from fieldwork, the gaps in the geoglyph borders are then closed in order to represent the most likely original form of the geoglyphs. This requires the manual edition and insertion of arcs and nodes in order to create closed polygons. Due to the somewhat limited functionality of ArcGIS 8.1 concerning object topology, vectors representing borders of two adjacent geoglyphs have to be marked twice, resulting in redundant data. This problem is expected to be solved in release ArcGIS 8.3. The editing process is accomplished in a two-dimensional coordinate system. The third dimension has to be added later by intersection of the objects with the DTM.

After the completion of the vectors, a polygon topology is created automatically using the "Clean" command in ArcGIS 8.1. This topology is stored in an arc.adf file that contains in a binary format the number of arcs and the arcs comprising a polygon. The resulting polygons shown in Figure 4 should not be confused with the geoglyphs. On the one hand, all geometrically possible polygons are created, including such areas enclosed by geoglyphs that are not themselves a geoglyph. On the other hand, due to the many overlaps between the geoglyphs, each polygon usually represents only a part of a given geoglyph, while it may belong at the same time to several overlapping geoglyphs.

Thus, the next step in the workflow is the revision of the polygon layer. For this step, ArcView 3.2 is presently used because of its comfortable editing functionality. It is planned in the future to accomplish this task in ArcGIS 8.3. Initially, all dispensable polygons that do not belong to any geoglyph are deleted. This is done manually by graphically selecting them, resulting in a shapefile with single polygons, as shown in Figure 5.

In a second stage, all polygons belonging to a given geoglyph are assigned an object ID that corresponds to their geoglyph number in the archaeological database. The resulting data is stored in a table. Where geoglyphs overlap, the corresponding polygons belong to two different archaeological objects. In such a case, the respective polygons have to be copied and stored as several objects in order to allow the assignation of several object IDs to a polygon. The result of this step is an assemblage of polygons, each of which carries an object ID that clearly indicates to which geoglyph it belongs.

In a final step, all geoglyphs with the same object ID are merged using ArcView's geoprocessing extension ("Dissolve" command) into a single geometric object using an aggregation by the attribute "ID". As a result, ESRI shapefiles and coverages, and their corresponding tables containing the objects are obtained. These now represent the real world geoglyphs, with their IDs as attributes, as indicated in Figure 6.

[2] Attribute data

The attribute data registered in the field was stored in a MS Access 2000 database. This data, exemplified in Figure 7, consists of partly standardized descriptions concerning, among other things, the form, construction technique, orientation, stratigraphy, and context of the geoglyphs.

The attributes are of different data types (text, numerical, Boolean) and some are constrained to a limited set of predefined values (e.g. the construction technique, form). The unique primary key of the table which identifies each record is the geoglyph number. The goal is to migrate the relational MS Access data to an object relational Oracle database. A prerequisite here is the storage of attributes according to the object-oriented data model (Lambers and Sauerbier, forthcoming), so a fully automated migration process is not possible. Another constraint is the non-compatibility between some of the MS Access data types and corresponding Oracle data types. For example, the Access data type MEMO (character field of unlimited size), used for textual descriptions, has no direct counterpart in Oracle. Long character fields in Oracle can be either up to 4 GB (CLOB, BLOB), with these being not compatible with ArcGIS 8.1, or up to 4000 Bytes (VARCHAR2), which restricts the length of textual descriptions. In this case it was decided to use VARCHAR2 since all geoglyph descriptions are shorter than 4000 Bytes.

The workflow for data migration starts with a conversion of the Access database (MDB) to an XML file which contains the table structure. For this step, the Oracle Migration Workbench, a free extension to Oracle 9i, is used (Fig.8).

In a second step, the table structure is converted from the XML file to an SQL-DDL script which contains the SQL statements for creating the table structure, constraints, primary and foreign keys, and references in an Oracle database. This step requires the setting of the corresponding data types, in parts manually, in a Migration Workbench menu. Afterwards, it is possible to create SQL statements for the data itself which can then be added to the Oracle database via the Oracle Migration Workbench or the SQLPLUS-Worksheet. The result is a database with a table structure based on the system independent conceptual data model, which has to be adapted and implemented iteratively.

[3] Data integration and user interface

With the spatial data stored in the ESRI file system and the attribute data stored in the Oracle database, a link between the two data sets has to be established that enables a graphical selection of objects and a display of the appropriate descriptive data. For this purpose, ArcView 3.2 offers the possibility of connecting to different DBMS, e.g. Microsoft SQL Server, IBM Informix/DB2 and Oracle via ArcSDE or ODBC (Open Data Base Connectivity). In this case, the ODBC interface has been employed to access the Oracle database tables for read only access in ArcView 3.2. The constraint for a correct natural join of two relations is a common attribute in both tables (Heuer and Saake 2000). Then, the tuples with equal common attribute values are attached, as per Figure 9. This common attribute is the object ID that has been assigned to each geometrical object and which corresponds to the geoglyph number, the primary key in the central table of the archaeological database. Having activated the database connection by using the "Add database table" menu in ArcView, SQL-queries on database tables can be performed. However, manipulation of database contents is not possible.

Conclusions and Future Tasks

At the end of the workflow described, we obtain as an intermediate step a read only, graphically accessible set of archaeological objects, in our case the geoglyphs, with all available data shown in a table under ArcView 3.2.

To complete the desired functionality, some further steps will have to be undertaken. Firstly, we want to achieve a read/write access from the GIS software to the database. For this reason, the most recent release of ArcGIS 8.3 will serve as an user interface in the future. It does not only offer the possibility of read/write-access to the database, but also allows the definition of topologies with certain constraints and rules for efficient error checking. Another aim is to store the geometrical data not in a file system but also in the database. For this purpose, Oracle provides the spatial cartridge (Sharma 2002) which allows the storage of geometrical objects independent of any GIS package's data structure. A further issue is the step from the system-independent, conceptual data model to the logical data model, along with its physical implementation in the database as well as the subsequent extension for the integration of additional data resulting from the analysis.

Acknowledgements

We are grateful for the generous financial support given by both the Swiss-Liechtenstein Foundation for Archaeological Research Abroad (SLSA, Zurich), for fieldwork, and ETH Zurich for analysis. We also would like to thank Prof. Clive Fraser for his contribution to the linguistic improvement of this paper.

1 The figures mentioned in the text are available on the accompanying CD only.

REFERENCES

BILL, R., 1999. Grundlagen der Geo-Informationssysteme, Vol. 1. Heidelberg.

GRÜN, A. and LAMBERS, K., 2003. The geoglyphs of Nasca: 3-D recording and analysis with modern digital technologies. In Acts of the XIVth UISPP conference, University of Liège, Belgium, 2-8 September 2001, section 1: theory and methods - general sessions and posters, Oxford (BAR Int. Ser. 1145):95-103.

GRÜN, A., SAUERBIER, M. and LAMBERS, K., 2003. Visualisation and GIS-based analysis of the Nasca geoglyphs. In Doerr, M. and Sarris, A. (eds.), The digital heritage of archaeology - proceedings of the 30th CAA conference, Heraklion, Crete, April 2002, Athens:161-167.

HEUER, A. and SAAKE, G., 2000. Datenbanken: Konzepte und Sprachen. 2nd revised ed., Bonn.

LAMBERS, K. and SAUERBIER, M., 2003, in press. A data model for a GIS-based analysis of the Nasca Lines at Palpa (Peru). International Archives of the Photogrammetry, Remote Sensing and Spatial Information Sciences. XIXth CIPA International Symposium, Antalya, Turkey.

REINDEL, M., LAMBERS, K. and GRÜN, A., 2003, in press. Photogrammetrische Dokumentation und archäologische Analyse der vorspanischen Geoglyphen von Palpa, Südperu / Documentación fotogramétrica y análisis arqueológico de los geoglifos prehispánicos de Palpa, costa sur del Perú. Beiträge zur Allgemeinen und Vergleichenden Archäologie 23.

SAUERBIER, M. and LAMBERS, K., 2003. A 3D model of the Nasca lines at Palpa, Peru. International Archives of the Photogrammetry, Remote Sensing and Spatial Information Sciences, International Workshop on Visualization and Animation of Reality-based 3D Models, Tarasp, Switzerland, Vol. XXXIV-5/W 10, CD-ROM.

SHARMA, J., 2002. Oracle Spatial. Technical White Paper. Oracle Corporation, USA (http://technet.oracle.com/products/spatial/pdf/9iR2_spatial_twp.pdf).

WHEATLEY, D. and GILLINGS, M., 2002. Spatial technology and archaeology: the archaeological applications of GIS. London.

Some Problems in Archaeological Excavation 3D Modelling

Juan A. Barceló
Univerasitat autonoma de barcelona, Spain
jbarcelo@seneca.uab.es

Oriol Vicente
Univerasitat autonoma de barcelona, Spain
oriol.Vicente@uab.es

See the CD for the extended version

Abstract

Photogrammetry has been used in archaeology for the recording of complex structures. Therefore, the use of those techniques is more frequent in medieval and classic archaeologies. In those cases the presence of buildings and architectonics makes profitable going beyond the traditional paper and pencil approach. For prehistoric sites, traditional hand drawing is still the most profitable approach. Elements and structures to be drawn are geometrically quite simple, and the problems are more on the complexity of stratigraphic and sedimentary aspects, than in buildings, walls or floors. In this paper, we explain how to use digital photography in prehistoric excavations, how to modify those pictures to adequately represent the archaeological record, and how to build geometric models from photographs. Our final goal is to build a geometric and dynamic model of the site, in order to explain not its architectonic complexity, but taphonomy and the site formation process.

Archaeological Excavation and the Definition of Archaeological Space

We can define archaeological space as a sequence of finite states of a temporal trajectory, where an entity (ground surface) is modified successively, by accumulating things on it, by deforming a previous accumulation (for instance, by spreading) or by direct physical modification (building, excavation). Archaeological sites should be considered as the result of successive and overlapping modification steps (Barceló et al. 2003).

Natural and human process modify physical space, and as a result we are able to distinguish phase or modification steps, which can be used as analytical units. A phase is a homogenous region in space delimited by a well-defined discontinuity or boundary. A well defined boundary is an abrupt change in some spatial values. We may define an interfacial boundary or interface, when two phases are in mutual contact, that is, when two neighbouring regions in space have different probabilities for the same formation processes.

A wall, a pit, a garbage accumulation are phases or distinct regions of the archaeological space, which can be defined not only in terms of their own properties, but also in terms of the differences with neighbouring phases. We need to distinguish where observed discontinuities or boundaries begin and end, that is what are the proper borders of an occupation floor, or the original shape of a pit, where pottery sherds are accumulated, or where an animal carcass has been broken into bones. Therefore, archaeological excavation cannot be reduced to the mere unearthing of artefacts and ruins, but an exhaustive documentation of an archaeological space in terms of a finite set of spatial variables. The purpose is to characterize observed discontinuities in terms of distinct components or relevant units with uniform value of shape, size, texture, composition. However, a phase cannot be defined only in terms of their boundaries (Barceló et al. 2003). They should be analyzed as the presence/absence of some qualitative spatial variable, that is a feature which has positive value if it is present, and negative value in case of absence. Observed discontinuities between phases can also be expressed in terms of quantitative variables. Quantitative variables exhibit a variation in value throughout spatial regions. Variables such as geomechanical properties, mineral grades, material accumulations, soil morphological features, or any other property of sedimentary/depositional units and archaeological contexts, can be sampled or measured in terms of real, numerical values.

Archaeological Information Sources: Working with Variables and Coordinates

Archaeological information sources can be reduced to three basic data types: variables, characteristics and coordinates. They are properties of the subsurface that can be samples and measured in terms of real, numerical values. In contrast, characteristics are observable qualities of the archaeological space that have a finite number of possible descriptive values, and uniform value within finite, irregular volumes. Characteristic values are associated with discrete archaeological areas with distinct boundaries (a wall, an occupation floor, a pit, etc.).

The common feature of the archaeological information sources is that every variable or characteristic value is associated with a location and an extent (point, line, area, surface, volume) which in turn are defined by an implicit data geometry.

The usual way to define dependent variables in an archaeological model is by using observable characteristics. For instance, consider a texture classification or material identifier, with the following values:

Limestone, quartz, granite, pottery, charcoal, clay,…
or
Wall, occupation floor, brown sediment, red sediment, black sediment,…

or

Wall1, wall 2, wall 3, Floor 1, Floor 2, Floor 3, Pit 1, Pit 2,...

Although archaeology deals necessary with 'time', this is one of the less tractable independent variables. Even in the case of C14 dates, we do not have scalar values, but irregular intervals, which constitute mathematically complex structures. The easy way to solve this problem is by using ordinal values:

Period 1, Period 2, Period 3

Note that we are not equating stratigraphic depth with 'time'. In some cases, the temporal evolution of the site can be correlated with stratigraphic ordination, but not necessary.

Archaeological visualization is the graphical expression of relationships between observed and sampled values distributed in space. To adequately visualize these complex data structures we must consider a semi-infinite continuum made up of discrete, irregular, discontinuous volumes which in turn control the spatial variation of archaeological features. Therefore, an archaeological site should be described in terms of a volumetric information, that is, a group of data that describe a solid object from a three-dimensional space. Volumetric data occupies a volume of space. For example, when data is collected by excavation, there are data points spanning the height, width, and thickness of the archaeological element and a data value representing the type of material, sediment, structure, bone accumulation, etc. at each point.

Volumetric Data Acquisition

The usual way is to acquire separately coordinates, and after excavation, characteristic regions are built by joining characteristic points with lines, areas or surfaces. The archaeologist has an external database with a qualitative description of information related to each point, and then is able to create geometric shapes and structure by linking already existing points. By using polygons connecting points, by interpolating parametric surfaces or volumetric primitives. For instance, an occupation floor is acquired by calculating the 3D coordinates of some characteristic points along the contour, and then fitting a polygon to those coordinates.

The problem with this approach is that it is a modellization process, and not a real sampling procedure. Archaeologist is projecting what he/she things to know about the nature of a finite and very reduced series of coordinates, and focusing only on specific characteristics. Walls, and built structures can be easily built using standard geometric fitting tools, but then we are forgetting all information about interfacial boundaries. A wall is a characteristic of the archaeological space, but the sediment covering the wall, or the accumulation of stones around it, are also important characteristics which should be sampled.

We suggest to use a different approach to sample archaeological data. We need coordinates and characteristics to be acquired simultaneously, and we need a huge quantity of data points to approximate to scalar fields. The only way is through photographs.

A photograph is a spatial pattern of different luminance values. It is not a surrogate for reality, but a device for capturing some initial input (luminance perception) which should be translated into observed data. Given that spatial discontinuities are the building blocks of archaeological discontinuities, and they can be analysed in terms of texture variation, photographs can be used as a model of texture discontinuities.

Of course, we need more than a single photograph. 3D data sampling presupposes that a series of cross-sectional images, representing some volume which was regularly sampled at some constant interval, exists in digital form. An 'image stack' is a display of multiple spatially or temporally related images in a single window. The images that make up a stack are called slices. Stacks can be viewed from different perspectives, treating the layers of the stack as another spatial dimension.

This works only if certain conditions are satisfied. First, the individual files must be in the same format and bit depth. Second, they must have similar structure, i.e., same numeric type, number and sizes of dimensions, etc (Fig.1).

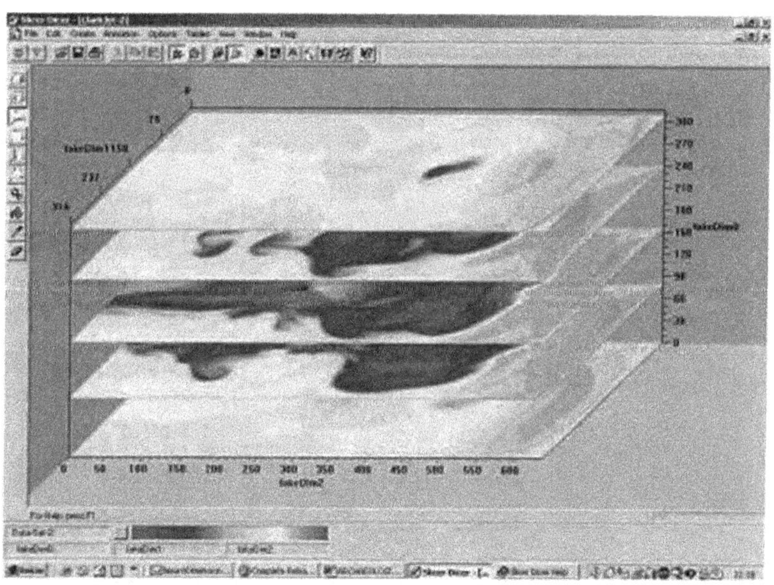

Figure 1 a stack of images sampled at different intervals. Simulated data using the SlicerDicer software from Pixotec,Inc.

Each image or 'slice' in a given dataset is made up of a number of picture elements or pixels. The distance between any two consecutive pixel centres in any slice within a dataset represents a real world distance referred to as the interpixel distance. Similarly, the distance between any two consecutive slices represents some constant real world depth with which the volume was sampled. This constant depth is referred to as the interpixel distance.

A series of cross-sectional digital images of this type is referred to as a 'volumetric dataset' or simply as a dataset. Such a data set is represented by a series of photographs, each containing a similar n-dimensional data array. Collectively, these files are interpreted as a single array of n+1 dimensions. For instance, we have 5 image files, each containing a 100 x 200 x 200 data array. By opening them all at once, a computer can read these data as if they come from a single 5 x 100 x 200 x 200 array, where 5 is the number of slices.

Processing a volumetric dataset begins by stacking the slices of a given dataset in computer memory according to the 'interpixel' and 'interslice' distances so that the data exists in a "virtual" coordinate space which accurately reflects the real world dimensions of the originally sampled volume. The next step is to create additional slices to be inserted between the dataset's actual slices so that the entire volume, as it exists in computer memory, is represented as 'one solid block of data'. The number of slices needed to fill in the blanks is based on the dataset's interpixel and interslice spacing and the slices needed are created through interpolation.

In this case, spatial values should be defined on regular, rectangular grids. Such data take the form of n-dimensional Cartesian arrays.

Once a dataset exists in computer memory as a solid block of data, the pixels in each slice take on an additional dimension. In effect, the pixels become volume pixels or 'voxels'. Once loaded into memory, a volume can be translated and rotated and a rendering of the dataset can be obtained (Fig.2).

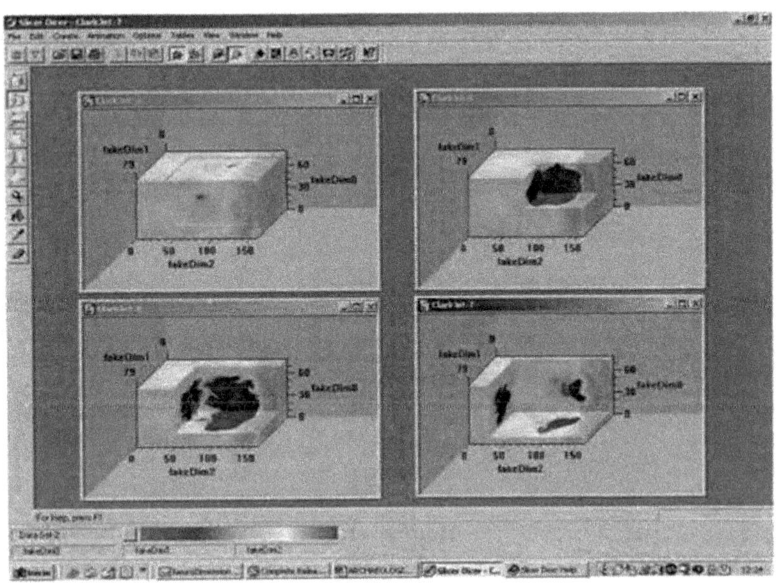

Figure 2 a rendered volumetric data set. Simulated data using the SlicerDicer software from Pixotec, Inc.

Technical Considerations before Building an Image Stack

If we need to overly photographs to 'slice' a 3D reality (the archaeological site), it is necessary that all slices be spatially related. Of course, sharing attributes on sample points only works if all scalar fields were sampled at the same locations. To make comparable all photographs in the stack, we have to geo-reference them. This step can be done with 'photogrammetrical' methods, where a image is modified introducing the real coordinates, and deforming it to be adjusted to a real scale.

Georeferenciation implies that discontinuities observed in the photographs should be positioned in space. The x ,y and z axes of the representation are aligned with the east, north and azimuth axes of the projection being used. Each pixel in the photo and each geometric primitive in the vectorial representation is geo-coded with its associated coordinate in the map projection. We introduce in the model a new variable: location. Some control points should be measured independently and then transferred to a database with a reference to the photo. Those control points will be used to 'register' the image and substitute pixels using specific geo-registration algorithms.

Also, the georeferenciation implies the rectification of perceptual errors, focal angle or focal distortion produced by the position of the archaeologist. In the case of a photograph a preliminary rectification transforms the pixels of the photo by interpolation of luminance wave length according to x and y axes.

However, even after geo-rectification photographs are always a misleading representation of reality. Scale variations are caused by the natural point-to-point variations in the elevation of the terrain being photographed. Scale variations are also caused by the varying distances of objects out from the principle point of the camera, as it is a perspective projection. Photographs are a 2D contrast map of luminance reflections on a 3D real surface. In other words, a 2D photograph is a deformed representation of reality. If it is deformed, a sequence of images within a stack would not be a right volumetric data set, because the spatial distribution of variables and archaeological characteristics sampled at one layer has nothing to do with data sampled at other layers.

Consequently, slices within an image stack cannot be considered a true input of volumetric data if we do not remove scale variations from any image. Once these variations in scale are removed from a photo, the photo becomes a true image map of the ground, where "map" is defined as a constant scale representation of a portion of the Earth's surface.

Limitations of this Approach

However, in most archaeological cases we cannot use this procedure.

There is a temptation to consider the process of building an image stack as a suggestion for excavating imposing artificial layers. Nevertheless, modern archaeology abhors the use of artificial layer slicing, because it makes impossible the correct reconstruction of stratigraphic sequence. We always follow observable discontinuities, that is, what we often call 'natural layers'. As a result, slices in an archaeological volumetric data set are not planes, but complex data arrays.

We should remember that archaeological characteristics and all variables describing the archaeological space are intrinsically four dimensional. Stratigraphic sequence is, in fact, a measure of temporal modification, which should be always taken into account.

Only when considering a stack of 4D data arrays we can represent properly archaeological of interfacial phases (contact surfaces between spatial discontinuities) at different time steps. For instance, Figure 3 is a 5D representation of an archaeological time step. Here colour (grey level) is used to represent different W values at different x,y,z coordinates. Let W' be the value at a position in the array defined by t = t', x=x', y = y', and z = z'. This datum will be rendered in the data view as a coloured pixel. The colour is defined by a data-to-colour mapping, or colour table, and the position of the pixel in the window is defined by a data-to-view coordinate mapping. If we had 3 files, each containing, for instance a 10 x 100 x 200 x 200 data array, we can integrate all data into a single volumetric set with 3 x 10 x 100 x 200 x 200 array, where 3 is the number of temporal steps (slices), 10 the number of values of the W characteristic, and 100 x 200 x 300 the dimensionality of the 3D grid where spatial values vary .

Figure 3 a 5Dimensional representation of an archaeological site

If we use z to represent height of ground level, then image rectification as explained previously does not work. We have here scale variations far greater than when considering only camera placement and focal angle. The only solution is to create a 3D model with textures, were textures can be used to build discontinuities, and then to locate characteristic values according to real topography. In this case, we can excavate following natural layers, and building 4D models (spatial coordinates plus textures) for each time step, identified also in terms of texture or qualitative discontinuity.

Conclusions

When visualizing archaeological spatial data, the exact values of the data are not as important as the relationship between values. Data visualization is used to gain insight into the data set, and expose relationships between values that might not be apparent in the raw data. As a result, intuitive, but less exact, representations of data values are often used.

Regularly gridded data are not very easy to take during field work (see an alternative method in Barceló et al. 2003). The problem is that spatial values defined on unstructured grids, which often take the form of tables with X-Y-Z-Value columns, cannot be directly visualized with common volumetric data visualizing software. It is often necessary to resample them onto a regular grid. One way of obtaining this rectangular grid is by creating images stacks.

However many real-world data sets are irregularly sampled. There may be a strong temptation to resample the data into a regular grid. This approach can be problematic in many cases, since the data may not change linearly between grid points. False relationships can be created in the resampled data.

This is the case in archaeological site modelling.

FINAL NOTE: A much more technical explanation of this procedure, and a suggestion for archaeological multidimensional field data format appears in the CD-ROM

REFERENCES

BARCELÓ, J.A., DE CASTRO, O., TRAVET, D. and VICENTE, O., 2003. A 3D Model of an Archaeological Excavation. In Doerr, M. and Sarris, A. (eds.), The Digital Heritage of Archaeology. Proceedings of the 30th Computer Applications and Quantitative Methods in Archaeology Conference, Heraklion, Crete. Hellenic Ministry of Culture, Archive of Monuments and Publications.

THE ORTHOGRAPHIC APPROXIMATION - A SIMPLE GEOMETRICAL MODEL FOR AVOIDING PERSPECTIVE ERROR IN CONSTRUCTING PHOTOMOSAICS

GEOFF AVERN
UNITED KINGDOM

ABSTRACT

With appropriate apparatus, ortho-photomosaics created on the desktop computer from near-range vertical photographs can be used to plan excavations more quickly than by conventional drawing techniques. However, a good fit of adjacent photographs is impossible when there is excessive perspective error, and the software "stretches" we might make to correct gross problems of alignment of features will often create error elsewhere in the image. A simple mathematical model, the Orthographic Approximation, provides a simple means of calculating the metric parameters which will give a result which is tolerably close to being orthographic, making the creation of an ortho-photomosaic a simple task.

INTRODUCTION

Since the advent of photography and the ready availability of film cameras, there have been many archaeologists who have sought to speed up the recording process by replacing plan drawing with ortho-photomosaics made from near-range vertical photographs taken from atop tripods or similar apparatus (see Sterud and Pratt 1975, for a useful history of these apparatus and their use, also Poulter and Kerslake 1997, Gruel, Buchsenschutz, Alliot and Murgalé 1993). Yet despite this, the technique remains infrequently used. The author originally presumed that this was due, in part at least, to the time and labour spent in developing and printing the photographs, and piecing together the mosaic, and so set out to test the notion that the process could be made much faster and simpler using today's affordable graphics software and digital cameras. It soon became apparent that there was another important issue.

The primary focus of the fore mentioned papers is usually a detailed description of the construction and use of some apparatus for holding the camera on high, the purpose of which is obviously to ameliorate the problem of perspective error, yet these authors pay little or no attention to ascertaining whether excessive perspective distortion is still present in their photographic results or not. Certainly, it was this author's experience that, despite photographing from elevations of 5m or 7m, considerable perspective error was still present in the photographs, making the task of creating a photomosaic a far from straight-forward task. This turned our attention from camera and software to an investigation of perspective error through the geometry of near-range photography. The results of this investigation provide a model for quantifying, or rather, approximating perspective error. This allows us to then define a region within the photograph where perspective error is acceptably small or, to put it another way, a region which is tolerably close to being orthographic. This model we call the Orthographic Approximation.

DEFINING THE QUESTION

In attempting to construct an ortho-photomosaic from near-range vertical photographs we are attempting to place a series of perspective views into a formal grid or loose framework to produce a mosaic image which is close to being an orthographic projection. In doing so the problem of perspective distortion within individual photographs must be addressed; firstly, from a purely functional stance, because of the probable difficulty of aligning features which overlap photographs or mosaic pieces, and secondly, because some or more of the remaining information in the photograph or mosaic piece may be similarly, if less conspicuously, affected by perspective distortion. Typically, attempts to minimise perspective distortion entail both photographing from a height and using only the central portion of the resulting photographs.

Thus far we have talked of the "problem", i.e. perspective distortion, and have perhaps implied moving from a bad situation towards something better. Let us turn this around and start from a "point of perfection", that is the only truly orthographic point in a photograph which lies at the point where the photographic axis intersects the terrain. A second point, very close to the first, might not technically be orthographic but it is almost certainly close enough for our requirements. The further we move away from the photographic axis, the more perspective distortion is introduced and the further we can consider to be from an "orthographic situation". The question of where to draw the line between useable and non-usable parts of a photograph now becomes "How far from the photographic axis can we move before the perspective error becomes too large?" To answer this we will seek a means of quantifying, or at least approximating, perspective error and go on to investigate its relationship to topographic relief, the height of the camera and the size of that part of the photographic field which yields a good orthographic approximation.

THE BASIC GEOMETRY

Figure 1 shows the situation of a vertical photograph of a horizontal surface containing two substantial topographical irregularities. Note three values; I which is half the photographic angle, d which is the distance between the camera lens and the centre of the field, and *f* which is half the photographic field width.

We can assign some further values in the regions of the topographical irregularities. We let the vertical heights of the irregularities be h and note that the angle I is formed between the photographic projection and a vertical line through point x_0. In the resulting photograph, the point x_0 (the corner of the irregularity) will appear to lie at x_1, introducing an apparent lateral translocation of the true x and y coordinates, an error due to perspective distortion which we label e.

In figure 1 we can identify three similar right-angled triangles, sharing the angle I. From these triangles we can form the equation;

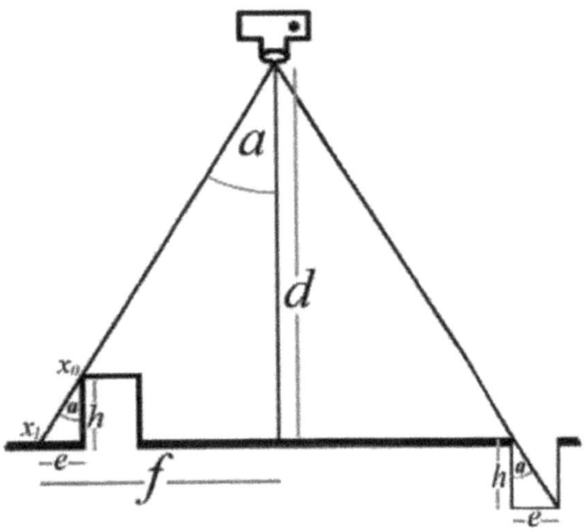

Figure 1 Basic geometry of vertical photography

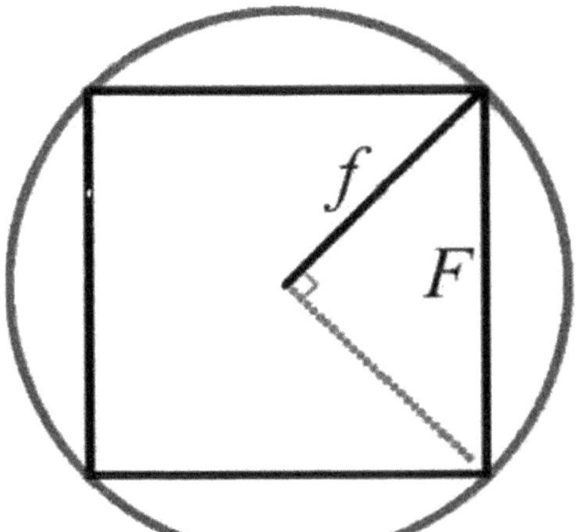

Figure 2 Calculation of F, the length of the sides of the largest square which will fit in a circle of radius f

Equation 1: $f/d = e/h$

This very simple equation relates the vertical topography of the terrain, the photographic geometry and the lateral perspective error, which is very useful for us. For a given (measured) amount of vertical relief in the terrain we want to record, and knowing the maximum height of the photographic apparatus, we can choose a value for e which represents the maximum perspective error that we are prepared to accept and the equation will give us f, which is the radius of a circle centred on the photographic axis, within which perspective error will be less than or equal to our chosen limit.[1]

Modifying the Equation for Practical Use

With a little modification, Equation 1 becomes somewhat more practical. When planning the photography of the site or working with photographs on computer, most people will find it easier to work with square portions of a photograph rather than circles. To this end, we calculate the size of the largest square which is encompassed by this circle Fig.2).

From this diagram, the right-angled isosceles triangle allows us to define

$F = \sqrt{(2f)}$, or in another form,

$f = \sqrt{(F^2/2)}$, which can be substituted into Equation 1.[2]

Equation 2: $$F = \sqrt{2\left(\frac{2e.d}{H}\right)^2}$$

The term F gives us a square that we use both on the terrain and in the photograph. On the terrain it is both the area covered by any individual photograph, and the distance between photographic centres, allowing a grid to be set out (or at least a plan for the photography) and the calculation of the number of photographs required for the job. In the photograph, F (scaled down to that of the photograph) defines the region of the photograph to be used, it allows us to create a software mask of that size for easy cutting, and it allows us to create a gridded blank sheet (grid size=F) for easy pasting of pieces to create the mosaic.

Calculations in the Field

Any three of the four variables in Equation 2 can be assigned values which will yield a value for the fourth, allowing us to answer the following questions;

- "Is there too much topographical relief in this terrain for the technique?"
- "How high must I postion the camera?"
- "How many photographs will I need to take?"

And, if any of the above questions do not yield satisfactory answers, it is possible to plug in practical or achievable values for each to learn just how large an error could result and whether you may judge it still small enough to warrant proceeding with the exercise.

Easy Reference Tables

Clearly, knowing the maximum height of the available apparatus on which we mount our camera, and having a consistent notion of what is an acceptable error, it is possible to use Equation 2 to draw up simple reference tables for use in the field.

The following is a table for use of an apparatus which holds the camera at height 5m, and chosen acceptable error of +/-1cm, to calculate the useable field within our photographs.

Topo. Relief (*H*) in cm.	5	10	20	30	40
Useable Field (*F*) in m.	2.83	1.41	0.71	0.47	0.35

A similarly useful table might be for using an apparatus giving maximum $d = 5$m and a photographic field width $F = 0.5$m to determine how large our error will be.

Topo. Relief (H) in cm.	5	10	20	30	40
Error (+/-e) in cm.	0.18	0.35	0.71	1.06	1.41

Case Study: Le Cheslé, Belgium.

Le Cheslé is a VIth-Vth century BCE (late Hallstatt/early La Tène) fortification of the éperon-barre type, on the Ourthe River, near the village of Bérisménil in the Provence of Luxembourg, southern Belgium (Fig.3). Professor P. Bonenfant (Service des Fouilles, Université Libre de Bruxelles) has been excavating the main barrage wall of the fort (Fig.3) since 1994.

Tranchée A was photographed in June 2000 using an Olympus Camedia C-2000Z digital camera (one of the few digital cameras to have an infra-red remote shutter operator) mounted at a height of 5 m on the Statif. The Statif (Fig.4) is an apparatus conceived by Yves Glotz of Mons (Belgium), designed by engineers at the Université Libre de Bruxelles and fabricated at the Polytechnic at Mons in 1984 to facilitate near-range vertical photography in linear sequences. The base of the Statif is a large, horizontal aluminium beam (5 m long, 12 cm square section) standing on 2 pairs of adjustable-length legs. A carriage, which slides along the length of the beam, has an inverted "L"-shaped aluminium arm (3.5 m high) clamped to it, on the end of which is a camera mount. The camera typically sits at a height of 5m to 7m and linear series of photographs are made by sliding the carriage across the beam. Parallel series of photographs may be acquired without necessarily needing to move the entire apparatus by altering the length of the horizontal section of the inverted "L"-shaped arm, and/or by rotating the arm 180° to photograph on the other side of the beam.

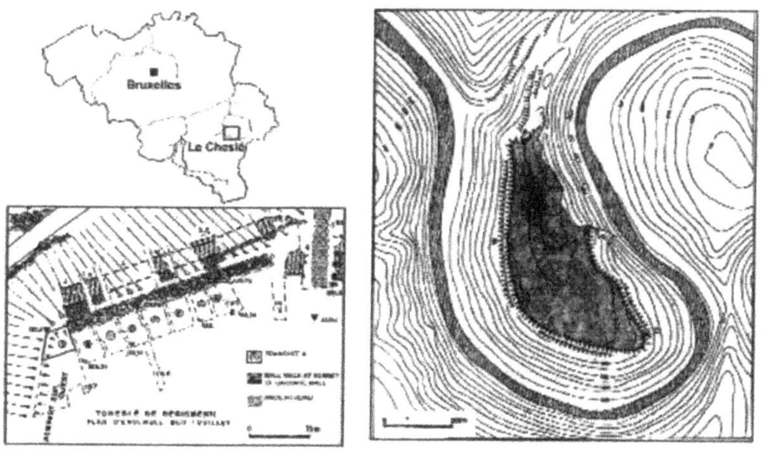

Figure 3 Locality of Le Cheslé. Lower left: Plan of excavations of the barrage wall (modified: after Bonenfant, 1995). Right: Topographic plan of Le Cheslé showing barrage wall, A (from Papeleux, 1988)

Figure 4 The Statif

Results

Photographs of Tranchée A taken from the Statif were cropped to a size calculated from Equation 2 and compiled into an ortho-photomosaic (Fig.5).

Having drawn Tranchée A by hand with a fellow student in the preceding year, I could compare the time taken to produce the final drawings against that of producing the photomosaic;

	Duration	*Labour (person days)*
1999 - Drawing by hand	*3.5 weeks*	*5 person-days*
2000 - Ortho-photo with Statif	*1 week*	*8 person-days*

Summary

By examining the geometry of vertical photography a simple equation was derived which allows us to make an approximate calculation of perspective error. The same equation can be expressed in a form which allows us to delimit a region around the photographic axis which is tolerably close to being orthographic. This "Orthographic Approximation" allows us to be predictive of "goodness of fit" at the initial photographic stage rather than reactive at the stage of constructing the photomosaic, with efficiencies accruing at both stages as demonstrated in the case study at Le Cheslé, Belgium.

Closing Note: While the geometry described above is obviously elementary in the fields of optics and photography, the author has not been able to find any reference in the archaeological literature to any similar considerations in the application of photography to field archaeology. Any claim of novel and original work that we might make in this respect we must, therefore, limit to its application in archaeology and to its dissemination amongst the archaeological community.

[1] In reality, the geometry of the situation described above is more complex, however, the influence of these factors is considered to be sufficiently small to be ignored in the derivation of the approximation of the situation.

[2] We also introduce the term H. In Equation 1, topographical relief, h, can be in either the positive or negative direction from the "average field height". Here, we let H = 2h where H is the total topographical relief, being the difference in measured height between the highest and lowest points of the terrain, not necessarily within just a single photograph but possibly across the entire area to be photographed.

Figure 5 Ortho-photomosaic of Tranchée A, Le Cheslé

REFERENCES

BONENFANT, P., 1995. Le Cheslé de Bérisménil. De La Meuse á l'Ardenne 20:122-124.

GRUEL, K., BUCHSENSCHUTZ, O., ALLIOT, J.-F. and MURGALÉ, H., 1993. Arkéoplan: a new tool for the archaeologist. In Andresen, J., Madsen, T. and Scollar, E. (eds.), Proceedings of the 20th Conference for Computer Applications in Archaeology at Aarhus University, Aarhus, Denmark, March 1992, Aarhus University Press:81-84.

PAPELEUX, J., 1988. Le Cheslé de Bérismenil (commune de La Roche) - Forteresses Celtiques en Wallonie. Archaeologicum Belgii Speculum XIV:15-30.

POULTER, A.G. and KERSLAKE, I., 1997. Vertical Photographic Site Recording: The "Holmes Boom". J. Field Arch. 24(2):221-232.

STERUD, E.L. and PRATT, P.P., 1975. Archaeological Intra-Site Recording with Photography. J. Field Arch. 2(1/2):151-167.

Photogrammetry and Archaeology: A Case Study in the Archaeological Site of Philippoi in N. Greece

SEE THE CD FOR THE EXTENDED VERSION

Olga Georgoula

Dimitrios Kaimaris

Georgios Karadedos

Petros Patias

The Aristotle University of Thessaloniki, Greece

Abstract

The present study belongs in the field of Geoinformation's applications, that are related with the documentation, management, protection and promotion of Archaeological sites. The value of these applications in National and Worldwide level is obvious, because, the essential tools for the continuous follow-up and management of Cultural Heritage are offered.

The monitoring of the remains of past human activities, especially when they are of complex form and cover large area, is more effective when it takes place from the air. The use of the aerial photography for surveying and documentation of archaeological sites is well known. The application of digital methods of processing, often of a large number of aerial photographs, gives the necessary outputs for the documentation of the historical sites. Of special interest is the case, where the aerial photography is used for the location of buried archaeological remains, which under the proper conditions, are best recognizable viewed from the air. On the other hand, the terrestrial photos are more appropriate for the documentation of man made constructions. The fusion of divers (aerial and terrestrial) data is a very interesting process because it offers an integrated product.

This project has as object the digital photogrammetric process of aerial and terrestrial photos as well and the production of suitable products aiming at:

- The documentation of an archaeological site
- The localisation of buried archaeological remains
- The application of virtual reality techniques aiming at the study and the analysis of archaeological landscape and of man-made constructions as well.

Region of study is the Archaeological site of Philippoi Kavala's in N. Greece.

Introduction

The present study belongs in the field of Geoinformation's applications that are related with the documentation, management, protection and promotion of archaeological sites. The value of these applications in national and worldwide level is obvious. Digital photogrammetry, one of the basic tools in this field of applications, is the method, which in comparison with other techniques like topographical survey, 3D scanning etc, offers the most advantages for the identification, understanding, interpretation and presentation of an archaeological site or a manmade construction. The use of the aerial photogrammetry for the monitoring of the remains of past human activities is well known (Doneus 2001, Patias, Karapostolou and Simeonidis 2002, Tozz and Harari 1984). The application of digital methods of processing, usually of a large number of aerial photographs, gives all the necessary outputs for this purpose, as maps, digital terrain models (DTM), orthophotomaps, 3D models etc. Especially, the orthophotomap, which combines the quantitative information of a map and the thematic information of a photograph, plays a significant role in archaeological research. On the other hand, the terrestrial photos are more appropriate for the documentation of manmade constructions. The fusion of different (aerial and terrestrial) data is a very interesting process because it offers a totally integrated product (Patias, Karapostolou and Simeonidis 2002).

Figure 1 Orthophotomap (zoom out) and the places of the traces that has been interpreted

Figure 2 Extract of the orthophotomap 1:5000

Figure 3 Traces of an unknown buried construction

Figure 4 Long trace in dark colour. Covered old stream?

This project has as object the digital photogrammetric process of aerial and terrestrial photos as well, aiming at 1) the compilation of the appropriate cartographic data for a future archaeological GIS for an archaeological site and its major area, 2) the location of buried archaeological remains and 3) the application of virtual reality techniques aiming at the study and the analysis of archaeological landscape and of manmade constructions as well. The paper refers the first result of a project that our group, in the Aristotle University of Thessaloniki, is working on. Our group is interdisciplinary and is consisted of Photogrammetrists and an Architect-Archaeologist. The area of study is the archaeological site of Philippoi Kavala's in N. Greece, where excavations are taken place for the past 90 years. During the excavations, characteristic monuments of the Hellenistic, Roman and Early Christian period were discovered, like the theatre of Philippoi, a small part of Via Egnatia and the city wall. The project is divided in two parts. The first concerns the archaeological site of Philippoi and the major area of it, where is the place of the battle of Philippoi at 42 BC and the second one is about the theatre, inside the city wall.

Figure 5 Straight trace of 6m width in light colour. An old road?

THE ARCHAEOLOGICAL SITE AND THE MAJOR AREA OF PHILIPPOI

The data that has been used was of different form and they came from different sources, as photogrammetric data, cartographic data and historical and archaeological data as well.

Photogrammetric data: 4 aerial photos in B/W, scanned at 1200dpi, scale 1:15,000, September 1996.

Cartographic data: 11 cadastre maps 1:5,000, dated from 1926 to 1960, from where information about the changes of the landscape and the fields has been drew off.

Historical sources: very important information has been drew off from the historian Appianos (IV:105-115), who describes in detail the landscape of the major area of city of Philippoi and he gives very interesting information about the battle of Philippoi.

Finally, information from the archaeological excavations has been used. All available information is processed and cross-evaluated. The photogrammetric process in the Digital Phtogrammetric Workstation included: a bundle adjustment (7 control points), production of the Digital Terrain Model and production of an orthophotomap at 1:5,000 scale (Figs.1, 2) and the compilation of a vector mp at the same scale (Kaimaris 2002).

During the photogrammetric process, the current images were examinated and a lot of traces have been found and they were interpreted as well. In figures 3, 4, 5 and 6, few examples are shown. In this point it must be mentioned the location of a very interesting trace (Fig.3). On the top of a small hill, the evident traces of a buried construction 53 x 47 m, unknown till now even for the archaeologists, are shown. In the next step the traces have been relocated on the orthophotomap (Fig.1). In this way it was easy to have a total view of the whole area with the traces and consequently to correlate various traces and to extract additional conclusions empowering our initial estimations. For example, the spatial arrange-

Figure 6 Straight trace of 7.8m width. A construction?

Figure 7 Orthophotomap and cadastre maps of 1928, 1932 and 1940

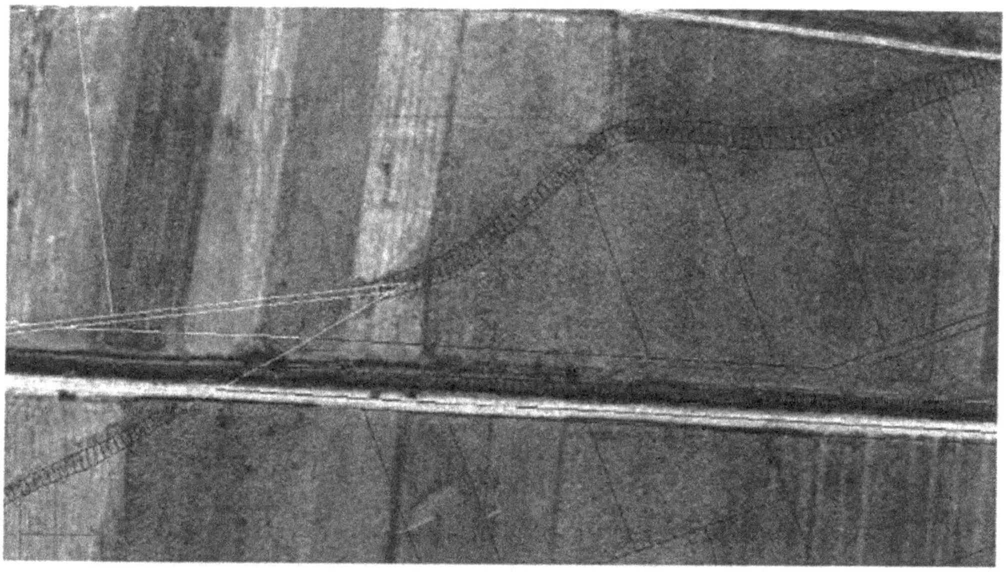

Figure 8 The branch of Gagitis river as is presented on the cadastre maps

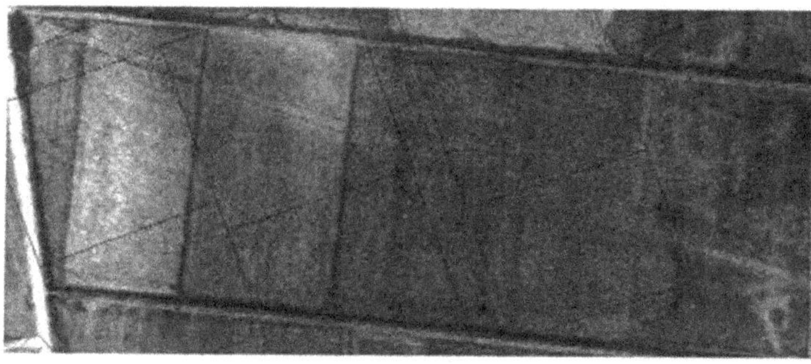

Figure 9 The cadastre map of 1940

Figure 10 The cadastre maps of 1940 and 1960

ment of 4 separated traces in connection with the fact that these traces had common characteristics as shape, size and tone led us to suppose that they are parts of a long straight element, a "construction" of 6m width, maybe a road. After that, the attention was focused on the research for the estimation of the chronological period that these traces appeared above the ground. For this purpose old cadastre maps (1928, 1938, 1940, 1957, 1960) has been processed and used. The cartographic process involved: scanning, transformations-georeferencing to the same cartographic system as the orthophotomap, merging of all cadastre maps in one map file and finally superimposition of the merged file on the othophotomap. On these maps the old geomorphology (old streams, old ownership boundaries etc), which does not exist, is shown (Fig.7). In some cases the initial estimation became true, like in the case of the branch of Gagitis river (Fig.8), which is shown in the cadastre maps of 1932 and 1940. Today is covered, leaving only its traces in the photos (Fig.4). In other cases, the cadastre maps did not help the research at all, because nothing was shown on them, leading to the conclusion that the traces belong to "something" before the time of maps (Figs.9, 10). In these cases significant role was given to historical sources (Appianos IV:105-115) and archaeological excavations.

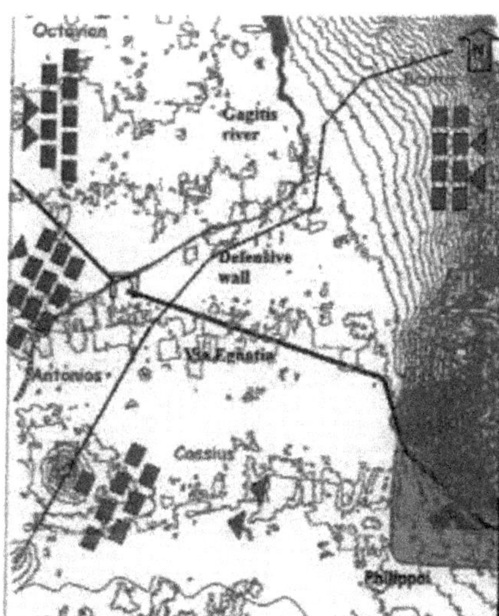

Figure 11 Via Egnatia and defence wall in the plain of Philippoi

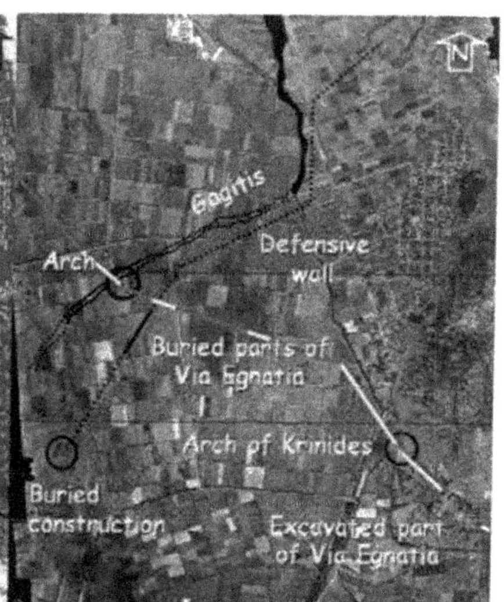

Figure 12 a historical map of the possible landscape in the time of battle of Philippoi (42BC)

In his manner it was possible to conclude that some of the traces belong to buried parts of Via Egnatia (Fig.5, 9, 11), and another trace to a part of a fortification wall (Fig.6, 10, 11), constructed by the democrats before the battle of Philippoi (42BC). After that a reproduction of a historical map of the possible landscape of that time and of the battle of Philippoi (Appianos IV:105-115) has been done (Fig.12). The combination of the photogrammetric products as the DTM and the orthophotomap has been used for the creation of an image-draped model. Using the appropriate software and process a fly-through video sequence around the model of the archaeological site has been created as an AVI file format, which anyone can use to fly virtually around (Fig.13). Additionally, MicrostationSE provides the ability to export the textured model of a CAD model in a VRML format file (Bakourou et al. 2002).

The Theatre

In the theatre of Philippoi the last years extended anastylosis-restoration works are in progress. These works are based on the result of the study of the historical and constructional phases of the theatre and the eastern city wall (Fig.14a, 14b), which is based to the archaeological excavations which are

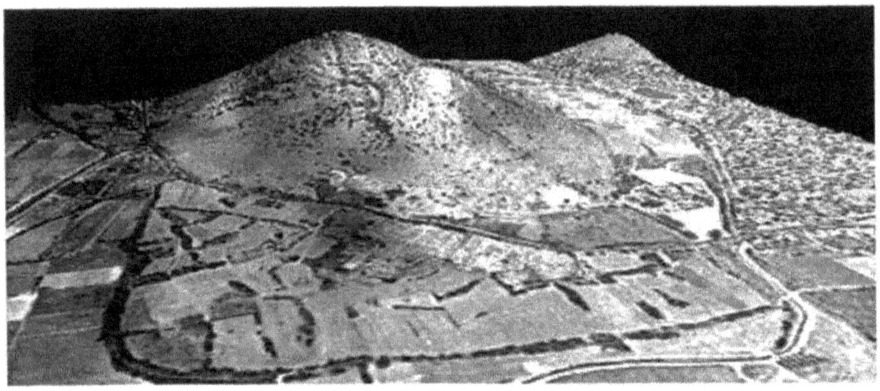

Figure 13 a view of the image-draped model of the archaeological site of Philippoi

continuing. The aim of our project is: 1) to surveying with the aid of photogrammetry the current status of the theatre, 2) to produce accurate 3D models of the different historical and constructional phases of it and 3) to proceed with visualization and virtual reality techniques that provide a better inspection of the theatre in study. Till now, the photogrammetric process of the terrestrial photos of the theatre has been finished and the construction of the 3D models is in progress (Fig.15).

Conclusions

Photogrammetry can give more than simple maps. It gives suitable products aiming at the documentation of an archaeological site, the location of buried remains of past human activities, the application of virtual reality techniques aiming at the study and analysis of archaeological landscape and of manmade constructions as well etc. These first encouraging results enforce us to continue the research in this area. In the near future we will process satellite images B/W and multiband of very high geometrical resolution for an extended area of Philippoi.

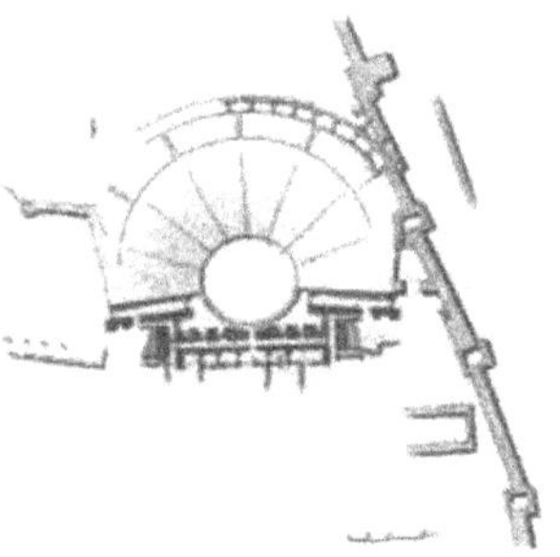

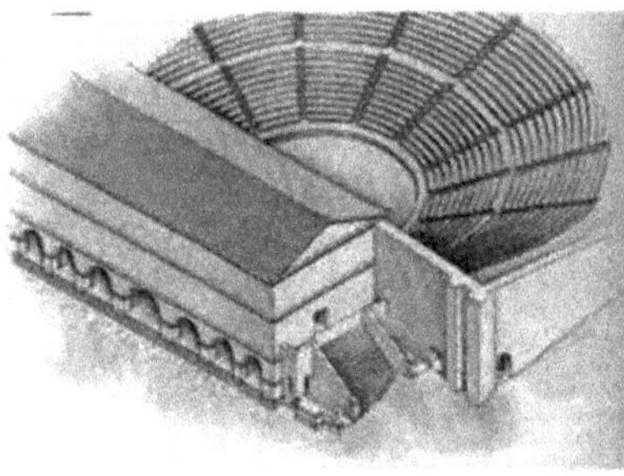

Figure 14a (left) Historical and constructional phases of the theatre
Figure 14b (right)

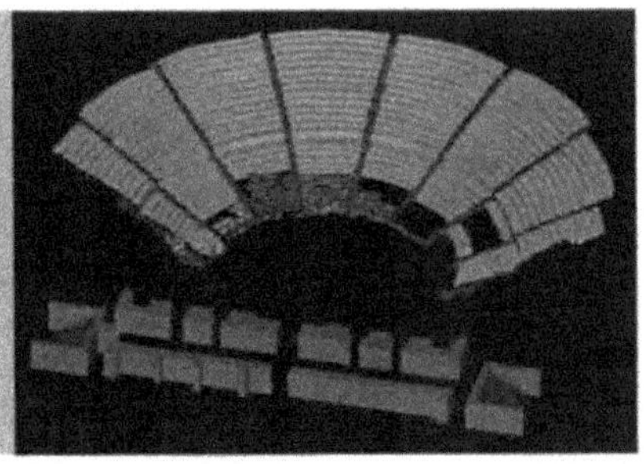

Figure 15 3D model of the theatre

REFERENCES

APPIANOS. IV:105-115.

BAKOUROU, E., TSIOUKAS, V., KATZOURAKI, I., STYLIANIDIS, E., PAPADIMITRIOU, K. and PATIAS, P., 2002. The promotion of Cultural Heritage through Internet using advanced audio-visual information: the Venetian castles of Peloponissos. Proceedings of the ISPRS Commission V Symposium, Close-Range Imaging, Long-Range Vision, Corfu, Greece, 2-6 September 2002, Vol. XXXIV, P5, Com. V:298-301.

DONEUS, M., 2001. The impact of vertical photographs on analysis of archaeological landscapes, Archaeological Prospection in Austria. Proceedings of the 4th International Conference on Archaeological Prospection, Vienna, 19-23 September 2001:94-96.

KAIMARIS, D., 2002. Photogrammetry-Photointerpretation: tools for the Archaeological research. Postgraduate diploma thesis, The Aristotle University of Thessaloniki (in Greek).

PATIAS, P., KARAPOSTOLOU, G. and SIMEONIDIS, P., 2002. Documentation and Visualization of Historical City centers: A multi-sensor approach for a new technological paradigm. Proceedings of the ISPRS Commission V Symposium, Close-Range Imaging, Long-Range Vision, Corfu, Greece, 2-6 September 2002, Vol.. XXXIV, P5, Com. V:393-399.

TOZZI, P. and HARARI, M., 1984. Eraclea Veneta, Imagine di una citta sepolta. Editore Compania Generale Aerea, Italy.

Statistic and Quantitative Methods

Establishing Optimal Core Sampling Strategies: Theory, Simulation and Practical Implications

Philip Verhagen
RAAP Archeologisch Adviesbureau BV,
Zeeburgerdijk 54, 1094 AE Amsterdam, The Netherlands

Adrie Tol
RAAP Archeologisch Adviesbureau BV,
Zeeburgerdijk 54, 1094 AE Amsterdam, The Netherlands

Abstract

Archaeological core sampling is an important surveying tool in the Netherlands. It is used widely to determine the archaeological content and value of the soil record. With the implementation of the Valletta Treaty the demand for this type of survey will even become higher, as it is a relatively cheap tool for site detection. Unfortunately, there is little documentation on the effectiveness of existing core sampling strategies for detecting and identifying specific site types. The paper will focus on the possibility of establishing optimal core sampling strategies for different site types, in particular through the use of simulations to predict the expected cost and benefit of each individual strategy. The simulations will be compared with the actual results of excavations of different site types in order to define the success rate of core sampling when it comes to determining content, size and value of an archaeological find spot.

Introduction

Archaeological core sampling is an important surveying tool in the Netherlands. It is used widely to determine the archaeological content and value of the soil record. Unfortunately, there is little documentation on the effectiveness of existing core sampling strategies for detecting and identifying specific site types.

The Senter agency of the Dutch Ministry of Economic Affairs is co-ordinating a programme of subsidized research projects aiming at promoting the use and development of technological innovations in public archaeology. Within this programme, the development of new, non-destructive ways of prospection is an important issue. In late 2001, RAAP Archeologisch Adviesbureau BV was asked to carry out a study within this programme on the effectivity of core sampling as a prospection technique. The project, that is currently near completion, has tried to gather information on this aspect and provide an assessment of the strengths and weaknesses of core sampling as a prospection technique. This paper focuses on the possibility of establishing optimal core sampling strategies for different site types, in particular through the use of simulation to predict the expected costs and benefits of each individual strategy, using the example of the excavation of Zutphen-Ooijershoek (province of Gelderland, The Netherlands).

Core Sampling: The Basics

Core sampling is not often used outside the Netherlands for archaeological prospection, although it is widely known as a geological survey technique. In areas where a strong accumulation of fluvial or marine sediments is found, core sampling is the only technique available that will provide a quick and cheap assessment of the local stratigraphy. Core sampling is still largely performed by means of manual labour, even though mechanical alternatives are currently being developed. Two basic types of equipment can be used. The auger has a diameter of 7 cm (sometimes 15 cm). It is screwed into the ground and takes small cores per sample (about 15 cm long). It is best suited for dry and sandy soils, and is not frequently used at depths below 2 meters. The gouge has a standard diameter of 3 or 5 cm and is driven with force into humid clayey soils or peat. The core obtained is 1 meter long. This type of core sampling can reach depths of 7 meters or even more, by extending the gouge with metal rods.

Given the fact that the Netherlands are covered by large areas of Holocene fluvial and marine sediments, it is not surprising that core sampling is also considered an appropriate tool for archaeological prospection. In many areas, there is no other way to obtain sufficient information on the (possible) presence of archaeological remains. In fact, its use has resulted in the discovery of some very important archaeological sites, like those found in the alignment of the Betuweroute railway, that runs straight through the river basin of Rhine and Meuse (Asmussen and Exaltus 1993, Asmussen 1994).

Statistical Background

The probability of discovering an archaeological site by means of any method of 'small unit sampling' (other possible methods are test pit sampling and machine trenching) is given by the following equation:

$$P = I \times D$$

where
P = discovery probability;
I = intersection probability; and
D = detection probability.

The intersection probability describes the relationship between the size of the object to be found and the distance between the sampling points. It can be determined using the following equation (Drew 1979):

$$I = A / (i \times s)$$

where
A = the area of the object;
i = the distance between the sampling points in a row; and
s = the distance between the rows.

This equation does not take into account the form and position of the objects. Krakker et al. (1983) have demonstrated that the optimal layout for a sampling grid is an equilateral triangular grid. In this case, the distance between rows s equals ½ i 3. For a standard core sampling survey, with sampling points every 50 meters, this equates to a distance between rows of 43.3 meters. The maximum diameter of a circular object that can be missed by such a grid layout is equal to s + (i2 / 4s), or 57.73 meters in the case of a standard grid (Kintigh 1988).

For elongated (elliptical) objects, the mean intersection probability is the same as for circular objects, but the probability distribution is different, and they may therefore slip through the net more easily (Gilbert 1987). However, when looking for elliptical objects, it is not necessarily useful to change the layout of the grid. Drew (1979) stated on theoretical grounds that using a rhomboid instead of an equilateral triangular grid is only effective when the orientation of the objects is more or less known. However, simulations carried out by ourselves show that there is a small positive effect of finding extremely elongated objects by using a rhomboid grid, even when the orientations are not known.

The detection probability for archaeological artefacts is given by the following equation (Stone 1981, Krakker et al. 1983):

$$D = 1 - e^{-A \times d \times W}$$

where
e = the base of natural logarithms (2.711828);
A = the area of the sampling unit;
d = the density of artefacts per area unit; and
W = the observation probability.

This equation describes a Poisson-distribution, that is appropriate for rare objects that are not very likely to be encountered in a sample. Artefact density determines whether a site may be detected or not, but the observation technique chosen determines whether an artefact will actually be observed. Very little data is available on the effects of sieving versus visual inspection, or of choosing a different sieving mesh. Groenewoudt (1994) showed that about 75 % of the flints found at the site of the Ittersumerbroek excavation were smaller than 4 mm, so choosing a smaller sieving mesh may drastically increase the amount of observed artefacts.

Very little data is available on the actual artefact densities encountered on archaeological sites in the Netherlands. Mean artefact density estimates are given by some authors. Groenewoudt (1994) for example estimated the mean artefact densities for Iron Age en Roman settlements at more than 120 per m2, an estimate obtained by extrapolating data from core samples. It should be noted that the actual detection probability of such a density is not very high when using a standard 7 cm auger (about 37%). For a selection of 79 Stone Age sites from NW Europe (kindly put at our disposal by dr. Willem-Jan Hogestijn) the mean artefact density is 140.4 per m2, but 70.9% of these sites have densities below 50 per m2. In the recent excavation of the Mesolithic site of the Hoge Vaart by Hogestijn and Peeters (2001), mean flint densities of only 18 and 16 per m2 were registered when sieving with a 2 mm mesh. Groenewoudt (2002) also mentions an example of a site with a mean density of only 6.4 artefacts per m2 (sieved with a 4 mm mesh); the site actually contained two house plans.

The observation method used is obviously very important in this respect. Core sampling is based on very small sampling units, the samples are usually thoroughly described, and the soil is sieved with a 1 mm mesh to obtain as many artefacts and other archaeological indicators as possible. Archaeological features are not usually recognized in core samples. During excavations, or even in machine trenching surveys, the features are of primary concern, and artefacts are usually only collected and described if they have diagnostic value. Given the already enormous amounts of artefacts collected in this way (e.g. almost 40,000 in the Malburg excavation; Oudhof et al. 2000), it is very understandable that a full count of all artefacts present per feature or quadrat is not performed. However, this implies that it is impossible to obtain reliable data on the spatial distribution of artefact densities.

Only a few examples could be found of sites that had been consistently sieved for artefacts in quadrats, and all of these concerned small excavated areas with relatively low artefact densities. Simulations performed on these data showed that these sites will be very difficult to discover by means of standard core sampling survey.

Establishing an Optimal Core Sampling Strategy: The Case of Zutphen-Ooijershoek

The Mesolithic site of Zutphen-Ooijershoek[1], for example, was sieved with a 3 mm mesh in 50 by 50 cm quadrats. The resulting flint counts ranged from 0 to 179, resulting in a mean artefact density of 66 per m2, on a total excavated area of 246.75 m2. A strong clustering of the flints was evident; in about two-thirds of the excavated area, the artefact density was below average. For purposes of comparison, the centre of the site was analysed separately from the periphery (see Table 1). The probability of finding the site using standard core sampling strategies was approached by simulating 1,000 hypothetical surveys of the site, using different parameters for grid size and sample diameter. In this way, the costs and benefits of each strategy can be compared. The probabilities given in table 1 should however not be seen as real probabilities of finding the site, as the effect of the observation method chosen has not been incorporated in the simulation runs.

It turns out that for this particular site, increasing the sample volume is a more cost-effective strategy than increasing the density of the sampling grid. However, it should be taken into account that taking a larger sample volume is a course of action that can only be applied once, as augers with a larger diameter than 15 cm are not available.

Conclusions

The results of the simulations, as well as theoretical considerations, point to the conclusion that core sampling is not a

ZUTPHEN-OOIJERSHOEK	centre	periphery	total	cost factor
mean artefact density per m^2	165.84	21.04	66.08	
area in m^2	76.75	170.00	246.75	
discovery probability 7 cm auger				
40 x 50 m	1.6%	0.1%	3.1%	1
20 x 25 m	6.2%	2.8%	7.7%	4
10 x 12.5 m	22.4%	9.2%	33.6%	16
6 x 6.25 m	64.8%	28.5%	73.4%	64
discovery probability 15 cm auger				
40 x 50 m	3.6%	2.3%	5.3%	2
20 x 25 m	11.1%	9.4%	19.1%	8
10 x 12.5 m	43.5%	34.7%	63.8%	32

Table 1 Comparison of the costs of different core sampling strategies for Zutphen-Ooijershoek, based on simulation results. The centre of the site is the area where artefact density is above average. An increase in grid density means a four-fold increase in number of samples, an increase in auger diameter implies a two-fold increase in time needed to take, sieve and describe a sample

very effective technique to discover small archaeological sites when they have a low density of artefacts. Even without the availability of much representative data on artefact densities from excavations, it can be suspected that especially Stone Age (and other briefly occupied) sites run this risk.

However, artefacts are not the only category of indicators looked for and registered in a core sampling survey. In fact, three classes of indicators are registered. The first of these are non-archaeological, like soil type and lithology which can serve as predictors of possible site locations. Secondly, there are (semi-)archaeological indicators with a higher detection probability than artefacts, like charcoal or occupation layers. Even if these indicators are not hard evidence of an archaeological site in the sense that artefacts are, they are almost certainly evidence of human occupation very near to the sampled location. Only in third instance 'real' archaeological indicators come into play, as the final corroboration that we are dealing with an archaeological site. It is only when geomorphological, pedological and archaeological 'predictors' are either absent or too small in size for detection in a standard core sampling survey that low density artefact scatters are likely to escape detection, as there will be no apparent reason to 'zoom in' on a specific location.

The absence of reliable data on the density and spatial distribution of indicators for different types of archaeological sites in the Netherlands makes it difficult to design site-specific prospection strategies. These data can only be obtained by registering the same data in excavations as during core sampling, and will need to be collected in a systematic way during future excavations and trenching campaigns. However, at the moment this is not happening in Dutch public archaeology, also because core sampling and excavation are often carried out by different commercial parties, that may not perceive the mutual benefit that can be obtained from investing time and money in this type of work. It is therefore hoped that the current project will provide the necessary impetus to actually start the comparative research needed for further improvement of archaeological prospection strategies in the Netherlands.

[1] The data of the Zutphen-Ooijershoek excavation was kindly put at our disposal by drs. Jos Deeben, Rijksdienst voor het Oudheidkundig Bodemonderzoek, Amersfoort.

REFERENCES

ASMUSSEN, P.S.G., 1994. Archeologische Begeleiding Betuweroute. Deel C: Waardering van de vindplaatsen. RAAP-rapport 86, Stichting RAAP, Amsterdam.

ASMUSSEN, P.S.G., and EXALTUS, R.P., 1993. Archeologische Begeleiding Betuweroute. Deel B: Inventarisatie. Deel C (gedeeltelijk): Waardering. RAAP-rapport 76, Stichting RAAP, Amsterdam.

DREW, L.J., 1979. Pattern Drilling Exploration: Optimum Pattern Types and Hole Spacings When Searching for Elliptical Shaped Targets. Mathematical Geology 11:223-254.

GILBERT, R.O., 1987. Statistical Methods for Environmental Pollution Monitoring. Van Nostrand Reinhold Company, New York.

GROENEWOUDT, B.J., 1994. Prospectie, waardering en selectie van archeologische vindplaatsen: een beleidsgerichte verkenning van middelen en mogelijkheden. Nederlandse Archeologische Rapporten 17, Rijksdienst voor het Oudheidkundig Bodemonderzoek, Amersfoort.

GROENEWOUDT, B.J., 2002. Sieving Plaggen Soils; extracting Historical Information from a Man-made Soil. Berichten van de Rijksdienst voor het Oudheidkundig Bodemonderzoek 45:125-154.

HOGESTIJN, J.W.H. and PEETERS, J.H.M., 2001. De mesolithische en vroeg-neolithische vindplaats Hoge Vaart-A27 (Flevoland). Rapportages Archeologische Monumentenzorg 79, Rijksdienst voor het Oudheidkundig Bodemonderzoek, Amersfoort.

KINTIGH, K.W., 1988. The effectiveness of subsurface testing: a simulation approach. American Antiquity 53:686-707.

KRAKKER, J.J., SHOTT, M.J. and WELCH, P.D., 1983. Design and evaluation of shovel-test sampling in regional archaeological survey. Journal of Field Archaeology 10:469-480.

OUDHOF, J.W.M., DIJKSTRA, J. and VERHOEVEN, A.A.A. (eds.), 2000. Archeologie in de Betuweroute: 'Huis Malburg' van spoor tot spoor: een middeleeuwse nederzetting in Kerk-Avezaath. Rapportage Archeologische Monumentenzorg 81, Rijksdienst voor het Oudheidkundig Bodemonderzoek, Amersfoort.

STONE, G.D., 1981. On artefact density and shovel probes. Current Anthropology 22:182-183.

The Problem of Determining Sex and Species of ancient Animals and their Height in Withers Based on Metapodium

Evgeniy Begovatov
Kazan State University, Russia

Aida Petrenko
Kazan State University, Russia

Abstract

The tasks of reconstruction of a whole object from its fragments are frequent in archaeological research. Usually a part of them is solved with the help of multivariate statistic analysis and another part needs different methods and approaches. This paper deals with questions of reconstruction. The problem of determination of cattle's sex and height in withers on the base of the metapodials. The program was worked out by application of discriminant and regression analyses. Osteological data on the ancient and medieval assemblages found in Eastern and Western Europe were studied by Prof. A. Petrenko.

Animal bones make up the most substantial material in the archaeology. But its fragmentary keeping in hand is the cause for the great importance of the reconstruction problem for the archaeological data.

In the number of works this problem is treated in the aspect of the sex and species determination for ancient animals as well as their height in withers estimate based on the metapodium.

The work of V. Tsalkin (1960) contains the measurements for the Kalmyk breed cattle and for some samples from the archaeological monuments in the Eastern Europe. G. Nobis (1954) put forward the idea to determine the animal sex on the base of the ratio of upper epiphysis x2 and the metapodium length x1 (in percentage). According to his measurements of the metacarpus the value of the ratio x2/x1 for cows, oxen and bulls varies over the intervals (26.5-29.7), (29.0-31.9), (30.6- 39.8), respectively.

V. Tsalkin's experiments showed that the use of these intervals brought to the wrong results in determining of the sex of the Kalmyk breed. The probable explanation is that G. Nobis has developed his methodics using the cattle bones, which were essentially different from the Kalmyk breed.

To decrease the possible errors, V. Tsalkin proposed to use the correlation graphs for the values of the absolute size (metapodium length x1) and the relative size (the ratio of diaphysis width x4 to the length x1).

The modern methods of mathematical statistics allow to use the data to the full extent, thus to decrease the errors when working with the cattle bones of various populations.

In this paper the methods of discriminant analysis are used to determine the cattle sex on the base of the metapodium materials, which were found at the archaeological monuments of the Eastern and Western Europe.
Measurements for the Kalmyk breed metapodium are used as the main training sample. They contain the measurements of the metapodium (the metacarpus and the metatarsus) width of the upper epiphysis x_2 and the lower epiphysis x_3, and the least width of diaphysis x_4. The measurements of other cattle species including materials of Julius- Kuhn Sammlung by courtesy of Dr. H.H. Muller and Dr. M. Teichert were used to correct the discriminant functions.

To solve this problem the absolute and the relative combinations of these four measurements were taken into consideration.

First of all, the index $J=100*(x_2/x_1)$ proposed by G. Nobis was used to determine the sex by the metacarpus method. The calculations are resulted in the following classification table

	Size	Cow	Ox	Bull
Cow	59	54	5	0
Ox	13	1	11	1
Bull	10	0	2	8

Percentage of the cases correctly classified is 90.24 %. A little of the cases of the better result was obtained by using an index $I = 100*(x_2 + x_3)/ x_1$. Classification table is as follows

	Size	Cow	Ox	Bull
Cow	59	56	3	0
Ox	13	0	12	1
Bull	10	0	2	8

Percentage of the cases correctly classified is 92.68 %. Greatest percentage of the cases correctly classified was obtained by the use of four coefficients x_1, x_2, x_4 and index I. Classification table is as follows

	Size	Cow	Ox	Bull
Cow	59	56	3	0
Ox	13	0	12	1
Bull	10	0	2	8

Percentage of the cases correctly classified is 96.34 %. The cattle of the ancient Europe is differ from the one of the Kalmyk breed. Moreover, they had differences at the conditions to the cattle breeding. Therefore it is necessary to make an adjustment of the discriminant functions obtained.

For this goal the mixture histograms concerning the index I for the cows, oxen and bulls of the Kalmyk cattle (Fig.1) were constructed.

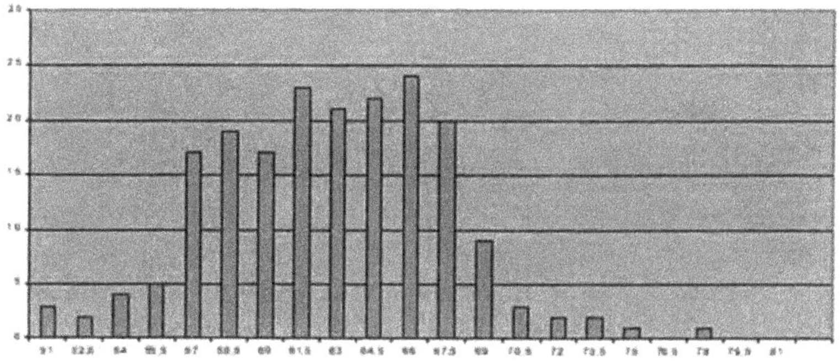

Figure 1 The histograms on index I of the Kalmyk cattle

This histogram contains three top values in the range of the index value: "58", "63" and "70", which are connected with an average value I for the cows (57.7), oxen (63.5) and bulls (69.2).

We assume that the index I have gaussian distribution for cows, oxen and bulls. Then the above expectations and the corresponding variances allow us to find the values of the index I separating the cows, oxen and bulls. These values are 60.8 and 65.8.

The same histograms were drawn for various monuments of Eastern and Western Europe (the part from them is indicated).

The above three top values are clearly recognized at some histograms (see Figure 3 for the cattle of the Manching monument (Durr 1961, Begatov 1994)). At the other histograms (see Figure 2 for cattle of the Bilyarsk monument of Volga Bulgaria XI - beginning XIII century (Begatov 1994)) they are diffused.

The consideration of these histograms allows us to assume that the distance between the top values with regard to definite sex is not practically changed from one monument to another, and that the value of the shift is determined by the position of the left maximum (for most of the cows).

The analysis of histograms leads to the conditional division of the cows into the species: the small size "forest" and the large size "steppe".

The separating point of an index I for the cows and the oxen of the steppe" breed is 60.8, the one for the "forest" breed is 58. The points separating the oxen and the bulls of the "steppe" and "forest" breeds have the values 65.8 and 63.0, respectively.

The separating point of an index J for the cows and the oxen of the "steppe" breed is 30.2, the one for the "bulls" breed is 28.5. The points separating the oxen and the bulls of the "steppe" and "forest" breeds are 33.3 and 31.5, respectively.

Generally, it is necessary to define a displacement of the left maximum of the index histogram with respect to the left maximum for the Kalmyk cattle. This displacement will define an appropriate displacement of the separating points.

The further researches using the measurements of the metapodium of the ancient and modern cattle allow us to verify the distributions of the sexes for various archaeology monuments and to correct the coefficients of the discriminant function.

The discriminant functions for the cows, oxen and bulls are as follows:

$$F\ cow = 12.51x_1 - 26.04x_2 - 15.33x_4 + 44.28I - 2489$$

$$F\ ox = 12.21x_1 - 26.05x_2 - 15.51x_4 + 45.67I - 2580$$

$$F\ bull = 12.15x_1 - 26.21x_2 - 15.23x_4 + 46.69I - 2635$$

They allow us to determine the cattle sex by the single metacarpus. When the monument has a great number of the cattle metacarpus, it is recommend to draw histograms for an index I to ensure the better results.

The problem of determining the sex by metatarsus was considered in this work too. This method is less reliable than the one by V. Tsalkin.

Now researchers use one-dimensional homogeneous model of regression $h=bx_1$ to define the height in withers by the metapodium length x_1. Here the coefficient of the regression depends on a sex of an animal. Some best result can be received, if it can be taken multivariate, homogeneous or non-uniform regression model. For the definition coefficient of the regression we used papers of V. Tsalkin (1960), J. Boessneck (1956), J. Matolcsi (1970), where the calculations have been made for metacarpus and metatarsus. They have shown that the mistake in the determination of height has the identical order to the cases of the known metacarpus and metatarsus. However, it is better to determine the sex by the metacarpus, than by the metatarsus. As a result, it is better to define the height by the metacarpus.

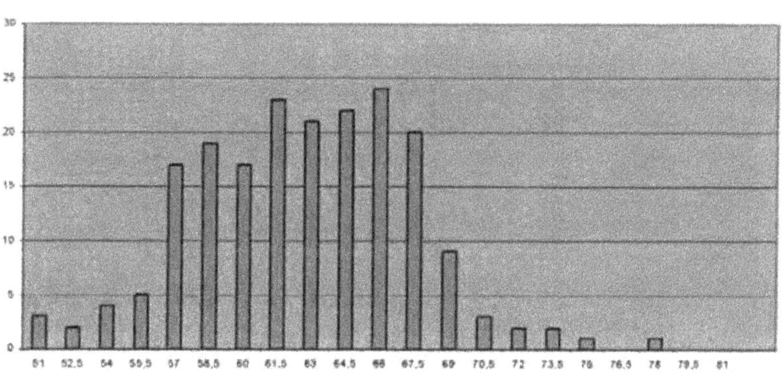

Figure 2 The histograms on index I for cattle of the Bilyarsk monument

The following formulas are received for the calcu-

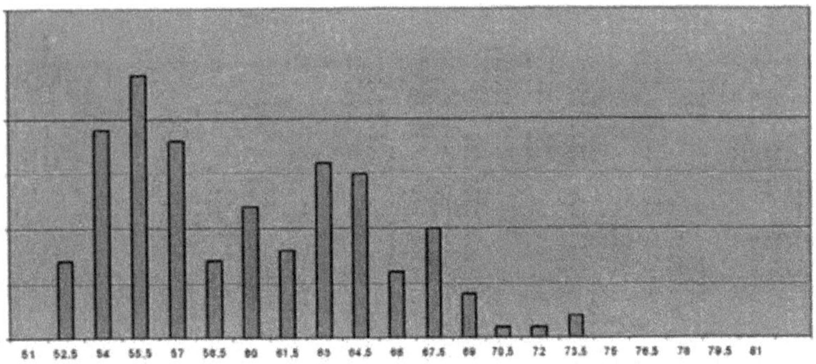

Figure 3 The histograms on index I for cattle of the Manching monument

lation of the height in withers on the base of the metacarpus length at homogeneous (h_h), non-uniform (h_n), V. Tsalkin's (h_t) models.

For cows
$h_h= 0.410x_1+ 0.066x_2+ 0.409x_3 +0.321x_4;$
$h_n= 0.143x_1+ 0.393x_2+ 0.133x_4 -0.524I +97.4;$
$h_t=0.598x_1.$

For oxen
$h_h= 0.815x_1- 1.297x_2+ 0.256x_3+0.732x_4;$
$h_n= 0.557x_1- 1.137x_2+ 0.463x_4 +1.1010I +5.6;$
$h_t=0.613x_1.$

For bulls
$h_h= 0.680x_1+ 0.029x_2- 0.777x_3 +0.995x_4;$
$h_n= -0.157x_1+ 1.180x_2+ 0.582x_4 -1.612I +165.6;$
$h_t=0.624x_1.$

A question about the preference between the homogeneous and non- uniform models remains open. This question is important because the calculation of the height in withers depending on metapodium length can strongly different from one to another for the homogeneous and non-uniform models.

The authors thank Dr. H. Muller and Dr. M. Teihert for their materials' provision.

The work is executed with financial support of Russian Fund for Basic Research.

REFERENCES

BEGOVATOV, E.A. and PETRENKO, A.G., 1994. Die Bestimmung des Geschlechtes und der Höhe in Holke des großen Hornviehs in der Archäologie. Kazan.

BOESSNECK, J., 1956. Ein Beitrag zur Errechnung der Widerristhöhe nach Metapodienmassen bei Rindern. Z. Tierzucht. Züchtungsbiol. 68:75-90.

DURR, G., 1961. Neue Funde des Rindes aus dem keltischen Oppidum von Manching. Studien an vor- und frühgeschichtlichen Tierresten Bauerns. XII, München.

FOCK, I., 1966. Metrische Untersuchungen an Metapodien einiger europäischer Rinderrassen. Diss. München.

MATOLCSI, J., 1970. Historische Erforschung der Körpergröße des Rindes auf Grund von ungarischen Knochenmaterialen. Z. Tierzucht. Züchtungsbiol. 87:89-137.

NOBIS, G., 1954. Zur Kenntis der Ur- Frühgeschichten Rinder und Nord- und Mitteldeutschlandes. Z. Tierzucht. Züchtungsbiol. 63:155-194.

TSALKIN, V. I., 1960. Die Veränderlichkeit metapodium und ihre Bedeutung für das Studium des großen Hornviehs des Altertumes. Den Bericht der Moskauer Gesellschaft der Prüfer der Natur. Die Abteilung der Biologie. T.XV.:109-126.

Group and Conquer - A Method for Displaying Large Stratigraphic Data Sets

Irmela Herzog
Rheinisches Amt für Bodendenkmalpflege, Bonn, Germany

Abstract

When dealing with a large stratigraphic data set, it is difficult to get a general idea of the excavation's main features and their chronological sequence because the Harris diagram will become very big. During data analysis, archaeologists combine individual stratigraphic units to form groups: For example, all the postholes belonging to a building are allocated to a group. Often lower level groups are aggregated to create higher level groups, resulting in an overview display of the main features of an excavation.

This paper will present a new computer program which does not only provide various methods for defining and checking stratigraphic units and their relationships but is also capable of working with a hierarchy of groups. An automatic Harris matrix layout algorithm was developed visualizing stratigraphic relationships by means of orthogonal lines and allowing to either expand or contract user-defined groups as required. Furthermore, phases can be integrated into the layout concept as well.

When dealing with a large stratigraphic data set, it is difficult to get a general idea of the excavation's main features and their chronological sequence because the Harris diagram will become very big. It is hard to understand a diagram consisting of hundreds of equally sized boxes, if no hints are given how to structure them. Large Harris diagrams are seldom published owing to the high costs, and they cannot be displayed properly on a web page. In addition, multilinear or floating sequences (Harris 1984:128) pose a problem, i.e. the network of stratigraphic relationships may be displayed in a large number of diagrams reflecting different chronological sequences.

These problems can be overcome in part by grouping stratigraphic units (Roskams 2001:258). During post-excavation analysis, archaeologists combine individual stratigraphic units to form groups: for example, all the postholes belonging to a building are allocated to a group. Often lower level groups are aggregated to create higher-level groups, resulting in an overview display of the main features of an excavation. British archaeologists call these overview displays group sequence diagrams (Hammer 2000:167-168). According to Hammer, a group summary including dating is vital for understanding a site. Clark (2000:157) notes that the approach which "may be described as the 'Grouping Hierarchy' system ... is now familiar and widely used, largely by default".

Though stratigraphic diagrams showing groups as rectangular boxes have been used for decades (an example from 1980 can be found in Hammond 1993:146-147), no computer program is yet available that handles both groups and stratigraphic data. The only exceptions are graph editors, which allow the user to drag the stratigraphic unit boxes manually so that all the boxes forming a group are within a rectangular area. But these programs do not support any consistency checks, nor do they allow the user to contract or expand a group.

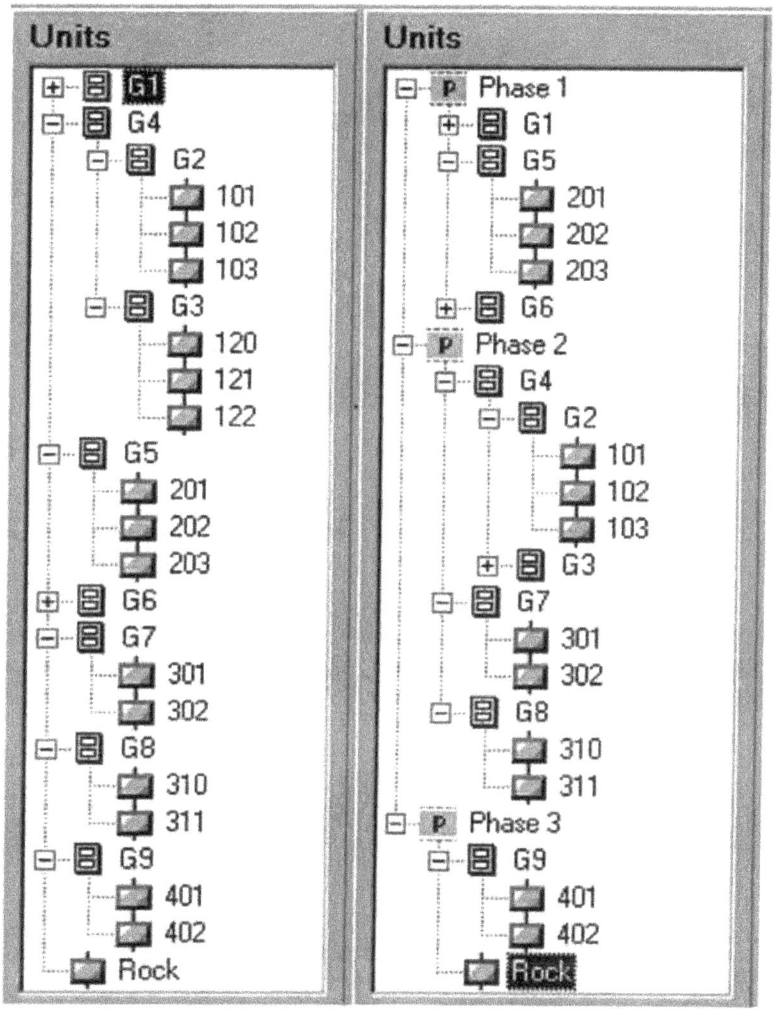

Figure 1 This figure shows two group hierarchy diagrams created by Stratify. On the right, phases are included in the hierarchy. Some of the groups are contracted, indicated by a "+"-sign to the left of the group name, i.e. the members are not shown. Different symbols are used for phases, groups, and contexts

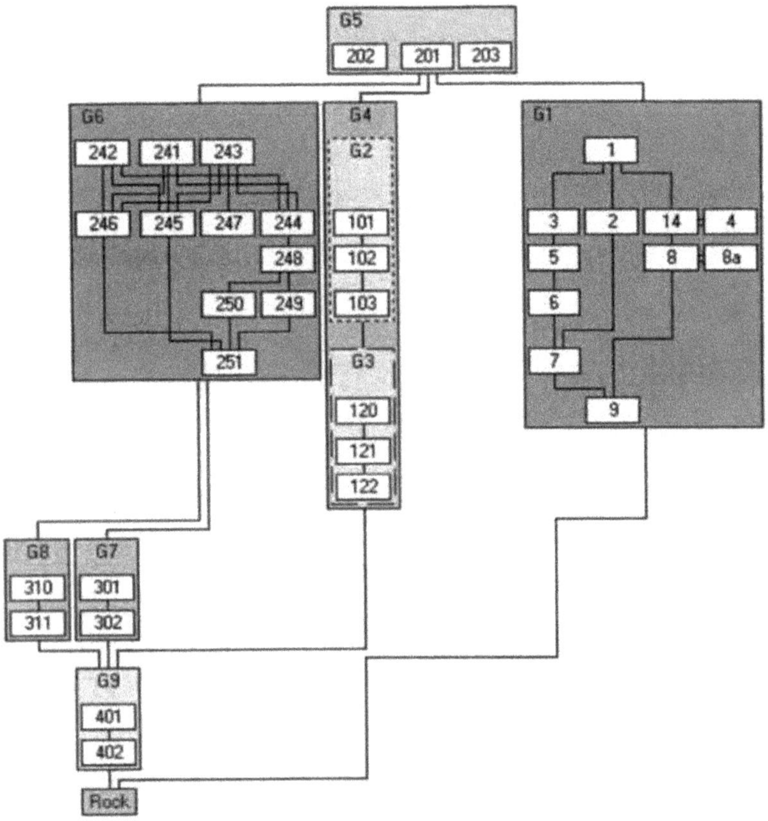

Figure 2 Example of a group layout created by Stratify

This paper presents features of the new computer program 'Stratify' (see also Herzog 2002), which not only provides various methods for defining and checking stratigraphic units (also called contexts in this paper) and their relationships, but is also capable of working with a hierarchy of groups and with phases. 'Stratify' allows the user to define 'earlier than', 'later than', 'contemporary with', 'equal to', and 'part of' relationships. The hierarchy of groups is created using 'part of' relationships: each stratigraphic unit may be part of one and only one group, which may be part of another group, and so on. Andresen and Madsen (1992:50) argue that data structures for recording archaeological excavation data should be able to deal with multiple functions of a single stratigraphic unit: for example, a wall may have been part of two buildings. This means that a context may be part of several groups. In practice, the proportion of contexts belonging to two or even more groups is fairly small. It is recommended to split such a context into several virtual contexts each belonging to a different group and to connect the contexts by 'contemporary with' relationships.

Stratify can show the group hierarchy in much the same way as a directory structure is displayed in the Windows Explorer, allowing the user to expand and contract branches of the hierarchy (Fig.1). This group hierarchy plays an important role for group layout and the checks that are necessary as a prerequisite. It is checked that at least one group is present, that groups are not part of themselves, that each group contains at least one context or another group, and that equal contexts belong to one group or to no group at all.

After a data set has passed these checks successfully, Stratify deduces the relationships between the groups from the relationships between members of these groups: for example, if a context, which is part of group 1, is later than a context belonging to group 2, then group 1 is later than group 2. When contradictions are found, i.e. group 1 is both later than and earlier than group 2, there are two possibilities: with the strict group chronology option, Stratify will show an error message and will not proceed with the layout. Otherwise, the two groups will be set as contemporary. Even if no cycle is detected in the context relationship network, a cycle may be present in the group chronology deduced from these relationships. In this case no group layout can be created. Some of the deduced relationships between the groups may be redundant, and these relationships are deleted (for detailed explanations of cycles and redundant relationships see Herzog 1993:207-208).

The group layout starts with the highest level groups in the group hierarchy, then the groups within these top level groups are laid out and so on, until the lowest level, the stratigraphic units, is reached. The algorithm is based on an idea published by Sugiyama and Misue in 1991 (a summary of this paper is given in Kaufmann and Wagner 2001:210-215): Imagine that the history of the stratigraphic units is written down in a book with several chapters that are numbered. The top-level groups correspond to chapter numbers 1, 2, 3 and so on, the groups within these top-level groups are assigned chapter numbers 1.1, 1.2, 1.3, etc., if they are part of a chapter 1 group. Finally, each stratigraphic unit receives a chapter number. For example, in Figure 2, context 103 is assigned the number 2.1.3: it is the third context in group G2, this group comes first in group G4, and G4 is in the second 'chapter' of the top-level group layout. In contrast to a book, a chapter number may occur several times. This happens whenever two groups or two contexts are positioned on the same depth level in the Harris diagram. With the chapter numbers the vertical sequence of the contexts and the groups is determined. If the nesting levels of the groups differ, a compaction procedure reduces the white space within each group, so that for example, in Figure 2, contexts 249 and 103 are positioned on the same depth level, though their chapter numbers differ (249 has chapter number 2.4, i.e. 249 is part of group G6 which is in the second 'chapter' of the top-level group layout, and within group G6, context 249 is on the fourth depth level).

This method may not only be used to establish the vertical sequence of the contexts and groups, but may be applied similarly to determine the horizontal order. The algorithm used for laying out the top-level groups and the contents of each group mainly draws on methods suggested by Sander (1996b). The 'earlier than' and 'later than' relationships are depicted as orthogonal lines that connect the appropriate context or group boxes. Some extra effort is required to ensure that any two horizontal relationship line segments do not share a point and, similarly, that vertical line segments never touch.

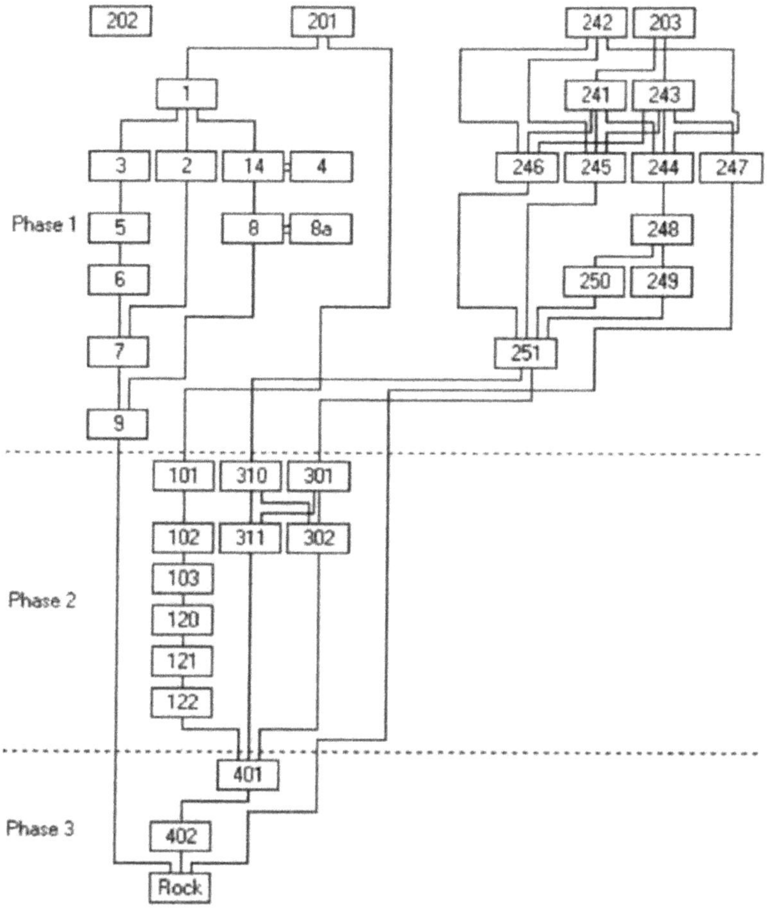

Figure 3 Example showing the data set of Fig.2 with phases but without groups

The result of the group layout algorithm is a Harris diagram of groups, whereby each group can be considered as a Harris diagram of groups and contexts and so on, until the lowest level in the hierarchy is reached. Contrary to a standard Harris diagram, the grouped Harris diagram consists of boxes whose sizes differ. The layout spaces the depth levels based on the height of the tallest group in the level. In the same way, each column width is dependent on the maximum group width in this column. This may force groups on subsequent levels to be placed further away from their predecessors than necessary. However, it is the only way to show contemporary relationships between groups.

Furthermore, with this layout strategy it is fairly easy to implement a contraction and expansion feature: if the user decides to hide the details of a group, i.e. to contract the group to a so-called black box (Paulisch 1993:77), then the only change is that the members of this group and their relationships are ignored during layout, but the chapter numbers remain unchanged for the rest of the groups and contexts. The layout method outlined in this paper creates a 'grey-box' display of the Harris diagram, i.e. the relationships between contexts belonging to different groups are not shown. A 'white-box' display, which does include all relationships, requires a more sophisticated layout strategy; the appropriate technical term in graph theory is 'compound digraph' (Sugiyama and Misue 1991, Sander 1996a).

The program 'Stratify' allows users to create a simple Harris diagram, a layout with phases (Fig.3), one with groups, or one with groups and phases. The procedures used for laying out a Harris diagram with phases bear some similarity to group layout. The chronological relationships of the phases are deduced from the relationships between contexts belonging to different phases. The aim is to create a consecutive chronological sequence of the phases. If there is no unique chronological sequence of the phases, the user will be asked to sort the phases. When combining groups and phases in a Harris diagram, each group must belong to one phase only. This is checked by creating a group hierarchy including phases, i.e. the phases are the top-level groups in the hierarchy.

The program can be downloaded at www.stratify.privat.t-online.de.

REFERENCES

ANDRESEN, J., and MADSEN, T., 1992. Data Structures for Excavation Recording. A Case of complex Information Management. In Larsen, C.U. (ed.), Sites & Monuments, National Archaeological Records, Kopenhagen:49-67.

CLARK, P., 2000. Post-Excavation Analysis: moving from the context to the phase. In Roskams, S. (ed.), Interpreting Stratigraphy. Site evaluation, recording procedures and stratigraphic analysis, BAR Int. Ser. 910, Oxford:157-159.

HAMMER, F., 2000. From Field Record to Publication with a Minimum Archive Report. In Roskams, S. (ed.), Interpreting Stratigraphy. Site evaluation, recording procedures and stratigraphic analysis, BAR Int. Ser. 910, Oxford:161-172.

HAMMOND, N., 1993. Matrices and Maya archaeology. In Harris, E. et al. (eds.), Practices of archaeological stratigraphy, London/San Diego:104-121.

HARRIS, E.C., 1984. The analysis of multilinear stratigraphic sequences. Scottish Arch. Review 3 (2):127-133.

HERZOG, I., 1993. Computer-aided Harris Matrix generation. In Harris, E. et. al. (eds.), Practices of archaeological stratigraphy, London/San Diego:201-217.

HERZOG, I., 2002. Possibilities for Analysing Stratigraphic Data, CD Workshop Archäologie und Computer Wien.

KAUFMANN, M. and WAGNER, D. (eds.), 2001. Drawing Graphs. Methods and Models, Berlin/Heidelberg.

PAULISCH, F.N., 1993. The Design of an Extendible Graph Editor. Lecture Notes in Computer Science 704, Berlin/Heidelberg.

ROSKAMS, S., 2001. Excavation. Cambridge Manuals in Archaeology, Cambridge.

SANDER, G., 1996a. Layout of Compound Directed Graphs. Technical Report A/03/96, Universität des Saarlandes, Saarbrücken.

SANDER, G., 1996b. A Fast Heuristic for Hierarchical Manhattan Layout. In Brandenburg, F. J. (ed.), Graph Drawing Proceedings (GD '95), Berlin/Heidelberg/New York:447-458.

SUGIYAMA, K. and MISUE, K., 1991. Visualization of Structural Information: Automatic Drawing of Compound Digraphs. IEEE Transactions on Systems, Man, and Cybernetics, Vol. 21, No. 4:876-892.

CUTTING OR SCRAPPING?
USING NEURAL NETWORKS TO DISTINGUISH KINEMATICS IN USE-WEAR ANALYSIS

JUAN A. BARCELÓ
DIVISIÓ DE PREHISTORIA. FACULTAT DE LLETRES. UNIVERSITAT AUTÒNOMA DE BARCELONA
E-08193 BELLATERRA. SPAIN
JBARCELO@SENECA.UAB.ES

JORDI PIJOAN-LÓPEZ
DIVISIÓ DE PREHISTORIA. FACULTAT DE LLETRES. UNIVERSITAT AUTÒNOMA DE BARCELONA
E-08193 BELLATERRA. SPAIN
JORDIPL31770@YAHOO.ES

SEE THE CD FOR THE EXTENDED VERSION

ABSTRACT

In this paper, we apply supervised neural networks (Backprop. learning algorithm) to the classical problem of statistical hypothesis testing. Processing experimental use wear in lithics we have found some contra-intuitive results using standard tests, which can be solved using the non-linear discriminant power of Neural Networks. Specifically when archaeological data do not fit parametric distributions, Supervised Learning algorithms appear as an alternative approach. Our particular case study is a set of digital images of experimental data showing use wear as a result of work actions. We have used replicated lithic tools in order to find similarities between use wear identified in experimental data. Previous studies shown that there is not an single discrimination rule to associate cause (kinematics) and effect (wear).

DESCRIBING USE WEAR AS TEXTURE PATTERN

Archaeologists studying lithic remains usually wish to determine whether or not these stones have been used as tools and how they were used. The best way to do this is through the analysis of macro- and microscopic traces of wear generated by the use of the tool.

The main assumption is that the surface of artefacts have specific features because of the way they have been made, or the way they have been used. Tools are made of solid materials and have rigid bodies which resist stress.

As any other physical entity, objects have surfaces, which can be defined in terms of their size, shape, composition and location. Texture can be defined then as the pattern of variability within this surface of those basic properties (Pijoan et al. 1999, Barceló et al. 2001, Adán et al. 2003).

In the case of tools, given that use and production make important alterations in surface features, we can use texture information to understand how the object was made and/or used (human work). Texture variations due to human work are evident, and vary according to the following causal factors:

- Movement: longitudinal (cut), transversal (scrape),...
- Worked Material: (wood, bone, shell, fur, etc.) the effects of its physical properties (hardness, wetness, porosity, plasticity, etc.) on the tool activity surface

We usually represent textures using images. What we are looking in that image is the patterning of luminance values across all pixels. Images have texture (luminance variation), which can be used to represent the variation of the object surface properties (surface texture). The texture of different images should allow us to discriminate between image groups with some characteristic pattern of luminance variation (Adán et al. 2003).

Texture is then described as the relationships of luminance values in one pixel with luminance values in neighbouring pixels (Pijoan et al. 1999, Barceló et al. 2001, Russ 1995, Fontoura and Marcondes 2001). These values can be modelled as forming a set of regions, consisting in many small subregions, each with a rather uniform set of luminance values. In our case, these values are defined as grey levels. A group of related pixels can be considered as a texture minimal unit, sometimes called ´texel -texture´ element- Texture patterning in an image should be described as associations between ´texels´ .

We define luminance discontinuities (region in an image) as ´texels´, if a set of local statistics or other local properties of the average density function are constant, slowly varying, or approximately periodic. Our goal is to segment those texture elements, in order to be able to study their variability in shape and spatial location .

´Texels´ may be geometrically described and measured or they can be Cidentified subjectively in the microscope image as texture primitives; the researcher ´sees´ stries, polished areas, scars, particles, undifferentiated background. Even the ´intensity´ of a trace has also been determined subjectively, introducing attributes like ´poor´, ´high´, ´developed´, ´greasy´, etc. However, we should calculate their formal and relational properties, using their variables of shape, size, composition, and location.

USING A NEURAL NETWORK

We have designed a neural network to use a quantitative description of use-wear texture to distinguish between lithic use (movement and worked material). In our PEDRA system (´pedra´ means stone in Catalan language), we wanted to distinguish those features produced by the movement of a lithic tool done on an specific material, from the macro and microscopic traces characteristic of the lithic surface alone. The idea was to calculate a non-linear discrimination rule for texel parameters, that is, how to distinguish texels generated because of longitudinal movement (cutting), from texels generated during transversal movement (scrapping).

As Input data, we have used the following texel measurements:

Shape:

- Elongation
- Circularity-Thinness
- Quadrature

- Ratio Compactness/Thinness
- Compactness, measured through two equations
- Irregularity
- Rectangularity, measured through two equations
- Ratio Perimeter/Elongation
- Feret diameter
- Minimum rectangularity

Composition:

- Mean, mean of luminance
- SD, standard deviation of luminance
- Mode, mode of luminance
- Min, minimum luminance value

Size:

- Area
- Major axis
- Major axis perpendicular to the major axis
- Perimeter

Location:

- Angle (orientation of each texel's major axis in relation to the edge of the tool)

The output units correspond to classes we want to learn from training data. We want to verify whether the shape, composition and size features have some variability degree related to work kynematics. Consequently, we have only used two outputs:

LONGITUDINAL MOVEMENT (cutting)
TRANSVERSAL MOVEMENT (scrapping)

Input vectors are quantitative values, which have been normalized to the range 1 to -1. Output values are binary.

We have used the standard backpropagation algorithm for supervised training (Bishop 1995, Kulkarni 2001, Principe et al. 2000). As a supervised training, we need a subset of well known output-input patterns, that is experimental data, where the origin of texels are known. For this experiment, we have processed 10 images from 6 tools. 3 tools were submitted to longitudinal work (cutting wood), and other 3 tools were submitted to transversal work (scrapping wood). Three different images from the first tool from each set were taken, and one additional photograph from each other tool. Texels measurements were calculated using the NIH Image software and some additional programs for calculating ratios. All data was and stored in a spreadsheet where each row contains measurements for a discrete texel. In this way we can compare within-image texel variation, within tool texel variation, and between tools and functions texel variation.

Different topologies were examined using 70 % of data (more than 650 texels) for training, and the resulting network was tested with the remaining 30 % which were not used for training. The best network had an input layer with 18 units (all shape/size/location variables, without the luminance intensity measures), 1 hidden layer with 24 units.

PEDRA 7: 18 Inputs (Shape/Location only variables), 1 hidden layer (24 units)

TRAINING RESULTS

75.46% right longitudinal classification, 24.54 % misclassified
58.3% right transversal classification, 41.7% misclassified

TRESTING RESULTS

68.59% right longitudinal classification, 31.41% misclassified
54.23% right transversal classification, 45.77% misclassified

We have created another neural classifier (Pedra13), using this time 6 inputs (the most relevant variables: ANGLE, ELONGATION, CIRCULARITY, RECTANGULARITY (index A), RATIO PERIMETER/ELONGATION and COMPACTNESS (index B) The network was configured with 1 hidden layer made of 13 units. In this case, we have obtained the best results in all series of experiments (Testing: 73% of correct longitudinal classifications).

To sum up, we have measured some degree of relationship, specifically when we analyse longitudinal movement. And this non-linear relationship explains (in average) more than 70 % of total variance. Neural Networks reveal that LONGITUDINAL action (cutting) is easier to identify than TRANSVERSAL movement (scrapping) when controlling all other elements in the experiment (raw material, worked material, time, intensity of work).

How good are Neural Network Results?

We have interpreted the network's output as an estimate of the likelihood that a given pattern belongs to the LONGITUDINAL or to the TRANSVERSAL class. In order to definitely assign a class from the outputs, the network must decide if the outputs are reasonably close to 0.0 and 1.0. If they are not, the class is regarded as undecided. This highlights the intrinsically probabilistic nature of the use wear classification problem. However, the interpretation of output values as intensities, does not mean that we can use them as probability estimations. Only by using probabilities instead of intensities, we can build optimal classifiers which have the potential to create arbitrary discriminant functions that separate data clusters according to the a posteriori probability. To understand the network results and evaluate the network as an optimal classifier or not, we should work with numerical outputs as a posteriori probabilities of the class given the data. In so doing we can minimize classification error and calculate the best one can hope for.

Up to now we have measured the performance of our neural network using classification errors. However, archaeological classification is never an 'all or nothing' type of problem, therefore it is important to evaluate how close we are to the desired LONGITUDINAL or TRANSVERSAL.

The hypothesis to test is that ANGLE values should be the only ones allowing differentiation between longitudinal and transversal action, because it is the only feature related to the direction of the energy flow, which sharps the texel. Theoretically we should imagine that an angle between 45 and 90 degrees should correspond to transversal movement, whereas an angle between 0 and 45° corresponds to longitudinal movement. Ideally, scrapping is a transversal movement and angle values should be around 90 degrees. Cutting is a longitudinal movement, and its angle values should be around 0°. In the middle (45°) we should imagine an indeterminacy area.

In our experimental data, texels observed in longitudinally processed tools have a mean of 42.4 and a standard deviation of 25.7. More than 50% of longitudinally generated data are below 45°.

Transversally processed tools have a mean around 53.6, and a standard deviation of 24.3, and 25% of data are higher than 75°. Although distributions are not normally distributed, mean difference is statistically significant according Student t test and other non parametric tests. The problem lies in the number of outliers. Means and medians can be different, but some outliers seem to attenuate the differences (Pijoan et al. 2002, Toselli et al. 2002).

Some texels from longitudinal tools have angle values greater than expected, whereas some texels from transversally processed tools have angle values lesser than expected.

Those statistical results can be compared with the neural network output values, to test its classification performance. We have used the best network (Pedra13), using a subset of input variables, where the influence of angle inputs on output probability estimates are easier to explain, and coincide with previous statistical results. The diagram below shows, schematically, both outputs (longitudinal, transversal), as a function of angle values input (this is a 1-D representation of the n-D input) (Fig.1).

Here, the probability estimate for longitudinal movement based on angle values is above the 0.5 threshold for the range of 'parallel' texels (angle between 10°-55°), and below that threshold for orthogonal angles, related with transversal movement (higher than 55°). The opposite is true for transversal outputs, with below the threshold results for the parallel range, and around the threshold for higher angle values. In any case, transversal output never goes beyond 0.6.

Conclusions

Our neural networks models can be described only partially as optimal classifiers. Neural computing shows the specific non linear relation between texture and work movement. The specificity of this relationship may arose from the fact that between images variation depends on within-images variation. That is, not all texels from an image have the same features, nor the different pictures from the same tool have similar texels.

Luminance maps may be not entirely related to movement, because there are three sources of luminance variations:

- one of them is generated by the object's surface and it should be explained in terms of the original texture of the object's surface before processing,
- a second one other is also identifiable with the object's surface, but it should be considered as the result of modifications on the surface generated by work activities (cutting, scrapping, etc.) and worked material,
- the third one is grey level variations in the image which are not related to the object's surface, but to the image acquisition process (photography). Furthermore, we should also distinguish luminance variations produced during the perceptual stage as a consequence of microscope functioning.

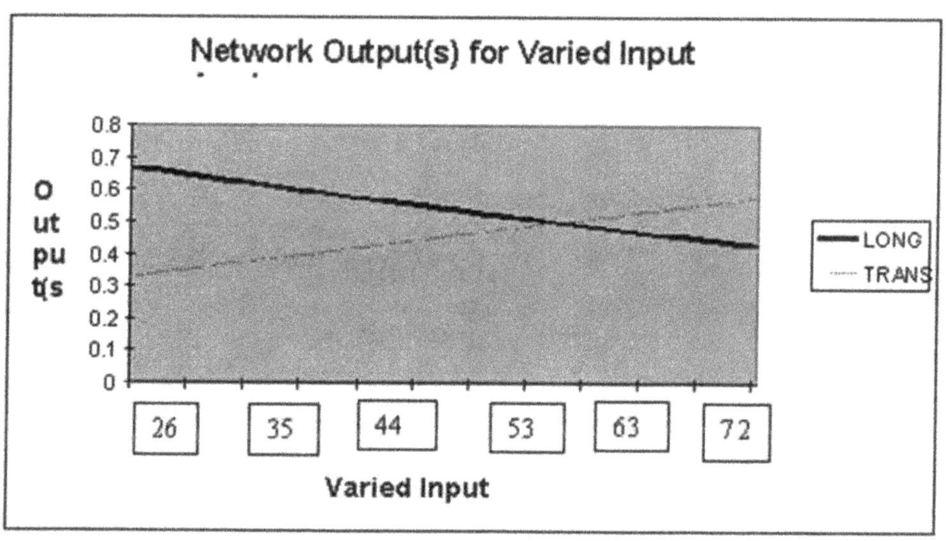

Figure 1 Output-Input Diagram for ANGLE input. Pedra 13 Network: 6 inputs, 1 hidden layer with 11 units

Statistical analysis proofs that the quantity of texels within an image and their size (area in pixels) are not normally distributed. All that points to the fact that not all texels in the sample are good indicators for movement: some of them and some of their features have been generated through specific kinematics, but other can be properties of the original surface of the flint tool, or even reflection consequences during microscopic image acquisition (see Barceló et al. 2001, Pijoan et al. 2002).

Texture is a phenomenon generated by a dynamic process. Consequently, the direction of the energy flux produced by movement determines the shape of texture elements. If the movement is longitudinal, then the energy generated by this movement will tend to create elongated texels, and their orientation according to the original movement is clearer. Transversal movement is much more irregular, and consequently energy flux is less focused at a single direction. The consequence is a higher dispersion and variability of texel shapes: elongated, and circular texels appear together.

References

ADAN, M., BARCELÓ, J.A., PIJOAN-LOPEZ, J., PIQUÉ, R. and TOSELLI, A., 2003. Spatial Statistics in Archaeological Texture Analysis. In Doerr, M. and Sarris, A. (eds.), The digital Heritage of Archaeology. Computer Applications and Quantitative Methods in Archaeology. Heleneic Ministry of Culture, Archive of Monuments and Publications.

BARCELO, J.A., PIJOAN, J. and VICENTE, O., 2001. Image Quantification as Archaeological Description. In Stancic, Z. and Veljanovski, T. (eds.), Computing Archaeology for Understanding the Past (BAR Int. Ser. S931):69-78.

BISHOP, C.M., 1995. Neural Networks for Pattern Recognition. Oxford University Press.

FONTOURA COSTA, L. and MARCONDES CESAR, R., 2001. Shape Analysis and Classification. CRC Press, Boca Raton/London/New York/Washington, D.C.

KULKARNI, A.D., 2001. Computer Vision and Fuzzy-Neural Systems. Prentice Hall.

PIJOAN, J., BARCELO, J.,A, BRIZ, I., PIQUÉ, R. and VILA, A., 1999, in press. Image quantification in use wear analysis. In Computer Applications in Archaeology '99, Dublin.

PIJOAN, J., BARCELO, J.A., CLEMENTE, I. and VILA, A., 2002. Variabilidad estadística en imágenes digitalizadas de rastros de uso: Resultados preliminares. In Análisis Funcional: su aportación al estudio de Sociedades Prehistóricas 1er Congreso de Análisis Funcional de España y Portugal, British Archaeological Reports, Archeopress S1073., Oxford:55-64.

PRINCIPE, J.C., EULIANO, N.R. and LEFEBVRE, W.C., 2000. Neural and Adaptive Systems - Fundamentals through Simulations. John Wiley Publ., New York.

RUSS, J.C., 1995. The Image Processing Handbook. 2nd ed., CRC Press, Boca Raton/London/New York/Washington, D.C.

TOSELLI, A., PIJOAN, J. and BARCELO, J.A., 2002. La descripción de las trazas de uso en materias primas no silíceas: resultados preliminares de un análisis estadístico descriptivo. In Análisis Funcional: su aportación al estudio de Sociedades Prehistóricas 1er Congreso de Análisis Funcional de España y Portugal, British Archaeological Reports, Archeopress S1073, Oxford:65-79.

Automated Classification of Stone Projectile Points in a Neural Network

E.S. Lohse
Department of Anthropology
Idaho State University, USA

C. Schou
Technology Innovation Center
Idaho State University, USA

R. Schlader
Technology Innovation Center
Idaho State University, USA

D. Sammons
College of Education
Idaho State University, USA

See the CD for the extended version

Abstract

We have built a working prototype for online classification of stone projectile points in a neural network. The initial application involves specimens drawn from the North American Pacific Northwest cultural area. The computing system environment for hardware is designed for I386 architecture. Software is coded in VB.net and C# for the DLLs. The current database design is not software specific; however, it requires a robust relational database server. The auto-classification system consists of three stages. Stage 1 is the classification system, with software that allows users to submit images of artifacts or actual specimens that are digitized by lab staff. Stage 1 generates projectile point classifications with specimens assigned to recognized types and is a .NET standalone application. Stage 2 consists of release of a typological descriptive report to system users, including a full image inventory of submitted and classified specimens with attached statistical probabilities of type assignment. Stage 3 is a web-based application hosted on the Technology Innovation Center system that serves as the educational system for public access and study. This paper presents the practical difficulties and successes encountered in automating stone projectile point classification in a neural network, which offers potential for development of a creative, thinking classification system and a rich, accessible, secure reference database.

Introduction

In 1985, Lohse produced a standard classification of stone projectile points from the Columbia Plateau region of North America. He used measurements taken from a collection of 600 points broken into eighteen identified types to classify over 1600 stone projectile points spanning a period of 8000 years (Lohse 1985). This classification was performed as a discriminant analysis. All specimens were photocopied, traced by hand, outline landmarks plotted, and then recorded on a digitizer. Type classifications were made with high confidence levels. All assignments were then checked against classifications made by experts, and were found to be quite accurate.

The process by which the 1985 typology was produced was tedious and long. With improvements in computer technology, especially imaging and data analysis, it is now possible to create a system that can automatically accomplish what took several researchers weeks to accomplish twenty years ago. In addition, there is now a demand from go vernment agencies, researchers, and archaeological contractors to complete typological assignments of points quickly, accurately, and inexpensively, with the data maintained and available over a long period of time and for a large geographical region. Accordingly, the Archaeological AutoClassification System (AACS) was developed (Lohse et al. 2003).

The initial research problem before the AACS is the typing of stone projectile points, but it can eventually be applied to other archaeological and non-archaeological object sets for which recognition and typing is important.

Archaeological Auto-Classification System Working as Virtual Analyst

Our current activity for the AACS Project has focused on obtaining authentic data to fill our database. By "authentic data," we mean information about projectile points from well-dated stratigraphic contexts. This allows us to produce a "clean" set of data that exactly reproduces the classification produced by Lohse (1985). The AACS first incorporated information from the same projectile points as those used by Lohse in the 1985 study.

Neural Networks and Archaeological Classification

The "brains" behind the AACS is a neural network, nicknamed SIGGI. Through training, SIGGI can replicate the actions of archaeological experts in identifying types. SIGGI has implemented a projectile point classification system, which replicates the results developed by Lohse (1985). There are several steps in training SIGGI and in making sure that SIGGI produces valid results. First, SIGGI must be given the necessary information and then be allowed to train itself on that information. Next, SIGGI must move beyond the initial training set of data and be able to classify new and original data sets correctly. Finally, SIGGI's thinking processes must be evaluated for quality, clarity and accuracy, not only to assess the validity of SIGGI's typology but also to gain insights into how human typologists think and work.

Feeding SIGGI

SIGGI's brain is an artificial neural network (ANN). An ANN is composed of a series of artificial neurons, or nodes, which are connected by a series of edges that possess an attribute known as a weight. The artificial neuron produced reflects the structure of a biological neuron (Buckland 2002). Like a biological brain, the ANN is composed of the intelligent combination of artificial neurons and weights. A typical ANN consists of three layers: an input layer, a hidden layers, and an output layer. The input layer consists of a set of data items that the ANN acts upon. The output layer is the set of valid responses that the ANN produces. Our input layer consists of information pulled from an image of a projectile point, while the output layers consists of the information set and classification of that point. The hidden layer is simply an intermediate stage in the actual analysis where the data are manipulated (Buckland 2002).

Weights are set values used by the ANN to modify the inputs. The key to having an intelligent ANN is identifying the proper combination of weights. Combination requires a two-step process. The first step has each artificial neuron performing a summation of all the inputs modified by the weight of the path that the specific input took to reach the neuron. This summed value is then fed through the second step or activation function. The activation function is a mathematical equation that evaluates the summed value of the inputs to the neuron and decides whether or not the neuron will fire. The activation function calculates the value passed to the next stage of the process. Typically, there are two types of activation function: a step function and a sigmoid function. SIGGI uses the sigmoid function (hence the name, "SIGGI"). The sigmoid function uses an equation to produce a value within a given range of values, rather than the all-or-nothing approach of the step function (see Lohse et al. 2003).

Once the activation of one layer is complete, the next layer performs the same calculations until the data reaches the output layer. In the output layer, the neurons pass the final values back to the program, and the process ends. These raw values themselves may then be further modified or may be used as they are, depending on what the ANN is directed to perform. In our case, we take the values, pass them through a normalization functions, which produces a normalized distribution across the domain of the outputs. This identifies a set of variables and establishes a set of statistical probabilities. The output of the ANN is the probability of inclusion in each projectile point type, rather than simply a set of activations of neurons (Buckland 2002).

Training SIGGI

There are two ways to train an ANN: supervised and unsupervised training. SIGGI utilized supervised training, since supervised training required implementing only one AI algorithm. A typical supervised training runs feedback from a set of predefined input that modifies the weights by a set value that is fed back into the network in reverse, or back propagation (Buckland 2002). Back propagation feeds the data set through the network and calculates the error between the expected values and the actual values returned from the ANN. The error value is then fed through a set of equations at each layer in the network. The equation for that layer then modifies the weights that feed into each layer so that the next that set of inputs is fed into the ANN, the ANN is more likely to produce the desired values. This process fits a line to the data set so that a predictive model is generated. In essence, the ANN is conducting a discriminant analysis.

The variables on which the ANN conducts its discriminant analysis are created through a process of image input and manipulation. Images of projectile points are converted to grayscale and then to black-and-white bitmaps. The backgrounds are removed and the edge distortions are smoothed. Finally, an outline of the point is created and submitted to a "tokenizer," a function of the ANN in which the outline is converted into a series of tokens (line segments with connecting end points). The distance and direction of line segment are packages into a vector, and these vectors are the tokens for input into the ANN. The accuracy of the system is dependent upon the number of tokens generated. The original discriminant analysis (Lohse 1985) generated eighteen line segments, but this was an insufficient number for SIGGI to successfully discriminant among point types. The final version of SIGGI establishes 100 tokens on each image, providing sufficient resolution to separate out the different types.

The Archaeological AutoClassification System and SIGGI On-line

SIGGI initially trained on the same set of projectile points used by Lohse (1985) in his original discriminant analysis. SIGGI has been able to successfully replicate that study. It is a working prototype for online classification of stone projectile points by a neural network. Our initial application uses projectile point specimens drawn from the Pacific Northwest cultural area, but the system is extensible to other cultural areas and other artifact classes. The ultimate goal of the AACS, and SIGGI's typology, is to create a web site on which researchers can obtain valid, reliable classifications of projectile points they have recovered or hold in collections, conveniently, on-line, and without the expense of either shipping points to the experts or bringing experts to the points.

The autoclassification system consists of three interrelated stages. Stage 1 is the classification system, with software that will allow users to submit images of artifacts or actual specimens that are digitized by lab staff. This stage generates projectile points classifications with specimens assigned to recognized types. Stage 2 consists of release of a typological descriptive report to system users, including a full image inventory of submitted and classified specimens with attached statistical probabilities of type assignment. Stage 3 is a web-based application hat will be an educational venue for public access and study.

Through these stages, we envision four different classes of on-line users of the AACS: government agencies, researchers, Native Americans, and the general public (Lohse et al. 2003). The look, options, and access of the AACS will be different for each of these four user groups. While some of the functions will be common to all four groups, some groups have very specifics interests and agenda for the information contained in the AACS. Government agencies will be most interested in having the ACCS analyze and classify projectile points which the agencies submits, and in having the AACS store data about the points securely and confidentially. Agencies will also request the AACS to produce summative temporal and spatial information about the projectile points. Academic researchers will have similar interests, with the added function of requesting images and descriptive information about specific projectile points from the database. Native American tribes could also interact with the AACS for similar reasons, but will be especially concerned that the information in the AACS databases remain secure and confidential, and most importantly, that locational information not be released. The fourth type of user, the general public, will be able to access the AACS for educational purposes, using the site like a key (what kind of point did I find?) or for general information (where are these points found? How old are they?). The AACS must be able to provide complete and valid information even to the lay-users, so that the complexity of their questions is limited only by their own knowledge and not by shortcomings or constraints of the AACS.

Conclusions

Our neural agent, SIGGI, has been trained to accurately reflect the thinking of archaeological typologists. SIGGI produces valid types in an automatic online environment, a confidential validated database that preserves information as images and as text fields, and insi ghts into how analysts think. Because SIGGI effectively learns, analysis of how SIGGI makes decisions and manipulates data can lead to new insights regarding redefinition of types and definition of new types, and may well revised how archaeologists consider doing typology. Input from archaeologists interacting with SIGGI will be included in the data being considered and will be reflected in types assigned and their evaluation. SIGGI will correspond with archaeological consultants in real time, responding to their inquiries and helping to enhance the operation of their own classification systems.

References

BUCKLAND, M., 2002. AI Techniques for Game Programming. Premier Press, Cincinnati, Ohio.

LOHSE, E.S., 1985. Rufus Woods Lake Projectile Point Chronology. In Campbell, S. (ed.), Summary of Results: Chief Joseph Dam Cultural Resources Project, Washington:317-364. Report to the U.S. Army Corps of Engineers. Office of Public Archaeology, University of Washington, Seattle.

LOHSE, E.S., SCHOU, C., STRICKLAND, A.W., SAMMONS, D. and SCHLADER, R., 2003. Information Data Archives: Management, Research and Information Distribution in a Secure, Controlled Environment. Paper presented at the World Archaeological Conference, Washington, D.C. (http://imnh.isu.edu/wac5/, June 20, 2003).

VOLUMES OF HISTORY: VOLUME CALCULATION FROM 3D SECTIONS AT THE TELL LEILAN CITY GATES EXCAVATION

CRISTIANO PUTZOLU
ITALY
CPUTZOLU@FREEMAIL.IT

ANDREW MC CARTHY
DEPARTMENT OF ARCHAEOLOGY
UNIVERSITY OF EDINBURGH
EPHPHILON@AOL.COM

LAUREN RISTVET
KING'S COLLEGE
UNIVERSITY OF CAMBRIDGE
LMR26@CAM.AC.UK

SEE THE CD FOR THE EXTENDED VERSION

ABSTRACT

Our work focuses on the problem of obtaining volumetric data from an excavation.
Last May in Tell Leilan (Syria) a 9x1x3 m area was excavated, in which 23 of the 47 lots recognized was entirely floated and sieved. We decided to use an ETS survey of the sections enclosing the area to calculate the volume of each lot. Once georeferenced we put them together in the same 3D Scene of ArcView 3D Analyst, linking each lot with the excavation DB.
In this way, each section maintains its original spatial relationship and the user has infinite points of view of the area. The link between the tables of attributes of the 4 sections enables us, with a simple query, to isolate a certain lot in each section. Once saved in a new shape file, the result of the query (at least 2 polygons) becomes the base for our volume calculation. Each polygon is divided in CAD into 2 polylines (upper and lower half) and 2 different TINs are generated from both "upper" and "lower" polylines. Finally with the "Cut Fill" tool of 3D Analyst we calculated the space between the 2 TINs, the lot volume.

INTRODUCTION[1]

What is the degree of approximation that we can accept in a work related to volumetric data from an excavation? Answering this is not easy because, until laser scanners decrease in price, we will continue to have to interpolate data to obtain surface and volume measurements. As a result, we need to fix a standard for the digital survey of surfaces that will be processed in 3 D managing environments in order to help surveyors gather the data needed for IT operations and to encourage excavators to document their fieldwork with more precision. An Electronic Total Station survey allows enough precision, if we take care to fit our surveyed points grid to the different slopes of the stratigraphic units, increasing the number of points where the surface is discontinuous and decreasing it where the ground flattens. Using a surveyed point grid variable from 120 to 5 points/sqm. (i.e. from 1 point each 10 cm to 1 each 50 cm) according to the sloping of the soil, ensures that the final digital surfaces will be of a high quality. Recently I surveyed the entire excavation of the Garamantian site of Fehwet, Libya (ca. 100 sqm.) by total station using such a standard and am working on the digital reconstruction of the stratigraphic units with good results. But what if there is no chance to respect this standard? In this paper we will describe our experience in the rescue excavation of the City Gate complex in Tell Leilan, Syria where, due to logistical problems, (the distance of the city gates from the main focus of the project, the acropolis, and the inability of the surveyor to be in both places at once) we could not survey the surfaces of the stratigraphic units, but instead surveyed the sections enclosing the excavated area. (C.P.)

THE ARCHAEOLOGICAL CONTEXT

Bulldozer damage from road construction in 2001 damaged the east side of the City Gate, and provided the opportunity to study a 24 meter span, north-south, through the Leilan City Wall The phasing and stratigraphy at CG is defined by the excavation of a small slice of this central occupation area, 9 meters long, 1 meter wide (Fig.1). Prior to excavation, this area was divided into stratigraphic units, which were carefully floated and dry-sieved to recover botanical and small remains and sampled for soil micromorphological investigations. The current paper grew out of our need to provide accurate volumetric measurements to contextualize the archaeobotanical and soil-micromorphological data.

The initial stage of this fortification system (ca. 2600 BC) was characterized by a set of earthworks, into which a wall was cut, to form a 7m wide, 3 m tall northern wall. At 2200 BC, another 1 meter wide buttressed wall was built up against this area. 20 meters to the south, a similar type of construction can be observed, which probably reveals the southern limits of these fortifications.

Between these two massive walls lay the administrative quarter of the city gate complex. Excavation revealed 9 phases of this complex, dating from the mid-third to the early second millennium. Phase 1, founded on virgin soil, dates to the building of the wall complex, while phases 2-3 (Leilan IIId-IIa) see the conversion of this area into administrative spaces divided with pisé and mudbrick walls. In the next phase, 4, a large northern wall was constructed and a fine floor was laid, although the area's function remained administrative. Phase 5 witnessed the construction of an enormous, 4 meter wide fireplace in the north, and a baked-brick platform to the west and south, which may have formed the surface of the road entering the gate. Following Phase 5, this area fell into a short period of disuse. In Phases 6 and 7, at approximately 2300 BC the city gate area was rebuilt. A new northern defensive wall and a new southern dividing wall were constructed during phase 6, and three, successive plaster floors were laid in phase 7. The subsequent phase, 8, witnessed an approximately 300 year abandonment, when the area was allowed to fill in with a meter-thick dust deposit. Finally, phase 9 dates to the resettlement of Tell Leilan in the early second millenium B.C., when a final set of earthworks was built. (L.R.)

VOLUME CALCULATION: FROM SECTIONS TO VOLUMETRIC DATA

In a previous paper written with Paolo Carafa and Sabatino Laurenza[2] we obtained volumetric data by creating pairs of TINs for each stratigraphic unit with ESRI ArcView 3D Analyst and calculating the space between them with the Cut Fill tool. In that case, the TINs were obtained by processing surface points surveyed by ETS, here we decided to create the TINs from the 3D polylines of the sections.

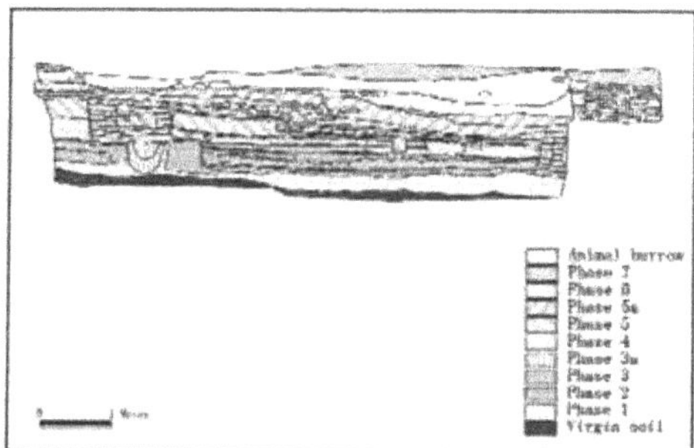

Figure 1 Tell Leilan City Gate excavation: Western section with different phases

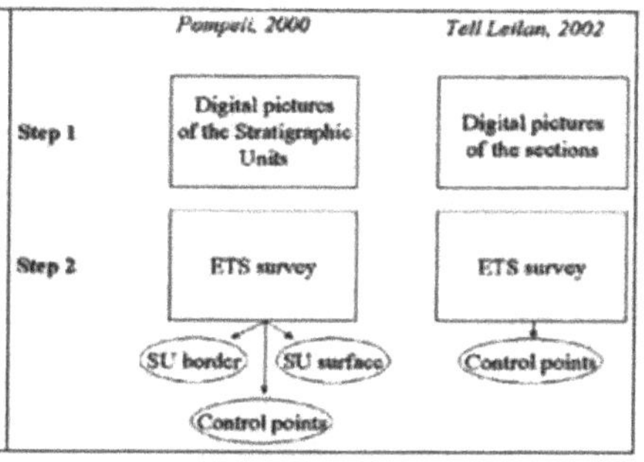

Figure 2 Fieldwork operations: Pompeii 2000 vs Tell Leilan 2002

The idea was that, by assuming that the surfaces between the western and eastern sections (separated by just one meter) had a regular slope, the sections' upper and lower profile of each stratigraphic unit was enough for the interpolation of the final surfaces. Although this resulted in a lack of precision of the resulting TINs, this strategy allowed us to limit the presence of the surveyor to only a few hours.

After fixing a grid of control points, we took digital pictures of the sections, taking care that each image was positioned between four control points, the coordinates of which was surveyed by the ETS. These two steps (digital photographing of the sections and ETS surveying of the control points) were the only fieldwork done by the surveyor. As our approach was different from Pompeii 2000 we also had to rethink our surveying strategy and post-processing operations: Figure 2 shows the sequence of six steps from the field to the volumetric datum for Pompeii, while Figure 3 shows six slightly different steps adapted to the new situation.

Image rectification and mosaic. In order to correct the digital images and to merge them together we first processed them in Rollcimetric MSR. This software commonly used for photogrammetry can resample the pixels in raster images according to the coordinates of at least four control points and, when there are different images with common control points, process the input photographs into a single orthorectified mosaic. The final output of MSR was a set of five scaled and orthorectified pictures representing our five sections.

'Vectorizing sections in a GIS environment.' The MSR output images were then imported into ESRI ArcView 3.2a[3] and the different Stratigraphic Units (lots) were drawn. Since the limits between lots were not always clear we processed the images in MSR at a very high resolution (1 pixel/1 mm) and checked the resulting printed image against the excavation. This enabled us to draw the profile of the different lots with a single graphic element in our GIS environment. Each lot and each feature within a lot (like a brick or stone) was drawn by a polyline and identified in the table of attributes by a Lot Id and/or a Feature Id. At the end of the process we obtained five ArcView Themes each of which grouped the graphic representation of one section and a linked table of attributes indicating the Lot Id of each polyline. The vectors were already scaled but they lacked georeferenciation and internal spatial relationships. The next step was locating our sections in 'real world' space.

'Georeferencing sections in a CAD environment.' We exported our vectors into Autodesk Autocad to give them a position in the 3D space that reflected their real world spatial relationships. For this purpose, we used the 3D align tool, assigning the coordinates of just three control points to each section. As they were already scaled, this gave them the correct position in the Cartesian space. We performed the georeferenciation in a local Cartesian space instead of a real world coordinate system because working with xx,xxx coordinates is definitely easier than working with xxx.xxx,xx ones!

'Creating TINs.' Once reimported into ArcView, the sections were all visible in the same 3D Scene. In order to speed up the selection of each lot in the different sections we linked the five tables of attributes with a simple database (with Lot ID, phase and period). The links between tables enabled us to iso-

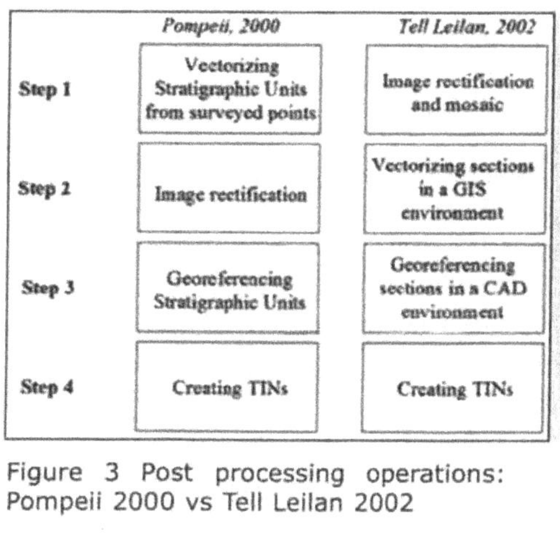

Figure 3 Post processing operations: Pompeii 2000 vs Tell Leilan 2002

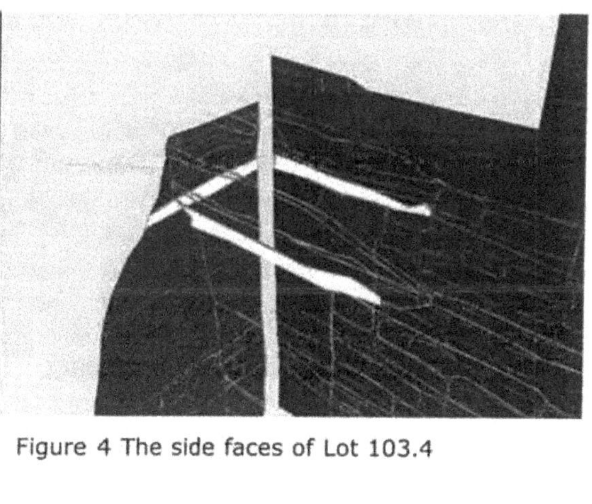

Figure 4 The side faces of Lot 103.4

Figure 5 The 3D reconstruction of Lot 103.4

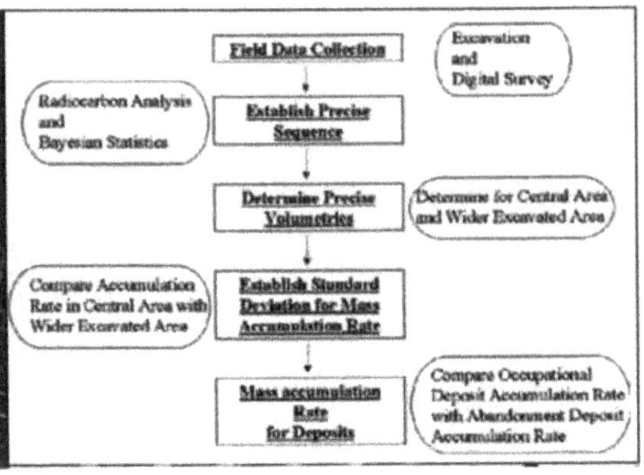

Figure 6 Site formation process analysys

late the 34 lots that were visible on both the N/S sections. A table with the 34 selected Lots was exported and linked with the five tables of attributes. In short we made a simple GIS where the representation of the lots in the five sections of the excavation was the only graphic element and an elementary relational database provided the alphanumeric data.

Once selected in the 'Lot table' a stratigraphic unit was automatically selected in each section, allowing us to isolate it in separate shapefiles. These shapefiles, which could be considered the side faces of the lot solid (Fig.4), were exported into Autocad to draw the top and bottom surface (upper and lower face of the Lot solid). The Autocad Object Snap Commands helped us create 3D polylines (one for the top and one for the bottom) which 3D Analyst needed to create TINs (Fig.5). Back in ArcView, the created TINs gave us two surfaces to use with 3D Analyst Cut Fill tool. This tool calculates the volume between a before surface and an after surface. In terms of 'archaeological' fieldwork the before surface is the top of a Stratigraphic Unit, while the after surface is what we see after its removal (i.e. the top of the units under it): in between we have its volume. The Cut Fill tool generated a table with the values of area and volume, and, after performing these operations for all 34 Lots, we could summarize the results in one table with the area, volume and average width (volume/area) of each Lot. (C.P.)

Site Formation Process Analysis: Diachronic Accumulation Rate

We plan to use the volumetric data to investigate site formation processes. The precise volume data recorded, alongside the impressive material culture and radiocarbon information retrieved from the excavation allow us a rare opportunity to explore the way in which a 'tell' site forms, both in occupational and abandonment contexts, by contrasting the accumulation rate from the abandonment layer (phase 8) with the data from phases 1-5 for which we have multiple radiocarbon dates. This analysis would be virtually inconceivable without the data in electronic form. The algorithms that can be created will allow us to model how the tell formed. The explanation of the process (Fig.6) is as follows:

'Field Data Collection: Excavation and Digital Survey.' First, careful excavation of the site is digitally recorded to produce a 3D model of the central section. Essential-ly, this gives us a 'core sample,' similar to units used in paleoecology. This is necessary in order to interpolate the results to obtain a standard deviation for the Mass Accumu-lation Rate.

'Establish Precise Sequence: Radiocarbon Analysis and Bayesian Statistics.' Second, multiple radiocarbon samples must be taken from key stratigraphic units and precisely analysed in order to establish a tight dating sequence. This sequence can be further refined by the use of Bayesian statistics, which essentially omits radiocarbon ranges that are stratigraphically impossible.

'Determine Precise Volumetrics: Determine for Central Area and Wider Excavated Area.' The next step is to determine precise volumes for each of the stratigraphic layers in the central 'core' section, as well as for a sequence that is stratigraphically identical to the central section, but spatially separated. This aspect of our analysis is still underway.

'Establish Standard Deviation for Mass Accumulation Rate: Compare Accumulation Rate in Central Area with Wider Excavated Area.' Using the volumetrics and the dating sequence, we can establish a Standard Deviation for the Mass Accumulation Rate in the excavation area. This can be done by creating a statistical algorithm, which calculates a rate of accumulation for occupational deposits at Operation CG.

'Mass Accumulation Rate for Deposits: Compare Occupational Deposit Accumulation Rate with Abandonment Deposit Accumulation Rate.' Finally, at Tell Leilan and other sites in the region, there is evidence for large scale abandonment of settlements at around 2200 BC (Weiss et al. 1993, Weiss 2000). This is referred to as the Habur Hiatus and is probably linked to an abrupt climate change. Using abandonment volumetric data from the Hiatus Phase at Operation CG, and Hiatus data from other excavations at Tell Leilan, it may be possible to create a site-wide Mass Accumulation Rate for the Habur Hiatus. We therefore can juxtapose occupational mass accumulation rate to abandonment accumulation rate. This method of interpretation can be used as a model to study site formation processes in general, and can be compared/contrasted to other contemporary occupations and abandonments at other sites.

A study of the mass accumulation rate of Tell Leilan Operation CG would be an important attempt at modelling the rate of site formation. Very rarely are we afforded such precise dates and a corresponding precision of volumetrics. In this circumstance, therefore, we feel that it is valid to adopt what is essentially a paleoecological model of accumulation rate into our analysis of cultural deposits. Furthermore, the abandonment phase at the City Gate operation accumulated an equally impressive amount of debris through presumably non-cultural means. The study of the site formation process will be an invaluable interpretative device. In conclusion, the 3D recording and GIS analysis speaks volumes about the history of the City Gate complex at Tell Leilan, and the integration of this computer application and more traditional means of analysis may serve as a model for future excavations of Tell sites. (A.M.C.)

[1] The authors would also like to thank Prof. Harvey Weiss (Yale University), the director of the Tell Leilan project for allowing us to excavate at Operation CG, The Syrian Directorate-General of Antiquities for aiding this project at every turn, Dr. Dominique de Moulins (University College London) for her work on the archaeobotany, Dr. Thomas Guilderson (Center for Accelerator Mass Spectometry, Lawrence Livermore National Laboratory) for his superb work on the radiocarbon dates, and our friends and colleagues from Tell Leilan, Tell Barham, and Qahtaniyeh for their outstanding excavation ability and invaluable assistance. Andrew McCarthy would like to thank colleagues at the University of Edinburgh, especially Dr. Mike Church, who has been very helpful in coming up with and discussing the practical details of the mass accumulation rate.

[2] Carafa P. Laurenza S. Putzolu C. 2002.

[3] The same could have been done in ArcGIS 8.2 (and now also in ArcGIS 8.3) but I was still using ArcView 3.2a at that time.

REFERENCES

AVERN, G., 2002. Reconstructions of the Excavations of Two Iron Age Chariot Burials from Belgium. In Bürenhult, G. (ed.), Archaeological Informatics: Pushing the Envelope (BAR International Series 1016).

BARCELÒ, J.A., FORTE, M. and SANDERS, D.A. (eds.), 2000. Virtual Reality in Archaeology (BAR International Series 843).

CARAFA, P., LAURENZA, S. and PUTZOLU, P., 2002. Stratigraphic Excavation from the Field to the Computer: The Pompeii Prototype. In Niccolucci, F. (ed.), Virtual Archaeology (BAR International Series 1075).

LAURENZA, S., PUTZOLU, C., 2002. From Stratigraphic Unit to the Mouse: a GIS Based System for the Excavation of Historical Complex - The Case Study of Pompeii. In Bürenhult, G. (ed.), Archaeological Informatics: Pushing the Envelope (BAR International Series 1016).

REILLY, P., 1989. Data Visualization in Archaeology. IBM Systems Journal 28 (4).

VITI, S., 2002. 3D Reconstruction of Prehistoric Living Floor Using Autocad - A Case Study: Grotta del Cavallo (Lecce, Italy). In Niccolucci, F. (ed.), Virtual Archaeology (BAR International Series 1075).

WEISS, H., 2000. Beyond the Younger Dryas: Collapse as adaptation to abrupt climate change in ancient West Asia and the Eastern Mediterranean. In Bawden, G. and Reycraft, R.M., Environmental disaster and the archaeology of human response, Maxwell Museum of Anthropology, Anthropological Papers 7, University of New Mexico Press, Albuquerque.

WEISS, H., COURTY, M.-A., WETTERSTROM, W., GUICHARD, F., SENIOR, L., MEADOW, R. and CURNOW, A., 1993. The genesis and collapse of third millennium north Mesopotamian civilization. Science 261.

ZHUKOVSKY, M., 2002. Handling Digital 3-D Record of Archaeological Excavation Data. In Bürenhult, G. (ed.), Archaeological Informatics: Pushing the Envelope (BAR International Series 1016).

An Overlapping Cluster Scheme

KAZUMASA OZAWA
DEPARTMENT OF ENGINEERING INFORMATICS,
OSAKA ELECTRO-COMMUNICATION UNIVERSITY,
NEYAGAWA, OSAKA, 572-8530, JAPAN

ABSTRACT

Classification plays an important role in understanding a set of findings in archaeology. Some of traditional cluster schemes have been employed for classification not only in archaeology, but also in other areas. Almost all of traditional cluster schemes provide non-overlapping partition of a set as seen in dendrograms. In archaeology, such non-overlapping partition might cause misunderstanding of a set of findings. Suppose a situation that three classes A, B, and C of a family of artefacts have already been defined. When an artefact belonging to the family is newly found, we immediately face at a problem to classify it into which class among A, B, and C. Sometimes, it might happen that we could find no clear attributes of the artefact to be classified into a single class, either A or B. In such case, a reasonable solution is that the artefact should be classified into two classes A and B; i.e. kind of overlapping scheme should be introduced. This paper theoretically presents an overlapping cluster scheme to provide overlapping classes of a set of findings. A consideration on an artificially defined set of patterns is also made to illustrate the proposed scheme.

INTRODUCTION

Clustering has been carried out in many fields. In one word, clustering is aimed at understanding of a set of individuals; i.e. a set of artefacts or findings in archaeology. Understanding of a set of individuals could be followed by abstracting new knowledge, building models and theories, structuring the set or defining new types. Some of traditional cluster schemes have been employed for understanding of a given set of individuals not only in archaeology, but also in other fields. Almost all of traditional cluster schemes provide non-overlapping partition of a set as seen in tree-diagrams (Jardin 1971, Van Ryzin 1977). In archaeology, such non-overlapping partition might cause misunderstanding of a set of findings. Suppose an artefact be newly found. Then we immediately face at a problem to classify it into which of existing types defined previously. It might sometimes happen that we could find no clear evidence to classify the artefact into a single type. In such case, one of reasonable solutions should be to classify the artefact into more than two types; i.e. kind of overlapping cluster scheme should be introduced (Jardin 1971, Matula 1977, Ozawa 1985). This paper presents a theoretical base of an overlapping cluster scheme that could be employed in archaeology.

TWO TYPES OF CLUSTER SCHEMES

A cluster is a subset of a set. Here a set means a collection of individuals such as artefacts or findings. Mathematical description of a set X and its n clusters v_1 to vn that cover X are written by

$$X = v_1 \cup v_2 \cup \ldots \cup v_n$$

Note that all clusters $v_1, \ldots, v_n$ are not needed to separate with each other: It is no problem that clusters contain the same individuals in common, i.e. clusters are overlapped. In contrast, if all clusters separate with each other, set X is to be partitioned into non-overlapped n clusters.

In theory, we can have two types of cluster schemes; one is non-overlapping and another overlapping. The first type is very conventional. In fact, almost all of popular cluster schemes have been non-overlapping. On the other hand, we face at some practical problems that would not be fit for such non-overlapping schemes. The real world has sometimes overlapping structures: For instance, there are many people who have plural nationalities. They belong to more than two clusters, in terms of nationality, of a set of individuals. We also have similar problems in archaeology, which should be handled by the overlapping cluster scheme.

On the other hand, another approach to the clustering problem has been well known as fuzzy clustering (Zadeh 1977). In short, with fuzzy clustering, membership grades of plural clusters are given to an individual. Then it looks to provide kind of overlapping clustering. However, when making a practical decision which cluster an individual strongly belongs to, i.e. when determining a threshold value to cut

Name	Attribute A	Attribute B
1	4	10
2	2	7
3	4	5
4	6	8
5	11	7
6	16	8
7	18	10
8	19	7

Table 1 Example set of eight individuals

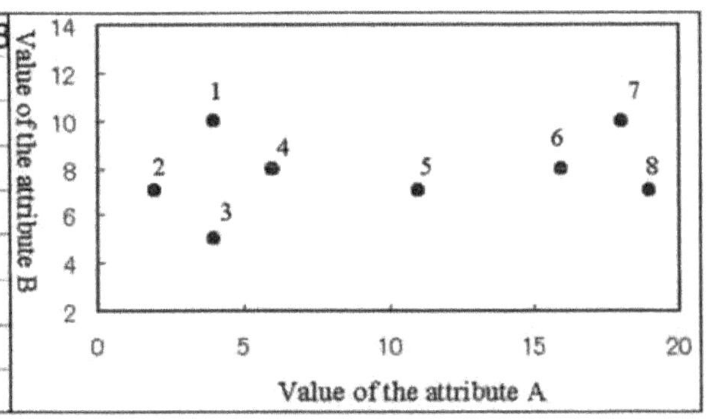

Figure 1 Distribution map of the example set of eight individuals

membership grades, you will find there nothing but non-overlapping clustering. In theory, overlapping clustering is given by excluding the transitive merging of clusters, described later, in either case of fuzzy or non-fuzzy clustering.

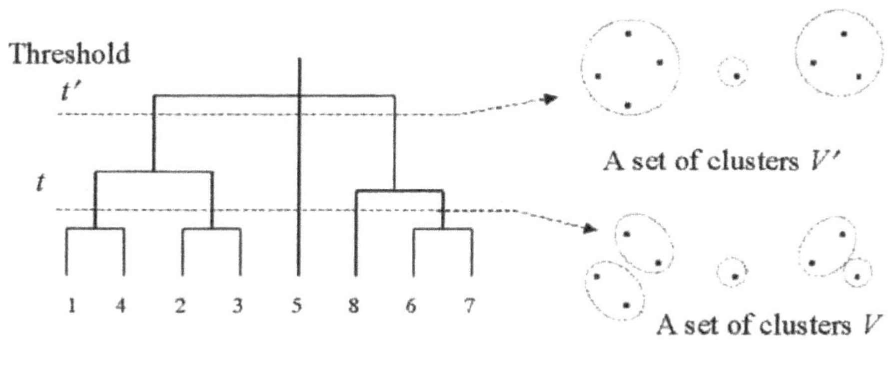

Figure 2 Tree-diagram representation of non-overlapping clustering of the example set

QUANTITATIVE ATTRIBUTES AND DISTANCE MEASURES

To discuss quantitative clustering, an example set of eight individuals named 1, 2,..., 8 is shown in Table 1, which is associated with two quantitative attributes A and B. The table presents two values of A and B for all individuals. Figure 1 presents a two-dimensional distribution map of the eight individuals. By visual inspection on the map, we can have three possible ways of clustering: The first is non-overlapping clustering such that we have three clusters {1, 2, 3, 4}, {5} and {6, 7, 8}. The second is also non-overlapping; we have two clusters such as {1, 2, 3, 4, 5} and {6, 7, 8}, or {1, 2, 3, 4} and {5, 6, 7, 8}. In this case, individual 5 is merged into either of two big clusters in the left hand and right hand sides. The third is overlapping clustering such that individual 5 belongs to both two big clusters; i.e. we have {1, 2, 3, 4, 5} and {5, 6, 7, 8}. To discuss this theoretically, we have to evaluate distances or dissimilarities between individuals. Let d(*i*, *j*) be the distance between two individuals *i* and *j*. And, ai is value of attribute A for individual i. Similarly b_j is a value of B for j. A matrix of Euclidean distances between eight individuals in the example set is given as follows:

$$D=\begin{bmatrix} 0 & 3.6 & 5.0 & 2.8 & 7.6 & 12.2 & 14.0 & 15.3 \\ 3.6 & 0 & 2.8 & 4.1 & 9.0 & 14.0 & 16.3 & 17.0 \\ 5.0 & 2.8 & 0 & 3.6 & 7.3 & 12.4 & 14.9 & 15.1 \\ 2.8 & 4.1 & 3.6 & 0 & 5.1 & 10.0 & 12.2 & 13.0 \\ 7.6 & 9.0 & 7.3 & 5.1 & 0 & 5.1 & 7.7 & 8.0 \\ 12.2 & 14.0 & 12.4 & 10.0 & 5.1 & 0 & 2.8 & 3.1 \\ 14.0 & 16.3 & 14.9 & 12.2 & 7.7 & 2.8 & 0 & 3.1 \\ 15.3 & 17.0 & 15.1 & 13.0 & 8.0 & 3.1 & 3.1 & 0 \end{bmatrix}$$

Where distance d(*i*, *j*) between *i* and *j* is written at the cross of *i*-th row and *j*-th column.

Here we introduce the between-cluster distance. Suppose there be two clusters *v* and *v'*. Then between-cluster distance of *v* and *v'*, written m(*v*, *v'*), has been given in several ways: One is the *single link* that defines it as the distance between the nearest two individuals each of which belongs to either cluster. Symbolically, we have $m(v,v') = \min_{i \in v-v',\, i' \in v'-v} d(i,$

Another is the complete link that defines it as the distance between the farthest individuals. Other definitions could also been employed.

PROCEDURES FOR MERGING CLUSTERS

Clustering can be regarded as a process of merging clusters. Here we discuss two types of merging procedures; i.e. transitive merging and strictly linked merging. We have the following definition of the transitive merging that is closely associated with non-overlapping clustering:

Definition 1 (*Transitive merging*) Now suppose there be three clusters *v*, *v'* and *v''*. For a given threshold value *t*, if the distance between *v* and *v'* is smaller than *t* and the distance between *v'* and *v''* is also smaller than *t*, then let the distance between *v* and *v''* be smaller than *t* automatically and merge the three clusters *v*, *v'* and *v''* into one cluster.

Metaphorically speaking, transitive merging means that if you are my friend and he is your friend, then he should automatically be my friend. This is obviously unrealistic. But almost all existing clustering schemes are based on this transitive theory associated with non-overlapping property. Another type of merging procedure is the strictly linked merging defined as follows:

Definition 2 (Strictly linked merging) For a given threshold value *t*, if and only if any pair of clusters selected from the three clusters *v*, *v'* and *v''* are closer than t, then they are merged into one cluster. We call such a situation that every pair of clusters is close as 'strictly linked'.

In other words, definition 2 says that if and only if you are my friend, he is your friend and he is my friend, then we all are friends. Note that such strictly linked merging allows clusters to overlap.

Figure 2 presents a well-known tree-diagram to illustrate a non-overlapping clustering process by transitive merging. At a stage on threshold *t*, five clusters are given. At the next stage, on threshold *t'*, they are merged into three clusters. The central cluster is a singleton that contains only single individual. The three clusters are merged into only one cluster at the final stage. In this example, single link has been employed for between-cluster distance. Other between-cluster distances would offer different tree-diagrams. On the other hand, when you introduce the strictly linked merging, you could meet with overlapped clusters. Figure 3 shows a process of strictly linked merging of clusters. In this case, we begin with an initial set of singletons and get to the final stage by increasing the threshold value step-by-step. At the third stage, we can meet with two overlapped clusters {1, 2, 3, 4, 5} and {5, 6, 7, 8}.

CONCLUSION

In this paper, we have discussed an overlapping cluster scheme associated with strictly linked merging. For simplicity, we have limited our discussion within an example set of eight individuals. Nevertheless, our results will have good generality in clustering every set of individuals in archaeology. It should be noted that the conventional transitive merging never offer overlapped clusters. In archaeology, overlapping clustering has never been employed for practical data analysis. The author is hoping it will be applied to understanding of sets of archaeological findings.

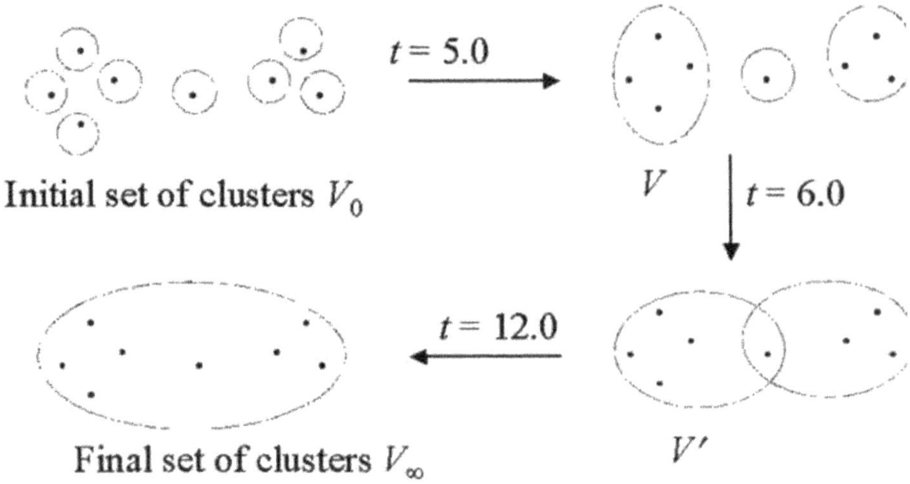

Figure 3 A process of strictly linked merging of clusters

REFERENCES

JARDIN, N. and SIBSON, R., 1971. Mathematical Taxonomy. Wiley.

MATULA, D.W., 1971. Graph Theoretic Techniques for Cluster Analysis Algorithms. In Van Ryzin, J. (ed.), Classification and Clustering, Academic Press.

OZAWA, K., 1985. A Stratificational Overlapping Cluster Scheme. Pattern Recognition Vol.18, Nos.3/4:279-286.

VAN RYZIN, J., 1977. Classification and Clustering. Academic Press.

ZADEH, L.A., 1971. Fuzzy Sets and Their Application to Pattern Classification and Clustering Analysis. In Van Ryzin, J., (ed.), Classification and Clustering, Academic Press.

Assessing Cluster Validity in the Study of Eneolithic Copper Artifacts from Romania

M. Kadar
"1 Decembrie 1918" University of Alba Iulia, Romania

I. Ileana
"1 Decembrie 1918" University of Alba Iulia, Romania

L. Marina
"1 Decembrie 1918" University of Alba Iulia, Romania

SEE THE CD FOR THE EXTENDED VERSION

Abstract

High purity copper artifacts dated to the Eneolithic period from Romania have been grouped by cluster analysis in order to establish the geological nature of raw materials used by the early metalworking in this area. The chemical composition in major and trace elements has been determined by ICP-AES. Trace elements used as markers for native copper have been analyzed and grouped in order to distinguish between made techniques of the heavy implements.
Here we discuss some aspects regarding the number of "real" clusters and methods of validation. A subset of samples of known origin that was "a priori" supposed to group together have been introduced in order to monitor the clustering. Additionally, principal component analysis has been used to supplement the cluster analysis.

Introduction

An important and still debated issue is the use of native copper in the early stages of the human prehistory and its influence to the development of the metallurgy world wide. Scholars have been studied ways of distinguishing between objects made of native copper from those made of smelted copper, some methodology being elaborated by combining chemical composition analysis with metallographic investigation (Maddin et al. 1980, Rapp 1982, Hancock 1991, Wayman and Duke 1999).

In order to understand the inception of metallurgy as one of the most important events of the prehistory, it is critical to recognize and to describe native copper in the unmelted state and after melting, to distinguish and to set up markers for several ore types and to assess the distribution of impurities remained concentrated in the copper objects.

In this archaeo-metallurgical project, 100 copper and bronze artifacts belonging to several museums from Transylvania have been investigated by ICP-AES and ICP-MS in the framework of project financed by the European Community - Access to Research Infrastructures Action of the Improving Human Potential Programme, Contract HPRI-CT-1999-00008 awarded to Prof. B.J. Wood from the EU Geochemical Facility, University of Bristol, UK. For this study have been chosen 31 artifacts of high purity concentration in copper and another 9 samples of native copper in unmelted and melted status.

Statistical Approach

Two main goals have been set up in this research:

1. What is the geographical component in the classification?
2. What are the major type of copper artifacts and their chief characterizations and best representations?

As software, has been used Clustan Graphics package elaborated by D.Wishart and SPSS 8.0.

The method was hierarchical agglomerative clustering analyses. Experimental matrix is consisted of 40 cases containing copper objects with 98-100% Cu and 5 variables. The variables chosen are those impurities which are considered to be markers for different ore types: Co, Fe, Sb, Ag and As. Standardization of data has been achieved by Z-score, dissimilarities between all pairs of individuals have been measured by squared Euclidian distance defined as:

$$d^2_{ih} = \frac{\sum_j (x_{ij} - x_{hj})^2}{v}$$

Where for each variable j, xij and xhj are the values in cases i and h and the summation is over all v variables. This method has been selected as the main interest was to find clusters that are relatively homogenous with respect to all variables.

Objects have been clustered by Ward's method or the method of increase in sum of squares. Increase in sum of squares assumes that the cases can be represented by points in Euclidian space and requires a proximity matrix of Squared Euclidian Distances.

$$E_p = \frac{\sum_{i \varepsilon p} \sum_j (x_{ij} - \mu_{pj})^2}{v}$$

$$I_{p\cup q} = E_{p\cup q} - E_p - E_q$$

The Euclidian Sum of Squares *Ep* for a cluster *p* is the sum of the Squared Euclidian Distances between all members of cluster *p* and is represented by the formula above, where for each variable *j*, *xij* is the value in case *i*, μ_{pj} is the mean in cluster *p* and *v* is the number of variables. The total Euclidian Sum of Squares over all clusters for a given classification is $E = \Sigma_p E_p$. Increase in Sum of Squares combines two clusters *p* and *q* which result in the least increase I_{puq} in *E* that is for which $I_{puq} = E_{puq} - E_p - E_q$ is minimum. Increase in Sum of

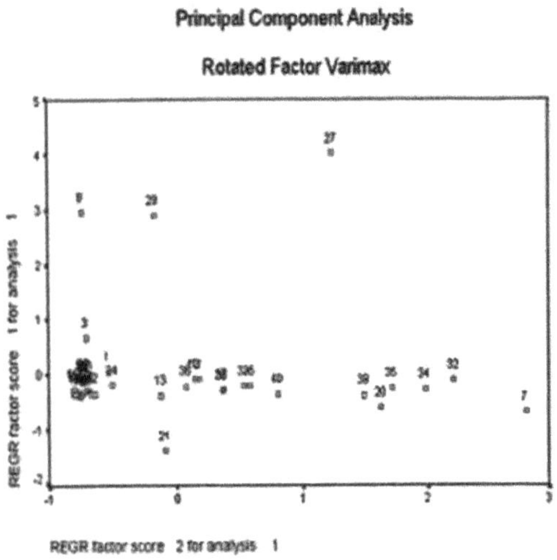

Figure 1 Principal component analyses by SPSS 8.0

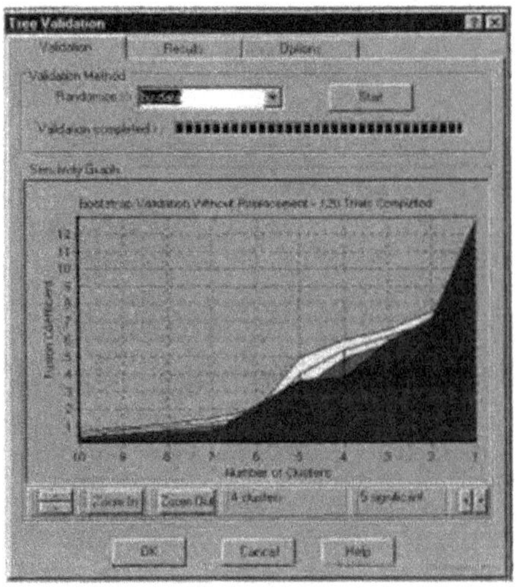

Figure 3 Tree Validation option in Clustan Graphics

Squares assumes that the cases can be represented by points in Euclidian space and requires a proximity matrix of Squared Euclidian Distances. *Clustan Graphics* can convert similarities to dissimilarities and this conversion is invoked automatically if Increase in Sum of Squares is selected with a similarity matrix (Wishart 2003:29).

Methods of Validation Applied

The real number of clusters is difficult to assess from the typical output of the dendrogram. The success of archaeological cluster analyses is connected with the ability to reproduce known archaeological groups. In the case of the provenancing studies it is confirmed that archaeologically-defined groups are chemically distinct in some cases (Baxter 1994:164).

A feed-back procedure is represented by the introduction of individuals known to group together, as a subset of the total number of cases studied. Native copper material has been introduced to monitor the clustering. Cases 32-40 are the samples of native copper, which have grouped together .

Graphical procedure by principal component analyses has been used to supplement cluster analysis. Clear structure on the component plot serve to confirm the reality of structure and offers additional information, if the cluster analysis is partitioning the data rather than identifying distinct clusters (Fig.1). Absence of structure on the plots does not however, imply that there are no clusters, since these may well exist in a higher number of dimensions than are capable of the representations on the plots.

Heuristic stopping rules for determining the number of clusters have been implemented by D. Wishart in Clustan Graphics.
The real number of clusters in this case has been obtained by applying the Best Cut option and the significance tests to the series of fusion levels in the current tree. By choosing significance test from the list a table shows the proposed cluster partitions in the current tree, which are significant. The corresponding partition of the tree is shaded (Fig.3) and the partition is saved as current cluster model.

Another procedure which can be used in Clustan Graphics is the Tree Validation to test the best number of clusters in the hierarchical classification. It compares the tree obtained for the data set with the family of trees generated by random permutation of the same data or the associated proximity matrix. A distribution is obtained for the set of trees from the randomly permuted data and a confidence interval is constructed about the mean. The tree for the given data is then compared with the confidence interval and significant departures from random are identified. Validate tree seeks to reject the underlying hypothesis that the data are randomly distributed. It searches for tree sections that correspond to the greatest departure from randomness, and in trials when random data evaluated it reassuringly reported no significant clusters (Wishart 2003:46).

Randomize can be done by data or by proximities with or without replacements.

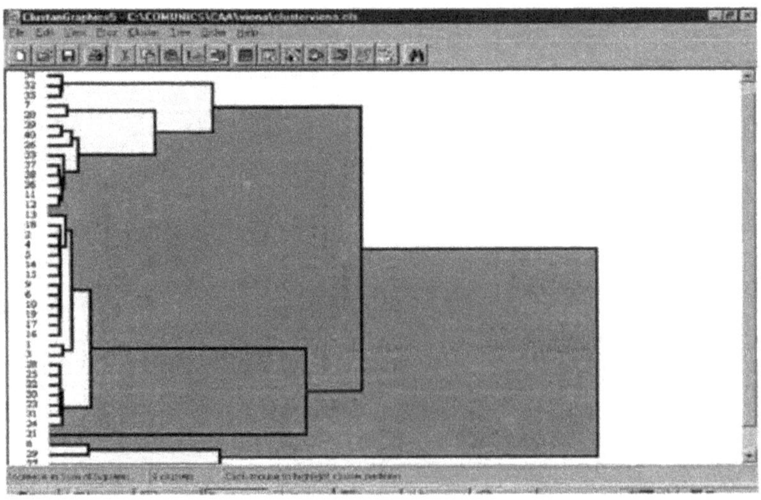

Figure 2 Dendrogram obtained by Tree validation

Conclusions

In this case study, satisfactory results have been obtained by combining principal component analyses with the clustering and validation methods implemented in Clustan Graphics.

As proposed by several scholars there is no generally applicable solution to determining the appropriate number of clusters, each case should be analyzed and checked by other additional methods. Reliance on the form of the dendrogram is unsafe therefore methods combining informal and subjective criteria based on subject expertise should be taken into consideration when dealing with statistical analyses of archaeological data, especially of chemical composition data.

Total Variance Explained

	Initial Eigen values			Extraction Sums of Squared Loadings			Rotation Sums of Squared Loadings		
Component	Total	% of Variance	Cumulative %	Total	% of Variance	Cumulative %	Total	% of Variance	Cumulative %
1	2.005	40.092	40.092	2.005	40.092	40.092	1.651	33.014	33.014
2	1.350	27.003	67.096	1.350	27.003	67.096	1.251	25.015	58.029
3	.748	14.962	82.058	.748	14.962	82.058	1.201	24.029	82.058
4	.666	13.327	95.385						
5	.231	4.615		100.000					

Extraction Method: Principal Component Analysis

Component Matrix

	Component		
	1	2	2
CO	.534	-.487	.548
FE	.489	.596	.441
SB	.913	-.106	-.129
AG	-7.546E-02	.848	8.029E-02
AS	.800	.161	-.480

Extraction Method: Principal Component Analysis.
a 3 components extracted

Rotated Component Matrix

	Component		
	1	2	2
CO	.135	-1.795E-02	.897
FE	.214	.833	.221
SB	..836	7.175E-02	.398
AG	--5.887E-02	.735	-.433
AS	.940	.101	-4.421E-02

Extraction Method: Principal Component Analysis.
Rotation Method: Varimax with Kaiser Normalization.
a Rotation converged in 5 iterations

REFERENCES

BAXTER, M.J., 1994. Exploratory multivariate analysis in Archaeology. Edinburgh.

HANCOCK, R.G.V., PAVLISH, L.A., FARQUHAR, R.M., SALLOUM, R., FOX, W.A. and WILSON G.C., 1991. Distinguishing European trade copper and north-eastern North American native copper. Archaeometry 33:69-86.

MADDIN, R., MUHLY, J.D. and WHEELER, T.S., 1980. Distinguishing artifacts made of native copper. Journal of archaeological Science 7:211-225.

RAPP, Jr.G., 1982. Native copper and the beginning of smelting: chemical studies. In Early Metallurgy in Cyprus, 4000-500 B.C., Nicosia:33-40.

WAYMAN, L.M., and DUKE, M.J.M., 1999. The effects of melting on native copper - The Beginnings of Metallurgy. Proceedings of the International Conference, Bochum 1995:55-67.

WISHART, D., 2003. Clustan Graphics Primer: A Guide to Cluster Analysis Second Edition.

Strange Attractors in the High Mountains

Espen Uleberg
University Museum of Cultural Heritage, University of Oslo

Abstract

A number of sites in the Norwegian high mountains are included in this study. Dateable artefacts and C14-datings can yield a relatively precise time for an activity on a site but do not supply information on the length of time involved. The allocations demonstrate the various interpretations of the landscape. Hunters and pastoralists have placed their sites differently in the landscape; dependent on how they interpreted their surroundings.
Some sites have been visited on a number of occasions. A certain site must have had specific traits that made it preferable for a hunter or a pastoralist. It is not possible for us to comprehend these preferences even though they are the main variables in the system. The strange attractor is a representation of the system and the attractor is reflected by the distribution of sites in the landscape. The concept of a strange attractor is used to focus on movement and flexibility within the system.

Stone Age sites are remains from movement and action in a continuous landscape. The static, dotted site map tends to conceal this, and leads to see the Stone Age landscape filled with a fixed number of closed units. The density of artefacts at any given place is in relation to the number of events that took place there. The varying density in the artefact distribution shows centres and less used areas. This shift from centre to periphery and movement in the landscape can be inferred from the static plot maps through a concept from chaos theory, namely the strange attractor.

The study area is in the Lærdal Mountains, between 1,100 and 1,500 m above sea level (Fig.1). The distance to the coast is around 250 km to the east and only 50 km to the west. The area is above the treeline which is around 900 m a.s.l., The only large game in the area is the reindeer, and scatters of quartzite flakes on the surface gives an impression that not much has changed since the Stone Age. The traditional interpretation of the sites is that they were used by hunter/gatherers living most of the year on the coast or in other lower lying areas, acting as specialised reindeer hunters in the mountains in the autumn (Johansen 1978).

The treeline is lower now than in the Stone Age. The effect of a colder climate and isostatic uplift is increased by human impact, especially the shieling in the 19th century. The existence of an earlier debated birch forest above the pine tree limit has been demonstrated through the finds of birch tree trunks in the moors. At 8700-8500 BP the pine treeline was probably around 1,200 m a.s.l. and the birch forest may have been as high as 1,400-1,450 m a.s.l. The tree limit stayed at this level till 5000 BP, and the subsequent decline was slow. (Aas og Faarlund 1995). Almost all of the Stone Age sites were therefore situated in the forest; some of them even in the pine forest. This suggests that moose and deer have been more important than the reindeer.

The last remains of the ice cover were melted not long after 9000 BP, and the oldest site in the Lærdal Mountains is dated to 8510 ± 110 BP. There are no organic material preserved at the sites, but bones from Stone Age sites in the nearby Hardangervidda area show the presence of reindeer in the mountains in Southern Norway from 8500 BP onwards. Bones from moose and possible deer in addition to bird (Indrelid 1994:237-40), show that the subsistence was more varied, and supports the idea that the sites were not situated above the treeline.

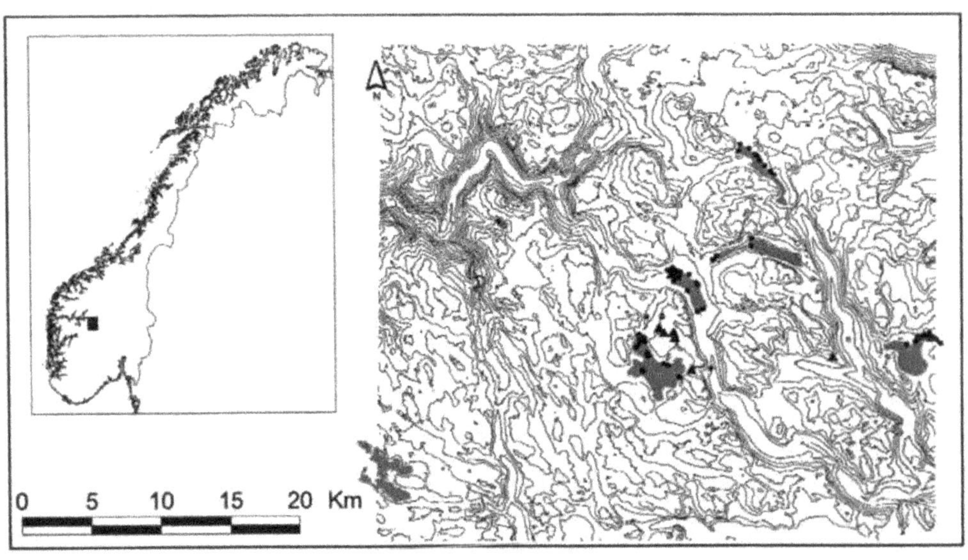

Figure 1

C14-results combined with dateable artefacts from several sites show repeated visits. The lack of cultural layers and frost perturbation makes it hardly possible to document any time sequence at the sites. What can be said with certainty is that the lakesides have attracted activity at several points in time.

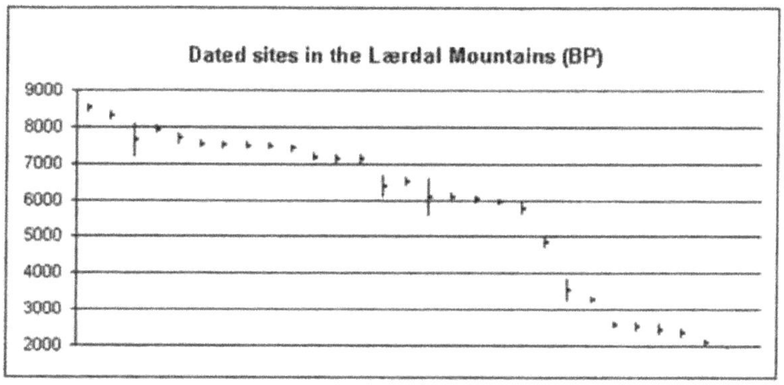

Figure 2

Two C14-datings or two dateable artefacts show two separate events, and it is not possible to say anything about what may have happened between these two points in time. A combination of a series of C14-datings from the high mountains shows longer periods with no activity at all. There is a noticeable gap in the period 6000 - 3500 BP, the last part of the Mesolithic period and the early Bronze Age (Fig.2). Dateable artefacts from this period indicate that the area was used, but the activity in the Neolithic and Early Bronze Age is not visible in the C14-samples. One explanation could be the use of fossil fuel. The tree species was not determined for the C14 samples from The Lærdal Mountains, so that old tree trunks, pine found in the bogs, might have been used as firewood. On the other hand, the same hiatus is visible in other high mountain areas where the dated material was twigs and branches (Bang-Andersen 1986), corroborating that the mountain areas in Southern Norway were less used in this period.

The hiatus has been explained by bad climate - that more precipitation followed by ice-covered ground in the winter would give worse conditions for the reindeer. When the reindeer population was reduced, the mountain areas would not be attractive for hunters (Moe et al. 1978). However, there does not seem to be sufficient data to support the idea that a climatic change caused the virtual abandonment of the interior areas. Longer and not simultaneous periods of no or less activity in different parts of Southern Norway implies that it was not only the natural environment but a cultural change that decided this development (Uleberg 2003).

Non-Linear Systems

The culture landscape is dependent on a beholder, but it is not necessary that the landscape be physically restructured. The landscape is structured through its description. It is the user of the landscape who, on basis of the cultural and historical context, ascribes importance to certain elements. The same landscape has differing contents for people with differing backgrounds (Meløe 1989). The subsistence pattern will outline what the individual sees and looks for in a landscape. The economical basis and individual experience decide the description of the cultural landscape. In this way, a description of a group's ecological niche will also be a description of the group's cultural landscape.

In this way, it is possible to have many landscapes within one geographical unit. These different, coexisting, landscapes are dependent on a subject to exist. The cultural landscape becomes a dynamic unit where changes happen in local environments and over short time spans (Welinder 1992:107). The system is dependent on individual decisions, and initial conditions and small variations can be decisive for the development. It is possible to use a model that presupposes that the system is changing. Small changes and initial conditions can be given great importance in non-linear systems theory. The nonlinear systems theory can show that unpredictable random variation can be produced in a deterministic system. The system can contain quick changes and shift between chaos and order. The apparently random variation is connected to orderly recurring patterns, where the system can change states from order to chaos and order can re-emerge from chaos (Spencer-Wood 2000:115-116).

One way of delineating the underlying dynamics of a system is by producing the attractor of the system. Simple systems converge towards simple attractors. A pendulum is a simple system. It moves back and forth, and stops eventually over a point. It has a point attractor. Other systems can have more complex attractors like a circle or a thorus. Chaotic systems, nonlinear systems, also converge towards an attractor, but this attractor is even more complex, and is called a strange attractor. The path is deterministic, but totally aperiodical. The movement converges, but two paths that are close to each other will diverge and follow quite different developments (McGlade 1995:119-120).

A strange attractor cannot be described beforehand. One way of describing a strange attractor is by studying the Poincaré section, which is a slice through the attractor. From an m-dimensional attractor, this can be obtained from the intersections of a continuous trajectory with an (m-i)-dimensional surface in the phase space. If the system is periodic, with period n, the section will contain a sequence of n dots repeating indefinitely. If the evolution is chaotic, the set will be a collection of points with no obvious repetition. The Poincaré section can be obtained by sampling the system occasionally, and not necessarily continuously (Tsonis 1992:83). It will not be possible to describe a model that can be tested, but it gives another way of thinking about the material. Important variables in the model are landscape and find density. Chronology gets less attention, so that finds from all periods can be incorporated in the same model.

Thinking of the artefact distribution as a strange attractor helps to move away from the site map's scatter plot to an understanding of interaction and movement. The small variations in the system can be visible in the pattern made by the artefacts in the landscape. In this way the archaeological recordings can be seen as a Poincaré set where the system is sampled occasionally. Stone Age men and women moved in a continuous landscape, performing activities, some of which resulted in physical remains, and some of these are preserved and documented by archaeologists. Each artefact is from an event in prehistory. At sites with one or only a few artefacts, there have been one or a few events. At sites with many artefacts and constructions, there have been several events, perhaps in a short period of time, perhaps spread out over a longer time span. There is a connection between the density of artefacts and the importance of the specific site, the place. Areas with high artefact density have been visited often, areas with lower density more seldom. This approach reduces the need to explain each site on its own, and strengthens the notion that all remains are parts of the same basic system.

The Lærdal Sites

There is a large difference in site density around the lakes in the Lærdal mountain area. Lake Eldrevatn has by far the largest number of sites. The Lakes Kvevotni and Flævatn have none. The other surveyed lakes have several sites. When the area is seen as a whole, there is no clear chronological difference among the lakes. All that have dateable sites have finds or C14-datings from all periods from the Mesolithic to the Bronze Age. The possible differences between the lakes Kvevotni and Flævatn and the other sites can give information about what people saw as a good allocation. The two lakes were earlier two series of smaller lakes and rivers. The lowest of these lakes were at 1,458 m a.s.l. The birch tree line was probably at 1,400-1,450 m a.s.l. These lakes have therefore always been above the treeline. All the other surveyed lakes were below the treeline. This suggests that a good allocation was a place near a lake below the treeline.

The density of sites around Lake Eldrevatn combined with C14-datings and dateable artefacts show that this has been an attractive area in several periods. The lake has attracted activity throughout the Stone Age. In addition to the definable sites, some with constructions, there are scattered finds along most of the lakeside. Sometimes exactly the same place has been used several times, although there does not seem to be anything that makes that particular place special. Some of the sites around the lake have finds from all periods, while an area in the West and in the Southeast have only finds from the Mesolithic period. This suggests that a change in subsistence is visible in the artefact pattern when seen at this scale.

A good allocation is determined by the reading of the landscape. The hunter looks for good hunting grounds, and the pastoralist looks for good pastures. Human activity at a place can give a positive feedback. The place will be prepared for living there. Human activity also leads to a concentration of debris, which again can give better grazing condition, and increase the possibility that a certain place can be seen as a good place to settle. The difference between the early hunter-gatherer sites and the Late Neolithic pastoral sites in the Lærdal Mountains is not very clear. The larger sites are all placed near the lakes that were used for shieling, even in modern times (Uleberg 2003). In the nearby Nyset-Steggje mountains areas, where lakes are not dominating the landscape in the same way, a shift in allocation from the early hunter sites to the pastoral sites of the late Neolithic is more clear (Prescott 1995). The situation changes when the scale is changed from the whole mountain area to one single lake. The mapping of artefacts along the lakeside makes it possible to a discern a shift in the reading of the landscape from the hunters of the Mesolithic to the pastoralists of the Neolithic period.

REFERENCES

AAS, B. and FAARLUND, T. 1995. Skoggrenseutviklingen i Norge. Lotte Selsing (red.), Kilder for klimadata i Norden fortrinnsvis i perioden 1860-1993:89-100.

BANG-ANDERSEN, S. 1986. Veden de fant - bålene de brant. Vedanatomianalyse som metode til rekonstruksjon av nærmiljøet rundt steinalderboplasser i høgfjellet. Viking XLIX:15-30.

INDRELID, S. 1994. Fangstfolk og bønder i fjellet. Bidrag til Hardangerviddas førhistorie 8500-2500 før nåtid. Universitetets Oldsaksamlings Skrifter Ny rekke Nr. 17.

JOHANSEN, A.B. 1978. Høyfjellsfunn ved Lærdalsvassdraget. Universitetsforlaget.

MCGLADE, J. 1995. Archaeology and the ecodynamics of human-modified landscapes. Antiquity 69:113-32.

MELØE, J. 1989. The Two Landscapes of Northern Norway. Inquiry, 31:401-17.

MOE, D., INDRELID, S. and KJOS-HANSSEN, O. 1978. Environment and Early Man. Norwegian Archaeological Review Vol. 9, No. 1:32-36.

PRESCOTT, C. 1995. Aspects of Early Pastoralism in Sogn, Norway. Acta archaologica Vol. 66:163-189.

SPENCER-WOOD, S.M. 2000. Streange Attractors. Feminist Theory, Nonlinear Systems Theory and Their Implications for Aarchaeological Theory. In Schiffer, M.B. (ed.), Social theory in archaeology, University of Uta Press, Salt Lake City:112-125.

TSONIS, A.A. 1992. Chaos. From Theory to Applications. Plenum Press, New York/London.

ULEBERG, E. 2003. Fra punkt til område. Steinbrukende tid i fjellet.

WELINDER, S. 1992. Människor och landskap. Aun 15 Societas Archaeologica Upsaliensis, Uppsala.

Aoristic Analysis: Seeds of a New Approach to Mapping Archaeological Distributions through Time

Ian Johnson
Senior Research Fellow, Archaeology,
University of Sydney NSW 2006
johnson@acl.arts.usyd.edu.au
http://www.timemap.net

Abstract

Aoristic analysis generates a temporal probability distribution from a spatially located population of temporally imprecise events. It is used, for example, in analysing burglaries, where only the terminus post quem and terminus ante quem of events are known, but not their actual time of occurrence. Chronological determination of artefact dating based on typology and/or attributes shares similar characteristics, in that we often know the range of dates over which the artefact could have been produced, but do not know its date within that range.
This paper considers the potential of aoristic analysis to provide probabilistic statements about the dating of assemblages, and its potential for extracting dating information from the full range of material, not just chronologically specific items which may occur rarely. The ability to combine narrowly dated characteristic artefacts with long-lived generic forms, should particularly suit mixed assemblages, such as those encountered in archaeological surveys.
The use of aoristic analysis is illustrated through an application to chronological identification of pottery from the Australian Paliochora-Kythera Archaeological Survey (APKAS), which shows how it might be used to map the changing pattern of artefact distribution through time.
It is also proposed that the concept of distributing probabilities through time might be extended further to cover spatially distributed assemblages of features, such as urban settlements, where there are multiple, potentially conflicting, sources of dating information. In this way we may be able to build accurate if fuzzy pictures of spatial distribution through time in place of representations which hide our uncertainties under a false appearance of precision.

Introduction

One of the fundamental problems in analysing the results of archaeological field surveys and in studying prehistoric settlement patterns, is the difficult question of 'contemporaneity'. Unfortunately - at least for archaeologists - Pompeii-style events are the extreme exception, and most of our evidence consists of palimpsests of the products of overlapping activities built up over years, centuries or millennia. Without a knowledge of the contemporaneity, or otherwise, of archaeological structures and discarded materials, discussion of landscape use and settlement patterning - so fundamental to much archaeological interpretation - often depends on convoluted arguments or, in some cases, simply conjecture.

For typical 'fieldwalking' surveys, a proportion of the material observed can be allocated to date ranges (generally determined - more, or less, narrowly - by pottery styles or stone tool technology). We generally make the assumption that there is not too much 'fossil' deposition, and that material deposition equates with activities (which can of course include secondary redistribution through e.g. manuring). For archaeological sites within a landscape or built structures in settlements, the dating imprecision is further complicated by their potential for long-term use and re-use.

The challenge is to work back from a loosely dated assemblage of objects - whether artefacts, structures or sites - to some estimation of the pattern of occupation in the past, and if possible to an understanding of the changes in occupation pattern, since it is on this information that so many archaeological theses are built.

APKAS: A Case Study

In the Australian Paliochora-Kythera Archaeological Survey (Johnson and Wilson 2003, Coroneos et al., in press), our attempts to quantify settlement patterns from intensive field survey results - in order to examine postulated depopulation of the island - have been frustrated by small numbers of precisely datable artefacts and the limited scale of the survey. Even with larger samples, the interpretation of settlement patterns from archaeological field surveys in the Mediterranean, as elsewhere, is often subjective, since we do not have appropriate tools for quantifying and analysing temporal spread, least of all when this is combined with spatial distribution.

The specific result we require from the Kythera survey is quantification of the change in settlement pattern to test the hypothesis of periodic abandonment of marginal environments - most specifically around the period of the sack of Paliochora (AD1537) - and to determine whether the 'fossil' distribution of churches (more than 80 for less than 1,000 present-day inhabitants, many isolated from villages or other habitation) reflects the former distribution of activities.

The Problem with Time

In my original proposal of the TimeMap methodology at the 1997 CAA Conference (Johnson 1999) I sought ways of explicitly recording the temporal signature of spatially located archaeological and, more generally, historical phenomena. These included the recording of 'snapshots' at known times, defining 'terminus post and ante quem' dates, and defining spread of uncertainty in dating as a through-time function.

The fundamental problems faced in satisfactorily recording historical phenomena include: 1. dating, even of individual events or objects, is generally only to within a loosely defined time range; 2. cultural areas - however defined - are not static objects which appear full-formed and disappear instantly; 3. cultural phenomena are generally not spatially uniform as demanded by most vector GIS data models, and we often do not know much about their internal pattern; and 4. our knowledge of spatio-temporal objects is typically anecdotal 'snapshots' of their extent at more, or less, accurately known times.

There is a difficult balance to be struck between adequately capturing the information available, and making the recording of such information unrepeatable and impossibly onerous. To my knowledge no-one has proposed a generalised method appropriate to archaeological material, although much research has been carried out - generally with an emphasis on contemporary data - into temporal databases (e.g. Tansel et al. 1993) and, to an increasing degree, spatio-temporal databases (e.g. Yuan 2001), and there are numerous domain bibliographies (e.g. Roddick et al. 2001).

Aoristic Analysis

Aoristic analysis (Ratcliffe 2000) is a method which is applied to the analysis of crime incidents. Archaeological events - such as the deposition of an artefact or the building, use and abandonment of a structure - share some of the characteristics of crime incidents: we normally know their location with some accuracy, but can only locate them temporally between a terminus post quem (the last known pre-incident moment or, for artefacts, the start of the known period of production of the artefact) and a terminus ante quem (the moment when the incident was discovered or, for artefacts, the likely end of production and/or deposition). Aoristic analysis has the potential to address some of the problems encountered in analysing spatio-temporal archaeological data by providing us a means of dealing with the variation in precision of dating of different types of artefact, from 'diagnostic' items to generic pottery or waste flakes.

In aoristic analysis, the probability of each event is distributed evenly across the time span during which it might have occurred. The probability of all events is the summed probability of individual events; events with tight temporal definition contribute more to the total probability over their range than do loosely defined events.

Figure 1 illustrates the way in which five objects/events of different temporal extents contribute to the overall probability. Note that, although the probability of an object/event occurring in a given time interval is higher in the second peak than in the first, the sum of probabilities (represented by the area under the graph) for the first peak is approximately three times that of the second.

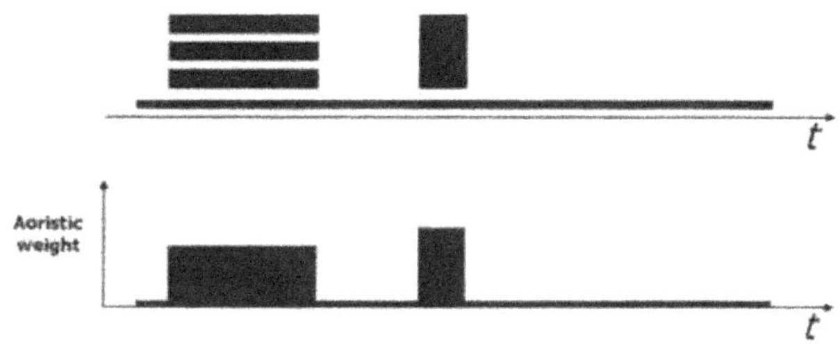

Figure 1 Contributions of different events/objects to aoristic weighting (areas of rectangles in upper diagram are equal, representing 1 object)

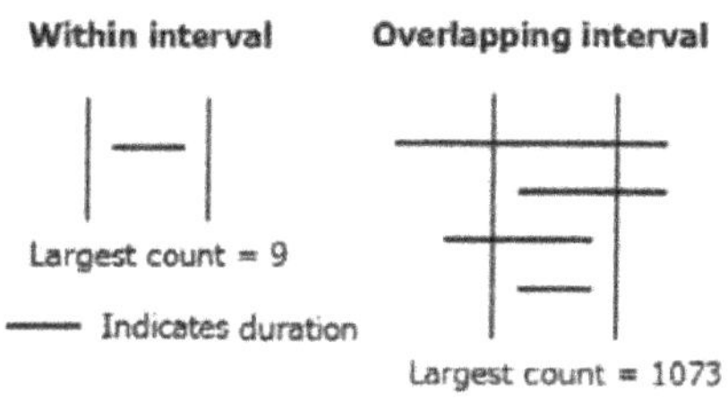

Figure 2 Alternative methods of determining object counts for a specified time interval (2a - left. diagnostic objects; 2b - right. all potential objects)

The aoristic approach of distributing probability across the possible temporal range of a crime incident, can be applied to archaeological material by distributing artefact or feature weighting across a time period identified on stylistic or other dating grounds. Where the time range is sharply defined (e.g. short-lived and easily recognised wares), the weighting is high, creating local evidence peaks - a 1950s building is stronger evidence for activity in the 1950s than a generic neo-classical building is evidence for activity in the 1850s (or indeed, in any other decade). Non-uniform distribution of weighting can also be applied if there is an 'a priori' reason such as additional knowledge about particular artefacts or periods from other sources.

There are of course alternatives methods for allocating objects/events to time intervals. Classically one would count the number of artefacts which fall within each time interval (Fig.2a), and the intervals themselves would be of different durations defined differentially in terms of perceived 'periods' recognisable from the artefacts. As well as being a slightly circular argument, this method does not allow for different degrees of temporal resolution in temporally overlapping artefacts. The problem is that if the interval is defined narrowly enough to provide temporal discrimination, very few objects/events will fall in the interval and the bulk of less 'diagnostic' objects will not contribute to the result.

On the other hand, if we count the number of objects which might have existed in a particular interval (Fig.2b), i.e. all those whose time range overlaps the interval in question, the

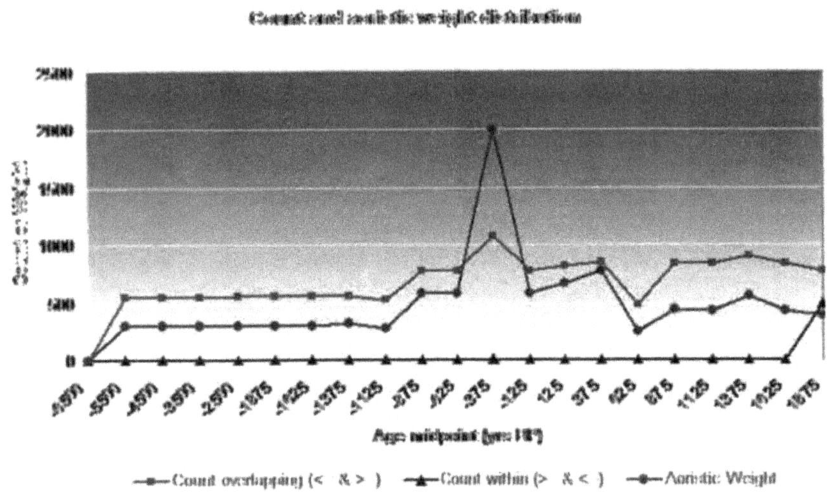

Figure 3 Kythera preliminary data: temporal discrimination of aoristic weights (circles) compared with distribution of temporal marker artefacts (triangles)

contribution of tightly dated 'diagnostic' pieces is swamped by the mass of undiagnostic material of indeterminate date.

Aoristic analysis steers a middle ground by attempting to weight material according to the strength of information that it contributes, and thus maximise the use of available information to define through-time variation from sparse data.

Aoristic Analysis and Spatial Distributions

The ability to make use of all available information is particularly critical when one wishes to look at the spatial distribution of activities or occupation. While there may be enough diagnostic pieces to describe the overall distribution of occupation through time for a study area, the division of a study area into spatial units imposes the need for a much larger sample if we are to define shifting activity foci. By combining the information from 'type specimens' with more loosely defined finds categories, aoristic analysis increases the effective sample size by sacrificing temporal precision for temporal pattern.

I have applied the aoristic approach, with some limited success, to preliminary data from the Kythera survey. The data available represents only a small part of the survey and reveals the need to systematically apply identification to all the survey material collected in previous seasons. We hope to complete identification of the backlog of collected material during the 2004 field season, allowing it to be incorporated into the analysis.

Figure 3 shows the distribution of aoristic weight (circle markers, highest peak) compared with a simple count of artefacts which have time ranges overlapping each interval (square markers, lower peak) and a count of artefacts distinctive enough to be identified to a time interval (triangle markers, merging with the X axis). The aoristic weight emphasises narrowly dated artefact types without ignoring the 'background' of less diagnostic pieces.

Method

We divide the overall time frame of the material recorded into equal intervals. In the case of Kythera, for example, we have a time range from a notional 10,000 BCE to 2,000 CE, which we have divided into arbitrary 100 year intervals. The artefact is allocated a weight, calculated as:

Aoristic Weight per interval = Interval Size / Time span for artefact type

for each interval which lies within the possible time span of the artefact. The weight is thus highest for artefacts with a limited potential time span ('diagnostic' artefacts) as opposed to generic artefacts which could have been produced and/or deposited over a wide range of time. For example, an artefact which can be identified to a 200 year time span will have a weight of 0.5 in each of two 100 year intervals. An artefact which might have been produced anywhere within a 2,000 year time span will have a weight of 0.05 in each of twenty 100 year intervals.

The time span for each artefact is determined from its identification to a 'chronotype' (see Given et al. 1999; examples of chronotypes are given below). Chronotypes are assigned a period, ranging from quite specific dating (e.g. Early Helladic IIIb) to a very wide range (e.g. Pre-Ceramic). The chronotype identifier is used as a foreign key to a table of periods, which in turn provides the time span for the chronotype, and hence the artefact. For example:

Narrow dating:

> Chronotype (allocated): Amphora, Smyrna Jar
> Period (derived from chronotype): Mediaeval, late
> Dating (derived from period): 1200 - 1537 CE

Broad dating:

> Chronotype (allocated): Pithos (undifferentiated)
> Period (derived from chronotype): Ceramic age
> Dating (derived from period): 6300 BCE - AD2000

The problem with equal weighting of time intervals is that historical time does not operate to a uniform scale. The upheaval of the industrial revolution is below the resolution of dating for the later Palaeolithic. Like an expanding universe, time draws out as one goes further from the here-and-now, artefacts become scarcer and the information available becomes thinner and less diagnostic.

To use equal time intervals and weightings based on them would seriously skew the analysis in favour of recent periods, where artefacts can be placed within a few years or decades. Even quite generic modern artefacts will achieve weightings of 0.5 - 1.0, whereas the most distinctive Palaeolithic artefacts would rarely achieve a weighting of 0.1.

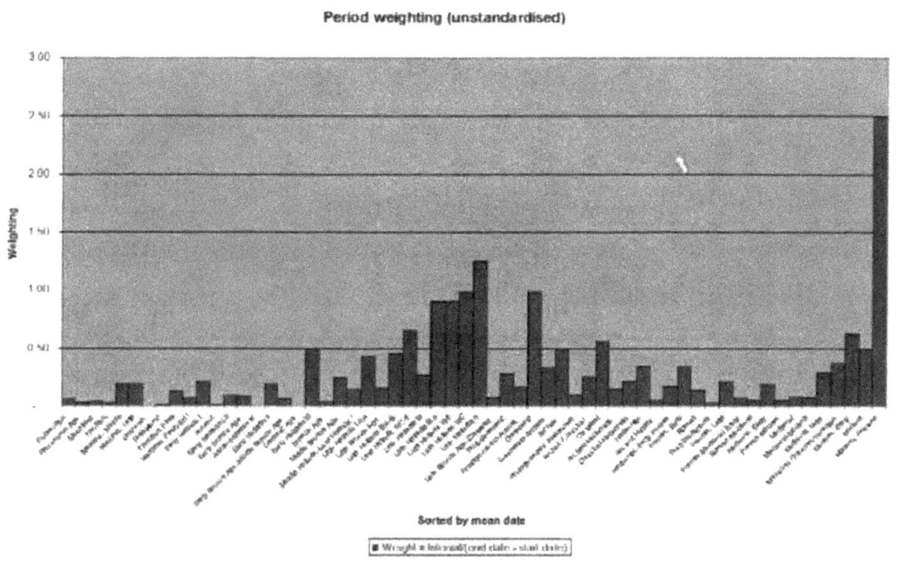

Figures 4a, 4b The effect of standardizing aoristic weights for pottery identification periods by median age of period. Horizontal axis: ranking by median age. (4a - unstandardised, 4b - standardized)

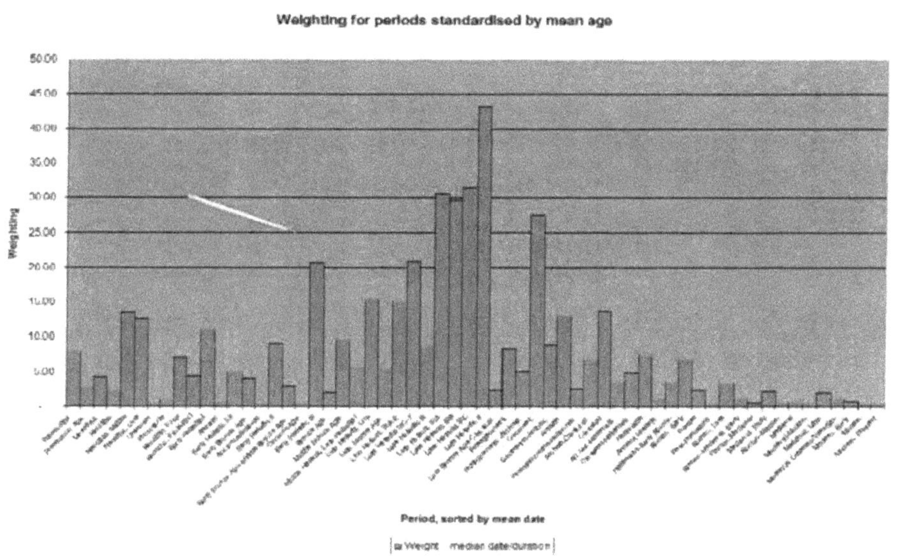

In order to counteract this tendency I used a simple correction which consisted in dividing the median date of the period, rather than the size of the intervals, by the length of the time span. Standardisation by age highlights 'better than expected' dating discrimination, such as the various phases of the Late Helladic (central peak of Figure 4b), each lasting only a couple of hundred years. However the values no longer reflect the total of probabilities (0 through 1) for the artefacts observed, but are instead greatly amplified.

A better scheme would have been to increase the size of intervals in proportion to age and to allocate probabilities equally between intervals irrespective of absolute duration. The results of the correction are shown in Figure 4b, which show the substantial emphasis created for narrowly dated prehistoric periods.

The preliminary analysis of Kythera was carried out using simple MapInfo SQL and MapBasic instructions to generate a series of calculated fields and apply thematic shading. The Finds, Chronotypes and Periods tables were joined to generate time ranges and aoristic weights for each find, and the weights were then summed for each survey unit for each time interval. Finally the survey units were shaded on the map according to the sum of the aoristic weights for the time period in question.

Two preliminary maps are shown in Figure 5 to illustrate the very provisional results of this analysis. It is worth noting that the despite use of the aoristic technique, the discrimination provided by more diagnostic pieces has been largely swamped by the effect of undifferentiated coarse pottery, which could date to anywhere in the last 8,000 years. In future analyses we will remove this hugely overwhelming body of material.

Future Directions

The aoristic approach discussed in this paper accommodates a single method of dating - in this case temporal ranges derived from (primarily) pottery identification. It cannot at first sight accommodate multiple sources of (potentially conflicting) dating which might be encountered, for example, in analysing an assemblage of structures in an urban context. However the allocation of probabilities across time ranges has potential as a means of tackling more complex problems.

For the pre-industrial city of Angkor (Pottier 1999, Fletcher 2001), systematic recording of more than 2000 Angkorean period settlement structures (temples, tanks, house mounds, canals and reservoirs) derived from analysis of airborne radar (Evans 2002), is uninformative on contemporaneous settlement patterns or settlement pattern change through time. However there are many sources of dating information, including inscriptions, feature style, surface finds, coring, stratigraphic relationships and spatial relationships. Using these sources we aim to build a probabilistic web of dating evidence which will combine aoristic analysis of datable artefacts with building styles, stratigraphic relationships of features and spatial associations between features.

Critical to the success of this approach will be the incorporation of stratigraphic relationships between features, defined through observation of their intersection and spatial relationships. Stratigraphic relationships have been widely exploited in archaeological excavations - explicitly identified and analysed since the 1970s using the Harris Matrix (Harris

1979) - and play an implicit role in the analysis of rock art (e.g. Taçon et al. 1996), but have not to the best of our knowledge been systematically applied to spatially distributed archaeological remains at a settlement level.

The eventual aim of these analyses is to generate time series of maps, and to use these maps to answer significant questions about the history of settlement patterns and issues such as depopulation, environmental or economic change. Because the dating evidence is necessarily expressed in probabilistic terms, these maps will not simply indicate a cut-and-dried picture of the layout and density of occupation through time, but will provide a probabilistic statement about what features might or might not have existed at each point in time, or the probable frequency of deposition events through time. The information in these maps may be less hard-edged than we would like, but this simply reflects our real knowledge of the data.

My aim in experimenting with aoristic analysis is to develop a methodology for more reliable assessment of archaeological field survey results and settlement patterning evidence, based not on selective interpretation of observations but on statistical analysis of all available chronological data. This methodology may fundamentally influence the ways in which we interpret palimpsests of archaeological features.

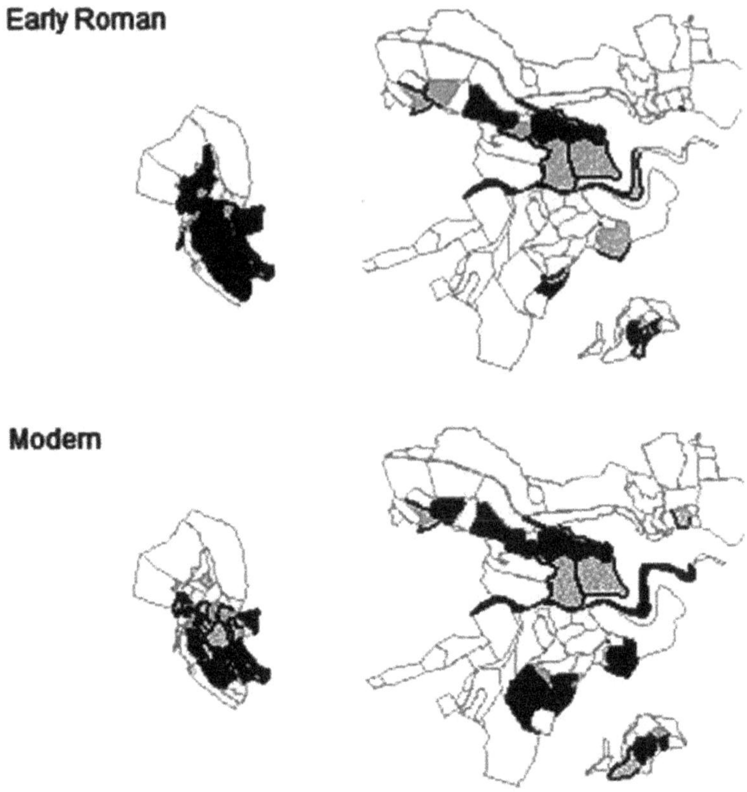

Figure 5: Shift in aoristic weightings between Early Roman and Modern periods: Vythoulas area (left) and Paliochora area (right)

REFERENCES

CORONEOS, C., DIACOPOULOS, L., GREGORY, T., JOHNSON, I., NOLLER, J., PASPALAS, S. and WILSON, A., in press. In The Australian Paliochora-Kythera Archaeological Survey Field Seasons 1999-2000, Mediterranean Archaeology 15.

EVANS, D., 2002. Pixels, Ponds and People: Urban Form at Angkor from Radar Imaging. BA Honours thesis, University of Sydney.

FLETCHER, R., 2001. A.R. Davis Memorial Lecture: Seeing Angkor, new views of an old city. Journal of the Oriental Society of Australia 32-33:1-25.

GIVEN, M., KNAPP, A.B., MEYER, N., KASSIAIDOU, V., NOLLER, J., GREGORY, T.E., WELLS, L., URWIN, N. and WRIGHT, H., 1999. The Sydney Cyprus Survey Project: an Interdisciplinary Investigation of Long-Term Change in the North Central Troodos, Cyprus. Journal of Field Archaeology 26:19-39.

HARRIS, E., 1979. Principles of Archaeological Stratigraphy. Academic Press, London.

JOHNSON, I., 1999. Mapping the fourth dimension: the TimeMap project. In Dingwall, L., Exon, S., Gaffney, V., Laflin, S. and van Leusen, M. (eds.), Archaeology in the Age of the Internet, British Archaeological Reports International Series 750:21pp.

JOHNSON, I., and WILSON, A., 2003. Making The Most Of Maps: Field Survey On The Island of Kythera. Journal of GIS in Archaeology 1:19-89.

POTTIER, C., 1999. Carte Archéologique de la Région d'Angkor, Zone Sud. Ph.D thesis, 3 vols. Universite Paris III - Sorbonne Nouvelle, UFR Orient et Monde Arabe.

RATCLIFFE, J.H., 2000. Aoristic Analysis: the spatial interpretation of unspecific temporal events. Int. J. Geographical Information Science 14(7):669-679.

RODDICK, J.F., HORNSBY, K. and SPILIOPOULOU, M., 2001. An Updated Temporal, Spatial and Spatio-Temporal Data Mining and Knowledge Discovery Research Bibliography (http://www.cs.flinders.edu.au/People/John_Roddick/STDM Papers/, 13 June 2001).

TAÇON, P., WILSON, M. and CHIPPINDALE, C., 1996. Birth of the Rainbow Serpent in Arnhem Land rock art and oral history. Archaeology in Oceania 32:103-24.

TANSEL, A., CLIFFORD, J., GADIA, S., JAJODIA, S., SEGEV, A. and SNODGRASS, R., 1993. Temporal Databases: Theory, Design and Implementation. Benjamin-Cummings, Redwood City, California.

YUAN, M., 2001. Representing Complex Geographic Phenomena in GIS. Cartography and Geographic Information Science 28:83-96.

Quantifying the State of the Art

Christian Mayer
Federal Commission on Historical Monuments,
Vienna, Austria

Abstract

Databases of archaeological sites provide a huge amount of information on the archaeological landscape. This information does not only refer to the location of a site but also to exactness of dating, number of reports and their temporal distribution thus providing an important source for assessing the state of the art in an archaeological perspective.
To analyse this information, each site is scored according to formal aspects (location, dating etc.) as prescribed by the data model of the database to which the information is stored and weightened against the maximum possible score of a site.
The spatial distribution of the scores is then analysed by means of Geostatistics, spatial autocorrelation and Geographically Weightened Regression (GWR) to provide an overview over the knowledge of the archaeological landscape under investigation.
The paper will present a study of about 25,000 sites from about 9,000 geographical units in Austria where information is taken from the National Archaeological Survey conducted by the Federal Commission on Historical Monuments, Vienna. Aim of the study is to explain the local differences in knowledge of Austria's archaeological landscape such determining regions of poor archaeological provision.

To quantify the state of the art is an important task, since this is the only way to point out deficits in the general set-up of a science. The usual approach to obtain such quantification is to rate publications by rank of the journals in which they are published. Certainly, that approach will give a general picture of the quality of scientific work done so far. But as far as archaeology is concerned, a spatial concept is of interest, that provides an idea how well an archaeological landscape is known. In this paper an alternative approach will be put forward that uses a database of an archaeological survey, actually its data model, to develop a measure for the state of the art of an archaeological landscape.

The database used for this purpose is the National Archaeological Survey (NAS) run by the Austrian Federal Commission on Historical Monuments. Its data model (MAYER 1996, 2002) discerns between topographical places and sites, where the latter contains - among others - information on dating, location, cultural affiliation and class of site (like settlement, burial place, deposit and stray finds) grouped by class of site and dating.

score	0	1	2	3	4
location	unknown	cadastrial unit	field name	coordinates	parcel of land
dating	unknown	period 1 (prehistoric,..)	period 2 (Iron Age)	period 3 (Early Iron Age)	phase (Ha C,..)
cultural affiliation	unknown	known			
class of site	unknown	contex unknown	settlement, burial place	specification of the latter	

Table 1 Scores

Basically, an entry into the database sets off with a record of any archaeological item, either find object or locality and this information is stored according to its available evidence. The data model formalises this evidence into a number of elements such giving an idea how well an archaeological item is known. In fact the terminology by which the evidence is transformed into a record is hierarchically stratified. The more evidence is available, the higher is the level in the hierarchy of the term applied. This provides us with a concept of exactness for each of the 4 major elements mentioned above (Tab.1): Assigning a score to each level of hierarchy in the terminology gives a measure of the amount of knowledge about each element. The sum of these scores serves as a measure of the amount of knowledge about the item as a whole. Since more than one site may be found at the same topographical place, the scores of each site are summed and the sum is divided by the number of sites at a place.

As an illustration of some methodological aspects of the approach introduced here, scores of prehistoric sites will be used that stem from the parts of Austria already covered by the NAS. The study area comprises 44,613.58 km^2 that is 53.18% of the Austrian territory. 5,640 out of 10,082 mapable places are considered. A full account of the results obtained from the analysis of all data will be given by a mapping project and be published in the Fundberichte aus Österreich.

The distribution of the sum of scores is of major importance for the methodological aspects of this approach to quantify the state of art. Firstly, the scores used here are discrete, meaning that only 13 distinct values can be attributed. Of course, a finer terminological grid would produce a greater variety in the values, therefore the distribution of scores is a distribution in blocked form. Secondly, this distribution gives an idea how we can rate the state of the art as good or bad: It is clear that a better state of the art than found in the working area would have a higher average score. If the state of the art is homogenous through the working area, the standard deviation would be small. Further, if the average of scores gets close to the maximum obtainable score, the distribution may be skewed to the right. Of course, life would be easier if the scores would be normally distributed which they are not. Therefore, we cannot use the three first moments of a normal distribution (average, standard deviation and skewness) but at least we can replace them by their robust counterparts.

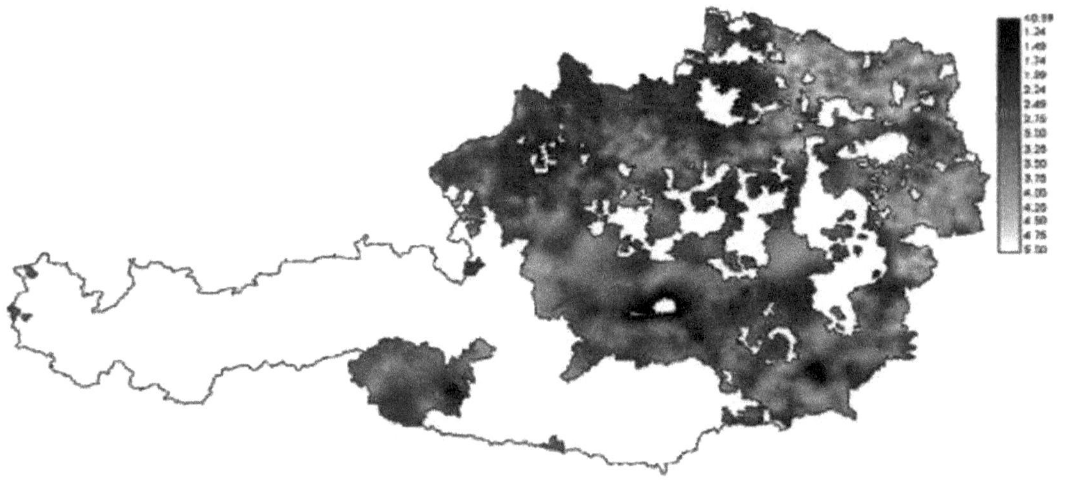

Figure 1 State of the Art in Austrian Prehistoric Archaeology

By virtue of the problem, we are dealing with a spatial phenomenon. Therefore, the overall distribution of scores calculated from the working area does not describe the state art in a region satisfactorily. Applying krigeing (Cressie 1993:29) as a method of spatial analysis to the data, we obtain a spatial picture of the general situation. But this picture is as informative as deceptive. In fact the cross validation coefficient of the underlying model is only about 0.45. Although a highly significant value, the determination coefficient is only about 20%. Of course, one wonders where the rest of the variation has gone. Looking at the variogram we see that the variation within the data sets off on an already high value at low distances and the spatial dependency is measurable only to rather small distances (about 1.8 km). This means, that good knowledge of one site does not necessarily guarantee a good knowledge of a site nearby. What about exactness of location, class of site, dating and cultural affiliation ? Exactness of location as well as class of site showed simply no spatial dependency. This is because the scores of these two attributes of the sites have very low standard deviations, the exactness of location due to the excellent archive work done with the archaeological survey, class of site because most of the sites are known only by surface finds. The map shown in Figure1 shows thespatial distribution of the sum of cultural affiliation and exactness of dating. The other variables are ignored for being of less importance. The correlation between estimated and actual data is only about 0.5 for reasons discussed already. Clearly "good knowledge" is quite rare.

Of course, one is interested to know, by what effects the scores are influenced. For instance, one would like to know, whether the vicinity of a museum influences the state of art in its vicinity. Typically, this question is transformed into a regression model by using the scores as the dependent variable and the distance to the nearest museum as the regressors. To illustrate this, the location of about 370 museums were collected and the distances to each mapable site calculated. To provide a better picture, the museums were classified to their body of responsibility being run by a private person, a local authority or a local society as the first group, the federal states as second group and the Republic of Austria as the third group. A classical regression model using these variables give a general correlation of only 0.125. Again this result is deceptive since classical regression does not consider the local dependency between the variables. As an alternative, geographically weighted regression has been applied (Fotheringham et al. 2002). Applying this method to the data, we obtain a correlation coefficient of 0.61. Variable selection detects the federal museums to be of little interest in this respect, since they are all situated at Vienna. Leaving them out we obtain a correlation coefficient of 0.60, museums run by the federal states and local museums produce a correlation coefficient of 0.57 and 0.58 respectively. Looking at these figures a second time the results seem paradox: One has to expect, that the closer a site is to a museum, the higher the knowledge of the site would be. In that case, the correlation coefficients would be negative, which they are not. In a map of the local correlation coefficients there are areas with very large negative values that are indeed close to museums. But again, this is not a general phenomenon since there are quite a lot of museums that have obviously no influence on the knowledge of their surrounding archaeological landscape. Further, one should not mix high negative correlation coefficients with good knowledge of the region. When the scores are multiplied with the sign of the correlation coefficient, it becomes obvious, how little the presence of a museum influences the knowledge of the surrounding archaeological landscape. Obviously, whether a museum influences the state of art in its vicinity depends on the museum's activity but not simply on its existence.

Undoubtedly, the quantification of the state of art in archaeology by the proposed approach has produced very interesting insight into Austrian archaeology, although results presented are not necessary good news.

REFERENCES

CRESSIE, N.A.C, 1993. Statistical for Spatial Data. Wiley.

FOTHERINGHAM, A.St., BRUNSDON, Ch. and CHARLTON, M., 2002. Geographically weighted regression, Wiley.

MAYER, Chr., 1996. Fundstellenbezogene Daten in der Abteilung für Bodendenkmale des Bundesdekmalamts, Fundberichte aus Österreich 35:321.

MAYER, Chr., 2002. Some aspects of SMR management in Austria. In Mapping the future of the past 37.

MACHINE LEARNING APPLIED TO GEO-ARCHAEOLOGICAL SOIL DATA

O. S. FARRINGTON
SCHOOL OF MATHEMATICAL AND COMPUTER SCIENCES, HERIOT-WATT UNIVERSITY, EDINBURGH, UK

N. K. TAYLOR
SCHOOL OF MATHEMATICAL AND COMPUTER SCIENCES, HERIOT-WATT UNIVERSITY, EDINBURGH, UK

ABSTRACT

'Decision tree learning' is one of the most versatile and practical methods of inductive inference developed in applied artificial intelligence. The 'Machine Learning' technique selected for this study was the ID3 algorithm, and represents a new application for this simple classification building technique.
The data, to which the ID3 algorithm was applied, consists of a collection of data sets relating to specific soil analytical procedures (e.g. heavy mineralogy, bulk geochemistry, fabric analysis, etc.) originally collected to address the dark earth conundrum.
In a previous study, PCO analysis proved adept at identifying contributions from individual variables such that it was possible to identify the geo-archaeological components that together constituted the 'character profile' of a typical dark earth.
In this study the classification building abilities of the ID3 algorithms, have been used to explore the possibility of establishing a classification (based on soil data) of not only dark earth's but other archaeological deposits characteristic of urban contexts.

INTRODUCTION

A unique database of geo-archaeological soil data was put together several years ago from the results of analyses of soil samples collected from selected urban archaeological deposits at excavation. The database at present consists of a collection of multi-variate ordination plots: the product of the application of principal coordinate analysis (PCO) to soil data from a uniquely comprehensive range of (standard) soil analytical procedures that were applied to a large collection of predominantly urban archaeological soil samples. It has only recently become possible to resume work on this data, which originally formed the basis for a University of London doctoral thesis, soon to be completed.

In the intervening period, the first author has been introduced to some of the newer techniques of Artificial Intelligence, and became aware of their potential suitability for further mining of the database. During preliminary assessment of the principal coordinate plots, it was noted that the urban archaeological deposits studied appeared to be made up of two components; one derived from the local geology and the other an 'anthropogenic' component. This is a new observation for an archaeological deposit. However new studies now in progress (located in the Department of Environmental Science, University of Stirling, under D.A. Davidson), and based on rather different types of archaeological soils than we intend to use, have recently reported a similar finding: that the archaeological soil ('plaggen' soil) may similarly have an 'anthropogenic' component.

We intend to examine in greater detail the supposition that archaeological deposits are in general made up of material from the local geology and an anthropogenic component. We are particularly interested in the composition of the anthropogenic fraction in our urban soils and its stability between different types of archaeological deposit. To this end the geo-archaeological database of urban archaeological material should prove invaluable. This very large archaeological soil database is probably the largest and most comprehensive dataset available of different types of archaeological soils. We have selected in the first instance to work on the mineralogy of the fine sand fractions of a collection of archaeological deposits that included dark earths, occupation deposits, cultivation, and other putative deposit types including comparative material from the local surficial geologies of the archaeological samples in the study because PCO plots show a clear separation between local geology and 'anthropogenic source material' in the form of a number of species of 'authigenic' minerals.

Initially we will look to use various artificial intelligence data-mining techniques, possibly in conjunction with more multivariate ordination statistical methods.

This new study will form the basis of a short research project in the School of Mathematical and Computer Sciences at Heriot-Watt University, Edinburgh.

THE DATABASE

The data to be used for data-mining concerns the mineralogy of the fine sand fraction of a range of urban archaeological deposits, summarised in a pair-wise multi-variate ordination plot between PC1 and PC2. This plot demonstrated a clear separation between minerals derived from the local geology and a putative anthropogenic component.

Several of the archaeological deposit samples were observed to be rich in authigenic minerals, such as the synthetic olivines (fayalite). Their presence in these deposits is easily predicted to represent the result of anthropogenic activities, such as metal processing, which produced these synthetic minerals.

The Proposed Data-Mining Methods

The Artificial Intelligence techniques which appear initially to be of most value are the Induction Tree techniques, of which the simplest is the ID3 Algorithm (Quinlan 1986), and Artificial Neural Networks, such as the Multi-Layer Perceptron (Rumelhart et al. 1986).

The Archaeological Context

The soil database was prepared from soil samples collected from urban archaeological sequences between 1984 and 1988, specifically for the purpose of investigating the composition provenance and depositional history of the enigmatic urban dark earth horizons.

The sampling programme adopted was fully comprehensive and involved collection not only of dark earths, but all other 'types' of archaeological deposit available in the same profile or near vicinity. Samples of the local geological substratum were also routinely collected.

A preliminary assessment of the data concerning soil samples taken from urban archaeological deposits (primarily dark earths) can be found in Farrington 1989, or Farrington and Bateman 1992.

Until recently, archaeologists have taken little interest in the routine analysis of the < 2 mm fraction of an archaeological deposit.

On the other hand, geomorphologists have over the past 100 years developed numerous methods of investigation of the <2mm fractions of soils and sediments. Quaternary soil scientists in particular have developed methods of identifying depositional processes and provenances of landscapes, which are especially suited to materials common in late Quaternary (Cenozoic sub-era) landscapes. As such these were deemed suitable for the investigation of the depositional history and provenance and their contribution to former urban landscape development.

The Adoption of Soil Science Analytical Procedures

This is the first attempt at a study based entirely on soil samples from archaeological deposits, using standard soil science analytical procedures. All soil analytical work was carried out in the Soils Division, Rothamsted Experimental Station, Harpenden, Hertfordshire. At the conclusion of the pilot study an internal report was sent to the sponsors; the Museum of London (Farrington 1983).

In 1984 a much more ambitious project was initiated as the basis for a doctoral research project shared jointly between the Department of Geography, Birkbeck College, University of London, and the Soils Division, Rothamsted. The topic was the application of soil analysis to an archaeologically and historically important archaeological deposit: Dark Earth.

A full list of all the analytical procedures used is inappropriate here, but will appear in the doctoral thesis.

When the plots were drawn, preliminary observations showed that the archaeological deposits which had been sampled were derived from two different fractions; one was the local geological substratum and the other was an anthropogenic component.

It is the aim of the present study to take as its starting point this observation and to examine more closely the distribution of minerals between the putative anthropogenic and geological components to identify the boundaries between the two.

This proposition is suitable for the two Artificial Intelligence procedures (ID3/MLP), as both test slightly different aspects of the problem.

Preliminary Statistical Analysis (Multivariate Ordinations)

A considerable volume of geo-archaeological data was summarised using the multivariate ordination procedure, know as principal coordinate analysis contained in the statistical pakkage Genstat developed at Rothamsted. Genstat Version 5 release 4.5 was used to produce scatter-grams based on pairwise plots of the first four principal co-ordinates. These were then visually inspected for groups, patterns or trends. Gower Similarity Coefficients were also calculated for all pair-wise combinations of samples and minimum spanning trees, linking samples by maximum Gower Similarity Coefficient, were superimposed on selected PCO scatter-grams.

It was observed that in general the first two principal co-ordinates provided the best separations, both with and without the addition of the minimum spanning trees. One database, that of the mineralogical data, was selected as the basis of this study. The plot of PC1 against PC2 produced a clear visual separation between archaeological deposits according to locality and anthropogenic content.

When plotted on a scatter-gram, the first two components resulting from the PCO clearly clustered the samples according to location. This is readily discernible visually, but the clusters are not linearly separable. I.e. one cannot draw straight lines on the scatter-gram which separate the clusters and so their boundaries remain unknown. Such data is well suited to processing with a Multi-Layer Perceptron (MLP).

Artificial Neural Networks

An MLP is an Artificial Neural Network (ANN) which can be trained by example to perform a non-linear mapping from a set of input values to one or more output values, i.e. it can discover an "optimal" set of curves, which do separate the clusters in a scatter-gram. Subsequent samples can then be presented to the MLP for classification into the cluster which they best fit.

One problem with ANNs is their 'black-box nature'. It is very difficult to ascertain exactly what they have learnt during training. It is somewhat easier to extract rules from symbolic learning systems.

Inductive Decision Tree Methods (ID3)

One very popular symbolic learning system, ID3, is capable of inducing the rules which determine where the clusters will appear on a scatter-gram. This system will, therefore, not only be able to classify new samples into appropriate clusters, it will also provide information indicating why the clusters were chosen. Symbolic systems generally perform better with discrete data values rather than continuous variables and tend to fail less gracefully than ANN approaches.

Machine learning techniques such as these offer great benefits in the analysis and interpretation of geo-archaeological soil data and the authors are currently investigating the potential of a number of algorithms.

Acknowledgements

Maria Chamberlain provided many helpful suggestions during the drafting of this paper.

Jill Stirling acted as scribe to Orpah Farrington during the drafting of this paper to circumvent some of the difficulties of Dyslexia.
All soil samples from the archaeological deposits in the database used in this study were obtained with permission.

Training in the methods of examination of the mineral components of archaeological soil samples was received from John Catt. All analytical work was carried out in the Mineralogy Laboratory Suite of the Soils Division, Rothamsted Experimental Station, Harpenden, Hertfordshire.

Much helpful discussion on the interpretation of Principal Co-ordinate Analysis plots was provided by John Catt, and by the late James Rayner of the Statistics Division, Rothamsted.

Orpah Farrington is in receipt of a Research Student Scholarship from Heriot-Watt University.

REFERENCES

DERCON, G., DAVIDSON, D.A., SIMPSON, I.A., DALSGAARD, K. and SPECK, T., 2003. The Nature and Formation on Anthrosols in NW Europe: a comparison between 3 countries. Extended Abstracts Second International Conference on Soils and Archaeology, Pisa:21-24.

DIGBY, P.G.N and KEMPTON, R.A., 1991. Multivariate analysis of ecological communities: population and community biology series. 3.5 Principal coordinates analysis:83-93. Chapman & Hall, London.

DORAN, J.E. and HODSON, F.R., 1975. Mathematics and computers in archaeology. Edinburgh University Press, Edinburgh.

DUCKE, B., 2003. Archaeological predictive modelling in intelligent network structures. In Doerr, M. and Sarris, A. (eds.), CAA2002: The Digital Heritage of Archaeology: Computer Applications and Quantitative Methods in Archaeology: Proceedings of the 30th Conference, Heraklion, Greece, April 2002. Archive of Monuments and Publications Hellenic Ministry of Culture:267-273.

FARRINGTON, O.S., 1983. The dark earth project. Unpublished Museum of London internal report.

FARRINGTON, O.S., 1989. Dark earth in Northwest Europe: application of geoanalytical techniques to a historically important archaeological deposit (abstract). 28th Int. Geol. Congr. Abst. 1:473.

FARRINGTON, O.S., 2001, unpubl. Machine learning (ID3), and artificial neural networks (MLP): A comparison of the properties of two different types of classification system applied to the same dataset: A selected subset of letters of the Latin Alphabet. MSc class report, Heriot-Watt University.

FARRINGTON, O.S. and BATEMAN, R.M., 1992. A holistic approach to the analysis of archaeological deposits, illustrated using a late Roman urban sequence from Northwest Europe. Mat. Res. Soc. Symp. Proc. Vol. 267:179-192.

GOWER, J.C, 1966. Some distance properties of latent roots and vector methods used in multivariate analysis. Biometrika 53:325-338.

GOWER, J.C., 1971. A general coefficient of similarity and some of its properties. Biometrics 27:857-872.

GOWER, J.C. and DIGBY, P.G.N., 1981. Expressing complex relationships in two dimensions. In Barnett, V. (ed), Interpreting multivariate data, Wiley, Chichester:83-118.

GOWER, J.C. and ROSS, G.J.S., 1969. Minimum spanning trees and single linkage cluster analysis. J. R. Stat. Soc. C 18:54-64.

MITCHELL, T., 1997. Machine learning. McGraw-Hill, NewYork.

QUINLAN, J.R., 1986. Induction of decision trees. Machine Learning 1(1):81-106.

RUMELHART, D.E., HINTON, G.E. and WILLIAMS, R.J., 1986. Learning internal represaentatioons by error propagation. In Rumelhart and McClelland (eds.), Parallel Distributed Processing Vol. 1, MIT Press, Cambridge Massachusetts:318-362.

SHENNAN, S., 1988. Quantifying archaeology. Edinburgh University Press, Edinburgh.

THOMAS, J., SIMPSON, I.A., DAVIDSON, D.A. and GAULD, J.H., 2003. GIS mapping of anthropogenic soils in Scotland: investigating the location and vulnerability of Scottish 'plaggen soil' deposits. Extended Abstracts Second International Conference on Soils and Archaeology, Pisa:127-128.

WILSON, C.E., DAVIDSON, D.A. and CRESSER, M.S., 2003. Multi-element soil analysis as an aid to archaeological interpretation. Extended Abstracts Second International Conference on Soils and Archaeology, Pisa:78-79.

Adding a New Dimension to Pollenanalysis: Human Impact in Space and Time

Jutta Lechterbeck
Institut für Geologie der Universität zu Köln
Zülpicher Str. 49a, 50674 Köln, Germany
jutta.lechterbeck@uni-tuebingen.de

Abstract

The high resolution palynological record of the annually laminated sediments of Lake Steisslingen (SW-Germany) was subject to correspondence analysis (CA). The results of the CA showed that a single gradient dominated the pollen distribution. Further investigations allowed identifying this gradient as increasing human impact from the Neolithic to the Middle Age. The eigenvalues of each pollensample were plotted against time. The resulting curve proved to be a proxy for human impact. This proxy could be tested against the archaeological record of the region. A tool for presenting complex stratigraphical data in a single curve was thus found.

This tool was applied to two other pollen profiles of the region, which proved likewise to be dominated by human impact. The next step will be to display the data spatially for different time slices and thus show the differences in the intensity of human impact in a certain region at a given time. This project is planned as a contribution to LUCIFS.

Introduction

Pollen analysis is a conventional and established method to reconstruct former vegetation. Pollen contain the male gametes, i.e. reproductive cells, of flowering plants. They are emitted in great numbers, transported and deposited in natural archives, such as lakes and mires. From those deposits they are recovered, determined and counted. The composition of the pollen assemblage is used for the reconstruction of the contemporary vegetation cover.

Naturally, the question after the pollen source area is one crucial to pollen analysis: How large is the area represented by a pollen assemblage?

The question was primarily addressed by determining how far pollen can be transported. It was generally agreed, that a given archive reflects the vegetation of the surrounding landscape depending on the openness of the area and the size of the archive (cf. Faegri and Iversen 1989).

More recent studies employing data recovered from pollen traps, however, showed that the role of long distance transport was generally overestimated (for example Damrath 1998).

If the modelling of vegetation of entire landscapes is concerned, the question "How far can pollen travel?" is not adequate. The appropriate question should be: "How large is the area which allows to reproduce the pollen record adequately?"

The same approach underlies simulation studies and much of the theoretical work already done by Odgaard and Rasmussen, who come to the conclusion that a diameter of 2 to 5 km accounts for over 90 % of variation in the pollen record (Odgaard and Rasmussen 2000). In this paper I will show how this radius can be confirmed with the help of archaeological and pollenanalytical data and correspondence analysis. With help of correspondence analysis a main dimension of explanation can be extracted from the pollen data and identified with a certain factor, for example human impact.

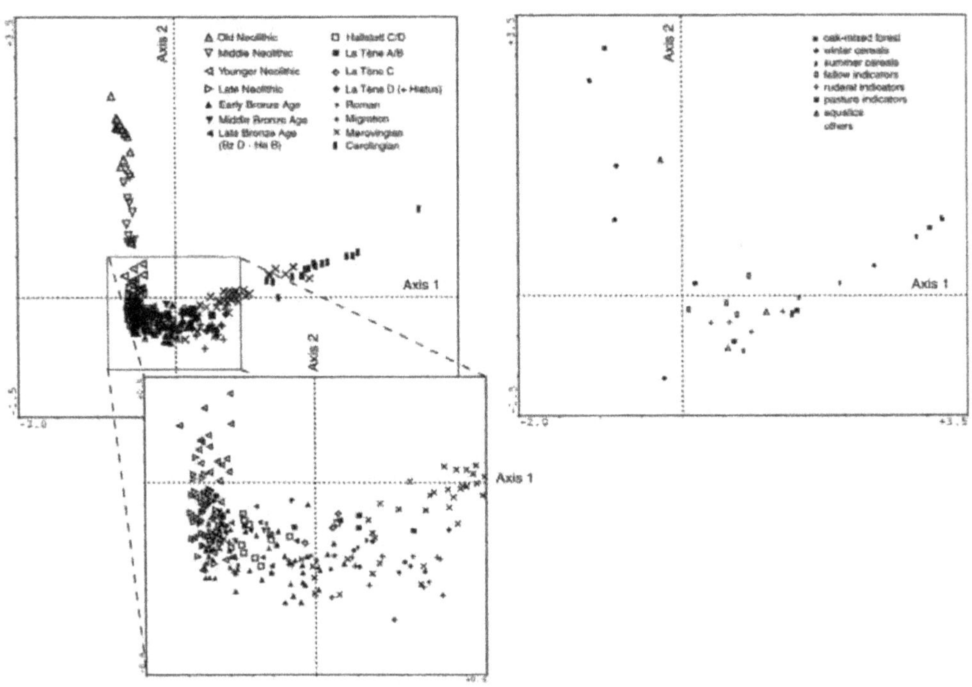

Figure 1 Lake Steisslingen. Correspondence analysis. Plot of sample and species scores in the plane of the first two eigenvectors

Basics of Pollen Analysis

The structure of pollen data is that of a contingency table, where the rows are the samples and the columns contain the taxa; the cells contain counts.

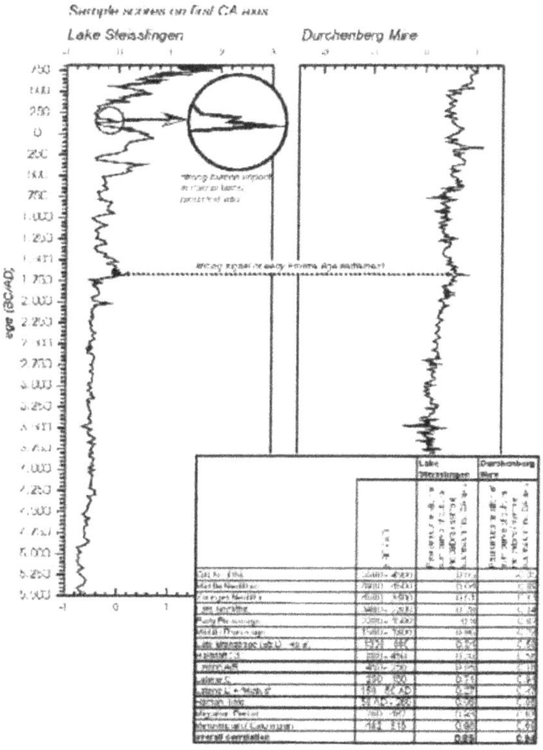

Figure 2 Lake Steisslingen and Durchenberg Mire. Sample scores plotted against time. Table of Pearson correlations between sample scores and sum of cultural indicators

Figure 3 Archaeological map and location of Lake Steisslingen and Durchenberg Mire

Usually pollen data are displayed as a pollen diagram: The percentage values of every taxon in every sample are calculated and represented in stratigraphic order. The resulting curves reflect the changes in the pollen composition through time. Actually, the pollen profile shows changes in the fossil record which are induced by several factors. The most important factor is vegetational changes but others like transportation or selective destruction of pollen grains have to be considered likewise.

Human impact in the pollen diagram can be determined in two ways: Positively by so called cultural indicators, which are pollen of cultivated plants, - such as cereals - , and of plants which are tolerant to human impact. And negatively by the disappearance of pollen from plants which do not tolerate human impact. The counts of cultural indicator pollen can be summarized in a single percentage curve. This curve already gives a kind of measure for the intensity of human impact through time.

Case Study: Lake Steisslingen

Lake Steisslingen is a little hardwater lake in south western Germany near Lake Constance. It contains sediments from 16,000 years ago until today, most of them annually layered. This annual lamination allows the establishment of a very detailed chronology by counting the single layers, analogue to tree rings.

Lake Steisslingen was subject to an interdisciplinary project and one part of it was a high resolution pollen analysis. The Lake Steisslingen pollen profile consists of more than 600 samples. In each at least 500 pollen grains were counted. The data matrix and the resulting pollen diagram are quite large, so it was obvious to apply dimension reducing multivariate statistical methods. Based on the works of (Birks et al. 1988) and (Kalis and Zimmermann 1997) correspondence analysis was chosen.

Here, the results of a reduced dataset are presented covering the time span between probable first human impact in the Neolithic and the early medieval. The data matrix contains all samples and all taxa except single finds.

The plot of sample scores (Fig.1) in the plane of the first two eigenvectors forms a parabola, the higher axes show likewise a polynomial relationship. This means that a single natural gradient accounts for the pollendistribution. The parabola is ecologically meaningful evidence, proving a mathematical continuity of a clearly prevailing dimension (Wartenberg et al. 1987). In archaeological seriation, this dimension is usually equalled with time. In Steisslingen this is not the case: On first sight one could assume that the samples are roughly ordered according to their age. But a closer look reveals that they follow another gradient: one arm of the parabola is built of the relatively undisturbed forest plant communities of the Neolithic and the other arm is formed by the open landscape as seen in medieval times. The parabola represents an ecological continuum. The underlying natural gradient -the main dimension of explanation- is an ecological one.

The scores on the first CA-axis for the samples can be plotted against time, the resulting curve (Fig.2) shows the behaviour of the main dimension of explanation through time. In the case of Lake Steisslingen the curve of the eigenvaluesolution ("eigenvaluecurve") correlates from the early bronze age on strongly with the curve of the cultural indicators. The two exceptions are correlated with sedimentological events.

From the strong correlation of the eigenvaluecurve with the cultural indicators it is concluded that the main dimension of explanation in this dataset is in fact increasing human impact from the Neolithic to the middle ages. The eigenvaluecurve is a kind of proxy for human impact and it has the advantage that it contains more information than the sum curve of the cultural indicators.

To compare this curve with the actual archaeological situation, a map and an inventory of the archaeological finds in the surrounding of the lake (Kerig and Lechterbeck 2000) was made. The area chosen for the mapping was an arbitrary 5 km radius around the lake (Fig.3). According to historic and ethnographic parallels, such a distance of one hour's walk is thought to be the range of the frequently used landscape around a settlement (Higgs and Vita-Finzi 1972, Higgs 1975). The pollenprofile as well as the eigenvaluecurve were compared with the archaeological map, but as these results are already published or under publication (Kerig and Lechterbeck 2000, 2003), only one striking example should be given: A little peak in the roman times (Fig.2) indicates roman impact in strength not visible in the pollen diagram. Before mapping a roman villa was predicted on the basis of the eigenvaluecurve, but could at first not be found in the course of the mapping. After mapping a post-it-note was found with a hint that the documentation of roman finds from Steisslingen was filed by accident under another parish name. It turned out to be a little cemetery of the kind regularly situated in the vicinity of roman villae. One terra sigillata sherd could be dated to exactly the period the peak of the eigenvaluecurve covers. This filing accident serves as an experiment: The hypothesis "roman villa" stated by the evidence of the curve could be tested against the actual archaeological situation. Thus the curve can serve for predicting unknown archaeological evidence.

With the eigenvaluecurve a tool to characterise the main dimension of explanation of pollen diagrams is found. In Lake Steisslingen this is human impact. The idea, that this is the case in all long settled landscapes was tempting.

Durchenbergried

The western Lake Constance region is such a long settled landscape and has furthermore several well-dated, high resolution pollen profiles. The next such profile comes from the Durchenberg Mire (Rösch 1990), about 5 km away from Lake Steisslingen, just outside the 5 km radius of the archaeological map (Fig.3). The Durchenberg Mire contains sediments from the Late Glacial until today. The Holocene deposits consist mainly of peats. The stratigraphy is dated by a series of radiocarbon dates. Analogue to Lake Steisslingen, a CA with the pollen data was carried out (Fig.2). Here, the eigenvaluecurve and the cultural indicators do not correlate much. Obviously, in the Durchenberg Mire human impact is not the main dimension of explanation or at least not in the same way, cultural indicators are. In both profiles however, the early bronze age settlement which lies in between the two localities is reflected clearly.

Conclusions

The established and conventional idea, that pollenprofiles are representative for large areas can be rejected by the combination of applying the established and conventional methods of pollen analysis and archaeological mapping. Furthermore, multivariate methods could show that the main dimension of explanation changes already at short distances. This result is in agreement with the simulation studies of Odgaard and Rasmussen.

Pollen analysis is thus a more sensitive method than hitherto thought. More and more detailed information can be gained from pollenprofiles. They deliver reliable data for the reconstruction of landscapes, but this demands a dense network of pollenprofiles.

The eigenvaluecurve is a tool to present complex pollen data in a single curve. The next step is to display the data spatially for different time slices and map the differences in the main dimension of explanation in a certain region at a given time. This is planned as a contribution to the LUCIFS project (Land Use and Climate Interactions in Fluvial Systems).

Acknowledgements

Thanks to M. Rösch, who freely provided the raw data of the Durchenberg Mire profile and to K.H. Lenz, who helped with the dating of the terra sigillata sherd.

REFERENCES

BIRKS, H.J.B., LINE, J.M. and PERSSON, T., 1988. Quantitative Estimation of Human Impact on Cultural Landscape Development. In Birks, H.H., Birks, H.J.B., Kaland, P.E. and Moe, D. (eds.), The Cultural Landscape - Past, Present and Future. Cambridge University Press, Cambridge:229-240.

DAMRATH, K., 1998. An attempt to quantify land surfaces cleared of forests by means of pollen analysis in southern Central Europe. In Gaillard, M.J. and Berglund, B. (eds.), Quantification of land surfaces cleared from forests during the Holocene: modern pollen/vegetation/landscape relationships as an aid to the interpretation of fossil pollen data. Paläoklimaforschung 27, Fischer, Stuttgart:17-30.

FAEGRI, K. and IVERSEN, J., 1989. Textbook of Pollenanalysis. Chichester:328p.

HIGGS, E.S., 1975. Palaeoeconomy - being the second volume of Papers in Economic Prehistory by members and associates of the British Academy Major Research Project in the Early History of Agriculture. Cambridge University Press, Cambridge:244pp.

HIGGS, E.S. and VITA-FINZI, C., 1972. Prehistoric Economies: A Territorial Approach. In Higgs, E.S. (ed.), Papers in Economic Prehistory: Studies by Members and Associates of the British Academy Major Research Project in the Early History of Agriculture. Cambridge University Press, Cambridge:27-36.

KALIS, J. and ZIMMERMANN, A., 1997. Anthropogene Einflüsse auf die Umwelt - Eine kanonische Korrespondenzanalyse von prähistorischen Pollenspektren. In Müller, J. and Zimmermann, A. (eds.), Archäologie und Korrespondenzanalyse - Beispiele, Fragen, Perspektiven. Internationale Archäologie 23, Leidorf, Espelkamp:179-182.

KERIG, T. and LECHTERBECK, J., 2000. 16000 Jahre unter dem See - Zum historischen Quellenwert der Steißlinger Laminite. Hegau:57, 85-109.

KERIG, T. and LECHTERBECK, J., 2003, in press. Laminated sediments, human impact, and a multivariate approach: A case study in linking palynology and archaeology (Steisslingen, South-West Germany). Quaternary International.

ODGAARD, B.V. and RASMUSSEN, P., 2000. Origin and temporal development of macro-scale vegetation patterns in the cultural landscape of Denmark. Journal of Ecology 88:733-748.

RÖSCH, M., 1990. Vegetationsgeschichtliche Untersuchungen im Durchenbergried. In Landesdenkmalamt Baden-Württemberg (ed.), Siedlungsarchäologie im Alpenvorland II. Forschungen und Berichte zur Vor- und Frühgeschichte in Baden-Württemberg 37. Theiss, Stuttgart:9-56.

WARTENBERG, D., FERSON, S., ROHLF, F.J., 1987. Putting things in order: A critique of detrended correspondence analysis. The American Naturalist 129/3:434-448.

Toward a Cultural Measure of Time: Remarks on Phasing CA-Derived Petrie-Matrices and the Use of DCA

Tim Kerig

Universität zu Köln - University of Cologne
Weyertal 125, 50923 Köln, Germany
kerigtim@hotmail.com

Abstract

Correspondence Analysis (CA) offers a proven algorithm for seriation. The eigenvalue-solution can be displayed in a coordinate system for showing the well known arch or parabola effect. There are serious obstacles for the use of these eigenvalues -or the differences between them- as arguments for phasing Petrie-matrices. Detrendet Correspondence Analysis (DCA) is an alternative algorithm, widely used in ecology, but not yet for seriation. DCA allows one to obtain an eigenvalue-solution different from that of CA, designed especially for eliminating the parabola effect. Displayed in a coordinate system the DCA eigenvalues allow a sounder procedure of phasing by measuring equally scaled distances. An example from the early Neolithic Mid-European Linearbandkeramik will show that these distances can be used as a measure of culturally represented time: The scaling of the DCA-solution leads to a general scale as an independent tool for quantifying changes in material culture by explicit archaeological means.

Relative chronology represents time - without any scale. Without external information we simply cannot measure the time depth between two strata in a stratigraphic context or between two graves in a seriation. We cannot even make a qualified guess. Here I will introduce a culturalistic concept of time and its formalized representation that allows an objective scaling of relative chronologies.

Seriation is the arrangement of combinations in terms of their elements along a gradient. Besides the strict definition (a matrix with the consecutive-1's property) I use the term Petrie-matrix in a broader sense: diagonalising a table with types of artefacts in the columns and finds containing the artefacts in the rows should come as close as possible to a matrix with a consecutive and diagonal block of numbers (cf. Kendall 1971).

Correspondence analysis (CA) is today the most common method for such a rearrangement (for example Baxter 1994:100-39, Müller and Zimmermann 1997). Before showing the advantages of an alternative algorithm, Detrended Correspondence Analysis (overview: Jongman 1995), it might be helpful to remember the principles of CA: The first step in CA is the conversion of the types counts into proportional data. A matrix of chi-squared distances is calculated. From this matrix eigenvectors are extracted by orthogonal regression, performed by singular value decomposition. There are as many eigenvectors computed as the smaller number of variables either in rows or columns, minus one. For the variables -both finds and types- the same principal component space is calculated. The first eigenvector accounts for most of the information (inertia) in the dataset; the second is situated to describe the variation of the remainder, and so forth. Let every eigenvector be an axis in an orthogonal coordinate system; hence the position of the variables -finds as well as types- can be determined by coordinates on the eigenvectors. The values on the first eigenvector - or first CA-axis - are of higher explanatory value than the values of all other single eigenvectors. If this main dimension of explanation shows characteristics that are assumed to be also the characteristics of our concept of relative chronology, then we can use the main dimension as gradient in seriation. One can display the existence of a dominant, single, and natural gradient by using the well-known Guttman effect (Greenacre 1984:226-232) for the so called parabola- or horseshoe-test: Here the values on the first eigenvector are plotted against the values on the second. A parabola forms, proving the existence of a single natural gradient. The seriation, the diagonalisation of the original contingency table, is quite simple: We have already identified the main dimension of explanation and equated it with time. The values on the first vector are - somehow!- representing time. The variables, the combinations in the rows as well as the types in the columns, are rearranged according to the ranks of their values.

In daily archaeological practice there is often a need for using phases rather than such ranks, e.g. if you want to date a handful of sherds from a surface collection. For such cases it is useful to determine typological phases. Phases may be defined as periods of hypothetical contemporaneity. There are different ways of phasing a rank-ordered Petrie-matrix: One may take the range of single dominant types for defining phases. This may be useful if there is already an established typological sequence and one wants to integrate the results of the seriation. There are also some examples in the literature of CA-derived Petrie-matrices, phased by Cluster-analysis of the CA-values. As the CA-solution is described as a set of coordinates, it is tempting to use these values as criteria for phasing. By using the coordinates in such a way one takes for granted that the computed CA-space allows the representation of an adequate concept of time. One would assume that the same absolute value in differences between the coordinates represent the same duration - that is not the case. The computed CA-space is not adequate for representations of our concept of absolute chronology. We assumed the CA-axes to be completely uncorrelated and we've therefore chosen an orthogonal coordinate system. There is no information about the absolute duration of anything at all. Expecting a chronology "a little bit more absolute" would be expecting the computer to do magic - doing the transubstantiation from relative to absolute chronology.

Detrended correspondence analysis (DCA) is an heuristic offspring of CA, especially designed to handle the problems in the relationship between the first and the second axis and to adjust the distortions already described. Software solutions are available in different Statistical Software Packages, such as CANOCO and the famous Bonn package WinBASP.

There are two principal methods of DCA: Detrending by polynomials and by segments. The first is more stable and considered more elegant. Nevertheless, detrending by segments gives a better first impression of the method. Imagine the solution of a CA displayed in a scatter plot. Now the first axis is divided into an arbitrary number of segments. These segments are moved in such a way that their mean on the second axis is approximately zero. Higher axes are treated in the same way with respect to each of the existing axes. In fact DCA by segments is not a method of rectifying CA-solutions; it is an algorithm of its own. Actually it is a CA that uses detrending instead of orthogonalisation.

DCA being without orthogonalisation offers the opportunity to go a step further: It is possible to scale the first axis in terms of the combination of variables, for example of artefacts of different types in graves. The appearance and disappearance of types along the first axis would be even - the rate of appearance along the time-representing gradient is held constant. Let the occurrence of artefacts of the same type be of normal distribution. As the standard deviation divides such curves in standardized areas, one can describe the phases in the use of artefacts of the same type in terms of standard deviation. By using the relation between the standard deviation and the variance - the first is the square root of the latter - one can equalize the curves' width: The variance of the optima of the types in a find should be an estimate of the average squared standard deviation of the types. For scaling one has to equalize the variances of the types-optima-in-the-finds as nearly as possible. Technically the first DCA-axis is divided in segments, and these segments are expanded if the finds in it show a low variance of the types' optima. Conversely, if the variance of the optima of the types is high, the segments are contracted. It is the positioning of the types that is rearranged. The finds are positioned in such a way that their coordinates are weighted means of the coordinates of the types that occur in them. As a standard scaling, the variance of the optima of the types in finds is set to 1. This method of scaling was introduced into ecology by Gauch and Whittaker 1972. They called the resulting unit Z. In 1980 Hill and Gauch described the scaling in combination with DCA. They named the unit standard deviation, sd.

Types situated at a distance larger than 4 sd will practically not occur together. The standardization allowed ecological sciences to describe the development of diversity along a gradient.

The use of standard deviations as units offers a wide range of possibilities. Distances between types or between finds can now be measured, and it will be possible to estimate contemporarity in terms of probability. Another important technical advantage is that one can define the length of the first axis as the range of the finds curves - if the range is too narrow we should check whether seriation is adequate. Finally, it allows the use of sd phasing: 1.) Perform the CA for identifying the main dimension of explanation; 2.) do the horseshoe-test for showing the qualities of the main dimension, and then 3.) calculate DCA and 4.) the scaling. If the first axis is of an adequate length, 5.) order the matrix using values as ranks, and 6.) finally show phases in units of standard deviations.

I will demonstrate the approach with an example (Fig.1): My data-set contains early Neolithic Bandkeramik from the region between the rivers Rhine, Main, and Weser. The contingency table contains settlement pits in the rows and types of ceramic decoration in the columns. The horseshoe-test shows the dominance of a single gradient. The ranks of the values of CA and DCA are the same and they remain practically the same after scaling. Earlier investigations made clear that on the first axis time is represented. On the second axis regional styles are displayed. The new scaling of the time axis confirmed the cultural sequence established by Meier-Arendt (1966). Meier-Arendt's sequence was recently attacked

Figure 1 Early Neolithic Bandkeramik from the region between the rivers Rhine, Main and Weser. The types of ceramic decoration are shown in the plane of the first two DCA axes, scaled in sd units. The smaller box shows the pits as calculated in the same way and with the phases of Meier-Arendt inserted

(Kneipp 1998) because of its purely impressionistic foundation. Sd-units show the accuracy of Meier-Arendt's sense of style and at the same time they make it useable in an objectified way. Just an impression: Compare the marked groups of ornaments - the standing triangles and the rhomboids. Both groups are practically of the same relative age, but they are found in different regions. And please pay attention to the overall pattern. Displayed is the growing diversity from the very first Mid-European ceramic into regional groups. After a century of research, DCA allowed us for the first time to detect and show simultaneously the stylistic sequence on the first axis and the spatio-stylistic development of the neolithic groupings on the second (for further detail, Kerig, in press).

Scaling of relative chronologies in units of standard deviations has several attractive advantages on a technical level. Furthermore, scaling of relative sequences is a search for cultural traits. Imagine a large graveyard dated by seriation. The measurement allows us to compare and to scale intervals of change in different classes of artefacts. It must be meaningful, if, e.g., weaponry changes ten times faster than jewellery during a shortage of resources. It seems as if we could obtain a kind of an independent archaeological measurement for artefactual change - something like an archaeological unit.

References

BAXTER, M.J., 1994. Exploratory multivariate analysis in Archaeology. Edinburgh.

GAUCH, H.G. and WHITTAKER, R.H., 1972. Comparison of ordination techniques. Ecology 53:868-875.

GREENACRE, M.J., 1984. Theory and Applications of Correspondence Analysis. London.

HILL, M.O. and GAUCH, H.G., 1980. Detrended correspondence analysis: an improved ordination technique. Vegetatio 42:.47-58.

JONGMAN, R.H.G., 1995. Ordination. In Jongman, R.H.G., ter Braak, C.J.F. and van Tongeren, O.F.R. (eds.), Data analysis in community and landscape ecology, Cambridge:91-173.

KENDALL, D.G., 1971. Seriation from abundance matrices. In Hodson, F.R., Kendall, D.G. and Tautu, P. (eds.), Mathematics in the Archaeological and Historical Sciences, Edinburgh:215-252.

KERIG, T., 2002, in press. Zur relativen Chronologie der westdeutschen Bandkeramik. Paper presented at the Symposium "Die Bandkeramik im 21. Jahrhundert", Brauweiler, 16-20 September 2002.

KNEIPP, J., 1998. Bandkeramik zwischen Rhein, Weser und Main. Studien zu Stil und Chronologie der Keramik. UPA 47, Frankfurt.

MEIER-ARENDT, W., 1966. Die bandkeramische Kultur im Untermaingebiet. Amt Bodendenkmalpfl. Darmstadt 3, Bonn.

MÜLLER, J. and ZIMMERMANN, A. (eds.), 1997. Archäologie und Korrespondenzanalyse. Beispiele, Fragen, Perspektiven. Internat. Arch. 23, Espelkamp.

Tesselations and Triangulations - Understanding Early Neolithic Social Networks

Erich Claßen
Institut für Ur- und Frühgeschichte
Universität zu Köln, Germany

Andreas Zimmermann
Institut für Ur- und Frühgeschichte
Universität zu Köln, Germany

See the CD for the extended version

Abstract

Different mathematical-statistical methods can be used in the analysis of the spatial distribution of archaeological data. On the one hand, we will consider Delauney-Triangulations, which connect neighbouring points of a distribution so that compact triangles are built, and on the other, we shall be looking at the use of tessellations in such an analysis. Around the points of a given distribution, polygons are formed; the junctions of the polygons are found in the zones of lowest point density.
Delauney-Triangulations of Early Neolithic settlements in the Rhineland, of the so-called Bandkeramik Culture, have shown that settlements were subjected to a standardised spacing in the landscape. Different distance-classes have been observed, uncovering different structures within the Bandkeramik settlement system. Within settlement groups, determined by triangulation, settlement hierarchies can be recognised by the evaluation of archaeological material. It is possible to distinguish big "central" settlements, and secondary settlements that were, more or less, dependent on these.
Tesselations make the possible hinterland visible, and allows us to make estimates regarding the economic necessities of individual settlements or settlement groups.
The combination of geometrically possible, geographically probable and archaeologically testable contacts is used to decipher past communication networks.

Introduction[1]

This paper deals with the application of tessellations and triangulations to landscape archaeology. In out presentation given at the conference we showed how these methods can be applied to cave art (Claßen and Zimmermann 2003).

However, in the following, tessellation and triangulation are used to analyse settlements of the so called Bandkeramik. We will also deal with a more generalized variant of triangulation which we would like to call the 'calculation of second degree neighbourhoods'. All these methods can help us in understanding the distribution of archaeological sites or artefacts.

'Delauney triangulations' connect neighbouring points of a distribution so that compact triangles are built, this means that a quadrilateral is subdivided into two triangles whereby the angles should be as obtuse as possible. No line may cross another so that only 'natural' neighbours are connected. In the result we see a conclusive definition of neighbourhoods (Ripley 1981).

Tessellations, synonymous are the terms 'Voronoi diagrams', 'Dirichlet cells' or 'Thiessen polygons' form polygons around the points of a given distribution. This method is closely related to triangulations as the edges of the polygons divide the lines of the triangulation perpendicularly in half (Fig.1). The vertices of the lines of polygons are found in the zones of lowest point density (Okabe, Boots and Sugihara 1992).

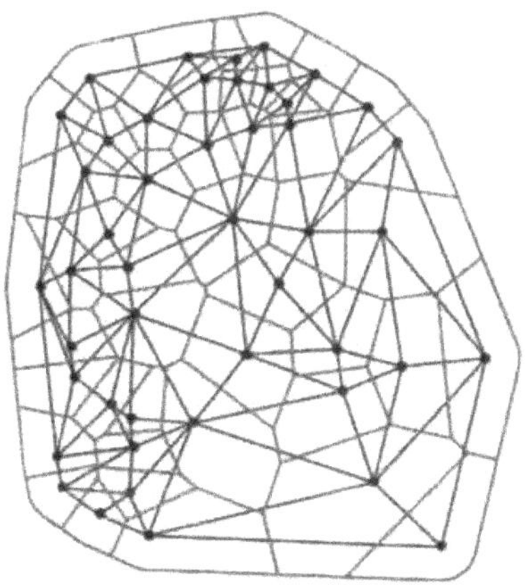

Figure 1 Schematic Triangulation (black) and Thiessen Polygons (red)

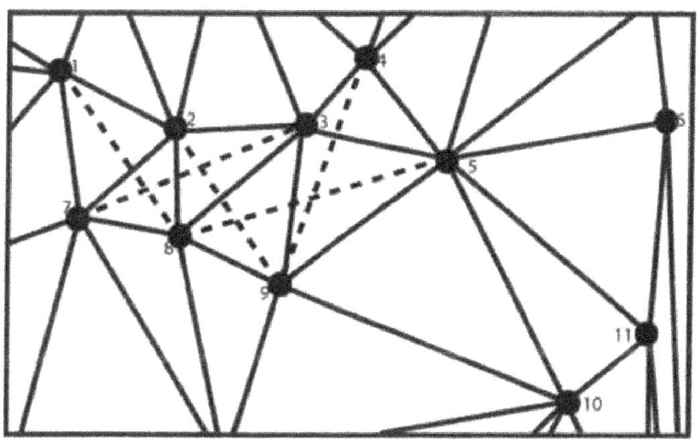

Figure 2 Schematic Triangulation of 'natural' neighbours (continuous lines) and 'neighbours of a second degree' (dotted lines)

The aim of the calculation of second degree neighbourhoods is to visualize additional structures in more complex distributions. In figure 2 these points are second degree neighbours which are connected by two lines in the triangulation of 'natural' neighbours. For example points 2 and 9 are neighbours to a second degree because the 'natural' neighbours, points 3 and 8, connect to one another.

The combination of possible geographical neighbourhoods, together with the archaeological data from the sites, should lead to an evaluation of the different possibilities in the sense of more or less probable past social networks.

Concerning the analysis of social networks, sociologists and social anthropologists have developed a broad range of methods (methods and theory: Wassermann and Faust 1994; ethnological applications: Schweizer 1996). In the coming years we aim to transmit these methods of social network analysis on to archaeological data. However, in this paper the attempt is made to deliver an insight on some archaeological data that may be seen as reflecting the social networks of the Early Neolithic.

As previously mentioned, we are dealing with the Early Neolithic of Central Europe (e.g. Lüning 1988). Bandkeramik is the term used to describe the culture of the first farmers in Central Europe. The name is derived from the typical ornamentation found on the vessels. Such pottery is found from the Carpathian mountains to the Parisian Basin and dates back to the time between 5500 and 4900 cal. BC. Important innovations from this time are farming and stokkbreeding, a sedentary way of life and the production of pottery.

At about 5300 cal. BC the farmers of the Bandkeramik arrived in the Rhineland. The settlements are situated mostly along small rivulets and consist of individual or several simultaneous farmsteads (e. g. Lüning and Stehli 1989, Stehli 1994). On the basis of the settlement size, the settlement duration, the supply with flint material and the peculiarities of the vessel decoration, we can reconstruct settlement groups. These settlement groups consist of a main settlement, several secondary settlements and/or single farmstead settlements.

In our definition a main settlement comprises up to ten farmsteads and was inhabited continuously for a long period of time. These settlements played a certain role in the procurement, processing and distribution of specific goods such as flint.

Secondary settlements are middle-sized hamlets, with three to four simultaneous farmsteads which were, to an extent dependent on the main settlements but which also developed their own traditions within time.

Single farmstead settlements were presumably more dependent on the main settlements.

Triangulations of Early Neolithic Settlements

The distribution map of Early Neolithic sites in the Rhineland shows clusters of sites and gaps in the distribution which are regarded as being meaningful (Lüning 1982:13-23, Fig.3 and 6).

One aim of the triangulation is to quantify the distances between neighbouring sites. In doing so it may be possible to realize certain regular distances, which can perhaps be explained by the past social relations.

On the one hand it is probable that less room between the individual settlements led to social stress owing to limited resources. On the other hand, if the settlements were situated further apart, it would have been difficult to keep contact and to work on common tasks. If a society strives for a balanced relation between communication and aggression, this leads to the establishment of certain rules as to where settlements could be founded. These possible rules can be reconstructed by analysing the distances between past settlements (Zimmermann 1992:108).

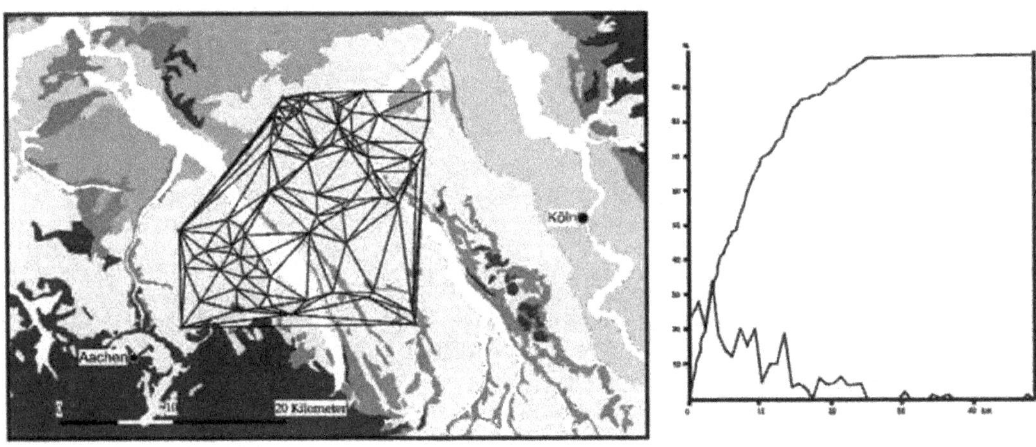

Figure 3 Triangulation of Early Neolithic sites in the lignite exploitation area west of Cologne, Germany. Map after Modderman 1970, Taf. 1, and a cumulative graph of the distances in the triangulation

In the Bandkeramik of the Rhineland regular distances between the settlements are visible because some ranges occur more frequent than others (Fig.3).

When looking at the triangulation we first recognize a maximum distance of about 1.5 km, which seems to be the maximum distance between two contemporaneous sites of one settlement group (Zimmermann 1992).
A peak at 3 km indicates the regular spacing between different settlement groups. This distance class corresponds with parallel running rivulets in the area under study. On the Aldenhovener Platte, where most of the settlements are fully examined, the distance between main settlements - with a high percentage of primary flint processing - is about 2 to 3 km (Zimmermann 1995:92).

Here we can see that archaeological and geographical information match. The settlement hierarchy developed on archaeological arguments - with main settlements, secondary settlements and single farmstead settlements - is supported by the regular geographical arrangement of main settlements.

The distances between 8 to 10 km and 13 to 15 km form the next groups. For the structures registered here, no sufficient explanations are available at the moment. It is possible that these groups mark the distances between settlement units on a higher level which may be connected with different natural landscape units.

The tessellations give an impression of the available land belonging to a settlement group. How far this land was used, and if this may have led to conflicts on resources, will be an aspect of further studies. The possible conflicts between neighbouring communities should appear in the archaeological database as apparent dissimilarities.

CALCULATION OF NEIGHBOURHOODS OF A SECOND DEGREE AND EARLY NEOLITHIC SOCIAL NETWORKS

At this point we have to consider triangulations of a second degree as they show the more complex and diverse relations. The possible relations of Early Neolithic settlements on the Aldenhovener Platte are shown in figure 5. In a next step we try to prove these possible connections using the archaeological data because the existence of some relations - which seem to be geometrically possible - are less realistic.

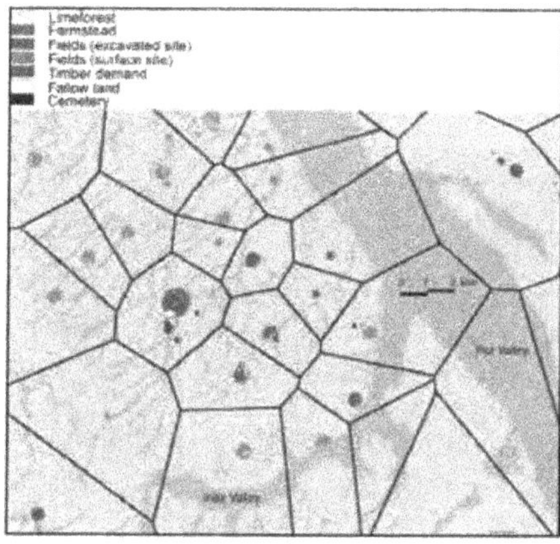

Figure 4 Tesselation of Early Neolithic settlement groups on the Aldenhovener Platte, Germany

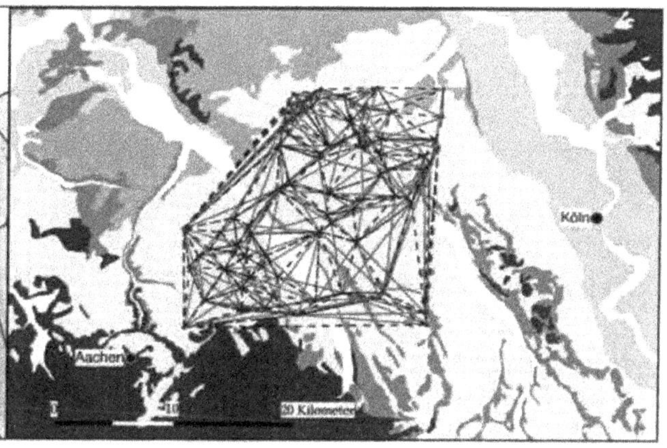

Figure 5 Triangulation of Early Neolithic sites on the Aldenhovener Platte, Germany. Continuous lines: 'natural' neighbours; dotted lines: 'neighbours of a second degree'. Map after Modderman 1970, Taf. 1

One way of checking the triangulation is via the very diverse pottery decoration (Frirdich 1994, Kolhoff 1999, Krahn-Schigiol 1999). However, as yet, this possibility has not been studied in detail.

Similarities in the pottery decoration of neighbouring settlement groups in one settlement phase show that some 'natural' neighbours don not share certain decorations, whilst some neighbours of a second degree do have a similar decorative spectrum (Fig.6).

TESSELATIONS OF EARLY NEOLITHIC SETTLEMENT GROUPS

If we calculate a tessellation of Early Neolithic settlement groups in the Rhineland the catchment area of each settlement group is displayed .

Figure 4 shows the estimated economic necessities for one settlement group in a single settlement phase with a duration of 25 years. The demand for farmsteads, fields, fallow land and timber acquisition has been calculated. The hinterland of the settlement- and field-area was needed for grazing the stock. The fact that there was no grassland to feed the animals led to grazing in the surrounding woodland, for which much more space was needed. In other settlement territories of the Bandkeramik land use models show that the whole possible hinterland was required for the upkeep of a population of livestock (Ebersbach and Schade this volume).

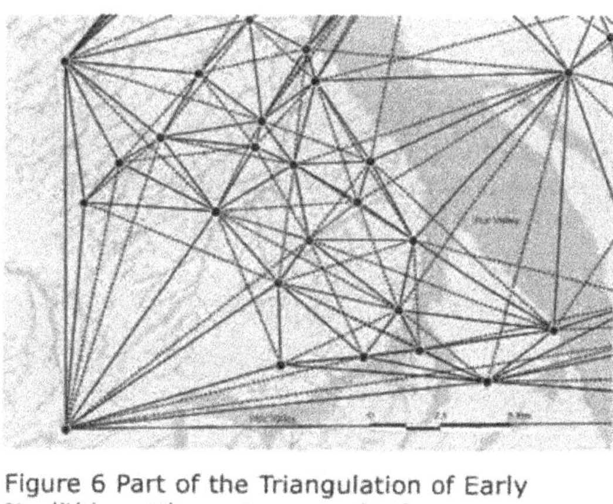

Figure 6 Part of the Triangulation of Early Neolithic settlement groups in the Rhineland. The red lines indicate similarities in the pottery decoration between different settlement groups in phase XII (after Kolhoff 1999, 107). Continuous lines: 'natural' neighbours; dotted lines: 'neighbours of a second degree'

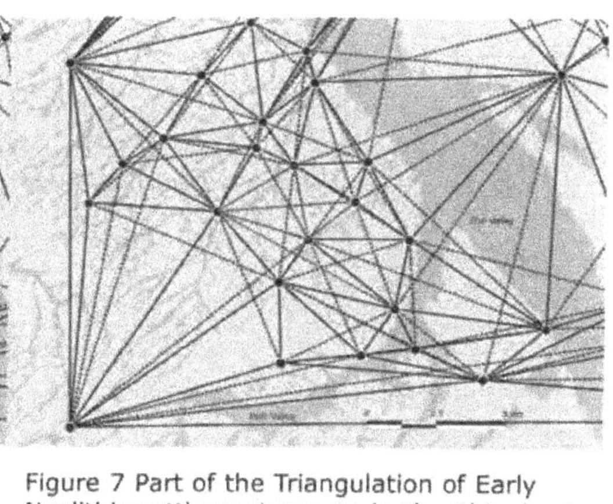

Figure 7 Part of the Triangulation of Early Neolithic settlement groups in the Rhineland. The red lines indicate similarities in the pottery decoration between different settlement groups in phase XIII (after Kolhoff 1999, 107).
Continuous lines: 'natural' neighbours; dotted lines: 'neighbours of a second degree'

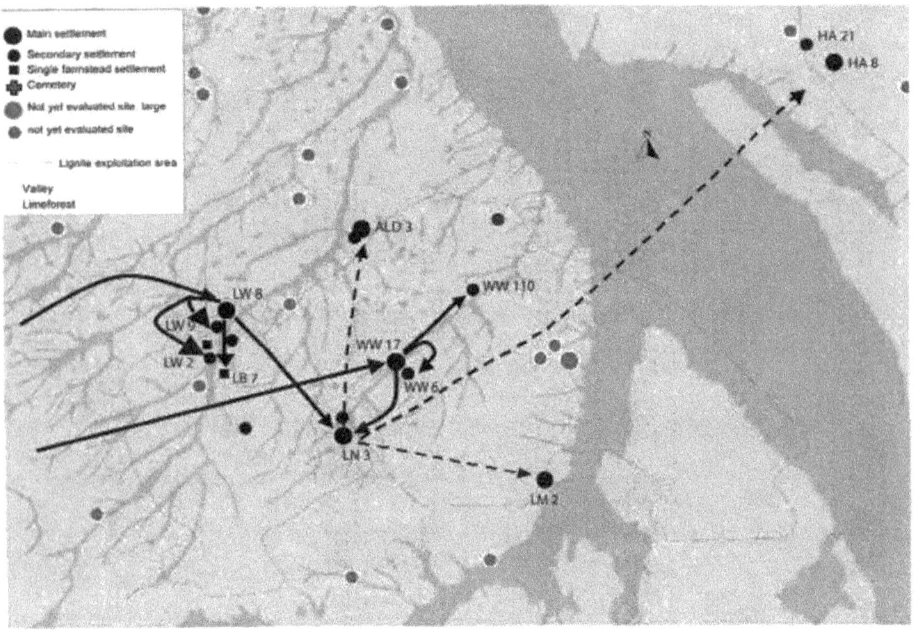

Figure 8 Simplified presentation of the flint raw material exchange on the Aldenhovener Platte, Germany

In the next settlement phase this picture changes slightly. Most 'natural' neighbours now have certain decorations in common, and even second-degree neighbourhoods located at greater distances from one another show similarities (Fig.7). Nevertheless, some settlement groups still appear to have had less contact with certain others.

Another source for the analysis of Early Neolithic social networks is the stone artefact raw material. On the Aldenhovener Platte a rather complex system of raw material exchange can be reconstructed. This region was mainly supplied with raw material from the west, which was won in the vicinity of Rijckholt.

To reconstruct the distribution of artefacts from one settlement to another criteria such as the amount of production waste, blades and final products are essential (Zimmermann 1982; 1995). They point to the production intensity and to the position of a settlement in the distribution system. The general raw material distribution network can be described as follows (Fig.8):

Firstly we can identify the supply of two main settlements with Rijckholt material, the reason being the high amount of primary processing that took place at the sites (LW8, WW17; Zimmermann 1988, Krahn-Schigiol 1999). This is shown by a high proportion of blanks with cortex which is an indication of the processing of flint nodules brought into these settlements.

These main settlements passed on cores, blades and final products to their neighbours in the settlement group (LW2, LW9, LB7, WW6; Langenbrink 1996, Gaffrey 1994, Krahn-Schigiol 1999).

One main settlement also supplied a smaller settlement that did not belong to the same settlement group (WW110; Bollig 2000).

In a further main settlement even more artefacts were produced in comparison to the two settlements already mentioned, although not from nodules but from prepared cores (LN3; Krahn-Schigiol 1999). This means that this settlement was supplied by one or both of the first mentioned main settlements with primary processing of nodules.

This main settlement probably gave parts of the produced artefacts to other settlement groups (ALD 3, LM 2, HA8 and HA 21; Deutmann 1997, Langenbrink 1992, Bender 1992, Hohmeyer 1997, Reepmeyer 2002).

The first network analysis of the raw material distribution network, which is not to be discussed in detail here, was conducted using the program package 'UCINET IV' (Borgatti, Freeman and Everett 1999) and considers the changes in the system from the middle Bandkeramik to the later Bandkeramik (Reepmeyer 2002:73-104).

Two settlements seem to change their position within the network, which means that a settlement declared as a producer and forwarder of raw material for the middle Bandkeramik became a receiver settlement in the later period.

If we look at the whole network, its structure does not change over time. The ratings for prominence show that the relations within the network can be described as being distributed rather uniformly. To keep it simple, all actors have nearly the same number of relations.

The network was therefore egalitarian.

There were enough settlements with a high production rate, so that each receiver settlement could interact with another producer settlement in a possible conflict situation.

This egalitarian network structure prevented the development of a more differentiated hierarchical settlement structure. This seems to change in the Middle Neolithic of the Rhineland, but this would be a subject for a further paper.

We are still in the investigation process, and of course our picture of the Bandkeramik settlements is not complete, However, we hope to specify the results presented here within a few years.

1 We would like to thank Lee Clare (Cologne) for the proof reading of this text.

REFERENCES

BENDER, W., 1992. Steingeräte. In Lüning, J. and Stehli, P. (eds.), Der bandkeramische Siedlungsplatz Lamersdorf 2, Gemeinde Inden, Kreis Düren, Rheinische Ausgrabungen 37:175-191.

BOLLIG, L., 2000, unpubl. Die Gesteinsartefakte des bandkeramischen Fundplatzes Weisweiler 110 (Kreis Düren). Master Thesis, Köln.

BORGATTI, S.P., EVERETT, M.G. and FREEMAN, L.C., 1999. UCINET IV. Analytic Technologies.

CLASSEN, E. and ZIMMERMANN, A., 2003. Räumliche Statistik, Soziale Netzwerkanalyse und Raumverständnis. In Pastoors, A. and Weniger, G.-C. (eds.), Höhlenkunst und Raum: Archäologische und architektonische Perspektiven (Cave art and space: Archaeological and architectural persepectives, Wissenschaftliche Schriften des Neanderthal Museums 3:91-103.

DEUTMANN, K.-H., 1997. Feuersteinmaterial. In Boelicke, U., Deutmann, K.-H., Lüning, J. and Schmidt, F.W.V., Der bandkeramische Siedlungsplatz Aldenhoven 3, Kr. Düren, Studien zur neolithischen Besiedlung der Aldenhovener Platte und ihrer Umgebung, Rheinische Ausgrabungen 43:83-127.

FRIRDICH, Chr., 1994. Kulturgeschichtliche Betrachtungen zur Bandkeramik im Merzbachtal. In Lüning, J. and Stehli, P. (eds.), Die Bandkeramik im Merzbachtal auf der Aldenhovener Platte, Rheinische Ausgrabungen 36:207-393.

GAFFREY, J., 1994. Die Steininventare der bandkeramischen Siedlungsplätze Laurenzberg 7, Langweiler 16 und Laurenzberg 8. In Lüning, J. and Stehli, P. (eds.), Die Bandkeramik im Merzbachtal auf der Aldenhovener Platte, Rheinische Ausgrabungen 36:395-531.

HOHMEYER, M., 1997. Ausgewählte lithische Inventare des bandkeramischen Siedlungsplatzes Hambach 8, Gem. Jülich, Kr. Düren. In Lüning, J. (ed.), Studien zur neolithischen Besiedlung der Aldenhovener Platte und ihrer Umgebung, Rheinische Ausgrabungen 43:229-318.

KOLHOFF, C., 1999, unpubl. Die Keramik des bandkeramischen Fundplatzes Weisweiler 110. Master thesis, Köln.

KRAHN-SCHIGIOL, Chr., 1999, unpubl. Vier bandkeramische Siedlungsplätze im Tal des Schlangengraben auf der Aldenhovener Platte im Rheinland. Ph. D. thesis, Frankfurt/Main.

LANGENBRINK, B., 1992. Rohmaterialien und Grundformen des Steinmaterials. In Lüning, J. and Stehli, P. (eds.), Der bandkeramische Siedlungsplatz Lamersdorf 2, Gemeinde Inden, Kreis Düren, Rheinische Ausgrabungen 37:134-174.

LANGENBRINK, B., 1996. Zu den Steinen der bandkeramischen Siedlungen Langweiler 2 und Langweiler 9. Bonner Jahrbuch 196:397-420.

LÜNING, J., 1982. Siedlung und Siedlungslandschaft in bandkeramischer und Rössener Zeit. Offa 39:9-33.

LÜNING, J., 1988. Frühe Bauern in Mitteleuropa im 6. und 5. Jahrtausend v. Chr. Jahrbuch des Römisch-Germanischen Zentralmuseums Mainz 35/1:27-96.

LÜNING, J. and STEHLI, P., 1989. Die Bandkeramik in Mitteleuropa: von der Natur- zur Kulturlandschaft. In Lüning, J. (ed.), Siedlungen der Steinzeit - Haus, Festung und Kult, Spektrum der Wissenschaft - Verständliche Forschung:110-120.

MODDERMAN, P.J.R., 1970. Linearbandkeramik aus Elsloo und Stein. Analecta Praehistorica Leidensia III, 's-Gravenhage.

OKABE, A., BOOTS, B. and SUGIHARA, K., 1992. Spatial Tessellations - Concepts and Applications of Voronoi Diagrams. New York/Brisbane/Toronto/Singapore.

REEPMEYER, Ch., 2002, unpubl. Das Steininventar des bandkeramischen Fundplatzes Hambach 21. Master thesis, Köln.

RIPLEY, B.D., 1981. Spatial Statistics. New York/Brisbane/Toronto/Singapore.

SCHWEIZER, Th., 1996. Muster sozialer Ordnung - Netzwerkanalyse als Fundament der Sozialethnologie. Berlin.

WASSERMAN, S. and FAUST, K., 1994. Social Network Analysis: Methods and Applications. Cambridge.

ZIMMERMANN, A., 1982. Zur Organisation der Herstellung von Feuersteinartefakten in bandkeramischen Siedlungen. In Pavuk, J. (ed.), Siedlungen der Kultur mit Linearkeramik in Europa. Internationales Kolloquium Nové Vozokany, 17-20 November 1981:319-322.

ZIMMERMANN, A., 1988. Steine. In Boelicke, U., von Brandt, D., Lüning, J., Stehli, P. and Zimmermann, A., Der bandkeramische Siedlungsplatz Langweiler 8, Gemeinde Aldenhoven, Kreis Düren. Rheinische Ausgrabungen 28:569-787.

ZIMMERMANN, A., 1992. Tesselierung und Triangulation als Techniken zur Bestimmung archäologischer Funddichten. Archäologische Informationen 15:107-112.

ZIMMERMANN, A., 1995. Austauschsysteme von Silexartefakten in der Bandkeramik Mitteleuropas. Universitätsforschungen zur prähistorischen Archäologie 26. Habelt, Bonn.

ESTABLISHING OPTIMAL CORE SAMPLING STRATEGIES: THEORY, SIMULATION AND PRACTICAL IMPLICATIONS

AVSHALOM KARASIK
INSTITUTE OF ARCHAEOLOGY, THE HEBREW UNIVERSITY, JERUSALEM, ISRAEL
DEPARTMENT OF PHYSICS OF COMPLEX SYSTEMS, THE WEIZMANN INSTITUTE, REHOVOT, ISRAEL

ILAN SHARON
INSTITUTE OF ARCHAEOLOGY, THE HEBREW UNIVERSITY, JERUSALEM, ISRAEL

UZY SMILANSKY
DEPARTMENT OF PHYSICS OF COMPLEX SYSTEMS, THE WEIZMANN INSTITUTE, REHOVOT, ISRAEL

AYELET GILBOA
DEPARTMENT OF ARCHAEOLOGY, THE UNIVERSITY OF HAIFA, HAIFA, ISRAEL

ABSTRACT

We describe a new method by which ceramics are classified and compared in terms of the curvature functions which are associated with each profile (cross section). The theoretical and practical aspects of the method are illustrated in terms of a test case of archaeological interest, where curvature analysis was successfully applied.

INTRODUCTION

Pottery is the most frequent find in typical excavations. Archaeologists who study ceramic assemblages seek to identify in them distinctive patterns that can be used for a better understanding of the material culture. Amongst the several attributes, which comprise the typological vocabulary, shape attributes are the most commonly used. Traditional shape descriptions and classifications, however, rely on intuitive, often vague characterizations, which are hard to quantify. Several attempts to harness computers and mathematical tools in pottery analysis have appeared in the proceedings of the CAA conferences (for instance: Hall and Laflin 1985, Goodson 1989, Liming et al. 1989, Durham et al. 1995 and 1996, Schurmans et al. 2001, Kampel et al. 2002), or books e.g., Orton (1980), Orton et al. (1993) and Whallon and Brown (1982). Here, we classify ceramic profiles by their curvature function - which gives the curvature as a function of the arc length along the profile (Mokhtarian and Boder 2003 for previous use of the same function). This function contains the entire information about the profile, and therefore it is a convenient basis for a detailed, objective and comprehensive typology of ceramics, which can be implemented in computer algorithms. The new aspects of our method and its advantages will be pointed out in the sequel.

THEORETICAL CONSIDERATIONS

We summarize here the main definitions and relations which are necessary for the subsequent presentation. Its implementation in practice will be described in the next section. The profile of a ceramic artifact is a curve in R^2 which is not necessarily closed, since we treat pot-shreds and whole vessels on the same footing. Let s be the arc length measured along the profile from the point which defines the top of the rim, and let $\theta(s)$ denote the direction of the tangent at s, so that $\theta(0)=\pi$. The curvature $\kappa(s)$ is the rate of change of $\theta(s)$ at s: $\kappa(s)=d\theta(s)/ds$. Given the curvature function $\kappa(s)$, the curve is *completely and uniquely* specified and can be reconstructed by integration. The expression of the two-dimensional profile by a single function, which emphasizes the features of archaeological relevance (highly curved sections such as the vessels rim or base appear as peaks, see (Fig. 1.d)), is one of the main advantage of the representation of curves in terms of their curvature functions.

In order to classify and compare profiles of vessels in a given assemblage, we standardize the description of the assemblage by scaling the arc length and the curvature of each profile by the rim radius, which is assumed to be provided. We define the distance $d_{\alpha\beta}$ between two curvature functions $\kappa_\alpha(s)$ and $\kappa_\beta(s)$ as $d_{\alpha\beta}^2=\int(\kappa_\alpha(s)-\kappa_\beta(s))^2\omega(s)ds$, where $\omega(s)$ is a non-negative, finitely supported, normalized *weight function*. In the present application we take $\omega(s)$ to be the characteristic function of the interval $[s_{\min},s_{\max}]$ which focuses the attention to the rim of the vessels. Alternative choices of $\omega(s)$ allow the archaeologist to introduce her/his judgment concerning the relative importance of the various features in the compared potsherds. The *scalar product* is defined by $\langle\alpha|\beta\rangle = \int \kappa_\alpha(s)\kappa_\beta(s)\omega(s)ds$, and the profiles correlation is $C_{\alpha,\beta} \equiv \langle\alpha|\beta\rangle / \sqrt{\langle\alpha|\alpha\rangle\langle\beta|\beta\rangle}$. These definitions allow us to treat the space of curvature functions as a Hilbert space, which is of great help in the analysis to be described below.

The profiles are classified by their affinity to some prototypes prescribed by the Archaeologist. Often, this is done by selecting a few representative subgroups of profiles which manifest in the best way the attributes that distinguish each of the prototypes. Taking the average of the curvature functions in each of the defining sets, we obtain the curvature functions of virtual profiles which represent the prototypes in the subsequent analysis (Fig.2). Let us assume that D predetermined prototypes were selected, and are represented by the curvatures functions $\kappa_{A_1}(s), \kappa_{A_2}(s), \ldots, \kappa_{A_D}(s)$. We must require that the prototype curvature functions are linearly independent (but not *necessarily* orthogonal), so that the D x D matrix $A_{r,s} = \langle A_r | A_s \rangle$ is regular ($\det A \neq 0$).

Assume that a given assemblage, which consists of M profiles (with curvature functions $\kappa_m(s)$, $m=1,\ldots,M$) are to be sorted according to their affinity to the set of D prototypes defined above. The coordinates $x_1(m), x_2(m), \ldots, x_D(m)$ which express $\kappa_m(s)$ in the basis provided by the prototypes, are computed for each fragment in the assemblage. Thus, $\kappa_m(s) = \sum_{j=1}^{D} x_j(m) \cdot \kappa_{A_j}(s) + \lambda_m(s)$, where $\lambda_m(s)$ stands for that component of $\kappa_m(s)$ which is *orthogonal* to the subspace spanned by the prototypes. $\lambda_m(s)$ must be included, since the prototypes do not span the entire space of curvature functions. To obtain the coordinates $x_i(m)$ we take the scalar product of $\kappa_m(s)$ as written above with each of the prototype curvatures. Since by definition, $\lambda_m(s)$ is orthogonal to the $\kappa_m(s)$, we obtain a set of D linear equations, with the explicit solution.

$$x_i(m) = \sum_{j=1}^{D} (A^{-1})_{i,j} \langle A_j | m \rangle$$

Each profile is now represented by a point $x(m)$ whose distribution in the D dimensional space gives a direct measure of the partition of the assemblage to types. Once the coordinate vectors $x(m)$ are known, we can easily compute the magnitude of the vectors $\lambda_m(s)$ which we denote by $\|\lambda_m(s)\|$. The assumption that the assemblage can be sorted by the prototypes $\kappa_{A_1}, \kappa_{A_2}, \ldots, \kappa_{A_D}$ is justified only if the *"quality factors"* defined by $\varsigma_m = \frac{\|\lambda_m\|^2}{\langle m|m\rangle} = 1 - \sum_{i=1}^{D} x_i(m) \frac{\langle m|A_i\rangle}{\langle m|m\rangle}$, are small (less than 1/2). Clearly, ς_m vanishes if $\kappa_m(s)$ is in the space spanned by the prototypes, and it equals one (which is its maximal value) if $\kappa_m(s)$ is orthogonal to all the prototypes. In summary, the quantities which we derived give us information on the affinity of the studied profiles to the prototypes, and, at the same time, they test the degree by which the prototypes account for all the features encountered in the assemblage.

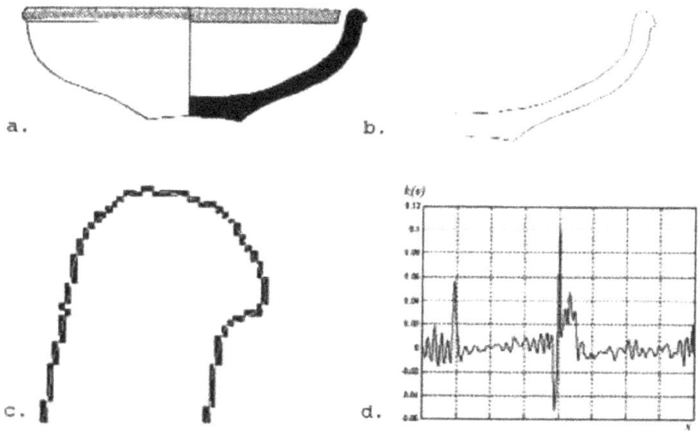

Figure 1 The steps leading from a scanned drawing to the curvature function. a. A scanned drawing of a bowl. b. The pixelized profile. c. Enlarged detail of the pixelized profile and the interpolated curve. d. The curvature function

The above construction can be applied in the frequently met situations, where one likes to quantify the observation that 'types' gradually change over time, space, or some other variable of archaeological interest. In such cases, it is convenient to demonstrate the transition between the types along a linear axis. For this purpose, one can try to analyze the assemblage in terms of only two prototypes *(D=2)*, and associate to each profile an interpolation parameter $\mu(m)$ which indicates the transition between the prototypes. We define the interpolating parameter by $\mu(m) = x_2(m)/[x_1(m) + x_2(m)]$, which takes the values 0 or 1 when the profile is identical to one of the prototypes. Such analysis is justified only if the corresponding quality factors ς_m are small.

Finally, an important advantage of the formal construction which was proposed above is that its application is independent of the form by which the curvature functions are represented. The curvature functions can be expressed as continuous or discrete functions in coordinate or in Fourier space. This freedom will be used in the subsequent description of the implementation of the method in practice.

Practical Implementation

[1] Data acquisition.

The data analyzed in the present work consists of linear drawings printed in archaeological reports. The various steps taken to extract the information and transform it to digital form are shown in (Fig. 1.a-c). The drawings are scanned, and the boundary of the cross section is identified and stored as a discrete set of (x,y) coordinates. The distance s traversed along the curve (measured with respect to the highest point) is computed, and the coordinate values at equal s intervals are obtained by interpolation, and are stored. In this form, the data still contains irrelevant 'noise', which is due to the finite resolution of the scanning device, imperfections of the origi-

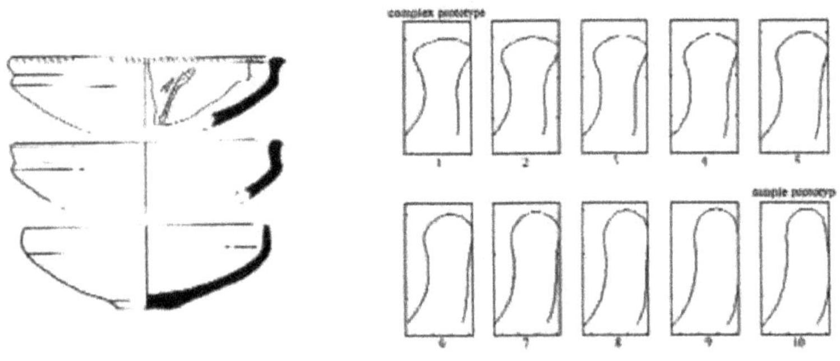

Figure 2 Three Dor bowls illustrating the typological transition. From top to bottom - the sub-types dominant in Iron Ia, Iron Ib and Iron II (left). The same transition is modeled by virtual rim profiles, constructed by interpolation between the two extreme prototypes (right)

nal drawings, etc. These features are filtered out to yield the smooth, yet accurate presentation of the original boundary as describe below. The whole procedure does not take more then a few seconds using MATLAB software.

[2] Smoothing, and the computation of the curvature function

The smoothing of the data alluded to above, is most conveniently performed in Fourier space: $\hat{X}(n)=\sum_{m=1}^{N} X(m)\cdot e^{-i\frac{2\pi}{N}(m-1)(n-1)}$
where $X(m)$ stands for the original discrete coordinate vector, and N is the number of points along the curve.

The smoothing is implemented by attenuating the irrelevant high-frequency fluctuations of the coordinated vectors. This is achieved by a filter function which suppresses the n'th Fourier component by a factor $\left(1+\exp\left[(n-n_0)/d\right]\right)^{-1}$ where n_0 is the filter range, and d is the transition width.

Transforming back to real space, we get the smooth coordinate functions, which are the basis of the curvature computations and the subsequent analysis. We found it necessary to suppress features which have a wavelength smaller than 8 pixels, however some times one can suppress even lower frequencies, without deteriorating the quality of the curves.

The first and second derivatives, which are needed for the curvature calculations, are also computed in Fourier space, and then transformed back to real space so the curvature is expressed as a function of s. Using the explicit formula $\kappa(s)=(x'(s)\cdot y''(s)-y'(s)\cdot x''(s))/([x'(s)]^2+[y'(s)]^2)$ the curvature is finally computed at a discrete set of equidistant points on the boundary (Fig.1d).

[3] Prototypes and Typology

To illustrate our method in action, we show here a few results obtained in the analysis of an assemblage consisting of 86 Iron-age bowls from the Tel-Dor (Israel). The bowls are all of the same general type (simple carinated bowls) and span a chronological range of one century or less. Manual analysis (Gilboa 2001; Gilboa and Sharon in press) suggested a typological development (albeit morphologically and quantitatively gradual) from "complex" rim type (dominant in the Iron-Ia stratum) to the "simple" rim type (abundant in the Iron I/II and Iron-IIa strata), as shown in (Fig.2). To test this hypothesis, we defined "complex" and "simple" prototypes (number 1 and 10 in (Fig.2), respectively), and analyzed the assemblage in the plane spanned by these profiles. The distribution of the bowls in the 2-dimensional complex-simple plane is shown in (Fig.3b,c), where the points representing profiles from different strata. In spite of the large (expected) overlaps between the distributions from the various strata, the increase of the mean values of μ is clear (Fig.3a). Notice that the μ values are increasing over time and that the ς values are consistently low, justifying *a-posteriori* the analysis of the assemblage in terms of two prototypes.

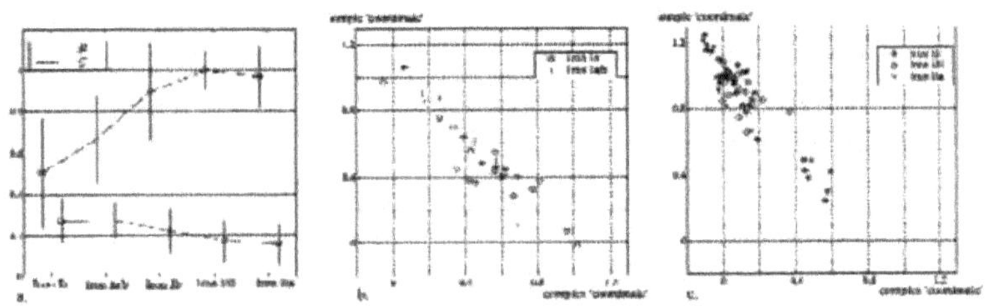

Figure 3 Representative results from the analysis of Dor bowls. a. The mean values and the standard deviations of and by periods. The lines through the points are to distinguish between the two functions which are shown on the same plot. b. The distribution in the complex-simple plane of the points corresponding to bowls from the Iron Ia and Iron Ia/b strata. c. Same as b. for the Iron Ib, Iron I/II and Iron IIa strata.

Detailed reports of typological analysis of assemblages from other Iron Age sites in the Southern Levant are in preparation. To summarize, the curvature analysis and the Hilbert space approach, enables a classification of ceramics based on rather minute stylistic details at a level and scope which was rarely attempted before. Its application is rather straight forward and it provided archaeologically relevant information.

ACKNOWLEDGEMENTS

This research is supported by a BIKURA grant from the Israel Science Foundation, and by the Kimmel Center for Archaeological sciences, in the Weizmann Institute of Science.

REFERENCES

GILBOA, A., 2001. Southern Phoenicia during Iron Age IIa in the light of the Tel Dor excavations. Ph.D. Thesis. The Hebrew University of Jerusalem.

GILBOA, A. and SHARON, I., in press. An archaeological contribution to the early Iron Age chronological debate: Alternative chronologies for Phoenicia and their effects on the Levant, Cyprus and Greece. Bulletin of the American Schools of Oriental Research.

MOKHTARIAN, F. and BODER, M., 2003. Curvature Scale Space representation, and MPEG-7 standardization. In Computational imaging and vision, vol. 25, Kluwer academic Publishers, Dordrecht, The Netherlands.

ORTON, C., 1980. Mathematics in Archaeology. Cambridge. University Press.

ORTON, C., TYERS, P., and VINCE, A., 1993. Pottery in Archaeology. Cambridge: University Press.

WHALLON, R. and BROWN, J.A. (eds.), 1982. Essays on archaeological typology. Center for American Archaeology Press.

Survey and Geographical Positioning Systems

MAPPING OUR HERITAGE

M. FARJAS
E. U. I. T. TOPOGRÁFICA,
UNIVERSIDAD POLITÉCNICA DE MADRID (UPM), SPAIN

M. ALONSO
E.T.S. ARQUITECTURA ,
UNIVERSIDAD POLITÉCNICA DE MADRID (UPM), SPAIN

CARPA RESEARCH TEAM
UNIVERSIDAD POLITÉCNICA DE MADRID (UPM), SPAIN

SEE THE CD FOR THE EXTENDED VERSION AND THE FIGURES

ABSTRACT

In this paper we present the results obtained during the search for cartographic methods, which can assist in the interpretation and dissemination of knowledge regarding Archaeology and our Heritage. Basing their research on techniques developed specifically for geodesic and cartographic engineering and architecture, the Polytechnic University of Madrid's Research Team LAR, are assessing how new tools applicable to their specialist can be used in an innovative way to capture and process information in the fields of archaeology and heritage.

To investigate different methods of portraying monumental edifices, we made use of new data capturing equipment in the Monastery of El Escorial, Spain for which traditional photogrammetrical methodologies were compared, taking advantage of the least square adjustment capability provided by the independent grids.

To investigate procedures for the treatment of cartographic data a study was undertaken, to develop a thematic cartography that combines the scientific results with a graphical presentation. The site used for this exercise was the archaeological site at Tuqueibah, in the Emirate of Sharjah, UAE.

All of the work referred to above strives to unite the disciplines of engineering and archaeology, using case studies to demonstrate the enormous potential that such interdisciplinary co-operation can offer.

INTRODUCTION

The research team CARPA (Cartography in Heritage and Archaeology) of the Universidad Politécnica de Madrid (UPM) was established in 1996 to explore new ways in which the cartographical sciences and archaeology could co-operate. The investigation adopts a dual approach. On the one hand we attempt to provide the archaeologists with basic cartography at different scales. On the other hand an attempt is made to provide cartographical documents which assist in exhibiting the information in paper format for static displays, or in digital format in the case of dynamic displays. In both cases our aim is to ensure that the topography, geodesy and related sciences serve to enhance the display material, and in this way expand the scope for application of our specialism.

We will present the results from three of our projects:

- Tuqueibah (U.A.E.)
- Escorial (Spain)
- Cibeles (Spain)

CARTOGRAPHY IN ARCHAEOLOGY: TUQUEIBAH

The overall objective of our research is to determine ways in which new engineering technology, and especially cartography, can be applied to Archaeology, to study processes and analyse proposals for the optimisation of archaeological data capture, treatment and display. The area used for the study is the whole of the Emirate of Sharjah (Fig.1,2).

In the Emirate are founds the archaeological sites of Tell Abraq, Al Manakh, Muweileh, Al Hamriyah, Mleiha, Jebel Al Emaileh, Jebel Falla, Jebel Al Buhais, Al Tuqueibah, Al Madam, Wadi Al Hilu, Wadi Shi, Khor Kalba y Khor Fakkan. One archeological team has been assigned to each of these excavation sites. This case study deals written the archaeological remains of Tuqueibah. This archaeological site has a total area of more than 50 hectares, in which the urban remains of the first settlement in the Emirate, dating back about 3000 BC. The archaeological team responsible for this excavation belongs to the Universidad Autónoma of Madrid (Fig.3).

The site was mapped at a scale of 1:1,000, covering the full 50 hectares of the archaeological area. The limited time available to collect field data, as well not being able to make use of aerial photogrammetry, we decided to make use of static GPS methods to quickly establish the basic grid and the RTK for the site. The equipment used and resulting grid areas follows: (Fig.4,5)

More than 20,000 points were read, and processing. This large amount of data involved detailed hand calculations to accurately reproduce the smooth curvatures of the dunes. Special care was needed in representing the vegetation, as this is an indicator of the stability of the dune over time, and a key indicator of orientation. This work was different to what we were used to in mapping more familiar shapes and materials in our own country, and it was essential not to miss any paints. (Fig.6)

The results obtained from the cartographical analysis at the Tuqueibah site encouraged us to sep new methods to improve the final image. The mapping contained all the necessary detail and satisfied the required tolerances but the results was not clear to the end-user. We projected images over the bakkground and manipulated relief detail using perspective views, colour and shading was enhanced using Adobe Photoshop. A range of colours was developed and a coloured version produced, and the resulting image to a process of Gaussian methods and filtering. The image obtained was as follows (Fig.7).

The image was superimposed an the highlighting the relief and facilitating interpretation (Fig.8,9).

Cartography in Heritage: El Escorial

The purpose of this exercise was to obtain a detailed mapping of the Basilica dome of the Monasterio de El Escorial in order to analyse the geometry. We carried out the mapping using topographical techniques that used one total-station without reflective prisms. Before undertaking this exercise the precise dimensions of the dome were not known. Only the lantern had been surveyed previously for the reconstruction of the needle in 1948.

We used a scale of 1:200 for the mapping, which is adequate for this type of work. Because the interior was an enclosed space and at a significant height above ground, the fundamental problem was to obtain coordinates for inaccessible points (Fig.10).

The equipment used was a Leica TCR305 survey station, an loan from Leica for this project. This equipment allowed us to measure points at a distance without a prism reflector, making use of a visible laser (LR). It also has an infrared distance meter for measuring with a prism (IR). In measuring distances it has an accuracy of 3mm + 3 ppm. The technical specifications for the equipment states that the range is dependant the same extent on the observed surface colour and surrounding light. The best results being obtained with white surfaces and in little background lighting.

From the points studied, 3 were chosen as survey stations for the mapping of the dome interior. The first of these was situated beneath the centre of the dome on the basilica floor, the second in the chorus left below the midday organ tribune and the third at the access to the cornice of the dome, 45 m above the basilica floor (Fig.11,12).

Our readings were taken from points situated at a distance of about 58 m fro points on a spherical surface, the angle of incidence of the laser on the surface being roughly orthogonal and therefore most favourable. We were aware that these readings were being taken at the practical limit for this station. The remaining points on the dome, being closer to the station, were measured without difficulty.

To carry out the mapping of the dome exterior, we positioned ourselves in the middle of the basilica bell towers, below the arches opposite the dome. From these two points we were able to view a little more than half of the dome exterior. We decided to base our work on readings from common points, and use Helmert 3D transformation techniques (Fig.13).

The first station we set up in the south tower, the clock tower. We started by taking 5 points which were used as a link to the second station on the north tower. We made sketches based on these readings in CD and CI. After that we attempted to define the cornice of the square dome base, for which 3 vertices were visible. We were able to measure the two closest vertices without problem, but not so with the furthest one. The distance was less than for previous readings but the complications stemmed form the fait that the view angle was acute. We guessed, and later continued, the importance of the angle of incidence on the suitability of the station.

The readings of the sternal surface of the dome from the clock tower hended to be complemented by corresponding readings from the north tower. We decided to take theses readings at dusle, with less light, as recommended by the manufacturer, to check whether under these conditions we would get better readings from the station which would allow us to obtain readings for the uppermost points on the dome (Fig.14).

Based on the data obtained in this way we were able to map the inside of the dome. The readings taken on the external surface had provide sufficient data on more than half of the dome. On the upper part, due to the acuteness of the view angle with respect to the surface, the laser station could not read. We could have used either bisection from the two towers or reach the surface of the dome armed with prisms. We decided on the later of the two options (Fig.15).

On this occasion we made use of a prism. We sere able to count on the help of Beatriz Mateios and David Dominguez, two E.U.I.T. Survey students and member of GEMAT, who climbed up the external face of the dome and fixed two safety cables around the perimeter. Using safety harnesses, secured to the cables and the lantern pedestal, they descended the external surface of the dome carrying mini-prisms. We took the equipment in non-prism mode, the laser distance meter. In this way we took co-ordinates for the missing points.

The time taken to obtain the readings was 7 working days, based on half days. The effective working time was 28 hours. We spent 19 hours on the inside of the dome where we took 3,250 points. The external surface required 9 hours to take 1,250 readings. After taking readings, as described above, we had the coordinates for 4,500 points on the dome. They were grouped according to survey station and in each case referenced to a local system (Fig.16).

The first step was to transform the data so that all the points for the external surface of the dome were referenced to a single co-ordinate system, and all those of the interior to another. After referencing all the points for the interior to a single co-ordinate system, we began to analyse the dome shape. The number of points strictly necessary to determine the centres and radii of the two spheres is 6. We had readings of 154 points. For each of the points we developed an equation, giving us a system with 154 equations and six unknowns.

To solve the equations we started with approximate values for the sphere parameters. The results were used to adjust the coordinates for the centres and radii of the spheres. An iterative process was used until the variations in the coordinates for the centres and spheres parameters was within acceptable limits. The calculations were carried out using a computer program that we developed for this purpose.

The following step was to rotate all the points for the dome interior about its own axis to locate them as a single vertical plane, with the help of an analysis program developed for this purpose. The ASCII file, with all the rotated coordinates, was converted into DXF format for use with AutoCad. In this way we obtained a point representation of the different dome meridian sections. We therefore had at our disposal the plan layouts and meridian sections handed to generate the dome. In other words we had the necessary data to plot the interior of the dome (Fig.17).

We followed the same procedure for the exterior of the dome as for the interior. The first step was to refer the points, taken from the 3 survey stations, to a single system of coordinates. We carried out a Helmert 3D transformation and calculated the parameters from common points, making least square adjustments (Fig.18,19).

Moreover, we imposed an assumption that the axis of the dome was over the centre of the enclosing circumference formed by the leutern butresses, for which we had coordinates for 8 points, the centre of the circumference coinciding with edge of the cornise at the base, for which we had coordinates of 15 points, and the centre of the circumference of the cylindrical from around the balustrade, for which we had coordinates of at least fifteen points.

For this we obtained a set of 80 equations and 7 unknowns, which we solved using least squares adjustments. The process was similar to that used for the dome interior. The calculations were carried out using a computer program developed for the purpose.

Having fixed the external dome axis, we rotated the points to locate them as the same vertical plane. We then converted the ASCII file with the rotated coordinates into .dxf format for use with AutoCad, and obtained layouts for the different meridian sections of the dome exterior. On the other hand, we also checked that the data for the principal elements of the radial and plan layouts for the dome interior and exterior agreed, these elements being the widten of ledges distances between them and windows opening. Having done this we had the data necessary to plot the dome exterior, the radial and plane layout and meridian sections.

One the internal and external dome surfaces were drawn, we had to related the one to he other to complete the section. The first thing to make the centrelines coincide, which we did graphically with AutoCad. Then we had to locate the forms in their correct position relative to each other on the common axis. We calculated the correct displacement using differences between levels of the seven horizontal and the horizontal drum corresponding to the window bulltresses, which enerved their location at the same level, resulting in a standard deviation of 8.6 mm (Fig.20).

The mapping of the dome described above, was done using two point scatters: one for the exterior and one for the interior; points which we rotated to locate them in a meridian plane. It was necessary to translate these rows of points into line drawings. At the end of the process to define the dome section, we had perfectly identified the external and internal location of each and all of the points (Fig.21,22,23).

Cartography in Future: Cibeles

The manufacturer Leica Geosystem offered us to use a Cyrax 2500 instrument to test its data acquisitions capacity. We chose the Plaza Cibeles in Madrid as our site (Fig.24).

This instrument has a laser scanner for topographical, engineering and construction applications. Points can be surveyed to an accuracy of within ± 4 mm and angles within ± 60 micro-radians. Its capacity for data capture reaches 1,000 points per second, achieving an internal accuracy between points of 1,2 mm (Fig.25,26).

We carried out a data gathering test on the Cibeles statue. The team handed only fifteen minutes to read the scatter of points (Fig.27).

Having seen the way in which the data gathering was achieved and the capacity of this new instrument, we were challenged by the question: What benefit does this new instrument being to Archaeological and Heritage site? In our experience the Leica TC 305 instrument (used in direct mode without prisms) allowed us to work with discrete points interconnected by means of adjusting by quadratic minimus. The Cyrax instrument allowed us to go further along this lines, but now with almost continuous sequence of points. We found the challenge of analysing and experiencing two achieve results using two different methodologies.

Conclusion

In the 'Cartography in Archaeology and Heritage' Team we aim to apply new methods of information technology to archaeology. By recording the geometric relationship between data points, we try to assist in display. In this way, using digital models we enhance the reference site through three dimensional representation without prejudice to our ability to produce more conventional cartographic models. We are actually working towards introducing interactivity in spatial models, and at the same time integrate the progress of the excavation over time in the exhibit.

Once we started to work with archaeological teams we learned about the enormous difference between scientific research and experimentation, and the use of technology as applied to the Social Sciences. In engineering it is possible to study, analyse and improve techniques and methods, but it is not normal to apply them to a different field, and especially to archaeology. It has been an experience in modelling and prototypes that perhaps has allowed us to confirm that we are learning archaeology and trying to discover ways to ensure that cartography becomes an indispensable tool for archaeologists.

GPS Survey: 3D Data for Easy Digital Terrain Modelling

SEE THE CD FOR THE EXTENDED VERSION

Robert Shaw
The Discovery Programme
34 Fitzwilliam Place, Dublin, Ireland

Abstract

Advances in GPS survey technology have made it relatively quick and easy for archaeological surveyors to gather high volumes of precise geo-referenced 3D data. At the same time the software to manipulate and display this data in the form of 3D models has become less technically demanding and more accessible. As a result its application in archaeology is becoming widespread. The aim of this paper is to look at some practical examples of these models and consider how they compare with the traditional methods of recording and depicting archaeological sites.

Introduction

The Discovery Programme, established in 1991, is a state-funded archaeological research company with the remit to enhance the understanding of Ireland's past. Projects undertaken have included research on the North Munster region, the Western Stone Forts, the Ballyhoura Hills and extensive study at Hill of Tara. Specialists from many disciplines are employed, archivists, environmentalists, excavation archaeologists, historians, GIS analysts and surveyors.

In 2001 The Discovery Programme began two new research projects, The Lake Settlement of Ireland, and The Archaeology of Medieval Rural Settlement in Ireland, both requiring extensive field survey programmes. No survey work had been undertaken for over a year as previous projects were in the report writing stage so a re-assessment of the survey department's equipment needs was undertaken. 3D modelling had been found to be a successful way of recording archaeological landscapes and sites, but that data acquisition using a total station was often too time consuming.

Global Positioning System (GPS) had been considered for the previous research projects but at that time it was decided that the technology was developing at such a rate that investment should be postponed until a later stage (Masterson 1999:131-146). An evaluation of the current technology in GPS indicated that development had 'stabilized' sufficiently and reached the level where we could invest in a system with a degree of security that it would not be radically superseded in the lifetime of the projects (6 years).

GPS Systems

GPS is a term that most people have encountered today through its navigation uses in recreational activities such as mountaineering and sailing, or from news reports of its role in guiding missiles during military conflict. Most archaeologists have probably read articles outlining its impact in surveying but with conflicting claims of its precision and cost. We need to clarify what we mean by GPS and how it is applied in the context of archaeological survey (Wheatley and Gillings 2002).

Table 1 shows how GPS can be broadly grouped in three types based on how the measurement is made, and how the system is operated. It's beyond the scope of this paper to go into a full explanation of these processes, but it is important to appreciate the impact this has on the precision achievable and the cost of a system.

GPS Type	Navigation	GIS	Survey grade
Measurement	Code	Code	Carrier phase
Operation	Autonomous	Differential	Differential
Accuracy	10-20m	1-5m	<10mm
Cost	250 euro	5000 euro	40000 euro

Table 1 Broad categories of GPS survey

A navigation system costing Euro 250,- is a GPS, has many valid applications in archaeology, e.g. reconnaissance and prospective walking, but it is not a precise surveying instrument. Our surveying requirements; accurate control points, detailed site surveys, precise height data, mean that only the survey grade systems are appropriate.

[1] System Specification

Trimble 5700 GPS Specification		
Measurement accuracy		
Static survey (control)	Plan:	5mm + 0.5ppm.
	Height:	5mm + 1ppm.
Kinematic survey (detail)	Plan:	10mm + 1ppm.
	Height:	20mm + 1ppm.
Physical specification		
Water resistance:	IPX7 for submersion to 1m.	
Shock resistance:	1m drop onto concrete.	
Climatic conditions:	-40 to +80°C, 100% humidity.	
Battery power:	7 - 10 hours continuous use.	
Weight:	Complete rover pole <4kg.	

Table 2 Specification of the Trimble 5700 GPS

High quality GPS systems are available from all the major survey equipment manufacturers but with a cost in the region of Euro 40,000,- the selection criteria was important - balancing performance, durability, and system compatibility with cost. The environment in which it would be operating demanded that it be rugged and water resistant, and this was one of our major considerations. Other factors were ease of integration with our existing total stations and the quality of processing software. After evaluating a number of options the Trimble 5700 system was chosen. Table 2 summarizes the specifications of the system.

[2] Real Time Kinematic (RTK) Surveying

GPS surveys are undertaken in the field in a RTK mode. To do this requires a pair of survey grade receivers operating in the same local area. One of them remains set up gathering data on a fixed point, known as the base station, while the other is moved around the site, the rover. Many of the errors that are inherent in the GPS system are the result of distortions as the signal passes through the ionosphere. These will be approximately the same in a local area, as the signals will have travelled through almost identical part of the ionosphere. As we know the precise position of the base station we can calculate the difference between the known and GPS calculated position at any instance - known as the differential correction. A real-time system broadcasts this correction to the rover via a radio link and the position is updated instantly. In this way the points we are surveying are measured to high accuracies relative to the base station. More than one rover GPS can be operated from one base station, allowing productivity to be effectively doubled by hiring and extra roving unit.

[3] Reference Systems

RTK surveys result in precise positions relative to a base station but a requirement of our work is to position surveys in the national mapping reference system of Ireland, the Irish Grid. To achieve this the continuously operating reference system (CORS) provided by the Ordnance Survey Ireland is used. This system records GPS data continuously at a network of stations around Ireland which is then made available, currently free of charge, through a website (www.osi.ie). GPS data gathered from our base station receiver is processed with data downloaded from the CORS stations and a precise Irish Grid coordinate obtained. All our RTK survey data can then be transformed into precise Irish Grid coordinates.

[4] Total Station Integration

GPS depends on receiving radio signals from satellites in space and therefore the receiver needs to have a clear view to the sky. However we often encounter sites that are all or partly obscured by trees, thick vegetation or tall buildings. In these cases we need to integrate total station surveys into our survey. The Trimble 5700 system operates with a TSCe controller unit that can be connected directly to a total station. Using control points set out using GPS total station data can be seamlessly integrated into the survey job. The controller has a graphic display that shows all the points surveyed so gaps can be easily identified and filled.

[5] Software and Processing

Whether surveyed using GPS or Total Station each point is described using a feature code. The processing software, Trimble Geomatics Office (TGO) converts these codes into lines, symbols or areas depending on the symbology defined in our feature code library. This vector graphic forms the basic survey of the site from which GIS layers and maps can be extracted. Although often viewed as 2D plans all the points automatically retain their full 3D description (x,y,z) and this is what makes GPS such a powerful survey method when it comes to modelling surfaces.

3D Surface Models

GPS was the key to creating 3D models of archaeological sites and landscapes because it allows us to gather large quantities of high quality data efficiently. Even with experienced operators total stations record points at such a slow rate that extensive modelling becomes prohibitively time consuming. Photogrammetric methods seem very attractive but have the problem of acquiring suitable large-scale aerial photography to achieve high accuracy results. This can only be done with purpose-flown photography that is expensive, and involve high performance high cost processing systems needing a high level of operator expertise.

Figure 1 Ringfort, Tulsk, Co. Roscommon

Although we refer to our DTMs as 3D models this is not strictly correct, as they contain no volume information. A more appropriate term 2.5D is often applied to these surface models (Wheatley and Gillings 2002).

[1] Point Sampling Strategies

Models are built from the network of surveyed points covering the landscape. The example around which this paper has been written is a medieval ring-fort site at Tulsk, Co. Roscommon, Ireland, see Figure 1.

Survey and Geographical Positioning Systems

The fort and immediate environment cover an area of approx. 10,000 m2. The form of the site is an awkward combination of steep slopes (height difference 6.8m) and the feint remains of features probably ploughed out over the centuries.

TGO processing software has shape-file as an output option and this is added as a dataset to a new map in ArcView. The software extensions 3D Analyst and Spatial Analyst are core tools in the model construction processes.

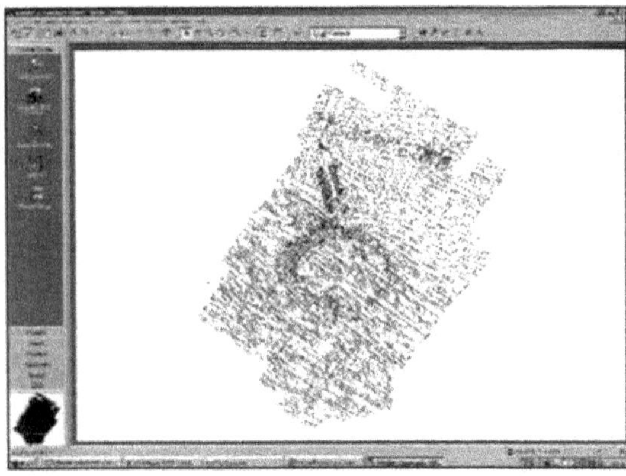

Figure 2 GPS points surveyed for the Tulsk model

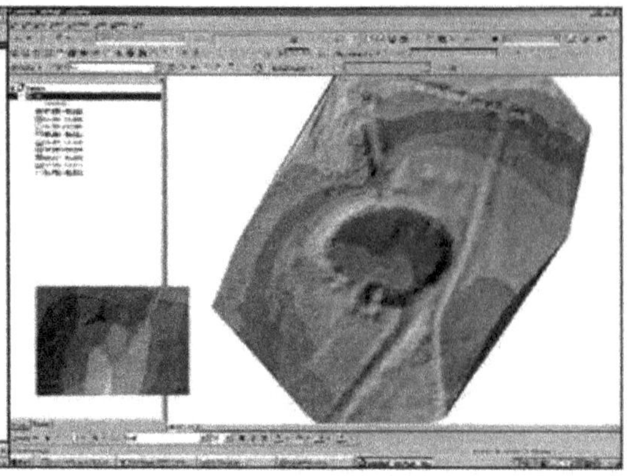

Figure 3 TIN model created in ArcView

The first stage is to build a basic Triangulated Irregular Network (TIN). The functionality of ArcView is such that this is a simple process with little inputs to consider. The shapefile of points is our only input, the resultant TIN automatically output, see Figure 3.

Vector break-lines can be added and the TIN built to respect them to varying degrees, but experiments on the earthworks sites undertaken so far has found these to be unsuitable. Tops and bottoms of slopes on the ground are rarely hard lines and all the break-line options overly distort the modelled slope if included. Our best models were achieved by including the points making up the break-lines without treating them as hard vector boundaries.

The sampling strategy employed is a combination of recognised spatial sampling strategies. The first stage is a modified systematic grid of points with approx. 1m spacing. It is considered modified because we increase the density of survey points where the surface is most complex. This is a subjective decision by the surveyor who has to assess whether the 1m grid is sufficient to reflect the subtleties of the surface.

A second stratified stage adds points along the obvious tops and bottoms of slope (break-lines) to ensure that they are included in the model.

This strategy has evolved from experiments with different approaches and seems appropriate to most of the sites encountered in the project so far. Importantly, it gives the surveyor flexibility to adjust the density of points to reflect changes in the terrain.

Figure 2 shows the result of the GPS survey of this site; two surveyors measured 8,750 points in two and a half days with two rover GPS units. This included a detailed reconnaissance of the site and the establishment of the base station.

[2] Model Construction

GIS analysis is a vital process in the archaeological research undertaken at the Discovery Programme, and ArcView 8.1 is the GIS software used. It includes a range of 3D modelling options and given our familiarity with ArcView as a whole we chose it as our model building software.

[3] Raster Models

Although the TIN gives an effective 3D model of the surface it has limitations. The faceted appearance limits its use as an illustration of a site, and its vector structure makes it less versatile for the further analysis vital to effective data interrogation.

The solution is to interpolate the point data to create a raster map using the spatial analyst tools in ArcView. This process results in a raster map in which the interpolated heights for each cell are stored. Both the algorithm for the interpolation, in most cases we use a spline, and the cell size (which determines the coarseness of the image) are selected depending on

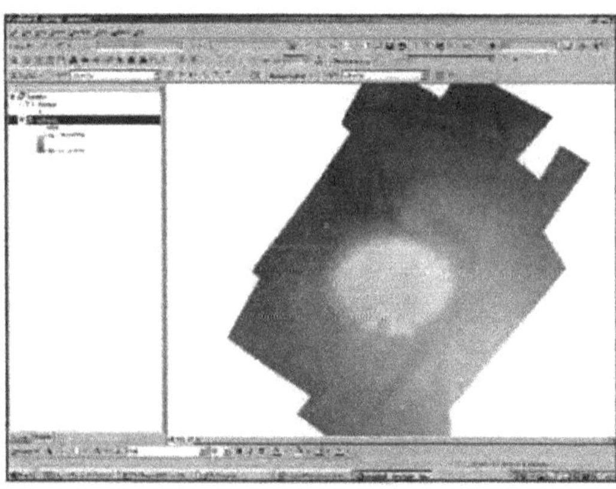

Figure 4 The spline model created in ArcView

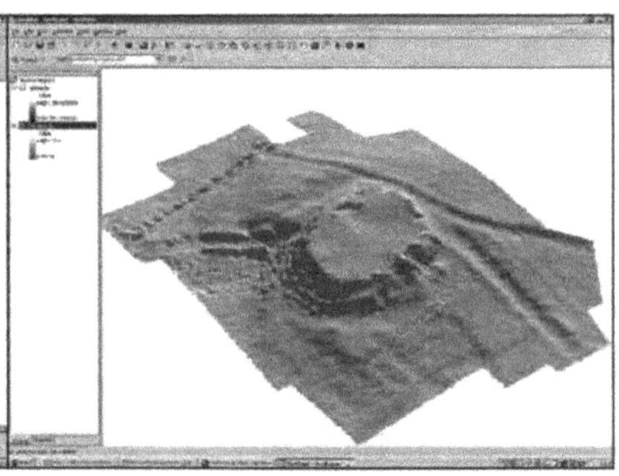

Figure 5 Hill-shade analysis draped over spline model, viewed in ArcScene

the nature of the site. The result is the basic rasterized surface model shown in Figure 4.

This rasterization process has to be approached with a high degree of caution. ArcView has a three of different algorithms for interpolating data, (nearest neighbour, kriging, and spline) and for each of these a range of parameters that can be adjusted. In the desire to produce more visually pleasing models it has to be recognised that the integrity of the data can easily be undermined. Interpolation in most cases does not totally respect the survey points that were surveyed using a carefully chosen sampling strategy. If the surface model then forms the basis for further analysis it's critical we are aware how the model was created. For each model created metadata should be included so the user can assess its validity.

With the data now in a raster form a range of spatial analysis tools can be used to further refine our model of the surface.

[4] Surface Analysis

The ArcScene image in Figure 5 shows how some basic surface analysis tools can be used to improve the quality of our model. It shows a hillshade model draped over the spline model with the height exaggeration by a factor two.

The azimuth and elevation of the light source (the effective 'sun' position) have been chosen to emphasise features not obvious when walking on the ground. In this example of this are the faint cultivation remains at the rear of the site.

Theses hill-shade models are the standard end product of our survey and should result in a largely objective model of the true land surface.

However, we can go further with surface analysis to perform more sophisticated interrogation our data. Figure 6 shows the 'synthetic' landscape that can been created when the slope angle values are calculated and draped over the existing surface Added emphasis has been given to the feint line of the fosse to the rear of the fort.

With such a system, and access to user-friendly manipulation software such as ArcView, it is entirely possible for non-specialist surveyors to consider surface modelling as a standard approach to recording archaeological sites.

But is this largely objective approach an appropriate way of recording and representing our surveys? To consider this we need the look more closely at the traditional approach to recording sites.

Traditional Approaches

The ring-fort at Tulsk was re-surveyed using the plane table survey method to allow an evaluation of the different techniques.

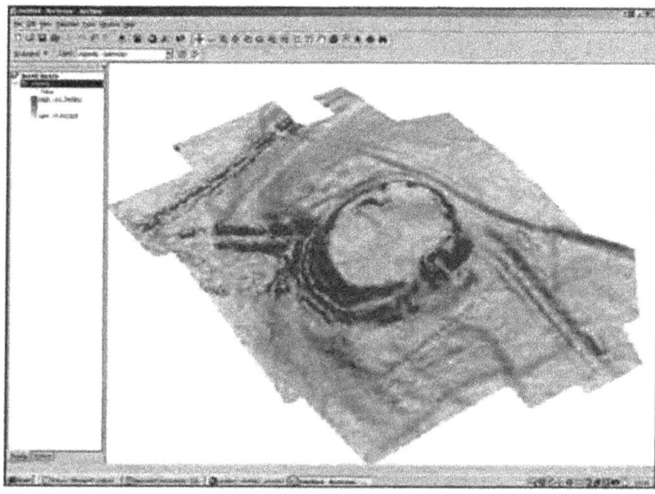

Figure 6 Slope analysis draped over spline model, viewed in ArcScene

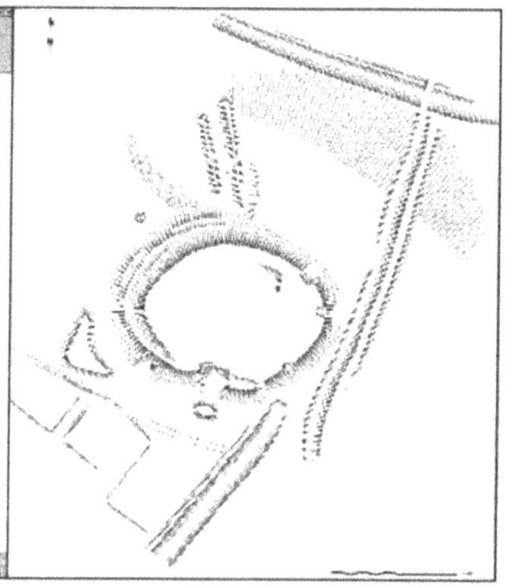

Figure 7: Hachure plan of the Tulsk Ringfort

[1] Hachure Plan

The process of drawing hachures immediately revealed the different level of interpretation and the subjectivity inherent in this type of survey (Tanaka 1950:444-456). How the different slopes interacted and related to each other had to be understood before they could be depicted. The process of defining tops and bottoms for the rows of hachures forced decisions to be made about phasing and relationships that could be largely ignored when surveying points for the 3D model.

In constructing the drawing compromises had to be made; which elements were selected for inclusion on the plan, how we depicted the long slight slopes and the short steep slopes. The process was translating our interpretation of the site to a graphical depiction on the paper. This was the key difference with 3D modelling - the emphasis on interpretation. The final drawing (Fig.7) shows the result of our interpretation of the site and the way this was depicted. It is important to acknowledge that this is a personal process and different surveyors with different levels of experience may produce very different plans.

As a method of portraying our understanding and interpretation of a site it has a great deal of merit, but an important part of any archaeological survey is to create a record of the site. Is a personal interpretation appropriate to be our record of what is there? Features omitted from the plan because we do not recognise their significance within the framework of our current knowledge base could turn out to be vitally important. We would have no idea they were there with our hachure plan but with a 3D surface the features are included even if we have failed to register their significance.

The more these two approaches were examined the less convinced we became that they were in fact comparable products. Our 3D surfaces appeared to be a good solution to the recording problem and an aid to analysing the site, but limited in presenting this understanding. How could this be resolved?

[2] Automatic Hachure Programmes

One approach that has been subject of much research in the UK[1] has been the development of an automatic programme to generate hachures from the surveyed top and bottom of slope lines. This development is achieving good results with most hachures being resolved without the need for manual editing. As our surveys include the top and bottom lines, integral to creating our models, hachures could be output automatically. Taking the process a stage further the hachures could then be draped over the 3D model to present the interpretative element on the basic site record (Kennelly and Kimerling 2000:78-81).

However, to interpret hachures requires a level of understanding and experience that cannot be taken for granted, so alternative approaches were examined (Dykes, Moore and Wood 1999).

[3] Virtual Tours

A straightforward export from ArcScene is a VRML file based on the TIN model. Free shareware makes this an easy way of allowing the model to be viewed. We are currently researching how we can program in VRML to take this a stage further and attach narratives to the model at the important waypoints. This could take the form of 'pop-up' text boxes and images or possibly a recorded commentary (Fisher 1997).

The tour would allow the interpretation to be presented in a logical sequence and ensure that the user is shown the important parts of the model. Different levels of narrative could be applied depending on the age or understanding of the user.

The obvious criticism of this is that it depends on access to the technology to view it. However as the illustrations throughout this paper show it is quite possible to represent excerpts and 'flat' images of the 3D world if needed.

Alternative Data Sources

Working with GPS on the ring-fort site the team felt restricted in how far we could extend the survey given our time constraints. It was highly unlikely that the hill-fort was isolated in the landscape and the potential of significant finds in the neighbouring fields was high. But needing to sample points at 1m intervals meant that ground survey was not a viable way to extend the survey.

[1] Balloon aerial photography

One option has been to re-examine digital photogrammetric systems (Pollesfeys et al. 2002).

To eliminate the high costs associated with conventional acquisition of aerial photographs we have been investigating the possibilities of a balloon-based system. A metric digital camera suspended from a modern survey balloon has been proved to be a successful way of capturing images.

The latest software (e.g. Erdas Imagine) includes functions for automatically extracting surface models from stereo pairs using limited ground control and pixel matching. Such a system would not only extend our modelling capabilities but would also open up other archaeological possibilities e.g. modelling of excavations, or terrestrial photogrammetry.

[2] 3D Laser Scanning Systems

Terrestrial applications of this technology have proved a successful method of acquiring data for surface modelling. Our continuing research intends to test the cost effectiveness of such systems to see whether their productivity can offset the high equipment costs.

Conclusions

Advances in GPS survey technology combined with access to user-friendly modelling software opened up the potential benefits of 3D modelling to our research programme. It allowed us to challenge the conventional approaches to monument recording, and in evaluating these forced us to assess the merits of our 3D models.

The need to create both a definite record and an interpretative result from our surveys indicates the need to continue our research into VRML tours.

Ironically, as the latter part of this paper demonstrates GPS may not be the best solution for gathering the data for 3D models. Having introduced us to the techniques it may be that GPS will only play a minor role providing the control points for other methods.

1 Work undertaken by Edinburgh University funded by the Ordnance Survey (Nicolas Regnauld, William A. Mackaness & Glen Hart: Automated Relief Representation for Visualisation of Archaeological Monuments and other Anthropogenic Forms. Paper presented at GISRUK 2001, University of Glamorgan) and subsequent development jointly funded by English Heritage, RCAHMS and RCAHMW. Work undertaken by Edinburgh University funded by English Heritage, OS, RCAHMS, RCHMW.

REFERENCES

DYKES, J., MOORE, K. and WOOD, J., 1999. Virtual environments for student fieldwork using networked components. International Journal of Geographical Information Science Vol. 13, no. 4:397-416.

FISHER, P. et al., 1997. The Virtual Field Course: An educational application of GIS. Geological Teaching and Learning Network Conference, Fieldwork in the Geosciences, University of Leicester, 12 November 1997.

KENNELLY, P. and KIMERLING, A.J., 2000. Desktop Hachure Maps from Digital Elevation Models. Cartographic Perspectives no. 37:78-81.

MASTERSON, B., 1999. Archaeological Applications of Modern Survey Techniques. Discovery Programme Reports Vol. 5:131-146.

POLLESFEYS, M., KOCH, R., VERGAUWEN, M., VAN GOOL, L., ESAT-PSI, K. and LEUVEN, U., 2002. An Automatic Method for Acquiring 3D Models From Photographs: Applications to an Archaeological Site.

TANAKA, K., 1950. The Relief Contour Method of Representing Topography on Maps. Geographical Review Vol. 40:444-456.

WHEATLEY, D. and GILLINGS, M., 2002. Spatial Technology and Archaeology: The Archaeological Applications of GIS. Taylor and Francis, London/New York.

The Arminghall Henge in Space and Time: How Virtual Reality Contributes to Research on its Orientation

Willem Beex

ARC-bv and National Museum for Antiquities, The Netherlands

John Peterson

School of Information Systems, University of East Anglia, Norwich NR4 7TJ, UK

Abstract

Neolithic and Bronze Age henges may be related to particular solar events, but confirmation of this requires an accurate determination of relative orientations. In the case of the Arminghall henge in Norfolk, England, the direction of midwinter sunset at that time was thought to have had an important influence, but our computer based methods reveal flaws in this proposal. A spreadsheet relating archaeoastronomical data to a horizon profile (created manually) and the orientation of the henge (derived from rectified aerial photography) allowed us to plot a simplified view of the sunset, but a virtual reconstruction, based on a digital terrain model and the excavation plan, gives a more accurate, useful and convincing visualisation. From this we can suggest that the henge is positioned so that from it the setting sun is visible "rolling down" the side of the most prominent nearby hill. However, it is clear that the axis of symmetry of the central wooden structure never pointed to this setting point, but towards the top of the hill. Virtual reality reconstruction thus allows alternative theories to be explored, and it may also provide a stimulus to further investigation of the landscape setting of similar structures.

The Arminghall Henge and its Environment

The Arminghall Henge is one of East Anglia's most significant prehistoric field monuments (Ashwin and Bates 2000:230). It lies near the junction of the rivers Yare and Tas, less than 4km south of the centre of Norwich (Fig.1). It is the focus of a large group of monuments, mainly dated to the Bronze Age (Fig.2). Itwas discovered from the air in 1928 and partially excavated in 1935. Two almost circular ditches define it. The inner is larger and has an gap opening towards the south-west. The outer ditch may have had a similar gap. The excavator suggested that there was a bank, about 2m high, between the two ditches. Within the inner ditch there was a setting of large wooden posts arranged in a horseshoe also opening to the south-west.

The contours of the local landscape are not very pronounced, with relative elevations of up to only 40m. Nevertheless, because the Henge is so low, one can see hills from it. The most prominent of these is Chapel Hill, which lies on the end of a spur between the two rivers. This hill lies, again, south-west of the monument. Trevor Ashwin suggests (Ashwin and Bates 2000:233) that the alignment of the Henge might be directed towards both midwinter sunset and the summit of Chapel Hill, although they are only approximately in the same direction. John North (1996:Fig.154) gives us a reconstruction which also shows the axis of symmetry of the horseshoe pointing to the gap in the inner ditch but, because of a miscalculation, it is not reliable and does not suggest a link with Chapel Hill.

An accurate figure for the Henge orientation, agreed among independent observers, was needed. So the first aerial photograph (published in 1935) was rectified to the Ordnance Survey base map by staff of Norfolk Landscape archaeology. One of the staff, Martin Horlock, and one of the authors then independently used drawing software to place a line on the rectified photograph which corresponded, in their judgment, to the axis of the horseshoe post setting (Fig.3). The difference between these estimates of orientation was 0.15°. Accordingly we may conclude that the Henge axis defined by the mirror symmetry is at 40.5° to OS north (to the nearest 0.5°). Since geographic north is 2.63° west of OS north at the Henge, this is equivalent to an azimuth of about 223° in relation to geographic north.

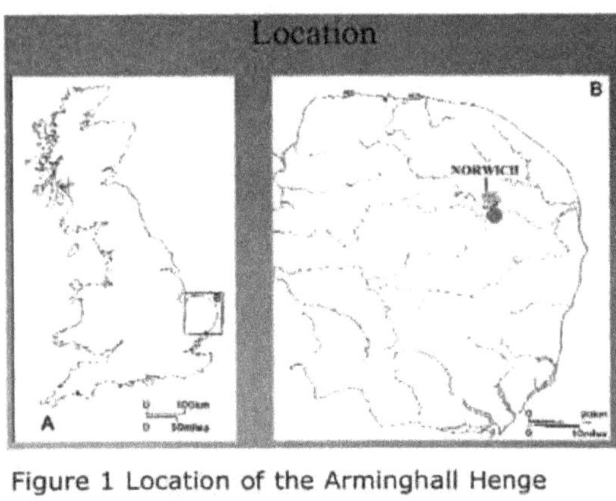

Figure 1 Location of the Arminghall Henge

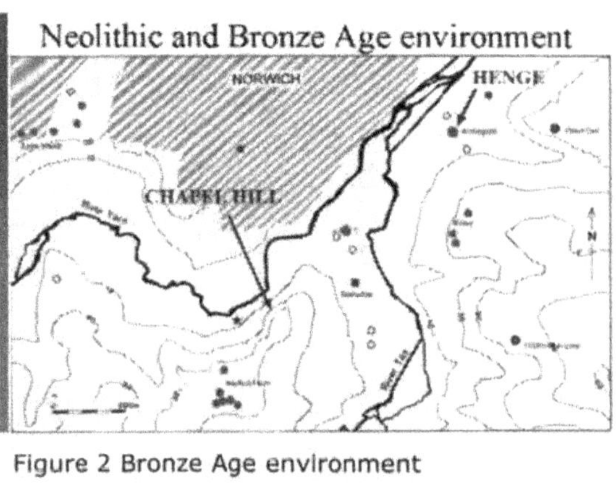

Figure 2 Bronze Age environment

The Relationship of the Henge to Midwinter Sunset in 4,000 BP

In the period 5,000 to 4,000 BP the last flash of the setting sun, making allowance for a horizon altitude of 0.8°, was at about 228°. This is 5° from the henge orientation so it is difficult to see a connection between the two unless we consider features of wider landscape, adopting the approach taken by (Freeman and Freeman 2001) in the case of Stonehenge. Using their approach, horizon profiles were constructed by drawing rays on the OS map from the centre of the henge and measuring the heights of points along those rays. Calculation then gave the (angular) altitudes, and hence the maximum altitude, along those rays.

well known case of Croagh Patrick (see http://www.carrowkeel.com/sites/croaghpatrick/reek2.html).

This suggests that the site of the Henge was chosen, perhaps many years before the Henge was built, because it was a place from which such a thing could be seen. It may be one reason why this Henge is so low lying. Only from such positions can hills be seen.

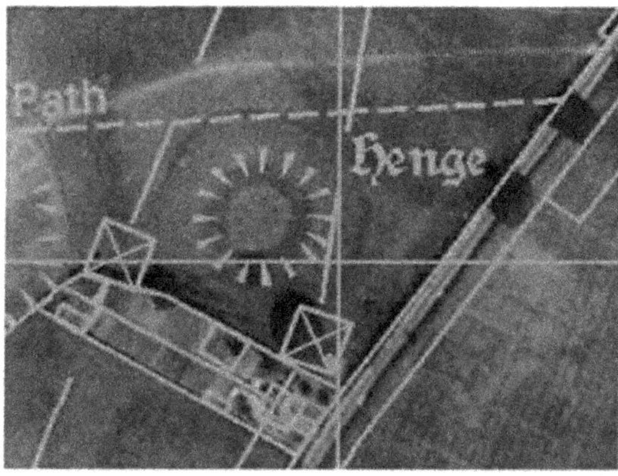

Figure 3 Rectified Aerial Photograph and base map

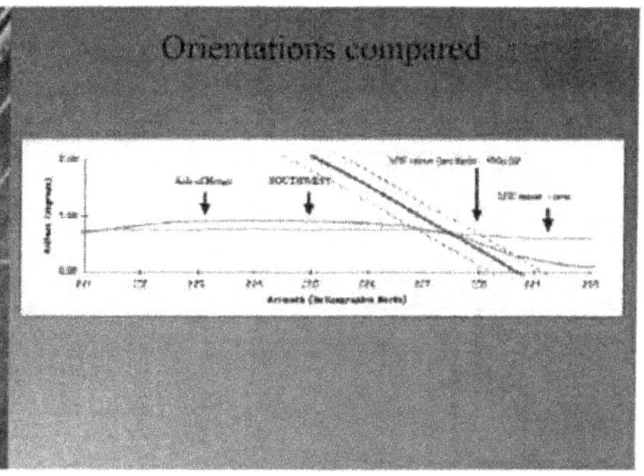

Figure 4 Orientations compared

Virtual Reconstruction of the Landscape

A more accurate model of the environment of the Henge required sufficiently accurate data on the elevation of the land surface within a radius of about five kilometres. Modern techniques such as photogrammetry were of little use, even if detailed aerial photographs had been available; half of the surrounding landscape is covered by modern day Norwich. Therefore the data had to come from contour lines at five-metre interval on 1:10, 000 OS maps. These maps have some disadvantage for areas with small differences in altitude, such as the flood plain of the rivers[1], but provide enough information for the reconstruction of Chapel Hill, whose prominence as seen from the henge is a key feature of the landscape.

A modified form of the Freemans' calculations gives, for midwinter sunset, a relationship between the apparent altitude of the sun's centre, h (for $h < 4°$) , and its azimuth. They are approximately related by a straight line

$$h = \frac{\text{Cos } Q \text{ Cos } l + \text{Sin } e}{\text{Sin } l} + r$$

Where
Q is the azimuth of the sun's centre,
l is the observer's latitude,
e is the obliquity of the ecliptic and
r is the increase in apparent altitude due to refraction.

A spreadsheet performing these calculations was validated by checking its results against those for Stonehenge given by John North. By trial and error it was established that a value for the refraction of 0.0081 radians gave the best fit of spreadsheet values to those published.

The spreadsheet values for the sun's path were drawn as a graph; the horizon profiles were also included. The date could be altered to affect the obliquity of the ecliptic and hence the sun's path. This showed that in 5,000 BP disc of the setting midwinter sun would have seemed to pass down the right hand (northern) side of Chapel Hill and be "eaten" by the angle between it and the far horizon (Fig.4). Like the Freemans, we tried to observe this on the ground, making an allowance for the change of about 1.5° in sunset azimuth over 5,000 years. This could not be done from the monument itself because a high hedge obscured the view. However, a line was drawn on the map from the Henge centre, at the appropriate angle. From the point where this line intersects a road, nearer to Chapel Hill, the "roll down" of the sun could be observed near midwinter, 2002. Such an effect can be compared to the

The maps were digitised in a CAD-application including their attribute data. The accuracy of this digitising was somewhat less than three metres which, for our purpose, is well within the maximum allowed tolerance[2]. The spatial data was then transferred to Surfer[3], which is probably the best software for creating a Digital Terrain Model (DTM). Its algorithms and parameters allowing for controlled and detailed processing of contour lines. They can be input as exact data and smoothed. For this the search radius, the distance over which data points are allowed to influence each other, is important. For example, the valley of the river Yare must not be smoothed because of measurements taken from a hill two kilometres away. Anisotropy, the irregularity of the landscape, must also be considered. In simple terms, there are two rivers cutting through the hills, which have generated the overall shape of the hillsides.

Experience showed which methods and parameters were most appropriate in this case. The commonly used Triangulation with Linear Interpolation (TIN) approach is not well suited for interpreting this type of data. A DTM produced in this way for the Arminghall Henge area produces a rather terrace-shaped landscape which does not convincingly reproduce the actual shape of the one significant hill. It gives it a flat top, even if smoothing is allowed for. Kriging seems more appropriate. The default application of this algorithm in

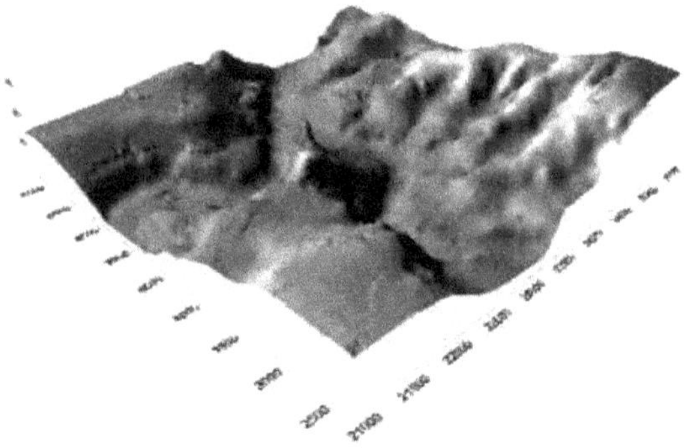

Figure 5 DEM created in Surfer 8

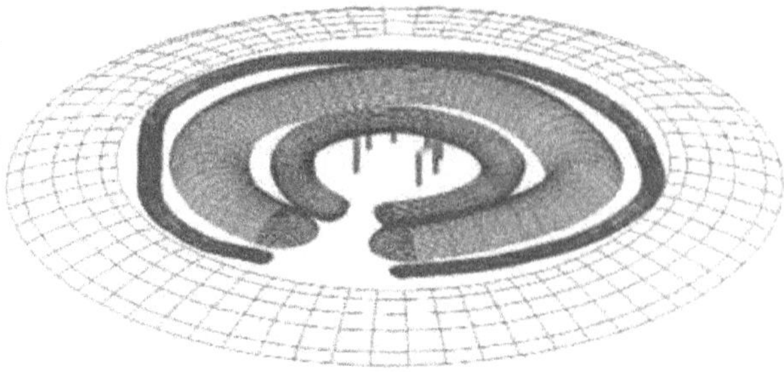

Figure 6 Building the henge

Surfer produces a model of the environment of Arminghall Henge which compares favourably with visual observation and photographs of the actual landscape. However, contour lines at five metre interval still lacked some detail needed to give a true profile to Chapel Hill. Adding valid additional data rectified this. The other problem was the tendency of the Kriging method to insert irregularities if the distance between two contour lines is more than ten times the size of the DTM mesh. This problem could be solved by using a custom-made variogram, despite the problems associated with creating this, about which the Surfer manual is quite explicit. The DTM of the Arminghall Henge environment was then prepared for modelling and animation (Fig.5). This required a custom-made LISP-routine in AutoCAD, capable of reading Surfer grid-files.

Meanwhile, using a CAD-program, the henge itself was reconstructed from the excavation plans, while correcting the north to the real world situation (Fig.6). When the basic model was considered to be correct, it was imported into the rendering and animation program. The trajectory of the sun 4,000 years ago was then recreated. Using positions calculated by astronomical software[4], a series of sightlines was plotted, identifying vertices of an arc from pre-sunrise till a few hours after sunset. A bezier-curve through these vertices, allowed for a smooth path for a virtual sun to follow.

With all the different elements available in the 3d-modelling application, textures, daylight effects, animation and post-production were added. Since midwinter sunset was to be simulated no summer textures had to be added to the ground surface and we could assume that at this particular midwinter it had snowed. An appropriate texture, onto which a "bump-map" was placed with a fractal algorithm, produced a snowy landscape (Fig.7).

The most significant part of creating the animation involved adjustment of the background sky and the overall lighting. This took some trial and error - reducing ambient light and smoothly inserting a dusk background sky and other effects - to obtain a convincing view of sunset (Fig.8).

The appearance of the sun itself was done by adding lens-flare effects and the midwinter 2002 photograph provided good reference information, especially for the apparent size of the sun in relation to the horizon. Changing the background from a blue into a dusk-coloured sky was done by using a gradient bitmap, a few times larger than the intended area of the background. This bitmap was then slowly moved upward during the animation, giving the impression of an atmosphere turning from blue into pink/orange.

The Value of VR Reconstruction

The VR reconstruction provides confirmation of the phenomenon modelled by the spreadsheet approach, but it has the great advantage that additional information can be presented that is not easily available in graphics and plans . This makes the results of the research on the orientation of the Henge much more understandable. Those to whom we are trying to communicate our ideas can be more confident that they are soundly based.

Figure 7 Adding textures and shades

Figure 8 Creating the sunset

The DTM also provides data which could can be used for further research. For example, there is another feature with a double circular ditch lying at the foot of Chapel Hill on its northern side. This may be another henge. Was it positioned so that a midsummer sunrise event was visible to the Northeast, over the hill on the south side of the Yare valley?

The situation of other similar monuments might also be explored in the same way. The so-called "Seahenge", which is so low that it is now in the tidal zone, may be just part of a group of monuments which lay north-east of the most prominent headland in north-west Norfolk. From this site the headland may have been visible in the direction of the setting midwinter sun, just as at Arminghall.

[1] The floodplains of rivers in the area are now flat, but we expect that 4,000 years BP sea level would have been lower. Our model does not attempt to take account of this, since it has no influence on sunset positions.
[2] When using a puck to digitise, accuracy in calibration of a map should be within 0.05 %.
[3] Surfer is product of Golden Software, Colorado. For the reconstruction of the landscape around Arminghall Henge, the newest version 8 was used.
[4] "Cartes du ciel" was used in this case.

REFERENCES

ASHWIN, T., and BATES, S., 2000. Excavations on the Norwich Southern Bypass, 1989, Part 1. East Anglian Archaeology 91. Norfolk Museums Service, Gressenhall.

FREEMAN, G. R., and FREEMAN, P. J., 2001. Observational Archaeoastronomy at Stonehenge: Winter Solstice Sun Rise and Set Lines Accurate to 0.2° in 4000 BP. In Pilon, J-L., Kirby, M. W. and Thériault, C. (eds.), A Collection of Papers Presented at the 33rd Annual Meeting of the Canadian Archaeological Association:200-219. (The Ontario Archaeological Society Inc., http://collection.nlc-bnc.ca/100/200/300/ont_archaeol_soc/annual_meeting_caa/33rd/freeman1.pdf).

NORTH, J., 1996. Stonehenge: Neolithic Man and the Cosmos. HarperCollins, London.

Argumentation paths in Information Infrastructure of the Archaeological Virtual Realities

Isto Vatanen
Information Studies, Åbo Akademi University
Cultural History, University of Turku
isto.vatanen@{abo.fi, utu.fi}

See the CD for the extended version

Abstract

As discussed recently (e.g. Vatanen in CAA 2002), documentation of digital multi dimensional data, is a acute considerable challenge for virtual archaeology. Being able to document precisely the archaeological work done also in virtual reality is essential for the scientific credibility of the discipline and the virtual archaeological methodology. This paper discusses some of the practicalities of implementing an efficient and yet cognitively unburdening system for the documentation of virtual realities, opening insights into what an archaeological record could be in a true digital form.
The essential difficulty of virtual spaces are the formal perspectives of the virtual. A strong emphasis is placed on the multi dimensionality and multi mediality of the virtual worlds. The rather more nature like form of a virtual reality, or simulation as has been recently argued, requires new concepts and approaches to the idea of annotation, documentation and information retrieval. Therefore a satisfactory solution cannot be reached simply by implementing linearity of traditional computing strategies.
Theoretical basis of the study is to conceptualise the tasks of making, reading and interpreting information in the virtual realities, as dynamic infrastructures of co-existing processes ending up as a semi-real organisation with inherent cognitive capabilities and intelligence. As the key point of the paper a theory on a plausible documentation strategy is presented with a reference to a practical case study on a documentation tool application.

Introduction

Virtual archaeology suffers from a paradox of lacking and an overload of resources. From a researcher's point there is seldom enough information on a particular site or subject to make satisfactory virtual representations of the past. From an information management point of view, virtual archaeology produces information and archivables in such amounts that their efficient storage and occasional retrieval becomes a critical issue. The key for managing both, the non-existent information and as a consequence, approximations and inconsistencies, and the existing data in the virtual models, is to develop functional measures for efficient and communicative documentation.

This article discusses documentation and usage issues of virtual reality documents in archaeology from the information management point of view. The focus is on defining theoretical premises on how and based on which grounds, a VR is a feasible research tool and a medium of communication in archaeology. The issue is conceptualised by using argumentation (Freeman 1991, Kircz 1991, Sillince 1992) and infrastructures (Star and Ruhleder 1994, Star 1995) approaches. From this discussion a proposal of "argument paths" is derived for an approach of enhancing usability of the archaeological virtual realities as publications and information storages. Special emphasis is placed on supporting the functionality of the document by documentation strategies.

Archaeology and Virtual Reality

During the period from the mid 1990's to the present, archaeological and cultural heritage applications using different multidimensional visualisation techniques have rather clearly established themselves as a one standard procedure and concept in the archaeology computing and cultural heritage sectors. A quick survey over recent literature gives an impression that the techniques have penetrated rather widely the archaeological work procedures. The use of the virtual reality technologies does however, vary considerably from one expedition and excavation to another, and the dissemination of the skills and experiments seem to be still geographically somewhat uneven, depending on institution and country. In Europe, the EU development, research and technology programmes have lead to a number of advanced projects and development in low-resource countries, but the actual involvement seems to depend considerably on participants personal skills and interest.[1]

Then what is an archaeological VR? An important aspect of a virtual reality is in its non-linear structure. Virtual realities do resemble, in this respect, hyper-texts, which are also descendants of the digital era making non-linear, a standard of publication. Another aspect draws from the question of what is a virtual and what is a reality. Thinking of the meaning of the term virtual reality, such a construct representing a historical subject, can hardly be an accurately measured theory-free three-dimensional sibling of cross-sections and plans, despite this approach was rather strongly argued by Gaiani (1999).

Unlike a multi-dimensional visual documentation technique to which Gaiani essentially refers to by using the term, a proper virtual reality should on my consideration defined rather as a dynamic projection of an assumed past reality following a suggestion by Barceló (2001). According to the term, a virtual reality ought to be a virtual interpretation of essentially real phenomena. A multi-dimensional document can be a representation of recent field documentation work, it can be such a document complemented with assumptions, hypotheses and theories, or it can be a simulation of the past reality only when it can be a virtual reality. An important aspect not to bypass is still that as a simulation a VR is still a theoretical composition, not a reality, nor belonging essentially to reality (Barceló 2001, Millette 2003).

A rather nihilistic view on virtual archaeology and archaeological virtual realities,[2] is functional especially because of the challenges regarding the illusion of false completeness easily summoned by photorealistic presentation (Eiteljorg 2001). As a plan drawn by an excavator is never completely theory-free, a virtual reality of the past as a cumulative of multiple archaeological and historical sources is in absolute and in relative terms considerably far less theory-free. Therefore an essential value in creating a virtual reality, or "writing space" (ref. Bolter 1990 and 1991), as might be said, and in studying it, or similarly "reading the space", is in the educated argument of the archaeologist responsible for the virtual reality. The value of an excavation report is in reporting minute details, but the value of an archaeological virtual reality is in distinguishing the essentials from the secondary information.

Multiple Dimensions of Archaeology

Different uses of virtual archaeological applications do raise a question on degrees and consequences of usage. Many of the possible practical applications of VR in archaeology proposed by Maurizio Forte (2000:250-251) have by now, if not had by then, been realised in various projects, and a number of new uses for the techniques have been introduced since then. Another question on the extents of virtual archaeology as a phenomenon presented in the same article (Forte 2000:250), has received, on the contrary, considerably less attention.[3] Despite the diversity of projects and definitions of, and the definitions used of, virtual reality applications, the overall aims of VR in archaeology have remained more or less the same from the beginning. As a provocative generalisation might perhaps be argued that most of the researchers being aware of the conceptual problems regarding the visualisations in general, do wash their hands of the problem by emphasising their scientific and scholarly aims over providing aesthetic images. It is highly questionable whether the rationale and the significance of VA for the archaeology as a scientific discipline, have been explained or adapted to the practical applications consistently.

The thing in the VA is not in creating a spatio-temporal simulation of a past structure or phenomenon. The thing is how this practical achievement does progress scientific research and in general, increase understanding of the past. Barceló (2001) discusses the question insisting on the imperative of implementing a true interactivity by placing the focus on the communication aspect in the virtual, before the user experience can enter to a virtual reality proper. Emphasising communication may be grounded also on the derivative of the most commonly presented explicit motivations for creating VA applications, which is by no doubt disseminating knowledge of the past in the sense of communication (Forte 1997:12). The communication, as referred here, has to be understood as a rather inclusive concept to consist of communication of a scientist to him/herself, communication within the scientific community and communication of science to the general public. In this sense even the personal research motive of creating virtual realities may be seen as a communication where a scholar arranges source data in a manner that the data communicates its essential information content back to the scholar him/herself.

Aspects of the communicativeness of the archaeological virtual realities have been recognised as a grave practical problem during recent years from the documentation point of view which is a familiar one for archaeology. The fear of not being able to deliver information on the accuracy of archaeological data and the premises on different interpretations to the audience has lead to a discussion on practical ways of documenting the models properly in scientific sense (e.g. Ryan 2001, Pekkola 2002, Niccolucci and Cantone 2003, Vatanen 2003). As noted already (Vatanen 2003), the debate has been focused, from a quite atomistic point of view, on rather mechanical issues of inserting this-and-that kind of metadata to the documents, instead of discussing also how and for what purpose the VR is actually expected to be used, and how the practical usability could be supported.[4]

Barceló refers to Hollan and Stornetta article of 1992 "Beyond being there" which discusses the problems of assuming that traditional face-to-face communication functions on computerised environment. This basic observation is indeed confirmed as a correct one by a number of usability trials (e.g. Chen and Czerwinski, M. 1997, Westerman and Cribbin 2000, Büscher et al. 2001). The complete picture is however by no means an uncomplicated one, and a functional usability springs from an amalgam of adapting both traditional and novel communicative techniques and approaches. Complementing the earlier observations, more recent research on the premises of human-computer interaction (HCI) and especially on the computer supported collaborative work (CSCW) has suggested a number of functional approaches to tackle the problem.

A rather widely discussed paper "Steps towards an Ecology of Infrastructure" by Susan Lee Star and Karen Ruhleder (1994) examined a collaborative computerised environment called Worm Community System (WCS) used by biologists studying genetics. The authors conceptualised the focus of their research by using a concept of an infrastructure to make visible the choices, politics and relations traditionally invisible in the structures. Star and Ruhleder view on infrastructure is to understand it as a relation between different entities than an entity, or a thing, itself. The concept has been referred to especially in literature on (management) information and knowledge systems, CSCW and classification (Bowker and Star 1994, Monteiro et al. 1994 etc.).

The point where genetics, collaborative work and infrastructure meet archaeological virtual realities, is at possessing a common aim of making something collectively known and used. Even though a VR is not a work group, strategies learned in and applied by CSCW research provide useful ground for enhancing the communicative capabilities of archaeological virtual realities. Archaeological VR is a dynamic (in the sense of historic processes, and the transitions in the archaeological stratum and the model, see Barceló 2001:224-229) multi-actor (the historic creators and users of the present material cultural remains, excavators, researchers, public audience) environment where different identifiable and hidden needs, ideas, politics, possibilities and states of willingness do coexist.

Infrastructures and Beyond

The conceptualisation of virtual realities through the glasses of infrastructures does not as such imply anything practical. Infrastructures become operational, through the reflection of an essential characteristics of an infrastructure to an archaeological VR, namely that the relations between the entities of information (Star and Ruhleder 1994), and especially in the political nature of the mentioned relations (Star 1995).

Documentation of virtual realities as a task of attaching descriptors into entities, is from a scientific point of view rather trivial operation. Intensive work for functional specifications and standards, and research for their premises, usability and use will be definitely needed. As the Semantic Web Activity of W3 Consortium (www.semanticweb.org and www.w3.org/2001/sw) or the compilation of CIDOC Conceptual Reference Model version 1.0 and 2.0, the amount of work not apparent challenges is not to be underestimated. The problem, of what these descriptors ought to be and how they should be structured, persists. However, even when a proper description of semantics and ontology of an archaeological VR data will be ready to use in the future, the resulting formal description is rather obscure from the human perspective even if supplied with a set of productivity tools.[5] From that point of view, discussing relations and usability issues, is rather essential even in this stage.

A notion of the obscureness of documentation data, upon which the argument in this article is largely based on, is by no means a novel one. Kircz (1991b) questions the efficiency of content driven data indexes of which the meta-data schemes are a good example. He suggested that argumentation rather than content driven approach would benefit considerably the efficiency of information seeking tasks. Sillince (1992) broadens the proposal to form an arguer search programme for theoretically every kind of academic articles. The use of rhetoric in non-linear documents is further discussed by Juby (1996) and Carter (2003).[6]

The problem with the approach presented in the original Kircz model, is its rather mechanistic and mono-disciplinary nature, which restricts its usability in scientific disciplines with different kinds of publishing cultures from the physics. The same has been noted to apply to other discourse-level documentation schemes. Schemes tend to be rather domain-specific which makes them somewhat unsatisfying for more general usage (Teufel et al. 1999). Restricting the documentation to a rigid framework of discourse would, even within one discipline, cause difficulties in implementing paradigm changes and new methodologies, not to mention extra-disciplinary references. The Sillince (1992) arguer programme does offer a basis for a functional framework of documenting archaeological virtual realities. The programme requires however some refinement to cover other forms of scientific publishing apart from the article, which is formally rather unlike the virtual realities despite the same communicative aims.

As a contribution to the form issue, Kircz discusses the question of new layout of scientific writing in a somewhat more recent article (Kircz 1998). He argues that as a product of historical evolution, the article could be replaced by a format more suitable for the electronic environment. It is plausible to assume that a scientific publication breaking the traditional form of an article, could have prospects to establish itself especially in hyper-text and in virtual reality environments. The Kircz suggestion (1998) on documents consisting of modules is not without some difficulties related especially to the physics-articles centred structural proposal, and, in the end, the rather conservative approach to breaking the form of a traditional article.[7] In this regard supplementing the discussion on possible future forms of scientific publication may profit of a contribution offered by the infrastructures theory even in general. For the debate on the use of virtual realities as scientific publications, the link is a rather natural one. Regarding the qualities of infrastructures may be argued, that the theory illustrates rather well the dynamics of virtual realities. Infrastructures can reflect effectively the aspects that a meaning, not the data matters, the central role of an argument, and the political nature of the structure and the relations.

Welding the argumentational and infrastructural approaches together do present a number of practical difficulties. As recognised by Star (1995) regarding the information structures,[8] every formalised set is in a sense political and therefore a structure presents only a relative truth despite its appearance. Inconsistencies have thus to be not only accepted but also expressed in a meaningful manner. Including the political dimension into a VR seems thus inevitable. As suggested earlier, conceptual models can be used to describe data, describe data about data, and relations between different data elements. In representing politics and inconsistencies though, the formal, or even fuzzy, logic of models tends to be weak. In this sense the Star and Ruhleder (1994) view on infrastructures could provide a plausible functional framework for defining the premises of a documentation scheme, and the argument theory, could complement this with the methods and techniques of expressing the content.

The problem with politics in scientific communication is though persisting despite the acceptance of the politics as an existing reality in the structure of the communication, and in the explicit and implicit interpretations made by the author. The politics should be transparent in a meaning that opinions, their nature and their degree have to be able to be distinguished from each other. This presents two important demands on the practical realisation of the infrastructure: (1) that the practical infrastructures consist of co-existing, interrelating, controversial, conflicting and incomplete structures that still complete each other. (2) Infrastructures have to be able to be evaluated, not by absolute values, but on relative, explicitly political scale that relates itself to relative yet established categories.

Infrastructures of Paths

Making a practical tool out of the infrastructures is far from being a straightforward task. Both the communication of arguments, and the dissemination of the political, are rather non-graphic processes that are difficult to represent in a multi-dimensional environment. Communicating an argument to an audience is according to the classic rhetoric theory, a progressive task consisting of cycles of motivation and presentation. The strategy is still available in a non-linear universe, though not readily exploitable by using conventional methods.

Carter (2003) discusses the aspects of rhetoric functioning in hyper-texts which in their structural alignment resemble virtual realities as already in the beginning. Different tactics of expressing arguments in non-linear documents presented by Carter can be utilised in documenting the virtual realities. An essential difference between the two does though exist. Thinking of the purpose and place of the documentation, in a hyper-text the argument is in the document, and in the virtual reality the argument is built on the document.

Of the argumentative tactics presented in the literature, the Freeman argument analysis model seems to offer the most functional basis for implementation of the VR documentation (Freeman 1991). Freeman extends argument diagram presented by Toulmin (1958) to cope with larger arguments consisting of a set of sub-arguments. An illustrative visual metaphor of the argument structure could be a path that leads the audience, occasionally not exactly straight but often bending, crossing and uniting with others, and passing through a number of stages to a final conclusion. The path metaphor is drawn as a practical proposition for coding arguments functionality into the virtual realities.

A path is defined in this context as a basically linear structure which guides the audience through the essential entities of the VR. Depending on the level of ontological information attached to the virtual reality document, an argument path can be constructed as a flat list of references to different relations and data objects in the ontological data (level A). This presupposes however the presence of motivation -type ontologies (because_that, because_of, due_to, derived_from), and support for expressing and tolerating inconsistencies, probabilities and assumptions within the ontology. A less developed variant of argumentation paths (level B) can be accomplished as a list of references to specially written objects containing the sub-arguments. Level C represents the trivial case of an external document with references to the VR describing the argument.

A path is required to possess certain qualities:

- A path is a list of references (nodes) representing sub-arguments, and forming the complete end-argument.
- A node may be descendant of one or more nodes of one or more paths.
- A node may be an ancestor of one or more nodes of one or more paths.
- A node may belong to one or more argument paths.
- To a single virtual reality document may refer unlimited number (including zero) of paths.
- Each path and each node has to have an owner (a person i.e. a scientist) explicitly intellectually responsible for the (sub)-argument.
- Arguments and sub-arguments of different paths may contradict, support or not interfere with each other.
- Sub-arguments within one path may not be contradictory with each other or with the end-argument of the same path.
- The path is supposed to communicate the sub-arguments and the end-arguments. Path becomes invalid in case this requirement is not met.
- In levels B and C the means to describe the arguments and the sub-arguments should not be explicitly restricted to any particular format or media of presentation. Descriptions can be (short) scientific articles, but any other form of scientific communication has to be considered acceptable.

An outline example of a hypothetic argument path could thus resemble the following pseudoformula:

1. TELL node_name call_name_of_the_entity_the_virtual_reality_represents
2. TELL node_name who_formulated_the_argument
3. TELL node_name call_name_or_id_for_the_argument
4. TELL node_name when_the_argument_was_formulated_and_revised
5. TELL node_name sub_argument_1_and_refer_to_motivations
6. TELL node_name sub_argument_2_and_refer_to_motivations
7. ...

Below, as a more practical example, the formula is applied to a argument made on a virtual reality constructed on a Roman republican era anchor:

1. TELL "2" NAME ! "La Giraglia wreck anchor" (Marlier, Sabine: La Giraglia, a dolia wreck of the 1st century BC from Corsica,France: study of its hull remains. IJNA (2002) 31.2: 161-171).
2. TELL "3" ARGUMENT_BY "Isto Vatanen <ivatanen@abo.fi>"
3. TELL "4" ARGUMENT_FORMULATED "2003/03/04"
4. TELL "8" ARGUMENT_REVISED "2003/04/16"

5. TELL "6" SUBARGUMENT !(reference to four moulded astragals) !(reference to those identification to Venus coup)! (reference to Marlier 2002, 169)
6. TELL "8" SUBARGUMENT !(reference to source 2)
7. TELL "8" SUBARGUMENT !(reference to decoration) FOUND_AT ((reference to site location) and (reference to a location information of source 2)
8. TELL "FINAL" ENDARGUMENT !(reference to decoration) CARRIES_ATTRIBUTE (likely not uncommon in Roman world) AND CARRIES_ATTRIBUTE (during the late republican times)

The given example is intentionally simplified to illustrate the basic capabilities of the argumentation paths -approach.[9] With the same basic framework it is possible to refer to considerably more complicated arguments including ones referring to dynamic components in the model, but also more simple argumentation diagrams of the Freeman categories (Freeman 1991).

CONCLUSIONS

The question on how to document effectively the genre of documents referred to as "virtual realities", is not a problem of virtual archaeology own. The demand for efficient management of information stored in similar documents is critical also in a number of other applications. Mutual benefits should and could thus be found in interdisciplinary efforts.

Virtual archaeology has though considerable contributions to offer regarding the general effort on the future of scientific publication, multidimensional documents and virtual realities. The rather unique variety of source data in multiple formats, the temporal span, scientific uncertainties and inconsistencies, and especially rather well established methods for coping these inconsistencies make archaeology an interesting case from the more general point of view.

The proposed method for virtual archaeology documentation has to be considered as an approach. The recent research in documentation and presentation of non-linear information has clearly shown that even though the general aims of the different actors are relatively consistent, the supportive methods for an effective use of information differ considerably. Argument paths intend to contribute on the issue of difficult comprehensibility in the content-based documentation. Further research is definitely needed on descriptors and the data upon which the argument paths are to be built. The most critical question is though the direction to which the usage patterns of the virtual realities are evolving: what for and for what purposes the virtual is used in reality.

1 Indicating, not definite survey was conducted by examining a random selection of recent archaeological publications, the BRITARCH-mailing list archives and European Union projects database at Cordis-server (www.cordis.lu).

2 Term "virtual archaeology" is used of a discipline working with archaeological virtual realities, in practise mostly visual, but essentially anything digital and virtual in academic sense of the meaning. (C.f. Reilly 1991; Forte 1997; Renfrew 1997)

3 Few papers and articles (e.g. Barceló 2001; Jablonka et al. 2003) tackle the question, but the general debate and questioning of the "what is being done" has remained almost non-existent.

4 The need for broader understanding of the potential position of VR in archaeology does not imply that metadata research is insensible as emphasised already in Vatanen 2003, 73-74.

5 In the sense discussed in this article, is important to note that the on-going semantics and ontology -projects aim to introduce and empower machine-readability of human-readable documents, not to enhance the structure of documents from the human reader point of view.

6 The general argumentation theory used in the presented texts is strongly influenced by the New Rhetoric of Burke and Perelman, (see Kircz, Sillince and Juby) and the treatise of Habermas (1994) that sees argumentation as a focus of a society.

7 Kircz proposal actually resembles the Carter (2003) tactic 1 of writing hyper-texts with a minimal number of nodes.

8 The question on the relative nature of accuracy of information and interpretations is not a novelty for archaeologists or historians, but is reflected rather poorly in information processing and storage systems. See e.g. Skeates 2000: 104-107; Sidiropoulos & Sideris 2003:65 on issues relating to uncertainty in virtual archaeology.

9 A small scale Java application framework "VRDocTOOL" has been prepared for practical documentation and consistency tests.

REFERENCES

BARCELÓ, J., 2001.Virtual Reality for archaeological explanation: Beyond "picturesque" reconstruction. Archeologia e Calcolatori 12.

BOLTER, J.D., 1990. The Writing Space: The Computer, Hypertext, and the History of Writing.

BOLTER, J.D., 1991. Writing Space.

BÜSCHER, M., MOGENSEN, P. and SHAPIRO, D., 2001. Spaces of practice. Proceedings of the Seventh European Conference on Computer Supported Cooperative Work, 16-20 September 2001, Bonn, Germany.

CARTER, L., 2003. Argument in hypertext: Writing strategies and the problem of order in a nonsequential world. Computers and Composition 20:3-22.

CHEN, C. and CZERWINSKI, M., 1997. Spatial ability and visual navigation: An empirical study. New Reviews of Hypermedia and Multimedia 3:67-89.

FORTE, M., 1997. Introduction. Virtual Archaeology. Great Discoveries Brought to Life Through Virtual Reality.

FORTE, M., 2000. About Virtual Archaeology: Disorders, Cognitive Interactions and Virtuality. Virtual Archaeology. Proceedings of Computer Applications and Quantitative Methods in Archaeology (CAA) Conference 1998, Barcelona, Spain.

FREEMAN, J.B., 1991. Dialectics and the macrostructure of arguments: A theory of argument structure.

GAIANI, M., 1999. Translating the Architecture of the Real Into the Virtual: Seven Years of Experimentation with Conservation and Representation at OFF (Officina Infografica, Faculty of Architehture, University of Ferrara). Canada-Italia Workshop. Heritage Applications of 3D Digital Imaging 1999.

HABERMAS, J., 1994. The theory of communicative action Vol 1.

JABLONKA, P., KIRCHNER, S. and SERANGELI, J., 2003. TroiaVR: a Virtual Reality Model of Troy and the Troad. CAA2002 The Digital Heritage of Archaeology. Computer Applications and Quantitative Methods in Archaeology. Proceedings of the 30th Conference, Heraklion, Crete, April 2002.

JUBY, D.L., 1997. Rhetoric in the Age of the World Wide Web (www.ou.edu/cas/english/agora/dianne.html, 19.6.2002).

KIRCZ, J.G., 1991. Rhetorical Structure of Scientific Articles: The Case for Argumentational Analysis in Information Retrieval. Journal of Documentation 47.4:354-372.

KIRCZ, J.G., 1998. Modularity: the next form of scientific information presentation?. Journal of Documentation 54.2:210-235.

MILLETTE, D.M., 1998. Textual Imaginations: Vitruvius in Archaeological [Re]Constructions. Assemblage 3.

RYAN, N., 2001. Virtual Reality for archaeological explanation: Beyond "picturesque" reconstruction. Archeologia e Calcolatori 12.

NICCOLUCCI, F. and CANTONE, F., 2003. Legend and Virtual Reconstruction: Porsenna's Mausoleum in X3D. CAA2002 The Digital Heritage of Archaeology. Computer Applications and Quantitative Methods in Archaeology. Proceedings of the 30th Conference, Heraklion, Crete, April 2002.

PEKKOLA, S., 2002. Critical Approach to 3D Virtual Realities for Group Work. Proceedings of The Second Nordic Conference on Human-Computer Interaction (NordiCHI2002). Aarhus, Denmark, October 19-23, 2002.

REILLY, P., 1991. Towards a Virtual Archaeology. CAA90 Computer Applications and Quantitative Methods in Archaeology 1990.

RENFREW, C., 1997. Foreword. Virtual Archaeology. Great Discoveries Brought to Life Through Virtual Reality.

SIDIROPOULOS, G. and SIDERIS, A., 2003. Requirements and Assumptions in Visualisation Process of Urban and Surrounding Areas (the Case Study of the Greek City in Time). CAA2002 The Digital Heritage of Archaeology. Computer Applications and Quantitative Methods in Archaeology. Proceedings of the 30th Conference, Heraklion, Crete, April 2002.

SILLINCE, J.A.A. and SAEEDI, M.H., 1999. Computer-mediated communication: problems and potentials of argumentation support systems. Decision Support Systems 26:287-306.

SKEATES, R., 2000. Debating the Archaeological Heritage.

STAR, S. and RUHLEDER, K., 1994. Steps Towards an Ecology of Infrastructure: Complex Problems in Design and Access for Large-Scale Collaborative Systems. CSCW '94, Proceedings of the Conference on Computer Supported Cooperative Work, October 22-26, 1994, Chapel Hill, NC, USA. 1994.

STAR, S., 1995. The Politics of Formal Representations: Wizards, Gurus, and Organizational complexity. Ecologies of Knowledge. Work and Politics in Science and Technology.

TEUFEL, S., CARLETTA, S. and MOENS, M., 1999. An annotation scheme for discourse-level argumentation in research. Proceedings of the Ninth Conference of the European Chapter of the Association for Computational Linguistics (EACL-99), Bergen.

TOULMIN, S.E., 1958. The uses of argument.

VATANEN, I., 2003. Deconstructing the (re)constructed: issues on annotation of the archaeological virtual realities. CAA2002 The Digital Heritage of Archaeology. Computer Applications and Quantitative Methods in Archaeology. Proceedings of the 30th Conference, Heraklion, Crete, April 2002.

WESTERMAN, S.J. and CRIBBIN, T., 2000. Mapping semantic information in virtual space: dimensions, variance and individual differences. International Journal of Human-Computer Studies 53:765-787.

Towards a Virtual 3D Reconstruction of a Rood-Screen from its Archaeological Fragments

Cédric Laugerotte
Systèmes Logiques et Numériques,
Université Libre de Bruxelles, Belgium
claugero@ulb.ac.be

Pierre Anagnostopoulos
Unité de Recherche en Histoire Médiévale,
Université Libre de Bruxelles, Belgium
pierreanagnostopoulos@hotmail.com

Alain Dierkens
Unité de Recherche en Histoire Médiévale,
Université Libre de Bruxelles, Belgium
adierken@ulb.ac.be

Nadine Warzée
Systèmes Logiques et Numériques,
Université Libre de Bruxelles, Belgium
Nadine.Warzee@ulb.ac.be

Abstract

The micro-architecture in the Netherlands (XVth - XVIth centuries) is being studied with a comparative method. More than ten thousand exceptional fragments of stone sculpture have been excavated at the Brussels main collegiate church. The study reveals that the fragments are part of a rood-screen and a carved tabernacle.
The purpose of this project is to develop a method in order to help the art historians and archaeologists to reconstruct archaeological objects. A first manual classification divides the non painted remains into three main groups: architectural, floral and figurative motives. We first focus on the architectural remains.
At present, several automatic computerized methods exist, but most of them apply to particular kinds of objects like potteries, plates... which are not true 3D common objects.
We propose an approach that virtually manipulates the digitalized fragments through 3D geometric primitives. Up to now, these primitives are planes, lines and circles describing characteristics present on the architecture. The use of these primitives, by putting them in correspondence and by respecting their continuity, reduces and simplifies the range of possible associations between fragments. Then, the result of the association performed by the computer is proposed to an expert who can validate it.

Introduction

In 1991, archaeological excavation at the Brussels Saints-Michel-et-Gudule collegiate church has revealed amounts of carved stone remains. These finds open the way to new interpretation on the production of the Brussels and Brabantine Sculpture. The authors are working on remains of gothic and renaissance stone sculpture from Brussels.

Cédric Laugerotte and Nadine Warzée (SLN, ULB) are developing 3D reconstruction methods in order to virtually rebuild micro-architecture monuments and propose a computer aided assembling and visualization tool for the work of archaeologists.

Pierre Anagnostopoulos and Alain Dierkens (URHM, ULB) are studying stone micro-architecture in the Low Countries during the XVth and XVIth centuries.

This collaborative work will provide the old rood-screen that stood in the Brussels main collegiate church with a closest reconstitution.

Rood-Screen

Evolved from early forms of roman "cancello" and Byzantine "iconostasis", the medieval rood-screen is a larger structure with a doorway and an high arcade. It stands at the entrance of the choir, facing the principal nave of the church. Rood-screens were once found in many churches in the Low Countries and the oldest surviving example is that of the Sint-Pieter collegiate church at Leuven, which has recently been restored (Leyrens 2000) (Fig.1). This rood-screen stands as a reference in stone micro-architecture in the XVth century Netherlands since only a few other rood-screens still remain today (Steppe 1952).

Figure 1 Rood-screen. Leuven Saint-Pieters collegiate church, 1450-1485

Figure 2 Rood-screen. Brou, Saint-Nicolas-de-Tolentin church, 1515-1530

Rood-screens have not only the physical function of dividing the ecclesiastical space; they also convey an emanation of the sacred (Jung 2000). Under the arcade, Brabantine rood-screens are open in the centre, flanked by solid walls in front of which stood altars. On the far side, facing the choir, it is connected to the choir stalls.

Above, rood-screens consist of openwork and an array of niches containing statues or religious scenes. Finally, they are crowned by a long openwork railing, as in the example from Brou (Fig.2). Steppe (1952) relates that during the mass or laic ceremonies the priest, the singers and also church officials could stand on the rood-screen. Rood-screens are good examples for demonstrating the evolution in micro-architecture in the Low Countries between 1450 and 1550, from the end of gothic period to the Renaissance (Kavaler 2000).

At Saints-Michel-et-Gudule church, Brussels, the rood-screen had probably been largely restored or partially remodelled around 1450 and was finally destroyed at the end of the XVIth century (Lefèvre 1942).

Archaeological Discovery at Brussels Main Collegiate Church

In the 1991 excavations under the floor of Saints-Michel-et-Gudule main collegiate church (Bonenfant 1998), 11600 stone sculpture fragments from the XVth and XVIth centuries were discovered. They represent the mixed remains of several monuments, most probably from around the choir, that were severely damaged by the Iconoclasts at the end of the XVIth century.

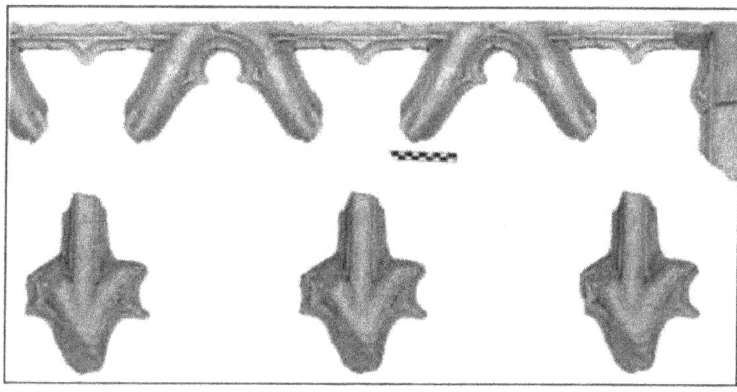

Figure 3 (left)Railing of the Rood-screen.
Brussels, Saints-Michel-et-Gudule, XVth century. Scale: 1cm
Figure 4 (right) Pinnacle. Brussels,
Saints-Michel-et-Gudule, XVth century. Scale: 1cm

In dealing with these remains, we are faced with five major problems which have formed the starting point of an original method of classification.

A global overview and simple classification is not feasible due to the great number of fragments. Also, the relatively small size of the remains reveal little of any larger structural or thematic elements or any continuous motifs. And, as several monuments have been identified among the fragments which are similar in style and shape (including a gothic tabernacle, a gothic rood-screen and a renaissance niche) it becomes extremely difficult to work with such an intricate group of sculpture. Furthermore, this micro-architecture has been carved from a soft white stone from Avesnes-le-sec, northern France, raising conservation priorities to ensure that these richly ornate sculptures do not suffer further damage.

Finally, there exist no XVIth century pictures, drawings nor prints which illustrate this gothic monument in the choir of the church. In order to arrive at a model close to the XVth century rood-screen in Saints-Michel-et-Gudule church, a comparative method is used to firstly classify the fragments, to understand their characteristics and, thence, by combining different motifs, provide a reconstruction of the whole architectural ornament.

Classification Method and Comparative Study: First Results

The classification method developed by one of the authors in a graduate thesis (Anagnostopoulos 2001) is currently used to enhance new results on the reconstitution of the rood-screen. The non-polychrome mouldings and floral elements are divided into 50 groups which are defined by their similarities in profile and size (Cabanot 1990). Combining together from two to six groups of mouldings can reveal the larger ornament of identifiable structures, such as the railing of the rood-screen (Fig.3) and a high pinnacle (Fig.4). The larger ornaments can be used to create new groups that can be matched together to form broader ensembles.

To facilitate this approach, digital photography was used, both to archive the fragments and to provide the means of creating a virtual 2D reconstitution of the rood-screen. With the help of a laser scanner, 3D reconstruction and 3D modelling is a key tool to virtually rebuild the rood-screen and other small scale monuments.

3D Tools

[1]Method

The approach we adopted to define a virtual environment for the manipulation of archaeological fragments consists in reproducing, on computer, the methodology already used by archaeologists. In essence, it consists in making successive steps of classification to isolate the sets of fragments displaying similar characteristics (size, colour, geometry, etc). Next, associations between fragments of the same set are attempted and tested on the basis of relations between these characteristics.

Most of the fragments used had already been classified by archaeologists according to their architectural form and/or floral ornamentation. Our technique of extracting characteristics and testing of assemblages by computer essentially replicates the manual method of the archaeologists.

[2] Set of validation

In order to validate the techniques of assemblage presented here, we first turned our attention to fragments from a niche which has been partially reconstructed and identified as a shell-like sculpture. The niche is "small architecture" (smaller than 1mx1m) and the number of available fragments is 24 (including fragments which may not belong to this niche). This limits the complexity of the tests while allowing us to focus on validation of the techniques. This choice was made for several reasons. First, details on these fragments are simple enough to identify points, lines, circles or planes. Secondly, the success of the technique can be judged by comparing our result against the manual assemblage of the fragments made by the archaeologists.

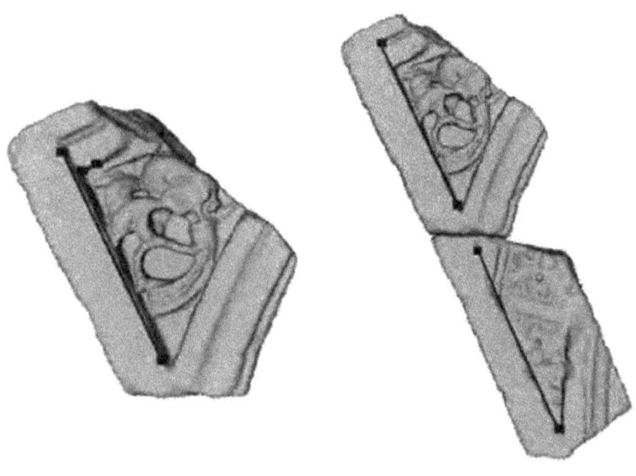

Figure 5 Extraction of features (left) and association between virtual fragments (right)

[3] Acquisition

It was felt that the technique demanded highly-accurate 3D models of the stone fragments since the level of detail on the fragments can be less than one millimetre. This, and the fragility of the fragments, suggested a non-contact acquisition technique and we chose a system based on the laser stripe triangulation. The device we used is the 3D scanner FastScanã commercialized by Polhemus.

[4] Extraction of features

The first step of our method consists in an interactive analysis of a fragment to select significant geometric features such as planar surfaces, ridges, valleys and axes of symmetry (Fig.5). The method used to isolate features are based on the analysis of differential parameters of the surface, such as normals (Thürmer 1998, Max 1999) and principal curvatures (Do Carmo 1976), deduced from the 3D mesh (Taubin 1995, Meyer 2002) and selected according the techniques of thresholding (Roessl 2000) and the analysis of local properties (Watanabe 2001). Then, for each feature, a simple geometric primitive (point, line, arc of circle) is associated. This primitive is obtained by approximation of the feature with classic methods of fitting and minimization (Eberly 2000, Thomas 1989).

[5] Assemblages

The association of different fragments of sculpture is possible through the manipulation of the geometric primitives which have previously been extracted. These primitives are selected manually and associated automatically by orienting and moving them in order to respect a continuity between them. For instance, two lines are oriented to be parallel and their extremities are identified (Fig.5). Concerning circles, they are oriented to make the tangents to the extremities parallel. Then, the extremities are identified. Moreover, the identification of the extremities can lead to situations where an overlap may occur between the fragments. To correct this effect, one of the fragments is moved in the direction of the geometric primitive used for the association till there is no more overlap. So, we get a configuration similar to a contact between the fragments.

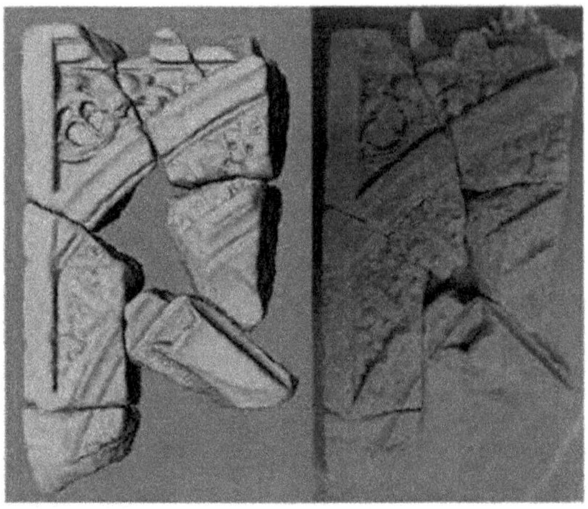

Figure 6 Reconstructed left top side of the niche.

Virtual Reconstruction of the Niche

[1] Assemblage of the fragments

After several associations between the 24 initial fragments through planes, lines and circles, the result was compared to the real assemblage on Figure 6. We notice that the virtual reconstruction is valid although there is one missing fragment due to the lack of information for a correspondence with its neighbours.

[2] Dimensions of the niche

From the assemblage we have obtained, it is possible to estimate the dimensions of the original niche (Fig.7).

Since the assemblage contains fragments from the upper and lower borders, it is possible to calculate the height of the niche as being 70 centimetres. In a similar manner, the depth of the niche is calculated to be 8.5 centimetres.

To estimate the width of the niche, we exploit the symmetry of the ornamentation appearing on the fragments. By identifying the point of intersection of the arcs leaving the base of the niche and the centres of the arcs of circle used for the assemblage, it is possible to estimate a centre of symmetry by averaging the previous points. Once this centre-point is determined, it is easy to measure the distance between it and the left side of the niche and thence determine the total width. The niche was estimated to be 1 meter wide.

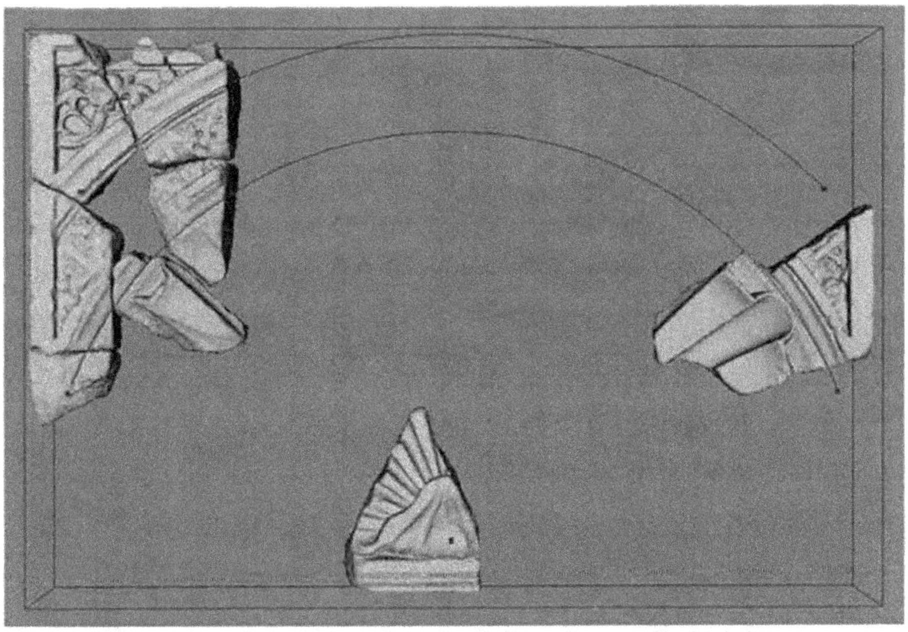

Figure 7 Dimensions of the niche. Virtually (left) and manually (right)

CONCLUSIONS AND FUTURE WORK

We have applied several techniques to develop a virtual environment to study, manipulate and assemble archaeological fragments into a virtual architectural reconstruction. The first method used photographs to archive and propose an initial 2D reconstitution. The second one is based on the analysis of 3D models of the fragments. These approaches were developed using "small architecture" as a study material for both simplicity and the ability to validate the approach by comparing "proposed" with "real" reconstructed assemblages. The results showed the efficiency of these tools to help us in the study of fragments and micro-architecture.

Thus far our work has focused on the reconstruction of a single niche. In the future, the tools and methods presented here will be used to further study the fragmentary micro-architecture of Brussels' main collegiate church, Saints-Michel-et-Gudule.

ACKNOWLEDGEMENTS

This work is partially granted by the "Région de Bruxelles-Capitale". Moreover, the authors are grateful to the reviewer for his comments. They also would like to thank deeply Dr. G. Avern for the final reading.

REFERENCES

ANAGNOSTOPOULOS, P., 2001. Architecture mobilière en pierre (1450-1550), Un matériel d'architecture ornementale bruxellois, Les fragments provenant des fouilles du chœur de la cathédrale Saints-Michel-et-Gudule de Bruxelles. Mémoire de licence, 3 vol., Université Libre de Bruxelles.

BONENFANT, P.-P. and FOURNY, M., 1998. Fouilles archéologiques à la cathédrale de Bruxelles 1987-1998. Annales de la Société Royale d'Archéologie de Bruxelles 62:225-257.

CABANOT, J., 1990. Constitution d'une banque de données sur les chapiteaux corinthiens et dérivés du corinthien: méthodes et perspectives. In L'Acanthe dans la sculpture monumentale de l'Antiquité à la Renaissance, Actes du colloque de la Sorbonne, October 1990, Paris, 1993.

DE RIDDER, P., BONENFANT, P.-P., BRAL, G.J., CLAES, H., BOLLAERTS, M., VANDEN BEMDEN, Y., DE CROMBRUGGHE, D., CAUDRON, J., HUYGENS, F., WEYTENS, R., VAN HOOF, E., FÉLIX, J.-P. and MAIRLOT, E., 2000. La cathédrale des Saints-Michel-et-Gudule. Editions Racine et Régie des Bâtiments, Bruxelles.

DO CARMO, M., 1976. Differential geometry of curves and surfaces. Prentice Hall.

EBERLY, D.H., 2000. 3D Game Engine Design. Morgan Kaufmann Publishers.

JUNG, J.E., 2000. Beyond the Barrier: The Unifying Role of the Choir-Screen in Gothic Churches. The Art Bulletin December, LXXXII, n°4:622-657.

KAVALER, E.M., 2000. Renaissance Gothic in the Netherlands: The Uses of Ornament. The Art Bulletin June, LXXXII, n°2:226-251.

LEFÈVRE, P., 1942. L'ancien Doxaal ou jubé de Sainte-Gudule à Bruxelles. Revue belge d'Archéologie et d'Histoire de l'Art XII, 1:31-48.

LEYRENS, I., 2000. Le jubé de la collégiale Saint-Pierre à Louvain. Bulletin de l'Institut Royal du Patrimoine Artistique Tome XXVII (1996-1998):303-310.

MAX, N., 1999. Weights for Computing Vertex Normals From Facet Normals. Journal of Graphics Tools 4(2):1-6.

MEYER, M., DESBRUN, M., SCHRÖDER, P. and BARR, A.H., 2002. Discrete Differential-Geometry Operators for Triangulated 2-Manifolds. Proceedings of the International Workshop on Visualization and Mathematics 2002 (Vismath 2002), May 2002.

POLHEMUS website (http://www.polhemus.com/).

ROESSL, C., KOBBELT, L. and SEIDEL, H-P., 2000. Extraction of feature lines on triangulated surfaces using morphological operators. Smart Graphics 2000, AAAI Spring Symposium, Stanford University:71-75.

STEPPE, J., 1952. Het Koordoksaal in de Nederlanden. Recueil de travaux d'Histoire et de Philosophie 3e série, 50e fascicule, Bibliothèque de l'Université de Louvain, Bruxelles.

TAUBIN, G., 1995. Estimating the Tensor of Curvature of a Surface from a Polyhedral Approximation. Fifth International Conference on Computer Vision (ICCV'95).

THOMAS, S.M. and CHAN, Y.T., 1989. A Simple Approach for the Estimation of Circular Arc Center and its Radius. Computer Vision, Graphics, and Image Processing 45:362-370.

THÜRMER, G. and WÜTHRICH, C.A., 1998. Computing Vertex Normals from Polygonal Facets. Journal of Graphics Tools Vol. 3, Number 1:43-46.

WATANABE, K. and BELYAEV, A., 2001. Detection of Salient Curvature Features on Polygonal Surfaces. Eurographics 2001. Computer Graphics Forum Vol. 20, No. 3:385-392.

On using State of the Art Computer Game Engines to Visualize Archaeological Structures in Interactive Teaching and Research

Martin Meister
Friedrich-Alexander-Universität Erlangen-Nürnberg, Germany
AERIA (Antikensammlung Erlangen Internet Archive)

Martin Boss
Friedrich-Alexander-Universität Erlangen-Nürnberg, Germany
AERIA (Antikensammlung Erlangen Internet Archive)

SEE THE CD FOR THE EXTENDED VERSION

Abstract

The AERIA (Antikensammlung Erlangen Internet Archive) Project is currently developing visualization platforms for architectural reconstructions of both greek and roman buildings.
The purpose of our work is to provide virtual interactive walkthroughs of structures of archaeological interest for both teaching and research, via the Internet. In our presentation we implement techniques provided by computer-game-engines. The objective is to create virtual worlds without the need to employ expensive CAD software. We demonstrate that game engines can easily be adapted to the needs of archaeological science.
This method was first introduced during an exhibition in the year 2000 in Rosenheim (Landesausstellung für Archäologie) "Die Römer zwischen Nordmeer und Alpen".
In our contribution we focus on the following reconstructions:
-Römerlager Marktbreit (Imperial Roman Fortress in Bavaria);
-Legrena 17 (a huge farmhouse in Attica during the classical period).
Both virtual worlds are based on a modification of the "Quake 2" render engine.
-The throne of Apollon Hyacynthios in Amyklai near Sparta.
This scenarion is based on a "Halflife" render engine.
The palace of Nestor in Pylos rendered with the "Morrowind" engine.

In the following pages we attempt to give an overview of our usage of computer games in archaeological visualization. During our experimentations with different games engines we came across very different scenarios in which such software can be used. We use games mainly for the rendering of reconstructions of ancient buildings with several target audiences ranging from school classes at public exhibitions, seminar sessions in the classroom with students of archaeology and public talks about latest research results. It is not within the scope of this article to give hints how to build an archaeologically sound model or to advise the reader on how to adjust his or her data to the different engines. However, beginning with early games with the need for highly specialized geometry files, we show that with newer games at least common graphics interchange formats are supported. We will give our view that it is possible nowadays to show almost any piece of archaeological 3D data with modern games and so avoid the need for expensive specialized solutions. The work presented in the following pages was carried out by both authors alone: all programming was done by Martin Meister, both the authors are responsible for graphics and design. All virtual worlds are based on 2D reconstruction drawings by Martin Boss.

The AERIA-Project originally started in 1994 as an electronic archive for the historic photo collection of the Erlangen Antikensammlung (collection of antiquities). Since then it has been expanded into a full scale virtual museum, with an online display of all parts (Collection of Greek and Roman Antiquities, Gallery of plaster-casts and photo-archive) of the collection.[1] Nevertheless, an online museum should not be restricted to simply mirroring the museum artefacts in terms of providing pictures or videofiles. The statement of 1857: "*Der Gewinn, welchen das unmittelbare Anschauen verspricht.*" remains valid today and cannot be substituted by looking at electronic images (Boss 1993).[2] Even the most extensive museum cannot recreate the atmosphere of an ancient building in operation. For example, the Ishtar-gate at the Pergamon Museum in Berlin gives an impression of the size of the museum rather than that of ancient Babylon. Here a virtual museum has genuine advantages, because it can show more than the actual state of preservation in virtual reconstruction. So in our case, the AERIA-Project at Erlangen can show the palace of Nestor in distant Greece without owning a single part of it for display.

From 1999 onwards AERIA has been developing visualization platforms for architectural reconstructions of both Greek and Roman buildings. First steps with still images and some short animated clips (reconstruction of geometric architecture at Perachora and Eretria[4]) were carried out with a 3D computer aided design[5] (CAD) package called LightWave[5]. However, as one cannot change the viewpoint or path, the disadvantage of prerendered views lies in their static nature.

The exhibition "Die Römer zwischen Nordmeer und Alpen" at Rosenheim (Wamser 2000) during the year 2000 offered the opportunity to experiment with real-time and interactive virtual 3D reconstructions of the Roman legionary fortress at Marktbreit and a wider spectrum of viewers. In view of the huge number of schoolclasses visiting the exhibition and the fact that most children today are familiar with computer games, a modification based on a first person action game or so-called shooter (FPS), Quake2[6], was developed.

At this point a definition of our notion of a (game) engine is apt. An engine is the core module of a software package that deals with the rendering of geometry from the memory of the computer or graphics card directly to the screen. It creates a transparent layer that wraps around the hardware and enables the user to process geometric shapes. In the game industry this division of, on the one hand, geometry and, on the other hand, the render engine existed from early on. Games manufacturers license graphics engines for their software and do not have to programme low-level graphics functionality at all. Instead, they concentrate on the nature of their game and the visual artwork, such as geometric shapes and textures. Consequently, the game community[7] is able to create modifications, changing the appearance and behaviour of these games with the aid of the game manufacturer but the engine cannot be altered. Our research is situated at exactly this point because we want to change the geometry displayed and hide most of the gaming aspects. Here, while a modification normally focuses on changing gameplay, textures and geometry, it was necessary for us to programme a modification module in the form of a dynamic link library (DLL) for Quake2 that made shooting and killing impossible, thus transforming the game into a visualization tool.

Figure 1 Fabrica at the Roman legionary fortress at Marktbreit; outside view

Figure 2 Principia at the Roman legionary fortress at Marktbreit; inside the central courtyard

During the above mentioned exhibition at Rosenheim the modified game (Quake2) offered four virtual spaces called maps, displaying various places within the camp (maingate, Fabrica, Principia and Praetorium), which could be explored in detail by free individual movement (Fig.1-2). In retrospect it proved to be a good addition to the real plaster-models which stood nearby. Additionally, the difference between virtual and physical reconstruction became very evident. While plaster models always stay the same scale, size and texture, the virtual model can be easily altered, individual viewpoints can be found and the materials can be changed by altering the texture mapping with the click of a button.

The Greek farmhouse at Legrena (Lohmann 1993) was the next model to be built (Fig.3-4). The purpose of this reconstruction, also written as a map for Quake2, was to test its usability as a more scientific visualization tool for teaching in university lectures. This unexcavated site was chosen because it is hardly visible today and its remains are unattractive for tourism as it lies almost forgotten on the road to the famous sights at Sunion. Nevertheless, the economic wealth of Athens in Classical times was based on such Attic farmsteads and oil mills. Using the limited abilities of the Quake2 render engine, it was possible to show the spatial relationships of the various buildings and to evaluate their functional relationship. The reconstruction of Legrena also shed new light on the controversial interpretation of the great tower bases found on such farms. Just like the Roman camp mentioned above, the textures provided with the original game were sufficient to give a basic impression of walls, roofs, wooden structures and even simple machinery such as the oil press. But it also became clear that this engine is not suited or built to display plants and trees. Therefore the garden lacks the crucial vegetation and its appearance in the map remains unfinished and bare.

Another weakness of the Quake2 engine lies in the limited number of possible surfaces that correspond to the binary space partitioning (BSP) algorithm employed. This algorithm most efficiently displays indoor geometry, which results in a restriction of the complexity of the building to be reconstruc-

Figure 3 Classical Greek Farmhouse at Legrena; outside view

Figure 4 Classical Greek Farmhouse at Legrena; the oilpress in the central courtyard

ted. Furthermore, poor performance and rendering flaws occur when attempting to build large outdoor maps.

Due to the given reasons, the next experiment was carried out using the Halflife[8] game engine, which is an enhanced and modified Quake2 engine released a few months later. Here, a modification can be kept simple (only geometry) because where Quake2 needed a modification (our specialized DLL) to prevent the shooting, in Halflife this part of scripting is less important because by default players are not equipped with any weapon. The site to be reconstructed was the so-called Palace of Nestor in Pylos, Greece (Blegen 1996)[9], and the subject to be investigated was the number of storeys at various locations throughout the building (Fig.5-6). In reality only the foundations of the palace are preserved and the remains of several staircases give a hint of further floors. During the 1950s the excavator Carl Blegen and his draftsman Piet de Jong used these remains in their reconstruction drawings and duplicated the layout of the ground level at the level above. But it is more likely that larger rooms, such as the central megaron and its anteroom, are higher than the narrow chambers in the wings. On the other hand, some parts of the palace such as later storage annexes certainly lakked an upper level. This is theoretically discussed on the AERIA-site with traditional 3D CAD imaging, whereas the virtual 3D reconstruction was carried out as proof of the concept. By freely walking through the building, the usefulness of our reconstruction becomes clear because it gives a plausible explanation of possible configurations of storeys. This experience is enhanced by an advantage of the BSP game engines: they feature physical simulation of light by means of radiosity algorithms that are precomputed and subsequently shaded on the textures. Realistically lit rooms were one of the main reasons for the submersive atmosphere of the reconstructed palace, even if the possibility to apply pseudorealistic textures for wall paintings for example was not applied and so the maximum detail was not exploited.

As a result, the virtual palace gives visitors a better visual clue than it might have done with simple CAD data viewers. When looking at Pylos in reality, its rather erratic and complex labyrinth of foundation-walls does not provide the amount of information that our reconstruction is able to give.

Nevertheless, with the main buildings of Nestor's palace the maximum complexity in Halflife was again reached, especially because the map was planned with huge outdoor areas that do not comply with the design of the rendering algorithm (BSP). Furthermore, detailed modelling of complex geometry, as occurs while constructing elaborate column-capitels or statues, is not possible with these earlier BSP-engines, because the coordinate space is quantized during a precomputed triangle generation step. As a result, only coarse structures can be visualized. This became an obstacle while creating a visual model of the Sanctuary of Apollon Hyakinthios at Amyklai, which is only preserved in a written description by the Roman writer Pausanias (Prückner 1992). This famous Thronos of Apollo could not be transferred into the gridded coordinate space of the early BSP games. As a consequence, our relatively fine model of the four main supports in the form of female Horai and Charites failed to render correctly (Fig.7). What is more, the work on the Thronos once again showed the weakness of the BSP engines when dealing with outdoor areas. Embedding a reconstructed building in its surrounding landscape is beyond its capabilities. In our case the top of the bronze-plated pole representing Apollo was about 49 feet high (15 metres) and is reported to have been visible from the Agora at Sparta, which is about 4 nautical miles (7 kilometres) away.

Figure 5 Mycenian Palace at Pylos; outside view

Figure 6 Mycenian Palace at Pylos; looking from rooftop into the central courtyard

Despite the disadvantages, game engines have proved their worth not only for entertaining school classes but also in scientific discussions and lectures held at our institute since 1999. The possibility to change one's viewpoint and examine the objects on one's own, help one to confront the issues at hand directly. In this way reconstructions can be closely evaluated and their plausibility can be questioned from all possible directions. For example, problems arising from the layout of roofs or the spatial relationships of buildings are more easily observed early on and changes can be dealt with more efficiently than can be done with traditional 2D line-drawings (Martin 1976). It is important to note that no third party was involved in the development of these visualizations. As soon as it was clear how the reconstruction would be laid out it was constructed with the required geometry editors by the authors themselves.

With the arrival of new rendering technology and newly developed graphic cards, new computer games have emerged. They can rely on fast growing hardware performance allowing them to transform ever more triangles with every new graphics board arriving from the manufacturers. Morrowind[10] is a recent game that uses the commercial

Figure 7 The sanctuary of Apollon Hyakinthios at Amyklai, general view rendered with Quake2 engine

Figure 8 The sanctuary of Apollon Hyakinthios at Amyklai, general view at sunset rendered with Morrowind

NetImmerse 3D engine[11], which incorporates both level of detail (LoD) and space partitioning techniques. Most important is the fact that complex CAD data, as produced by, for example, 3D Studio Max, can be imported into the game's editor by automated conversion. With this, the need to use specialized level-editors has become obsolete. For this reason, a second reconstruction of Amyklai was carried out using this engine (Fig.8). Morrowind now enables us to give a highly triangulated view of Amyklai as a site lying on a hill in a countryside full of vegetation.

Apart from the graphic enhancements, different kinds of computer games offer different advantages. FPS games like Quake2 or Halflife basically do subtractions of hit-points when a player fights his way through monsters or human controlled opponents. Apart from pushing a chair to the side or throwing a handgrenade, further interaction with the virtual world is limited. Role playing games (RPG), on the other hand, offer a lot of interaction with both virtual characters and objects: books can be read, objects can be picked up, replaced and examined. Virtual characters are not simply shot at but talked to and learned from. All this can be converted easily to suit archaeological needs. With the virtual reconstruction of Amyklai for example, a book can contain the various reconstruction drawings done so far since 1852, another book can give the original text of Pausanias about the Thronos at Amyklai. Or even better: one might meet Pausianias himself or chat with a fellow traveller acting as a tour guide on his virtual pilgrimage. Examining the altar, one might pick up one of the offerings lying there and examine it: a pop-up box provides all known scientific background in pictures and text. All this can be done without altering the nature of the original game. All that the developer needs in order to add such information is already provided. The only requirement is to learn the editing software of the game that is supplied along with it. There one can specify new objects, give them properties and leave a description. Furthermore, one can define the behaviour of virtual characters and automatically check artificial intelligence scripts for consistency with the logical dependencies defined.

However, this form of interaction with the virtual reconstruction is not the only one that computer games can offer.

While Morrowind is our most advanced example tested for scientific reconstruction, it is a single player game, while our FPS modifications offer a multiplayer mode. With Morrowind, having downloaded our modification from the internet, one can explore the building of Amyklai at home; the interaction remains restricted to the virtual characters and their predefined behaviour. With our reconstructions that use the Quake2 or Halflife engine, it is possible to lead whole schoolclasses through virtual buildings on a server at our institute via TCP/IP protocol over the internet. In this way we can monitor the log-on to a map and choose a modification of our control while forbidding weapons, shooting and death. In this way two visitors in different locations in the world are able to chat to each other as well as to ask questions to the scientists at Erlangen while exploring our reconstructions through the game built-in chat mode; interactivity is not restricted to computerized counterparts. Using multiplayer features a game becomes a platform of virtual communication with a scientific background.

In the future, we hope that advanced RPG games similar to Morrowind will provide similar multiplayer capabilities. In the year 2003 several promising games such as Halflife2 and Doom3 may prove useful for our needs depending on the context of use. There is certainly no doubt that the functionality that games already provide should be exploited by archaeologists. But there is also the need to carefully consider the purpose of the visualization. The amount of control and detail needed to convince researchers might be more than a game can provide, whereas the needs of public exhibitions and other educational displays may be less demanding. Overall such games are not only inexpensive but also provide the most up-to-date graphics and, in most cases, a good editing platform.

[1] all references and abbreviations comply with DAI (German Archaeological Institute) guideline in: AA 1997, 611-628; all hypertext-links were active, when this article was written (May/June 2003); http://www.phil.uni-erlangen.de/~p1altar/aeriahome.html.
[2] http://www.phil.uni-erlangen.de/~p1altar/geschichte_html/ geschichte.html#photosammlung.
[3] http://www.phil.uni-erlangen.de/~p1altar/ausstellung_html/lectures_html/perachora/perachora01.html.
[4] http://www.phil.uni-erlangen.de/~p1altar/forschungsstation.html.
[5] http://www.newtek.com.
[6] http://www.idsoftware.com/games/quake/quake2/.
[7] Fan communities for Quake2 and Halflife are found at http://www.planetquake.com and http://www.planethalflife.com.
[8] http://www.half-life.sierra.com.
[9] http://www.phil.uni-erlangen.de/~p1altar/ausstellung_html /lectures_html/pylos/pylos01.html.
[10] http://www.elderscrolls.com.
[11] http://www.ndl.com.
[12] unreleased, published by http://www.sierra.com.
[13] unreleased, published by http://www.idsoftware.com.

REFERENCES

BOSS, M., 1993. Die Antikensammlung der Friedrich-Alexander-Universität. In Die Friedrich-Alexander-Universität Erlangen-Nürnberg 1743-1993, Ausstellung im Stadtmuseum Erlangen:597ff.

BLEGEN, C., 1966. The Palace of Nestor at Pylos in Western Messenia.

LOHMANN, H., 1993. Atene:95ff, 144ff, 515f. Abb. 30, 36 Taf. 98.

MARTIN, R., 1976. RA:205ff.

PRÜCKNER, H., 1992. Ein Thron für Apollo. In Kotinos, Festschrift für E. Simon:123ff. Taf 21.1 mit Lit.; Pausanias, Periegesis III:18.9ff.

WAMSER, L. (ed.), 2000. Die Römer zwischen Alpen und Nordmeer. Katalog-Handbuch zur Landesausstellung des Freistaates Bayern, Rosenheim:435ff. Kat. Nr. 250.

THE REBIRTH OF A ROMAN FORUM - THE CASE STUDY OF THE FLAVIAN FORUM OF CONIMBRIGA

ALEXANDRINO JOSÉ MARQUES GONÇALVES
ESCOLA SUPERIOR DE TECNOLOGIA E GESTÃO
DO INSTITUTO POLITÉCNICO DE LEIRIA

ANTÓNIO JOSÉ NUNES MENDES
CENTRO DE INFORMÁTICA E SISTEMAS
DA UNIVERSIDADE DE COIMBRA

ABSTRACT

The Flavian Forum of Conimbriga was located on the highest point of the ancient Roman city of Conimbriga (one of the largest of the west of the Iberian Peninsula). The municipality built the Flavian Forum, dedicated to imperial cult, during the reign of Flavius Vespasianus. Due to its architecture and location within the city, the Forum was one of the most important places in most Roman cities.

This paper presents a virtual model of the Flavian Forum of Conimbriga. It is common to question the accuracy of this type of architectonic model. However, this virtual model was built in accordance with the reliable blueprints in possession of Museu Monográfico de Conimbriga, from the time of the excavations.

The basic use of VRML to elaborate this work, allowed us to fine-tune some of the more problematic points related with historical reconstructions in virtual format:
- Accessible to a wide public over the Internet;
- Easy to use;
- File size (always a problem): small, very small;
- Performance via optimization techniques.

You can make your own "visit" to this historical place in http://lsm.dei.uc.pt/forum.

INTRODUCTION

Beginning with the Middle Ages or perhaps before, questions related to the historical and cultural past, mainly at the level of archaeological findings, have exerted an enormous and mysterious allure on Man, who has come up with methods and forms to recreate such places.

Mainly in these past few years, the rise in number of projects involving this theme (historical reconstruction), has been real and effective due to two main factors: the increasing technological development that allows such designs to be accomplished more easily, with a larger impact; and the continuous and increasing interest of humanity for questions related to its cultural and architectural patrimony.

Some examples are Archeoguide (archeoguide.intranet.gr), which implemented a prototype of Augmented Reality in the archaeological area of Olympia in Greece or 3D Murale Project (www.brunel.ac.uk/project/murale), that in Sagalassos (Turkey) is developing and using 3D Multimedia tools to measure, reconstruct and visualise archaeological ruins in virtual reality.

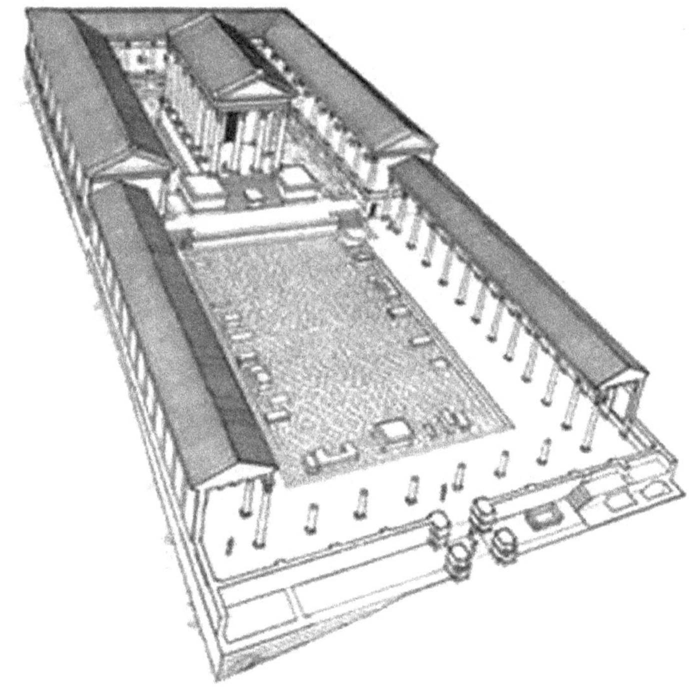

Figure 1 Image of the Flavian Forum of Conimbriga

THE FLAVIAN FORUM OF CONIMBRIGA

The twelve campaigns of excavations carried out in the ruins of Conimbriga by the French Archaeological Mission in Portugal and the Monographic Museum of Conimbriga, between 1964 and 1971, led, in particular, to the discovery of the foundations of the former Forum in this ancient Roman city. The Flavian Forum of Conimbriga was built between the years 75 and 80 AD by the municipality during the reign of Flavius Vespasianus, dedicated to the Imperial Cult.

Three monuments composed the Flavian Forum of Conimbriga: the temple, the porches of the temple and the porches of the Forum (the square). The temple of the imperial cult was raised on the highest point of the city, precisely in the middle of a sacred area. This location and its characteristics confer impressive dimensions, as was originally desired. This entire sacred area was surrounded by a porch whose function was essentially architectural. The square was the civic place par excellence. Its two lateral porches were placed at an inferior level in relation to the porches of the temple and the floor was now covered with a meticulously lined up pavement (Alarcão et al. 1994).

The Reconstruction

In 1899, Queen Amélia gave orders to initiate the archaeological excavations of Conimbriga, but only after 1930 were these excavations complemented and deepened in such a way that they show undeniable vestiges of a city constructed by the Roman civilization. The amount and quality of the archaeological findings presented an architectural and cultural capital of such impact that it was decided to let the general public have contact with this reality, significantly distinct from today's.

Any visitor of this space, when facing the scene of buildings in ruins, degraded floors or damaged columns, is unconsciously led to an imaginary time, by recreating, through the set of sensations assimilated at that moment, all the urban, social and architectural environment of those days. Given the distinct ability of abstraction and interpretation of each human being, even when based on drawings and blueprints, this task of recreating will always be full of imprecision, which can vary with the mood, age or even the cultural level of each person.

The work developed in this project, and now presented, had the main purpose to make it possible for everyone to pay an interactive visit, although virtual, to one of the most emblematic places in the cities of the ancient Roman Empire, the Forum. We recreated a three-dimensional model, as reliably as possible (from the number of steps of staircases to the number of columns, patterns of the time, colours, etc.) and at real scale, of the Flavian Forum of Conimbriga (Gonçalves 2002).

[1] The technology

Taking into account the characteristics of the project and the main goals, the technology to use would have the capacity to achieve good visual results but, at the same time, to supply the author with the necessary potentialities to carry out an efficient development. The choice was the Virtual Reality Modeling Language (VRML), a language that allows the creation of virtual environments, and that allows users, now converted into visitors, to explore the reconstructed spaces by simply moving the mouse.

Since VRML is the Web standard for publication of three-dimensional contents, the range of potential users will certainly grow when placing the reconstruction of this historical relic on any Web server, where anybody, anywhere will be able to easily take a trip in time and visit one of the monuments of the ancient Roman Empire.

[2] The working tool

To carry out this work, we use a tool that operates natively with the VRML, the Cosmo Worlds 2.0 from Silicon Graphics. This tool allowed us to perform a supported and gradual development of the project. Its characteristics and potentialities, where part of the responsible items for the final results.

[3] Historical accuracy

In these types of historical reconstructions, the question of scientific veracity of the architectural model to use is always argued and a motive for profound analysis. There are situations where the lack of consensus between investigators causes the presentation of more than one proposal for the model of the space or area to be recreated. In the case of the Flavian Forum of Conimbriga that did not happen. The excavations, concluded in 1971, culminated with the attainment of trustworthy blueprints of the near total of the architecture structure of the Forum.

This way, the blueprints elaborated during that period and gently yielded by the head of the Monographic Museum of Conimbriga, were our work base for "rebuilding" the Flavian Forum. The implementation of this work was, consequently, made solely with these blueprints, with the observation of the Forum ruins and the access to some of the removed fragments of its original construction.

We would like to refer that this work would not have been possible, without the direct contribution of the Monographic Museum of Conimbriga and mainly its Director, Dr. Virgílio Correia.

Main Characteristics

Here are some of the main characteristics of this work:

[1] Realistic model and up to scale

It has already been said that the modelling of the Forum would be true to the original building and would be materialized to the real scale, which allows for:

- The possibility to visit the Flavian Forum of Conimbriga in the perspective of an inhabitant of that period, and thus have a real perception of the colours and forms of the architecture of the ancient Roman Empire;
- Simultaneously, and from that same perspective (about 1.70m high), give the visitor the real magnitude and impressive surrounding area of the Forum: the high and imperial temple with its colossal columns; the beauty and symmetry of the porches of the temple, whose sections we can now visit, unlike many of the local inhabitants (Alarcão et al. 1994); and finally the enormous square of cult and knowledge, where the members of that society met for their discussions.

The realism of our construction, compared to the 1st century building was approved by the experts, especially by the Director of the Monographic Museum of Conimbriga.

[2] Interactivity

Another main characteristic of this work is related to its interactivity. We intended to create a virtual space where the visitor (the user) could have complete autonomy of movements in the area of the Forum, which means he can go where ever he wants, following the path he chose, by using an interaction device (keyboard or mouse).

[3] Guided Tour

Besides the total liberty of movements, we also created a version with a guided tour that shows all the monuments that constitute the Flavian Forum of Conimbriga, for eventual low levels of agility and dexterity when handling the devices that permit to simulate the movements in the three-dimensional space.

[4] Easy to use

Whoever decides to visit the virtual space of the Forum will be presented with another feature of the virtual worlds elaborated in VRML: it is very easy to use. A simple touch of the directional keyboard keys or a click and drag with the mouse, and any layperson in the matter will be able to move himself in the virtual space without any difficulty. All the movements are made in order to give the visitor a realistic perception of what would physically be happening to him, such as the notion of climbing stairs or falling off a high place.

[5] File dimension

This was, without a doubt, one of the main achievements of this work. Due to the optimisation techniques used in the elaboration of the project, we were able to obtain a very small dimension of the files needed to generate all the visualization. For example, the main, and only, VRML file has approximately 22Kb. It is possible to store many copies of this work in a small and almost obsolete floppy disk, and practically without losing any of the realism obtained.

[6] Performance

Still, due to the optimisations made in the elaboration of this work, and because this is a problematical item in the viewing of three-dimensional spaces in a low end Personal Computer (PC), the performance of this virtual space is fully satisfactory in any current domestic PC.

[7] Accessibility

The access to some of the projects, namely the ones that involve the use of Virtual Reality techniques, is normally restricted and limited because of the specificities used in its development and/or its visualization, which, in general, unable the presenting of these kinds of projects to the wide public, for example over the Internet. Our work tries to differ from that tendency.
The two last points presented in this paper ([5] and [6]) have contributed much to this fact. Due to the language used, the VRML, and the results of the optimisations made, all the care placed in the search of a good balance between realism, performance and files dimensions, resulted in a final file of reduced dimensions, easy to download and with a good performance in any recent computer. Therefore, any person using the Internet can easily access it. This work can be seen in: http://lsm.dei.uc.pt/forum/.

[8] Cost of the project

This type of work is not very expensive, financially speaking, in contrast to others with similar objectives, namely the time required to implement it or the tools to make it possible (in VRML), where, for instance, all the applications needed for visualization are available for free.

Even after its conclusion, it is possible to make improvements to the realism of the recreated space, without any alteration in the VRML file. For this, a simple improvement of the textures used in the scene, automatically updates the visualization.

Figure 2 View (in greyscale) of the main entrance of the Virtual Flavian Forum of Conimbriga

Conclusion

The last years have confirmed an increasing concern, mainly of great institutions such as UNESCO or the European Union, with the preservation, interpretation and spreading of the historical and cultural legacy of our ancestors. Such fact originated a new wave, where the use of new technologies, namely of Virtual Reality, acquires a larger role in the achievement of such intentions.

The Flavian Forum of Conimbriga supported the work presented in this paper, which aimed to recreate with accuracy, in a virtual way, this impressive but disappeared monument. The interactive three-dimensional representation enables a deeper contact, better than any drawing or written description. The work has, thus, a double function: pedagogical and social. Therefore it enables the "reading" of the past to some levels and it assures the transmission of culture to future generations. And since anyone, with no exception, can access it, this work contributes to a true democratisation of knowledge.

Thus, and for the accomplishment of this design, all the area belonging to the Flavian Forum was modelled three-dimensionally (with exception of the interior of the temple) to the real scale and always with special care to preserve its geometric characteristics, so that the historical precision, always essential in these situations, is unquestioned.

This work also intends to contribute, even if modestly, to the spreading, preservation - by preservation we also refer to the "reconstructions" by scale models, drawings or virtual representations of spaces forever lost - and better knowledge of a past, which is important for both the historical interest and the values that the Flavian Forum of Conimbriga represents.

References

ALARCÃO, J., ETIENNE, R. and GOLVIN, J.-C., 1994. A Maqueta do Centro Monumental Flaviano. Instituto Português de Museus/Museu Monográfico de Conimbriga.

GONÇALVES, A., 2002. Reconstrução de ambientes históricos utilizando VRML: o caso do Fórum Flaviano de Conimbriga. Dissertação de mestrado em Eng. Informática pela Universidade de Coimbra.

THE 3D DOCUMENTATION OF AL-JAHILI FORT IN AL-AIN, UNITED ARAB EMIRATES (UAE): A VIRTUAL REALITY APPLICATION

SEE THE CD FOR THE EXTENDED VERSION

IHAB NABIL ELKHAWAS
DEPARTMENT OF ARCHITECTURAL ENGINEERING,
COLLEGE OF ENGINEERING, UNITED ARAB EMIRATES UNIVERSITY
IELKHAWAS@LINK.NET

AMAL ABDULLAH AL QUBAISI
DEPARTMENT OF ARCHITECTURAL ENGINEERING,
COLLEGE OF ENGINEERING, UNITED ARAB EMIRATES UNIVERSITY
AMALQ@EMIRATES.NET.AE

ABSTRACT

This paper describes and documents the general procedure followed to create a complete 3D geometric model of the distinguished Jahili Fort located in Al-Ain, one of the major cities of the United Arab Emirates (UAE). The model was in-turn used for the development of a virtual reality 3D model that can be viewed and manipulated online.
The aim of this study is two folds. The first is to introduce an online easy-access tool for general public. This online tool would inform and increase the public awareness of their-own architectural heritage. The second fold, however, is to present a pilot example for architectural heritage documentation, recording, and online presentation. This pilot model can be applied in the future to other historical sites and buildings in the UAE. The initial results of the above mentioned effort are promising, and it should be further tested on several other buildings and sites.

INTRODUCTION

The conservation of historical buildings has always been one of the ways of preserving a country's identity and culture. Historical buildings have always been the pride of nations and their icons. In this paper, the authors are trying to highlight and present one of the most important cultural icons of the United Arab Emirates (UAE), namely, the distinguished Jahili fort. This fort is one of the most famous historical forts in the UAE. It was constructed by Sheikh Zayed Bin Khalifa Al Nahyan (Zayed the First 1855-1909) in 1898. The fort was built for defense purposes and to guard precious palm groves. It was also used as a residence for the local governor. The fort was constructed using mud-brick, stones, and palm-trees products. Al-Jahili fort was selected for this study because of its distinguished architectural and historical value, and also for its simple geometry and texture (Fig.1).

This work is a starting step in the effort of the 3D documentation of the architectural heritage in the UAE. It sets the stage for others to step-in and actively participate in this process. It also provides a working example to be generalized and followed by other teams. In the long run, this effort should cover the entire architectural heritage of the UAE and combine them in an online virtual library. This library would not only help in increasing the local cultural awareness, but would also help in introducing these cultural icons to the general public world wide.

LITERATURE REVIEW

The Internet is rich with VRML models that document architectural heritage around the world. In the process of preparing this study, the authors reviewed a large number of similar web-sites. Some of these sites are listed below as online references.

One of the best online examples would be the VRND project on Virtual Notre-Dame. This example is unique because of it provides very high level of details for the interior of the monument. Another example is the Abbey of Pomposa. The surveying method adopted in this project --that is far more advanced than that adopted by the authors in this study-- made it possible to document even the very small details of the building. But unlike the Notre-Dame monument, the Abbey is much smaller.

Most of these sites present VRML models and rendered images from the 3D models. In this study, however, the authors incorporated 3D modeling with real panoramic-views of the existing building and its surroundings. Thus, mixing the virtual and the real fort to enhance the user's experience and take him or her as close as possible to the real scenes of the existing fort.

DATA COLLECTION METHODOLOGY

There are three main pillars in the process of architectural heritage documentation. These three data collection modes include: (1) Written Histories; (2) Photo Documentation; and finally, (3) Site Surveying. The work progress in these three modes can be done in parallel in most cases. However, sometimes the history of the building can be the focus of the initial preparation for the documentation process, especially if the building has seen significant physical changes over its history. In the following sub-section, each of these modes will be separately discussed.

[1] Written Histories

In this data collection mode, all previously written history of the site were collected and studied to compile and verify a single story for the building. Interviews with previous users were also documented to further verify both the history of the building. They were also documented to verify the time period and the reason for each physical change that were done to the original building. The following briefly illustrates the history and the backgrounds of the building under investigation:

Figure 1 Al-Jahili fort: Simple geometry and texture

- Al-Jahili Fort has a great history. It was built on stages and witnessed many major events in Al-Ain. It started by building a massive round watchtower on a hilly location in Al-Jahili area. The aim was to safeguard the settlement from any intruder and to watch the water courses that were the essential means of living and survival in the oasis.
- The watchtower was built from mud bricks and it was inhabited by guards who used to watch the area day and night. A mosque was then built close to the watchtower for the use of the guards.
- Later on, in 1989 Sheikh Zayed Bin Khalifa (Zayed the First) built the rectangular fort. As he ruled the region during that period, the fort was used as his headquarter and as the ruling family residence.
- The fort was abandoned after the death of Sheikh Zayed the First since his son Sheikh Sultan moved to his fort at Al-Ain city centre (Sheikh Sultan Fort) and the watchtower was used as a storage facility.
- In the late fifties the fort was inhabited again but by soldiers as the area became a British Protectorate. The Oman Trucial Scouts --the representatives of the British Army in the region- - used the fort as their headquarters until the independence of the Emirates in 1970.
- Currently, the fort is the subject of a major restoration, preservation, and rehabilitation project. This project is sponsored and managed by Al-Ain Economic Development and Tourism Promotion Authority.

[2] Photo Documentation

In this data collection mode, digital images were used to photo-document the building with its different sections, feature, and architectural elements. These images also documents the physical state the building, including cracks, structural and finishing-materials damages, and the position of any missing ornaments or any other missing architectural element. They also record the current urban setting around the fort and its placement with regards to other buildings or urban elements. Digital Photos were meant to preserve visual information about the monument and provide a real visual feel of the fort structure, materials, texture, and its' current physical state.

[3] Surveying of Al-Jahili Fort

This data collection mode basically captures and records the actual dimensions of the fort to be later converted into two-dimensional drawings. In this case, manual surveying techniques were used to verify some old existing drawings from some previously done work. Following that verification process, all theses data were plotted in two-dimensional drawings for different plans, elevations, and sections. Finally, a three-dimensional AutoCAD model was constructed for the monument.

The Digital Presentation of the Fort: Web-based Documentation

After the initial recording and documentation of the monument, several levels of digital online presentation techniques were used to publish this effort to the public. The aim of this digital presentation is two folds. The first is to introduce an online easy-access tool for the general public. The second is to present a pilot example for architectural cultural heritage documentation, recording, and online presentation that can be applied in the future to other historical sites and buildings in the UAE. These different levels of digital online presentation techniques were all embedded and integrated into an Internet Web Site that includes the following:

[1] Web-based History Documentation

The history of the fort was embedded into a Web-based Document to be viewed by the general public. This web-text document includes all verified historical information as well as the verified stories of pervious users of the fort. It also includes some old photographs and postcards of the fort that show the original settings of the monument.

[2] Non-interactive Internet Based Picture Book System

Some of the previously mentioned captured images were used to create a Non-interactive Internet Based Picture Book System. In this system, the user can browse through a number of thumbnails that are categorized by subject, and then click to enlarge his/her choice for a closer look.

[3] Pseudo-3D-immersive Environments

Some of these captured images were also later used to create a number of Pseudo-3D-immersive Environments. Here, Pixmaker© was utilized to stitch a consequent set of images together utilizing the technology presented by Apple Quicktime. This process forms partial or full 360 degrees panoramic images of the monument site that can be displayed online. The user can literary look around the monument, and zoom in and out on demand.

[4] Nested Pseudo-3D-immersive Environment

Some of these Pseudo-3Ds were later nested together in a sequence of immersive images that take the user from the front gate into the main court and then to the fort's inner court through a series of Hotspots. The aim is to introduce a user friendly way to explore all the external details of the monument while sitting at home, in a library, or in a class room anywhere in the world. Pixmaker© was again utilized to accomplish this task following a preset storyboard. This sequence of pseudo images takes the user from a position in front of the main gate into the main court. It then simulates a 360 degrees panoramic view of the court interior from more than one vanishing point. The 3D-immersive presentation continues by taking the user inside the inner court of the building through the southern door; simulate a 360 degrees panoramic view of the court interior, and then takes the user out of the western door. Finally, from a vanishing point in front of the western door, a 360 degrees panoramic view of the main court interior that includes the main cylindrical tower is presented.

[5] The Production of a VRML-Based Virtual Model

As a mentioned earlier in Section 3.3, a three-dimensional AutoCAD model was constructed using both the survey and the digital images. In this step, however, the aim is to produce a VRML model of the monument. This process started by exporting the 3D model from AutoCAD to AutoDESK VIZ 4. Since the model was huge in size (55 MB), several steps has to done to successfully import the model. First, the number of polygons was tremendously decreased by replacing some parts of the model with regular primitives and through unifying Faces and Normals. Then, the model was broken down into several small files that were later imported one by one and merged into a single file in VIZ 4. Still, the model contained an enormous amount of faces (literally, millions.) These faces were further decreased using the optimization modifier currently embedded in VIZ 4. Finally, a VMRL model was exported from AutoDESK VIZ 4 using the "helper." This model was later embedded into a Web-page to be viewed and manipulated online.

Conclusions and Future Work

The initial results of the above mentioned effort are promising, and it should be further tested on several other buildings and sites. The following conclusions have materialized from the above illustrated experiences:

- Due to the monument present condition, the nested Pseudo 3D-immersive model is the best digital presentation of the monument. This is because of its simple geometry and the presence of the two wide courts that made the 360 degrees panoramic views possible.
- The virtual reality model would be a great tool for the online exploration of the historical stages and of the details of all the architectural elements of the fort. This may be the subject of a future study.
- Moreover, the virtual reality model would greatly help in the process of rehabilitation and reuse of the fort. The model can be utilized through out the design process and later on in evaluating different design proposals.
- The product of this research work was not publicly presented in the UAE or tested by users. It is our future goal, however, to test the output of this study on a number of randomly selected users. This process would help us in evaluating the users' response and modify our work accordingly.
- It is also our future goal to study the local cultural responses to the web-based heritage documentation system as an educational tool. The aim is to find out whether this process is a successful way of conveying the intended message to its' intended audience or not.

REFERENCES

ABBEY OF POMPOSA: VRML Representation (http://www.vit.iit.nrc.ca/Biris_Italy/pomposa/models.htm, January 2003).

THE VIRTUAL TOMB OF MENNA (Egypt) (http://www.doc.mmu.ac.uk/RESEARCH/virtual-museum/Menna/visit.html, January 2003).

VIRMLS Project; Virtual Open Air Museum (http://inf2.pira.co.uk/factsheets/inform/digicult/virmus.html, January 2003).

VRND: Notre-dame Cathedral a Real-time Virtual Reconstruction (http://www.vrndproject.com/)

A Cultural Heritage Dialogue
IT Support for Reflections on the Present and the Future

Susanne van Raalte
Chalmers University of Technology, Göteborg, Sweden

Rolf Källman
The Swedish National Heritage Board, Stockholm, Sweden

Tomas Wikström
School of Architecture, Lund University, Lund, Sweden

See the CD for the extended version

Abstract

A society's cultural heritage and the discussion surrounding such heritage form a possible and fruitful point of departure for reflections on the present and the future. Digital communication and visualization aids can, when properly used, provide a resource with which to develop and broaden discussions concerning the contents, usage, and significance of the cultural heritage.
To try to contribute to a broadening of the cultural heritage dialogue, with those who actually live in today's society, a church building is chosen as the object of study. The church enables a large number of opportunities for interpretation and experience and has many different meanings and values for different groups of people.
As a technical platform, a multi-user environment on the Internet, Active Worlds, is used. Within these virtual environment two interactive 3D-models of Old Uppsala Cathedral, the current church and the reconstructed 12th century cathedral, can be reached and explored. In the reconstructed model an imaginary mass on Saint Erik's Festive Day on May 18, 1180 is simulated. The models are closely integrated with contents on a website.
By actively involving two target groups, professionals involved in the church and its future (found in the ecclesiastic and conservation professions) and the school, to participate and interact with the virtual environment an evaluation of the intersections between the core of content and techniques and of possibilities for communication and dialogue has been made.

A project in collaboration between the Swedish National Heritage Board and Chalmers University of Technology

Our cultural heritage is not defined once and for all. What it means to you depends on who you are, what you know about it and from which angle you look upon it. To put the questions about the cultural heritage, its values and management, on the society's agenda it is important to establish and broaden the dialogue concerning these questions with the citizens.

In the project "A Cultural Heritage Dialogue"[1] we have tried to involve new and old target groups in dialogues based upon interactive visualizations. To reach a level where the cultural heritage concerns and engages as many as possible we believe it's important to go from one-way, top down communication, to a dialogue.

Few cultural environments can compete with churches in terms of claiming relationships to many disparate groups of people. The churches offer a large number of opportunities for interpretation and have different meanings and values for different groups of people. Therefore the parish church of Old Uppsala was selected the object of study.

Old Uppsala - The Place and the Church

Old Uppsala is one of Sweden's most famous and historically charged cultural environments. With its royal burial mounds, traces of a manor house and the former cathedral, the place itself speaks of the central importance it had in prehistoric and early medieval times.

In the year of 1164, Sweden was given permission to form its own diocese. The Archdiocese was situated in Old Uppsala in the heart of Uppland, a place that enjoyed great importance in pagan Svealand.

Figure 1 The parish church of Old Uppsala

Figure 2 The reconstructed 12th Century Cathedral

The cathedral was completed during the 12th century and was then some time during the first half of the 13th century, the cathedral burned down and the

Archdiocese was moved to East Aros (Uppsala). In 1270, the church in Old Uppsala was transformed into a smaller parish church.

Activeworlds and TimeDoc

As a technical platform, an online multi-user environment, Activeworlds[2] is used. It is a commercial system based on streaming technologies. The system hosts over a thousand collaborative 3D virtual worlds in its main universe. Within the browser users can enter and explore different worlds together.
The Activeworlds graphical interface (Fig.3) consists of three main windows: a graphics window with the current 3D world, a chat window and an interconnected standard web browser window. Entering a world a user can navigate and appear as an avatar in real-time 3D, click on objects to display the corresponding web page, teleport to a different area, chat with others etc. Activeworlds is one of the most popular and used online browsers. It stands out from the others due to its fast browser download, low system requirements and the possibilities for users to build within the world.

Within our world TimeDoc in the Activeworlds universe, two church models, the current Old Uppsala parish church and the reconstructed 12th century cathedral, can be explored.

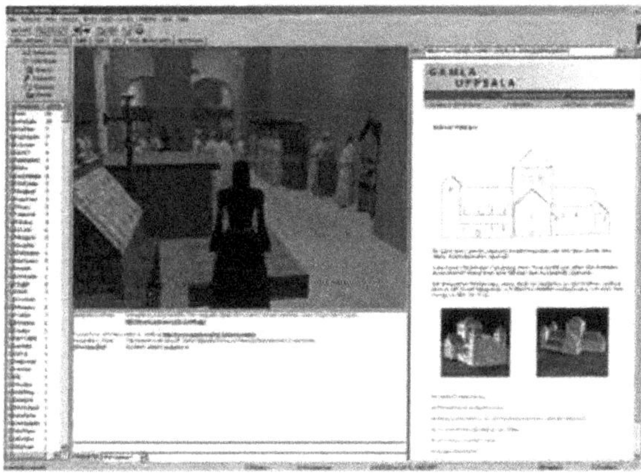

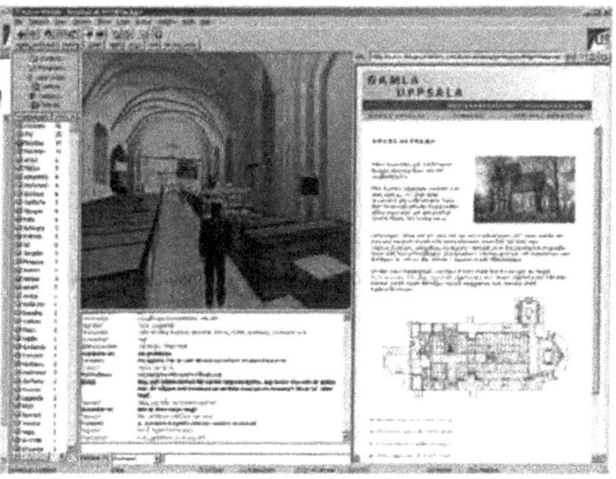

Figure 3 Attending the virtual mass, in TimeDoc, Active Worlds Figure 4 Chat going on in TimeDoc

In the parish church, which is represented in a photo-realistic way, you may click on certain objects to gain more information on the closely connected website.[3] There are also possible to see the planned changes in the church. If you do not want to explore the world by yourself there are two guidebots[4] that are programmed to show you around.

The second model represents the 12th century cathedral and is based on a scientific interpretation of the source material and available references. In this case, we have used full colour illustrations instead of photo-realism. It was important to make evident that this model is an interpretation of the source material. Via interactive solutions, visitors find out, how the source material was interpreted. Another reason was to pave the way for additions and alternative interpretations. In the reconstructed model, an imaginary mass on Saint Erik's Festive Day on May 18, 1180 is simulated.

Using the Virtual Models in Dialogues with Users

It is neither possible, nor desirable, to address all target groups simultaneously. Moreover, the issues that concern, or interest, different groups and individual persons are seldom the same. Therefore the information and dialogues must be based on who one wants to address and in which context.

The overall aim of the project is to establish a dialogue, where the virtual model of the church becomes a place for conversations not only about this specific church, but also of church buildings more generally, today and in the future. The point of departure for the project was the planned changes of the interior of Old Uppsala parish church. That has given us the privilege to work in close contact with the staff of the church. Together with the staff we were able to formulate the questions we wanted to discuss with the public. The planning process also gave us the opportunity to employ the virtual models and the ongoing dialogues as material in the discussions between the church and the regional heritage authorities.

We have also collaborated with the Tunaberg School in Uppsala, where a number of students from the 7th-9th grades have participated. The aim in this case was to explore the perspective of the user thirsting for knowledge. What do the students like to know? What interests them? And how should the information be presented in the virtual model? The students have contributed by doing their own research about persons and events associated with the present as well as the medieval church. The result of the students work is presented in TimeDoc.

Experiences

Based on a core content representing an important Swedish cultural environment, Old Uppsala, the intention of the project was to test in which ways it is possible to develop dialogues with different target groups using Virtual Reality technology.

The results of this project and the dialogues held with our target groups will provide an important foundation for continued discussions at the Swedish National Heritage Board.

What will the dialogues lead to in terms of expertise needed, and - perhaps - altered methods of work? The result of the dialogues, that still are running, may lead to the conclusion that experts on cultural heritage do not have all the competence and knowledge the citizens ask for. And perhaps that will lead to a change of positions. If, as a result of the dialogue with its target groups, the National Heritage Board has to transform its ways of working, this means that the strive to establish a way to deal with the heritage in a democratic way.
We were happy to find out that those, who had the opportunity to take part in events in TimeDoc, experienced that they actually were communicating in a "real" space. Of course the pupils, one of our target groups,, are quite used to spend time in three-dimensional virtual space, as they are playing Internet based online games. But still, we found it easy to discuss difficult questions with them; what does a church mean to them, what values do they see, and what are their opinions of what you can and cannot do with a church. The pupils willingly took part in what developed into a virtual excursion and very much enjoyed the dialogue.

Not only did we successfully conduct an interactive lesson in TimeDoc with the pupils of the Tunaberg School, on more than one occasion we arranged open houses and other presentations were experts participated. The general experience of these "seminars" was that the experts very quickly forgot the virtuality of the setting and became involved in discussions about the content. The three dimensional context evoked certain topics that concerned experts from different fields and led to re-evaluations of existing knowledge. For example, after some discussion, the reliquary with the bones of Saint Erik was moved to another location within the medieval church, to allow better access for pilgrims.

Finally, within the model of the parish church, the members of the parish of Old Uppsala had the opportunity to examine proposed changes sketched by the architect. The parish church model was demonstrated on meetings between parish members and involved experts. It was employed to demonstrate different ways to rearrange the interior. However, few of the parish members took the chance to explore TimeDoc on their own. The quite simple technology involved still seems to represent an obstacle for many potential users.
One important problem we have experienced so far concerns how to sort out the dialogue when the there are more than 4-5 active visitors present simultaneously. It becomes hard to figure out who is talking to whom and some people tended to become involved in two or three conversations at the same time. Another problem is that the number of objects that a certain area within Activeworlds may hold is limited. Thus, for example, some of the proposed changes of the parish church had to be shown in the separate web-window, which decreases interactivity.

Conclusions

This is the first time the National Heritage Board is involved in real-time discussions with its target groups in a Virtual Reality medium. The results of the project so far are convincing and speak for a continuation. A dialogue in a real-time simulated world provides the opportunity do discuss questions concerning cultural heritage with groups of people that are normally hard to reach through other media. And the multimedia environment provides new angels.

The project was financed by Stiftelsen Framtidens Kultur (The Culture of the Future Foundation) and with funds from the Swedish National Heritage Board's Research and Development grant.

1 See http://www.design.chalmers.se/kulturarvsdialog.

2 See http://www.activeworlds.com.

3 See http://www.design.chalmers.se/kulturarvsdialog/gamlauppsala.

4 Bots: short name for robot.

References

SCHROEDER, R., HUXOR, A. and SMITH, A., 2001. Activeworlds: geography and social interaction in virtual reality. Elsevier Science Ltd.

VAN RAALTE, S., 2001. Virtual Heritage -3d-Visualization Of Cultural Heritage Information. Institute of Conservation, Göteborg University.

WIKSTRÖM, T., 2001. Virtualitet och rumslighet - Om nygamla media och platser. Nordisk arkitekturforskning no.3.

Computer Games and Archaeological Reconstruction: The Low Cost VR

Michael Andrew Anderson
Department of Archaeology and Anthropology
Cambridge

Abstract

As computer game designers struggle to reproduce reality ever more faithfully for game players, they simultaneously provide archaeologists and researchers with an inexpensive and effective tool for real-time three-dimensional reconstructions of ancient sites. This paper presents the results of recent work examining the potentials of 3D game rendering engines for the production of educational and scientific visualisations of ancient sites. Since games are designed to run on personal computers with minimal system requirements often far below those required for many immersive real-time virtual reality environments they present a resource that can be accessed by many different users without great expense. In addition, most recent game engines come packaged with software designed for the creation of new 'levels' and features. With careful planning, these tools can be used to create non-game environments that allow users to experience an archaeological reconstruction by personally navigating it at their own pace. Following in the footsteps of recent work at the Martin Centre for Architecture, Cambridge, which used the QuakeTM engine to produce new architectural designs, the present work critically appraises the advantages, disadvantages and utility of this resource for the archaeological reconstruction of ancient Roman domestic environments.

VR and Computer Games

The value of virtual reality and computer visualisation in archaeology for scientific research, archaeological preservation, public access, documentation and publication has been the subject of considerable interest in recent years. Projects like the UCLA Cultural VR lab's Roman Forum project, the Troia VR project, and the CNR ITABC[1] have set the standards for realism, scientific examination, archaeological accuracy and transparent documentation in computer reconstruction and emersive virtual reality. However, while much effort has centred on improving the quality and scientific validity of computer-aided visualisation, their actual production has only been possible for large-budget projects with access to powerful computers and specialised interface equipment. While most projects do not have the funds necessary to work at this scale, they would nevertheless benefit from computer visualisation and reconstruction.

I suggest that the computer game industry has already provided archaeologists with tools that can be used to produce virtual reconstructions at a fraction of the cost of more 'academically recognised' software with only marginal losses in quality with minimal investment of time on low-end systems. Furthermore, the growth of the computer game industry will continue to provide ever more useful and realistic visualisation tools. Computer game platforms also have unique abilities that their more official cousins do not, such as inherent Internet functionality and multi-user access. While game engines cannot and should not replace high-end reconstruction, they do offer a greater spectrum of choice for individual projects and the opportunity to use the appropriate tool for the job.

The Project

As a component of my PhD research, I began a project in order to examine whether computer game engines could provide the tools necessary for archaeologists to create scientifically acceptable emersive 3D environments that would ultimately be useful for academic purposes as well as for public presentation. Since my research centres upon the use of Roman domestic space and on Pompeii in particular, I decided to reconstruct a single Pompeian house: that of P. Paquius Proculus (or C. Cuspius Pansa - Regio I, 7, 1), restoring it to its appearance just before the AD 79 eruption of Vesuvius. I already had a large database of digital photographs of this house including images of each side of every wall.[2] In many cases floors were also photographed, but the archive had not been taken with this project in mind. This meant that many of the photographs had to be rectified before they could be used as flat textures in the virtual environment. Nevertheless, I felt that this resource, in combination with accurate floor plans[3] would allow for a reasonably accurate reconstruction of the house. The project used a Compaq Presario 1700 Laptop, Pentium III 800MHz with 512MB RAM and an 8 MB graphics card for authoring and presentation.

Many different game engines were considered for the project. Because of similar work previously carried out by the Martin Centre for Architecture at the University of Cambridge, and also because of an amateur production of an accurate model of the Titanic with this engine, Quake 2 seemed at first to be a strong candidate.[4] However, from a cursory examination, both the interface and the graphics quality seemed to be inferior to other engines. Ultimately, I decided upon the Quake 3 engine because it offered a greater degree of control over lighting and 'radiosity' reflection properties as well as true curved objects. While it lacked the ability to create 'story line' scenes that might have been ultimately useful for presentation or teaching, I felt that greater accuracy of the model itself was more important.

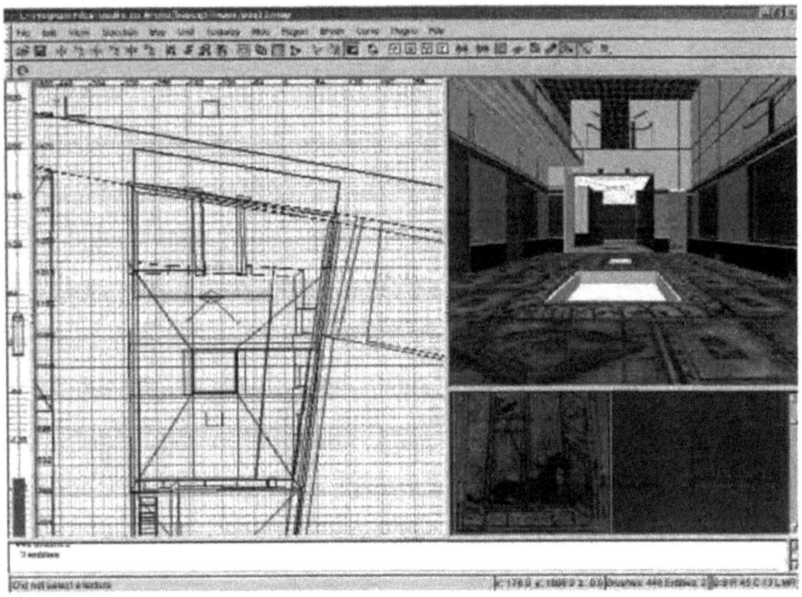

Figure 1

The project used a variety of software packages to achieve the greatest possible degree of processing convenience and accuracy. The primary editor used was Gtk-Radiant 1.2.11, which was freely available on the web.[5] The interface consists of three windows that present the map overview, an elevation window, and a low-resolution 3D view (Fig.1). However, while this editing package does place objects on grid of coordinates, there is no direct command line for accurate point entry such as is found in many CAD packages, and archaeologically accurate placement would have required a painstakingly slow approach of carefully placing each vertex by hand using the x, y and z coordinates displayed in the corner of the screen. To make matters worse, Gtk-Radiant does not support the DXF format either for import or export.

I found that a freeware beta version of Discreet's 3DStudio Max called Gmax: Tempest could import DXF files.[6] It provides a similar interface to that provided by Gtk-Radiant but I found that exporting maps for the Quake engine was not as straightforward. The beta status of Gmax:Tempest also meant that complete documentation was lacking and this hampered its use. The decision was therefore made to use Gmax:Tempest to produce a map containing the basic structure from DXF files, but to detail and finalise it in Gtk-Radiant.

Accordingly the floor plan of the house was digitised in Adobe Illustrator and saved as a two-dimensional DXF file, which was then scaled appropriately in AutoCAD. Afterward, the two dimensional DXF file was imported into Gmax:Tempest and the polygons were extruded to a standard height. This resulted in a set of 3D Studio Max polygons that were then capped and turned into Quake polygons. The result was saved as a map file in such a format that Gtk-Radiant would understand. After loading the new map file into Gtk-Radiant, it became apparent that the polygons would have to be replaced by hand, as miscommunication between the two programs meant that when maps were compiled rather serious errors occurred.

Finally, with an extruded floor plan consisting entirely of stable polygons, the true work of reconstruction and detailing could begin. Upper roofing configurations and wall heights had to be largely hypothetical, but this is not unusual in virtual reconstruction. Once the polygonal structure of the building was complete, textures had to be created in order to recreate their AD 79 appearance. Wherever possible, the actual preserved walls were used as templates for the textures. In cases where no information was known about the wall decoration, they were either left as plain white or decorated with a standard wall texture that could be easily identified as merely representative.

The most challenging stage of the production was the final rendering, which required setting light values for the sky and sun and working with the radiosity settings. Surprisingly, these did not function as would have seemed logical, and it was often necessary to use ridiculously high values in order to generate 'realistic' results.

Results

The final result of the project was the successful reconstruction of the house of P. Paquius Proculus as it may have appeared just before the AD 79 eruption of Vesuvius. The model compared favourably to photos taken within the currently preserved ruins, and the interface through the game engine provided a realistic experience of the house structure, size

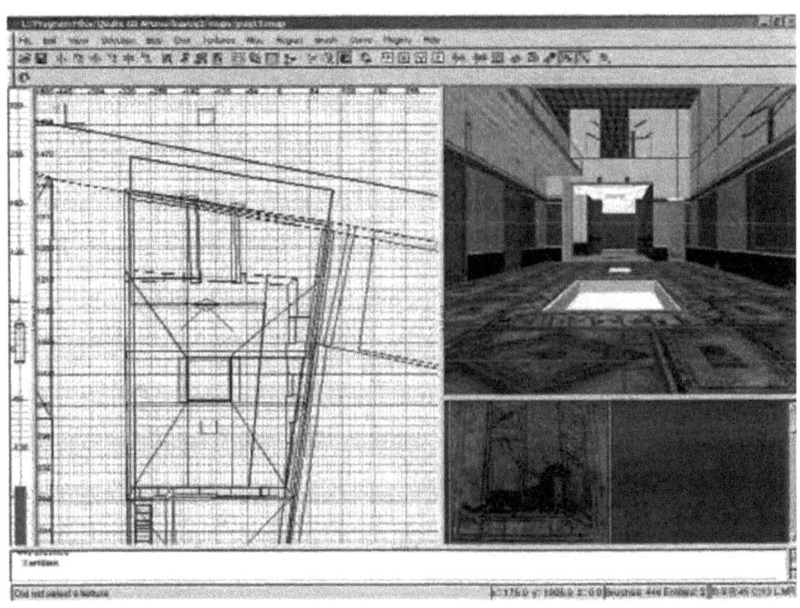

Figure 2

Figure 3

and phenomenology (Figs.2 and 3). The surrounding city environment was created in 'movie set' fashion so that the vistas from upper stories would be more accurate and the visualisation more complete. Surrounding landscapes were also supplied so that standing on the balcony of the house the Sorrentine peninsula is visible. All lighting in the model comes from the sky and reflectance within the house, and highlights the fact that some rooms were evidently quite dark even at noon.

LIMITATIONS

This project set out to assess whether computer game engines could be used for scientific archaeological VR and visualisations. While the final results certainly justify the use of computer game engines for archaeological reconstructions, it is apparent that they do suffer from significant limitations that will need to be overcome before the use of such engines can become widespread.

The most pressing issue for the use of computer game engines is access to the source code. Even in cases where the source code is 'open' and can be modified into completely different games certain aspects of the code remain occluded either because of 'black boxes' or more frequently because the information is not clearly documented but must be gleaned from endless online forums filled with extraneous material.[7] For instance, while it is possible to change almost all aspects of the Quake 3 engine mechanics the exact functioning of the radiosity calculation used for rendering Quake maps is not immediately transparent. This means that evaluating the accuracy of results is difficult, especially since the immediate effect of changes made during compilation is not always clear. This is made all the more frustrating by the fact that problems do exist with the compilation process that are infuriatingly hard to solve.

Most notable of these is the 'light leak' problem that exists at the edges of polygons. When the polygons are too thin (i.e. less than 16 units) light will leak between the join, no matter how the polygons are arranged. Indeed, this will occur even if they are not touching at all, and the problem is even worse if the polygons are not aligned with the main grid axes. This light does affect the overall radiosity calculations and can lead to some very inaccurate results. Clearly when the game was written, it was felt that thick, heavy constructions were the type of architecture likely to be built by game players. Unfortunately, since scientific visualisation aims at accuracy, one cannot simply thicken the walls to solve the problem. In fact, the only reliable way to solve this problem is to scale up the entire structure so that all polygons are more than 16 units thick. In turn the height of the player's viewpoint is too low to approximate reality.

Another major problem is that of the polygon dynamics themselves. The game engine has difficulties when polygons overlap in any way, and generally expects all joins to occur perfectly on the imposed grid system. Of course, when designing a game level of an imaginary place, then this is a simple matter of keeping the polygons relatively simple and angles sharp. However, it is much more complicated when using real archaeological data which does not always line up perfectly. This can lead to walls that suddenly disappear during the rendering process, or are visible from some locations within the model and not others. Such infuriating problems consumed many hours of the time and solutions to such problems were not immediately apparent or available on the Internet.

Finally, there is the problem of the scientific nature of the reconstruction. I have already suggested that lighting values are somewhat arbitrary. In fact, not a single measurement in the system corresponds to a real-world measure. Units on the grid are 'units' and although on-line there is a general belief that eight units equal one foot, in my experience this measure does not seem to be accurate. One can only be certain that objects in the current model are accurate with respect to each other since the floor plan was scaled correctly to begin with and all vertical projections were carefully performed.

Finally, there are certain limitations that are not related to the actual functioning of the software, but rather rest on the idea of using game engines at all. To begin with, there is the problem of using proprietary software for academic work. Distribution of the end product relies upon the end-user having access to specific software and licenses. Since the production of low-end virtual realities, particularly ones to be experienced over the Internet is one of the major goals of this project, it must also be considered whether requiring users to purchase proprietary software in order gain access to information is really in line with the goals of academia and democratisation of information.

Persistence and longevity of the data are also a problem because the vagaries of the market may suddenly cause a game to be no longer supported, or future operating systems may no longer enable it to run properly. Of course, this problem exists for using any proprietary software, and to a certain extent is a problem with digital data as a whole, but the situation is more acute with computer games due to the velocity of development. Certainly this will continue to be a problem for any digital reconstruction using a game engine.

With these limitations noted however, it should be pointed out that not a single one of them is insurmountable. The creation of an online academic community dedicated to the production of scientific VR and visualisations with game engines would soon lead to workable solutions. Such a consortium could produce its own compilers, or at least properly document those currently available. At the same time, additional features could be added such as a standard scale for light and size. New features such as ODBC compliant database interaction allowing documentation and transparency of reconstruction decisions are also a possibility.Such modified versions might present a more serious user interface specifically designed for the archaeological audience.

While the fast paced development of new game engines and computer abilities means that current work will need to be constantly migrated to newer and better systems, the result is that continually greater degrees of realism and emersion will be possible. So long as archaeology constantly remains just behind the crest of the wave of development, then the money spent on the computer gaming industry can be made to work for archaeology at the same time. The Quake 3 engine cost the current project about ten pounds sterling, and the original Quake has already been made freeware. As engines become outdated, they will become increasingly accessible to all. There could be no greater force for the democratisation of information.

Additional Abilities

It should also be pointed out that the use of computer game engines provides a host of advantages not provided by any other system. Most notable is the ease with which such material can be presented over the Internet. While the map files could be downloadable via the Internet for personal use, they could just as easily be hosted either on a LAN or the Internet. Because they are designed with a multi-user environment in mind, the games contain systems for communication between participants. I can envision such a reconstruction being visited by a set of experts from around the world, who could discuss in real time and give immediate feedback on the reconstruction they were experiencing. Additionally, experts could give guided tours for students or the interested public regardless of their physical location. Since certain game engines were also designed to present non-adversarial entities for interaction, this presents the possibility to 'repopulate' these ancient spaces with computer-controlled agents performing their daily tasks, or an expert 'tour guide' with audio explanations of various aspects of the reconstruction.

Future Directions

This project has critically assessed the use of computer game engines for the construction of scientific visualisation and virtual realities, and has found that while significant limitations do exist, that this avenue of work should provide useful and worthwhile results. It is particularly important with the rapidly shrinking budgets of academic projects that money be spent wisely. The computer gaming industry has funds and expertise with which archaeologists can never hope to compete, and we should not attempt to do so. Extending or adapting existing technology to our purposes is an excellent way to avoid reinventing the wheel. I have suggested ways in which future development should iron out problems that exist in the current games engine software. Since the gaming industry will continue to advance technology, future projects will be able to produce even more impressive results. We must continue to bridge the gap between tools and products designed for games and those that produce results acceptable to academia. As the gap between the highest quality academic VR projects and computer games closes, it will become even more important that standards, methods, tools and community be created to make these programs more accessible and acceptable to the archaeological community. It is the further goal of this project to begin work on these tasks.

1 cf. Forte, M., this volume.

2 By the kind permission of the Soprintendenza archeologica di Pompei, and with help from the Anglo-American Project in Pompeii.

3 Plan taken from Corpus Topographicum Pompeianum, 1986.

4 http://www.arct.cam.ac.uk/research/pubs/pdfs/rich00a.pdf, http://www.planetquake.com/titanic/, cf. also Meister, M., Boss, M., this volume.

5 GtkRadiant is available from: http://qeradiant.com/.

6 Gmax and Tempest are available from: http://www.discreet.com/products/gmax/.

7 Such as those found at: http://planetquake.com and http://www.quake3world.com.

References

VAN DER POEL, H., 1986. Corpus Topographicum Pompeianum. Rome.

Between Reconstruction and Reproduction. The Role of Virtual Models in Archaeological Research

SABINA VITI
DIPARTIMENTO DI SCIENZE DELL'ANTICHITÀ G.PASQUALI, UNIVERSITY OF FLORENCE, ITALY

SEE THE CD FOR THE EXTENDED VERSION

ABSTRACT

The recent years have witnessed a high amount of computer aided, virtual reconstructions of archaeological sites, objects or artefacts. This relatively new interest in archaeological research gives new theoretical inputs on how to elaborate archaeological data and how to apply and build, in a virtual model different interpretations. Most of the literature concerning this subject outlines the impact of virtual reconstruction on one hand in educational contexts, in museum environments or in archaeological sites - as informative tools for the visitors - and on the other hand focuses on the high potentiality in the field of archaeological research. In the two cases the final product, the virtual model, is often presented without a specification where the archaeological reality ends and the virtual reconstruction starts, or without giving different interpretative options.
The paper will discuss the difference between reconstruction (beyond the archaeological reality) and reproduction (the recreation of the archaeological scenario) of archaeological entities. Similar approaches but different concepts and goals. The methodological aspect will be developed by the analysis of different case-studies.

Introduction

Over the last five years the interest and attention focused on issues regarding virtual archaeology has produced a wide body of literature, which has investigated technical, theoretical and methodological matters.

Two main directions are recognizable. One concerns the impact of VR within the educational context, such as the museum environment or the archaeological site, with VR employed as a didactic tool and a means to popularize archaeology and illustrate cultural heritage (Roussou 2002). The other direction has led to the development of new epistemological approaches in the field of archaeological research (Reilly 1990, Forte and Beltrami 2000). Two different goals, two different ways of developing 3D reconstructions. To simplify this statement, it may be said that, in the first case, i.e. within the public framework, the product is meaningful only in its final realization. The purpose of the virtual restoration of a monument, usually monumental architecture, is principally the aesthetic and/or educational impact on the public at large or on the student, and it results in indistinct interpretations. In the second case, i.e. within the research framework, the chief aim of VR is to analyze the different processes, which allow us to reach the final result. It represents "...the manipulation of an archaeological interpretation" (Barcelò 2000).

In this paper we will discuss the use of VR as a research tool in archaeology by examining the distinction between two different underlying processes, virtual Reconstruction and virtual Reproduction.

Virtual Reconstruction

Virtual may be taken to mean "... something that can act as surrogate or replacement for an original" (Barcelò, Forte and Sanders 2000), "...the representation of some (not necessarily all) features of a concrete or abstract entity", (Barcelò 2000) or basically the re-creation of what cannot be visible in reality. Reconstruction can be interpreted as the integration of entities not present, totally or partially, in the archaeological scenario, moving beyond the archaeological reality to produce a possible scenario, following a deductive process:

- Start from presumed objective archaeological data (what is seen, recognized and documented);
- Proceed on the basis of different assumptions (for example the estimation of certain parameters like the height of a wall by the evaluation of the collapsed construction material);
- Arrive at interpretations that will determine the integrations (adding a second floor to the analyzed structure if the walls reach a certain supposed height).

The model can be read as a synthesis of these various processes.

Virtual Reproduction

I mean by Reproduction that process which leads to the modelling of what is present in the archaeological scenario.

In the prehistoric context, for example, often the lack of standing elements, posts, walls etc, as well as the general lack of "monumentality", restricts our ability to reach the reconstruction stage since any further interpretation of the structural features would tend to be artificially contrived especially when the data is not sufficient or since the data manipulation is not necessary.

In this case, we may take virtual to mean that what is seen and modelled represents the complexity of the entire state of the archaeological findings: what, during the different phases of digging was achieved partially is now visible in all its complexity.

We may define Virtual Reproduction as the overlay of the different excavation layers, the re-composition of all the elements, for example the stratigraphy referring to one living area over a period of time, is never seen contemporaneously during the field investigation because excavations by their very nature, destructive, constitute an unrepeatable experience.

Therefore the possibility of reiterating the exploration virtually, not as a digging simulation, (since by virtue of documentary constraints, absolute integrity is effectively impossible to maintain), but as a bringing to light of the different steps of the excavation, allows us to improve the interpretation process and to carry on different experimentations.

By means of the analysis of two case studies the distinction between the two experiences may be clarified.

Case Study

[1] The virtual reconstruction of Hut A in Maccarese Le Cerquete Fianello (Rome)

Within the sphere of a larger project, promoted by the Italian Institute of Pre-history and Protohistory, entitled "Computational analysis and data-treatment of habitation structures of the prehistory and protohistory in Italy" approved and financed by M.U.R.S.T (2000-2001), the analysis and the reconstruction of hut A in the Calcolithic site of Maccarese, Le cerquete Fianello, was performed ((Manfredini et al., in print, VVAA 2002).

The postholes found in the village of Maccarese and easily recognizable as pertaining to human habitation, fuelled an interest in the modelling of the foundation floor of Hut A in order to enhance the bi-dimensional graphics of the archaeological data acquired in the excavation phase and to further enhance in 3D the surface profile of the ground and the morphology of the single holes that held the structural supports of the dwelling.

The attempt to recreate this structure was made possible through the use of different data sources: the 2D traditional documentation (plans and contour lines), photographic documentation, ethnographic data and photos, anthracological data, spatial analyses.

The re-composition of all this data produced the reconstruction and provided us with hints on how to experiment with several elaborations to restore virtually the upper parts of the hut in different ways.

The position of the postholes is quite clear and gives us the shape of the hut itself: it is regularly ovular.

The surface modelling, based on the interpolation of the contour-lines (Surfer 6),was performed to analyze more precisely the morphology of the post-holes, allowing us to visualize them from any vantage point.

Several observations were made once the modelling was done; even though this information was embedded in the 2d documentation it did not have the same impact. In order to model the posts to their approximate dimensions, the calculation of the diameter was based on the measurements of the cross-section that cuts through the post-holes. Once the plans were acquired and vectorialised in a Cad program (AutoCad2002), the further steps were performed by the use of 3D StudioMax4.

The height (4.5/5 m), and the general aspect of the rough straw clusters of the hut was extrapolated by analyzing early twentieth century photos from the photographic archive of the area of Maccarese. Keeping in mind the limitations of an exact correlation between these structures, we recognize some interesting similarities: same plan, disposition of the posts, same soil and similar environment.

The study of the anthracological remains encountered in the interior of the structure revealed the presence of various tree species, most especially the oak (Quercus sp.) and the elm (Ulmus sp). If we compare other findings (Bietti Sestieri et al. 1998), it is reasonable to believe that the perimeter walls were made of oak and that the covering was constructed from lighter material such as elm. For this reason, a texture like that of oak bark was chosen for the manufacture of the support posts and a wood possessing the coloration of elm was used for the sloping posts of the roof, taking into account that some form of humidity resistance treatment of the wood altered its original texture.

The 2 different interpretations that are presented were tested. Even though there are many variables such as the humidity, the ecological constraints etc., the principles of static rules were taken under consideration. Several other variations were

Figure 1 First reconstruction hypothesis of of Hut A

Figure 2 Second reconstruction hypothesis of Hut A

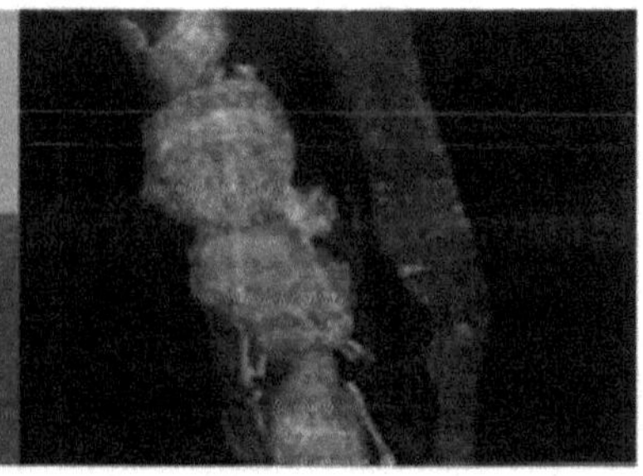

Figure 3 The use of opacity to enhance visualisation

tried once the basic model was built up and the one shown seems to be the most feasible (Fig.1, Fig.2).

In order to avoid "...the difficulty for the end user to distinguish the model's realism and the archaeological reality..." (Kanter 2000) and to create for this reason a transparent model (Ryan 2001), the rendering of the roofing was left opaque (medium transparent) to underline in a more evident way the uncertainty of the interpretation. The production of an animation allows us to show and visually explain as well, with minimal artifice (the lighting of the post inside the perimeter of the hut), where it is thought and verified, by effecting a spatial analysis of the general distribution of the archaeological material, that a passage existed between the two living areas recognizable in the structure. The process, which produced this animation, clarified certain points and during the construction it was possible to formulate different interpretations in a concrete way. From all the data collected we produce a graphic representation and the final animation represents the synthesis of all the accrued data and its interpretation.

It is important to point out that the interaction of the modeller with the model during the different steps of the creation process represents the topic phase of the research. Many complicated concerns are in fact dealt while the model is being created, (e.g. the calculation of the diameter of the posts, the problems related to the issue of static, the entries, the alignment of the beams etc.) ultimately making reconstruction viable. Thus it could be said that the model not only represents the ultimate result of the archaeological interpretation, but may also constitute the means to understanding the data itself.

[2] The virtual reproduction of grave 7 in Romito Cave (Papasidero, Cosenza)

The Epi-gravettian grave in Romito cave, (Martini et al., in print) is a good example by which to illustrate the concept of virtual reproduction.

In this case, the model was created through the use of different data sources. The contour lines permitted the modelling of the pit (AutoCad 2002) and by the acquisition of different cross sections of the stones, we performed their modelling (Rhino1.0). The photographic documentation provided the textures in order to achieve a photo realistic result and was also used to have a more precise view of the conservation state of the bones. The modelling of the skeleton is still on going and was carried out by the acquisition of the 2D documentation, the photos and anthropological data (3DStudioMax4).

In this instance it is not necessary to undertake a restoration of any missing feature because the grave appears in its integrity. The 3D reproduction of the pit and the filling in of the depression, including the covering architectural stone complex, allows us to reconstruct and underscore the dynamic anthropical events that generated the burial formation process: the excavation of the grave, the deposition of the body, the stone covering in its architectural complexity. The model provides us with the opportunity to visualize the burial-place with all its composite characteristics, even where it was not possible to appreciate it in the archaeological scenario; e.g. it is not possible to visually distinguish the respective position between the skeleton and the stones that directly cover it during the excavation stages, nor from the analysis of the traditional 2D documentation, whereas it lies within the scope of modelling to exploit various tactics to engineer visualization: the stones are rendered with a high opacity in such a way as to enable us to see the bones underneath and we can achieve a more comprehensible reading of their relationship (Fig.3).

It is possible to glean essential taphonomic data by observing the model. From the position and the rotation of the bones, it is feasible to determine in a significant way, how the overlapping stones acted upon the bones (as soon as the soft components of the interred body deteriorated) or how they splintered them (Fig.4).

Figure 4 Taphonomic analyses

The significance of this virtual reproduction is demonstrated via its capacity to explore iteratively the grave, the stone architecture, and the post-depositional event, and its value is confirmed in the power to provide this all-encompassing image not accessible previously to us solely by means of excavation.

Conclusion

When making a virtual reconstruction, all the interpretations developed during the modelling are represented graphically and explicitly. In the case illustrated here, the process being described may be defined as an action that extends from a specified perceived reality to a generalized conceived reality, in other words what we originally see in its fragmented state eventually becomes virtually reconstructed into a coherent meaningful form.

When a reconstruction is not possible or necessary, the virtual Reproduction helps and enhances the representation of the events that determined the archaeological state: for example, the post-depositional phenomena, or the stratigraphic relationships. In this instance, the process travels along a different route, one that begins with an diminished perception of reality and arrives at a more substantial or detailed reality.

Such "virtuality" still needs to be further developed and exploited in archaeology or more specifically in prehistory research.

The goal of either process (Virtual Reconstruction and Virtual Reproduction) is to create a tool for a better understanding of the past in a visual way. The strategies to achieve this goal follow a parallel course. Where their paths diverge is in the final result and in the practical applications that may arise.

ACKNOWLEDGEMENTS

The realisation of this paper was possible with the help and support of Joyce Myerson and Giovanni Puliti. I should also thank for their valuable technical contribution Arch. Riccardo Nigi, and Stefano Ricci (University of Siena).

REFERENCES

BARCELÒ, J.A., 2000. Vizualizing what might be: an introduction to virtual reality techniques in archaeology. Barcelò, J.A., FORTE, M. and SANDERS, D.H. (eds)., Virtual reality in Archaeology (BAR Int. Ser. 843):9-36.

BARCELÒ, J.A., FORTE, M. and SANDERS, D.H., 2000. The diversity of Archaeological virtual worlds.In Barcelò, J.A., FORTE, M. and SANDERS, D.H. (eds.),Virtual reality in Archaeology (BAR Int.Ser. 843):3-6.

BIETTI SESTIERI, A.M., DAMIANI, I., DE GROSSI MAZZORIN, J., DE SANTIS, A., RECCHIA, G. and REMOTTI, E., 1998. Un edificio della I età del Ferro nell'abitato di Fidene (Roma): posizione stratigrafica, caratteristiche strutturali, materiali. Proceedings of the XIII U.I.S.P.P. Conference, Forlì, 8-14 September 1996, Sections 10-12, IV:595-599.

FORTE, M. and BELTRAMI, R., 2000. A proposito di Virtual Archaeology: disordini, interazioni cognitive e virtualità. Archeologia e calcolatori 11:273-300.

KANTNER, J., 2000. Realism vs. reality: creating virtual recontrutions of prehistoric architecture. In Barcelò, J.A., Forte, M. and Sanders, D.H. (eds)., Virtual reality in Archaeology (BAR Int.Ser. 843):47-52.

MANFREDINI, A., CIRIACO, A., RINALDI, L., VITI, S., in print. Maccarese, Le Cerquete Fianello (Roma). Proceedings of the conference, Ferrara 26-27 November 2001. Analisi informatizzata e trattamento dei dati delle struttura di abitato della preistoria e protostoria italiana, Ferrara Novembre 2001, Origines IIPP, Pisa.

MARTINI, F., CASCIARRI, S., LEONINI, V., LO VETRO, D., MALLEGNI, F., MARTINO, G., NOTO, F., RICCI, S., RICCIARDI, S. and RICKARDS, O., in print. La nuova sepoltura epigravettiana "Romito 7" a Papasidero. Atti della XXXVII Riunione Scientifica dell'IIPP, Scalea 29, 9-4 October 2002, Scalea.

REILLY, P., 1990. Towards a virtual archaeology. In Lockyear, K. and Rahtz, S. (eds.), Computer Applications in Archaeology 1990 (BAR Int. Ser. 565):133-139.

ROUSSOU, M., 2002. Virtual Heritage: From the research Lab to the broad public. In Niccolucci, F. (ed.), Virtual Archaeology: Proceedings of the Vast Euroconference, Arezzo, Italy, November 2000 (BAR Int. Ser. 1075):93-100.

RYAN, N., 2001. Documenting an validating Virtual Archaeology. Archeologia e Calcolatori 12:245-253.

VVAA, 2002. Le dune, il lago, il mare: una comunità di villaggio dell'età del Rame a Maccarese. In Manfredini, A. (ed.), Origines, Firenze.

The Multimedia Room of the Scrovegni Chapel: A Virtual Heritage Project

Maurizio Forte
CNR ITABC, Roma, Italy
maurizio.forte@mlib.cnr.it

Sofia Pescarin
CNR ITABC, Roma, Italy
sofia.pescarin@mlib.cnr.it

Eva Pietroni
ARACNET, Roma, Italy
evapie@tin.it

Claudio Rufa
4ARACNET, Roma, Italy
clarufa@tin.it, www.aracnet.it

Dorina Bacilieri
srl, Bologna, Italy
info@arche-srl.it, www.arche-srl.it

Davide Borra
No Real, Torino, Italy
info@noreal.it, www.noreal.it

Abstract

The Scrovegni Chapel of Padova has been re-opened in March 2002, after a long closing period, due to necessary restoration activities on Giotto's frescos conducted by Italian ICR (Central Institute of Restoration). Giotto's masterpiece was given back to visitors and citizens, but with some restrictions: the chapel is solely visible for group of 20/25 people and for a limited period of 15 minutes. In order to get over this limitation, Padova Municipality decided to employ an empty and unused room, to be used as a multi-media centre for the visitors. The project was developed by Archè srl, under the technical direction of Sofia Pescarin, with the founding of an American Foundation (Wiegand Foundation). The multimedia room is planned to represent a cognitive space where the visitor can interact with the monument, with different approaches (active and passive fruition) through seven installations. The équipe of CNR ITABC (Institute of Technologies Applied to Cultural Heritage), directed by Maurizio Forte, was in charge of developing a Desktop Virtual Reality application (DVR) for the Chapel. With this application the visitor can interact with the frescos, through the 3D model of the Chapel (that is a three-dimensional, metrically reconstructed and interactive menu itself), and can access all the complex information, made immediately comprehensive by virtual reality and by the navigating metaphors. This application is one of the first examples of DVR system, made really accessible to a various public, who can use it, autonomously and without a guide, just discovering the VR world and the content.

Introduction

Because of serious problems of preservation, Giotto's paintings, in the Scrovegni Chapel in Padova, have been restored in the last two years. One year ago (March 2002) the works were finished and the chapel was finally given back to the visitors. Unfortunately the accessibility to the monument was limited to 20 people any time, while the visit cannot last more than 15 minutes. This is not enough, obviously, not only to understand this complex and wonderful piece of Italian 13th Century art, but also to have a complete view of the monuments and its frescos. Moreover it is almost impossible to see properly at least the 20 % of the paintings, because of the distance from the visitor.

When the municipality of Padua decided to face this problem of fruition, accessibility and comprehension, we proposed an integrated project that could be a possible solution: a system of seven multimedia and VR installations. The municipality completed the project by offering an unused space of 200 m2, an hypermedia room under the City Museum and by finding out a financial support, contacting an American foundation (Wiegand Foundation). The aim was to realise a "multimedia centre" in order to give the visitors appropriate tools to examine thoroughly, understand and appreciate Giotto's frescos, the Chapel and the 13th century background.

[1] Musealising the Virtual

"Without context there is no communication" (Bateson 1973), so only in an alphabetised space we identify rules of learning. Then, it is on the basis of these concepts that a territory called museum can become a place, a coded space. How is cyberspace being used to reconnect people in a new environment? What rules can be used in this digital ecosystem of virtual heritage? According to the cybernetic principle "the map is not the territory" (Bateson 1973), we have no information about the space without maps. The map is the visual grammar for getting information about and through a digital ecosystem. According to these premises it is possible to distinguish four domains of digital learning through maps (coded spaces) showing the relations between spaces and places:

1.Territory - Space - real. Territory is the space and it is in the Reality.
2.Map - place. The map is an alphabetised space, namely it is a place.
3.Map - virtual. In short, the map is the Virtual, in terms of virtual heritage.
4.Musealized area - Metaterritory (non-territory).

Finally, what is a museum? A musealised area can be identified in the middle between map and territory, therefore it is a meta-territory.

The main features of a virtual museum are mentioned below:

- Multimedia-interactive.
- Multidisciplinary.
- Multisensorial.
- Multi-dimensional (geometry of the museum and of the linked information).
- Multi-temporal (museum evolution).
- Connective (in the sense of spatialised information).
- Dynamic (information changes through the time).
- Meta-context (the virtual space is an informative space aimed
 to multiply quality and quantity of information).
- De-territorialised (museum information is referenced within an informative space, out of a physical territory);
- Meta-alphabetised (digital learning through complex metapzors);
- Cognitive. As structure-system the virtual museum becomes a cognitive space, therefore it is increased of significances according to the context it creates, namely the totality of information and association it includes.
- Inter-connected (with other real and virtual environments).
- Narrative. It can show not only objects, but events or activities (Forte and Franzoni 1998).

In this interpretation of virtual museum the concept of virtual heritage can be identified: virtual heritage is the perceived heritage, an information coded by a cognitive, inclusive interactive action of learning. In this sense the virtual museum can be even seen as "virtual model of museum", open museum, digital emission of an exhibited place. This kind of virtual musealisation can embrace contexts, works or objects, disseminated through space and time in a non-territory, a map coded by the Virtual and not by the reality.

In our case, the Scrovegni Chapel Project represents a pilot case of "itinerary of the virtual", an attempt to describe a communicative interchange and cultural assimilation throughout a virtual exhibit (five installations) and a physical musealisation (two installations). Therefore, the museum dialectics evolves between real and virtual, with different degrees of immersion and interaction. Musealising the virtual means to describe new spaces of composites geometries, or physical whether virtual; the physical interaction is with the environment, the virtual interaction is only within the multi-interactive space, making remote actions towards the context of digital information. That is what we see within the display systems is remote in comparison with the context of the digital information, but very close in comparison with the sensorial involvement: in this way we define the deterritorialisation of the space from real to virtual.

Figure 1 The Virtual Reality application. On the left: the main menu, on the right: the Cybermap

THE MULTIMEDIA ROOM

The Wiegand Multimedia Room is accessible to the public from the 25th of March 2003 (www.cappelladegliscrovegni.it, www.mlib.cnr.it/itabc/giotto). It has been planned and realised as a 'cognitive space' for the visitors. In this space it's possible to interact in different ways with the paintings, the monument and the city, through seven installations that bring the visitors across passive learning and active interaction phases. The entertainment aspects have been considered, but always in connection with a more complex scientific approach. This approach, obtained by words, voices, images, sounds, 3D worlds, etc. was reached even thanks to the strenuous work of the team composed by art historians, topographers, architects, graphic artists, communication and IT experts. Even the staging of the room was thought to stress the historical, didactical and technological aspects. Hardware systems, together with Software applications and Content, were considered all part of the same world. The installations are thought as content keeper but also as physical part of the virtual museum, the multimedia room (Fig.2).

The Multimedia room is not just a sequence of computers and monitors. To complete the exhibition, a few real spaces were created, such as the reconstruction of St. Anne's house (Fig.2, left). This model, realized such as it is painted in the scene "Announcement to St. Anne", is particularly useful in order to complete the explanation on Giotto's architecture. Moreover it has been recreated even a model, an example of an artist's workshop, with tools and material, while panels are placed, in another area, to describe the painting of a scene, through its different work stages and the fresco technique.

THE INSTALLATIONS

Apart the above mentioned installations, other five applications were realized, partly non-interactive, partly interactive, partly semi-immersive. The first Installation is an introductive DVD, a multivision that welcomes the public in the virtual museum (Fig.2, right). This is a non-interactive video, in five languages, realized with compositing techniques, using only graphic materials and 3d animations. The visitor feels to be involved in the story because it is made by Enrico Scrovegni himself, with his human vicissitude, his relation with Giotto, in a dramatic tone.

Figure 2 The Wiegand Multimedia Room

The second and third computers host two DVD interactive application, on Padua in the 14th Century and Everyday life, nature and architecture in Giotto's frescos (Fig.2, left). With these DVD applications the visitor can receive information on the context. These applications have been thought even to create a "space of experiences". In some cases, such in the multivision, this means that the visitor can use its own memories to find out a relation with passed history. The history of a monument and of its paintings can be explained, indeed, even in a narrative way, without loosing any scientific precision.

Focus of the virtual museum is the virtual reality system located towards the end of the room (Fig.1-2): the starting interface of the application is the entrance into the chapel, where the user is free to navigate through the global 3D space. This spatial dimension represents the first level of detail. At the beginning of the navigation no themes are activated and no elements can be selected. For querying the scenes and for opening other types of contents it is necessary to visualize the main menu. The main menu is composed by two sections (Fig.1, left). The first section can be defined VISIT (available for both levels of detail) and it is relative to the sequences of the virtual visit according to the available documentation and to the level of detail of visualization; the second section can be defined TOOLS and provides some typological interactions and utilities in order to interpret the space of the virtual chapel and the organization of the structured contents. Some of these tools have been just implemented, other ones will be added in a future development of the project. In the use of texture mapping we have provided for two levels of detail: a first intermediate level of detail for a comprehensive vision of the model and the scenes, and a second high level of detail where it is possible to appreciate every particular of each scene from a very near distance, simulating, in this way, a quality of visualization that is totally inaccessible during the real visit. For navigating through the model, the user can use the mouse (free rotation) and some keys of the keyboard (linear movements in all the directions).

CONCLUSION

The Wiegand Multimedia Room can be considered the first Italian Virtual Museum and one of the most important digital-virtual environments in Europe for the Cultural Heritage on the semantic level of the informative and communicational aspects. In the domain of the Virtual Heritage the risk is to enhance the amazing aesthetic features despite the informative/narrative feedback and cognition within the virtual worlds. How much information can I get from a virtual system? How does it communicate? How can we process this kind of interactive information? The importance of the virtual reality systems in the applications of cultural heritage should be oriented towards the capacity to change ways and approaches of learning. The Virtual communicates, the user learns and creates new information. Typically we define a linear learning, tools and actions, such as books, audio guides, catalogues and so on (in this case the communication is a linear sequence), and reticular learning VR systems where the user is immersed within reticules of information and visual data.

Psychological and cognitive learning in the museum cultural exhibition can be interpreted as the principle of communicating vases: the communicative flow of information crosses the museum's territory and the user's territory. The relations between Virtual and psychological reactions (action/reaction) are the basis of any cognitive processing: a good impact corresponds to a good memory, a good memory very probably will be a good tale. In a museum exhibition, for instance, in order to interpret an object, we compare old mental "maps" with our newer "maps": this mutual interaction can create an aesthetics of the fruition, a new context.

In this project we have embraced the philosophy of the ecological thinking for the Virtual reality applications, interpreting the virtual worlds such as ecosystems. According to this cybernetic approach, we consider these basic concepts in the virtual reality: feedback or retroaction, circularity, redundancy of information, difference, hierarchical levels of information, map, connectivity, metaphors, context, narrative visuality, aesthetic of the fruition. "Complex systems are those that include large numbers of components interacting in nonlinear ways, often leading to unexpected behaviour. Complexity sciences explore how parts are related to wholes, describing the interactions between environment, system, and observer. In common language one is reminded of the saying that 'the whole is greater than the sum of its parts.' Understanding art as a language, the historical development of art, and the creative process itself are all key areas in which art and complex systems have a common ground" (Levy and Galanter 2002).

References

ARNHEIM, R., 1969. Visual Thinking. University of California Press.

BATESON, G., 1973. Steps to an Ecology of Mind. Paladin Books.

BOWEN LOFTIN, R., 2002. Psychophysical Effects of Immersive Virtual Reality. In Proceedings of the IEEE Virtual Reality Conference 2002:145-148.

FORTE, M., 2000. About virtual archaeology: disorders, cognitive interactions and virtuality In Barcelo, J., Forte, M. and Sanders, D., 2000 (eds.), Virtual reality in archaeology, ArcheoPress, Oxford (BAR Int. Ser. S 843):247-263.

FORTE and FRANZONI, 1998. I musei virtuali - Comunicazione e metafore. In Sistemi Intelligenti.

FORTE, M., PIETRONI, E. and RUFA, C., 2002. Musealising the Virtual: the Virtual reality Project of the Scrovegni Chapel of Padua. In VSMM 2002, Proceedings of the Eighth International Conference on Virtual Systems and Multimedia, "Creative and digital culture", Gyeonggju, Korea, 25-27 September 2002:43-52.

FORTE, M., PIETRONI, E., RUFA, C., BIZZARRO, A., TILIA, A. and TILIA, S., 2001. DVR Pompei Project: Reconstructions of Archaeological Contexts in Desktop-OpenGL Environments. In Thwaites, H. and Addison, L. (eds.), VSMM 2001. Proceedings of the Seventh International Conference on Virtual Systems and Multimedia. Enhanced Realities: Augmented and Unplugged, 25-27 October 2001:77-85.

GREGORY, R.L. Eye and Brain: The Psychology of Seeing. Princeton University Press.

LEVY, P., 1997. Becoming Virtual: Reality in the Digital Age. Plenum, New York.

A Concept for 3D Mapping with Augmented Reality Technologies

SEE THE CD FOR THE EXTENDED VERSION

K. GRETHER
ARCTRON 3D-VERMESSUNG & SOFTWAREENTWICKLUNG GMBH
RINGSTRAßE 8, D-93177 ALTENTHANN
TEL.: +49 9408 85010, FAX: +49 9408 850121
INFO@ARCTRON.DE, HTTP://WWW.ARCTRON.DE

ABSTRACT

Augmented Reality (AR) is a new form of human/computer interaction (HCI), whereby computer data is superimposed onto real life photographs and all kinds of information can be displayed to the operator depending on the content.
In this article, a possible application of AR technology will be introduced using the example of the damage mapping carried out on the monumental sculpture "Bavaria" in Munich. Techniques such as AR have great potential in this area, especially when combined with a complete, laser scan recording of the 3D geometry of the object. Using this technique, the virtual object, whose precision depends on the quantity of measured points, can be superimposed with a 1:1 scale 3D photograph. Damage mapping can be carried out on the virtual object and on computer simultaneously, and then converted into a corresponding damage map. In this way, effective documentation, visualisation and structured data records are created, which can be managed in a content management system (CMS). The application is supported by 3D user interfaces, information filtering and automatic integration mechanisms.
We at ArcTron GmbH, are currently still in the development stage of this system. In the future, it will provide a 3D information system, specifically conceived for 3D documentation in archaeology and monument heritage at a variety of different levels.

Initial Proposal

The requirements for computer-supported, scientific classification of archaeological excavation features are computer-supported recordings and the integration of work carried out by various specialists in an information and communication system.

Existing, potential IT solutions for mastering this task have only been inadequately exploited until now. At ArcTron we are working on methods and techniques of 3D mapping and documentation. Our aim is to develop a total concept using existing object surveys, which will enable all recorded data, geometries and other documented descriptions to be efficiently collected, managed and represented.

Fields of Application in 3D Mapping

3D mapping is the structured collection and management of information in an information system. During data collection, the data is reduced down to that which will later be of use. The most important data comes from the recording of a point cloud and the wire mesh model generated from this. Other factual and general data is also collected (photographs, video sequences, sketches etc.).

All existing recording techniques (e.g. laser and stripe light scanning, total station surveying, photogrammetry, hand drafting, photography etc.) have certain technical merits. If the task formulation is adequate, these various advantages can be combined to produce the best-possible, most efficient result.

On the other hand, no single method alone can meet the ever-changing, diverse demands made by object recording. It is essential to combine available methods and technologies in order to avoid inefficiency.

Damage mapping	Archaeology/ Stratigraphy
- Segmentation of the object - Surveying damage - Documenting damage	- Finds/ Feature mapping - Finds/ Feature classification

Table 1 Main focuses of 3D mapping

The aim of our research is to examine how far 3D mapping can be realised using AR/VR technologies. Other main focuses of our research are:

- The evaluation of existing recording methods
- Assistance for recording and recording instruments/ mobile data collection
- Integration in a total system (CMS/ ArchaeoData/ Aspect3D)
- Classification of processed data (Organisation system)
- Modifying and adding information (context relative infor mation)
- Processing and monitoring data, attaching information to the 3D model (Ranze 1999)
- Structured storage of recorded data (Database)
- A modular tool for recording and logging the work stages
- Displaying information ("context awareness") (Rose 1995)
- Filtering information (Julier 2000)
- User friendly Power Management AR technology has proved itself to be useful in the following situations:
- Supporting communication and provision of filtered information (Julier 2000)

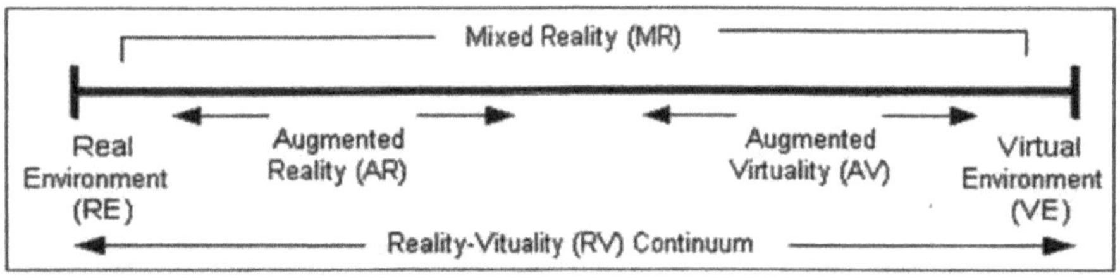

Figure 1 Definition of Augmented Reality (AR)

- Displaying information that is not directly accessible/visible (clearness)
- Direct, in situ collection and recording of object information (illustration)
- Generating and processing new information
- Structured data storage in a content management system (CMS)

AR & VR AND THE MAPPING PROCESS

[1] Technology

(Fig.1) An AR system generates a composite view for the user. In order to be able to see the view, it is necessary to use display equipment known as a Head Mounted Display (HMD). It displays a combination of a real image and a computer-generated image and enhances the picture with additional information. The superimposed image is continuously recalculated depending on the users' head movements.

1) Combines real and virtual space
2) Interactive in real time
3) Registered in 3-D

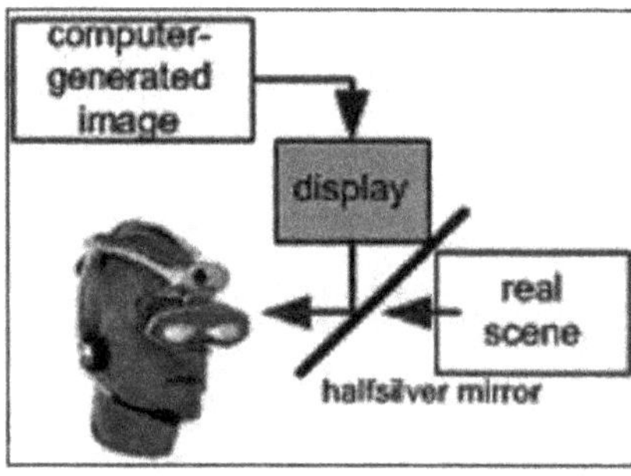

Figure 2

[2] See-through Head Mounted Display (st-HMD)

There are two different see-through HMD'S used in AR and VR systems, the optical see-through HMD (ost-HMD) and the video see-through HMD (vst-HMD).

The optical see-through HMD (Fig.2) combines a real and a computer-generated image by using an optical mechanism with a half-silvered mirror. In this way, the surroundings can still be directly observed. The Video see-through HMD (Fig.3) records and displays the surroundings to the user using a video camera. The software (Video composer) ensures that the video of the surroundings and the computer-generated images (constructed using a Scene generator) coincide.

[3] Tracking Systems

The position and orientation of the user's head must be known so that virtual objects can be correctly placed. Tracking is the method of measuring the position of bodies moving in 3D space. When location (3 positional coordinates) and orientation (3 independent angle coordinates) are measured at the same time, we call it a 6 Degrees of Freedom measurement (6dof).

It is important for the system to be able to react to the movements of the user in real time and deliver the calculated images without delay. The time lag is calculated from the time that elapses between the movement of an object and it is realisation through the computer.

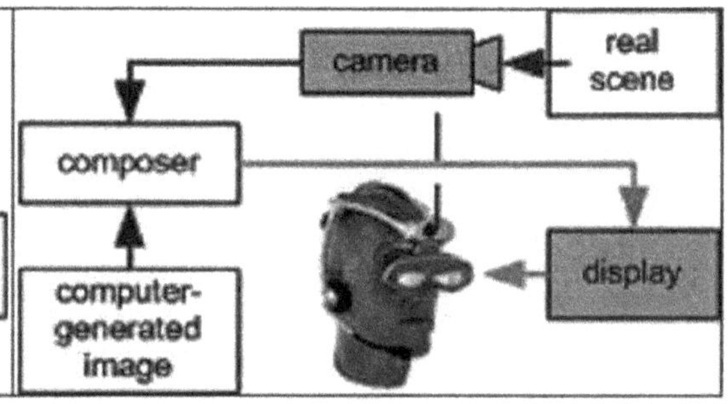

Figure 3

MEASUREMENT DATA COLLECTION IN THE FIELD

Basically, there are two approaches to data collection in the field:

[1] Examination of existing plan documents

Existing 2D drawings or 3D models form the basis of the documentation that can be examined, especially the geometry. The user is shown these in a see-through display.

The concept of using a combination of existing recording technology (laser scanners, stripe light scanners, total stations, photogrammetry, hand drafting) is an allusion to work carried out on a prototype system called CyliCon that was developed by Siemens (Navab 1999, 2000).

Figure 4

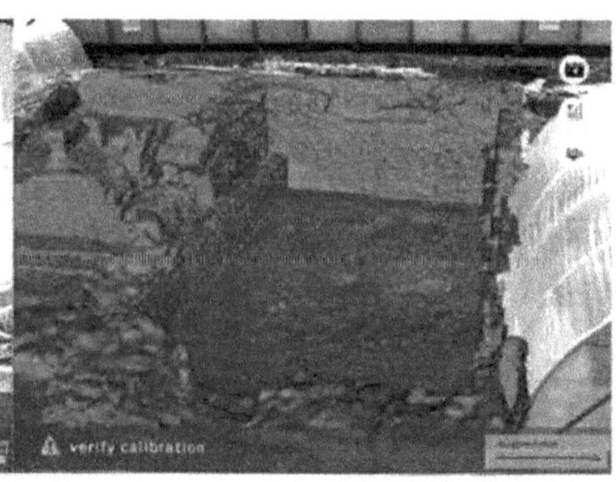

Figure 5

Furthermore, the focus of this work lies in the direct integration of data into the generated geometric model and it's storage in a Content Management System.

[2] Collecting measurement data

Stripe light scanners are employed for high-accuracy, 3D digitisation (to a tenth of a millimetre) and reconstruction as well as for documentation in the field of heritage (Schaich 2000). The plausibility of measurement data is checked by overlaying the generated model on the real object. The model can be visually compared to the real object in situ. The individual systems and the combination of these have not been extensively tested. In order for the systems to be constructively employed in our ever-changing working conditions, the interaction of the components must be examined and adjusted for each case for the following reasons:

- Every object is unique
- Operational possibilities outside of laboratory conditions
- Environmental influences
- Minimum calibration input

Figure 6

Figure 7

3D Mapping Scenarios

[1] Scenario 1 - Plausibility check

The system supports the user by displaying the data that has already been recorded (Fig.4-5). The plausibility of the data can be checked the same second that it is recorded and the data can then be linked to a suitable context. Additional information can be inserted at the correct position on the object.

- Tracing areas of damage with a finger (Fingertracking MIT)

- Recording visual details (digital photography/ damage documentation)
- Refining the mesh in order to document the exact positions of damage
- Voice memos for making notes and determining damage classes
- Virtual surveying tool for determining the amount of damage
- Arranging the objects in the 3D virtual illustration

[2] Scenario 2 -Excavation documentation

Recordings taken using an Laser scanner (LS system can produce exact documentation of an excavation. Automated integration mechanisms assist with the structured storage of information in the database and it's integration in the CMS (Fig.5).

- Finds, finds complexes etc. can be recorded on site using digital finds labels.
- Finds locations are directly incorporated into the 3D information system using finds symbols.
- Features can be described, classified and digitally recorded on site.
- Visual representation of the excavation is possible at any time.

[3] Scenario 3 - Magic Book Metaphor - Documentation

The "Magic Book" (ARToolkit) (Azuma 2001) metaphor is a new form of interpretation and leads to a new understanding of damage mapping. Using markers on the printed pages of documentation, the object in question or a section of the excavation can be viewed in 3D either via a web cam on a monitor, controlling a virtual camera in a special documentation application (Fig.6), or using a st-HMD for viewing the object directly on the documentation page (Fig.7).

- Visual monitoring of the section in question

Figure 8

Figure 9

- Easy access to the corresponding source data

FURTHER FIELDS OF APPLICATION

AR/VR as a resource for historical contexts

One of the best prospects for the use of AR/VR is offered by the constant trueness-to-scale (see Archeoguide (Fig.8-9), Reconstruction of historical buildings (Höllerer 1999, Stricker 2001). At a very early work stage, the user can view realistic images of the historical situation without the mental burden of the compilation procedure, as is the case when using conventional two dimensional graphic or screen based three dimensional representations.

CONCLUSION AND FUTURE WORKS

Future research will increasingly concentrate on system specification and prototypical implementations as well as the further development of data management and the database. AR offers us the opportunity to reduce the duration of the working process and to link information to the correct locations. An improvement of the information technology infrastructure would also reduce the cost of projects and maintenance and optimise the running of projects. The developments in this area correspond to the general situation, which is characterised by:

- a continuous increase inthe complexity of projects,
- optimising and improving the efficiency of the conception and planning processes,
- increasing specialisation of applications and among professional planners
- globalisation of the market (virtual organisations).

Thanks to the growing distribution and development of the basic technology, future projects may include the following:

- Integration of superimposition technology for quality assurance (data accuracy).
- Mapping instructions for complete, integrated and uniform documentation.
- Devolution of existing data (for future scientific evaluation, restoration).
- Transfer of captured data into superior information systems, e.g. GIS/WCS and georeferenced data.

REFERENCES

ARTOOLKIT (http://www.hitl.washington.edu/research/shared_space/, February 2003).

AZUMA, R., BAILLOT, Y. et al., 2001. Recent Advances in Augmented Reality. IEEE Computer Graphics and Applications 2001:34-47.

AZUMA, R., 1997. A Survey of Augmented Reality. Presence: Teleoperators and Virtual Environments Vol. 6, no. 4, August:355-385.

HÖLLERER, T. et al.,1999. Exploring MARS: Developing Indoor and Outdoor User Interfaces to a Mobile Augmented Reality System. Computers and Graphics Vol. 23, no. 6, December.

JULIER, S., LANZAGORTA, M. et al., 2000. Information Filtering for Mobile Augmented Reality. Proceedings of IEEE and ACM ISAR'2000, Munich, Germany, 5-6 October 2000.
MILGRAM, P., Colquhoun, H. Jr., 1999. A Taxonomy of Real and Virtual World Display Integration. Proceeding of ISMR, Yokohama, Japan.

NAVAB, N., CUBILLO et al., 1999. CyliCon: a software platform for the creation and update of virtual factories. Proceedings of the 7th IEEE Int'l. Conference on Emerging Technologies and Factory Automation, ETFA '99, Barcelona, Spain, Vol.1:459-63.

NAVAB, N., GENC, Y. and APPEL, M., 2000. Lines in one orthographic and two perspective views. Proceedings of the IEEE Int'l. Conf. on Computer Vision & Pattern Recognition, CVPR 2000, June 2000, IEEE Computing Society Press, Vol. 2:607-14.

RANZE, K.C., 1999. Real Component Help: a mobile multimedia help system (http://www.tzi.de/, 23 April 2001).

ROSE, E., BREEN, D. et al., 1995. Annotating Real World Objects Using Augmented Reality. Proceedings of Computer Graphics International ´95, Leeds, UK, June 1995.

SCHAICH, M., 2000. Computergestützte Grabungsdokumentation im Netzwerk aus elektronischer Vermessung, Fotogrammetrie, CAD und Datenbanken. In Harl, O. (ed.), Proceedings Workshop - Archäologie und Computer (CD), Vienna.

STRIKER, D. et al., 2001. Design and Development Issues for ARCHEOGUIDE: An Augmented Reality-based Cultural Heritage On-site Guide. EuroImage ICAV 3D Conference in Augmented Virtual Environments and Three-dimensional Imaging, Mykonos, Greece, 30 May - 1 June 2001.

The Aksum Project: A VR GIS for a 3D Inclusive Interaction with an Archaeological Landscape

Bonfigli M.E.
CINECA, via Manganelli 6 b, Casalecchio di Reno (BO), Italy
e.bonfigli@cineca.it

Forte M.
CNR ITABC, via Salaria Km 29,300, C.P.10, Monterotondo St. (Rome), Italy
maurizio.forte@mlib.cnr.it

Guidazzoli A.
1CINECA, via Manganelli 6 b, Casalecchio di Reno (BO), Italy
a.guidazzoli@cineca.it

Pescarin S.
CNR ITABC, via Salaria Km 29,300, C.P.10, Monterotondo St. (Rome), Italy
sofia.pescarin@mlib.cnr.it

Zane M.
Modena and Reggio Emilia University, Faculty of Engineering, Italy

Abstract

The Aksum Project is a joint research project between Boston University, IUO (Istituto Universitario Orientale of Naples, Italy), CNR-ITABC (Istituto per le Tecnologie Applicate ai Beni Culturali, of Rome, Italy) and CINECA Supercomputing Center (Visit Lab, of Bologna, Italy) supported by Agenzia 2000 CNR, aimed to the virtual reconstruction of the archaeological landscape throughout paleo-environmental data, GIS and remote sensing applications. The project started in 2001 and we expect to complete the first phase of software development within 2003.

In this context, after a long and complex archaeological fieldwork activity, a VR system consisting in the 3D inclusive interaction with the archaeological landscape has been created according to a multiplatform approach. In fact the application is developed for desktop virtual reality systems (inexpensive PCs) and for Virtual Theatres based on Unix-SGI systems. The application is planned using Java instead of C++, the language used for the main application. The choice of the Java language has been made for two reasons: 1) the perfect portability of Java application across different platforms, which represents a great help for the development of a multiplatform application, and 2) the availability of a powerful and complete library for the creation of GUIs.

We can define the final result a first fundamental step for the realization of a VR GIS with a specific interface dedicated to all the 3D data: layers, themes, movies, archives, DEMs, maps, etc, everything connected in a 3D space. The faculty to query and to interact the system in 3D opens new perspectives for the next generation of VR applications oriented towards the virtual reconstruction of archaeological landscapes or, better, "mindscapes"(perceptual condition to interpret the anthropic landscape).

Introduction

The Aksum Project is a joint research project between Boston University, IUO (Istituto Universitario Orientale of Naples), CINECA (Supercomputing Centre, Bologna), and CNR-ITABC (Istituto per le Tecnologie Applicate ai Beni Culturali, of Rome), supported by Agenzia 2000 CNR, aimed to the virtual reconstruction of the archaeological landscape throughout paleo-environmental data, GIS and remote sensing applications. Based on a very detailed mapping and 3D archives of graphic libraries (vegetation, buildings, villages, animals, effects etc.), the GIS project represents a very articulated context of information for testing the possibility to implement a post-GIS phase, that is the creation of a VR spatial system for archaeological landscapes.

In the field of the landscape archaeology the GIS can be considered the fundamental basis for VR applications in the context of the virtual reconstruction but it cannot satisfy all the conditions of a perceptual reconstruction: the perception of landscapes through the ancient minds, in other terms, a mindscape (Forte and Williams, in print). With this methodological approach we would like to discuss the potentialities of VR in the domain of geographic and ecological data; in fact, real time applications multiply the interpretative faculties thanks to the inclusive interaction within a 3D space. For this aim we have developed a multiplatform system, running on PC and on a Virtual Theatre and integrating Java language with C++. For the final users it will be very important to use the same graphic interface on both platforms and sharing the same spatial archives, such as metadata (html, xml, alphanumeric databases and so on).

Overview

Many commercial GIS include (i.e. ArcView, ArcInfo) in general a 3D extension, namely a 3D interface for visualising multilayered surfaces (vector and raster) but no one is aimed for an articulated 3D interaction with all the archives of spatial data (bi-dimensional and three-dimensional).

In the state of art of GIS applications, typically, the three-dimensional components are considered the basis only for a visualisation process and not for complex analyses and queries. So the "3D" in this context is, more or less, a scenic effect and not a real increased value. The main idea of this work is to project a VR geographic system, starting from a virtual reality application, namely, beginning "from the end", from the inclusive interaction in real time and not using a standard 2D interface. The challenge is to demonstrate that a VR GIS, based on desktop virtual reality systems, or, better, a VR interface for an interactive accessibility to a virtual landscape, multiplies the faculty to perceive the spatial data with a user-friendly interface: maybe it could be defined a perceptual GIS, more than a simple 2D or a 3D visual GIS. The faculty to query and to access to connective spatial information in 3D and in real time represents a new digital perspective for an advanced interpretation of cultural and natural landscapes (Raper 1989).

Figure 1 (left) Map and location of Aksum and a photo of the actual landscape with the typical Stele. (right) Remoted sensed image: Spot XS satellite image with GIS data overlaid

Then, the potentialities of this kind of system are relevant or in terms of scientific interpretation/visualisation, whether in terms of communicative/didactic aspects, for example in the case of a museum or a cultural exhibition. For example a final user could interact with the landscape with a "simple" virtual navigation or he could query the landscape within a 3D visualisation, facing up the problems of ecological components and of metadata.

THE AKSUM PROJECT

[1] Aksum project

The primary goal of the project is the reconstruction of the archaeological landscape of Aksum, in Ethiopia, on the basis of ethno-anthropological, paleo-environmental, GIS, remote sensing and spatial data and with a spatial - virtual - remote sensed and ecological approach (Bard, Coltorti, DiBlasi, Dramis and Fattovich 2000:65-86).

The amount of archaeological data regarding Aksum is quite relevant, because of the several investigations undertaken, in the past and in the last years, by IUO (Istituto Universitario Orientale, Naples), University of Padua (Department of Archaeology) and CNR ITABC (Rome). The whole project indeed includes systematic researches and excavations, in multidisciplinary fields such as paleobotany, archaeozoology, etnoarchaeology, ethnography, anthropology, geology, geomorphology, palinology, finally GIS, remote sensing and, finally, also virtual reality (Littmann, Krencker and von Lüpke 1913, Fattovich and Bard 1993:41-71, Phillipson 1998, Bard et al. 2003).

In the case of Aksum there was, first of all, a big problem with the cartographic base, since there was not any adequate map (just one map 1:25,000). Just a survey in Ethiopia, with DGPS equipment allowed to take enough points in order to build a detailed Digital Elevation Model. And then, thanks to Remote Sensing analysis, connected with data that come from fieldworks, it was possible:

- to underline a connections between soils, terrains and sites;
- to study the vegetation, calculating a vegetation index based on satellite imagery and some aerial photos of the 60's (this operation has shown the extension of eucalyptus's forests that are progressively hiding, and in some cases destroying, the visibility of the archaeological landscape);
- to study the distribution of the archaeological sites in the area;
- to obtain an archaeological interpretation and reconstruction of Aksum and, in particular, of the area of Bieta Giyorgis. For instance a remarkable interpretation comes from these remote sensing applications: an ancient path (dating from Pre Aksumite period) crossing the hill from the East is visible from the SPOT XS; it passes very close to a stone monolith located just in the middle of the hill. Our ipothesis is that this monolith-stele, that is not linked with a tomb, is to be interpreted as a boundary-landmark. Therefore two principal archaeological regions or districts can be distinguished, in the North and in the South, index of a relevant and organized power landscape.

The new methodological approach adopted, that put together GIS, Remote Sensing and Virtual Reality applications, indicates the relevance of this international archaeological project. With the co-operation of CNR ITABC (Rome, Italy), indeed, it was possible to specifically addressed the researches even towards a Virtual Reality spatial interaction to GIS, Remote Sensed data and to data acquired from fieldworks. The main aim, indeed, was to maintain georeferenced information in VR application. A first attempt was done building a Desktop Virtual Reality (DVR) software and interface, written in C++ , using Terravista and Vtree (thanks to a collaboration with Terrex Inc. and Tesscom). This application was presented to the public during an Exhibit on the History of cartography organized in Rome in 2002 (Fig.2).

Further developments were done in co-operation with CINECA Supercomputer Centre, in order to create a fully interactive and cross-platform application, that could be used in a Virtual Theatre, but also in a PC with 3D Graphic Card.

THE AKSUM VIRTUAL ARCHAEOLOGICAL LANDSCAPE

[1] User Requirements

The Aksum Virtual Landscape application was developed into two distinct versions: the first - the semi-immersive version - was designed to work in the Virtual Theatre at CINECA, the second - the desktop VR version - was designed for common personal computers. The two versions share the same functionalities, and they differ only for the way they are used and fruited. The semi-immersion version is more suitable for the exploration and the navigation of the scene, thanks to the more realistic rendering of the 3D view and the great sense of presence allowed by the Virtual Theatre. The desktop VR version is more suitable to enjoy the multimedia data and for object manipulation.

Figure 2 Vision Station and the DVR system written in C++ using VTree and Terravista

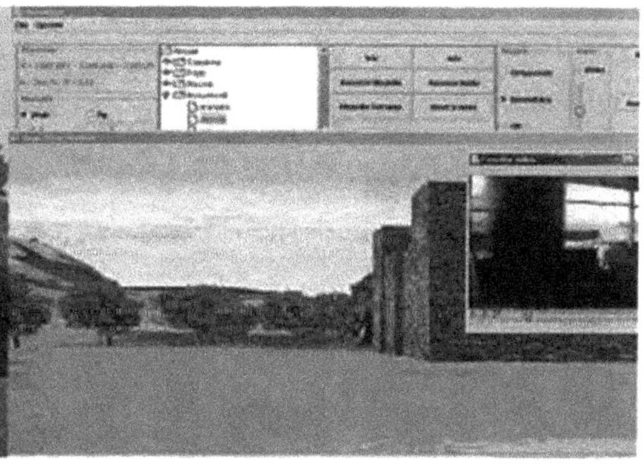

Figure 3 Multimedia data linked to 3D objects

The navigation and interaction techniques were designed in order to be user friendly, considering as a target for this application users without a specific experience and knowledge on VR applications. The design of the user interface is most critical for the semi-immersive version, in order to avoid the user to loose the sense of presence due to the a bad designed and hard-to-use interaction method.

To add information to the 3D reconstruction, multimedia data such as HTML pages and short movies realized by archaeologists were linked to the objects in the scene. In this way more data are provided with the visual reconstruction and the user can compare the virtual environment with the pictures and movies taken from the real world.

[2] Features and Implementation

Object Classification

The objects in the scene are organized into groups, (i.e. classes), in order to simplify their selection, search and manipulation and to link them with multimedia data and with the 3D models (properties). Each class represents a collection of objects which are quite identical (e.g. the Fences class) or share similar features (e.g. the Buildings class).

The properties can refer to the single objects and to the whole class: when a property is not defined for a particular object, it is inherited from the class the object belongs to. This way both classes of generic objects, which do not define different properties for each instance, and classes of more specified objects, with different properties for each element, can be handled. For instance, each member of the Buildings class has its own properties and model, while members of Fences class are not defined with specific information, and simply share the properties of the whole class.

When the application starts, classes are created and properties are defined reading data from a set of configuration files. These are short text files which define the properties of objects and classes using a simple syntax. Properties can be modified run-time using the commands disposed on the graphical interface and the changes made can be saved to the configuration files for further use of the application.

The Graphical Interface

The graphical user interface (GUI) carries out three tasks: allows the user to easily access the system control parameters, offers a method for selecting objects by name from a list that groups them by classes, integrates simple HTML browser and media player for the visualisation of multimedia data.

The GUI was developed as a little client application that sends commands for the main application, which behaves like a visualisation server. The only operations that are performed completely within the GUI code are the ones that concern with the visualisation of multimedia data.

The GUI has been developed using Java instead of C++, the language used for the main application. The choice of the Java language has been made for two reasons: 1) the perfect portability of Java application across different platforms, which represents a great help for the development of a multiplatform application, and 2) the availability of a powerful and complete library for the creation of GUIs (through the classes provided by Swing and AWT packages), which allows the design of graphical user interfaces that can run on different platforms without the need of the use of specific APIs.
The interaction between the main application, written using the C++ language, and the Java GUI has been obtained using the functions provided by JNI library. The Java Native Interface allows to insert fragments of platform-dependent native code into Java. The JNI library provides also the functions for embedding a Java Virtual Machine into native applications and for accessing Java classes and their methods from native code.

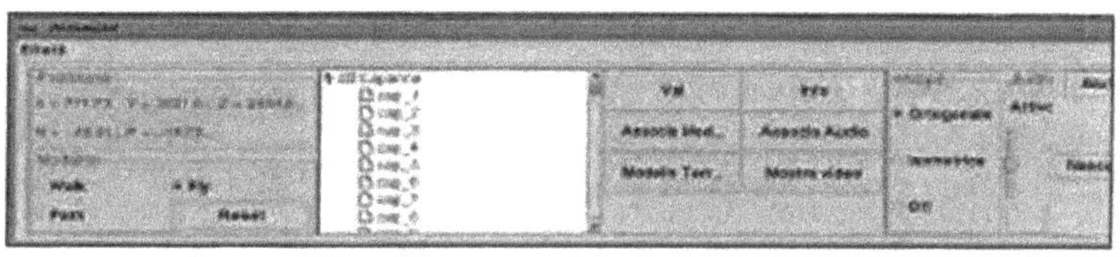

Figure 4 Graphical Interface

Conclusions

The Aksum Project represents a very interesting case study including very articulated series of spatial data coming from different sources of information (Forte and Williams, in print), bi-dimensional and three-dimensional. We have obtained a first VR reconstruction using GIS and remote sensing platform (ArcView, Er Mapper) and rendering all the models using Terravista 3.0. In this first process a real time visual navigation (without behaviours) has been produced basing the graphic performance on OpenGl technologies. In the second phase of work a specific graphic interface has been implemented in C++ and in Java in order to include different possibility to query the landscape according to the type of 3D spatial data: terrains, remote sensed data, movies, architectural/archaeological structures, paths, roads, thematic layers and navigation modes (walk, fly). In this way a preliminary VR and georeferenced informative system has been realized with all the links and the connections available in 3D. It is even important to remark that the application is multiplatform, namely is developed with the same GUI for immersive systems, such as virtual theatres, and for desktop computers, keeping the same graphic performances and connections. Additionally, a virtual timeline will be developed on the basis of the evolution of the archaeological-anthropological landscape.

Finally, on the basis of this approach, we could say that a new era of 3D VR systems for interaction and advanced visualisation of archaeological landscapes is in progress, integrating all the spatial properties of spatial maps and objects in a perceptual "place".

Acknowledgements

In the Aksum project many teams and researchers have collaborated and are still collaborating. A special thank is due to: R. Fattovich, S. Kay, A. Manzo, C. Perlingieri, R. Perlingieri, B. Trabassi and, as Institutions, to CNR and CINECA. For the software and hardware support we thank the companies Terrex, Tesscom and Elumens.

References

BARD, K.A. et al., 2003, in press. The Joint Archaeological Project at Bieta Giyorgis (Aksum, Ethiopia) of the Istituto Universitario Orientale, Naples (Italy), and Boston University, Boston (USA). In Proceedings of Boston Workshop: Results, Research Procedures and Preliminary Computer Application.

BARD, K.A., COLTORTI, M., DIBLASI, M., DRAMIS, F. and FATTOVICH, R., 2000. The Environmental History of Tigray (Northern Ethiopia) during the Holocene: A Preliminary Outline. The African Archaeological Review 17(2):65-86.

FATTOVICH, R. and BARD, K.A., 1993. Scavi archeologici nella zona di Aksum. C. Ona Enda Aboi Zaguè (Bieta Giyorgis), Rassegna di Studi Etiopici, 35:41-71.

FORTE and WILLIAMS, 2003, in press. The Reconstruction of Archaeological Landscapes through Digital Technologies. Proceedings of Boston Workshop, Boston, Massachusetts, USA, 1-3 November 2001.

LITTMANN, E., KRENCKER, D. and VON LÜPKE, T., 1913. Deutsche-Aksum Expedition I-IV, Druck und Verlag von Georg Reimer, Berlin.

PHILLIPSON, D.W., 1998. Ancient Ethiopia - Aksum: Its Antecedents and Successors. British Museum Press, London.

RAPER, J., 1989. Multidimensional GIS - Extending GIS In Space And Time. Taylor and Francis Ed.

The Citadel of Nimrud, Iraq: A Virtual Reality Interactive Model as a Resource for World Heritage Preservation[1]

Samuel M. Paley
The University at Buffalo
State University of New York

Donald H. Sanders
Learning Site, Inc.
Williamstown, Massachusetts

See the CD for the extended version

Abstract

A virtual reality reconstruction of the citadel at Nimrud (ancient Calah), Iraq, originally constructed by the late Assyrian King, Ashurnasirpal II (9th c. BC), with his palace and has now expanded to the Central Palace of Tiglath-pileser III (8th c. BC). Excavated by A. H. Layard 153 years ago, it continues to be researched by the British, Italians, Iraqis and Poles. Evidence of its ancient occupation is detailed enough to embark on a full-scale interactive VR model, working across the citadel, incorporating the evidence from all excavations. The models are being developed in a PC version at Learning Sites, Inc. (Williamstown, Massachusetts: http://www.learningsites.com) and a supercomputer model for Immersadesk™ and CAVE™ at the Center for Computational Research (University at Buffalo: http://www.classics.buffalo.edu/projects). The latter will be available for distance learning across Internet II. The finished models will be excellent tools that can be used to support claims for Cultural Heritage: monuments from Nimrud have been dispersed to dozens of museums and private collections in the last 150 years, and have been robbed further in recent decades.

Since 1998, I (S.M. Paley) have been the principal investigator of a virtual reality reconstruction of the citadel of Nimrud (ancient Kalhu/Calah), Iraq at the University at Buffalo's Virtual Reality Laboratory, Department of Mechanical and Aerospace Engineering and at Learning Sites Inc. in Williamstown, Massachusetts. UB and LSInc. are collaborating in the preparation of interactive VR models, together with a portfolio of 3D images for scholarly publication and education and public interest. Models are being built that can be used for distance learning among institutions with Immersadesk™ and CAVE™ technology and a PC version, for individual and classroom use.[2]

Ancient Nimrud is east of the present course of the Tigris River and north of the Tigris' intersection with the Greater Zab River. King Ashurnasirpal II (883-859 BC) founded its citadel on a ridge near the edge of the Tigris flood plain. The rest of the city was laid out to the east and north. Today there are rolling farmlands, partly in the flood plain and partly to the east. When the Tigris is within its banks, the land is fed by rain and canal water. The king planned palaces and temples on the site but, in his lifetime, never finished all that he planned. Successors added their own royal and religious buildings, rebuilding and re-designing the citadel as needed.

The evidence of ancient occupation on the site is detailed enough for us build an interactive VR model. Because of the deteriorating state of preservation and looting of the archaeological site as a result of sanctions and the second and third Gulf wars in 1991-2 and 2003, we think that one of the main resources for the preservation of the memory of Nimrud is a VR reconstruction with proper scholarly documentation available on the World Wide Web. Digital records and VR can both document the state of preservation and conservation at different periods of the site's modern history and the recovery of stolen objects from it. They can contribute to the future history of the site by creating a resource for study of the history and culture of this ancient Assyrian capital as well as help plan for the reinstitution of the site as a museum so that scholars could visit and continue excavation and tourists can visit. All this can be made possible by publishing all material in one digital place. Iraqi permission for this plan has been in part made possible by the inclusion of the site of Nimrud, with Nineveh, in 2002, on the World Monuments Watch's list of 100 most endangered world historical sites. A VR model for study with appropriate documentation would be a useful introduction, especially to such a large site: 200 hectares. The finished VR model and documentation will be used to support claims for cultural heritage preservation and conservation.

We began the project with Ashurnasirpal's palace, one of the best-preserved monuments on the site. We have over 150 years of various kinds of documentation for this building. We are expanding to include the temple area north of the palace and the area south of it, where two other buildings, one of Ashurnasirpal and one of his son, Shalmaneser III (late 9th c. BC), were found, and where the northern part of the very much destroyed palace of King Tiglath-pileser III (8th c. BC) was located. The web pages are organized to include the VR models with the reasoning behind the model's structures, the history of the excavations, and photographic documentation of the excavation, conservation, and the present state of preservation.

Nimrud is also a site from which, for over the last 150 or more years, cultural heritage has been dispersed to more than 75 museums and private collections across the world, not including what survives in the museums of Iraq, the site museum at Nimrud itself, and any stolen pieces, whose whereabouts are unknown. Digital documentation is one rather inexpensive way to document the site as it now exists and as we know it from excavation records, collect the material that has been taken from the site over time and bring everything into one virtual place so that the citadel's monuments can be studied as close to the "whole" as discovered by the archaeologists who excavated it. Also, broken monuments can be restored digitally to approximate the way they were found and with good scholarly surmise even reconstruct their original state. In the event of a solution to the present situation in Iraq, this information can help in the restoration and conservation of what survives and even become part of the exposition of the monument at the local site museum when tourists return. For people who would never have the chance to visit the site or never could, a virtual model would present to them as near a real experience as they would ever have. And students would have a rich resource to use in learning about the archaeology of Assyria.

Archaeological reconstruction is needed for nearly every building on the site. Just figuring out the physical relationships among the rooms of the Great Northern Courtyard of Ashurnasirpal's palace, the city and citadel wall and the temple area on the northwestern corner of the site is full of problems as the excavation publications differ in their various iterations. Talk about the need for good data. Scholarly surmise about the various reconstructions present us with formidable choices and difficult decisions to be made among them. The plan of a building may differ; the place of a road or street may alter. And yet we must make some decisions or our project cannot go on and its benefits, with all the caveats, will not be realized.

Figure 1 (Paley01_LSInc_Nimrud_massing-test_032703.jpg) Massing model of part of the Nimrud citadel, created by Learning Sites for study of the relationships among the many buildings at the northern end of the site; image copyright 2003 by Learning Sites, Inc; reproduced with permission

Looking at a model (Fig.1) of the northwestern corner of the Nimrud citadel, we are again reminded of the problems of reconciling the data, as the widths of the city wall and the heights of the palace floor paving, temple tower, temple floor paving and palace and temple walls become relevant. We must be and are sensitized to the problems of reconstructing an archaeological site in VR and not only because we must, from time to time, make informed guesses. We agree with other scholars that we have only a part of the story and that we are making our own interpretations. And we must also deal with and be cognizant of what has been said many times, recently by Alexander Stille, about the field of archaeology and ancient studies that "...our vision of the 'past' is heavily conditioned and distorted by what monuments happen to have survived the ravages of time."

If anything, digital archaeology and VR reconstruction can provide the means to collect and study vast amounts of data and organize material that can be retrieved and presented in ways that provide us with new possibilities to see, interpret and present. As information is added and opinions change, the VR model can be easily corrected and new data added to the database. New ways of seeing and interpreting the relationship between architecture and design in the Northwest Palace come as a result of customized research made available because it is possible to "walk" around the VR model quite freely. Standing outside doorways (Fig.2) and looking into the various rooms of the palace or stopping in doorways and looking around, or standing where the king or his courtiers or the representatives of foreign countries might have stood during the various ceremonies associated with the audience halls in the various wings of the official part of the palace, can elucidate the ways the architect/builder-contractor of the palace and the designer of the bas-relief must have worked together to arrive at their creative decisions. A room had to be so wide, the sculptures on the bas-relief arranged on the stone slabs in specific positions, the doorways so deep. It stands to reason that the position of the central doorway, which has been reconstructed from scholarly surmise and some evidence on the ground - mud brick traces, fragments of the doorway figures, parts of the threshold construction - and its relationship to the position of the images of the king across from it, behind the throne and in adjacent rooms, was at least in part the focus of the design of the whole throne room complex. These choices of place must have been part of the decisions made by artist and architect/builder as part of the planning, when they made other choices like that of the nature of the image of the king, which we know were presented for royal approval. It is implied in the claims by the king that he built the palaces. This sort of research is made easier and can be tested with VR.

Of course, the possibilities of new insights into the meaning of the palaces and the temples in the course of the reconstruction of the citadel in VR cannot be predicted. However, preservation though digital documentation of the existing remains can not only gives us a picture of the current state of the ruins, but can also document all known phases in the building's history, and thus can be a major resource for new research. Maybe the only way to correlate the vast amounts of data in meaningful ways for retrieval, analysis, and teaching, is digitally, allowing customized research or collaboration on the web among scholars at different web portals in immersive virtual re-creations of the site. There are also traditional distance education possibilities and certainly localized classroom uses. In the virtual environment, a student or scholar could do things like take measurements or change or manipulate the world for personal research, just as if one was at the site in different phases of its occupation and deterioration. Exposition in both conventional and site museum settings is possible. Links between various VR methods and the heritage preservation/documentation of the

Figure 2 (Paley02_LSInc_NWP_Entry-ED-to-B13.jpg) Rendering from the Learning Sites virtual reality model of the Northwest Palace, Nimrud, showing the view from the Great Northern Courtyard through the main central doorway into the Throne Room, toward Relief B-13; image copyright 2002 Learning Sites, Inc; reproduced with permission

Stille's recent work on the presentation of the past is fair warning of the many dangers to archaeological sites. Speaking of Egypt, he writes:

We are faced with a dramatic paradox: soon we may have virtual realities of the tombs of the Valley of the Kings in classrooms all around the world-a genuine democratic advance-but the tombs themselves, most of which looked almost new when they were discovered in the late nineteenth and early twentieth century, may no longer exist.

We hope of course that VR will not be the only way to remember Nimrud place that allow different kinds of investigation and learning opportunities are limitless. Perhaps this "brilliant new technology", as Stille has called it, will enhance the study of antiquity and make us more informed than we ever have been before.

There are many reasons why a project like this is important for world heritage at the present moment in our history. As I stated a moment ago, because of the political and military situation in Iraq today, it is nearly impossible to maintain - both police and conserve - all the archaeological sites in Iraq sufficiently well to preserve very many of them. Alexander

[1] A full bibliography can be found in the CD-ROM version of this paper.

[2] There are several colleagues involved in this project under my direction: Dr. Donald Sanders, President of Learning Sites, Inc.; Richard Sobolewski, architect and "Second Director" of the Polish excavations at Nimrud from 1974-76; Professor Thenkurussi Kesavadas, Director of the University at Buffalo Virtual Reality Lab and his graduate assistant, Mr. Youngseok Kim; Professor Alison Snyder, Department of Architecture, University of Oregon; Professor Stuart Shapiro. Department of Computer Engineering, University at Buffalo, a specialist in natural language interface; Dr. Julian Reade, British Museum (ret.) for the painted decorations that accompanied the bas-relief; and Dr. Elizabeth Hendrix, Materials Research Laboratory at the Massachusetts Institute of Technology, a specialist on paint and painted stone. For this paper Mr. Mark Altaweel, a graduate student in Archaeology at the Oriental Institute of the University of Chicago, has provided GIS data for the preparation of the terrain maps around Nimrud, rendered by Learning Sites. We hope that we can add to the list of contributors as funding grows.

REFERENCES

KIM, Y., 2001. A Real-Time Immersive Virtual Environment for an Archaeological Site: The Northwest Palace of Ashurnasir-pal II (883-859 B.C.). Thesis in partial fulfillment of the Master of Science, Department of Mechanical and Aerospace Engineering, The University at Buffalo.

SANDERS, D.H., 2002. Virtual Archaeology and Museums: Where Are the Exhibits? In Niccolucci, F. (ed.), Virtual Archaeology: Proceedings of the VAST Euroconference, Arezzo 24-25 November 2000 (BAR Int. Ser. 1075):187-194.

SNYDER, A.B. and PALEY, S.M., 2001. Experiencing an Ancient Assyrian Palace: Methods for a Reconstruction. In Jabri (ed.), W., Reinventing the Discourse - How digital tools help bridge and transform research, education and practice in architecture. Association for Computer-Aided Design in Architecture, Buffalo:62pp.

STILLE, A., 2002. The Future of the Past. Farrar, Straus and Giroux, NY.

CARPINIANAN: A VIRTUALIZED BYZANTINE CRYPT

V. VALZANOA

A. BANDIERA
SIBA COORDINATION, UNIVERSITY OF LECCE, ITALY
SIBA@SIBA2.UNILE.IT; HTTP://SIBA2.UNILE.IT/

J-A BERALDIN

M. PICARD

S.F. EL-HAKIM

G. GODIN
SIBA COORDINATION, UNIVERSITY OF LECCE, ITALY
SIBA@SIBA2.UNILE.IT; HTTP://SIBA2.UNILE.IT/

ABSTRACT

This paper presents the 3D modeling work that was accomplished in preparing a multimedia CDROM about the history of a Byzantine Crypt. An effective approach based upon high-resolution photo-realistic texture mapping onto 3D models generated from range images is used to present the spatial information about the Crypt. Usually, this information is presented as 2D images that are flat and do not show the three-dimensionality of an environment. Both a CDROM and a video animation have been created for this project.

INTRODUCTION

In recent years, high-resolution recording (as-built reality) of heritage sites has stimulated a lot of research in computer graphics and vision. If the only purpose is the generation of photo-realistic images for visualization, then purely image-based rendering techniques offer a general solution. However, if the goal is to analyze the works, to preserve and share a record of their geometry and appearance, then explicit shape information must be acquired. A 3D model contains a wealth of information that can be analyzed and enhanced (Godin 2002). Sites that must be closed for conservation reasons can still be studied and visited. Virtual restoration can be used to improve the legibility of textual information, without turning to interventions often traumatic for the original copy. Elements that were added over the years can be removed and the digital 3D model of a site can then be viewed in the correct historical context. When preparing a multimedia CDROM about the history associated to a particular historical or archaeological site, spatial information is usually presented using 2D images that are flat and don't show the three-dimensionality of that site or environment. An effective approach based upon photo-realistic texture mapping onto 3D models generated from high-resolution range images was used to present that spatial information about a site. The results are applied to the creation of a CDROM about the Byzantine Crypt known as the Crypt of Santa Cristina (see Fig.1), which is located in Carpignano (LE), Italy.

Figure 1 Byzantine Crypt, a) two outsideentrances, b) interior located underground

USING 3D INFORMATION FOR OUR MULTIMEDIA CDROM

When trying to present the history of a heritage site, the use of spatial information becomes very important in order to facilitate an understanding of that particular site. One can resort to hand drawn or computer generated isometric views, CAD models based on more of less reality and 3D models built from reality. The source of information includes among other things drawing/paintings, papers/digital photographs, or laser scanner data. Some of these provide dimensions directly but others need indirect ways to get scale and/or dimensions. We opted to represent our site (a Byzantine Crypt) using both photogrammetic techniques for the outside, and, for the inside dense 3D laser scanner information combined with high-resolution color images. The Crypt of Santa Cristina was excavated (rupestrian site) at around the 9th century c.e. and it is characterized by two entrances. One of them leads to the area that served as a cemetery and the other, as the church. Furthermore, irregularly shaded walls covered with a number of fairly well preserved frescoes characterize this Crypt. One of them, Christ (Pantocrator) and the Annunciation, is dated at 959 c.e. and is signed by Theophylact. The structure of the image is similar to images of the same period, present in Constantinople and in Cappadocia (Turkey). Mostly experts of the Byzantine period come from around the world to visit and study this site. During the course of history, a Baroque altar was added (1775 c.e.) along with three pillars that replaced one that collapsed. Many aspects of sensing and modeling must be understood before starting such a large project. The typical processing pipeline used for 3D modeling includes geometric modeling and appearance modeling. These are well documented in references (Godin 2002, Beraldin et al. 2002, Soucy et al. 1996, Bernardini et al. 2001, El-Hakim et al. 1998, Sequeira et al. 1999). Here, we summarize what was done in the case in point. The technical aspects of the project are described in Beraldin et al. (2002).

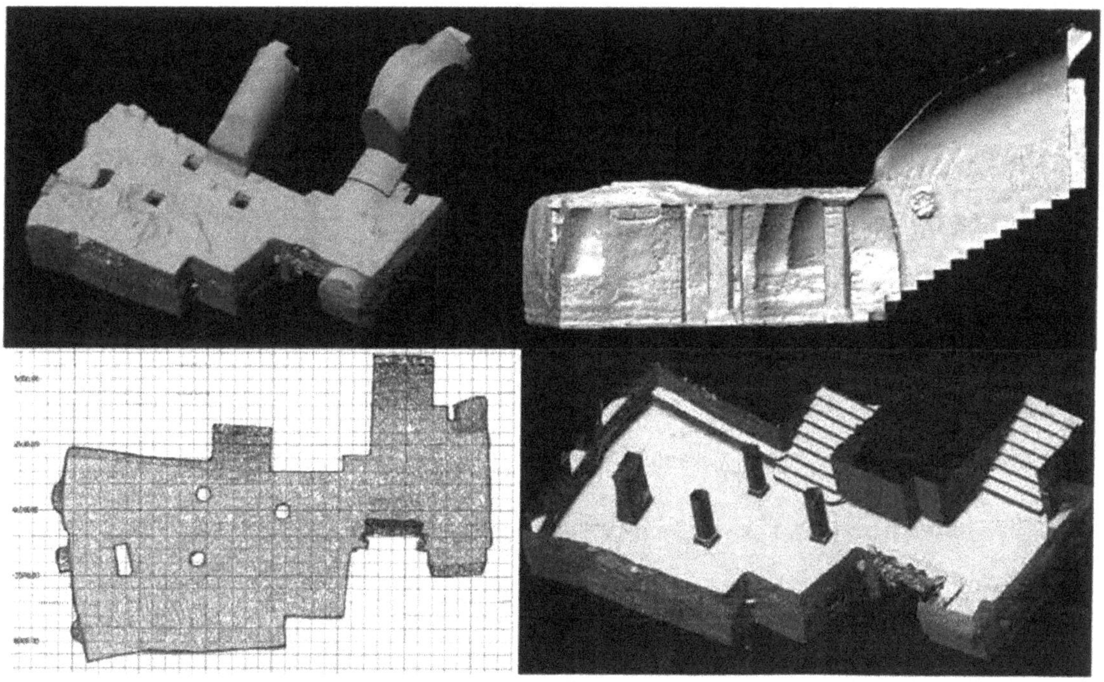

Figure 2 Complete 3D model of the Byzantine Crypt, a) view from outside shown with synthetic shading, b) a particular view of the stairs leading to the Crypt, c) floor plan generated from an orthographic view of the 3D model, d) section of 3D model showing the space inside the crypt

MODELING TECHNIQUES USED FOR THE BYZANTINE CRYPT

[1] Geometric Modeling

A model is a digital representation of the object or site on which one can perform operations. Acquiring dense 3D data of surfaces has been a hot topic of research in the last 20 years. Though not as mature as photography, 3D imaging is seeing new applications emerging every year. Numerous commercial systems are available to measure dense 3D data. The Byzantine Crypt is relatively large (16.5 m by 10 m by 2.5 m) and we wanted to model it with a fairly high spatial resolution. For these kinds of environments, there are not a lot of range cameras on the market that could provide us with the desired level of spatial resolution and low measurement uncertainty. This range of distances represents the transition between optical triangulation and time of flight technologies. In order to create a dense 3D model of the Byzantine Crypt, a MENSI SOISICTM-2000 SD was used. This triangulation-based laser scanner can acquire 3D images at a minimal distance of 0.8 m and at up to 10 m. The range uncertainty on a cooperative surface at 2.5 m is about 0.4 mm (1 sigma) and the data rate is 100-3D points per second. The spatial resolution of a model mesh is determined by the angular resolution of the laser inside the scanner. The lowest possible resolution with the Mensi scanner is 0.1 milli-radians. For instance at a distance of 2.5 m (standoff distance selected), this angular resolution corresponds to a spatial resolution of about 0.25 mm. To minimize the amount of time spent in the Crypt, a strategy for scanning was defined well before starting the work. In order to keep a quasi-constant spatial sampling on the surface of the walls, 3D vertical scans were used to build the 3D model. In theory we could have used the lowest angular resolution available with this scanner but in practice, this would have required very long scanning sessions. Therefore, it was decided to increase the sampling step on the mesh to 5 mm. This gave an average scan time per 3D image of about 80 minutes. As a result, surface details like small tool marks were not measured well at that resolution but the overall shape of the Crypt is excellent. Figure 2 present the complete 3D model (without color information) that would appear if one could see through the ground. From this model, a floor plan was created. The generation of the model was done with both the camera manufacturer's own software 3DipsosTM and PolyworksTM software from Innovmetric. The actual measurement uncertainty was found to be twice the value given for cooperative surfaces, i.e. 0.8 mm at a distance of 2.5 m.

[2] Appearance (Color Texture)

Appearance modeling includes methods like Image Perspective Techniques (IPT) and Reflectance Modeling (RM). IPT is concerned with direct mapping of photographs onto a 3D model and reflectance modeling (Bernardini et al. 2001, El-Hakim et al. 1998, Sequeira et al. 1999, Weinhaus and Devarjan 1997). Reflectance modeling is used to extract from the measured color and shape those physical properties of an object that are intrinsic to it and that determine its appearance when viewed with artificial lighting on a computer screen (Soucy et al. 1996). The true appearance of an object is the result of the interaction of light with material. The knowledge of such information is important in order to reproduce hypothetical lighting conditions under varying observation points. For this project, texture mapping using IPT was chosen. With current CCD and CMOS technology, access to high quality texture images for IPT is now within reach of everyone. Though the Mensi provides 2D images from its internal video camera, the resolution and color quality is not acceptable for our application. So a lens-interchangeable SLR-type 6-MegaPixel digital camera was used for the texture acquisition. Proper texturing of the 3D model requires special lighting fixtures in order to control illumination. Good uniformity of the illumination is essential in order to ease the virtual restoration tasks and improve the visual quality of a textured model. Flashtubes with a color temperature of about 5600 K were used to acquire textures.

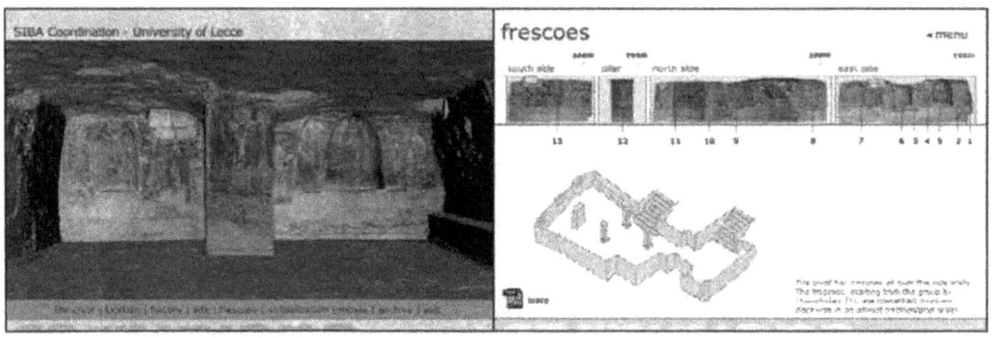

Figure 3 CDROM: a) entrance page, b) use of orthophotos generated from 3D model to navigate through the frescoes

CDROM AND VIDEO ANIMATION: CARPINIANA

Realistic estimate of the time to acquire the range images, build a 3D model and the determination of the required quality of model is very critical for such a project. For this site, with an average scan time on the walls of 80 minutes per 3D image, a total of 92 hours were spent in the Crypt. During that time, fifty 3D images were acquired for the Crypt along with thirty 3D images for the altar. The 3D model was created over a period of one month. The acquisition of the texture took 3 days and the actual mapping was done in 4 days.

A number of models with different levels of complexity were created from the original data. At the highest quality, the spatial resolution on the wall is about 5 mm and on the ceiling and floor, 15 mm. The range uncertainty is about 0.8 mm. We are currently working with 3 models: one 4.6 million-polygonal un-textured model (10 mm resolution) of the complete Crypt, a 12.8 million-polygon fully textured model (5 mm resolution) of one half of the Crypt (contains the two apses), and, a lighter textured model with 0.4 million polygons. These different models were further transformed in order to fit the format used in a CDROM. All of these representations are aimed at showing the three-dimensionality of the site and visualize artifact that are not visible in a typical visit to the site. A movie called "Carpiniana" showing a fly through of the Byzantine Crypt was also prepared. Snapshots of the CDROM and Video animation are shown on Figures 3 and 4 respectively.

CONCLUSIONS

The potential of modeling as-built reality in heritage opens-up applications such as virtual restoration or as an input to virtualized reality tours. As demonstrated with a Byzantine Crypt, a high degree of realism can be attained by those techniques and the context in which the artifacts were discovered or were used can be recreated. Real world acquisition and modeling is now possible. Technological advances are such that difficulties are

Figure 4 Movie: a) view of crypt without texture, b) view of main pillar with texture

more of a logistical nature than technological per se. Models of large objects, structures and environments are possible but as demonstrated here require the combination of a number of techniques. The problem we addressed in this paper is the use of 3D modeling to enhance the understanding of a heritage site that needs to be preserved and shown to more people in order to raise awareness and understanding of the Byzantine period. Both a CDROM and a video animation were created to fulfill these hopes.

ACKNOLEDGEMENTS

Contributors: E. Bandiera, F. Bergamo, M.C. Catamo, Sac. G. Colavero, D. Lucarella, R. Ingrosso, A. Marra, F. Melcarne, S. Nuccio, P. Pulli, A. Van Den Troost, U. Parrini, M. Biliotti, C. Sempi, V. Blasi, R. De Rinaldis and M. L. Blasi. Co-financing was provided by the Ministero dell'Istruzione, dell'Università e della Ricerca and the European Union (F.E.S.R.) for initiative 18 of the Piano Coordinato delle Università di Catania e Lecce.

REFERENCES

BERALDIN, J.-A., PICARD, M., EL-HAKIM, S.F., GODIN, G., VALZANO, V., BANDIERA, A. and LATOUCHE, C., 2002.Virtualizing a Byzantine Crypt by Combining High-resolution Textures with Laser Scanner 3D Data. Proceedings of 8th International Conference on Virtual Systems and MultiMedia, Gyeongju, Korea, 25-27 September 2002.

BERNARDINI, F., MARTIN, I.M. and RUSHMEIER, H., 2001. High-quality texture reconstruction from multiple scans. IEEE Transactions on Visualization and Computer Graphics 7(4):318-332.

EL-HAKIM, S.F., BRENNER, C. and ROTH, G., 1998. A multi-sensor approach to creating accurate virtual environments. ISPRS Journal of Photogrammetry & Remote Sensing 53:379-391.

GODIN, G. et al., 2002. Active Optical 3D Imaging for Heritage Applications. IEEE CG&A 22(5):24-36.

SEQUEIRA, V., NG, K., WOLFART, E., GONCALVES, J. and HOGG, D., 1999. Automated reconstruction of 3D models from real environments. ISPRS Journal of Photogrammetry & Remote Sensing 54:1-22.

SOUCY, M., GODIN, G., BARIBEAU, R., BLAIS, F. and RIOUX, M., 1996. Sensors and algorithms for the construction of digital 3-D colour models of real objects. Proceedings of the IEEE International Conference on Image Processing, Lausanne, Switzerland, 16-19 September 1996.

WEINHAUS, M. and DEVARJAN, V., 1997. Texture mapping 3D models of real-world scenes. ACM Computing Survey 29(4):325-365.

QUICKTIME VIRTUAL REALITY (QTVR) AND THE DOCUMENTATION OF ROCK ART LOCALITIES

CONNY MEISTER
UNIVERSITY OF CAPE TOWN
MEICON@WEB.DE

BASTIAN ASMUS
UNIVERSITY OF CAPE TOWN
BASTIAN@SCIENCE.UCT.AC.ZA

ABSTRACT

The Cederberg Rock Art Survey Project (CRASP) aims to record, and therefore preserve, South African rock art, particularly that of the Cederberg. The preservation of this unique world cultural heritage was attempted through the help of various methods and techniques, including traditional documentation, as well as a newly-introduced fully digital recording system. The visualisation of archaeological results for the public is becoming a predominant aspect of archaeology. This is primarily because successful funding often depends on the way results are presented, and also because we, as archaeologists, owe the publication and visualisation of the results to a broader spectrum of non-professionals. Therefore we feel that finding ways of visualising our results for the general public in order investigate and disseminate knowledge about the past, is one of the main principles of archaeology.

To present the unique rock art of South Africa within the landscape where these sites are found, we sought a way to amalgate the aspects of publication and visualisation. Quick Time Virtual Reality (QTVR) panoramas make it possible to achieve these two aims using a low-cost recording system. This article demonstrates the relevance of QTVR as a supplementary tool to the traditional way of documenting, preserving and visualising rock art, and also discusses the equipment required for the task.

SETTING AND BACKGROUND

The Cederberg is a hilly to mountainous area stretching about 200 km from north to south and 100 from east to west dominated by outcrops and cliffs of reddish sandstone, derived from the Table Mountain Formation. It is situated about 200 km north of Cape Town in the western Cape region of South Africa (Fig.1).

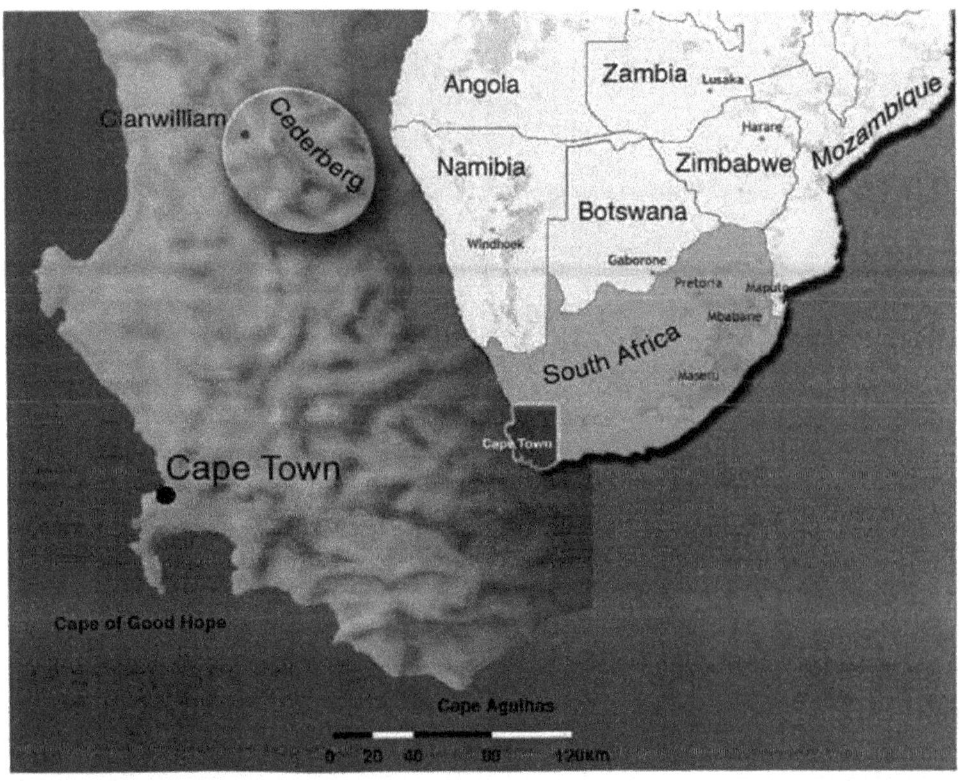

Figure 1 Southern Africa and the location of the Cederberg

The research project area covers 35 km2 and is located about 5km south of Clanwilliam stretching southward along the Olifantsriverdam and eastward into the tributary Rondegatvalley (Fig.2). To date, through a survey of the project area, we have identified and recorded 65 rock art sites, such as shelters, caves and painted boulders. Found within the Cape vegetation that consists mostly of drought-resisting sclerophyllous fynbos scrubs, these shelters and caves exhibit a great number of rock art paintings, mostly in the style of the fine-line tradition.

Generally speaking, the rock art of South Africa can be grouped into three distinct categories: fine-line paintings (which also might be referred to as representational art with a conceptual component which cannot be omitted (Lewis-Williams 1981)), handprints, and the colonial/historical period paintings. The fine-line paintings have also been referred to by other researchers as detailed representatio-

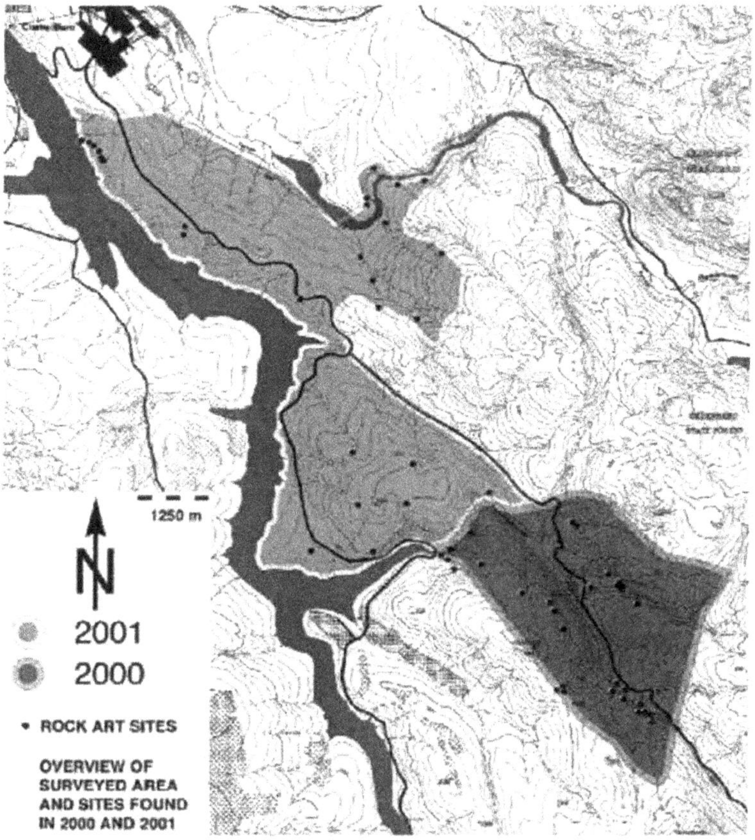

Figure 2 Surveyed area of the CRASP until 2003

nal paintings (Yates et al. 1993), finely detailed paintings or fine-line imagery (Manhire 1998) as well as fine painting or representational painting (Yates et al. 1994). These paintings were created by using a thin instrument, possibly a stick or a brush, applying the paint to elaborate, resulting in fairly naturalistic drawings of animals, humans, as well as objects of everyday use, such as bows, quivers, bags, nets and sticks (Asmus and Meister 2001, Yates et al. 1994). These occur in various colours, predominantly shades of red, yellow, white and black. Apart from this, the paintings are often painted in a bichrome and polychrome manner, especially those of humans and various animals, such as eland, elephant, hartebeest and others. (Asmus 2003, Asmus and Meister 2001, Yates et al. 1993).

While these paintings occur almost anywhere suitable rock surfaces can be found, handprints are primarily found in the western Cape. These prints "...are often grouped in clusters or are displayed in lines and seem to be secluded from other images. Multiple handprints and fine-lined paintings rarely occur in the same place, as handprint sites are ordinarily quite segregated" (Meister 2003). Due to these and other differences in repetitiveness, size, style, age, intra-site and general distribution of fine-line and historical paintings, handprints should be considered as a category of their own (Manhire 1998, Meister 2003).

Also in contrast to the detailed representational paintings are the crude finger painted colonial applications, which differ not only in quality, but also in subject matter. Most of these images are depictions of horses, cattle, other unidentifiable animals and humans with typical attributes of the European settlers, like brimmed hats, high-heeled shoes and guns, as well as images of ships, wagons and crude grid symbols (Yates et al. 1993). The crudeness of finger-applied artistical expressions, including graffiti, corresponds most closely with the printing of hands rather than with the fine-lined drawings of traditional images. It is possible, however, to find crude finger-painted signs, symbols or depictions of animals which are not necessarily attributed to the colonial style or context. For more information about the background of the rock art of southern Africa, see Willcox 1984, Yates et al. 1994, Lewis-Williams and Dowson 1999.

Recording Rock Art Digitally

The project set out to document rock art of the Cederberg digitally, linking a low-cost, low-tech recording system with a relational database in which to store and access the collected data. A handheld GPS-receiver to map the sites, and a digital camera, together with a conventional 35 mm camera and recording forms, were used to record the rock art images (Asmus and Meister 2001). The suitability of this relational database for rock art is demonstrated by Asmus (2003).

Nonetheless, we feel that these tools have their limitations in terms of their visualisation of the images and the culturally important landscape where they are found. This aspect has been neglected in previous rock art recordings in the area (Asmus 2003). Due to this, we saw the need to connect the rock art in context with its surroundings to make this unique human heritage available to everyone. At the same time the visual recording would shelter the fragile art from destruction, an inevitable process of the revelation of cultural monuments and heritage to the public. These intertwined aspects of documentation and touristical exposition gain more importance as exposure of archaeological sites to the public increases in the late 20th and early 21st century. Moreover funding of archaeological projects often depends on the publicity gained from such.

There are no safe guidelines for the protection of sites, but a project which combines research, education and preservation will not only offer scientific results for archaeologists and interrelated research colleagues, but also help to establish public awareness of the fragility of our heritage and its need for protection. Archaeology therefore needs to be informative and interesting to a layman, which is most easily accomplished visually. To secure the aspects of visualisation and information in the CRASP we used "3D"-panoramas, utilizing Quicktime Virtual Reality Authoring Studio®, which were generated within the rock art sites and the surrounding landscape.

Introducing Quicktime Virtual Reality (QTVR) Panoramas

QTVR-panoramas are composed of 12 to 16 single images, whose edges slightly overlap. The images were recorded on a

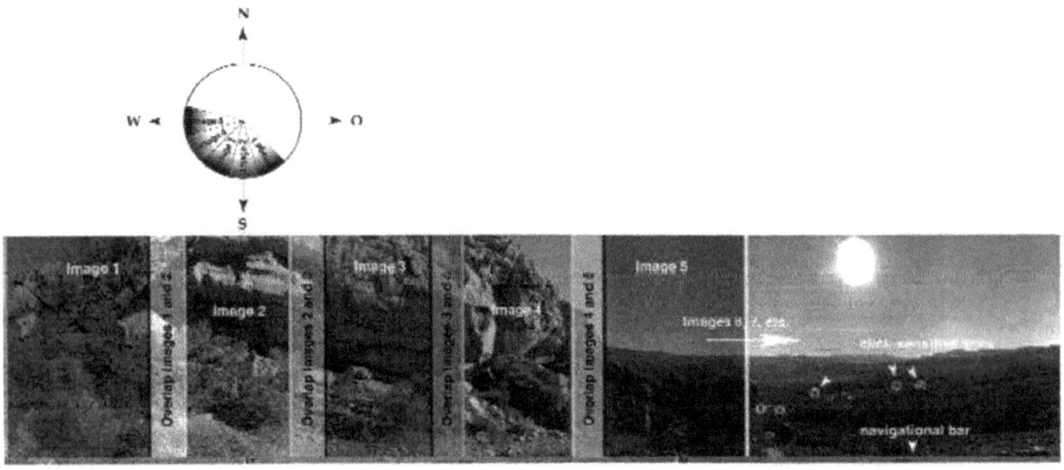

Figure 3 Part of a panorama, showing click-sensitive links (hotspots) and navigational bar (bottom). Also visible here are the aforementioned single images mended into a single panoramic image

levelled tripod. The overlap makes it possible for the Quicktime Authoring Studio application to mend all the images together into one large image file. The appearance of the resulting image-file was further modified with image-processing software. This allowed us to include indicators for other sites in the vicinity that can be seen from within the panorama (Fig.3). These "click-sensitive" areas or "hotspots" connect either one panorama with another, or refer to previously defined URL-links, to directly access rock art sites and images thereof. Furthermore, for orientation purposes we included a navigational bar to the mended image-file to indicate direction of view, such as northing and southing (Fig.3). The modified image-file was then flexed, warped and mended together at the ends, to create a cylindrical image in whose virtual centre the viewer stands. The final Quicktime®-movie, in which the "hotspots" function as interactive links, can be used with the Quicktime Player®.

Summary: Use of QTVR

The finished Quicktime Movies can be utilized as modules for Multimedia-CDs in combination with other multimedia applications or components, such as Flash or Internet-Browsers, which exploit web-based interactive links. This allows for a new range of visual exposure, as Quicktime Movies can be published on the internet, empowering many people with virtual access to the rock art within its environment. Such Multi-media-CDs (Parkington, Asmus and Meister 2003) and Internet sites provide substantial help to preserve sites, as laymen are educated and able to experience archaeological sites without destroying them through ignorance. Furthermore such applications are by far more informal and accessible for laymen, who are potential sponsors, than scientific publications. Apart from this, is it conceivable that local people will benefit from the visualisation by both selling the products (e.g. multimedia CDs) and learning about the value and need for their protection of the nearby heritage sites.

We therefore encourage archaeologists to utilize the media of visualisation and offer a affordable low-tech application in this article.

Acknowledgements

The project was financially supported by the Herbert-Quandt Stiftung der Varta AG, the Bund der Freunde der Universität Tübingen and the University of Cape Town. We are endeared to John Parkington, and the Clanwilliam Living Landscape, who gave us the opportunity to realize the Multimedia application of the Clanwilliam Rock Art Trail (Parkington, Asmus and Meister). Furthermore we would like to acknowledge the tremendous work of Tobias Tonner and Tony Manhire in surveying and recording, and especially Tobias Tonner for making invaluable comments and contributing ideas towards the projects aims. We also acknowledge Mary Anne Murray and Joellyn Palomaki for proof-reading this paper.

REFERENCES

ASMUS, B., 2003, unpubl. Painted images, landscape position and site characteristics in the Rondegatvalley, Western Cape, South Africa. MSc. Thesis, Department of Archaeology, University of Cape Town, Cape Town.

ASMUS, B. and MEISTER, C., 2001. Projekt zur Aufnahme von Felskunst in den Cederbergen, Südafrika. Archäologische Informationen 24,1:93-96.

LEWIS-WILLIAMS, J.D., 1981. Believing and seeing: Symbolic meanings in southern San rock paintings. London Academic Press, London.

LEWIS-WILLIAMS, J.D. and DOWSON, T.A., 1999. Images of Power. Southern Book Publishers (PTY) Ltd.

MANHIRE, A.H., 1998. The role of hand prints in the rock art of the south-western Cape. South African Archaeological Bulletin 53:98-108.

MEISTER, C., 2003, unpubl. Handprints of the Western Cape: Recording, measuring, identifying. MSc. thesis, Department of Archaeology, University of Cape Town, Cape Town.

PARKINGTON, ASMUS and MEISTER, 2003. The Mantis, the Eland and the Hunter - A virtual journey through rock art sites in the Warmhoek valley. CD-Rom.

WILLCOX, A.R., 1984. The rock Art of Africa. Croom Helm. London/Canberra.

YATES, R., MANHIRE, A.H. and PARKINGTON, J.E., 1993. Colonial era paintings in the rock art of the south-western Cape: Some preliminary observations. South African Archaeological, Goodwin Series 7:59-70.

YATES, R., MANHIRE, A.H. and PARKINGTON, J.E., 1994. Rock painting and history. In Dowson, T.A. and Lewis-Williams, J.D. (eds.), Contested Images, University of Witwatersrand University Press, Johannesburg:29-59.

Subterranean-Modelling
From Plan to Computer Model - Entering the 3rd Dimension

GERO STEFFENS
FACHBEREICH MONTANARCHÄOLOGIE
DEUTSCHES BERGBAU-MUSEUM
D - 44787 BOCHUM
GERO.STEFFENS@BERGBAUMUSEUM.DE

SEE THE CD FOR THE EXTENDED VERSION

ABSTRACT

The German Mining Museum is mainly working on the exploration and examination of ancient mining. There for it is important that a precise map is drawn up where the results of e.g. an excavation or the ancient working traces are documented. Sometimes there is already a 2D plan and one wants to "add" or improve more information to the plan. Very soon a plan becomes confusing e.g. when passages intersect on different levels or (interesting) parts are hidden by others in the respective projection.
In this article possible workflows will be shown, starting with the measuring of cross sections with various equipment (optical, electro optical). Further more different possibilities of how to orientate the single sections are discussed and the construction of the 3D model is revealed by using AutoCAD and 3D Studio MAX. Finally leading up to the following article, an example of an information system, built in 1999 will be demonstrated as well as some visualisations are examined.

Subterranean measurement, respectively subterranean documentation is a very special and difficult topic, which is to be handled of course with special (technical-) solutions. The very detailed requirements are pointed out and (our) solutions are described.

Mines in 3D

In mining archaeology it is often not enough to record the excavation exclusively, but also the mine surrounding it. All objects like graves, cellars, caves and mines, do not only reveal their information on the ground but also at the sidewalls and at the roof ridge. These irregular (surface) structures can not always be simplified in a plane (ground plan, section plan, etc.) therefore it is necessary to keep the information e.g. in a 3D model. The advantages of having a computer model in addition to photos, videos and plans must not be pointed out separately, but in mining archaeology you finally get the possibility of:

- calculating the volume (how much material was extracted)
- looking at your model from outside ("through the rock")
- constructing any section you wish
- preserving the information of roof, walls and ground in one "plan"

Having come to the conclusion to document your object (-structure), 3-dimensionaly, you must decide how the data is to be captured. Beside the well known and often used "laser scanning", we are glad to present an alternative method, the so called "profile scanning".

These two scanning methods have in common that simply spoken a laser beam measures the distance to any object in a range of 20 cm to 300 m with the speed of light.

Just pointing out some of them, the main differences between those scanning systems are, that a laserscanner is put up a tripod, providing range, horizontal- and vertical angle (3D) and signal amplitude, while the profile scanner is a 2D handheld system. 2D means that in addition to the distance only one angle is provided so you get a flat shape. Furthermore a laserscanner measures about 1,000 times as much points as the profile scanner. The operating range of most laserscan-

Figure 1 standard photography

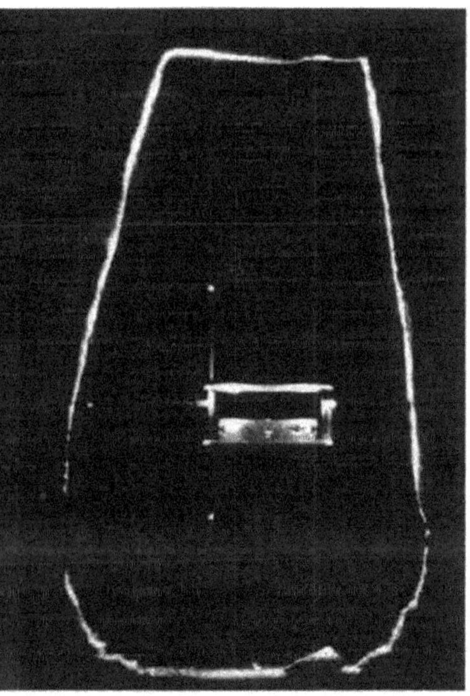

Figure 2 cross section of the same situation as figure 1

ners start at 2 m which is of course to much for narrow ancient mines.

MEASURING CROSS SECTIONS

The measuring of cross sections is suitable wherever you work in small spaces (e.g. between two walls) or in stretched structures like galleries. There are two main methods of measuring cross sections

1. an optical one[1]
2. an electro-optical one

The optical method is operating by taking pictures of a light-bar emitted with a flashlight. The light-bar is an app. 5cm width shape covering the walls (Fig.2).

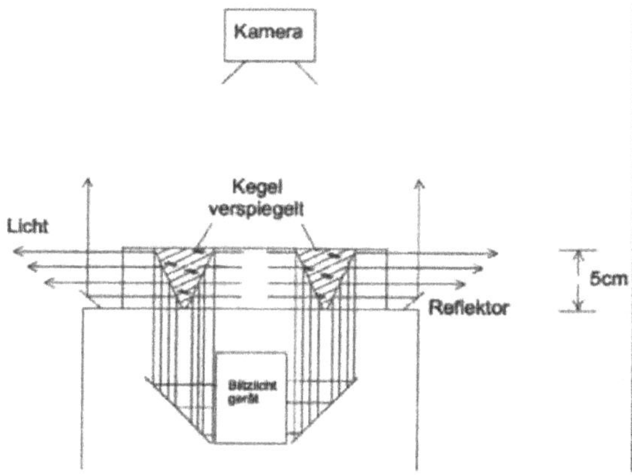

Figure 3 flashlight device

Figure 4 laser scanning device

The little light spots next to the device (visible in the second picture) are put up in a known distance, giving the scale while digitising the photo. There are several limitations to observe using this method. e.g. you always have to get far away enough from the cross section to take a picture of it, which can be limited by the available space and the focal length of your camera. This is the reason why (and because of other limitations) the Deutsches Bergbau-Museum Bochum in association with partners from speleologie[2] has developed a device which measures a cross section electro optically.

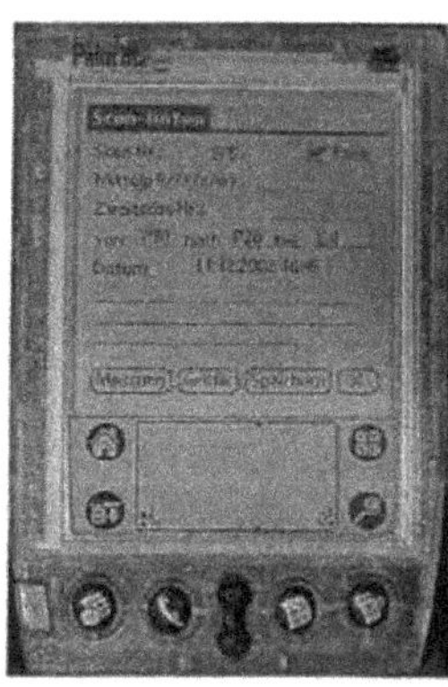

Figure 5 Palm input screen

Figure 6 Palm control screen

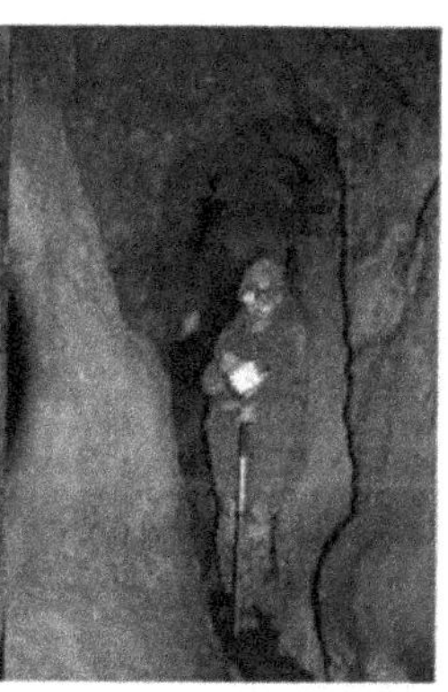

Figure 7 measurement in an ancient mine

THE HANDHELD 2D SCANNER

The frequency of the laser is 100 Hz. During 3 seconds the laser rotates about 360° with the result that a single cross section consists of 300 vertices.

The data is sent to a Palm IIIe handheld device where it is calculated and displayed. The user is now able to check the scan for errors, caused by e.g. less reflection or invisible parts of the section.

To built up a 3D model you need the position where each cross section was captured.

Therefore a measuring tape is put between two polygon points with known coordinates during the measurement. The particular position (read off the tape) of taking the scan is input manually into the palm device (Fig.5). Holding the scanner-device horizontally and parallel next to the tape the scan is started and 3 seconds later the cross section is displayed on the Palm.

After transferring the data to a PC, the computer can put the shape nearly in the right position in (virtual) 3D space. Nearly, because you slightly have to adjust the horizontal and vertical position. This is because you may not always have put the profile scanner right next to the tape, but due to local circumstances differed from this specification (but you always have to put up the device horizontally and parallel to the tape).

WHAT TO DO WITH THE DATA ?

The following figures show the recorded data first being reduced from 300 points to app. 30 points. The reduction of the data is very important to keep the amount of polygons of your 3D model as low as possible, and thus manageable. Second the spline has to be smoothed which gives the structure a much more natural appearance and also eliminates "jumps" between two vertices caused by the measurement error of the laser scanner. As you can see inFigure 8, the tape, which was put between two polygon points, can be identified as a peak in the blue shape. You also see the vertical and horizontal offset of the measuring tape "connecting" the point of origin of the profile scanner with the tape.

After positioning the shapes in 3D space with Auto CAD, we export the complete data set to 3D Studio MAX. A mesh is constructed based on all the single shapes. Taking measurements e.g. every 30 cm provides the opportunity to built up a very detailed 3D model. This sounds very easy, but of course there are

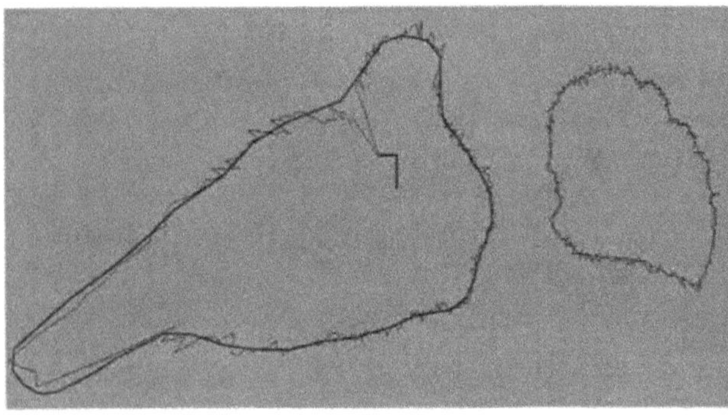

Figure 8 point reduction, smoothing and localisation of the tape

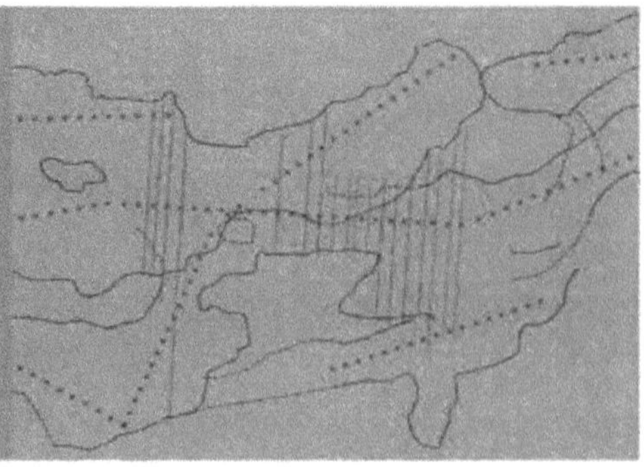

Figure 9 arranging the shapes with the help of a ground plan in AutoCAD

several "problems" which have to be solved during the construction:

WHERE IS THE RUB?

Especially in the area of conjunctions, not the complete cross section will be valid. Think (3 dimensionally!!!) of a Y-junction where the single incoming tubes are named A, B, C starting clockwise with the upper left one. Coming closer to the conjunction out of tube A, the laser will hit (on the left side) more and more parts of the B tube before A and B merge to tube C. Then you do the same with tube B (delete the parts witch hit tube A). This means that you have to delete those parts of a cross sections manually, which do not belong to the respective tube (gallery).

Now we have open cross sections and closed cross sections. If you come out of tube A or B, you first have to connect the closed cross sections with each other. In most cases there is no major problem of letting the computer do this automatically, but of course there are limitations of an automatism e.g. think of connecting two triangles, one of them upside dow .

There is at least one position in every tube where closed cross sections have to be connected with open ones (nearby the conjunction, remember that you opened them manually to get valid cross sections) Also the connection of two open cross sections can be difficult (e.g. two C-shaped sections one of them reverse). At the moment, those areas must be meshed manually because the computer can not decide automatically how to construct this.

Finally the meshed object is constructed, you can do "everything" you want to do with your 3D model. Adding textures and lights to the wire frame model, you can arrange an almost photo realistic 3D computer model. You can animate it (fly through the mine), or export it as e.g. VRML. Depending of the size of your model (and of course the capability of your computer), this gives you the possibility of walking through the mine in real-time and looking from any position, to everywhere you want.

FURTHER DEVELOPMENTS

At the moment we are thinking of different possibilities of capturing the position and orientation (6 degrees of freedom) of the profile-scanner as precise as possible (in relation to the local circumstances in a mine- dark, muddy, narrow). If the position (and orientation) of the scanner in relation to the tape is known (and digitally stored), it would be possible to arrange each single cross sections automatically in e.g. AutoCAD. There are many different trakking systems available, but for our field of application the inertial tracking system and the magnetic tracking system seem to be most suitable.

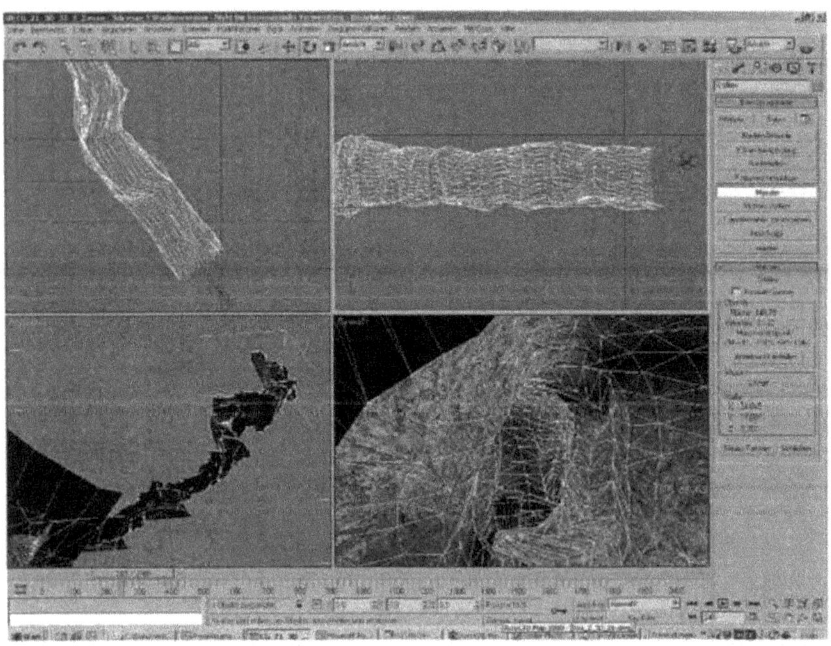

Figure 10 screenshot of a model of a mine in 3D Studio MAX

1 This optical light section device was developed and built in the Deutsches Bergbau-Museum in the early 80s.

2 Martin Melzer (Switzerland) from the HFG Karlsruhe did the programming of the Software for the Palm device and he developed the control unit. Martin Heller (Switzerland) has programmed the tool for converting the Palm -data to an AutoCAD readable dxf file.

Progressive 3D Modelling of the Theban Tomb 32

Zsolt Vasáros
University of Technology and Economics,
Faculty of Architecture

Gábor Divinyi
University of Technology and Economics,
Faculty of Architecture

Abstract

The presentation is a virtual reconstruction of the Theban tomb 32, which aims at modelling the topographical placing of the architectural mass and how the area was built in on a ridged surface. The densely built in areas of the preferred slopes of the Theban cemetery are visualised in space, as well as the layout survey - thereby helping the chronological classification of certain structures. The elaboration of detail in architecture, sculpture and painting tries to understand the structures better and to discover the planning-artistic concept from a new viewpoint. The 3D reconstructions of the fragmental structures throw light on excellent workmanship - and give away those awkward solutions which would not have been visible on 2D drawings. Alternative proposals for ambiguous details or for those which cannot be accounted for from the layout can be visualised in a 3D model. The find assemblage is of significance, the mass reconstruction of the interior scenes and use of space makes the visualisation of the interior complete. During archaeological excavations most of the contexts are not intact, thus real and virtual reconstruction of fragmental and incomplete states is unavoidable. Placing the finished models into the architectural environment is a testing of tomb-reconstruction, but at the same time it also provides a picture which rarely comes up in excavations.
Besides the presentation of the reconstruction we also wish to show how this is relevant in the assessment of scientific results, apart from offering an aesthetic outlook.

Nowadays, computerised representation of architectural structures and historical relics amounts to nothing new within the field of archaeology. The multifunctional application of the computer for the assessment of archaeological data brought forth novel results from a scientific perspective as well. Modern exhibitions, in situ presentations and popularising publications are an opening towards the wider audience, thereby creating an increasing interest on the part of the public towards archaeological results.

The aim of computer-aided planning and modelling is to expose and get acquainted with contemporary professional skills, creativity and innovative thinking, besides creating virtual reality displays. Thanks to ongoing software developments, modelling is increasingly successful in capturing the richness of surface detail and the materials used. The photo-realistic representation of virtual reality is a general expectation towards modelling.

The material to be presented is the architectural-archaeological reconstruction and virtual modelling of the spatial structure of the Theban tomb no. 32 and its close environment. Apart from representing the architectural units, we also wished to give an account of the chronology, that is, presenting the periods the environment in question was used - through the course of thousands of years. The tomb was constructed during the reign of Ramses II, in the period of the New Empire, as the final resting place of Djehutymes, a principal fulfilling various important positions. The excavation of the complex started in 1983, with the clearing of the underground areas. Excavation of the forecourt began in 1986, while the pyramid was uncovered in 1997-98. During the excavation of the outer court, several former burials and earlier structures came to light (Fig.1).

Computerised-digital processing started after the surveying and the architectural reconstruction had been completed. The survey drawings, aerial and site photographs, the find assemblage and analogies were our points of departure. The 2D processing of the drawings compiled for the presentation was

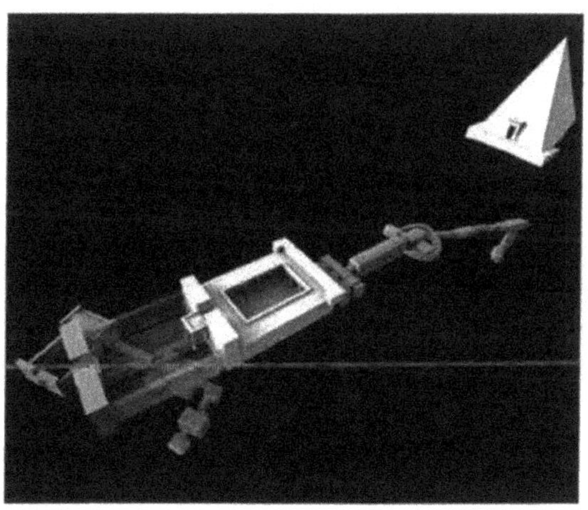

Figure 1

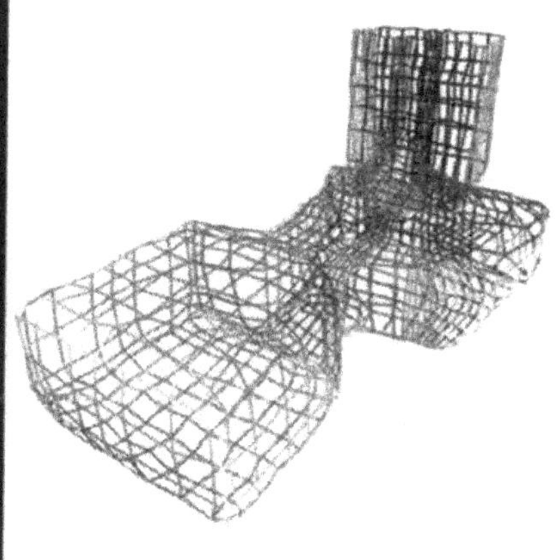

Figure 2

created by the Autodesk Architectural Desktop Autocad platform. For the 3D modelling, we used the Autodesk VIZ4 software, which is compatible with the 2D system. The structure of the model was determined by the chosen software; in this case this is a surface model created by bitmap-based shading, mapped onto a wireframe. The courts below the surface, the pyramid and the chambers which can be traced back to simple geometrical shapes were created by typical architectural modelling practices. The individual, built-in elements are composed of parametrically based solids.

The mapping of the organic and amorphous spaces and surfaces was created by polygon-based modelling. The sub-object level polygons, vertices and edges provide the tools of modelling.

To give an account of the modelling process; we would like to present the following stages and illustrations:

1. wireframe building (Fig.2)
2. built-in smoothing
3. bitmap-based shading
4. rendering

With the help of 3D modelling, exciting virtual models can be constructed, besides uncovering the structure of the tomb. Such an element of the tomb is this painted relief detail from the second chamber. The picture depicts the entrance of the tomb, which has a portical structure, with the pyramid above. During the architectural reconstruction, a construction very similar to the Deir el-Medine tomb pyramids came into mind. After several unsuccessful excavation seasons, the expedition discovered the pyramid nearly exactly over the tomb - though 35 metres higher. The unambiguous connection was provided by frescoes from the inside of the pyramid, bearing the name of Djehutymes. Although the pyramid is situated 70 metres away from the entrance and is located 15 metres higher, it shows the picture depicted in the tomb - as it can be seen on the CAD-design. The iconography - as well as the actual existence of such representations - can become the focus of study. As our example also shows, the strict representational conventions nevertheless allow for realistic depictions (Fig.3).

The reconstruction of the statues is a significant and spectacular part of virtual modelling. The inner court, facade and first chamber of the former complex contained twenty-four high reliefs. In addition, fully sculpted statues must have stood between the columns and in the inner areas. The life-size representations in the chambers that were carved into the rock managed to survive more-or-less completely, while those in the court got very damaged. However, by reconstructing the large mass of excavated stone fragments, first the faces and then by the process of proportioning the bodies themselves could be modelled. The detailed reconstruction of the spatial structure gives an account of the tense, dynamic characteristics of the contemporary space; while it also provides a system of differing visual displays in the course of time. The court model might help to locate the original positions of the thousands of stone fragments, though it might also be useful in the preparatory stage of in situ presentation (Fig.4).

During the excavation of the outer court, several earlier burials and Late Period structures were unearthed. These structures - taking the environment into consideration - are usually chambers that were carved from horizontal shafts. On the presentation an adobe structure is clearly visible, which - during its construction - ruined the forecourt of the tomb. The modelled "negative spaces" presented here shed light on the archaeological-architectural feature of the Theban cemetery, and characterise its structured aspect, even within this small segment as compared to the much larger whole. The actual cemetery extends over several square kilometres, the area that has been excavated by us is only a couple of thousand meters. The main purposes of the construction of the virtual model was to present the below-surface areas of the cemetery and to further the precise determination of chronology.

The environment of the tomb of Djehutymes was segmented by the construction of earlier graves, chambers carved out of rock and several courts. The rock quality of the region differs from layer to layer and the continuously widening gaps further render stone cutting more difficult. This is why excavations are dangerous, and 3D documentation ever important. The wall face of the chambers and facade follow the plains of larger rock movements; documenting these processes can provide us with valuable information regarding the strengthening of the rock walls and the consolidation of the remains. These factors made the preparatory planning stage and the actual execution all the more complicated even in ancient times. The strange axis changes observable on the ground plan structure also reflect the complexities of the architectural conception and the difficulties that had to be faced during

Figure 3

Figure 4

the realisation of the plan itself. One had to adopt to the circumstances and conform to the contemporary conventions regarding the construction of tombs. An interesting example of strict axis composition and creative planning can be seen in the connection of the tomb's courts and the parts that were carved into the rock. The neighbouring tomb which was constructed during the 18th dynasty and a most probably unfinished tomb on the western side determined the maximal width that could be carved out or constructed. Thus, the axis of the first chamber is not in line with the axis of the proportionally designed courts. In order to balance the anomaly, a facade with half-statues, a stele and a false door was constructed; one of its secondary axes is the actual entrance of the tomb. This apparent solution created a problem on the eastern side, the planned symmetrical composition could not be attained. The remains suggest the construction of a further, unknown element beside the entrance. Answers will be provided after the stone fragments have been processed.

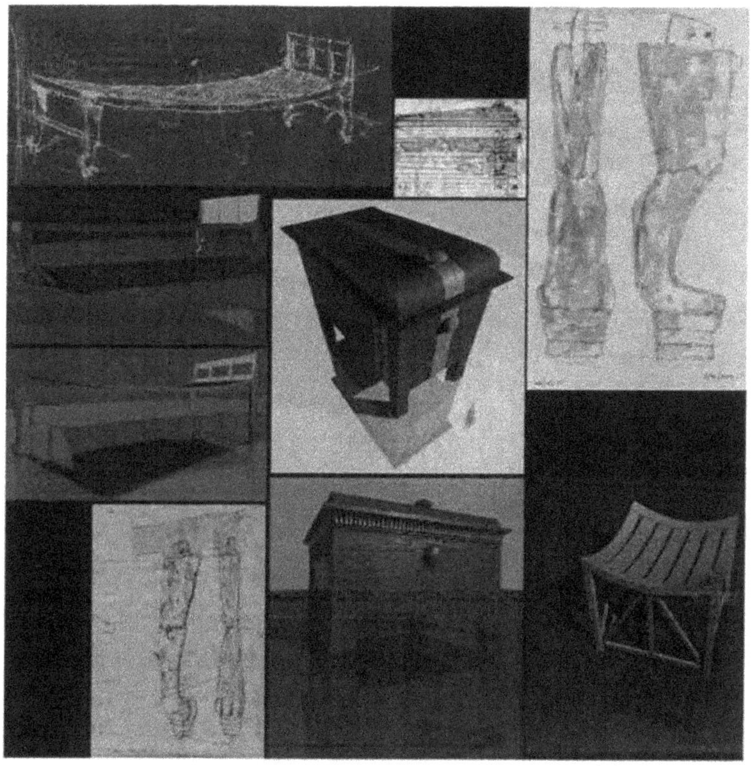

Figure 5

By the construction of the complex, we wished to model the topographical location and building of the architectural mass on this highly segmented landscape. The exceptionally dense building of the preferred slopes of the Theban cemetery is presented in space as well - besides the ground plan surveys; thereby helping to establish the chronological ordering of the specific structures. The finish of the architectural, sculptural and pictorial details aims at a higher-level understanding of the structure and the discovering of the planning-artistic conception from a new perspective. The 3D reconstruction of the fragmentally survived superstructures shed light on the contemporary master strokes of design and at the same time give away the awkward, less-successful solutions - which otherwise would not have been visible in a 2D model. 3D modelling can visualise alternative suggestions for details that cannot be decided straightforward or cannot be deduced from the ground plan structure, thereby presenting the best possible solution.

The mass reconstruction of the significant pieces of the find assemblage make the visualising of the interiors and presumed burials possible. During archaeological excavations, undamaged, complete contexts are very hard to find. Due to grave robbing and art collecting, the virtual reconstruction of the fragmented, incomplete state is highly desirable from a scientific point of view as well. Placing the finished model into the architectural environment can be seen as tomb-reconstruction, a trial of the burial and also provides a sight very rarely experienced in reality at archaeological excavations (Fig.5).

Besides presenting the reconstruction, we would also like to introduce those possibilities which might play an important role in the assessment of scientific results, besides offering an aesthetic experience.

References

ARNOLD, D., 1991. Building in Egypt: pharaonic stone masonry. New York/Oxford.

ARNOLD, D., 1994. Lexikon der ägyptischen Baukunst. München/Zürich.

EIGNER, D., 1984. Die monumentalen Grabbauten der Spätzeit in der thebanischen Nekropole. Wien.

GARDINER, A.H. and WEIGALL, A.E.P., 1913. A Topographical Catalogue of the Private Tombs of Thebes. London.

KÁKOSY, L., 1989. Dzsehutimesz sírja Thébában - Magyar ásatások Egyiptomban. Budapest.

KÁKOSY, L. and GAÁL, E., 1985. First Preliminary Report on the Hungarian Excavation in Thebes-West, Tomb 32. ActaArchHung 37:13-22.

KAMPP, F., 1996. Die thebanische Nekropole: zum Wandel des Grabgedankens von der XVIII. bis zur XX. Dynastie /Theben 13. Mainz am Rhein.

MACKAY, E., 1921. The Cutting and Preparation of Tomb-Chapels int he Theban Necropolis. JEA 7:154-168.

MANNICHE, L., 1987. The Tombs of the Nobles at Luxor. Cairo.

SPENCER, A.J., 1979. Brick Architecture in Ancient Egypt. Warminster.

Visualizations - A Critical Survey on Opportunities and Limits

Wolfgang Knoll
University Stuttgart
Department of Architecture and Urban Planning
Institute of Presentation and Design, Chair 1

Peter Braumann
University Stuttgart
Department of Architecture and Urban Planning
Institute of Presentation and Design, Chair 1

Ildikó Frels
University Stuttgart
Department of Architecture and Urban Planning
Institute of Presentation and Design, Chair 1

See the CD for the extended version

Abstract

New visualization technologies developed in the last 15 years have yielded a wide variety of new presentation and design tools. These tools are having seminal influence on the character of scholarship in many fields of visual studies. In archaeology for instance, in the area of reconstruction and visualization of archaeological objects, the choice of and appropriate use of available presentation media is freeing scholarship from former restraints imposed by technical problems and limitations, and from the formerly high costs of archaeological object documentation. Reflection on the effects of this new scholarly freedom has led to an important insight: The kind of presentation media chosen by a scholar highly influences the quality of an archaeological objects initial statement, and consequently highly determines subsequent interpretation of the objects character.

Visualizations - A Critical Survey on Opportunities and Limits

The development of new visualization technologies in the past 15 years has yielded a wide new variety of presentation and design tools. To those who work in the field of reconstruction and visualization of archaeological objects, opportunities and limits that were typical to this field have become dramatically less bounded by technical problems and by formerly high costs of rigorous documentation or of good presentation materials. The choice of available media and their appropriate use have become far more central issues in field.

Today the range of presentation forms at our disposal includes traditional pencil drawings, perfect CAD - Simulations, cardboard models or CNC supported models. Each of these techniques offers distinguishing features and qualities advantageous to producing particular visualizations. An essential change in paradigm had occurred: Today the kind of presentation media chosen by an author highly influences the quality of an archaeological objects initial statement, and consequently highly determines subsequent interpretation of the objects character.

Meaning of Visualization

The term "Visualization" is generally taken to refer to computer-aided presentation. But visualization ideally comprises more than merely creating perfect renderings with an electronic instrument.

A maxim of Paul Klee was that art never shows the apparent but always unveils the invisible. Visualization has the potential to do both: The visible is shown, but also coherences which are not obvious, and their interpretations, are potentially unveiled. Simplified, visualizing is showing what one thinks about an object.
Especially in the field of reconstruction of archaeological objects, the suggestive power of the presentation is often difficult to distinguish from the realistic contents.

Naturally visualization of archaeological contents relies on facts, but at the same time visualization is subject to a different set of terms and parameters which very much lie in the field of art and design, rather than the traditional archaeological language and limits which lie more in the fields of science, natural history, and sociology.

Figure 1 reconstruction - roman bath Baden-Baden 2003 © ArEP, Braumann, Frels, Knoll

Figure 2 reconstruction - San Giovanni dei Fiorentini 2001 © Idg1, Universität Stuttgart

Visualization - Manipulative or Cognitive Tool?

The starting point of every visualization is a scientific assumption or thesis, or archaeological on-site-findings. The Intention is to reconstruct buildings on a scientific basis. The distinction between the aspects of a presentation that are based on facts and the aspects that are speculative or interpretative must be made clear. An essential task of any good visualization is to make these distinctions abundantly clear.

If this separation of scientific report and free interpretation is clearly made, then a widespread prejudice that visualization is a manipulative tool can be weakened and replaced with an understanding of the cognitive use of visualization. The suggestive power of images or models must be used to clearly differentiate the scientific root from fictive speculations and interpretations.

Figure 3 reconstruction - roman bath Baden-Baden 2001 © ArEP, Braumann, Frels, Knoll

Another responsibility of visualization, no matter what method, analogous or digital, is to make an audience understand abstract contents that would otherwise be hard to imagine. Visualization serves as interface in communication of scientists and laypersons.

When a visualization enables a laypersons basic attraction of an archaeological object to flourish, the layperson is in a better position to develop a relationship to the history of the object, to scholarship about the object and to interpretations of the object.

Figure 4 reconstruction - roman thermal bath Badenweiler © ArEP, Frels, Huster, Knoll

Finally, visualization is able to contribute to archaeological understanding and findings by checking assumptions and by facilitating the development of new theories based on speculations developed through visualization techniques.

Visualization as a Result of Teamwork

Ideally, visualization is created in close teamwork of experts in archaeology, history of art, sociology or engineering. A dialogue between these experts and specialists in visualization generates potential solutions for questions and problems such as the following.

[1] Range of Application and Target Group:

- Reconstruction of buildings/objects that exist only rudimentarily in written texts, drawings etc.
- Construction of buildings/objects that never were built and only exist in planning, for example competitions, architectural utopias etc.
- Buildings with constricted access to public
- To record a state, a condition before the restoration of a building
- To envision aspects of: reuse, rededication and remodel, computer as a tool for decision support
- Scientific-work related: checking assumptions, defining problems, computer as cognitive tool?

[2] The Didactic Concept:

- The application of media and the interconnection of different media
- What shall for whom and how be shown, with what kind of media, techniques and methods and which objective is set?
- What is the learning effect for the viewer and the user?

[3] Levels of Consideration:

- Sociocultural level:
  historic environment, building history, society and its involvement, perception of the subject in due time and today, cultural context, such as: art, inventions or technical achievements at the time of origin
- Architectural level:
  architectural form, space, spatial composition, usage, geometry, construction, building technology, engineering structure, material, colour, light and illumination

Choice of Presentation

In choosing the most appropriate visualization between potential solutions to the questions and problems developed by teamwork each available presentation media should be tested. An analysis should be done of each potential visualization to determine its correspondence with the archaeological purpose. Examples of correspondences with archaeological purpose follow.

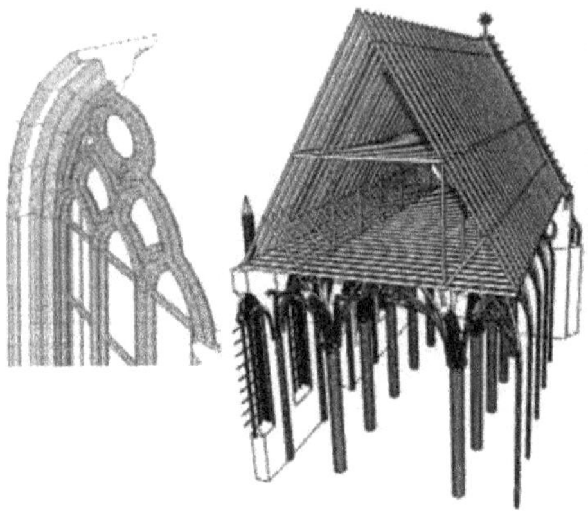

Figure 5 visualization - Heilig Kreuz Münster, Schwäbisch Gmünd 2001 © Dr. Zöller

Figure 6 visualization - Burg Falkenstein 2001 © Dr. Zöller

DRAWING

A Drawing, for example, offers the possibility of a graphically lively presentation. The particular drawing instrument, the expression and flow of the lines and the character of execution allow a high extent of individuality. In preliminary drawings, the process of finding forms can be revealed for instance through a series of lines which slowly condense to the final contour. Traces of work stay visible, the composition of a drawing is always understandable for the viewer. Still a drawing needs a skilled imagination to compose the single pieces into a spatial synthesis.

MODEL

The model is experienced visually and haptic. It shows the objects in their entity. Because of its three dimensions laypeople can generally more easily understand a model. The viewer is free to choose his or her position and point of view. The audience is able to look at volume, scale and proportion as a whole. This raises the comprehension for the observer. Since material and scale typically limit the degree of detailing in a model, its purpose is typically focussed on the characteristic marks of an object.

In addition to a CAD reconstruction of the project this model shows the essential parts of the building as: volume, roof, structure

COMPUTER

Working with Computer Aided Design, the question that arises within this context is, if presentations on the computer activate fantasy and imagination or rather reduce or even anticipate them. To what extend is the viewer confronted with an illusionary world, being feigned by virtual reality?

In the experience of reading, for example a novel or poetry, in the readers mind, pictures emerge, notions of the world described. The reader will generate a different scene, according to his constitution which may not at all match the images of other readers. The reader is asked to give something towards the text actively. If he does not do this or if the text does not claim it, it will stay irrelevant for him.

One is asked to get involved creatively, to invest fantasy and ideas, in all levels of perception, similar to experiencing architectural objects.

Looking at drawings, pictures and sculptures also requires this creative input. Watching a photorealistic Animation in the computer, only little imagination and fantasy is needed, respectively the creative process of getting involved creatively is restrained.

Figure 7 visualization - military historic museum, Dresden 2001 © Braumann, Frels, Huster

Architecturally complex buildings are to be read and interpreted in multiple ways. Spaces are not experienced just by looking and watching, but particularly by moving, listening and touching. At the computer the viewer only gets this possibility as much as it is provided in the software program. In addition, the display screen shows a two-dimensional image, space exists only seemingly.

Figure 8 wood model: teatro olympico, Vicenza
The Section was chosen to emphasize the relationship between the different functional areas of the theatre 2001 © Idg1, Universität Stuttgart

Figure 9 Reconstruction - roman thermal bath in Badenweiler, wood model. In addition to a CAD reconstruction of the project this model shows the essential parts of the building as: volume, roof, structure
2001© ArEP, Frels, Huster, Knoll

While watching a computer the viewer is bounded by the structure of the presentation. Even with interactive presentation media, the range of movement is predetermined and limited.

What is the capability that makes the computer in some respects more powerful than other media? Besides the great possibilities to develop forms and their coherences, the computer can show different levels of regard or build variants. The ability to simultaneously show different information gives the user the opportunity to depict time and to illustrate time-lap, for example the light on a building over the entire day.

The potential of computers can be found in their capacity to link different levels of examination together, for instance in the analysis of possible geometries, in the development of form and its coherences, or in the creation of virtual spaces, and most notably the demonstration of time.

Dangers and shortfalls of computer visualization lie within the shine of perfection and the compulsion to be concrete. As a helping tool the computer serves well in combination with analogue techniques. The greatest insight may be the result of overlapping computer and conventional analogue visualizations.

Conclusion

The most effective visualization for rich interpretation occurs when an interconnection of different media is possible, where different senses and sensibilities of the viewer are addressed. A palette for effective visualization includes drawing, photography, models, computer animation, computer networking of different information, sounds and tone, even samples of materials with different surfaces. Developers of visualization must persistently question what kind of media, what techniques, what vehicles of expression are most suitable for the insight being developed. For the realisation of data into a visualization, typographic, and graphic know-how is required, as well as the knowledge of composing a movie: directing camera, lights and time flow.

Ideally, those who deal with presentation and visualization of archaeological buildings and objects have some of the skills and abilities of an architect: skills with building techniques, construction, structure, materials, geometry and free planes geometry. But more essential than these skills are architectural abilities: a feeling for architectural forms, spatial rhythm and spatial qualities - qualifications that reside in an artistic creative domain.

A precondition for successful visualization is teamwork - scientists, archaeologists, art historians, sociologists, engineers, and experts in visualization working together. Within that context, visualization has potential to exert far deeper significance to archaeological scholarship than simply drawing, modelling or animating with a range of different visual instruments could provide. It has the potential to reshape aspects of the character or this scholarship and of human understanding.

Figure 10 reconstruction - roman thermal bath, Badenweiler, interior
2001 © ArEP, Frels, Huster, Knoll

Figure11 reconstruction - roman thermal bath Baden-Baden
2001 © ArEP, Braumann, Frels, Knoll, Rometsch

LATEST NEWS FROM VIENNA'S MEDIEVAL JEWISH QUARTER

M. GORIANY
STADTARCHÄOLOGIE, WIEN

D. SCHÖN
INDEPENDENT ARCHAEOLOGIST AND BUILDINGS RESEARCHER

ABSTRACT

This talk shows the possibilities which arise through the combination of scientific research results and computer technology. Recent buildings archaeology in the late medieval Jewish quarter of Vienna has produced results from around a dozen properties. Because of this work we can now reach general conclusions about late medieval property structure, the size of buildings and the accompanying infrastructure. A new picture of the medieval Jewish quarter has emerged, which includes important architectural detail.
With the house Judenplatz 8, the structure of a property from the second half of the thirteenth century will be examined. The 3D reconstruction of the building led to new discoveries about its construction. The application of computer technology threw up many new questions, but also helped to solve problems about the house. Starting with the reconstruction of this individual house, surrounding properties are filled with the shapes of the buildings that once stood there. By placing the house in its urban context in this way, a fresh picture emerges of the late medieval Jewish district. A method of demonstrating in the visualisation of the buildings the qualitative differences in the states of research of the individual buildings will also be demonstrated. The system is a modular one, allowing the integration of new results.

In the mid-nineties a new wave of research began in the area of Vienna's medieval Jewish quarter. It began with the excavation of the synagogue and the establishment of a museum in the neighbouring house Judenplatz No. 8. The museum necessitated a structural survey of the house and the results of this research fed into a 3D reconstruction of the building. Through computer technology, the building was resurrected as it appeared in the second half of the thirteenth century.

In the following years one of the authors was able to survey archaeologically a range of houses in the former Jewish quarter. This research has made it possible to reconstruct large parts of Vienna's Jewish quarter. Forgotten streets, several house plans and significant architectural details have come to light.

The most painstakingly analysed building in the Jewish district is the house Judenplatz No. 8. Between 1996 and 2000 this house was archaeologically analysed and documented from the cellar to the attic. The building's relative chronology was documented and analysed in a Harris matrix. Dendrochronological testing and written sources contributed to the absolute dating of the house's history.

Our research has shown: Property boundaries were laid down in this part of town during the twelfth century. The house in this case stretched originally to Wipplingerstraße, where the front was situated. The rear part of the plot, our part of the property, was originally a garden or yard area, in which several pits could be found, probably dug to dispose of refuse.

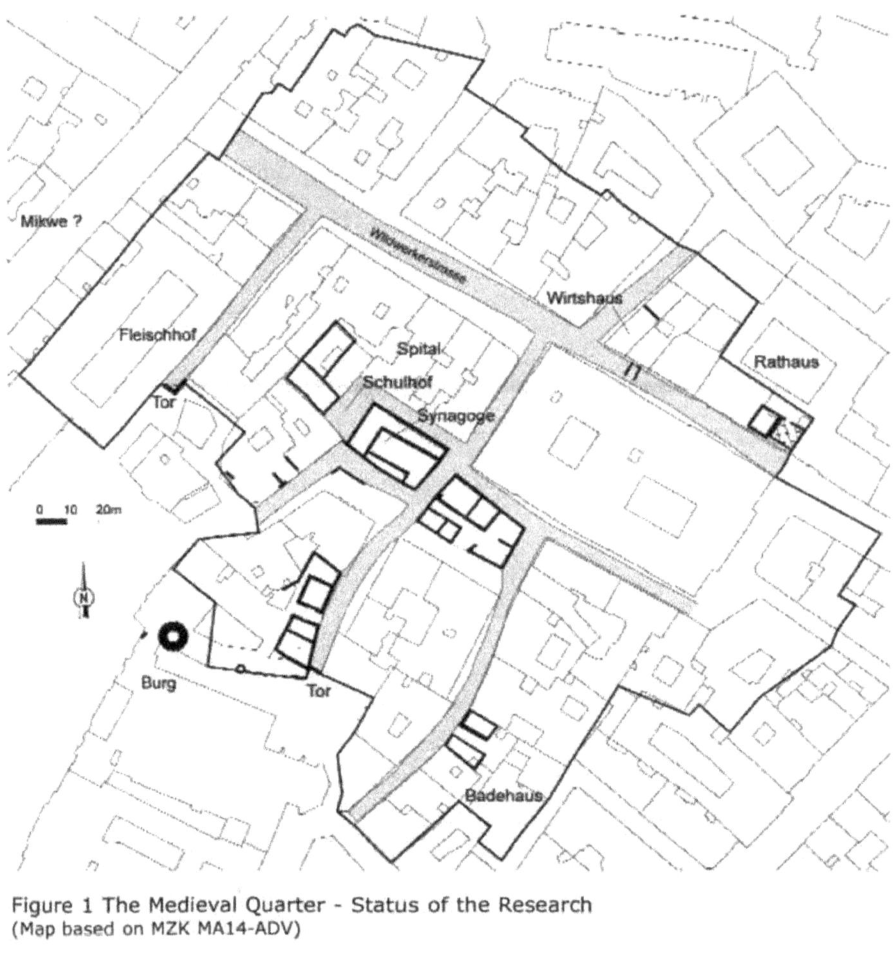

Figure 1 The Medieval Quarter - Status of the Research
(Map based on MZK MA14-ADV)

In the mid-thirteenth century the synagogue was built on the western half of today's square. At the same time our plot was reorganised. This reorganisation was sparked off by the construction of the synagogue, which became the new centre of the district. Two very large stone buildings were now built opposite the synagogue, the southern wing of our house was approximately the same size and height as the 'men's school' or the central room of the synagogue complex.

The documented wall sections, shown in dark grey in the rendering, allow the reconstruction of two wings, arranged in a right angle to each other. The lowest and highest points of the walls are taken from the deepest and highest surveyed parts of the buildings.

The southern building is externally 8 to 8.4 metres wide and approximately 18 metres long. The building reaches a height of 10.3 metres above the late medieval floor level.

The floor level is taken from a ledge in the eastern wall which marks the top of the foundations.

The level of the ground floor ceiling is taken from a horizontal ledge in the west wall on which the boards of the first upper storey floor rested. This ledge allows us to place the height of the ground floor at around 3 metres. In the south wall of today's first upper storey a 50 centimetre high horizontal scar, filled with brick was, discovered. This is the last trace of the first upper storey ceiling, left as the beams were ripped out, and fixes the first upper storey height at around 4 metres. 2.4 metres remain for a further storey.

The beams of the attic lay directly on the walls, the height of the roof's ridge line is unknown. The roof which we have reconstructed is based on roofs of the same period from Regensburg (above all the so-called 'Degginger House', Wahlenstraße No. 17).

Nothing can be said about the internal room structure of the house because of later alterations. No medieval windows were documented for the same reason. In the reconstruction, the similarly dated windows from Parisergasse No. 1, another Jewish house nearby, were used in the ground floor. Those in the first upper storey are larger, but based on the ground floor windows. The positioning of the windows is purely fictional.

The well-preserved western ground floor area had a door in the northern wall, but definitely no windows. The door's threshold is 50 centimetres higher than the ground floor of the southern wing, but no traces of stone steps remain. We suggest a semicircular wooden staircase at this point, an idea which is based on a staircase from the western room of the synagogue in its' third phase. That staircase however was a hollow quarter circle with stone steps filled with earth and covered by mortar .

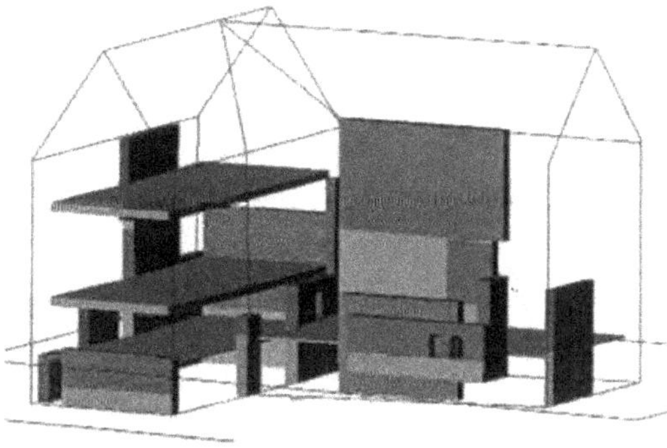

Figure 2 Original and Assumed Layer of the medival period

Figure 3 House No 8 reconstructed in its medieval phase

The western wing of the house was 16.5 metres long, approximately 10 metres wide and at least 9.5 metres high. Because of later alterations, neither the internal room structure nor the division of the storeys are known in this wing. However a small earth section in the courtyard of today's house revealed an internal mortar floor 50 centimetres higher than the ground floor of the southern wing. Proof that this level represents the height of the late medieval floor came from several layers of rendering on the north wall, which ended at the same height.

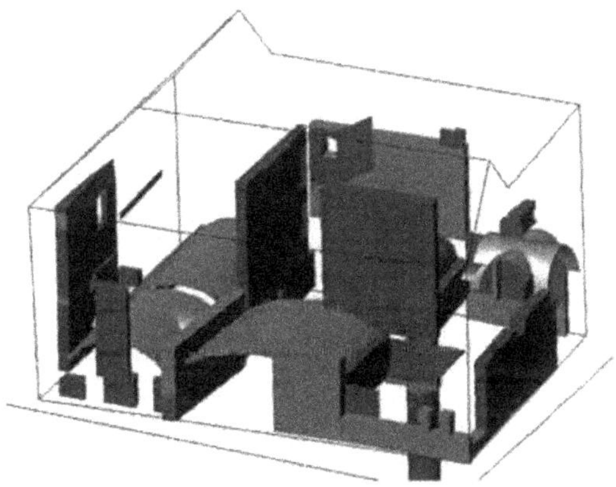

Figure 4 Original and Assumed Layer of the renaissance period

Figure 5 House No 8 reconstructed in its renaissance phase

In about 1528, a date revealed by written records, the late medieval house was converted to a considerable renaissance four-winged building around a small central courtyard.

THE MEDIEVAL QUARTER

Based on the results of the Harris-Matrix we tried to fill in the missing links in the cubature of the building using CAD.

Today the visitor to Judenplatz sees the house in its baroque appearance. We therefore decided to reconstruct not only the medieval phase of the house but also its renaissance appearance. The intense investigation of two years provided us with enough data for a reconstruction of the house in these different phases. We wanted to show not only a detailed rendered animation of Judenplatz No.8, but also the building within its urban context of the medieval quarter. Because so many buildings in the area of Judenplatz - the synagogue and the houses of the eastern part of the square - no longer exist today-- it is difficult to understand what the medieval quarter might have looked like. Our aim is to create an image in the visitor's head using only scientific data.

METHOD

Out of the over 4000 numbers of the house's matrix we extracted first only these applying to the buildings structure. Simplifying its layout every original structure was defined by:

- The height of its bottom edge
- The height of its upper edge
- Its width
- The position in the planimetry

Several interfaces such as the level of the floors were first marked in the building's coordinate system, but actually created later in the reconstruction phase. The input of the original layer gave us a first sight of the building's layout in the medieval and renaissanceperiods. This step allowed us to make corrections to the complex's general appearance.

Perhaps the most interesting step was the addition of the Assumed Layer. Because most parts of the Original Layer didn't start at the buildings ground floor but were somehow "flying" between the ground floor and the 2nd floor, we needed to unite every single structure to the buildings zero level. The assumed Layer gave us the chance to draw the heights and thickness of the vaulting by using the bases of the vaults and the position of the floor above. The different floors and the positions of some windows and doors were added. Most of the Interfaces were integrated into the structure at this stage of the work.

The last Layer, the Reconstruction, closed the cubature of the building, a roof was added and the layout and position of windows and entrances defined. Unfortunately, we found no original opening in the medieval phase on which we could build in terms of size and layout. References were found at Parisergasse No.1 , as mentioned above, and in Regensburg / Bavaria, a comparable medieval town.

Figure 6 House No.8 and the synagogue

Figure 7 The medieval quarter rendered in transparency

As well as the reconstruction of the single building Judenplatz No. 8, we have begun a 3D map of the whole Jewish Quarter. In several investigations over the years, we collected planimetric data on the position and size of medieval house remains. Unfortunately the remains are very incomplete. They don't allow us to reconstruct every single building of the quarter in detail. For this situation we created a kind of stereotype of the Viennese medieval house. The stereotype is adopted to the size and layout of the property. Rendered with transparency we aim to keep the difference between secure data, for example house No. 8, and assumed data, the stereotype house, absolutely clear. In addition to the building data we are able to reconstruct the streets of the quarter, which act as a unifying element in the incomplete form of representation of the medieval quarter.

This 3D Map serves as a base for further investigations into medieval Vienna.

REFERENCES

BUCHINGER, G., MITCHELL, P. and SCHÖN, D., 2002. Das Palais Collalto - vom Herzogshof und Judenhaus zum Adelspalast, ÖZKD 4/2002:402-419.

BUCHINGER, G., MITCHELL, P. and SCHÖN, D., 2002. Fensterformen des 13. Jahrhunderts an Wiener Profanbauten, ÖZKD 4/2002: 444-453

BUCHINGER, G. and SCHÖN, D., 2001/2002. Die Baugeschichte der Wiener Schranne nach bildlichen Quellen und archäologischen Befunden. Jahrbuch des Vereins für Geschichte der Stadt Wien 57/58: 301-334.

GAISBAUER, I., MITCHELL, P. and SCHÖN, D., 2003. Forschungen zum mittelalterlichen Wien. Neuansätze und Verpflichtungen zum Weiterdenken. In: Beiträge zur historischen Archäologie. Festschrift für Sabine Felgenhauer-Schmiedt zum 60. Geburtstag. Beiträge zur Mittelalterarchäologie in Österreich, Beiheft 6/2003: 125-140

MITCHELL, P. and SCHÖN, D., 2002. Brunnen und Latrinen in historischen Wiener Profanbauten, Österreichische Zeitschrift für Kunst und Denkmalpflege 4/2002. 474-480.

MITCHELL, P. and SCHÖN, D., 2002. Zur Struktur und Datierung des Mauerwerks in Wien, ÖZKD 4/2002: 462-473.

SCHÖN, D., 2003. Von mittelalterlichen Mauern, renaissancezeitlichen Fenstern und barocken Fußböden. Bauforschung im Haus Wien 1, Judenplatz 8. FWien 6/2003: 96-141.

SCHÖN, D. and GAISBAUER, I., 2000. ...und jenseits der Straße beginnt das Judenviertel. Zu mittelalterlichen Parzellenstrukturen in Wien 1, Kurrentgasse 4-8. FWien 3/2000: 62-74.

WAWRUSCHKA, C. and SCHÖN, D., 2002. Bauhistorische Untersuchung im Haus Zu den fünf Kronen, Beiträge zur Mittelalterarchäologie in Österreich 18/2002: 165-186.

The Charismatic Approach to Modelling and Rendering of Historic Urban Environments

D.B.Arnold
University of East Anglia, UK

A.M.Day
University of East Anglia, UK

D.W.Fellner
Inst. für Computergraphik Braunschweig, University of Technology, Germany

S.Havemann
Inst. für Computergraphik Braunschweig, University of Technology, Germany

See the CD for the Paper

Abstract

The Charismatic research project (www.charismatic-project.com) is about the production and the interactive rendering of complete reconstructions of populated Cultural Heritage Sites.

Both the re-creation and the 3D rendering of virtual reconstructions of city-sized historic environments are technically quite difficult. In order to accomplish them, a full range of approaches and tools were developed in the course of the project. Their underlying idea is to exploit domain knowledge, and also the semantics, of the types of objects to be re-created in order to make the modelling of virtual reconstructions less cost-intensive.

As an example, the tool-set includes techniques such as

- shells, i.e. deformable boxes that can be flexibly distorted, textured, and combined,
- their arrangement in a specially-tailored hierarchical scene-graph, and also
- multi-resolution data structures such as progressive meshes and subdivision surfaces for free-form ornaments attached to shells.

The same structural and semantic information useful for modelling is also exploited by the Charismatic renderer: Interactive display is optimised by applying appropriately chosen level-of-detail and culling techniques.

In summary, we present a tool-box of different techniques, designed to make the modelling and rendering of city-sized models feasible, covering as many aspects of this challenging task as possible.

The Electronic Cultural Atlas Initiative at the Enter the Past Conference, Vienna, April 2003

The Electronic Cultural Atlas Initiative (ECAI) (www.ecai.org) is a global consortium of scholars, librarians, curators and information technology professionals using emerging technologies in geographic information systems and digital libraries to display and share knowledge about history and culture. The work of ECAI is directed toward the integration of information technology research with humanities teaching and scholarship; providing expertise and training for faculty and graduate students in the humanities, while suggesting new research and development challenges for librarians, graduate students and faculty in information technology fields.

The ECAI strategies and technologies use the categories of time and space to discover, organize, integrate, visualize and use digital cultural resources. Time and space are organizing principles that can be used to integrate content on servers around the world regardless of differences in format, language and content. A focus on time and space also allows scholars and students in the humanities and information technology to take advantage of important emerging analytical approaches and develop new research questions. In order to illustrate these objectives, ECAI is conducting research; supporting the development of software, infrastructure and standards; creating content and exemplary Atlases; maintaining a data-sharing clearinghouse, and hosting conferences and workshops.

In April, 2003, ECAI was one of the sponsors of the "Enter the Past: the E-Way into the Four Dimensions of Cultural Heritage" conference in Vienna. Our goal was to initiate discussion between the well-developed field of GIS and archaeology represented by this conference, and ECAI's interdisciplinary digital Cultural Atlas community. ECAI sponsored five sessions at the conference in order to introduce the GIS and archaeology community to the methodology, technology and content that ECAI has developed. The five sessions were:

Introduction to ECAI

- The Intellectual Agenda: The Cultural Meaning of Time and Space
- When, Where and Why: Space and Time in Research
- Building the Infrastructure
- ECAI and Historical GIS: The State of the Art

The Introduction to ECAI session was chaired by Paul S. Ell. ECAI's Director Lewis Lancaster spoke on "The Role of ECAI in the Humanities and Social Sciences." This was followed by a talk entitled "Content, Research, and Publishing in ECAI: New Models for Digital Scholarship and Heritage Preservation" by Head of Collections Development Ruth Mostern. These two talks proposed the idea that the disciplines which have traditionally studied culture in universities need to begin to think rigorously about time and space as organizing categories for research, teaching, and the dissemination of research results. This ought to involve the use of emerging technologies for analyzing and disseminating information that is visual, complex and quantitative and thus poorly suited to traditional forms of communication and analysis. The next two talks introduced the audience to the infrastructure and map delivery technology that ECAI has developed to make this vision possible by integrating, displaying and publishing information about culture. The two talks were "The ECAI Infrastructure for Data Sharing" by Technical Director Jeanette Zerneke, and "TimeMap: An Interactive, Time-Enabled Web Mapping System for Internet Datasets Registered with ECAI" by Ian Johnson, developer of the TimeMap Project and the ECAI Metadata Clearinghouse. Figure 1 shows two screens from Dr. Johnson's talk, illustrating the TimeMap of World Heritage Sites. The final talk in the session was "Creating a Community and Building an Atlas: Sharing and Visualizing Cultural Data for Southeast Asia" by Southeast Asia Editor Caverlee Cary. Through an Atlas that uses the ECAI infrastructure, data from various fields, created for very different purposes by individual scholars or at the direction of major institutions, is recontextualized on a map-based interface. The Atlas allows users to select and display the datasets they require, and can be revised as new data layers are contributed or as existing ones are updated. Seeking, uniting, and preparing collaborators is a significant part of Atlas development. One mechanism that serves to further the growth of the Atlas on many levels is offering training workshops in ECAI methodology. This presentation also outlined the process of community building that has been important to ECAI.

The final ECAI session at the conference, "ECAI and Historical GIS: The State of the Art", was chaired by Lewis Lancaster. It illustrated the range and scope of historical and

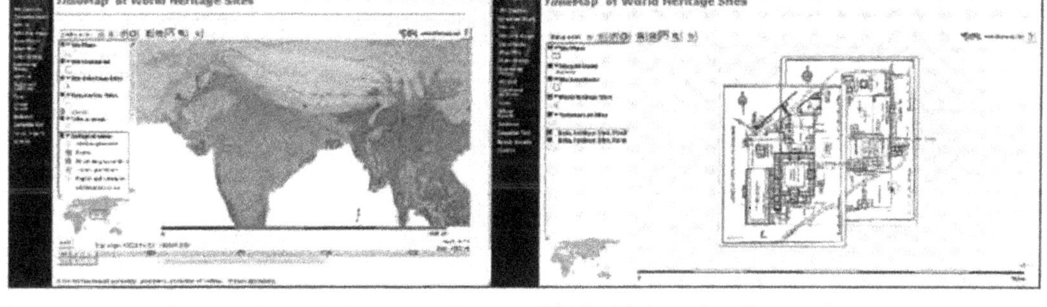

Figure 1 From "TimeMap: An Interactive, Time-enabled Web Mapping System for Internet Datasets Registered with ECAI"

cultural GIS developed by ECAI-affiliated scholars. The six projects represented on this panel were diverse in scale, approach and content. However, they all used historical spatial data derived from print sources, and all used time-variant maps in order to display results and facilitate analysis.

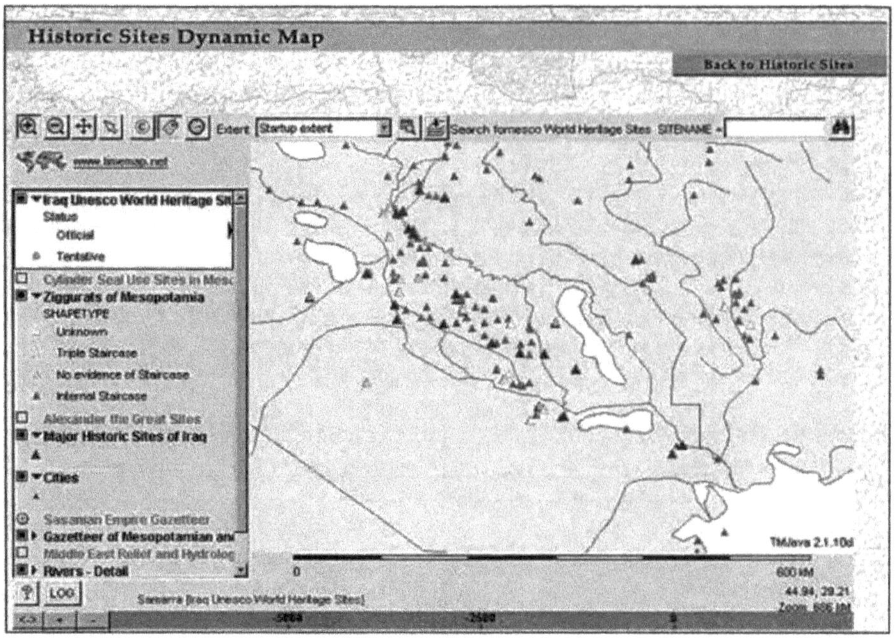

Figure 3 The Iraq Cultural Atlas

Four of the presentations concerned historical political geography. Irina Merzliakova, Merrick Lex Berman, and Alexei Karimov's, "From Tibet to Altai: Place Names of Western China on Old Russian Maps" illustrated the use of historical maps as a source of place names and features as well as images that could be incorporated into a GIS. Figure 2 is a screenshot of this project. Kim Heung-kyu, Choe Young-joon, Kim Jong-hyuk, Jung Chi-young, and Sin Hang-su presented "The Compilation of the Administrative Atlas of the Chosun Dynasty (1392-1910)." While their Atlas will eventually include information about literature, religion and other domains, this talk introduced the administrative atlas of the Chosun period, the geographical foundation of the Atlas. Merrick Lex Berman presented "Time - Differentiated Datasets for Chinese Historical Administrative Units." This presentation used ECAI TimeMap software to demonstrate browsing and searching of time-differentiated datasets. Ian Gregory and Paul Ell's "The Great Britain Historical GIS" introduced one of the best developed national GIS databases, containing a wide variety of census, vital registration and other statistical information published in Britain in the nineteenth and twentieth centuries, linked to a temporal GIS containing the changing administrative boundaries.

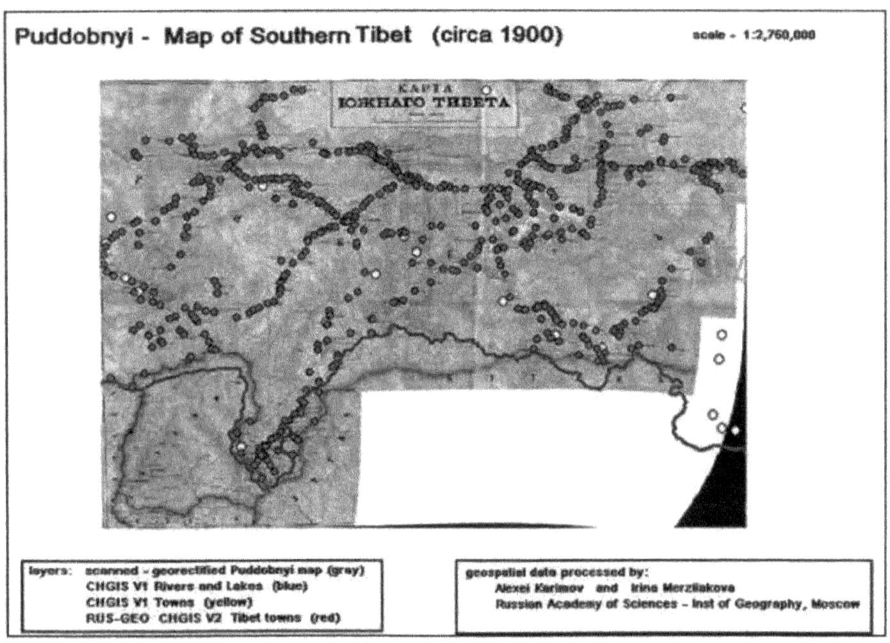

Figure 2 From "From Tibet to Altai: Place Names of Western China on Old Russian Maps"

The other two presentations addressed religious geography in a national context. Rain Simar's "Medieval German Shrines and the Reformation: A Geographic Information System" represented data from a groundbreaking article on the geography of German religion published more than 20 years ago. David Bodenhamer's "The North American Religion Atlas" presented a web-based, interactive attempt to work effectively with both the religious and geographical diversity that is a hallmark of the United States.

All of the presenters discussed issues pertaining to the creation and use of historical GIS. Berman, and Merzliakova et al., made particular note of standards development and the need for common standards in order to allow data integration among different systems. Another theme of the session was work with print sources to create a historical GIS. Kim et. al discussed their use of archival sources to create the first Choson atlas in any medium. Simar talked about the ramifications of moving from a print article into a digital environment. Merzliakova et. al presented methods for harvesting data from historical printed maps. Finally, Ell and Gregory, Bodenhamer, and Simar all addressed questions of analysis and scholarship. As Gregory and Ell pointed out, geographical analysis and data interpolation is needed in order to create comparable units; as they, Bodenhamer and Simar all noted, once a historical GIS is complete, it allows new approaches to visual and quantitative analysis in classroom and research settings.

The ECAI delegation at the Enter the Past Conference also helped to organize the response from the conference to the crisis of Iraqi cultural heritage. News about the looting and destruction of museum collections and archaeological sites was emerging as the conference occurred. As a result of meetings during the conference, ECAI has created a pilot Cultural Atlas of Iraq. This would not have been possible without the contacts and sessions held in Vienna in April. We invite all Enter the Past participants to log in to the Atlas at www.ecai.org/iraq. Figure 3 shows one of the pages from the Atlas.

The ECAI strand of the Conference reported on current activity in the humanities to interlink disparate digital cultural resources in space and time through the developing sub-discipline of historical GIS. The Initiative aims to break conventional boundaries in the humanities which tend to restrict scholars to the study of specific datasets in isolation using methodologies which largely ignore spatio-temporal issues. ECAI believes that through the combination of existing digital resources best use can be made of research materials. Once sources are related through space and time researchers have more information available upon which to develop their scholarship. The potential of temporal GIS goes much further however. It allows for the visualization of information in new ways. The facilities of GIS make exploratory spatial and temporal analysis possible. Sophisticated multivariate analytical techniques are conceivable such as Geographically Weighted Regression and multivariate modeling. Finally, longitudinal temporal analysis becomes relatively straightforward. The humanities stand on the brink of a methodological revolution, a revolution which ECAI aims to promote.

www.ingramcontent.com/pod-product-compliance
Lightning Source LLC
LaVergne TN
LVHW072246300926
849160LV00018B/1601
* 9 7 8 1 8 4 1 7 1 5 9 2 6 *